Building the Future Internet through FIRE

2016 FIRE Book: A Research and Experiment based Approach

RIVER PUBLISHERS SERIES IN INFORMATION SCIENCE AND TECHNOLOGY

Indexing: All books published in this series are submitted to Thomson Reuters Book Citation Index (BkCI), CrossRef and to Google Scholar.

The "River Publishers Series in Information Science and Technology" covers research which ushers the 21st Century into an Internet and multimedia era. Multimedia means the theory and application of filtering, coding, estimating, analyzing, detecting and recognizing, synthesizing, classifying, recording, and reproducing signals by digital and/or analog devices or techniques, while the scope of "signal" includes audio, video, speech, image, musical, multimedia, data/content, geophysical, sonar/radar, bio/medical, sensation, etc. Networking suggests transportation of such multimedia contents among nodes in communication and/or computer networks, to facilitate the ultimate Internet.

Theory, technologies, protocols and standards, applications/services, practice and implementation of wired/wireless networking are all within the scope of this series. Based on network and communication science, we further extend the scope for 21st Century life through the knowledge in robotics, machine learning, embedded systems, cognitive science, pattern recognition, quantum/biological/molecular computation and information processing, biology, ecology, social science and economics, user behaviors and interface, and applications to health and society advance.

Books published in the series include research monographs, edited volumes, handbooks and textbooks. The books provide professionals, researchers, educators, and advanced students in the field with an invaluable insight into the latest research and developments.

Topics covered in the series include, but are by no means restricted to the following:

- Communication/Computer Networking Technologies and Applications
- Queuing Theory
- Optimization
- Operation Research
- Stochastic Processes
- Information Theory
- Multimedia/Speech/Video Processing
- Computation and Information Processing
- Machine Intelligence
- Cognitive Science and Brian Science
- Embedded Systems
- Computer Architectures
- Reconfigurable Computing
- Cyber Security

For a list of other books in this series, visit www.riverpublishers.com

Building the Future Internet through FIRE

2016 FIRE Book: A Research and Experiment based Approach

Editors

Martin Serrano
National University of Ireland Galway
Ireland

Nikolaos Isaris
European Commission
Belgium

Hans Schaffers
Saxion University of Applied Sciences
Netherlands

John Domingue
Open University
United Kingdom

Michael Boniface
IT Innovation
United Kingdom

Thanasis Korakis
Polytechnic Institute of NYU
USA

Published, sold and distributed by:
River Publishers
Alsbjergvej 10
9260 Gistrup
Denmark

River Publishers
Lange Geer 44
2611 PW Delft
The Netherlands

Tel.: +45369953197
www.riverpublishers.com

ISBN: 978-87-93519-12-1 (Hardback)
978-87-93519-11-4 (Ebook)

Printed on acid-free paper.

Contents

PART III: Research PROJECTS and CASES Using Experimentation TESTBEDS

Dedications

"To those who believe that science is a lifestyle and not only job, for your time, for your efforts and for your trust on the edition of this book."

"To the community of researchers and technologists that believed that their efforts and experience can be collected in a book for the new generations."

Martin Serrano

Acknowledgements

The 2016 FIRE Book Editors' team acknowledge the great efforts and contributions from the FIRE community representing the flagship projects and the best practices from the different domain areas included in this book. The editors also would like to thank the European Commission for their support in the planning and preparation of this book. The content, recommendations, roadmaps and vision expressed in this book are those of the editors and contributors respectively, and do not necessarily represent those of the European Commission.

Martin Serrano
Nikolaos Isaris
Hans Schaffers
John Domingue
Michael Boniface
Thanasis Korakis

Editors Biography

Martin Serrano is Principal Investigator, Data Scientist and Research Director at the Internet of Things and Stream Processing Unit (UIoT) within the Insight Centre for Data Analytics Galway at the National University of Ireland. He has more than 15 years experience in industry and applied research within a wide range of successful EU (FP5-FP7/H2020) collaborative software projects, Irish National Projects (HEA PRTLI, SFI) and also Enterprise Ireland (EI) innovation projects. He is a recognized IoT expert on systems interoperability, big data management and End-to-End solutions architect with a strong background on applied semantic, services and network management.

Nikos Isaris is Deputy Head of the "Internet of Things (IoT)" Unit within the European Commission's Directorate-General for Communications Networks, Content and Technology (DG CONNECT). Before that, Nikos was Deputy Head in the Unit responsible for Future Internet Research and Experimentation in the same Directorate. He joined DG CONNECT in 2013 coming from the Directorate-General for Home Affairs, where he was Deputy Head in the Unit

"Large-scale IT systems and Biometrics" dealing with IT systems for the Schengen area. Before Nikos worked in the private banking and electronic banking sectors in his home country Greece.

Hans Schaffers is a professor on Digital Business Innovation at the research group Ambient Intelligence of Saxion's School of Creative Technology. Hans Schaffers has been the coordinator of various research and innovation projects in the area of context awareness, living labs and regional innovation, smart cities, and (currently) Future Internet. His main interest is to develop a multi-disciplinary, user-oriented and responsible approach for the development of intelligent applications in an increasingly connected world, with a focus on technical, human and social issues.

John Domingue is professor at and Director of the Knowledge Media Institute at the Open University in Milton Keynes, and researcher in the Semantic Web, Linked Data, Services and Education. Domingue started his academic career late 1980s as researcher at the Human Cognition Research Laboratory of the Open University. In 2008 he was appointed Professor at the Open University in Milton Keynes. Domingue was the Scientific Director of SOA4All and has worked in dozens of other research projects. He is chair of the Steering Committee for the European Semantic Web Conference Series and member of the Future Internet Symposium Series steering committee. He coordinates the Future Internet Service Offer Working Group within the Future Internet Assembly, and is on the editorial board of the Journal of Web Semantics.

Professor Michael Boniface is Technical Director of the IT Innovation Centre. He joined IT Innovation in 2000 after several years at Nortel Networks developing infrastructure to support telecommunications interoperability. His roles at IT Innovation include technical strategy of RTD across IT Innovation's project portfolio, technical leadership, and business development. He has over 16 years' experience of RTD into innovative distributed systems for science and industry using technologies such as new media, secure service-oriented architectures, semantic web, and software defined infrastructures.

Thanasis Korakis is currently a Research Assistant Professor in the Electrical and Computer Engineering Department of Polytechnic Institute of NYU. He is also affiliated with CATT and WICAT at Polytechnic. His research interests are in the field of wireless networks with emphasis on access layer protocols, cooperative networks, directional antennas, quality of service provisioning and network management. He leads the Wireless Implementation Testbed Laboratory (Witest Lab) of the Department of Electrical and Computer Engineering at Polytechnic Institute of NYU.

Foreword

2016 FIRE Book

Research and Experimentation: Past, Present and Future

The design and building on how the future Internet will looks like by 2020-2025 does not only imply technologist and scientific communities, but the society in general, it is a multidisciplinary activity where all the professionals from science, technology, sociology and arts have participation. Mainly because by doing it in this way there will be more opportunities that the future Internet will help for promoting more users engagement, addressing important societal challenges and facilitate companies in finding solutions and activate business markets.

Building the Future Internet is an important activity that will help to incentive the growth of ecosystems and in this line the research and development of experimental platforms will strongly benefit this activity. Particulalrly the future Internet research and experimentation activity looks at the scalability aspects of the technology and applications, best practices for large-scale deployments, infrastructure and facilities orchestration, pilots and testbeds federation alike empirical results.

At the forefront of the future technology and applications for the future Internet, user-driven experimentation and co-creation models are now driving the evolution not only of the device technology, Internet virtual infrastructures and middleware platforms but the Internet of End-to-End applications. Research and Experimentation must promote the growth of ecosystems that are supported by the research and development of experimentation platforms that promote users engagement and facilitate companies in finding solutions, activate business markets, and address important societal challenges.

The 2016 FIRE book focuses on the role, evolution and importance of research and experimentation on the Internet of Future based on testbed facilities and experimentally-driven research. The book presents results of on going and selected past flagship Future internet Research and Experimentation

(FIRE) projects, and addresses developments both in Europe as well as international collaboration. Among the themes to be addressed in the book chapters are the following:

- Results and impact of FIRE facility and experimentally driven research projects.
- The role of "experimentation" and "experimentally-driven research" in bringing advances in the Future Internet.
- Evolution of experimental facilities since 2007. From facilities oriented towards networking research and datasets towards addressing new application domains (e.g. media, education, etc).
- Experimental facilities and experimental research on the future Internet addressing real-life environments or ecosystems, including humans (e.g. in smart cities).
- The role of federation and interconnection between testbed facilities to accommodate a European-wide testbed infrastructure.
- Developments in experimentally driven research and innovation, making use of the testbed facilities and capabilities for industrial problems.
- International collaboration within FIRE (GENI in the US, Japan, Brazil, South Africa and other initiatives).
- Impact of FIRE on business and societal innovations. Importance of FIRE for SMEs research and innovation.
- Sustainability and business models of Future Internet facilities and facilities covered by other actors and initiatives.
- FIRE outlook and vision 2020, including application domains, collaboration with other initiatives and technology domains (IoT, smart Cities, 5G and other).

European experimental facilities are facing up the challenge to evolve towards a dynamic, sustainable and large-scale not only European but world-wide infrastructure, connecting and federating existing and next generation testbeds for emerging technologies. During several years one of the most representative initiative named Future Internet Research and Experimentation (FIRE), have focused on offering wide-scale testing and experimentation resources demanded by competitive research organisations, industry and SMEs alike to speed up the time-to-market for innovative technologies, services and solutions. As all in life FIRE, gradually re-defined its original focus on advanced networking technologies and service paradigms expanding towards new emerging areas of technological innovation such as Internet of Things, and to application domains and user environments such as for networked media and smart cities. This evolution raised the issue on how European experimental facilities could further evolve as core resources of an innovation ecosystem and act as accelerator platform for future Internet research, experimentation and innovation.

The research and experimentation landscape shaping the Future Internet is undergoing into a major transformation. Service and application developers (including SMEs) make use of advanced networking, communication and software concepts. Smart City initiatives and technology-intensive domains such as healthcare, manufacturing, e-government and financial services present new challenges to such developers. European-wide initiatives have also emerged where FIRE's experimental facilities may bring value added such as advanced networking (5G PPP), Big Data, Internet of Things and Cyber-Physical Systems. Traditional boundaries between facility developers, researchers and experimenters, and end users in vertical application domains start blurring, giving rise to experimentation and innovation-based platform ecosystems which bring together a wide range of stakeholders to collaborate on innovation opportunities driven by Future Internet technologies. Correspondingly the demands of experimenters and researchers serving those users and developers are changing, pushing for the development of new types of experimental facilities and experimentation methods and tools.

In this context the AmpliFIRE project [2], has provided a future vision concerning the potential of experimental testbed facilities and experimentally driven research for the coming decade. In this vision, FIRE's federated facilities fulfil a key role within the currently evolving innovation ecosystem for the Future Internet.

2016 has been a year with multiple changes and the Future Internet Research and Experimentation (FIRE) initiative at the DG Connect experimental facilities unit in the European Commission has not been the exception. Nevertheless the FIRE initiative with duration now for almost 10 years has been considered a key initiative within the Horizon 2020 program. Since the introduction of the FIRE initiative in the EC ICT Work programme FP7 Objective "The Network of the Future" back in 2007–2008[1], and along the subsequent programs the FIRE initiative continues being a critical pillar on the design of the Internet network and services.

The Future Internet Research and Experimentation Initiative addresses the evolutionary expectations that are being put upon the current Internet, by focusing on providing a research environment for investigating and experimentally validating highly innovative and revolutionary ideas with a vision on user acceptance and industrial market impact. In December 2015 at the 3rd edition of the FIRE Forum it was discussed the objective to create awareness about the evolution towards supporting a third dimension in FIRE and define the possible roadmap towards the innovation, by means of more open services interactions, interoperability, secure methods and mobile integrated services, user acceptance and validation while increasing Internet network capacity and preserving quality of networking services across FIRE facilities.

The other two related dimensions of FIRE[1] are: on one hand, "promotion of experimentally-driven long-term, visionary research on new paradigms and networking concepts and architectures for the future Internet" (Experimentation); and on the other hand, "building a large-scale experimentation facility supporting both medium- and long-term research on networks and services by gradually federating existing and new testbeds for emerging or future Internet technologies" including the emergent technologies, new paradigms and methodologies, to cope with the networks, services and applications demands in today's more integrated Internet of everything, virtualized networks and open information systems.

FIRE is now in a continuous evolution of the testbeds and facilities ecosystem, towards the achievement of the Horizon 2020 vision and beyond into the next Framework Programme and comprises the latest generation of FIRE resources and projects, which started with the H2020-ICT-2014 Call.

FIRE in its evolution is addressing the emergent technologies, paradigms and methodologies, to cope with networks, services and applications users demands & validation in today's more integrated experimentation as a service experience. Experimentation drives the evolution not only of the device technology, infrastructure and platforms but the Internet of End to End applications. Experimentation must promote the growth of ecosystems that are supported by the research and development of experimentation platforms that promote users engagement and facilitate companies in finding solutions, activate business markets, and address important societal challenges.

The 2016 FIRE Book Editors' team acknowledge the great efforts and contributions from the FIRE community and is proud and happy to bring to you this book.

[1]Future Internet Research and Experimentation: The FIRE Initiative DOI 10.1145/1273445.1273460

2016 FIRE Book Editors' team

List of Figures

List of Tables

PART I

The Next Generation Internet with FIRE

1

European Challenges for Experimental Facilities

Hans Schaffers[1], Thanasis Korakis[2], Congduc Pham[3], Abdur Rahim[4], Antonio Jara[5] and Martin Serrano[6]

[1]Saxion University of Applied Sciences, Netherlands
[2]University of Thessaly, Greece
[3]University of Pau, France
[4]CREATE-NET, Italy
[5]HES-SO, Switzerland
[6]Insight Centre for Data Analytics Galway, Ireland

1.1 Evolution of Experimentation Facilities into Open Innovation Ecosystems for the Future Internet

There have been considerable changes in FIRE as a consequence of the evolving vision and the needs and interests of the industrial and scientific communities. Originally established from a core of networking testbeds and aimed at investigating fundamental issues of networking infrastructure, FIRE's mission has changed to deliver widely reusable facilities for the Future Internet community, resulting in the current emphasis on federation. Figure 1.1 provides an overview of representative testbeds that forms the European federated ecosystem.

New domains are coalescing within Future Networks, such as the Internet of Things, Internet of Services, Cyber-Physical Systems, Big Data and other areas, giving rise to new research and innovation challenges and demands to experimentation facilities. Interactions with communities such as Smart Cities, Cloud computing and Internet of Things already brought new perspectives into FIRE's portfolio. To some extent this is visible in the new Work Programme 2016–2017, in particular in relation to Internet of Things, where FIRE testbeds are considered to support technology validation before deployment

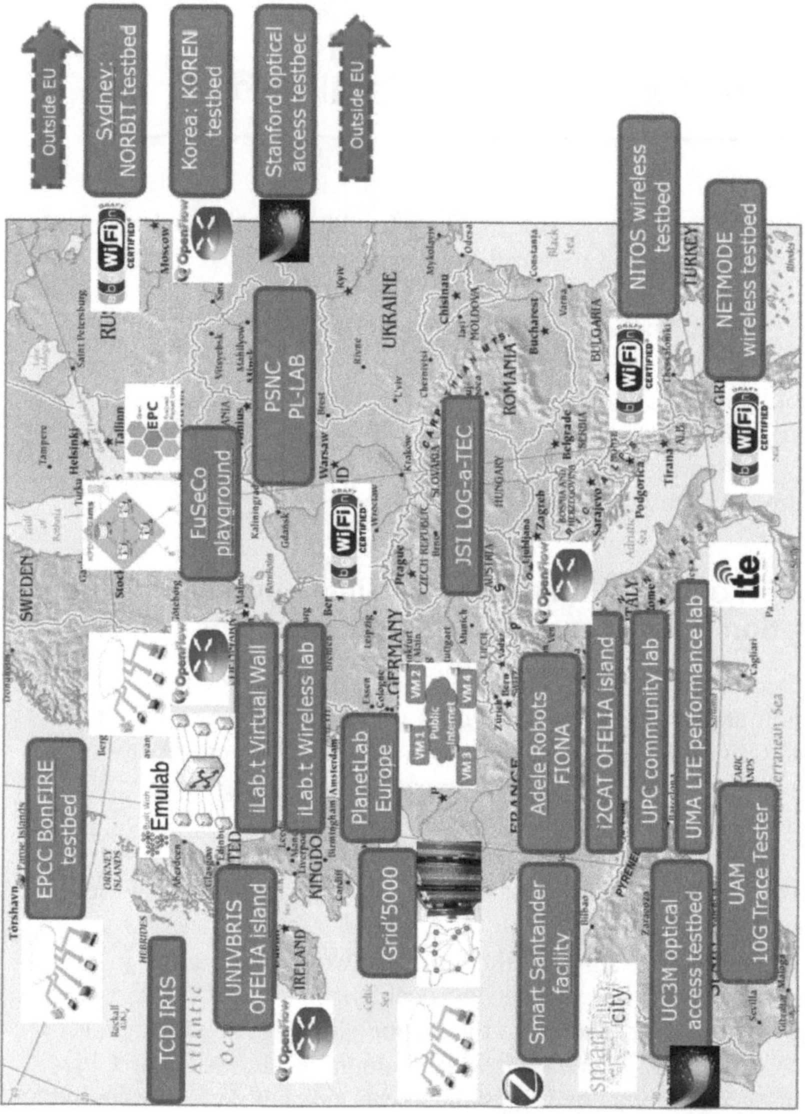

Figure 1.1 European federated testbeds ecosystem.

Source: FED4FIRE Project.

in field trials. AmpliFIRE identifies several key trends, such as the integration of a broad range of systems (cloud services, wireless sensor networks, content platforms, and mobile users) within Future Internet systems in large-scale, highly heterogeneous systems, to support increasingly connected and networked applications. This new emphasis calls for looser forms of federation of cross domain resources.

Whereas FIRE has become meaningful in the context of the Future Internet and its research community, FIRE also increasingly addresses the demand side of experimentation, the need to engage users and to support innovation processes. This way FIRE's evolution must find a balance between coherence and fragmentation in shaping the relation between facility building projects and research and experimentation – and increasingly innovation – projects. In this respect a specific development is how FIRE is increasingly shaped by new, flexible demand-oriented instruments such as Open Calls and Open Access, which demonstrates how customer "pull" is increasingly supplementing and balancing technology "push."

As experimenter needs and requirements are becoming more demanding, expectations are rising as regards how FIRE should anticipate the needs and requirements from SMEs, industry, Smart Cities, and from other initiatives in the scope of Future Internet such as Internet of Things and 5G. New types of service concepts for example *Experimentation-as-a-Service* aim at making experimentation more simple, efficient, reliable, repeatable and easier to use. These new concepts affect the methods and tools, the channels for offering services to new categories of users, and the collaborations to be established with infrastructure and service partners to deliver the services.

Thus it is expected that experimentation will increasingly be shaped by demand-pull factors in the period 2015–2020. These user demands will be based on four main trends:

- The Internet of Things: a global, connected network of product tags, sensors, actuators, and mobile devices that interact to form complex pervasive systems that autonomously pursue shared goals without direct user input. A typical application of this trend is automated retail stock control systems.
- The Internet of Services: internet/scaled service-oriented computing, such as cloud software (Software as a Service) or platforms (Platform as a Service).
- The Internet of Information: sharing all types of media, data and content across the Internet in ever increasing amounts and combining data to generate new content.

- The Internet of People: people to people networking, where users will become the centre of Internet technology—indeed the boundaries between systems and users will become increasingly blurred.

In order to contribute to these four fast moving areas, the FIRE ecosystem must grow in its technical capabilities. New networking protocols must be introduced and managed, both at the physical layer where every higher wireless bandwidth technologies are being offered, and in the software interfaces, which SDN (Software defined Networks) is opening up. Handling data at medium (giga to tera) to large (petabyte) scale is becoming a critical part of the applications that impact people's lives. Mining such data, combining information from separated archives, filtering and transmitting efficiently are key steps in modern applications, and the Internet testbeds of this decade will be used to develop and explore these tools.

Future Internet systems will integrate a broad range of systems such as cloud services, sensor networks and content platforms into large-scale heterogeneous systems-of-systems. There is a growing need for integration, for example integration of multi-purpose multi-application wireless sensor networks with large-scale data-processing, analysis, modelling and visualisation along with the integration of next generation human-computer interaction methods. This will lead to complex large-scale networked systems that integrate the four pillars: things, people, content and services. Common research themes include scalability solutions, interoperability, new software and service engineering methods, optimisation, energy-awareness and security, privacy and trust solutions. To validate the research themes, federated experimented facilities are required that are large-scale and highly heterogeneous. Testbeds that bridge the gap between infrastructure, applications and users and allow exploring the potential of large-scale systems which are built upon advanced networks, with real users and in realistic environments will be of considerable value. This will also require the development of new methodological perspectives for experimentation facilities, including how to experiment and innovate in a framework of collaboration among researchers, developers and users in real-life environments.

As we emphasize a focus on "complex smart systems of networked infrastructures and applications" within the experimentation, the unique and most valuable contribution of experimental facilities should be to "bridge" and "accelerate": create the testing, experimenting and innovation environment which enables linking networking research to business and societal impact. Testbeds and experiments are tools to address research and innovation in "complex smart systems", in different environments such as cities, manufacturing industry and data-intensive services sectors. In this way, experimentation widens its primary focus from testing and experimenting, building the facilities, tools and environments towards closing the gap from experiment to innovation for users and markets.

1.2 Support, Continuity and Sustainability: The NITOS Testbed Example

1.2.1 NITOS Future Internet Facility Overview

University of Thessaly operates NITOS Future Internet Facility [http://nitlab.inf.uth.gr/NITlab/index.php/nitos.html], which is an integrated facility with heterogeneous testbeds that focuses on supporting experimentation-based research in the area of wired and wireless networks. NITOS is remotely accessible and open to the research community 24/7. It has been used from hundreds of experimenters all over the world.

The main experimental components of NITOS are:

- A **wireless experimentation testbed**, which consists of 100 powerful nodes (some of them mobile), that feature multiple wireless interfaces and allow for experimentation with heterogeneous (Wi-Fi, WiMAX, LTE, Bluetooth) wireless technologies.
- A **Cloud infrastructure**, which consists of 7 HP blade servers and 2 rack-mounted ones providing 272 CPU cores, 800 Gb of Ram and 22 TB of storage capacity, in total. The network connectivity is established via the usage of an HP 5400 series modular Openflow switch, which provides 10 Gb Ethernet connectivity amongst the cluster's modules and 1 Gb amongst the cluster and GEANT.
- A **wireless sensor network testbed**, consisting of a controllable testbed deployed in UTH's offices, a city-scale sensor network deployed in Volos city and a city-scale mobile sensing infrastructure that relies on bicycles of volunteer users. All sensor platforms are custom, developed by UTH, supporting Arduino firmware and exploiting several wireless technologies for communication (ZigBee, Wi-Fi, LTE, Bluetooth, IR).
- A **Software Defined Radio (SDR)** testbed that consists of Universal Software Radio Peripheral (USRP) devices attached to the NITOS wireless nodes. USRPs allow the researcher to program a number of physical layer features (e.g. modulation), thereby enabling dedicated PHY layer or cross-layer research.

- A **Software Defined Networking (SDN)** testbed that consists of multiple OpenFlow technology enabled switches, connected to the NITOS nodes, thus enabling experimentation with switching and routing networking protocols. Experimentation using the OpenFlow technology can be combined with the wireless networking one, hence enabling the construction of more heterogeneous experimental scenarios (Figure 1.2).

The testbed is based on open-source software that allows the design and implementation of new algorithms, enabling new functionalities on the existing hardware. The control and management of the testbed is done using the cOntrol and Management Framework (OMF) open-source software. NITOS supports evaluation of protocols and applications under real world settings and is also designed to achieve reproducibility of experimentation.

1.2.2 NITOS Evolution and Growth

The NITOS Future Internet facility has been developed and constantly expanded through the participation in several EU-funded FIRE projects. During these projects, the testbed has been enhanced with diverse hardware and software components, aiming to provide cutting-edge experimentation services to the research community, in an open-access scheme and remotely accessible, as well as augmented with user friendly orchestration of experiments. Below, we provide a brief overview of the key projects that assisted in the NITOS development.

OneLab2 (https://onelab.eu/) started in 2008, was the FIRE project that laid the foundations of the NITOS experimental facility. OneLab2

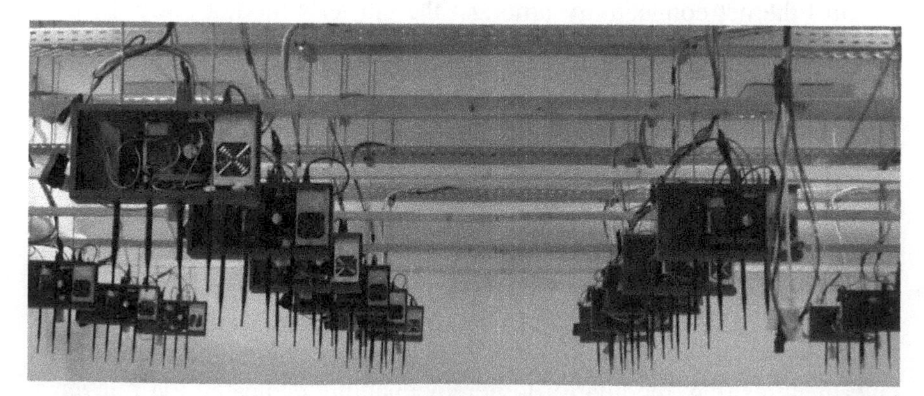

Figure 1.2 The NITOS Indoor deployment.

has developed one of the first pan-European experimental facilities, offering experimentation services involving both wired and wireless resources. During the project, the first tools for provisioning testbeds and conducting experiments were realized. Through OneLab2, NITOS was initially developed, operating with a small number of nodes, offering experimentation services involving open source WiFi networks and adopting the state-of-the-art OMF framework.

Following OneLab2, **OpenLab** (http://www.ict-openlab.eu) was one of the first projects to address testbed federation for both the control and experimental plane. By control we mean the way that the testbed resources are represented, reserved, provisioned and accessed, whereas by experimental we refer to conducting experiments over the testbeds. During OpenLab, NITOS testbed was extended with a large number of nodes and first steps towards federation were taken. In addition a WiMAX macroscale base station was installed, along with the respective end-clients, and a commercial LTE network was provisioned. Tools for enabling experimentation with a plethora of different components were implementing, by extending the OMF framework to support Wi-Fi, Wired, WiMAX and Software Defined Radio (SDR) components.

In **FIBRE** (http://www.fibre-ict.eu/) project, the first results of federation in Europe were extended in order to also cover Brazil. Moreover, focus was placed on Software Defined Networking (SDN), and its integration in the existing testbeds. Through FIBRE, NITOS was extended with OpenFlow enabled switches, and the extensions in the respective control and management tools for supporting them. In FIBRE, NITOS was one of the key European facilities, and following its paradigm, NITOS-like testbeds were installed at six different brazilian sites.

CONTENT was a project that investigated the integration and convergence of wireless resources, along with SDN-enabled wired and optical networks. During the project, NITOS was the key testbed where all the developments took place, and was extended with advanced frameworks for the configuration and management of the wireless resources. Aspects such as end-to-end network slicing, including both optical and wireless resources were examined, as well as network virtualization of the LTE and WiFi resources of the testbed.

NITOS is also one of the core wireless testbeds participating in the **Fed4FIRE** (http://www.fed4fire.eu/) project. NITOS has been developing for the project software dealing with the control plane federation of the testbeds (NITOS Broker), easing and unifying the federation of any NITOS-like testbed in Fed4FIRE.

In **CREW** (http://www.crew-project.eu/) NITOS testbed was extended with USRP devices for Software Defined Radio related research, whereas energy monitoring devices, with very high resolution were developed and installed at the testbed. These devices are able to measure the energy spent in the wireless transmissions in even a per packet basis, thus rendering them a valuable tool for energy minimization experimentally driven research.

In **SmartFIRE** (http://eukorea-fire.eu/) federation with South Korea was addressed. The project was coordinated by the NITOS team, and developed all the extensions in the testbed control and management frameworks that ease the federation of Korean testbed sites. The testbed was further expanded in terms of equipment, increasing the SDN capabilities and experiments that can be conducted.

Through the participation in **XIFI** (https://fi-xifi.eu/), NITOS was extended significantly with the integration of Cloud infrastructure in the testbed. The Cloud system is interconnected with the experimental resources of the testbed, thus enabling meaningful experiments including multiple technologies using Cloud processing and storage capabilities. Although the tools managing the Cloud infrastructure differed from the ones developed through FIRE projects, the NITOS team developed the appropriate drivers for their intercommunication.

Finally through **FLEX** (http://flex-project.eu) project, the testbed has been extended with commercial and open-source LTE infrastructure. NITOS team is coordinating the project, and is leading the development in all the control and management software for the LTE testbed components, as well as the uncontrolled and emulated mobility toolkits that are offered to experimenters.

After the completion of the aforementioned projects, NITOS has evolved into a truly heterogeneous Future Internet Facility providing a strong set of tools and hardware for experimental research. The tools that NITOS is offering are going beyond the existing 4G research and towards 5G, as the testbed is highly modular and can be tailored for supporting a very diverse set of experiments.

1.2.3 Facilitating User's Experience

The expertise of NITOS team on supporting experimenters, gained from the long experience on maintaining and managing the NITOS facility from 2008, led to the design and development of various tools and frameworks aiming at proactively assisting them and addressing possible issues before they arise.

Examples of such tools that have been designed, developed and extended in the context of the aforementioned EU-funded projects are the NITOS Portal (http://nitos.inf.uth.gr), the NITOS Documentation portal (http://nitlab.inf.uth.gr/doc/) and the NITOS Broker, which all targeted in operating, controling, managing and federating the facility to the most possible unobstructed way.

NITOS Portal

The NITOS Portal is the entry point for experimentation in NITOS Facility providing a wide range of web-based tools for discovering, reserving, controlling and monitoring testbed resources, including but not limited to the Scheduler, the Node Status tool, the Testbed Status tool, the Distance tool and the Spectrum Monitoring tool (Figure 1.3).

The **Scheduler** is a web-based tool that allows experimenters to discover and reserve resources from the testbed in order to conduct their experiments. Through this tool, experimenters are able to observe nodes' characteristics, filter them and finally reserve them based on their availability on time. They are also able to observe their current or future reservations in NITOS, in order to edit or cancel them. The **Node Status tool** allows a user to monitor and control the status (turn on/off and reset) of his/her reserved nodes and the **Distance tool** allows him/her to find out the physical distance between the nodes of the testbed. Finally the **Testbed Status tool** reports the functional state of each node of the three NITOS deployments together with their characteristics.

NITOS Documentation

NITOS provides a wide variety of use cases and tutorials online, on the Documentation portal of NITOS facility (http://nitlab.inf.uth.gr/doc). There is a basic tutorial with simple but detailed enough documentation, in order for every novice user to easily manage and configure NITOS resources and setup an experiment. In addition, for each of the specific testbeds that NITOS provides, for example the WiMAX or the LTE testbeds, there is a separate tutorial which guides users to the whole experimentation procedure. From the reservation of the proper resources to the configuration of them and the execution of the experiment. Finally, video tutorials can be found in the official YouTube channel of NITlab (https://www.youtube.com/channel/UCPfbZTgTk5gapcJbF85DI-w) for facilitating users during the experimentation process.

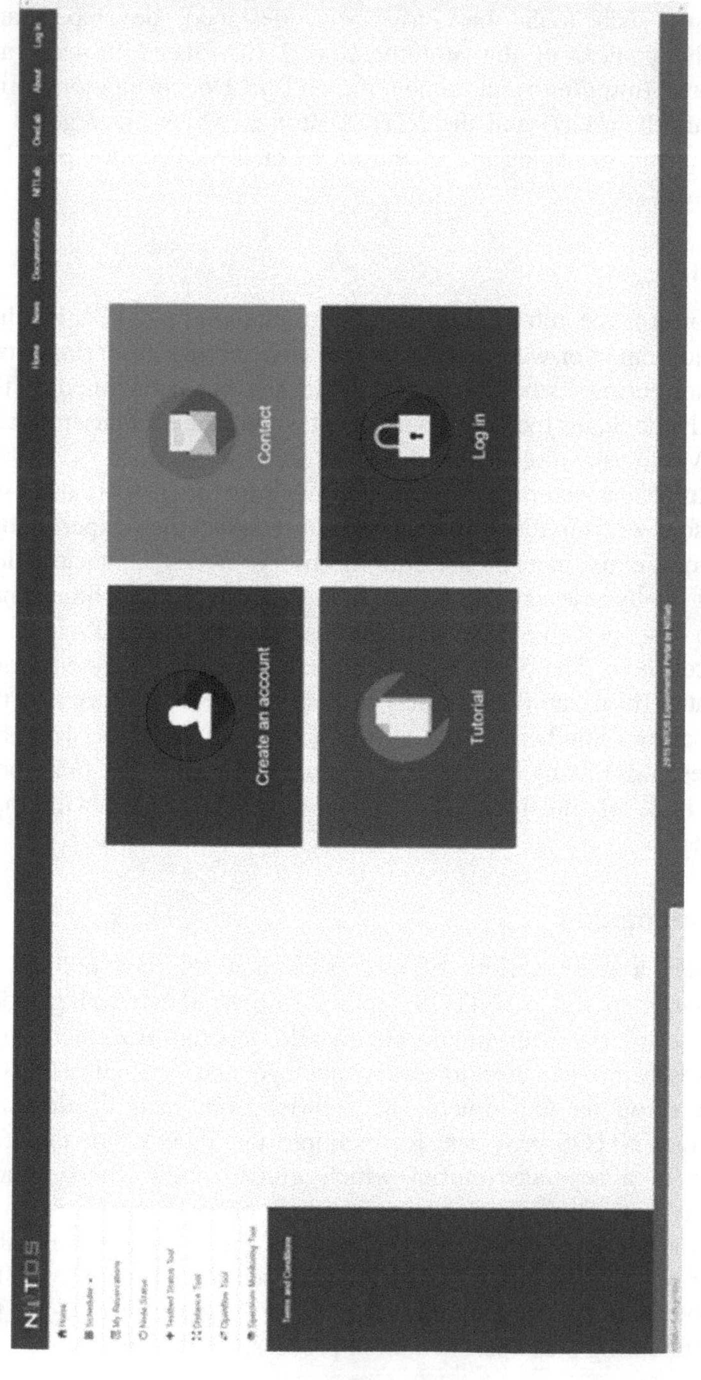

Figure 1.3 The NITOS testbed portal.

OMF Extensions

As mentioned before, the integration of a hardware extension in NITOS was constantly followed by the integration of this hardware to the control and management framework, namely the OMF [http://mytestbed.net/]. This way, all the heterogeneous hardware components were controllable through a single OMF script, enabling NITOS to effortlessly control every component, as well as combine diverse resources and design advanced experiment topologies.

In addition, the trend for the federation of experimental facilities in recent years, led to the design and implementation of the Broker entity [1] which is an OMF component responsible for controlling, managing and exposing properly the testbed's resources. It features all the necessary intertaces (XML-RPC, REST, FRCP [2], XMPP) for the federation of an OMF testbed with other heterogeneous facilities under the scope of SFA [3].

1.2.4 Exploitation of NITOS and Users Statistics

The NITOS facility attracts a large amount of research experimenters from all over the world, with a significant part coming from Industry and SMEs. More particular:

- Approximately 25% of the NITOS usage comes from Industry/SME.
- Approximately 75% of the NITOS usage comes from research institutions.

The distribution of the visitors based on their country is indicated in the following Figure 1.4:

Around 55% of the users are from EU countries, namely France, UK, Spain, Germany, Belgium, Italy and Greece, while 20% of them come from countries like US, Brazil, Australia, India, China, South Korea and Canada. Currently, NITOS counts around 500 subscribed experimenters who use the testbed in a daily basis.

Federation

The number of the NITOS users and the reservations for resources experienced significant increase upon the addition of the testbed in several federations, like OneLab [https://onelab.eu/] or the Fed4FIRE [http://www.fed4fire.eu/]. Currently NITOS is federated with facilities all over the world, including all the major EU facilities and testbeds in Brazil, South Korea and USA, providing heterogeneous resources to its users. This way, experimenters are able to form large-scale topologies including diverse resources, spanning from wireless nodes to OpenFlow switches, mobile robots, sensors and 4G equipment.

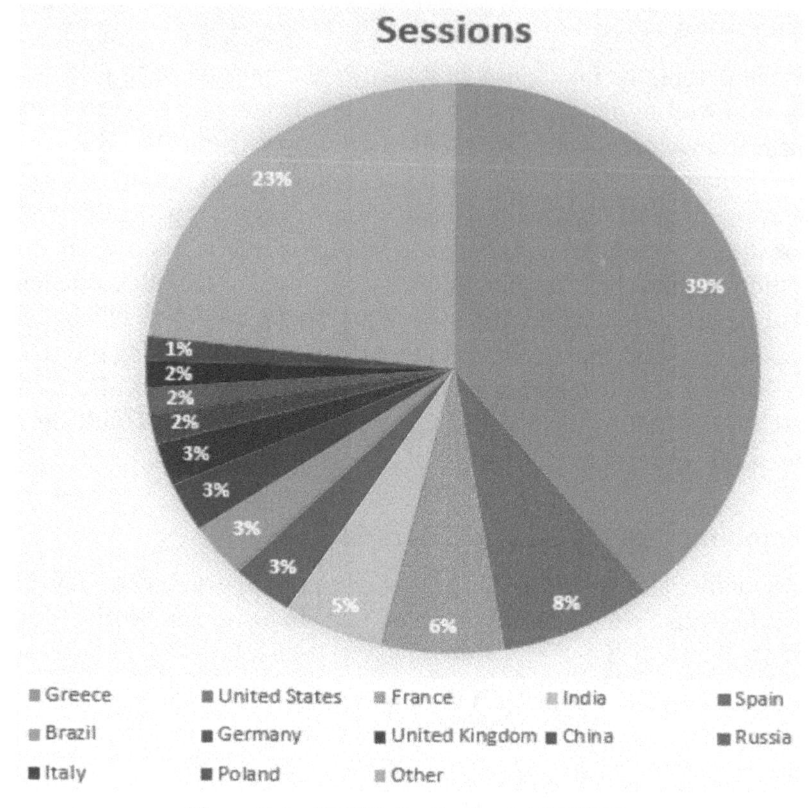

Figure 1.4 NITOS distribution in EUROPE.

Education

NITOS is deployed in Volos, Greece and specifically in University of Thessaly, thus it has very strong bonds with the University's community. During each semester, at least one course of the University is using NITOS. Students are conducting experiments using real resources, which enhance their overall knowledge on state-of-the-art wireless and wired network technologies and enables them to study and identify practical problems and solutions. In addition, NITOS is being frequently used in semester courses of the NYU Polytechnic School of Engineering.

Moreover towards the familiarization of the students with the testbed, Students Labs and "NITOS days" are often organized in the context of courses. These courses introduce NITOS portal and NITOS testbed to the participants, as well as other EU facilities and federations like OneLab, encouraging them to create accounts and use them for experimentation. Finally, there is a variety of master thesis and PhD dissertations that take advantage of the testbed, publish experimental results and disseminate experimentation-driven research.

1.2.5 References

[1] D. Stavropoulos, A. Dadoukis, T. Rakotoarivelo, M. Ott, T. Korakis and L. Tassiulas, "Design, Architecture and Implementation of a Resource Discovery, Reservation and Provisioning Framework for Testbeds", to be presented in WINMEE, Bombay 2015, India, May 25, 2015.
[2] W. Vandenberghe, B. Vermeulen, P. Demeester, A. Willner, S. Papavassiliou, A. Gavras, M. Sioutis et al. "Architecture for the heterogeneous federation of future internet experimentation facilities." In Future Network and Mobile Summit (*FutureNetworkSummit*), 2013, pp. 1–11. IEEE, 2013.
[3] Peterson, L., R. Ricci, A. Falk, and J. Chase. "Slice-based federation architecture (SFA)" Working draft, version 2 (2010).

1.3 Experimentation: Vision and Roadmap

In Europe there are several initiatives that seek into the Future for establishing an ecosystem for Experimentation and Innovation. FIRE (Future Internet Research and Experimentation) seeks a synergetic and value adding relationships with infrastructures and stakeholders. GÉANT/NRENs and the FI-PPP initiatives related to Internet of Things and Smart Cities seek for the interactions with large deployments and big number of users. EIT Digital, the new 5G-PPP and Big Data PPP initiatives and the evolving area of Cyber-Physical Systems aims for defining ecosystems for large deployments. For the future, it is foreseen a layered Future Internet infrastructural and service provision model, where a diversity of actors gather together and ensure interoperability for their resources and services such as provision of connectivity, access to testbed and experimentation facilities, offering of research and experimentation services, business support services and more. Bottom-up experimentation resources are part of this, such as crowd sourced or citizen/community-provided resources. Each layer is transparent and offers interoperability. Research networks (NRENs) and GÉANT are providing the backbone networks and connectivity to be used by FIRE facilities and facilities offered by other providers.

European testbeds ecosystem core objective is to provide and maintain sustainable, common facilities for Future Internet research and experimentation, and to provide customized experimentation and research services. In addition, given the relevance of experimentation resources for innovation, and given the potential value and synergies that experimentation facilities offers to other initiatives, testbeds assume a role in supporting experimentally-driven research and innovation of technological systems. For this to become reality FIRE and other initiatives related to the Future Internet, such as 5G, should ensure sharing and reusing experimentation resources. FIRE should also consider opening up to (other) public and private networks, providing customized facilities and services to a wide range of users and initiatives in both public and private spheres. Specifically FIRE's core activity and longer term orientation requires the ability to modernize and innovate the experimental infrastructure and service orientation for today's and tomorrow's innovation demands. Really innovative contributions may come from smaller, more aggressive and riskier projects. Large-scale EC initiatives such as the 5G PPP, Big Data PPP and regarding the Internet of Things should have an influence on their selection and justification. Early engagement and dialogue among concerned communities is essential to accomplish this goal.

1.3.1 Envisioning Evolution of Experimentation Facilities into the Future

For setting out a transition path from the current FIRE facilities towards FIRE's role within a "Future Internet Ecosystem", four alternatives for future development patterns which equally represents the spectrum of forces acting upon FIRE's evolution have been defined:

- **Competitive Testbed as a Service**: set of individually competing testbeds offering their facilities as a pay-per-use service.
- **Industrial cooperative**: become a resource where experimental infrastructures (testbeds) and Future Internet services are offered by co-operating commercial and non-commercial stakeholders.
- **Social Innovation ecosystem**: A collection of heterogeneous, dynamic and flexible resources offering a broad range of facilities e.g. service-based infrastructures, network infrastructure, Smart City testbeds, support to user centred living labs, and other.
- **Resource sharing collaboration**: federated infrastructures provide the next generation of testbeds, integrating different types of infrastructures within a common architecture.

These future scenarios aim at stretching our thinking about how experimentation must choose its operating points and desired evolution in relation to such forces. Simplifying the argument, Experimentation evolution proceeds along two dimensions.

One dimension ranges from a coherent, integrated portfolio of activities on the one side to individual independent projects (the traditional situation), selected solely for their scientific and engineering excellence, on the other. A second dimension reflects both the scale of funded projects and the size of the customer or end-user set that future projects will reach out to and be visible to, ranging from single entities to community initiatives.

Some particular lines of FIRE's future evolution can be sketched as follows in Figure 1.5. **In the short term,** FIRE's mission and unique value is to offer an efficient and effective federated platform of facilities as a common research and experimentation infrastructure related to the Future Internet that delivers innovative and customized experimentation capabilities and services not achievable in the commercial market. FIRE should expand its facility offers to a wider spectrum of technological innovations in EC programmes e.g. in relation to smart cyber-physical systems, smart networks and Internet architectures, advanced cloud infrastructure and services, 5G network infrastructure for the Future Internet, Internet of Things and platforms for connected smart objects. In this role, FIRE delivers experimental testing facilities and services at low cost, based upon federation, expertise and tool sharing, and offers all necessary expertise and services for experimentation on the Future Internet part of Horizon 2020 (Figure 1.5).

For the **medium term**, around 2018, FIRE's mission and added value is to support the Future Internet ecosystem in building, expanding and continuously innovating the testing and experimenting facilities and tools for Future Internet technological innovation. FIRE continuously includes novel cutting-edge facilities into this federation to expand its service portfolio targeting a range of customer needs in areas of technological innovation based on the Future Internet. FIRE assumes a key role in offering facilities and services for 5G. In addition FIRE deepens its role in experimentally-driven research and innovation for smart cyber-physical systems, cloud-based systems, and Big Data. This way FIRE could also support technological innovation in key sectors such as smart manufacturing and Smart Cities. FIRE will also include "opportunistic" experimentation resources, e.g. crowd sourced or citizen- or community-provided resources.

In this time frame, FIRE establishes cutting-edge networked media and possibly Big Data facilities relevant to research and technology demands

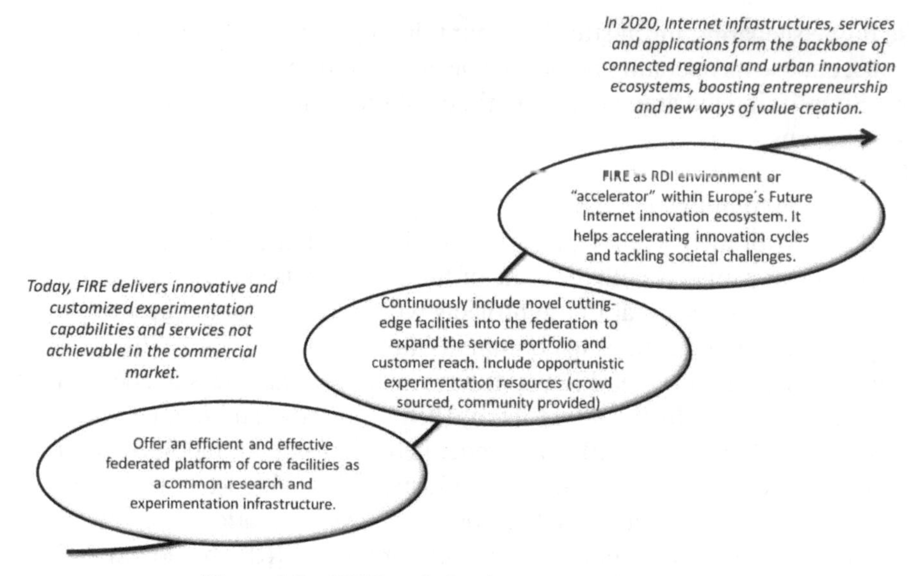

Figure 1.5 FIRE evolution longer term vision 2020.

to support industry and support the solving of societal challenges. Federation activities to support the operation of cross-facility experimentation are continued. A follow-up activity of Fed4FIRE is needed which also facilitates coordinated open calls for cross-FIRE experimentation using multiple testbeds. Additionally, a broker service is provided to attract new experimenters and support SMEs. This period ensures that openly accessible FIRE federations are aligned with 5G architectures that simplify cross-domain experimentation. Second, via the increased amount of resources dedicated to Open Calls, FIRE will create an Accelerator functionality to support product and service innovation of start-ups and SMEs. For this, FIRE will establish cooperation models with regional players and other initiatives. FIRE continues to implement professional practices and establishes a legal entity which can engage in contracts with other players and supports pay per use usage of testbeds.

For the **longer term**, by 2020, our expectation is that Internet infrastructures, services and applications form the backbone of connected regional and urban innovation ecosystems. People, SMEs and organisations collaborate seamlessly across borders to experiment on novel technologies, services and business models to boost entrepreneurship and new ways of value creation. In this context, FIRE's mission is to become the research, development and innovation environment, or "accelerator", within Europe's Future Internet innovation ecosystem, providing the facilities for research, early testing and experimentation for technological innovation based on the Future Internet. FIRE in cooperation with other initiatives drives research and innovation cycles for advanced Internet technologies that enable business and societal innovations and the creation of new business helping entrepreneurs to take novel ideas closer to market.

In this timeframe it is envisaged that FIRE continues to add new resources that match advanced experimenter demands (5G, large-scale data oriented testbeds, large-scale Internet of Things testbeds, cyber-physical systems) and offers services based on Experimentation as a service. The services evolve towards experiment-driven innovation. More and more FIRE focuses on the application domain of innovative large-scale smart systems. Implementing secure and trustworthy services becomes a key priority, also to attract industrial users. Responsive SME-tailored open calls are implemented, to attract SMEs. FIRE continues the accelerator activity by providing dedicated start-up accelerator funding. FIRE also takes new steps towards (partial) sustainability by experimenting with new funding models. Sustainable facilities are supported with continued minimum funding after project lifetime. FIRE community has achieved a high level of professional operation. FIRE contributes to establishing a network of Future Internet initiatives which works towards sharing resources, services, tools and knowledge and which is supported by the involved Commission Units.

Around 2020, FIRE thus may have evolved towards a core infrastructure for Europe's open lab for Future Internet research, development and innovation and FIRE has evolved into a technology accelerator within Europe's innovation ecosystem for the Future Internet. Clearly this implies that FIRE should achieve a considerable level of sustainability, possibly as (part of) the core infrastructure of a thriving platform ecosystem which creates technological innovations addressing business and societal challenges.

In summary, some of the key strategic objectives for FIRE proposed by AmpliFIRE are the following:

- For 2016: to increase its relevance and impact primarily for European wide technology research, but also to increase its global relevance.
- For 2018: to create substantial business and societal impact through addressing technological innovations related to societal challenges. To become a sustainable and open federation that allows experimentation on highly integrated Future Internet technologies; supporting networking and cloud pillars of the Net Futures community.

- For 2020: to become a research, development and innovation space that is attractive to both academic researchers, SME technology developers and industrial R&D companies, with emphasis on key European initiatives such as 5G, Big Data, Internet of Things and Cyber-Physical Systems domains.

1.3.2 Vision and Opportunities of OMA LwM2M/oneM2M and Its Role in the Monitoring and Deployment of Large Scale Unmanned Networks

OMA LwM2M improves existing functionality for device management and brings new features for the resource management tool through the provisioning of a standardized resources description based on OMA Objects. Homard platform acts as a horizontal application to enable the device management tool with the capabilities for remote firmware upgrade, remote maintenance, standard interface for subscription to events/data, access to statistics regarding communications/performance/status/devices health etc., and finally a standards description for the metadata of the nodes/devices (manufacturer, version, security, firmware etc.).

OMA LwM2M is a very relevant standard based on the experience and knowledge from the most validated and extended protocol for device management (firmware upgrade over the air, remote monitoring, remote reboot, maintenance etc.). In details, the operations offered by the device management platform Homard using OMA LwM2M protocol are:

- **Software Management**: enabling the installation, removal of applications, and retrieval of the inventory of software components already installed on the device and the most relevant firmware upgrade over the air.
- **Diagnostics and Monitoring**: enabling remote diagnostic and standardized object for the collection of the memory status, battery status, radio measures, QoS parameters, peripheral status and other relevant parameters for remote monitoring.
- **Connectivity and security**: allowing the configuration of bearers (WiFi, Bluetooth, cellular connectivity), proxies, list of authorized servers for remote firmware upgrade and also all the relevant parameters for enabling secure communication.
- **Device Capabilities**: allowing to the Management Authority to remotely enable and disable device peripherals like cameras, Bluetooth, USB, sensors (ultrasound, temperature, humidity, etc.) and other relevant peripherals from the nodes.

- **Lock and Wipe**: allowing to remotely lock and/or wipe the device, for instance when the device is lost (relevant for devices in open ocean, air etc.), or when the devices are stolen or sold. It enables the remote erase of personal/enterprise data when they are compromised.
- **Management Policy**: allowing the deployment on the device of policies which the client (node, device, sensor) can execute and enforce indepen dently under some specific conditions, i.e., if some events happen, then perform some operations.

In addition to the functionalities, OMA LwM2M defines the semantics for the management objects. These objects have been defined with other standards organizations such as oneM2M and IPSO Alliance, which cooperate with OMA to avoid fragmentation and duplication that enables the semantic integration with the Management Objects. OMA LWM2M provides service providers with a secure, scalable, application-independent IoT control platform that provides control and security across multiple industries.

Thereby, this extension will also enable the integration into other initiatives such as oneM2M[1], which is the major initiative being led by ETSI and all the members from 3 GPP to enable a worldwide architecture for Internet of Things. It has a special focus on Semantic Web and interoperability. Therefore, Homard via the integration of OMA LwM2M support and oneM2M interworking will enable the openness of the platform towards possible future expansion through the integration with other IoT-based testbed infrastructures.

In addition, OMA LwM2M promotes the integration of a wide range of IoT enabled with OMA LwM2M for standardized management and data modelling based on Web Objects. OMA LwM2M and IPSO Alliance/OMA Web objects provide the capabilities for remote management and cloud computing integration. In addition, the OMA LwM2M clients are being supported in C and Java for integrating other sensors/nodes.

It is well known that there are an important number of IoT protocols with different adoption rate competing in the market as a consequence of the diversity of application domains in combination with the continuously increasing number of devices. In this direction, **oneM2M is an open standard that is based on the collection of the practices from the state of the art in a common framework** rather than the introduction of new approaches. In this way, oneM2M is gradually covering interoperability gaps and addresses

[1]OMA LwM2M is a key component from oneM2M [6, 7], it is the official device management component for oneM2M and it enablers interworking of the devices with oneM2M-based architectures.

pending difficulties using the global experience of IoT technologies. Lead by ETSI and the other SDOs such as ARIB, ATIS, CCSA, TIA, TTA and TTC, the oneM2M standard is totally coherent and has integrated outcomes from IETF, IPSO Alliance, IEEE, W3C and OMA, presenting a strong acceptability and maturity. oneM2M provides a well-defined service layer architecture as well as specifications for integrating existing IoT-specific technologies and standards such as CoAP, MQTT and OMA LwM2M.

1.3.3 Large Deployments with Low-power, Long-range, Low-cost

Internet of Things (IoT) devices are typically envisioned as the fundamental building blocks in a large variety of smart digital ecosystems: smart cities, smart agriculture, logistics&transportation... to name a few. However, the deployment of such devices in a large scale is still held back by technical challenges such as short communication distances. Using the traditional telco mobile communication infrastructure is still very expensive (e.g. GSM/GPRS, 3G/4G) and not energy efficient for autonomous devices that must run on battery for months. During the last decade, low-power but short-range radio such as IEEE 802.15.4 radio have been considered by the WSN community with multi-hop routing to overcome the limited transmission range. While such short-range communications can eventually be realized on smart cities infrastructures where high node density with powering facility can be achieved, it can hardly be generalized for the large majority of surveillance applications that need to be deployed in isolated or rural environments.

Future 5G standards do have the IoT orientation but these technologies and standards are not ready yet while the demand is already high. Therefore, and independently from the mobile telecom industry, recent modulation techniques are developed to achieve much longer transmission distances to a gateway without relay nodes to reduce the deployment cost and complexity. Rapidly adopted by many Machine-to-Machine (M2M) and IoT actors the concept of Low-Power Wide Area Networks (LPWAN), operating at much lower bandwidth, is gaining incredible interest. In addition, from a business perspective, the entry threshold for companies is much smaller with LPWAN than with traditional cellular technologies.

Some LPWAN technologies such as Sigfox^{TM} are still operator-based. However, other technologies such as LoRa^{TM} proposed by Semtech radio manufacturer can be privately deployed and used. Although direct communications between devices are possible with some technologies, most of IoT applications follow the gateway-centric approach with mainly uplink traffic patterns. In the typical architecture for public large-scale LPWAN, data captured by end-devices are sent to gateways that will push data to well-identified network servers, see Figure 1.6. Then application servers managed by end-users could retrieve data from the network server. If encryption is used for confidentiality, the application server can be the place where data could be decrypted and presented to end-users.

The advantages of long-range transmission comes at the cost of stricter legal regulations as most of them operate in the sub-GHz, unlicensed bands (for both increased coverage and flexibility). In Europe, electromagnetic transmissions in the 863–870 MHz band used by Semtech's LoRa technology falls into the Short Range Devices (SRD) category. The ETSI EN300-220-1 document [1]\cite{etsi EN300-220-1} specifies for Europe various requirements for SRD devices, especially those on radio activity. Basically, a transmitter is constrained to 1% duty-cycle (i.e. 36 s/hour) in the general case. This duty cycle limit applies to the total transmission time (referred to as time-on-air or air-time), even if the transmitter can change to another channel. In most cases, however, the 36 s duty-cycle is largely enough to satisfy communication needs of deployed applications. Note that this duty-cycle limitation approach is also adopted in China in the 779–787 MHz Band. US regulations in the 902–928 MHz Band do not directly specify duty-cycle but rather a maximum transmission time per packet with frequency hopping requirements.

1.3.3.1 LoRa technology

Although SigFox technology can have longer range than LoRa (40 kms have been reported for Sigfox while LoRa is typically in the range of 10 to 20 kms) when taking deployment flexibility into account, LoRa technology,

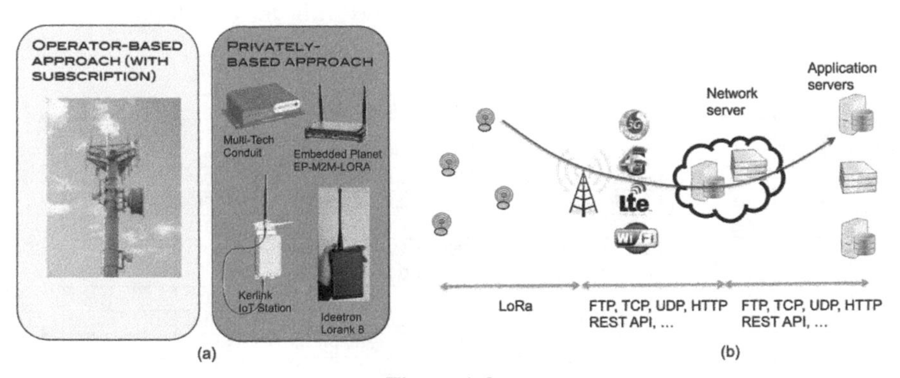

Figure 1.6

which can be privately deployed in a given area without any operator, has a clear advantage over Sigfox which coverage is entirely operator-managed.

Semtech's LoRa (LOng-RAnge) technology [2, 3]\cite{semtech-lora, Goursaud15} belongs to the spread spectrum approaches where data can be "spreaded" in both frequencies and time to increase robustness and range by increasing the receiver's sensitivity, which can be as low as −137 dBm in 868 MHz band or −148 dBm in the 433 MHz band. Throughput and range depend on the 3 main LoRa parameters: BW, CR and SF. BW is the physical bandwidth for RF modulation (e.g. 125 kHz). Larger signal bandwidth (currently up to 500 kHz) allows for higher effective data rate, thus reducing transmission time at the expense of reduced sensitivity. CR, the coding rate for forward error detection and correction. Such coding incurs a transmission overhead and the lower the coding rate, the higher the coding rate overhead ratio, e.g. with coding_rate = 4/(4+CR), the overhead ratio is 1.25 for CR = 1 which is the minimum value. Finally SF, the spreading factor, which can be set from 6 to 12. The lower the SF, the higher the data rate transmission but the lower the immunity to interference thus the smaller is the range. Figure 1.7 shows for various combinations of BW, CR and SF the time-on-air (ToA) of a LoRa transmission depending on the number of transmitted bytes. The maximum throughput is shown in the last column with a 255B payload. Modes 4 to 6 provide quite interesting trade-offs for longer range, higher data rate and immunity to interferences. Mode 1 provides the longest range.

1.3.3.2 LoRaWAN

Promoting the LoRa radio technology, the LoRa Alliance proposes a LoRaWAN [4]\cite{lorawan} specification for deploying large-scale, multi-gateways networks (star on star topology) and full network/application

				time on air in second for payload size of						
LoRa mode	BW	CR	SF	5 bytes	55 bytes	105 bytes	155 Bytes	205 Bytes	255 Bytes	max thr. for 255B in bps
1	125	4/5	12	0.95846	2.59686	4.23526	5.87366	7.51206	9.15046	223
2	250	4/5	12	0.47923	1.21651	1.87187	2.52723	3.26451	3.91987	520
3	125	4/5	10	0.28058	0.69018	1.09978	1.50938	1.91898	2.32858	876
4	500	4/5	12	0.23962	0.60826	0.93594	1.26362	1.63226	1.95994	1041
5	250	4/5	10	0.14029	0.34509	0.54989	0.75469	0.95949	1.16429	1752
6	500	4/5	11	0.11981	0.30413	0.50893	0.69325	0.87757	1.06189	1921
7	250	4/5	9	0.07014	0.18278	0.29542	0.40806	0.5207	0.63334	3221
8	500	4/5	9	0.03507	0.09139	0.14771	0.20403	0.26035	0.31667	6442
9	500	4/5	8	0.01754	0.05082	0.08154	0.11482	0.14554	0.17882	11408
10	500	4/5	7	0.00877	0.02797	0.04589	0.06381	0.08301	0.10093	20212

Figure 1.7

servers architecture as previously depicted in Figure 1.7. This specification defines the set of common channels for communications (10 in Europe), the packet format and Medium Access Control (MAC) commands that must be provided. In addition, LoRaWAN also defines so-called class A, B and C devices. Class A are bi-directional devices with each device's uplink transmission is followed by two short downlink receive windows for possible packets from the gateway. All LoRaWAN devices must at least implement Class A features. Class B and Class C devices are bi-directional devices with scheduled receive slots and bi-directional devices with maximal receive slots (nearly continuous listening) respectively. Class C devices consume a lot of power and few battery-operated applications can implement such behavior. Most of telemetry applications however use so-called Class A devices.

To optimize radio channel usage, Adaptive Data Rate (ADR) allows end-devices to use different spreading factor values depending on their distance to the gateway. By using a smaller spreading factor, the ToA is reduced therefore a larger amount of data can be sent within the 36 s of allowed transmission time.

When developed countries discuss about massive deployment of IoT using new LPWAN technologies, developing's countries are still far from being ready to enjoy the smallest benefit of it: lack of infrastructure, high cost of hardware, complexity in deployment, lack of technological eco-system and background, etc [5]\cite{IoT-newletter-zennaro}. For instance, in Sub-Saharan Africa about 64% of the population is living outside cities. The region will be predominantly rural for at least another generation. The majority of rural residents manage on less than few euros per day. Rural development is particularly imperative where half of the rural people are depend on the agriculture/micro and small farm business. For rural development, technologies have to support several key application sectors like water quality, agriculture, livestock farming, fish farming, etc.

Therefore, while the longer range provided by LPWAN is definitely an important dimension to decrease the cost of IoT, there are many other issues that must be addressed when considering deployment in developing countries: (a) Simplified deployment scenarios, (b) Cost of hardware and services and (c) Limit dependancy to proprietary infrastructures and provide local interaction models.

1.3.3.3 Simplified deployment scenarios

This typical LPWAN architecture depicted in Figure 1.6 can be greatly simplified for small, ad-hoc deployment scenarios such that those for agriculture/micro and small farm businesses, possibly in very remote areas.

Some LoRa and LoRaWAN community-based initiatives such as the one promoted TheThingNetwork^{TM} [6]\cite{TTN} may provide interesting solutions and feedbacks for dense environments such as cities but under simplified scenerios depicted in Figure 1.8 an even more adhoc and autonomous solution need to be proposed. In Figure 1.8, the gateway can directly push data to some end-user managed servers or public IoT-specific cloud platforms if properly configured.

Case A depicts a cellular-based and a WiFi Internet long-range gateway scenario. The Internet connection can be either privately owned or can rely on some community-based Internet access. Case B shows a no-Internet scenario where it is required that the gateway works in fully autonomous mode, capable of local interactions using standardized, consumer-market short-range technologies such as WiFi or Bluetooth.

Cost of Hardware and Services

The maturation of the IoT market is happening in many developed countries. While the cost of IoT devices can appear reasonable within developed countries standards, they are definitely still too expensive for very low-income sub-saharan ones. The cost argument, along with the statement that too integrated components are difficult to repair and/or replace definitely push for a Do-It-Yourself (DIY) and "off-the-shelves" design orientation. In addition,

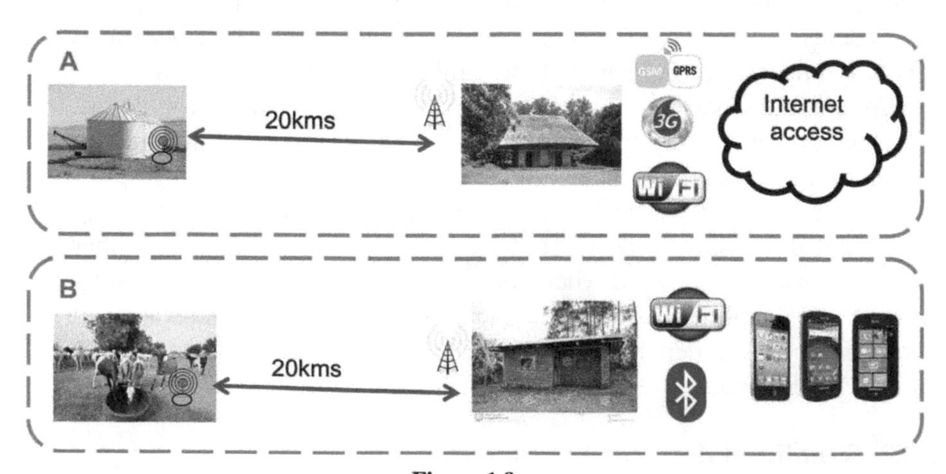

Figure 1.8

to be sustainable and able to reach previously mentioned rural environments, IoT initiatives in developing countries have to rely on an innovative and local business models. We envision mostly medium-size companies building their own "integrated" version of IoT for micro-small scale services. In this context, it is important to have dedicated efforts to design a viable exploitation model which may lead to the creation of small-scale innovative service companies.

The availability of low-cost, open-source hardware platforms such as Arduino-like boards is clearly an opportunity for building low-cost IoT devices from consumer market components. For instance, boards like Arduino Pro Mini based on an ATmega328 microcontroller offers an excellent price/performance/energy tradeoff and can provide a low-cost platform for generic sensing IoT with LoRa long-range transmission capability for a total of less than 15 euro. In addition to the cost argument such mass-market board greatly benefits from the support of a world-wide and active community of developers.

With the gateway-centric mode of LPWAN, commercial gateways are usually able to listen on several channels (e.g. LoRaWAN) and radio parameters simultaneously. For instance the LoRaWAN ADR mechanism may appear at first sight an interesting approach but it puts high complexity contraints on the gateway hardware as advanced concentrator radio chips, that alone cost more than a hundred euro, must be used. Besides, when a large number of IoT devices needs the longest range, the ADR mechanism provides only very small benefit.

Here, the approach can be different in the context of agriculture/micro and small farm business: simpler "single-connection" gateways can be built based on a simpler radio module, much like an end-device would be. Then, by using an embedded Linux platforms such as the Raspberry PI with high price/quality/reliability tradeoff, the cost of such gateway can be less than 45 euro.

Therefore, rather than providing large-scale deployment support, IoT platforms in developing countries need to focus on easy integration of low-cost "off-the-shelves" components with simple, open programming libraries and templates for easy appropriation and customization by third-parties. By taking an adhoc approach, complex and smarter mechanisms, such as advanced radio channel access to overcome the limitations of a low-cost gateway, can even be integrated as long as they remain transparent to the final developers.

Limit Dependancy to Proprietary Infrastructures and Provide Local Interaction Models

Data received on the gateway are usually pushed/uploaded to some Internet/cloud servers. It is important in the context of developing countries to be able to use a wide range of infrastructures and, if possible, at the lowest cost. Fortunately, along with the global IoT uptake, there is also a tremendous availability of sophisticated and public IoT clouds platforms and tools, offering an unprecedented level of diversity which contributes to limit dependency to proprietary infrastructures. Many of these platforms offer free accounts with limited features but that can already satisfy the needs of most agriculture/micro and small farm/village business models. It is therefore desirable to highly decouple the low-level gateway functionalities from the high-level data post-processing features, privileging high-level languages for the latter stage (e.g. Python) so that customizing data management tasks can be done in a few minutes, using standard tools, simple REST API interfaces and available public clouds.

In addition, with the lack or intermittent access to the Internet data should also be locally stored on the gateway which can directly be used as an end computer by just attaching a keyboard and a display. This solution perfectly suits low-income countries where many parts can be found in second markets. The gateway should also be able to interact with the end-user' smartphone to display captured data and notify users of important events without the need of Internet access as this situation can clearly happen in very remote areas, see case B in Figure 1.8.

Single-Connection Low-cost LoRa Gateway

Our LoRa gateway [7]\cite{pham-lcgw} could be qualified as "single connection" as it is built around an SX1272/76, much like an end-device would be. The low-cost gateway is based on a Raspberry PI (1B/1B+/2B/3B) which is both a low-cost (less than 30 euro) and a reliable embedded Linux platform. There are many SX1272/76 radio modules available and we currently tested with 6: the Libelium SX1272 LoRa, the HopeRF RFM92W & 95W, the Modtronix inAir9 & inAir9B, and the NiceRF SX1276. Most SPI LoRa modules are actually supported without modifications as reported by many users. In all cases, only a minimum soldering work is necessary to connect the required SPI pins of the radio to the corresponding pins on the Raspberry pin header as depicted in Figure 1.9. The total cost of the gateway can be less than 45 euro.

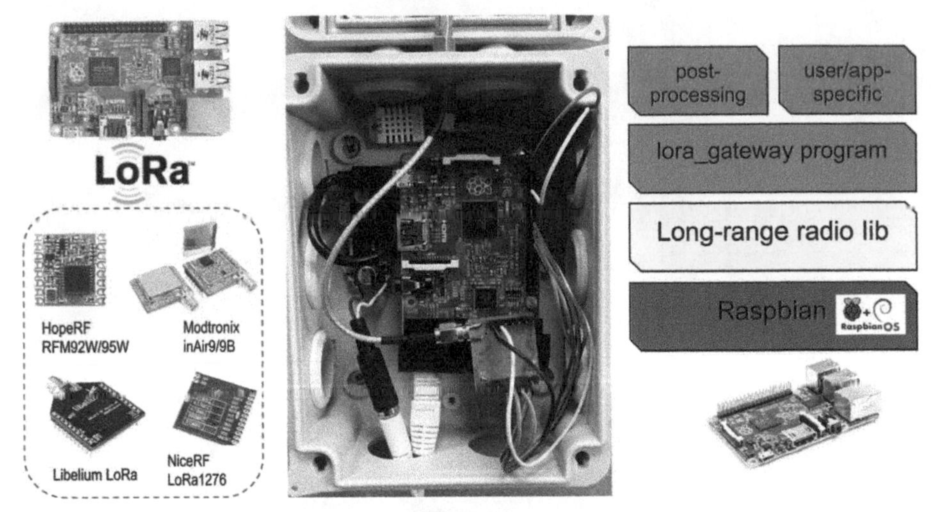

Figure 1.9

Together with the "off-the-shelves" component approach, the software stack is completely open-source: (a) the Raspberry runs a regular Raspian distribution, (b) our long range communication library is based on the SX1272 library written initially by Libelium and (c) the lora_gateway program is kept as simple as possible. We improved the original SX1272 library in various ways to provide enhanced radio channel access (CSMA-like with SIFS/DIFS) and support for both SX1272 and SX1276 chips.

We tested the gateway in various conditions for several months with a DHT22 sensor to monitor the temperature and humidity level inside the case. Our tests show that the low-cost gateway can be deployed in outdoor conditions with the appropriate casing. Although the gateway should be powered, its consumption is about 350mA for an RPIv3B with both WiFi and Bluetooth activated.

Post-Processing and Link with IoT Cloud Platforms

After compiling the lora_gateway program, the most simple way to start the gateway is in standalone mode as shown is Figure 1.10a. All packets received by the gateway is sent to the standard Unix-stdout stream.

Advanced data post-processing tasks are performed after the gateway stage by using Unix redirection of gateway's outputs as shown by the orange "post-processing" block in Figure 1.10b. We promote the usage of high-level language such as Python to implement all the data post-processing tasks

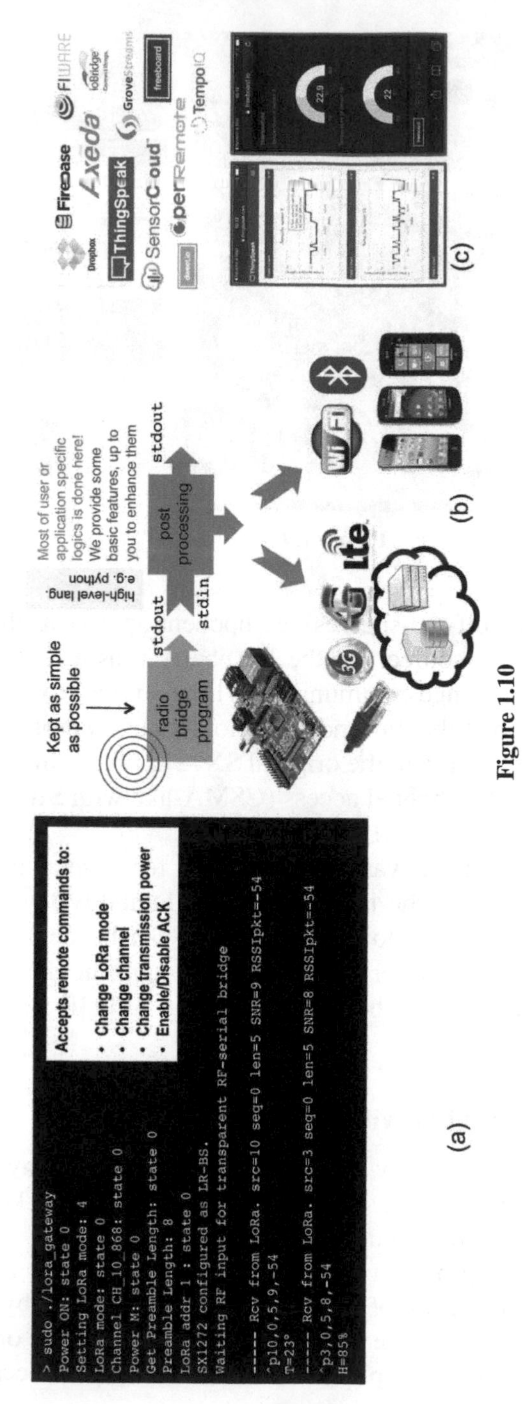

> sudo ./lora_gateway
Power ON: state 0
Setting LoRa mode: 4
LoRa mode: state 0
Channel CH_10_868: state 0
Power M: state 0
Get Preamble Length: state 0
Preamble Length: 8
LoRa addr 1 : state 0
SX1272 configured as LR-BS.
Waiting RF input for transparent RF-serial bridge
----- Rcv from LoRa. src=10 seq=0 len=5 SNR=9 RSSIpkt=-54
^p10,0,5,9,-54
T=23°
----- Rcv from LoRa. src=3 seq=0 len=5 SNR=8 RSSIpkt=-54
^p3,0,5,8,-54
H=85%
Accepts remote commands to:
Change LoRa mode
Change channel
Change transmission power
Enable/Disable ACK
(a)
Kept as simple as possible
radio bridge program
stdout
stdin
high-level lang. e.g. python
post processing
stdout
Most of user or application specific logics is done here! We provide some basic features, up to you to enhance them
3G
4G
lte
Wi Fi
(b)
Dropbox
Firebase
FIWARE
Xeda
ioBridge
ThingSpeak
GroveStreams
SensorCloud
freeboard
OpenRemote
TempoIQ
22.9
22
(c)

Figure 1.10

such as access to IoT cloud platforms and even advanced features such as AES encryption/decryption. Our gateway is distributed with a Python template that explains and shows how to upload data on various publicly available IoT cloud platforms. Examples include Dropbox™, Firebase™, ThingSpeak™, freeboard™, SensorCloud™, GrooveStream™ and FiWare™, as illustrated in Figure 1.10c.

This architecture clearly decouples the low-level gateway functionalities from the high-level post-processing features. By using high-level languages for post-processing, running and customizing data management tasks can be done in a few minutes. One of the main objectives of IoT in Africa being technology transfer to local developer communities, we believe the whole architecture and software stack are both robust and simple for either "out-of-the-box" utilization or quick appropriation & customization by third parties. For instance, a small farm can deploy in minutes the sensors and the gateway using a free account with ThingSpeak platform to visualize captured data in real-time.

Gateway Running Without Internet Access

Received data can be locally stored on the gateway and can be accessed and viewed by using the gateway as an end computer by just attaching a keyboard and a display. The gateway can also interact with the end-users' smartphone through WiFi or Bluetooth as depicted previously in Figure 1.8b. WiFi or Bluetooth dongles for Raspberry can be found at really low-cost and the smartphone can be used to display captured data and notify users of important events without the need of Internet access as this situation can clearly happen in very remote areas. Figure 1.11 shows our low-cost gateway running a MongoDB™ noSQL database and a web server with PHP/jQuery to display received data in graphs. An Android application using Bluetooth connectivity has also been developed to demonstrate these local interaction models.

Low-cost LoRa End-devices

Arduino boards are well-known in the microcontroller user community for their low-cost and simple-to-program features. These are clearly important issues to take into account in the context of developing countries, with the additional fact that due to their success, they can be acquired and purchased quite easily world-wide. There are various board types that can be used depending on the application and the deployment constraints. Our communication library supports most of Arduino boards as illustrated in Figure 1.12.

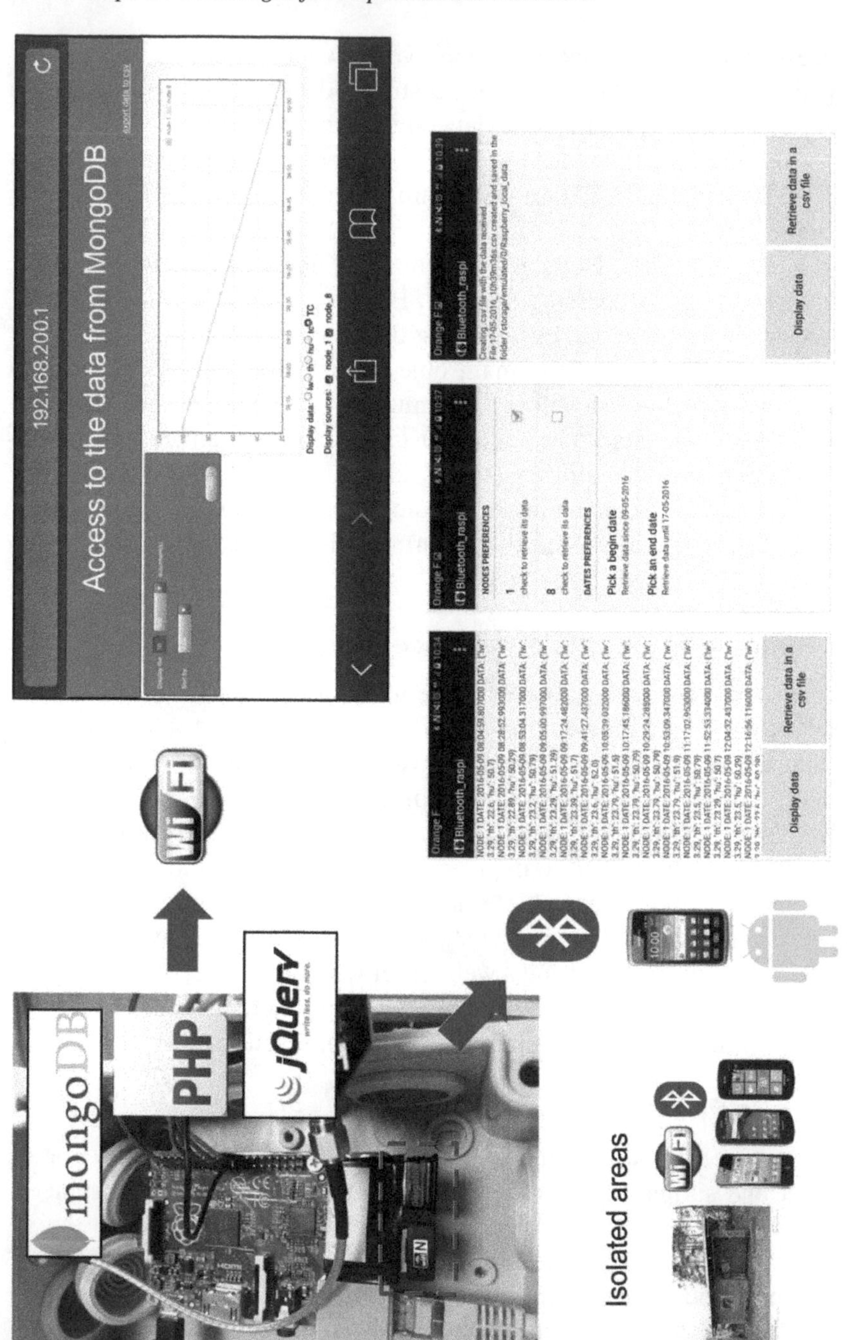

mongoDB
PHP
jQuery
write less, do more.
Isolated areas
192.168.200.1
Access to the data from MongoDB
Bluetooth_raspi
NODES PREFERENCES
DATES PREFERENCES
Pick a begin date
Pick an end date
Display data
Retrieve data in a csv file

Figure 1.11

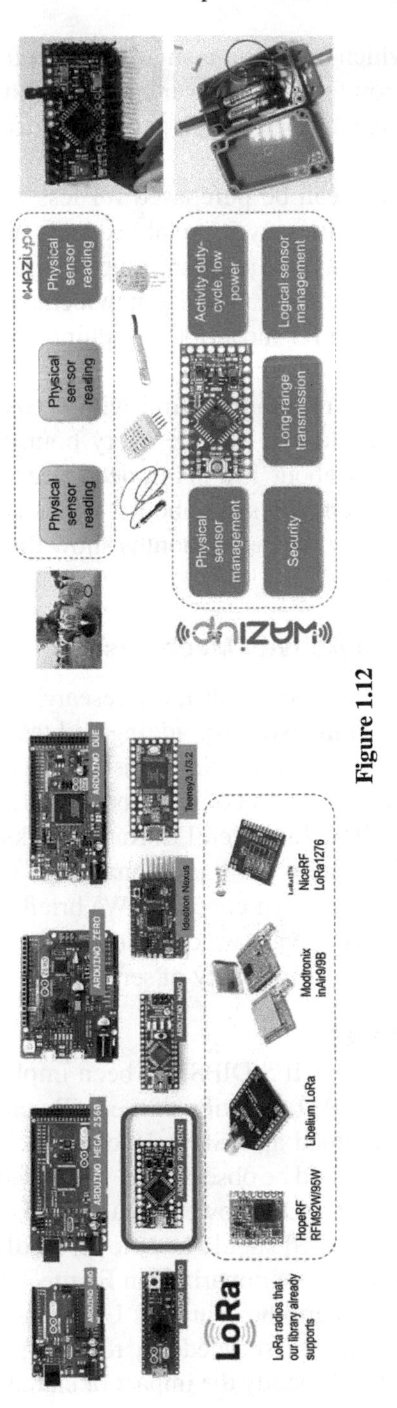
LoRa
LoRa radios that our library already supports
HopeRF RFM92W/95W
Libelium LoRa
Modtronix inAir9/9B
NiceRF LoRa1276
Ideetron Nexus
Teensy3.1/3.2
Physical sensor reading
Physical sensor reading
Physical sensor reading
Activity duty-cycle, low power
Logical sensor management
Physical sensor management
Long-range transmission
Security
WAZIUP

Figure 1.12

The Arduino Pro Mini, which comes in a small form factor and is available in a 3.3 v and 8 MHz version for lower power consumption, appears to be the development board of choice for providing a generic platform for sensing and long-range transmission.

Arduino Pro Mini clones can be purchased for less than 2 euro a piece from Chinese manufacturers with very acceptable quality and reliability level. Similar to the low-cost gateway, all programming libraries are open-source and we provide templates for quick and easy new behaviour customization and physical sensor integration for most of the Arduino board types.

For very low-power applications, deep-sleep mode are available in the example template to run an Arduino Pro Mini with 4 AA regular batteries. For instance, with a duty-cycle of 1 sample every hour, the board can run for almost a year, consuming about 146 uA in deep sleep mode and 93 mA when active and sending, which represents about 2 s of activity time. Our tests conducted continuously during the last 6 months show that the low-cost Pro Mini clones are very reliable.

Adding Advanced Radio Activity Mechanisms

The proposed framework leaves room for more research-oriented tasks as it actually provides a flexible framework for adding and testing new advanced features that are lacking in current LPWAN. For instance, while the LoRaWAN specifications may ease the deployment of LoRa networks by proposing some mitigation mechanisms to allow for several LoRa networks to coexist, it still remains a simple ALOHA system with additional tight radio activity time constraints without quality of service concerns. We briefly describe below 2 issues of long-range networks that are we currently study: improved channel access and activity time sharing for quality of service.

Improved channel access

A CSMA-like mechanism with SIFS/DIFS has been implemented using the Channel Activity Detection (CAD) function of the LoRa chip and can further be customized. A DIFS is defined as 3 SIFS. Prior to packet transmission a DIFS period free of activity should be observed. If "extended IFS" is activated then an additional number of CAD followed by a DIFS is required. If RSSI checking is activated then the RSSI should be below –90 dB for the packet to be transmitted. These features are summarized in Figure 1.13.

By running a background periodic source of LoRa packets, we observed that the improved channel access succeeds in reducing packet collisions. The current framework is used to study the impact of channel access methods

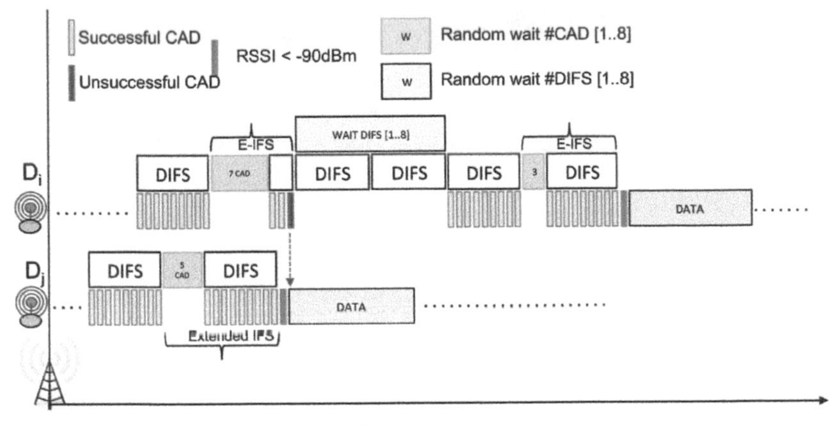

Figure 1.13

in a medium-size LoRa deployment when varying timer values due to the longer time-on-air.

Activity time sharing

We also propose and implement an exploratory activity time sharing mechanism for a pool of devices managed by a single organization [8]. We propose to overcome the tight 36 s/hour radio activity of a device by considering all the sensor's individual activity time in a shared/global manner. The approach we propose will allow a device that "exceptionally" needs to go beyond the activity time limitation to borrow some from other devices. A global view of the global activity time, G_{AT}, allowed per 1 hour cycle will be maintained at the gateway so that each device knows the potential activity time that it can use in a 1-hour cycle. Figure 1.14 shows how the deployed long-range devices Di sharing their activity time initially register (REG packet) with the gateway by indicating their local Remaining Activity Time l^{i}_{RAT0}, i.e. 36 s. The gateway stores all l^{i}_{RAT0} in a table, computes G_{AT} and broadcasts (INIT packet) both n (the number of devices) and G_{AT}. This feature is currently tested for providing better surveillance service guarantees.

Use Case: Fish Farming – Fish Pond Monitoring

With our WAZIUP partner Farmerline (http://farmerline.co/) we deployed a small number of our low-cost IoT sensor boards in a fish farm which operates several ponds of different sizes (http://www.kumahfarms.com/). This farm engages in pond culture and do both tilapia and catfish (Figure 1.15).

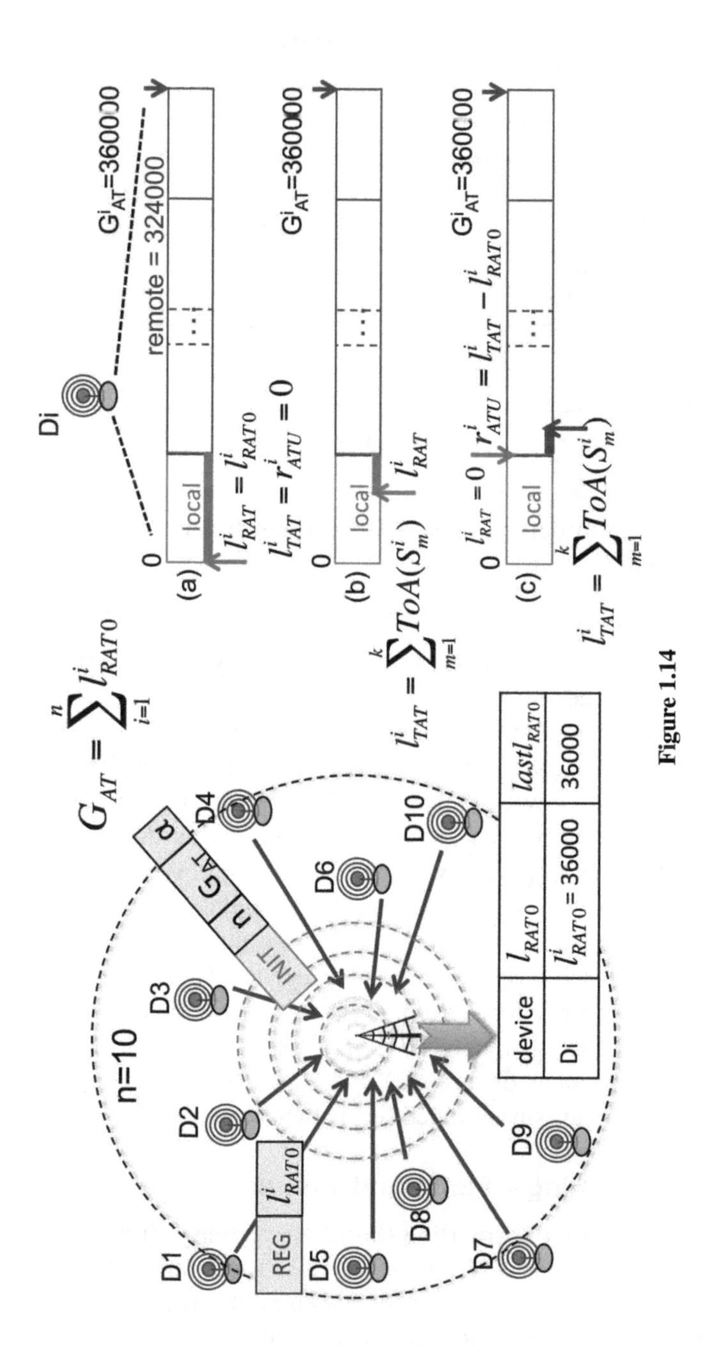
n=10
D1
D2
D3
D4
D5
D6
D7
D8
D9
D10
REG
INIT
n
α
$G_{AT} = \sum_{i=1}^{n} l^i_{RAT0}$
device
Di
l_{RAT0}
l^i_{RAT0} = 36000
$lastl_{RAT0}$
36000
Di
remote = 324000
G$^i_{AT}$=360000
local
(a)
(b)
(c)
0
$l^i_{RAT} = l^i_{RAT0}$
$l^i_{TAT} = r^i_{ATU} = 0$
l^i_{RAT}
$l^i_{TAT} = \sum_{m=1}^{k} ToA(S^i_m)$
$l^i_{RAT} = 0$
$r^i_{ATU} = l^i_{TAT} - l^i_{RAT0}$

Figure 1.14

Figure 1.15

Their main needs is to get water quality indicators such as temperature and dissolved oxygen. 3 sensors are connected to the generic sensor board: a DHT22 ambient air temperature and humidity sensor, a PT1000 sensor for water temperature and an AtlasScientific DO sensor for water dissolved oxygen level. Using the generic activity duty-cycle module, the board will periodically read values on the 3 connected physical sensors every 3 minutes for our test scenario. The concatenated message string format is as follows: "TC/27.35/HU/67.5/WT/23.47/DO/10.42" where TC and HU are for the air temperature and humidity level from the DHT22, WT for water temperature from the PT1000 and DO for dissolved oxygen level from the AtlasScientific DO sensor. However, at the time of writing, we didn't receive the DO sensor yet so the DO values are emulated.

The gateway is installed on one of the farm's building and can have Internet access. The post-processing stage simply takes the message string to separate it into a list of fields: ['TC', '27.35', 'HU', '67.5', 'WT', '23.47', 'DO', '10.42']. The gateway then pushes data to the GroveStream cloud (with free account) which provides a very flexible framework where it is possible to create several data streams (e.g. TC/HU/WT/DO) per component (the sensor node) in a dynamic manner. Figure 1.16 shows for the 3 deployed sensors their data streams with a focus on the DO stream from sensor 9.

Figure 1.16 also shows the no-Internet connectivity scenario as illustrated previously in Figure 1.6 : the gateway also stores data from the various sensors in its local MongoDB database and acts as a WiFi access point and web server to display the sensed value (here, screenshot from an Android smartphone).

With the generic sensor board, with ready-to-use duty-cycle and low-power building blocks, deploying and setting the whole system was easy and quick. Regarding the physical sensor reading, each environmental parameter is wrapped in a Sensor class object that can implement pin reading and specific data conversion tasks to provide a usable value through a virtual

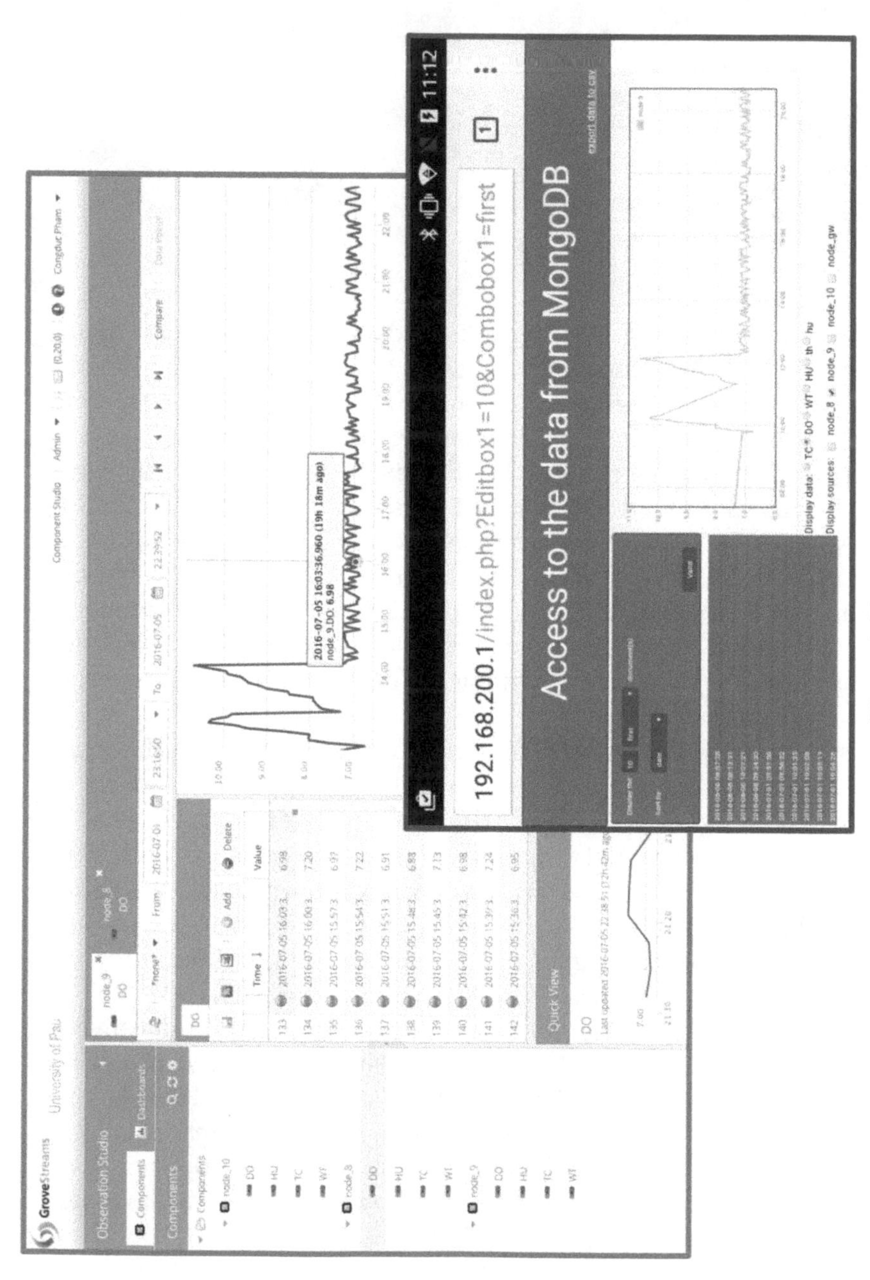
Access to the data from MongoDB
192.168.200.1/index.php?Editbox1=10&Combobox1=first
11:12
GroveStreams
University of Pau
Observation Studio
2016-07-05 16:03:36.960 (19h 18m ago)
node_9_DO: 6.98

Figure 1.16

get_value() method. For instance, the DHT22 sensor that provides 2 environmental parameters is represented by 2 different Sensor class objects. The sensor board will simply loop and call all get_value() methods of all connected sensors. At the gateway, the post-processing template written in Python can handle an arbitrary number of data streams therefore the whole post-processing stage was left unchanged for uploading data from the 3 physical sensors to our GroveStream cloud account.

1.4 Conclusions

FIRE has evolved into a diverse portfolio of experimental facilities, increasingly federated and supported by tools, and responding to the needs and demands of a large scientific experimenter community. Issues that require attention include the sustainability of facilities after projects' termination, the engagement of industry and SMEs, and the continued development of FIRE's ecosystem to remain relevant to changing research demands. A more strategic issue is to develop a full service approach addressing the gaps between ecosystem layers and addressing integration issues that are only now coming up in other Future Internet-funded projects. A related challenge is to expand the nature of FIRE's ecosystem from an offering of experimental facilities towards the creation of an ecosystem platform capable to attract market parties from different sides that benefit from mutual and complementary interests. Additionally, FIRE should anticipate the shifting focus of Future Internet innovation areas towards connecting users, sensor networks and heterogeneous systems, where data, heterogeneity and scale will determine future research and innovation in areas such as Big Data, and 5G and Internet of Things. Such demands lead to the need for FIRE to focus on testbeds, experimentation and innovation support in the area of "smart systems of networked infrastructures and applications".

To address the viewpoints identified by the FIRE community, the FIRE initiative should support actions that keep pace with the changing state-of-the-art in terms of technologies and services, able to deal with current and evolving experimenter demands. Such actions must be based upon a co-creation strategy, interacting directly with the experimenters, collecting their requirements and uncovering potential for extensions. FIRE must also collaborate globally with other experimental testbed initiatives to align with trends and share expertise and new facilities. Where major new technologies emerge, these should be funded as early as possible as new experimental facilities in the FIRE ecosystem.

This analysis leads to some recommendations regarding the future direction of FIRE, concisely summarized below.

- FIRE's strategic vision for 2020 is to be the Research, Development and Innovation environment for the Future Internet, creating business and societal impact and addressing societal challenges. Adding to FIRE's traditional core in networking technologies is shift of focus in moving upwards to experimenting and innovating on connected smart systems which are enabled by advanced networking technologies.
- FIRE must forcefully position the concept of experimental testbeds driving innovation at the core of the experimental large-scale trials of other Future Internet initiatives and of selected thematic domains of Horizon 2020. Relevant initiatives suitable for co-developing and exploiting testbed resources include the 5G-PPP, Internet of Things large-scale pilots, and e-Infrastructures.
- FIRE should help establish a network of open, shared experimental facilities and platforms in co-operation with other Future Internet initiatives. Experimental facilities should become easily accessible for any party or initiative developing innovative technologies, products and services building on Future Internet technologies. For this to happen, actions include the continuing federation of facilities to facilitate the sharing of tools and methods, and providing single access points and support cross-domain experimentation. Facilities also should employ recognized global standards. At the level of facilities, Open Access structures should be implemented as a fundamental requirement for any FIRE facility. To extend open facilities beyond FIRE, for example with 5G-PPP or Géant and NRENs, co-operation opportunities can be grounded in clear value propositions for example based on sharing technologies and experiment resources.
- FIRE should establish "technology accelerator" functionality, by itself or in co-operation with other Future Internet initiatives, to boost SME research and product innovation and facilitate start-up creation. The long-term goal of FIRE is to realize a sustainable, connected network of Internet experimentation facilities providing easy access for experimenters and innovators across Europe and globally, offering advanced experimentation and proof-of-concept testing. The number of SMEs and start-ups leveraging FIRE can be increased by offering professional highly supported facilities and services such as Experimentation-as-a Service, shortening learning time and decreasing time to market fort

experimentation. A brokering initiative should provide broker services across the FIRE portfolio or via exploitation partnerships. Additionally, community APIs should be offered to make FIRE resources more widely available.

- FIRE's core expertise and know-how must evolve: from offering facilities for testing networking technologies towards offering and co-developing the methodologies, tools and processes for research, experimentation and proof-of-concept testing of complex systems. FIRE should establish a lively knowledge community to create innovative methodologies and learn from practice.
- FIRE should ensure longer term sustainability building upon diversification, federation and professionalization. FIRE should support the transition from research and experimentation to innovation and adoption, and evolve from single area research and experiment facilities towards cross-technology, cross-area facilities which can support the combined effects and benefits of novel infrastructure technologies used together with emerging new service platforms enabling new classes of applications.
- FIRE should develop and implement a service provisioning approach aimed at customized fulfilment of a diverse range of user needs. Moving from offering tools and technologies, FIRE should offer a portfolio of customized services to address industry needs. FIRE should establish clear channels enabling interaction among providers, users and service exploitation by collaboration partners.

FIRE should become part of a broader Future Internet value network, by pursuing co-operation strategies at multiple levels. Cooperation covers different levels: federation and sharing of testbed facilities, access to and interconnection of resources, joint provision of service offerings, and partnering with actors in specific sectoral domains. In this FIRE should target both strong and loose ties opportunistic collaboration. Based on specific cases in joint projects, cooperation with 5G and Internet of Things domains could be strengthened.

Finally, FIRE should evolve towards an open access platform ecosystem. Platform ecosystem building is now seen critical to many networked industries as parties are brought together who establish mutually beneficial relations. Platforms bring together and enable direct interactions within a value network of customers, technology suppliers, developers, facility providers and others. Developer communities may use the FIRE facilities to directly work with business customers and facility providers. Orchestration of the FIRE platform ecosystem is an essential condition. Steps towards forming a platform ecosystem include the encouragement of federation, the setting up open access and open call structures, and the stimulation of developer activities.

The concept of Low-Power Wide Area Networks (LPWAN), operating at much lower bandwidth, is gaining incredible interest in the IoT domain. In this contribution we presented several important issues when considering deploying low-power, long-range IoT solutions for low-income developing countries: (a) Simplified deployment scenarios, (b) Cost of hardware and services and (c) Limit dependancy to proprietary infrastructures and provide local interaction models. We described our low-cost and open IoT platforms for rural developing countries applications that addressed these issues. Targeted for small to medium size deployment scenarios the platform also privileges quick appropriation and customization by third parties.

References

[1] ETSI, "Electromagnetic compatibility and radio spectrum matters (ERM); short range devices (SRD); radio equipment to be used in the 25 MHz to 1000 MHz frequency range with power levels ranging up to 500 mw; part 1." 2012.

[2] Semtech, "LoRa modulation basics. rev.2-05/2015," 2015.

[3] C. Goursaud and J. Gorce, "Dedicated networks for IoT : PHY/MAC state of the art and challenges," EAI Endorsed Transactions on Internet of Things, Vol. 1, No. 1, 2015.

[4] LoRaAlliance, "LoRaWAN specification, v1.01," 2015.

[5] M. Zennaro and A. Bagula, "IoT for development (IOT4D). In IEEE IoT newsletter." July 14, 2015.

[6] TheThingNetwork, "http://thethingsnetwork.org/," accessed 13/01/2016.

[7] C. Pham, "A DIY low-cost LoRa gateway. http://cpham.perso.univpau.fr/lora/rpigateway.html and https://github.com/congducpham/lowcost loragw," accessed Apr 29th, 2016.

[8] C. Pham, "QoS for Long-Range Wireless Sensors under Duty-Cycle Regu lations with Shared Activity Time Usage", to appear in ACM Transaction on Sensor Networks (TOSN).

2

Next Generation Internet Research and Experimentation

Martin Serrano[1], Michael Boniface[2], Monique Calisti[3], Hans Schaffers[4], John Domingue[5], Alexander Willner[6], Chiara Petrioli[7], Federico M. Facca[8], Ingrid Moerman[9], Johann M. Marquez-Barja[10], Josep Martrat[11], Levent Gurgen[12], Sebastien Ziegler[13], Serafim Kotrotsos[14], Sergi Figuerola Fernandez[15], Stathes Hadjiefthymiades[16], Susanne Kuehrer[17], Thanasis Korakis[18], Tim Walters[19] and Timur Friedman[20]

[1]Insight Centre for Data Analytics Galway, Ireland
[2]IT Innovation Centre, UK
[3]Martel Innovate, Switzerland
[4]Saxion University of Applied Sciences, Netherlands
[5]Open University, United Kingdom
[6]FhG-Fokus, Germany
[7]University of Rome 'La Sapienza', Italy
[8]Martel Innovate, Switzerland
[9]iMinds, Belgium
[10]Trinity College Dublin, Ireland
[11]Atos, Spain
[12]CEA, France
[13]Mandat International, Switzerland
[14]Incelligent
[15]i2CAT, Spain
[16]University of Athens, Greece
[17]EIT Digital
[18]University of Thessaly, Greece
[19]iMinds, Belgium
[20]LIP6 CNRS Computer Science Lab, UPMC Sorbonne Universités, France

2.1 Experimentation Facilities in H2020: Strategic Research and Innovation Agenda Contributions

The Internet as we know it today is a critical infrastructure composed by communication services and end-user applications transforming all aspects of our lives. Recent advances in technology and the inexorable shift towards everything connected are creating a data-driven society where productivity, knowledge, and experience are dependent on increasingly open, dynamic, interdependent and complex networked systems. The challenge for the Next Generation Internet (NGI) is to design and build enabling technologies, implement and deploy systems, to create opportunities considering increasing uncertainties and emergent systemic behaviours where humans and machines seamlessly cooperate.

Many initiatives investigated approaches for measuring, exploring and systematically re-designing the Internet, to be more open, efficient, scalable, reliable and trustworthy [FIWARE/FIPPP, CAPS, EINS, FIRE, GENI, US IGNITE, AKARI]. Yet, although no universal methodologies have emerged due to the continuously evolving interplay among technology, society and the economy such initiatives produce a richer awareness of the socio-economic and technological challenges and provide the foundation for new innovative ICT solutions.

The Internet has evolved to the point that today is a vast collection of technologies and systems and has no overall defined design path for its inherent expansion and neither shall the Next Generation Internet. The actual experience is telling us that the Internet evolves through widely adopted experimentation that engages active users and communities rather than through purely technological advances invented in closed laboratories. Individuals and companies use larger experiments as a way to build the knowledge and necessary insights to verify and validate theories and ideas, and as the basis for creating viable, acceptable and innovative solutions driving benefits to Internet ecosystems and their stakeholders. For example *"by the end of 2018, 90% of IT projects will be rooted in the principles of experimentation, speed, and quality"* [Forrester2015].

The actual evidence, based on practical industrial experiences is unambiguous:

- Facebook is a huge and wide ranging social experiment investigating broad topics such as the economics of privacy, appetite for disclosure of personal data, and role of intermediaries in content filtering including emotional effect [14].

- Google's Experiments Challenge and Showcases uses Android as an open platform to engage large participation from OSS communities in the creation of inspirational, distinctive and unique open source mobile applications [5].
- Ericsson uses experimentation to explore opportunities in enterprise ecosystem related to localised applications, global applications along with added value services supporting security, device management and mobile productivity [Ericsson15].
- Smart Cities and underlying programmable network infrastructures uses social experiments with citizens in applications as diverse as transport, energy and environmental management [18].
- Netflix uses an experimentation platform to ensure optimal streaming experience with high-quality video and minimal playback interruption to its customers by testing adaptive streaming and content delivery network algorithms across so called experimental groups involving Netflix engineers and Netflix members [NETFLIX2016].
- Experimentation plays a vital role in business growth at eBay by providing valuable insights and prediction on how users will react to changes made to the eBay website and applications. A/B testing is performed by running more than 5000 experiments per year on the eBay Experimentation Platform [eBay2015].
- Apple used experimentation extensively to explore smart watch ideas initially starting from primitives as simple as an iPhone with a Velcro strap [WIRED14].
- Many industries targeting large online communities (e.g. gaming) use open beta programmes to investigate features and experiences with end user and developer ecosystems, to gain initial market attraction, for example only, the recent Overwatch programme secured 10 million players [17].

These strategies demonstrate that many successful Internet technologies are now developed through experimentation ecosystems allowing creative and entrepreneurial individuals and companies to explore disruptive ideas, freely with large "live" user-driven communities.

Innovation also plays a dynamic role in the process of large experimentation adoption. Experiments are conducted with ecosystems using platforms and infrastructures (e.g. mobile platform, social network, smart spaces and physical wireless spaces) designed to foster innovation by considering value creation through openness, variation and adaptability. These strategies show an increasing need to structure and engage society and communities of users in the co-creation of solutions (one of the multiples forms for innovation) by bridging the gap between vision, experimentation and large-scale validation sufficiently to attain end-user (citizens or industry) investment, either in terms of time or money.

Addressing directly the demand for innovation, Europe must establish large-scale experimental ecosystems aligned with NGI architectures that are sustained beyond individual EU project investments, with full involvement of end-users (i.e. citizens and SMEs), since they provide applicability validation of outcomes. Ecosystems help in anticipating possible migration paths for technological developments, create opportunities for potentially disruptive innovations and discovery of new and emerging behaviours; as well as in assessing the socio/economic implications of new technological solutions at an early stage. In addition, experimentation is an effective way to build evidence for the robustness, reusability and effectiveness of emerging specifications and standards. Note that it is important to recognise that there is no such thing as a "failed experiment". Even if the findings point to a null hypothesis, learning what doesn't work is a necessary step to learning what does correctly. Discovering that a technology fails to perform, is not commercially viable or is not accepted by end users is a clear route to future research and innovation challenges for the NGI.

2.1.1 European Ecosystem Experimentation Impacts

Ecosystem experimentation and trials using open platforms are a major contributor to the success of European research and innovation programmes investigating the future of the Internet. Initiatives such as Future Internet Research and Experimentation (FIRE), the Community Awareness Platforms for Sustainability and Social Innovation (CAPS/CAPSSI), the Future Internet-Public Private Partnership (FI-PPP), the 5G-Public Private Partnership (5G-PPP), European Institute of Innovation & Technology (EIT) Digital, and the European Network of Living Labs (ENoLL) have all been delivering platforms and ecosystems that have advanced Internet-based technologies towards markets and society. Each flagship initiative has been designed to fulfil specific complementary socio-economic and technical objectives. For example, CAPS enables societal innovation through open platforms supporting new forms of social interaction, FI-PPP enables innovation through accelerator ecosystems building on the open platform FIWARE, whilst FIRE enables innovation through highly configurable technology infrastructures and services. In particular, selected FIRE examples show that significant long lasting European impacts can be delivered:

- ***SME competitiveness***: experimentation has enhanced 100's of companies' product and service offerings have benefited by validating performance, acceptance and viability using experimental platforms. Examples include: Televic Rail launching their SilverWolf passenger information product on more than 22,000 railcars following complex end-to-end networking performance tests; Evolaris GmbH launching Europe's 1st Smart Ski Goggles service in the Ski Amadé, Austria, Europe's 2nd largest ski area based on user-centric networked media experiments; Incelligent proactive network management products building on cognitive radio experiments, involving realistic conditions and actual testbeds leading to the company being selected as one of the 12 startups awarded to work with Intel, Cisco and Deutsche Telekom, through the next phase of their joint ChallengeUp! Program.
- ***Pioneering concepts***: experimentation has demonstrated groundbreaking results that the world has never seen before. Examples include: Open platforms to transforming the education of the next generation of Internet scientists and engineers through remote experimentation on top of FIRE facilities and open online courses supporting over 1,000s of students and more than 16 courses across several countries (e.g. Belgium, Greece, Ireland, Spain, Brazil and Mexico) by allowing the creation, sharing and re-use of learning resources based on real experiments and data, accessible anytime/anywhere learning [6]; The World's 1st mixed reality ski competition broadcast across European television (BBC, ORF, etc.) radio and online to a global audience of over 700 million [2]; the first generation of networked Internet of Things technologies for pervasively monitoring the underwater environments; validation of HBBTV technology in European broadcast events [10].
- ***Interoperability and standardisation***: experimentation has established evidence and contributed to the development of new international standards, many of those adopted by the market. Examples include: Licensed Shared Access (LSA) technology to maximize mobile network capacity in LTE (4G) communications presented to the ETSI TC Reconfigurable Radio Systems WG1; Transceiver API for a hardware-independent software interface to a Radio Front-Ends developed by Thales Communications and Security SAS standardised in Wireless Innovation Forum (WInnF); Contributions to standardisation fora (Wireless

Innovation Forum, ITU-R, ETSI, IEEE 802, IEEE P1900.6, DySPAN); Simplifying spectrum sensing measurements through a common data collection/storage format, based on the IEEE 1900.6 standard, enabling sharing of experiment descriptions, traces and data processing script for heterogeneous sensing hardware; Establishment of the W3C Federated Infrastructures Community Group to start the standardization of according semantic information models and facilitate collaboration with other groups such as the IEEE P2302 Working Group – Standard for Intercloud Interoperability and Federation (SIIF) – or the OneM2M Group on Management, Abstraction and Semantics (MAS).

- ***International collaboration***: experimentation has raised the global profile and reputation of European research and innovation initiatives. Examples include: Establishment of the Open-Multinet Forum to facilitate the international collaboration between FIRE and GENI and other members for harmonizing interfaces and information models; Global reconfigurable and software defined networks between Europe, Korea, Brazil, South Africa, Japan and US.
- ***Internet regulation and governance***: experimentation has delivered results driving the evolution of policies regulating networks and services; Examples include: interaction with national regulators (BIPT-Belgium, National Broadband Plan NBP – Ireland, BNetzA – Germany, ANFR – France, ARCEP – France, AKOS – Slovenia, Ofcom – UK); PlanetLab Europe supports the Data Transparency Lab (http://www.datatransparencylab.org/), an initiative of Telefónica I+D, together with Mozilla and MIT, to understand data policies around the world; Internet measurement testbeds are observing the efforts of network regulators around Europe as they implement the European Network Neutrality mandate.
- ***Productivity***: experimental platforms have delivered methodologies, tools and services to accelerate Internet research and innovation. Examples include: evaluation of novel concepts (5G, cognitive radio, optical networks, software-defined networks, terrestrial and underwater IoT, cloud) through pathways from laboratory to real-world settings (i.e. cities, regions and global); Easy access to different individual testbeds through a common portal with a comprehensive description of the and guidelines on how to access and use the federated testbeds; Increasing the reproducibility of experiments through experimentation descriptors linked to provisioning policies supported by benchmarking methodologies and tools to execute experiments, collect and compare results;

2.1.2 Drivers Transforming the Next Generation Internet Experimentation

The drivers expected to transform the NGI can be categorised into advances in intelligent spaces, autonomous cooperative machines and collective user experiences supported by key networking technologies are summarised as follow:

- **Intelligent Spaces**: enabling computers to take part in activities in which they never previously involved and facilitate people to interact with computers more naturally i.e. gesture, voice, movement, and context, etc. Internet of Things (IoT) enrich environments in which ICTs, sensor and actuator systems become embedded into physical objects, infrastructures, the surroundings in which we live and other application areas (e.g. smart cities, industrial/manufacturing plants, homes and buildings, automotive, agrifood, healthcare and entertainment, marine economy, etc.).
- **Autonomous Cooperative Machines**: intelligent self-driven machines (robots) that are able to sense their surrounding environment, reason intelligently about it, and take actions to perform tasks in cooperation with humans and other machines in a wide variety of situations on land, sea and air.
- **Collective User Experience**: human-centric technologies supporting enhanced user experience, participatory action (e.g. crowd sourcing), interaction (e.g. wearables, devices, presentation devices), and broader trends relevant to how socio-economic values (e.g. trust, privacy, agency, etc.) are identified, propagated and managed.
- **Key Networking Technologies**: physical and software-defined infrastructures that combine communications networks (wireless, wired, visible light, etc.), computing and storage (cloud, fog, etc.) technologies in support of different models of distributed computing underpinning applications in media, IoT, big data, commerce and the enterprise.

Within each category listed above, there are trends driving the need for experimentation that leads to the identification of Experimentation Challenge Areas that exhibit high degrees of uncertainty yet offer high potential for Next Generation Internet impact, as presented hereafter in this document.

2.1.2.1 Intelligent spaces

Internet of Things (IoT) is transforming every space in our daily professional and personal lives. IoT is one paradigm, different visions, and multidisciplinary activities [1] that much motivate this change. Today's Internet of Things is the world of everyday devices; 'things' working in collaboration, using mainly the Internet as a communication channel, to serve a specific goal or purpose for improving people' lives in the form of new services. In other words Internet of things has evolved from being simply technology protocols and devices to a multidisciplinary domain where devices, Internet technology, and people (via data and semantics) converge to create a complete ecosystem for business innovation, reusability and interoperability, without leaving aside the security and privacy implications.

The European and Global market for IoT is moving very fast towards industrial solutions, e.g. smart cities, smart citizens, homes, buildings this race is generating that IoT market applications have multiple shapes, from simple smart-x devices to complete ecosystems with a full value chain for devices, applications, toolkits and services. Making a retro-inspection and looking at this evolution and the role that Experimentation has played in this evolution, IoT have covered various phases in the evolution. IoT area has run a consolidation period in the technology, however yet the application side will run a long way to have big business markets and ecosystems deployed [3] and what is most important, the IoT users acceptance that will pay for services.

Wearable devices are the next evolution in the IoT horizon providing clear ways for user acceptance and further user-centric applications development. Wearable technology has been there since early 80's, however the limitation in technology and the high cost on materials and manufacturing caused wearable ecosystem(s) to lose acceptance and stop grow at that early stage. However in todays' technology and economic conditions where technology has evolved and manufacturing cost being reduced, Wearable Technology is the best channel for user acceptance and deployments in large user communities. Demands in technology & platforms (Supply Side) require further work to cope with interoperability, design and arts for user adoption, technology and management and business modelling. On the other hand from User & Community (Demand Side) it is required to pay attention in reliability of devices, cross-domain operation, cost reduction device reusability and anonymity and security of data.

Experimentation Challenges Areas for intelligent spaces may include:

- Engagement of large number of users/communities for co-creation, awareness and design constrains to improve user acceptability.
- Provisioning of large numbers of cooperative devices.
- Scale of data management associated with the scale of devices.
- Interoperability management considering the large array of "standards" that are emerging in the IoT space.

- Energy optimisation for low-powered chips, aligned with intelligence for smart devices and spaces.
- Security, anonymity and privacy because at intelligent spaces the amount of data that is produced is large and most of the time associated to users, by location, usage and ownership.
- Trust management mechanisms and methodologies for ensuring safe human acceptance/participation.

Next Generation Internet impacts are expected to include the:

- Acceptability for new innovative devices and technology that can change aspects of how we perceive aspects at work, live and home.
- Creation of communities for user acceptance and design including user personal identity and reflects the fashion trend of the users.
- Growth and matureness of particular areas, as result of the involvement of users in the process of validation and certification.

2.1.2.2 Cooperative autonomous machines

Autonomous machines operating in open environments on land, sea and air will cooperate to revolutionize applications in transport, agriculture, marine, energy and ecosystems dependent on high fidelity and real-time earth and environment observation and management. Local, regional, national and European initiatives are exploring how autonomous machines can become an integral part of the Internet infrastructure by bridging technical challenges (robotics, cyber physical systems, IoT, Future Internet) and dealing with social challenges of trustworthiness, dependability, security and border control.

Swarm robotics is here allowing collective behaviour by multi-robot systems consisting of boat/aircraft/ground vehicles. Miniaturization will be a continuous trend with nano- and micro-robotics (e.g., robotic implants). This leads to challenges in relation to human-robot coexistence and interaction (e.g., collective human-robot cooperation) along with machine simulation of human behaviour (e.g., reasoning, learning, feelings, and senses). In addition, current machines offer poor interaction with complex dynamic uncertain human-populated and natural environments.

Experimentation Challenges Areas include:

- Mixed human-robot environments (e.g., ITS environment where driverless vehicles can coexist with vehicles having human drivers).
- Heterogeneous mix of autonomous, manual and remotely operated machines.

- Machines operating in natural open and uncertain environments.
- Active security design, monitoring and mitigation in relation to emergent threats from deep learning intelligence machines and systemic dependencies.
- Paradigm shift within the Industrial Internet of Things domains towards Edge Computing, in which programmable, autonomous IoT end-devices can communicate with each other and continue to operate event without connectivity.
- 5G dense network infrastructures with Edge computing capabilities that are complemented with new M2M communications protocols/networks (i.e. NB-IoT).

Next Generation Internet impacts are expected to include:

- Systems that mix humans, machines and all ICT capabilities in ways that are acceptable to society.
- Operational models that optimize the use of distributed intelligence schemes (e.g., distributed AI reasoning, planning etc.).
- Methodologies and knowledge for investigating, developing and operating non-deterministic systems.
- Insights into the trade-offs between autonomy vs. predictability vs. security in cooperative machines.
- Insights into the evolution of legislation and regulatory policies.
- A digitalisation strategy for the industry 4.0 path supported by IoT emergence.

2.1.2.3 Collective human experience

Collective human experience is probably the major driver of Next Generation Internet as it dictates what the Internet is used for and its benefits to both individuals and the overall society. Internet participation is changing due to trends in open data, open and decentralised, shared hardware, knowledge networks, IoT and wearable technologies. Experiences are increasingly driven by participatory actions facilitated by decentralised and peer-to-peer community and open technologies, platforms and initiatives. Concepts such as decentralised network and software architectures, distributed ledger, block chains, open data, open networks, open democracy enable an active role of citizens rather than passive consumption of services and content. Internet participation is reaching, informing and involving communities of citizens, social enterprises, hackers, artists and students in multidisciplinary collaborative environments, as fostered by Internet Science and Digital Social Innovation communities, where creativity, social sciences and technology collide to create innovative solutions mindful of issues of trust, privacy and inclusion.

In addition, human-machine networks are emerging as collective structures where humans and machines interact to produce synergistic and often unique effects. In such networks humans and machines are both actors (Human to Machine – H2M and Machine to Machine – M2M) that raises important issues of "agency", to identify what actors are capable of and permitted to do. This is especially relevant to emerging machine intelligence where machines are capable of evolving intention based on sensing and learning about environments in which they operate. Facebook itself is purely a social machine as it supports Human to Human – H2H interaction whereas for example, precision agriculture with autonomous tractors, survey drones, and instrumented animals self-reporting health would be considered a H2M network.

Collective Awareness Platforms for Sustainability and Social Innovation (CAPSSI) are designing and piloting online platforms creating awareness of sustainability problems and offering collaborative solutions based on networks (of people, of ideas, of sensors), enabling new forms of sustainability and social innovation. These platforms provide strong ecosystems with thousands, or even millions of users, is built by mutual trust that interactive players are providing value to one another. The critical mass in the diffusion of innovations is "the point after which further diffusion becomes self-sustaining". The use of creativity in the innovation process through approaches such as "gamification" is a promising solution for keeping the critical mass of users engaged. The challenge is to identify innovative combinations of existing and emerging network technologies enabling new forms of Digital Social Innovation coming bottom-up from collective awareness, digital hyper-connectivity and collaborative tools.

The major underlying trends in this area include:

- Increasing self- and observer quantification and participation driving post broadcast networks with end user engagement in creative wide ranging processes.
- Increasing machine agency shifting beyond automation systems to situations post automata networks where autonomous machines increasingly evolve their own intentions and goals driven by increasingly high level

human defined policy constructs necessary to deal with the complexity of interaction.
- Increasing geographically localized interaction moving towards situations post "mega" mediator networks (interaction purely supported by Internet giants such as Google and Facebook).

Experimentation Challenges Areas include:

- Hyperlocal infrastructure, service and platform models.
- Deep "Me-as-a-Service" provisioning, orchestration and choreographies.
- Distribution of agency in networks, machines and people.
- Intention independent and transparent networking.
- Decentralized and distributed social networks, wikis, sensors, block chains value networks, driven by real-time human monitoring and observation sensor data streams.
- Accounting for the context through changing conditions.
- Experimenters' participatory involvement in collective awareness/intelli gence production.

Next Generation Internet impacts are expected to include:

- Operational models fostering localised ownership and control building on international standards.
- Multi-actor protocol/system design principles and methodologies for cooperating machines and people.
- Networking protocols robust to and adaptable to variations of outcomes and with transparent constraints.
- Participatory innovation and interaction models supporting collective intelligence production.
- Insights into the disruption of new value systems supported by emerging technologies such as block chains.
- Definition of new legislation to accommodate the entrance, and reduce barriers, of new technology, service and applications into daily lives of European citizens.
- Democratisation of the internet across new open and innovative services.
- Technology drivers that facilitate the emergence of new business models that may also operate under a collaborative economy based model. Thus, citizens and social impact is considered as a key driver for technology evolution.

2.1.2.4 Key networking technologies

Major initiatives such as the 5G-PPP are transforming wireless networking technologies and software defined infrastructures. 5G standardization is driving the activities for designing new protocols addressing diverse aspects of wireless networks and services.

Experimentation Challenges Areas include:

- Wireless investigations closer to real world ecosystems providing ways to demonstrate the applicability of experimental evidence to real-life application scenarios and to explore realistic coexistence/interference scenarios.
- Involve end devices: more flexible, compact, energy efficient radio platforms.
- End-to-end experimentation integrating radio – network – application/ services through co-design in early phases through multi-disciplinary research, development and innovation.
- Low-end vs. high-end flexible radio platforms considering new high end spectrum bands (e.g. cm and mmWave) in contrast to mobility scenarios with (very) large-scale experimentation standardisation of low-cost SDR.
- Massive (cooperative) MIMO aiming to reduce complexity & cost, and involve distributed, heterogeneous devices forming virtual antenna arrays.
- Multi-channel radio supporting multiple virtual Radio Access Technologies (RATs) running simultaneously in a single wireless node, supporting simultaneous operation of new-innovate (RATs) and traditional RATs.
- Over the air downloading of new RATs, live reprogramming of wireless device & synchronous instantiation of new RATs (adding/updating RATs) on a set of co-located wireless devices.
- SDR 'record-and-replay' building real world wireless environment (background scenarios), E.g. out-of-band transmissions (satellite, TV, aviation, etc.) to instantiate real-life scenario emulating many concurrent systems in real world.
- Co-design of the wireless access and the optical backhaul and backbone in an integrated manner, researching at the convergence point between optical and wireless networks (FUTEBOL) [15].
- NFV/VNF applications over the platforms employed by the testbeds can assist in building modular testbeds.

- New protocols based on existing technologies (e.g. beyond LTE for cellular communications, WiGig, etc.).
- New management architectures moving towards the orchestration of functionalities towards the extreme edges of the network to reduce latency, enhance reliability and ensure data sovereignty (Edge Computing).
- Complete slicing of network-topologies including available frontend and backend services such as EPC to setup separate management domains for various use cases that require partly orthogonal QoS parameters, such as IoT/M2M or CDN networks.
- Convergence of new 5G scenarios with new IoT capabilities and technologies.
- Architectures that reduce the limitations that TCP-IP have towards the expansion of Internet (i.e. mobility, addressing, etc.).

NGI impacts are expected to include:

- Evidence for performance, viability and acceptability of approaches and technologies for 5G. Proof of scalability of 5G able to cope with increasing network traffic demand, viability to migrate from legacy to 5G, coexistent of 5G and legacy.
- Evidence for robustness of networking standards.
- Homogeneous services across networks, information technologies, IoT devices and people.

2.2 Policy Recommendations for Next Generation Internet Experimentation

The drivers for the Next Generation Internet presented in this document i.e. Intelligent Spaces, Autonomous Cooperative Machines, Collective User Experience, Key Networking Technologies act as study areas that requires a dedicated consideration in policy support and European agenda reorganisation. The clear view in how the drivers are a priority for Europe, likewise the increasing convergence of Internet technologies and more involvement of the society drive the need to reconsider the design and scope of future initiatives. The following recommendations are designed to maximise the potential for Europe to create technological breakthroughs and deliver truly global impact towards Next Generation Internet Experimentation.

More Than Just Technology Networks: Successful Internet platforms deliver technology-enhanced ecosystems supporting large-scale efficient interactions between platform users. A technologically advanced platform without users will deliver no impact. Europe must focus on developing where networks of users and technology can coexist in ways that support sustainable growth of real life network and as a consequence drive demand for emerging information and communications network architectures.

Transparent and Accelerated Innovation Pathways: Industry and SMEs need clear routes to market for research and innovation activities. Platforms that deliver insight that cannot be adopted within applicable investment cycles are not relevant to business. Europe must establish experimental platforms with clear innovation pathways that deliver commercial opportunities whilst addressing contemporary/legacy constraints, market-driven interoperability/standardisation, and regulation.

Programmatic Consideration of Business and Technology Maturity: Large industry and SMEs have different capacity to invest, appetites for risk and rates of return. Europe must design and nurture current initiatives with a business and technical strategy that optimally aligns technology lifecycle phases with appropriate business engagement models for different stakeholders (Industry vs SMEs vs Research).

Quantifiably Large and Dynamic: Ecosystems must be sufficiently large and interactive to understand performance, acceptance and viability of platform technologies in real-world scenarios. Large-scale is often cited but rarely quantified. Europe must establish measurable criteria and tools for Next Generation Internet ecosystems (e.g. infrastructure, platforms, data, users, etc.) necessary to support research and pre-commercial activities ecosystems (i.e. up to city-scale), and mechanisms to rapidly scale networks towards market entry.

Nondeterministic Behaviour vs Replicability: Insights gained in one specific physical or virtual situation need to be applied in many global situations to maximise the return on investment. Computer science wants to deliver replicable experimentation however this is looking increasingly unachievable considering that networks are inherently non-deterministic and that open systems and real-life experiments only exacerbate uncertainties. Europe must foster the development of methods and tools supporting investigation into non-deterministic systems incorporating human and machine interaction in open environments that allow for insights to be replicated across the globe.

Next Generation Internet Technology and Investment Education: Learning about the potential of NGI technologies and business implications is essential for the next generation of entrepreneurs and SMEs in Europe and beyond. Unless innovators understand the ecosystem and technology potential sufficiently to convince investors (e.g. business units, venture capitalists, consumers, etc.) of the value proposition continuation funding and consequent impact will not be delivered. Europe must support platforms that educate the next generation entrepreneurs and technologists whilst supporting SMEs in the development of NGI business plans and provide ways to test the viability of solutions with potential investors.

Multidisciplinary Action: The interconnectedness of Next Generation Internet Experimentation systems means that multidisciplinary teams must work together through common objectives. Europe must support end-to-end experimentation driven by multidisciplinary teams from different technology domains (e.g. wireless networks, optical networks, cloud computing, IoT, data science) in relation to vertical sectors (healthcare, creative media, smart transport, marine industry, etc.) and horizontal social disciplines (e.g. psychology, law, sociology, arts).

Efficient and Usable Federations: Collaboration is often the most cost effective way to acquire capability, scale or reach necessary to achieve an objective. Yet the benefits of collaboration through federated platforms are limited by the barriers of interoperability, multi-stakeholder control, trust concerns and policy incompatibilities. Europe must support federated Experimentation-as-a-Service approaches where there are clear benefits to users of the federation and where techniques lower the barrier to experimentation and cost of maintaining federations through increased interoperability, usability, trustworthiness, and dynamics by contributing to or leading market accepted standardisation efforts.

2.3 References

[1] Luigi Atzori, Antonio Lera and Giacomo Morabito "The Internet of Things: A survey" Computer Networks: The International Journal of Computer and Telecommunications Networking archive, Volume 54, Issue 15, October, 2010, Pages 2787–2805, Elsevier North-Holland, Inc. New York, NY, USA.

[2] http://www.bbc.co.uk/news/technology-31145807

[3] Myriam Leggieri, Martin Serrano, Manfred Hauswirth "Data Modeling for Cloud-Based Internet-of-Things Systems" IEEE International Conference on Internet of Things (IEEE iThings 2012) France.

[4] http://www.ericsson.com/thinkingahead/the-networked-society-blog/2015/04/02/exciting-enterprise-eco-system-experiments-how-operators-can-find-the-next-growth-trajectory/

[5] https://www.androidexperiments.com/

[6] G. Jourjon, J. M. Marquez-Barja, T. Rakotoarivelo, A. Mikroyannidis, K. Lampropoulos, S. Denazis, C. Tranoris, D. Pareit, J. Domingue, L. A. DaSilva, and M. Ott, "FORGE toolkit: Leveraging distributed systems in eLearning platforms," IEEE Transactions on Emerging Topics in Computing. [Online]. Available: http://dx.doi.org/10.1109/tetc.2015.2511454

[7] FUSION Catalogue, March 2015, http://www.sme4fire.eu/documents/FUSION_Catalogue_web.pdf

[8] European Network of Internet Science D2.1.3: Repository of methodologies, design tools and use cases, http://www.internet-science.eu/publication/1268

[9] European Network of Internet Science, http://www.internet-science.eu/

[10] http://www.tvring.eu/new-tv-ring-cross-pilot-action-international-hbbtv-eurovision-song-contest/

[11] Future Internet Research and Experimentation, https://www.ict-fire.eu/

[12] https://www.fi-ppp.eu/

[13] https://ec.europa.eu/programmes/horizon2020/en/h2020-section/collective-awareness-platforms-sustainability-and-social-innovation-caps

[14] http://www.forbes.com/sites/kashmirhill/2014/06/28/facebook-manipulated-689003-users-emotions-for-science/#5700eb0c704d

[15] J. M. Marquez-Barja, M. Ruffini, N. Kaminski, N. Marchetti, L. Doyle, and L. DaSilva. Decoupling Resource Ownership From Service Provisioning to Enable Ephemeral Converged Networks (ECNs). In 25th European Conference on Networks and Communications EUCNC [To appear], 2016.

[16] http://www.wired.com/2015/04/the-apple-watch/

[17] https://playoverwatch.com/en-us/blog/20119622

[18] http://cityofthefuture-upm.com/smart-city-platform-at-madrid-moncloa-university-campus/

2.4 Experimentation Facilities Evolution towards Ecosystems for Open Innovation in the Internet of Future

2.4.1 Changes in the FIRE Portfolio

The FIRE **demand side** is changing as well, with changes in experimenter demands and requirements, and higher expectations as regards how FIRE should anticipate the needs and requirements from SMEs, industry, Smart Cities, and from other initiatives in the scope of Future Internet such as Internet of Things and 5G. Within FIRE this is also anticipated by **new types of service concepts**, for example Experimentation-as-a-Service. These new concepts affect the methods and tools, the channels for offering services to new categories of users, and the collaborations to be established with infrastructure and service partners to deliver the services.

2.4.2 Technological Innovation and Demand Pull

In response to the envisaged changes in the FIRE landscape, AmpliFIRE has identified new research directions based on interviews, literature surveys and leading conferences, and highlighted what the FIRE research, facilities and community may look like in the future [1]. We found that funded Open Calls and STREPs, and unfunded Open Access opportunities, which are increasingly aligned with the main FIRE experimental facilities, are influencing FIRE's evolution from the demand side, by showing customer "pull" supplementing and even replacing technology "push." Thus it is expected that FIRE, which has been technology-driven, will increasingly be shaped by demand-pull factors in the period 2015–2020. These user demands will be based on four main trends:

- The Internet of Things: a global, connected network of product tags, sensors, actuators, and mobile devices that interact to form complex pervasive systems that autonomously pursue shared goals without direct user input. A typical application of this trend is automated retail stock control systems.
- The Internet of Services: internet/scaled service-oriented computing, such as cloud software (Software as a Service) or platforms (Platform as a Service).
- The Internet of Information: sharing all types of media, data and content across the Internet in ever increasing amounts and combining data to generate new content.

- The Internet of People: people to people networking, where users will become the centre of Internet technology—indeed the boundaries between systems and users will become increasingly blurred.

In order to contribute to these four fast moving areas, the FIRE ecosystem must grow in its technical capabilities. New networking protocols must be introduced and managed, both at the physical layer where every higher wireless bandwidth technologies are being offered, and in the software interfaces, which SDN is opening up. Handling data at medium (giga to tera) to large (petabyte) scale is becoming a critical part of the applications that impact people's lives. Mining such data, combining information from separated archives, filtering and transmitting efficiently are key steps in modern applications, and the Internet testbeds of this decade will be used to develop and explore these tools.

Future Internet systems will integrate a broad range of systems (cloud services, sensor networks, content platforms, etc.) in large-scale heterogeneous systems-of-systems. There is a growing need for integration e.g. integration of multi-purpose multi-application wireless sensor networks with large-scale data-processing, analysis, modelling and visualisation along with the integration of next generation human-computer interaction methods. This will lead to complex large-scale systems that integrate the four pillars: things, people, content and services. Common research themes include scalability solutions, interoperability, new software engineering methods, optimisation, energy-awareness, and security, privacy and trust. To validate the research themes, federated experimented facilities are required that are large-scale and highly heterogeneous. Testbeds that bridge the gap between infrastructure, applications and users and allow exploring the potential of large-scale systems which are built upon advanced networks, with real users and in realistic environments will be of considerable value. This will also require the development of new methodological perspectives for FIRE [8].

As we emphasize focusing on "smart systems of networked applications" within the FIRE programme, the unique and most valuable contribution of FIRE should be to "bridge" and "accelerate": create the testing, experimenting and innovation environment which enables linking networking research to business and societal impact. FIRE's testbeds and experiments are tools to address research and innovation in "complex smart systems", in different environments such as cities, manufacturing industry and data-intensive services sectors [9]. In this way, FIRE widens its primary focus from testing and experimenting, building the facilities, tools and environments towards closing the gap from experiment to innovation for users and markets.

2.4.3 Positioning of FIRE

This leads to the issue of how to **position FIRE in relation to other initiatives** in the Future Internet landscape. FIRE is one among a number of initiatives in the Future Internet research and innovation ecosystem. FIRE seeks a synergetic and value adding relationship with other initiatives and players such as GÉANT/NRENs and the FI-PPP initiatives related to Internet of Things and Smart Cities, EIT Digital, the new 5G-PPP and Big Data PPP initiatives, the evolving area of Cyber-Physical Systems, and other. For the future, we foresee a **layered Future Internet infrastructural and service provision** model, where a diversity of actors gather together and ensure interoperability for their resources and services such as provision of connectivity, access to testbed and experimentation facilities, offering of research and experimentation services, business support services and more. Bottom-up experimentation resources are part of this, such as crowd sourced or citizen- or community-provided resources. Each layer is transparent and offers interoperability. Research networks (NRENs) and GÉANT are providing the backbone networks and connectivity to be used by FIRE facilities and facilities offered by other providers.

In this setting, FIRE's core activity is to provide and maintain sustainable, common facilities for Future Internet research and experimentation, and to provide customized experimentation and research services. However, given the **relevance of experimentation resources for innovation**, and given the potential value and synergies which FIRE offers to other initiatives, FIRE should assume a role in supporting experimentally-driven **research and innovation of technological systems**. For this to become reality FIRE and other initiatives should ensure cooperation and FIRE should also consider opening up to (other) public and private networks, providing customized facilities and services to a wide range of users and initiatives in both public and private spheres. FIRE's core activity and longer term orientation requires the ability to modernize and innovate the experimental infrastructure and service orientation for today's and tomorrow's innovation demands.

2.4.4 Bridging the Gaps between Demands and Service Offer

The gaps between the technologies presently offered in FIRE as testbeds, and the gaps between the layers in which its communities have formed are large. For example, the gaps between wired and wireless networking, between networking researchers and cloud application developers, and between both sorts of developers and end user input all require bridges that exist today only as research efforts (an example is the Fed4FIRE project). Developing future scenarios and identifying prospective user requirements are useful tools to shape and drive those bridging activities and chart the most direct paths from the present fragmented FIRE portfolio of testbeds, which are either hardware or user-oriented, to the goals of Horizon 2020. This requires a sustained effort to articulate how the technical goals of the present FIRE activities can be lifted, channelled and amplified to support the societal goals represented in Horizon 2020. This places requirements on the FIRE community which, as engineering teams with an often academic focus, will need to collaborate with different types of communities and actors. The FIRE community needs to clarify and justify such requirements and identify new instruments and relationships with business and SMEs that can draw upon FIRE's strengths. For this, we must expose the gaps and identify the communities that need to be engaged or created. This helps to create the "pull" that can make FIRE effective as 2020 approaches, and assist the individual projects as they provide the "push."

2.4.5 Testbed-as-a-Service

Increasingly, experimenters, developers and innovators expect to find the tools and services they need and the infrastructure in which they will do measurements and develop applications packaged in groups that allow easier access and more rapid development. The catch phrase "X as a service" (XaaS) captures these expectations. Today's infrastructures, even with the strides made towards federation and provision of powerful standard enablers, are still far from the desired shape presented in Figure 2.1. The Testbed as a Service concept (all of Figure 2.1) consists of as many as three connected layers and two value-added offerings, each of which needs to offer standard APIs and be easily adapted to multiple purposes over both long and short term.

Infrastructure available as a service benefits from the federation accomplishments of Fed4FIRE and GENI using the model of slices, and the technologies around SFA and OMF or NEPI for access to infrastructure, acquisition of reservations for resources, dispatch of experiments and capture of their results. But there is much more to be done to make these tools available to a broader audience, reduce the training requirement and learning curve. There are common elements now standardized in the OpenFlow community to make the interface to more flexible and powerful networking infrastructures itself more flexible, but these only begin to explore the ways in which the communications infrastructure can be more responsive to application

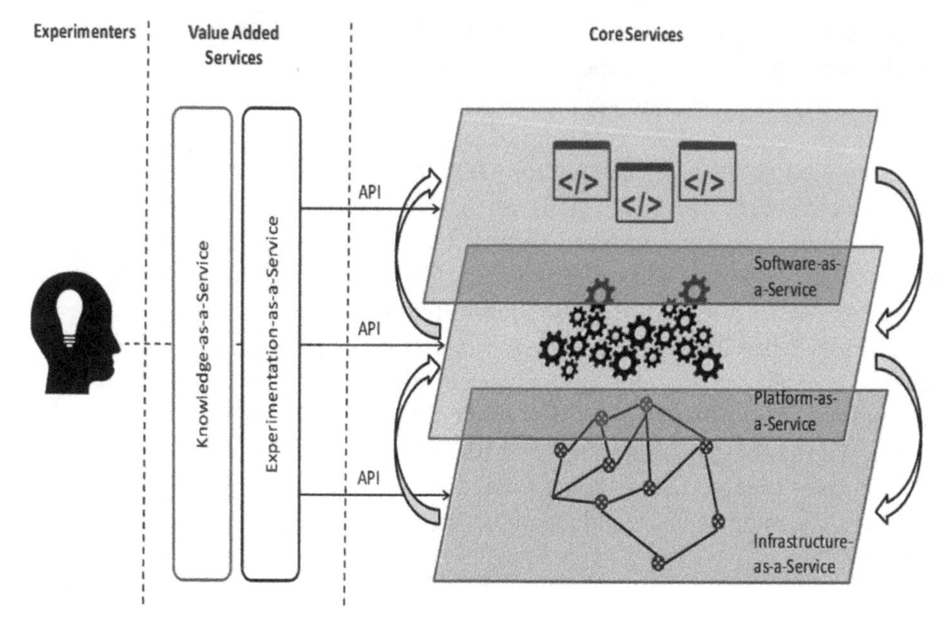

Figure 2.1 X as a Service [7].

requirements. While standard building blocks such as OpenStack exist, there are strong pressures to enhance these. FIRE can make a critical difference in evaluating the platform components proposed for extending this service concept and understanding their value. Also needed are studies of the possible options at the interfaces and their codification into APIs between the layers, and to implement services to support new demands from users more interested in the results of an experiment rather than performing their experiments themselves.

Data curation, archiving, and tools for access of experimental data, learning from experimental data, and extracting useful information using sparse sampling and other complexity techniques will be key components of Knowledge-as-a-Service. While much research in these "big data" areas is being done already in academia and in industry, FIRE with its rich trove of experimental data from Smart Cities projects, can make a contribution. Focusing on the environmental data that sensor-rich cities collect might be a good strategy, avoiding the sensitivities around healthcare data and the proprietary nature of most commercial and market activity data. Also, "big data" studies do not as a rule involve truly vast amounts of data, or require access to data centers on the largest commercial scales.

Benefiting from these opportunities requires a foundation of adaptable infrastructure, wired and wireless, software-defined, more open than ever before. The FIRE projects have made great strides in federating different kinds of facilities and exposing their novel capabilities to experimenters and end users. To meet the new demands and support the expansion to become an Internet of Things, Services, Information and People, FIRE will provide testing facilities and research environments richer than the commercial world or individual research laboratories can provide.

2.4.6 Future Scenarios for FIRE

For setting out a transition path from the current FIRE facilities towards such a "FIRE Ecosystem", AmpliFIRE identifies two key uncertainty dimensions and in that space of outcomes proposes four alternative future development patterns for FIRE (illustrated in Figure 2.2):

1. ***Competitive Testbed as a Service***: FIRE as a set of individually competing testbeds offering their facilities as a pay-per-use service.
2. ***Industrial cooperative***: FIRE becomes a resource where experimental infrastructures (testbeds) and Future Internet services are offered by co-operating commercial and non-commercial stakeholders.
3. ***Social Innovation ecosystem***: FIRE as a collection of heterogeneous, dynamic and flexible resources offering a broad range of facilities e.g. service-based infrastructures, network infrastructure, Smart City testbeds, support to user centred living labs, and other.
4. ***Resource sharing collaboration***: federated infrastructures provide the next generation of testbeds, integrating different types of infrastructures within a common architecture.

These scenarios are aimed at stretching our thinking, but FIRE must choose its operating points and **desired evolution** along these two axes. The vertical axis ranges from a coherent, integrated portfolio of FIRE activities at bottom (a natural foundation) up to individual independent projects (the traditional situation), selected solely for their scientific and engineering excellence. The horizontal line reflects both the scale of the funded projects and the size of the customer or end-user set that future FIRE projects will reach out to and be visible to. Clearly FIRE must be open to good ideas at multiple points along the scale of size. For the larger efforts, which need to engage a broad cross-section of the engineering community or the end users, the impact can be enormous.

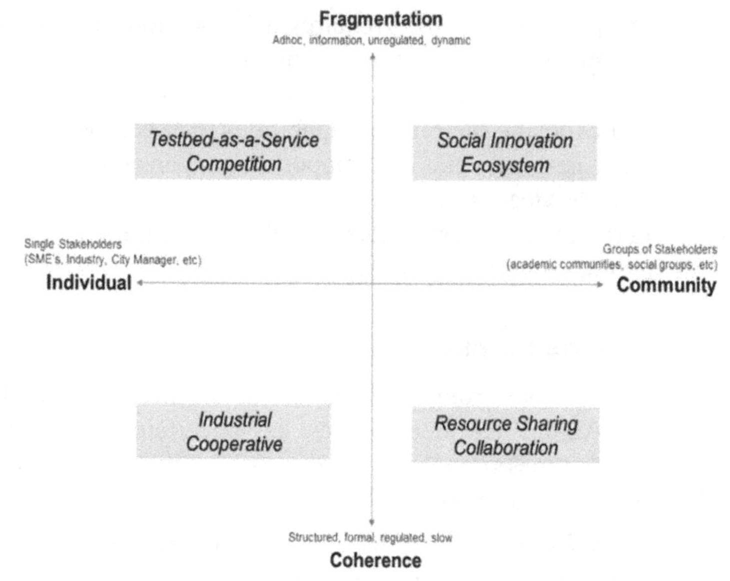

Figure 2.2 FIRE scenarios for 2020 [1].

Really innovative contributions may come from smaller, more aggressive, riskier projects. Large-scale EC initiatives such as FI-PPP, 5G-PPP, Big Data PPP and around Internet of Things (AIOTI) should have an influence on their selection and justification. Early engagement is essential to accomplish this.

2.5 FIRE Vision and Mission in H2020

FIRE's current mission and unique value is to offer an efficient and effective federated platform of core facilities as a common research and experimentation infrastructure related to the Future Internet; this delivers innovative and customized experimentation capabilities and services not achievable in the commercial market. FIRE should expand its facility offers to a wider spectrum of technological innovations in EC programmes e.g. in relation to smart cyber-physical systems, smart networks and Internet architectures, advanced cloud infrastructure and services, 5G network infrastructure for the Future Internet, Internet of Things and platforms for connected smart objects. In this role, FIRE delivers experimental testing facilities and services at low cost, based upon federation, expertise and tool sharing, and offers all necessary expertise and services for experimentation on the Future Internet part of Horizon 2020. For the **medium term**, FIRE's mission and added value is to support the Future

Internet ecosystem in building, expanding and continuously innovating the testing and experimenting facilities and tools for Future Internet technological innovation. FIRE continuously includes novel cutting-edge facilities into this federation to expand its service portfolio targeting a range of customer needs in areas of technological innovation based on the Future Internet. FIRE assumes a key role in offering facilities and services for 5G. In addition FIRE deepens its role in experimentally-driven research and innovation for smart cyber-physical systems, cloud-based systems, and Big Data. This way FIRE could also support technological innovation in key sectors such as smart manufacturing and Smart Cities. FIRE will also include "opportunistic" experimentation resources, e.g., crowd sourced or citizen- or community- provided resources.

For the **longer term**, our expectation is that Internet infrastructures, services and applications form the backbone of connected regional and urban innovation ecosystems. People, SMEs and organisations collaborate seamlessly across borders to experiment on novel technologies, services and business models to boost entrepreneurship and new ways of value creation. In this context, FIRE's mission is to become the research, development and innovation environment, or "accelerator", within Europe's Future Internet innovation ecosystem, providing the facilities for research, early testing and experimentation for technological innovation based on the Future Internet. FIRE in cooperation with other initiatives drives research and innovation cycles for advanced Internet technologies that enable business and societal innovations and the creation of new business helping entrepreneurs to take novel ideas closer to market.

In 2020, FIRE is Europe's open lab for Future Internet research, development and innovation. FIRE is the technology accelerator within Europe's Future Internet innovation ecosystem. FIRE is sustainable, part of a thriving platform ecosystem, and creates substantial business and societal impact through driving technological innovation addressing business and societal challenges.

2.6 From Vision to Strategic Objectives

The role of the FIRE vision and mission statement is to inspire for the future, answering the question "Why FIRE?" and "Where to go?" Within the context of uncertainties surrounding FIRE's longer term future, the actual evolution of FIRE is shaped by the range of scenarios and by the planning and implementation decisions that are being taken within the EC and within FIRE and related initiatives. For example, the Fed4FIRE project to create a high-level framework is driving coherence in technology, operations and governance across many of the FIRE facilities. There are also interesting implications regarding collaboration of FIRE facilities with related programs such as Future Internet PPP and possibly the Big Data PPP which are more oriented towards business innovation than FIRE. Testbeds participating in these initiatives may have to operate in more than one scenario, requiring them to adapt new operational models, legal contexts and technical implementations.

To structure the process of identifying future directions, FIRE should agree on strategic objectives for its mid- and longer term evolution. Technical objectives oriented towards FIRE's core activity are a necessity but they are not sufficient on their own as FIRE also needs strategic positioning in terms of how it achieves sustainable value creation activity and how it positions and interacts with other major initiatives.

2.6.1 Strategic Objectives

We identified the overall strategic objective for FIRE as to become a sustainable environment for research, development and innovation in the Future Internet, supporting researchers and the community to tackle important problems, and acting as an accelerator for industry and entrepreneurs to take novel ideas closer to market. Figure 2.3 visualises the potential strategies that could be employed to achieve these objectives in a high-level roadmap.

The key strategic objectives for FIRE will be:

- For 2016: to increase its relevance and impact primarily for European wide technology research, but will also increase its global relevance.
- For 2018: to create substantial business and societal impact through addressing technological innovations related to societal challenges.
- For 2018: to become a sustainable and open federation that allows experimentation on highly integrated Future Internet technologies; supporting networking and cloud pillars of the Net Futures community.
- For 2020: to become the RDI environment space that is attractive to both academic researchers, SME technology developers, and industrial R&D companies with emphasis on key European initiatives such as 5G, Big Data, IoT and Cyber-Physical Systems domains.

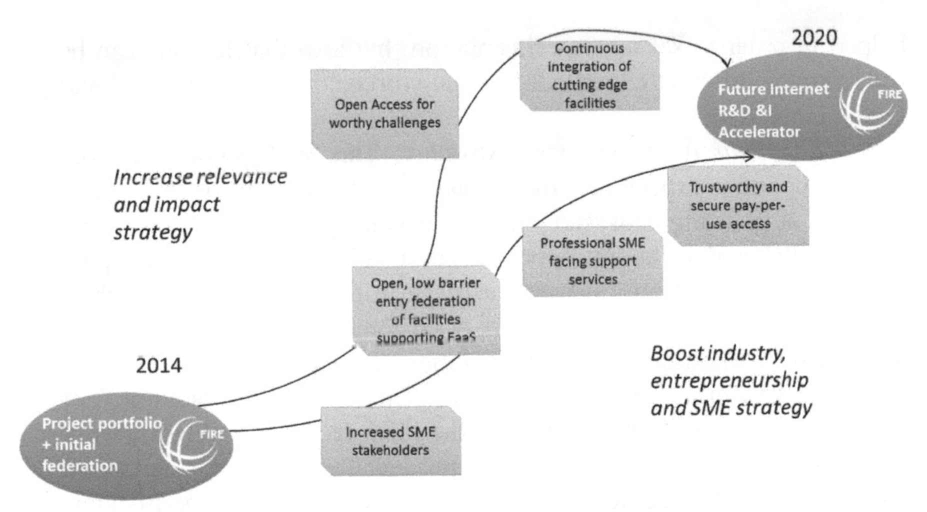

Figure 2.3 Overall strategic direction of FIRE [2].

2.6.2 FIRE's Enablers

AmpliFIRE's report on FIRE Strategy [2] provides a detailed elaboration of strategic directions for FIRE's "enablers": the domains of service offering, facilities and federation, EC programme relations, ecosystem development, and collaboration. Below we concisely address some of the main points.

Service offering. On the shorter term, FIRE's service offer strategy must ensure that FIRE remains relevant and meet current and future experimenter demands and be driven by demand [5, 7]. FIRE should also promote common tools and methodologies to perform experiments. FIRE's offer in the next years will transform towards a service-oriented framework where the concept of Experimentation as a Service is central. The model presented in Figure 2.1 depicts how facilities or federations can offer a service to experimenters. The lowest layer is the infrastructure, the actual physical machines. In the middle is the platform layer, able to control the infrastructures in a more organized manner, making use of predefined APIs, such as software-defined networks. On the topmost layer, software can be run as a service, giving experimenters access to applications. Crossing these layers, two services can be defines. One is experimentation as a service, where experimentation is offered in a customized approach with less or no concern about the infrastructure, platform or services behind the scene; just knowing that it is available and can be accessed is in most cases enough. The Fed4FIRE project serves as an example. Additionally a final step could be knowledge as a service, where experimenters are helped in order to set up experimentation, but also that lessons can be learned from the different experiments (what worked, what didn't work) and can be disseminated.

User and community ecosystem strategy. This will become a more and more important aspect of FIRE strategy and future business model. The concept of platform ecosystem and multi-sided platforms is potentially relevant for FIRE and opens new opportunities. Unlike a value chain or supply chain, a (multi-sided) platform-based activity brings together and enables direct interactions within a value network of customers, suppliers, developers and other actors. The range of FIRE facilities and services can be seen as constituting a platform ecosystem facilitating multi-sided interactions. For example, developer communities may use the FIRE facilities to directly work with business customers on technology and product development, whereas the current FIRE service model focuses on giving researchers and experimenters access to FIRE facilities[1]. The issue is then to what extent the current FIRE ecosystem realizes its opportunities and what the strategic options are to extend the current FIRE model to a platform-based ecosystem model.

Collaboration strategy. Given FIRE's positioning in the wider Future Internet ecosystem collaboration in the shorter and longer term is essential and must be grounded in clear value propositions [10]. To reach the next phase FIRE should target both strong ties and loose ties collaboration. By strong ties we refer to relationships that have developed throughout many years, while loose ties collaboration is represented by more dynamic relationships. Both are of equal importance. By close collaboration between different actors within the FIRE value-network we can capitalize on sharing of testbed resources, and foster FIRE to become more dynamic and user-driven to attract and serve a wider base of partners. This also includes a complex prosumer exchange value-network structure where providers of testbed assets also can be users and vice versa. In existing FIRE collaborations these prosumer structures can be found as strong elements for sustainability beyond the lifetime of a project and foster long-term relationships. Also the framework for cooperation must support flexible forms and easier entry into collaborations as well as to sustain beyond the lifetime of a project.

As FIRE is positioned in an environment of continuous change also FIRE collaboration relations will evolve and new relationships and partners

[1]In [3], AmpliFIRE discusses broadening the Future Internet user base by providing experimenter solutions, offering APIs that match community practices (BonFIRE, Experimedia).

will emerge finding new opportunities for win–win by collaboration but also defining new demands for being part of the FIRE value-network. In this context FIRE needs evolution in several domains and even to reflect on its position in being a "research and experimentation environment" as this is being more attractive for research partners than other actors. How can FIRE also serve stakeholders with specific interest in the development of new services and products for the Future Internet with a commercial purpose? These stakeholders are mainly representatives from industry and their requirements on collaboration models might differ from the existing more research oriented. To increase their attraction for FIRE collaboration the FIRE value-network should be extended by complementary partners to the traditional ICT actors, e.g. customers and users. But can FIRE really fit all? FIRE will remain interested to cooperate with core initiatives within the landscape of Future Internet research, innovation and experimentation, like 5G-PPP, FI-PPP, Internet of Things, Smart Cities, Big Data, which requires FIRE to show a clear position on its offerings and uniqueness. Some examples:

- **5G-PPP**: FIRE experimental facilities could potentially be of use for the 5G-PPP. Fed4FIRE offers a large number of federated facilities across Europe of which most are potentially important for 5G testing (including cellular networks, WiFi and sensor based networks, cognitive radio networks, but also SDN and cloud facilities). CREW offers open access to wireless testbed islands and advanced cognitive radio components as well as support services.
- **FI-PPP**: integration of relevant FIRE facilities in XiFi's federated nodes infrastructure, especially physical computing/storage facilities and back-end infrastructures such as sensor/IoT networks used by applications and services to run experiments on top of them.
- **GÉANT/NRENS**: cooperation in terms of connectivity is ongoing in several FIRE projects. Other opportunities could include extending GÉANT service offerings to include testbed as a service. Some related activities are going on in federation of testbeds, and experiment management towards Experimentation-as-a-Service (Fed4FIRE), and resource control and experiment orchestration and monitoring (OpenLab, FLEX, CREW). FIRE projects might extend their use of GÉANT/NREN resources and FIRE and GÉANT may cooperate in services and resources. FIRE may leverage GÉANT facilities and improve GÉANT services adding services such as testbed access. FIRE and GÉANT can also collaborate on SDN/Networking Protocols & Management.

- In relation to **Smart Cities**, and technological innovations in the domain of **Internet of Things** and **cloud computing** of high relevance for city innovation, FIRE moves further into this direction in projects such as OrganiCity and FIESTA. Next steps would be to establish project-oriented discussions and explore opportunities for common calls with key organisations in this area.

In order to develop FIRE collaboration opportunities for the future the ability in realization and implementation of concrete collaboration models will be essential. To do so collaborating partners must be able to define what is the goal of collaboration, what is the win-win and what are the assets used to enable collaboration and to establish an exchange structure for the collaboration as well as models for governance. Therefore we should ask ourselves *Who* is the formal body to interact with and to formalize collaboration? Finally, realizing the FIRE collaboration vision beyond 2020 requires to be linked with and to influence what FIRE partners today and in the future define as the strategic directions of FIRE and what partners want it to be to be attractive for collaboration.

Portfolio management. There is an inevitable problem of getting coherence with a selection of projects chosen individually for their excellence by mostly academic referees. Incentives added in the past include asking projects to present evidence of a relationship with existing FIRE projects (easy to do towards the end of a Framework Programme, not so easy at the outset of one, but FIRE's continuity may alleviate this). This results in project groupings which allow more varied approaches still focused on a single infrastructure technology or bringing a single technology closer to end users.

One suggestion that has been raised in recent years is finding ways in which the FIRE programme can provide some of the assistance and even direction that is offered to start-up companies. This may involve management attention and involvement in changing project directions that were difficult to achieve under FP7 and may have become impossible in Horizon 2020. Nonetheless, we present in this review the suggestion that a support action focused on achieving earlier and better exploitation might be considered and describe how it could work, and what problems it would solve.

Managing innovation and exploitation needs attention and could be addressed more systematically. Today, many projects end after the first demonstrations are presented. Exploitation may be planned, but it lies in the future, if it happens at all. Project structures, as specified in future calls could, by the middle period of 2016–2018, require that some projects have their capabilities demonstrated and external interfaces ready for the first full review.

These projects could then report progress on external utilization and exploitation by the end of the project. Although not all projects will, or should, achieve this, we can imagine seeing identification of partners and a pathway to commercialization by the end of a FIRE project.

Future sustainability. Sustainability of the FIRE ecosystem has been raised as a concern in many of the interviews we conducted after issuing the draft FIRE radar vision document. Users want to see one or a few components of a FIRE testbed sustained (or successfully evolving) and the ultimate responsibility lies with the institutions in which these components reside. If only one institution is involved, as is the case with iMinds in Belgium, a member of OpenLab, Fed4FIRE and CREW, then sustainability of the several component testbeds that iMinds supports (the Virtual Wall, the W.iLab.t and others) is addressed through the institution pursuing multiple means of support. In the case of iMinds, all modes seem to be open – EC funding, regional support and industrial partnerships have all contributed. For testbeds whose components are distributed over multiple institutions, projects like BonFIRE and OFELIA have created informal consortia which continue beyond any single EC integrated project, and link only the key partners. Typically these consortia intend to offer something like Open Access or similar lightweight short-term involvement in their testbed's use, and will explore multiple sources of funding to make this happen. Accounting systems to allow fairly precise allocation of costs to the different uses that result are being created as they will be needed downstream in this model. Finally, the OneLab Foundation is an actual legal entity that has been created to manage the activities of the PlanetLab Europe, NITOS, and FIT-IoT Lille testbeds using the network operating center (NOC) and federation toolkit that has been created under OpenLab and Fed4FIRE.

2.7 FIRE Roadmap towards 2020

2.7.1 Milestones

The FIRE Roadmap of milestones is shown in Table 2.1 [3]. It essentially pinpoints milestones for FIRE to deliver within the framework of roadmap solutions. For example, "before 2016, open access will be a requirement of a FIRE testbed". The table is split into three phases: i) 2014–16, ii) 2016–2018, iii) 2018–2020 that identify the milestones and decision points of the roadmap. These phases are then broken down into a common template of solutions within layers of the FIRE ecosystem:

Table 2.1 FIRE Roadmap Milestones 2014–2020

	2014–2016	2016–2018	2018–2020
FIRE Resources Solutions	Testbeds will be established in the domain of software services (2016) Gradual implementation of converged federation (2016)	Cutting-edge FIRE testbeds are established in key areas such as 5G, IoT, Big Data (2016–2017) A converged set of resources is aligned with 5G architectures (2017–2018)	Continuing to establish cutting-edge FIRE testbeds in key areas such as 5G, IoT, Big Data (2018–2020) A converged set of resources is aligned with 5G architectures (2018–2020)
FIRE Services and Access Solutions	Open Access is implemented as a requirement (2015–2016) Projects are funded that develop services supporting reproducibility (M16) EaaS solutions will get harmonized and interoperable (2016–2017) All FIRE Open Access projects get integrated into one single portal for offering coherent package of services (2015–2016)	Mechanisms are set in place that support cross-facility experimentation through a central experimentation facility (2016) A FIRE Broker initiative is implemented providing broker services across the FIRE portfolio (2017)	Implementation of a new financing model to ensure sustainability of resources (2019) FIRE legal entity enables pay-per-use services (2018–2019) FIRE facilities implement secure and trustworthy resources capabilities (2019)
FIRE Experimenters Solutions	Alignment of EC units leads to cross-domain access to facilities and services (2016–2017) FIRE is made accessible to wider communities by offering community APIs (2015–2016)	Alignment of FIRE and 5G in terms of facilities, services and experimentation actions (2016–2017) Introduction of accelerator functionality for "technology accelerator"	SMEs are key target group of FIRE, with Open Calls specifically dedicated to SMEs (2018) Professionalisation of FIRE services marketing Introduction of startup funding as part of "full-service accelerator"
FIRE Framing conditions solutions	Professionalization of FIRE's internal organization (2015) Collaboration agreements in place between FIRE and large initiatives such as 5G PPP (2015)	A Network of Future Internet Initiatives is established (2016–2017) Cross-initiative collaboration in the Future Internet domain is implemented to enable seamless interconnection	FIRE, within NFII, is operating as legal entity to ensure sustainability and professionalisation

- The FIRE *resources* layer considers the role of the testbeds made available through FIRE i.e. whose development is funded in part by the FIRE programme. These represent an important element in achieving objectives through making the right experimental facilities available, sustaining facilities, and ensuring provision meets user demands.
- The FIRE *service and access* layer considers the services provided to the user to allow them to perform experiments; these can be experimental services to perform and monitor experiments (set up experiment, report on results, etc.), services to utilise facilities directly (SLA management, security, resource management), and central services managing the FIRE offering (e.g. a FIRE portal). Also the mechanisms employed to allow users to access and make use of the testbed are considered e.g. fully open access, open calls, policy based access, etc.
- The FIRE *Experimenter* layer considers the consumer, i.e. the overall FIRE user base who utilise the available FIRE testbed resources. Solutions in this layer will implement changes in the user base, e.g. changing from a traditional academic community in Europe, to a more global community, and/or more industry and SME users.
- FIRE *framing conditions* solutions address the activities concerning the ecosystem conditions and the activities carried out to operate FIRE, and also integrate FIRE with wider initiatives.

Phase I: 2014–2016

In this period, partly covering the new Work Programme 2016–2017, we expect continued and intensified attention to funding facilities that increase impact and relevance by balancing Future Internet pillars. Testbeds in the domain of software and services are prioritized. Cutting-edge testbeds should be added in key areas 5G, IoT, Big Data and Cyber-Physical systems. Loosely-coupled FIRE federation will be continued in order to simplify cross-domain experimentation. In order to increase the experimental use of facilities, FIRE-funded facilities will be required to offer open access Also, ease of use and repeatability and reproducibility of experiments must be improved by promoting Experimentation-as-a-Service concepts. Both actions aim at simplifying cross-domain experimentation. The main priority regarding experimenter solutions is to increase the user base and actual use of facilities, by making FIRE accessible to the larger Future Internet community, by offering community APIs and establishing interoperability. The FI-PPP and GENI are prominent initiatives in this time period. Also, common experimentation standards across initiatives will be required, such as cloud and IoT APIs. Strategic alignment and collaboration between FIRE and other EC programmes (DG CONNECT and wider) needs to be pursued, e.g. preparing for joint calls and stimulating interactions among the Unit priority areas. FIRE as a community needs to start working towards a credible level of organisation to prepare for sustainability and professional service offers.

Phase II: 2016–2018

In this period, FIRE establishes cutting-edge Big Data facilities relevant to research and technology demands to support industry and support the solving of societal challenges. Federation activities to support the operation of cross-facility experimentation are continued. A follow-up activity of Fed4FIRE is needed which also facilitates coordinated open calls for cross-FIRE experimentation using multiple testbeds. Additionally, a broker service is provided to attract new experimenters and support SMEs. This period ensures that openly accessible FIRE federations are aligned with 5G architectures that simplify cross-domain experimentation. Second, via the increased amount of resources dedicated to Open Calls, FIRE will create an Accelerator functionality to support product and service innovation of start-ups and SMEs. For this, FIRE will establish a cooperation with regional players and other initiatives. FIRE continues to implement professional practices and establishes a legal entity which can engage in contracts with other players and supports pay per use usage of testbeds.

Phase III: 2018–2020

FIRE continues to add new resources that match advanced experimenter demands (5G, large-scale data oriented testbeds, large-scale IoT testbeds, cyber-physical systems) and offers services based on Experimentation-as-a-service. The services evolve towards experiment-driven innovation. More and more FIRE focuses on the application domain of innovative large-scale smart systems. Implementing secure and trustworthy services becomes a key priority, also to attract industrial users. Responsive SME-tailored open calls are implemented, to attract SMEs. FIRE continues the Accelerator activity by providing dedicated start-up accelerator funding. FIRE takes new steps towards (partial) sustainability by experimenting with new funding models. Sustainable facilities are supported with continued minimum funding after project lifetime. FIRE community has achieved a high level of professional operation. FIRE contributes to establishing a network of Future Internet initiatives which works towards sharing resources, services, tools and knowledge and which is supported by the involved Commission Units.

2.7.2 Towards Implementation – Resolving the Gaps

Setting out a vision, strategy and roadmap must go hand in hand with being aware about the gaps that need to be resolved. Two categories can be distinguished: 1) gaps with respect to the current FIRE offerings, and 2) gaps with respect to the FIRE vision. The current FIRE offering has evolved from individual projects, many of which had specific project objectives to build testbeds on which to make experiments, but were not expected to federate with others, be open for researchers outside of the project consortium, or continue after the end of the project contract timeframe. The fact that these features are now increasingly being offered is a result of earlier gap analyses by FIRE stakeholders and actions taken by the EC to address the issues incrementally in successive Calls for Proposals. The assessment of FIRE's relevance for Future Internet experimenters is, however, a continuous process; new technologies, devices and protocols emerge and new ways of improving the experience for both experimenters and testbed providers are identified. AmpliFIRE's Portfolio Capability Analysis [4] lists some of the main gaps with respect to the current FIRE offering that have been identified by experimenters (or potential experimenters). In many cases, these gaps reflect the increasing interest being shown in the FIRE facilities by SMEs and industry organisations, as opposed to the traditional users, who are largely from the academic community.

Many of the gaps, in particular those associated with the usage of FIRE testbeds by a higher number of SMEs and industrial organisations, are common to the needs for FIRE testbeds identified by the reports on FIRE Vision [1] and FIRE Future Structure and Evolution [2]. However, we have identified additional requirements, related to 1) the concept of FIRE becoming the common European Experimentation Infrastructure incorporating FIRE testbeds with ESFRI, FI-PPP, CIP ICT-PSP, GEANT; and 2) the transitioning of the more mature FIRE facilities towards business innovation and education platforms within (for example) the EIT Digital context. In general terms – whilst FIRE has been strong, historically on networking topics – more effort needs to be placed now on service aspects and extending expertise into the commercial area. Testbed-as-a-Service, Experimentation-as-a-Service, Knowledge-as-a-Service, and all of the functions and tools that underpin these concepts become increasingly important. We propose the following actions to address the identified gaps:

- Common FIRE tools should be built for TaaS, EaaS and KaaS, rather than each project developing their own.

- One FIRE portal should exist, through which the resources of all FIRE projects can be accessed by experimenters as a single entity.
- There should be a more coordinated approach to FIRE collaboration (e.g. with respect to support for the FI-PPP, 5G-PPP, Big Data PPP etc.), rather than the ad-hoc mechanisms applied today.
- For addressing the sustainability issue, an independent stakeholder alliance funding mechanism to manage the European common platform should be considered.

2.8 Main Conclusions and Recommendations

FIRE has evolved into a diverse portfolio of experimental facilities, increasingly federated and supported by tools, and responding to the needs and demands of a large scientific experimenter community. Issues that require attention include the sustainability of facilities after projects' termination, the engagement of industry and SMEs, and the further development of FIRE's ecosystem. A more strategic issue is to develop a full service approach addressing the gaps between ecosystem layers and addressing integration issues that are only now coming up in other Future Internet-funded projects. A related challenge is to expand the nature of FIRE's ecosystem from an offering of experimental facilities towards the creation of an ecosystem platform capable to attract market parties from different sides that benefit from mutual and complementary interests. Additionally, FIRE should anticipate the shifting focus of Future Internet innovation areas towards connecting users, sensor networks and heterogeneous systems, where data, heterogeneity and scale will determine future research and innovation in areas such as Big Data, and 5G and IoT [9]. Such demands lead to the need for FIRE to focus on testbeds, experimentation and innovation support in the area of "***smart systems of networked infrastructures and applications***".

To address the viewpoints identified by the FIRE community, the FIRE initiative should support actions that keep pace with the changing state-of-the-art in terms of technologies and services, able to deal with current and evolving experimenter demands. Such actions must be based upon a co-creation strategy, interacting directly with the experimenters, collecting their requirements and uncovering potential for extensions. FIRE must also collaborate globally with other experimental testbed initiatives to align with trends and share expertise and new facilities. Where major new technologies emerge, these should be funded as early as possible as new experimental facilities in the FIRE ecosystem.

This analysis leads to conclusions and recommendations regarding the future direction of FIRE. The following is a concise summary of conclusions and recommendations, grouped in three areas: (1) the vision and positioning of FIRE, (2) the strategic challenges, and (3) the action plans. These conclusions and recommendations have been elaborated in more detail in the AmpliFIRE D1.2 report [11].

2.8.1 FIRE Vision and Positioning

- FIRE's strategic vision for 2020 is to be the Research, Development and Innovation (RDI) environment for the Future Internet, creating business and societal impact and addressing societal challenges. Adding to FIRE's traditional core in networking technologies is shift of focus in moving upwards to experimenting and innovating on connected smart systems which are enabled by advanced networking technologies.
- FIRE must forcefully position the concept of experimental testbeds driving innovation at the core of the experimental large-scale trials of other Future Internet initiatives and of selected thematic domains of Horizon 2020. Relevant initiatives suitable for co-developing and exploiting testbed resources include the 5G-PPP, Internet of Things large-scale pilots, and e-Infrastructures.

2.8.2 Strategic Challenges for Evolution of FIRE

- FIRE should help establish a network of open, shared experimental facilities and platforms in co-operation with other Future Internet initiatives. Experimental facilities should become easily accessible for any party or initiative developing innovative technologies, products and services.
- FIRE should establish a "technology accelerator" functionality, by itself or in co-operation with other Future Internet initiatives, to boost SME research and innovation and start-up creation. A brokering initiative should provide broker services across the FIRE portfolio or via exploitation partnerships. Community APIs should be offered to make FIRE resources more widely available.
- FIRE's core expertise and know-how must evolve: from offering facilities for testing networking technologies towards offering and co-developing the methodologies, tools and processes for research, experimentation and proof-of-concept testing of complex systems. FIRE should establish a

lively knowledge community to innovate methodologies and learn from practice.

- FIRE should ensure longer term sustainability building upon diversification, federation and professionalization. FIRE should support the transition from research and experimentation to innovation and adoption, and evolve from singe area research and experiment facilities towards cross-technology, cross-area facilities which can support the combined effects and benefits of novel infrastructure technologies used together with emerging new service platforms enabling new classes of applications.
- FIRE should develop and implement a service provisioning approach aimed at customized fulfilment of a diverse range of user needs. Moving from offering tools and technologies FIRE should offer a portfolio of customized services to address industry needs. FIRE should establish clear channels enabling interaction among providers, users and service exploitation by collaboration partners.
- FIRE should become part of a broad Future Internet value network, by pursuing co-operation strategies at multiple levels. Cooperation covers different levels: federation and sharing of testbed facilities, access to and interconnection of resources, joint provision of service offerings, and partnering with actors in specific sectoral domains. In this FIRE should target both strong ties and loose ties opportunistic collaboration. Based on specific cases in joint projects, cooperation with 5G and IoT domains could be strengthened [10].
- FIRE should evolve towards an open access platform ecosystem. Platform ecosystem building is now seen critical to many networked industries as parties are brought together who establish mutually beneficial relations. Platforms bring together and enable direct interactions within a value network of customers, technology suppliers, developers, facility providers and others. Developer communities may use the FIRE facilities to directly work with business customers and facility providers. Orchestration of the FIRE platform ecosystem is an essential condition.

2.8.3 Action Plans to Realize the Strategic Directions

- The ongoing development towards federation of testbeds should be strongly supported; it is a key requirement now and in the future. We have proposed several actions to accomplish this goal, which is taken up in the Work Programme 2016–2017.

- FIRE should strengthen the activities aimed at wider exploitation of its testbed resources by increasing the scope and number of experiments and experimenters using FIRE facilities.
- FIRE should increase the number of projects and experiments that lead to resolving societal challenges. Bring end user communities to the FIRE community to stimulate innovation for the social good. Promote open source community building methods such as hackhatons and open source code.
- FIRE should initiate actions to leverage its resources to start-ups and SMEs.
- FIRE should initiate activities aimed at decreasing the time to market for experimenters.
- FIRE should maintain and strengthen its relevance for the researcher community.
- The potential capability of FIRE facilities and resources for regional development, to support technology development and product and service innovation, should be exploited.
- FIRE should expand its range of facilities to also address research and innovations in sectors where "networked, smart systems" are crucial for innovation.
- FIRE facilities are to be exploited for standardisation activities (proof-of-concept).
- FIRE should selectively engage in international co-operation, based on reciprocal and result oriented actions.
- Create co-operation across Future Internet related initiatives and stimulate alignment of EC units.
- FIRE should establish a professionally coordinated community to lead its development toward 2020.

2.9 Final Remarks

As explained in Section 2.2's vision and mission statement for FIRE and detailed in Sections 2.3–2.4, we foresee a further development of FIRE's mission and value offer. One particular challenge is to expand the nature of the FIRE's ecosystem, from offering facilities to mostly experimenters in academic research institutes towards a **wider spectrum of actors** in a growing FIRE ecosystem, including large businesses and SMEs, developer communities, and other initiatives or programmes. FIRE will continue to offer an efficient and effective federated platform of core facilities as a common research and experimentation infrastructure related to the Future

Internet; this delivers innovative and customized experimentation capabilities and services not achievable in the commercial market. FIRE will expand its facility offers to a wider range of technological developments in EC programmes e.g. in relation to smart cyber-physical systems, smart networks and Internet architectures advanced cloud infrastructure and services, 5G network infrastructure for the Future Internet, Internet of Things and platforms for connected smart objects. FIRE delivers experimental testing facilities at low costs based upon federation, expertise and tool sharing, offering all necessary expertise and services for experimentation on the Future Internet part of H2020. In the longer term, FIRE's mission is to be the research, development and innovation environment, or "accelerator" within Europe's Future Internet innovation ecosystem, providing the facilities for research, early testing and experimentation of innovative technologies and solutions, by accelerating Future Internet technology-induced innovation cycles resulting in advanced applications and business support leading to the creation of new market opportunities. The overall strategic objective for FIRE is to become a sustainable 'R&D&I lab'-like facility for research in the Future Internet; supporting researchers and the community to tackle important problems, and acting as an accelerator for industry and entrepreneurs to take novel ideas closer to market.

The strategy to realize this future role is multidimensional and AmpliFIRE jointly with the FIRE community and the Commission have been working towards the definition of a set of strategic objectives aimed at 2020, and a range of activities to realize the 2020 objectives.

The strategy includes the following key recommendations:

- Establish an easily accessible network of open and shared experimental facilities and platforms and create partnerships with other Future Internet initiatives to realize this.
- Target industry and SME innovators by establishing an "accelerator" functionality, starting with creating a market interface aimed at aligning demands and offers.
- Increase the number of experiments and experimenters using FIRE, attracting new user/stakeholder groups such as large ICT companies, developer companies, SME innovators, Smart Cities and regions, and other EC programmes.
- Target business innovator needs related to accelerating product and service innovation and go-to-market, addressing the needs and demands of companies in different stages of their development lifecycle. Work together with innovation intermediaries.

References to AmpliFIRE Reports and White Papers

[1] AmpliFIRE D1.1: FIRE Vision and Scenarios 2020. Final report, May 2015.
[2] AmpliFIRE D1.2: FIRE Future Structure and Evolution. Interim report, June 2014.
[3] AmpliFIRE D1.3: FIRE Ecosystem Progress Report. FIRE Roadmap towards 2020. Final report, June 2015.
[4] AmpliFIRE D2.1: FIRE Portfolio Capability Analysis. Final report, June 2015.
[5] AmpliFIRE D2.2: Overview of Experimenter Requirements. Final report, February 2015.
[6] AmpliFIRE D3.1: FIRE Collaboration Models. Final report, April 2015.
[7] AmpliFIRE D3.2: FIRE Service Offer Portfolio. Final report, March 2015.
[8] AmpliFIRE White Paper: Experimental Methodology Challenges. June 2015.
[9] AmpliFIRE White Paper: Potential of FIRE as Accelerator for New Areas of Technological Innovation. June, 2015.
[10] AmpliFIRE White Paper: FIRE Value Proposition. September 2014.
[11] AmpliFIRE D1.2: FIRE Future Structure and Evolution. Conclusions and Recommendations for FIRE's Future. Final report, February 2015.

This section makes reference to the AmpliFIRE Final Summary Report, December 2015, presents a synthesis of results of the AmpliFIRE Support Action, funded by the European Commission (Grant Agreement 318550).
Editor: Hans Schaffers, AmpliFIRE coordinator, Aalto University School of Business, Centre for Knowledge and Innovation Research (CKIR), E-mail: hans.schaffers@aalto.fi
Contributors: Michael Boniface (IT Innovation), Stefan Bouckaert (iMinds), Monique Calisti (Martel), Diana Chronér (LTU), Paul Grace (IT Innovation), Jeaneth Johansson (LTU), Scott Kirkpatrick (HUJI), Timo Lahnalampi (Martel), Jacques Magen (InterInnov), Malin Malmström (LTU), Santiago Martiñez Garcia (Telefónica R&D), Bram Naudts (iMinds), Michael Nilsson (LTU), Jan van Ooteghem (iMinds), Martin Potts (Martel), Géraldine Quetin (InterInnov), Sathya Rao (Martel), Mikko Riepula (Aalto University), Annika Sällström (LTU), Hans Schaffers (Aalto University).

Project Officer European Commission, Experimental Platforms Unit E.4: Nikolaos Isaris.
The AmpliFIRE consortium gratefully acknowledges the results of discussions organised during the AmpliFIRE's lifetime, especially in the context of the FIRE Board and FIRE Forum events and dedicated workshops.
For more information, see AmpliFIRE's website http://www.ict-fire.eu/home/amplifire.html
AmpliFIRE reports and White Papers are available for downloading at this website.

PART II

Experimentation FACILITIES Best Practices and Flagship Projects

3

Fed4FIRE – The Largest Federation of Testbeds in Europe

Piet Demeester[1], Peter Van Daele[1], Tim Wauters[1] and Halid Hrasnica[2]

[1]iMinds, Belgium
[2]Eurescom GmbH, Germany

3.1 Introduction

The Fed4FIRE[1] project has established a European Federation of experimentation facilities and testbeds and developed necessary technical and operational federation framework enabling the federation operation. With its 23 tesbeds, the Fed4FIRE represents the largest federation of testbeds in Europe which allows remote testing in different areas of interests; wireless, wireline, open flow, cloud, etc. Various user friendly tools established by the Fed4FIRE project enable remotely usage of the federated testbeds by experimenters who can combine different federation resources, independently on their location, and configure it as it is needed to perform the experiment.

The main idea behind the Fed4FIRE Federation of testbeds is to enable easy and efficient usage of already available experimental resources by the entire research and innovation community in broad area of Future Internet and Communications Technologies (ICT) as well as various vertical application sectors applying the ICT, such as Energy, Health, Automotive, Transport, Media, etc. To ensure it, the Fed4FIRE project worked on establishing the federation of testbed for benefit of both testbed providers and experimenters by taking into consideration their particular requirements and interests.

Until now, more than 50 experiments have been using the Fed4FIRE experimental facilities and tools. Part of them took opportunity of seven Open

[1]Fed4FIRE is an Integrating Project under the European Union's Seventh Framework Programme (FP7) addressing the work programme topic Future Internet Research and Experimentation. It started in October 2012 and has been running for 51 months, until the end of 2016 – http://www.fed4fire.eu/

Calls for Experiments organized by Fed4FIRE project in last three years. Other experimenters used the Fed4FIRE Open Access mechanism which allows free of charge access to the experimental facilities and support for setting up the experiments from Fed4FIRE team.

The Fed4FIRE experimenters had opportunity to experience all advantages of the Fed4FIRE tools, to configure and successfully execute planned experiments. The feedback received from the experimenters on usability of Fed4FIRE facilities and tools was very positive. Moreover, the most of the performed experiments would be even not possible without provision of the Fed4FIRE federation and its experimental facilities. Thus, the Fed4FIRE facilities helped the experimenters to further explore their research and business development based on results gathered from the experiments.

This chapter is organized as follows; In Section 3.2, overall needs for the federated experimentation facilities and scope of a federation of testbeds as well as Fed4FIRE approach to establish a testbed federation, including currently involved testbeds, have been elaborated. Common framework for establishing large-scale federation of testbeds, including its architecture, federation tools, and specific requirements for the involved testbeds are presented in Section 3.3, followed by discussion on experiments performed in Fed4FIRE and related added value for both experimenters and the federation, including support provided to various types of experiments performed by different type of organizations, in Section 3.4. The federation operation models and possible structures are presented in Section 3.5, where related sustainability issues are considered as well. The chapter is concluded with a brief summary of main Fed4FIRE achievements (Section 3.6).

3.2 Federated Experimentation Facilities

3.2.1 Requirements from Industry and Research

The Future Internet experimentation require a broad availability of facilities offering testing resources which apply the latest developed networking solutions and computing technologies, including testbeds established by the most relevant actual and recent research activities across Europe and worldwide. The researchers and developers from both industry and academic environments need to be able to perform experimental research by using the up-to-date testbeds as efficient as possible, to cope with nowadays' trends of a very fast development and implementation of innovative services and applications. Moreover, for the efficient experimental research and development of complex Future Internet solutions and systems, possibility to use combinations of different testing resources simultaneously is also extremely important. As the different testing resources are geographically distributed, a significant requirement on the Future Internet experimentation facilities is to be accessible and configurable from remote locations.

In order to meet the mentioned requirements, the future experimental facilities have to ensure the following:

- Simple, efficient, and cost effective experimental processes considering requirements and constraints of both experimenters and facility owners.
- Common frameworks that will be widely adopted by different experimentation facilities and used by different experimenter communities, and
- Increased trustworthiness and efficiency of the experimental facilities, including a sustainable environment for the needed testbeds continuously ensuring their updates in accordance with actual experimenters needs.

A specific requirement of the academic communities, such as universities and research centers is support for long-term research and the related scientific activities. On the other hand, the industry stakeholders, in particular SMEs, are interested to test systems and solutions under investigation for specific operational scenarios, directly aiming at exploitation of innovative products and services and establishing short-term close-to-market solutions. Of course, in lots of cases, interests of both industry and academia are overlapping, in particular in medium-term and applied research. Furthermore, there are joint undertakings by industry and academia in the research and innovation activities, including knowledge transfer, where interests of both communities are merging into common requirements towards the future experimental facilities.

However, contrary to the all research and industry requirements discussed above, the existing testbeds in Europe, which also apply for rest of the world, have been created to support experimentation in specific domain, targeting a narrow set of technology, and are usually a limited number of potential users and experimenters. The testbeds are implemented by various initiatives; e.g. EU or national research project, individually established partnerships among academia and industry, private investments in industry environments, publicly funded universities and research institutions, etc. Accordingly, all the individual testbeds are using different frameworks and tools to set-up and execute experiments creating of course a big disadvantage for experimenters, who need to get familiar with the different experimentation tools every time they use different testbeds. Furthermore, only a limited number of testbeds can be combined with other testing facilities placed in different locations and do not foresee remote configuration of the experiments and their execution.

Further important aspects of having appropriate experimental facilities is their maintenance to ensure that the testbeds are always ready to be used and are updated in accordance with the newest technological developments and trends. To ensure it, it is necessary to establish a common testbed framework supporting the testbed owners and operators to cope with this requirement within a kind of sustainable environment by involving both the experimenters and the testbed providers.

3.2.2 Establishing Fed4FIRE Federation of Testbesd

Fed4FIRE project defined its objectives along the broad requirements of the industry and research community on the Future Internet experimental research. Accordingly, establishment of a sustainable large-scale federation of testbeds has been identified as the main Fed4fFIRE project goal.

On the first instance, the federation of testbeds has to be established for benefits of both experimenters and testbed providers (Figure 3.1) and to enable easy usage of experimental resources available in the federation for a broad range of experimenters as well as to allow testbeds to easily join the federation and offer their testing and experimental services.

To ensure it, Fed4FIRE has been working on definition nand implementation of a federation framework, which includes a set of federation tools ensuring the following:

- Easy discover of testing resources in the federation by the experimenters
- Easy set-up and configuration of the experiments, by combining various experimental resources available in the federation

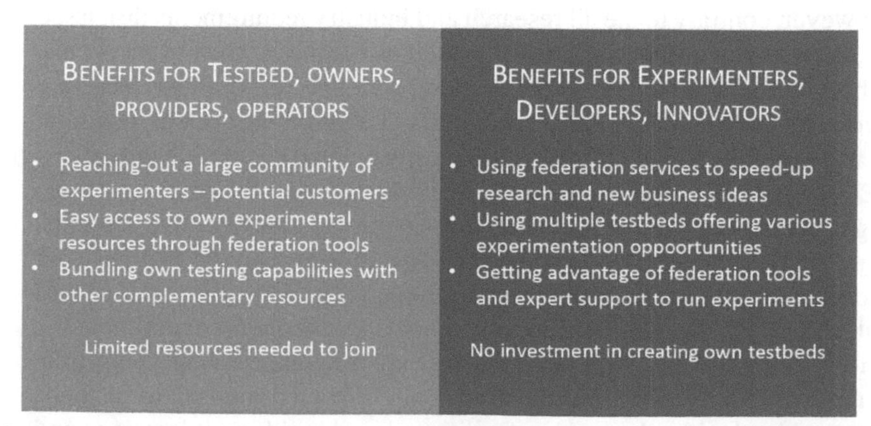

Figure 3.1 Benefits for experiments to use and for testbeds to join the federation of testbeds – overview.

- Experiment execution, including experiment scheduling, monitoring, and gathering testing results

The Fed4FIRE project worked on establishment of the federation framework and tools in several development cycles. Between the cycles, Fed4FIRE offered its experimental facilities to a wide range of users to gather feedback on their usage, which was then taken into account while improving and upgrading the common framework and the experimentation tools. Furthermore, Fed4FIRE started with a number of testbeds involved and over the project life time further testbeds joined, so that the Fed4FIRE federation offer has been significantly enlarged and experience from joining process of the new testbeds has been gathered to improve the overall framework and the related tools.

3.2.3 Experimentation Facilities in Fed4FIRE

Fed4FIRE established a federation of 23 testbeds encompassing different technologies and stretching over Europe (Figure 3.2), also with connections outside Europe, and its represents the largest federation of testbeds in Europe

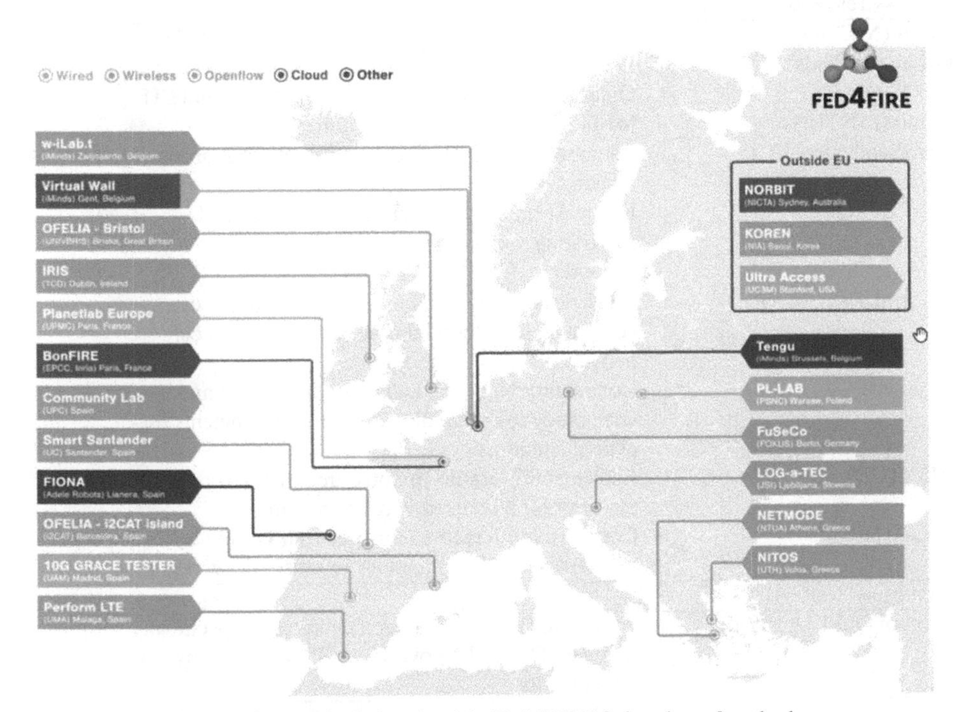

Figure 3.2 Testbeds involved in Fed4FIRE federation of testbeds.

and probably also world-wide. The federation involves testbeds focused on wired and wireless communications as well as open flow and cloud based technologies, including further specific testbeds (Table 3.1). The Fed4FIRE federation is open for new testbeds which are willing to join and is expected to grow further in the future.

Table 3.1 Brief description of Fed4FIRE facilities per testbed category

Wired Testbeds:	
Virtual Wall (iMinds)	Emulation environment with 100 nodes interconnected via a non-blocking 1.5 Tb/s Ethernet switch and a display wall for experiment visualization
PlanetLab Europe (UPMC)	European arm of the global PlanetLab system, providing access to Internet-connected Linux virtual machines world-wide
Ultra Access (UC3M, Stanford)	Next Generation of Optical Access research testbed
10G Trace Tester (UAM)	10 Gbps Trace Reproduction Testbed for Testing Software-Defined Networks
PL-LAB (PSNC)	Distributed laboratory in Poland focusing on Parallel Internet paradigms
Wireless Testbeds:	
Norbit (NICTA)	Indoor Wi-Fi testbed located in Sydney, Australia
w-iLab.t (iMinds)	For Wi-Fi and sensor networking experimentation
NITOS (UTH)	Outdoor testbed featuring Wi-Fi, WiMAX, and LTE
Netmode (NTUA)	Wi-Fi testbed with indoor facilities
SmartSantander (UC)	Large scale smart city deployment in the Spanish city of Santander
FuSeCo (FOKUS)	Future Seamless Communication Playground, integrating various state of the art wireless broadband networks
PerformLTE (UMA)	Realistic environment composed of radio access equipment, commercial user equipment, and core networks connected to Internet
C-Lab (UPC)	Community Network Lab involving people and technology to create digital social environments for experimentation
IRIS (TCD)	Implementing Radio In Software, a virtual computation platform for advanced wireless research
LOG-a-TEC (JSI)	Cognitive radio testbed for spectrum sensing in TV whitespaces and applications in sensor networks
Open Flow Testbeds:	
UBristol OFELIA island	Testbed for Future Internet technologies, specifically Software Defined Networking (SDN)/OpenFlow and virtualization

Table 3.1 Continued

i2CAT OFELIA island	Testbed for Future Internet technologies, specifically Software Defined Networking (SDN)/OpenFlow and virtualization
Koren (NIA)	High-speed research network in Korea interconnecting six nodes with OpenFlow and DCN switchess
NITOS (UTH)	Outdoor testbed featuring Wi-Fi, WiMAX, and LTE
Cloud Computing Testbeds:	
BonFIRE (EPCC, Inria)	Multi-cloud testbed for services experimentation
Virtual Wall (iMinds)	Emulation environment with 100 nodes interconnected via a non-blocking 1.5 Tb/s Ethernet switch and a display wall for experiment visualization
Other Technologies:	
FIONA (Adele Robots)	Cloud platform for creating, improving and using virtual robots
Tengu (iMinds)	Big data analysis (iMinds)

3.3 Framework for Large-scale Federation of Testbeds

3.3.1 Framework Architecture and Tools

3.3.1.1 Experiment lifecycle

The Fed4FIRE architecture has been built taking requirements from various stakeholders into account, including testbed and service providers and experimenters, with sustainability in mind and aiming to support as many actions from the experiment lifecycle as possible. The experiment lifecycle covers a number of functionalities summarized in Table 3.2.

Table 3.2 Functionalities of Fed4FIRE lifecycle

Function		Description
Resource discovery		Finding available resources across all testbeds, and acquiring the necessary information to match required specifications.
Resource specification		Specification of the resources required during the experiment, including compute, network, storage and software libraries.
Resource reservation		Allocation of a time slot in which exclusive access and control of particular resources is granted.
Resource provisioning	Direct (API)	Instantiation of specific resources directly through the testbed API, responsibility of the experimenter to select individual resources.

(*Continued*)

Table 3.2 Continued

Function		Description
	Orchestrated	Instantiation of resources through a functional component, which automatically chooses resources that best fit the experimenter's requirements.
Experiment control		Control of the testbed resources and experimenter scripts during experiment execution through predefined or real-time interactions and commands.
Monitoring	Facility monitoring	Instrumentation of resources to supervise the behavior and performance of testbeds, allowing system administrators or first level support operators to verify that testbeds performance.
	Infrastructure monitoring	Instrumentation by the testbed itself of resources to collect data on the behavior and performance of services, technologies, and protocols.
Measuring	Experiment measuring	Collection of experimental data generated by frameworks or services that the experimenter can deploy on its own.
Permanent storage		Storage of experiment related information beyond the experiment lifetime, such as experiment description, disk images and measurements.
Resource release		Release of experiment resources after deletion or expiration the experiment.

3.3.1.2 Resource discovery, specification, reservation and provisioning

3.3.1.2.1 *Architectural components*

Figure 3.3 details the part of the architecture responsible for resource discovery, specification, reservation and provisioning, from the viewpoints of the federator, the testbed provider, the experimenter and actors outside of the federation.

At the federator side, the following components are located: the portal (central starting place for new experimenters), the member and slice authority (registration), the aggregate manager (AM) directory (overview of the contact information of the AMs of all available testbeds available in the federation), the documentation center (http://doc.fed4fire.eu), the authority directory (authentication/authorization between experimenters and testbeds, supported through specific experimenter properties included in the experimenter's certificate, signed by an authority), the service directory (federation and application services), the reservation broker (for both instant and future reservations).

At the testbed side, the resources (virtual or physical nodes) are located, as well as the testbed management component (AM, responsible for discovery, reservation and provisioning of local resources through any desired software

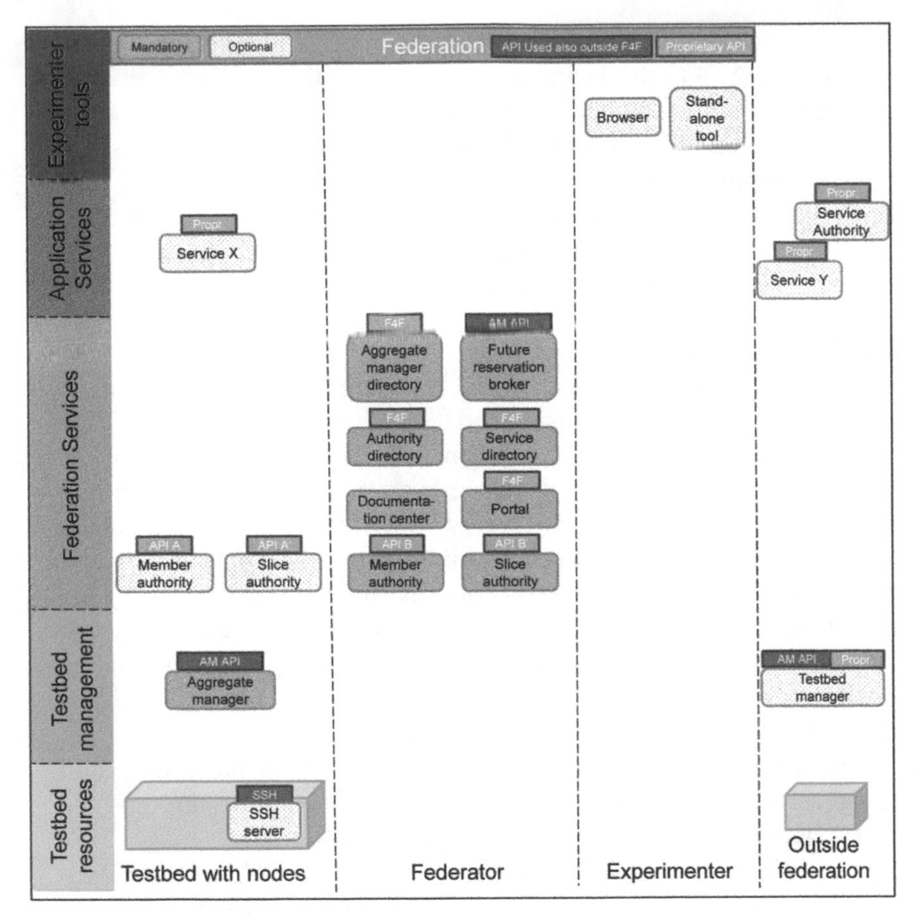

Figure 3.3 Fed4FIRE architecture components.

framework), an optional authority (member and slice) and optional application services (abstracting the underlying technical details of the provided services, relying on X.509 certificates for authentication and authorization).

At the experimenter side, we find the toolset to facilitate experimentation, such as a browser to access the hosted tools (portal, future reservation broker, documentation center, application services, etc.) and stand-alone tools to handle testbed resources (Omni, SFI, NEPI, jFed, etc.).

Outside of the federation, relevant components include the resources of testbeds that are not part of the federation, the testbed manager to handle these resources, any application services on top of resources in- or outside of the federation, and services authorities.

Several aspects of this architecture originate from the Slice-based Federation Architecture (SFA)[2]: the Aggregate Manager API, the member authorities and the slice authorities. A slice bundles resources belonging together in an experiment or a series of similar experiments, over multiple testbeds. A sliver is the part of that slice which contains resources of a single testbed. One uses an RSpec (Resource Specification) on a single testbed to define the sliver on the testbed. The RSpec and thus the sliver can contain multiple resources. The GENI AM API details can be found at the documentation website[3].

3.3.1.3 Other functionality

Similar architecture diagrams are available for monitoring and measurement, experiment control, SLA management and reputation services.

For monitoring, the following components can be distinguished at the testbed side: (1) facility monitoring (to see if the testbed is up and running) that exports an Open Measurement Library (OML) stream to the federator's central OML server, (2) infrastructure monitoring (to collect data on behavior and performance of local services, technologies, and protocols, as well as on resources from a specific experiment), (3) the OML measurement library (for measuring specific experiment metrics), an optional OML server (the endpoint of a monitoring or measurement OML stream that stores that in a database) and (4) an optional measurement service with proprietary interface. The federator then offers the FLS dashboard to give a real-time view on the facilities' health status, the central OML server for FLS data, nightly login testing and the (optional) data broker for experiment data from OML streams.

For experiment control, the testbed provides (1) an SSH server on each resource, (2) a resource controller that invokes actions through the Federated Resource Control Protocol (FRCP), (3) an Advanced Message Queuing Protocol (AMQP) server to communicate the FRCP messages, (4) the Policy Decision Point (PDP) that enables authorization and (5) the experiment control server to execute the experiment's control scenario.

Related to SLAs, the SLA management module at each testbed is responsible for supervising the agreement metrics and processes all relevant measurements from the monitoring system. The SLA collector acts as a broker between these modules and the client tools, such as the SLA front-end tool provided in the Portal, and gathers warnings and experimenter-specific evaluations. The SLA dashboard allows testbed providers to view the status of active SLAs on their facilities.

---

[2]http://groups.geni.net/geni/attachment/wiki/SliceFedArch/SFA2.0.pdf

[3]https://fed4fire-testbeds.ilabt.iminds.be/asciidoc/federation-am-api.html

The architecture further supports layer two connectivity between testbeds, service composition (through YourEPM), speaks-for credentials for trust chain relationships, ontology-based resource selection and first level support (FLS) monitoring.

3.3.2 Federating Experimentation Facilities

In order to support the federation of experimentation facilities, we define different classes of testbeds and different types of federation.

3.3.2.1 Classes of testbeds

A testbed is a combination of hardware and testbed management software. We make a difference between two classes of testbeds which could join the federation or be compatible with Fed4FIRE: (1) type A, which includes testbeds with resources that can be controlled through SSH, FRCP or Openflow, and (2) type B, which are accessible through service APIs only. Type A testbeds have the ability to share resources between different users, shared over time or in parallel (through multiplexing or slicing) and support the concept of credentials and dedicated access (e.g. through SSH). Type B testbeds offer a particular service with a (proprietary or standard) API and support the concept of credentials.

As an example, the Virtual Wall which provides physical or virtual machines with SSH access is type A, while SmartSantander, providing a proprietary REST API to fetch the measurement results, is a type B testbed.

3.3.2.2 Types of federation

Three types of federation are defined: (1) association, (2) light federation and (3) advanced federation. Associated testbeds are not technically federated, but are mentioned on the Fed4FIRE website with a link to the testbed specific documentation. These testbeds have to organize their own support.

Light federation is the same for type A and type B testbeds. The testbeds need to provide support for Fed4FIRE credentials in a client based SSL API, maintain specific documentation for experimenters (on a webpage maintained by the testbed), adhere to the policy that everyone with a valid Fed4FIRE certificate can execute the basic experiment that is document without extra approval, provide facility monitoring and ensure a public IPv4 address for connectivity to the API server. The Fed4FIRE federation in turn offers test credentials for testing the federation, information on enabling PKCS12 authentication, a central monitor dashboard, at least one client tool exporting PKCS12 credentials from the X.509 certificate, at least one authority to provide credentials, a central documentation website linking to all testbeds and central support (google group and NOC) for first help and single point of contact. This light federation makes it possible to have an easy way to federate with Fed4FIRE and as such testbeds can easily join a very ad-hoc and dynamic way for a short period of time.

For advanced federation, type A and type B testbeds are treated differently. Type A testbeds need to provide support for GENI AMv2 or AMv3 (or later versions), maintain specific documentation (on a webpage maintained by the testbed), adhere to the policy that everyone with a valid Fed4FIRE certificate can execute the basic experiment that is document without extra approval, provide facility monitoring through the GENI AM API and ensure a public IPv4 address for the AM and a public IPv4 or IPv6 address for SSH login to the testbeds resources, and offer basic support on the testbed functionalities towards experimenters. In turn, the Fed4FIRE federation offers testing tools for the AM API, nightly testing of the federation functionality, a central monitor dashboard, at least one client tool having support for all federated infrastructure testbeds, at least one authority to provide credentials, an SSH gateway (to bridge e.g. to IPv6, VPNs, etc.), a central documentation linking to all testbeds and central support (google group and NOC) for first help and single point of contact.

Advanced federation for type B testbeds can be supported through service orchestration on the 'YourEPM' (Your Experiment Process Model) tool which is designed to provide high level service orchestration for experimenters, based on open standards such as BPMN (Business Process Model and Notation) and BPEL (Business Process Execution Language). YourEPM presents a web GUI that automatically obtains information on available services from the service directory that collects service descriptions from the specific URL provided by each testbed. The communication with the services from YourEPM is ensured using general wrappers to specific technologies (i.e. REST, SFA). This tool can also be integrated with the jFed tool to extend the orchestration to include testbed resources. In order for YourEPM to use application services available in the federation, type B testbeds which want to have an advanced federation with Fed4FIRE have to provide a description of the service API in RAML, so that the tool can invoke it automatically.

3.3.2.3 Workflow for federation

Figure 3.4 highlights the typical workflow for a new testbed to be federated, starting with the existing documentation on how experimenters can use already federated testbeds.

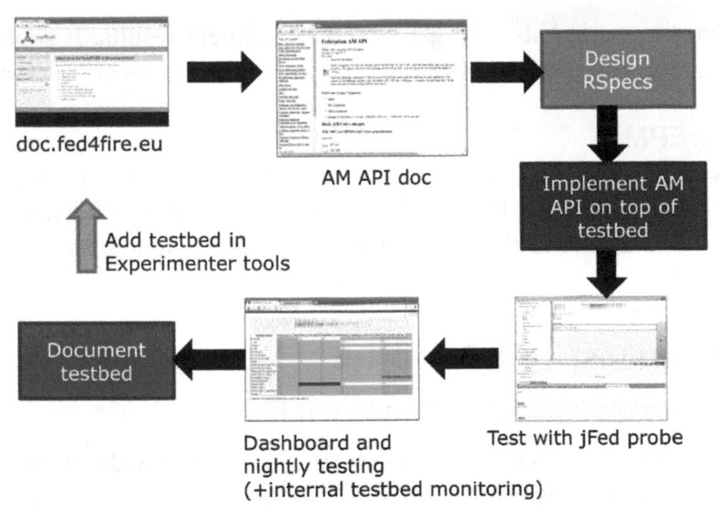

Figure 3.4 Workflow for testbeds joining the federation.

3.3.3 Federation Tools

3.3.3.1 Portal

The Fed4FIRE portal[4] is the central starting place for new experimenters and provides the testbed and tools directory, links to the project website and to the First Level Support service, support for the registration of new users. Furthermore, it acts as an experimentation tool for discovery, reservation and provisioning of resources and as a bridge to experiment control tools. It is powered by MySlice software[5].

3.3.3.2 jFed

jFed[6] is a java-based framework to support experimenters to provision and manage experiments, to assist testbed developers in testing their API implementations and to perform extensive full-automated tests of the testbed APIs and testbeds, in which the complete workflow of an experiment is followed.

3.3.3.3 NEPI

NEPI[7], the Network Experimentation Programming Interface, is a life-cycle management tool for network experiments, that helps to design, deploy and

[4]https://portal.fed4fire.eu
[5]http://myslice.info
[6]http://jfed.iminds.be
[7]http://nepi.inria.fr

control network experiments, and gather the experiment results. It supports design and control through the federated resource control protocol FRCP.

3.3.3.4 YourEPM

YourEPM is an Experiment Process Manager that allows high level application service orchestration in the federation. It connects experiment owners, testbed facilities and federator central coordination with both automated and manual processes for experiment planning, execution and analysis.

3.4 Federated Testing in Fed4FIRE

3.4.1 Overview of Experiments on Fed4FIRE

Fed4FIRE offers its testbeds for use and experimentation to a wide community and to all interested parties. This is offered through a system of either Open Calls by which selected proposals received financial support to carry out the experiments or through a system of Open Access by which any interested party can set up and run an experiment on the facility. Since its initial set up as a federation, Fed4FIRE has supported over 50 experiments through its Open Calls, out of over 150 submitted proposals, which were oriented towards SMEs, industry, academic or research parties (Figure 3.5).

Utilization of the federation testbeds used by different experiments accepted in the Open Calls is presented in Figure 3.6 (colors indicate type of the testbeds used according to testbed overview from Figure 3.2).

3.4.2 Complexity of the Fed4FIRE Experiments

One measure which can be used to indicate the complexity of the experiment which is run on the Fed4FIRE facilities is the number of testbeds in use.

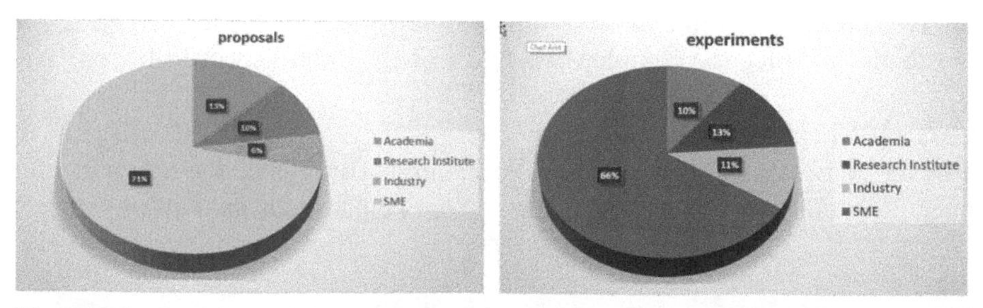

Figure 3.5 Overview of the proposals and accepted experiments through the open call mechanism.

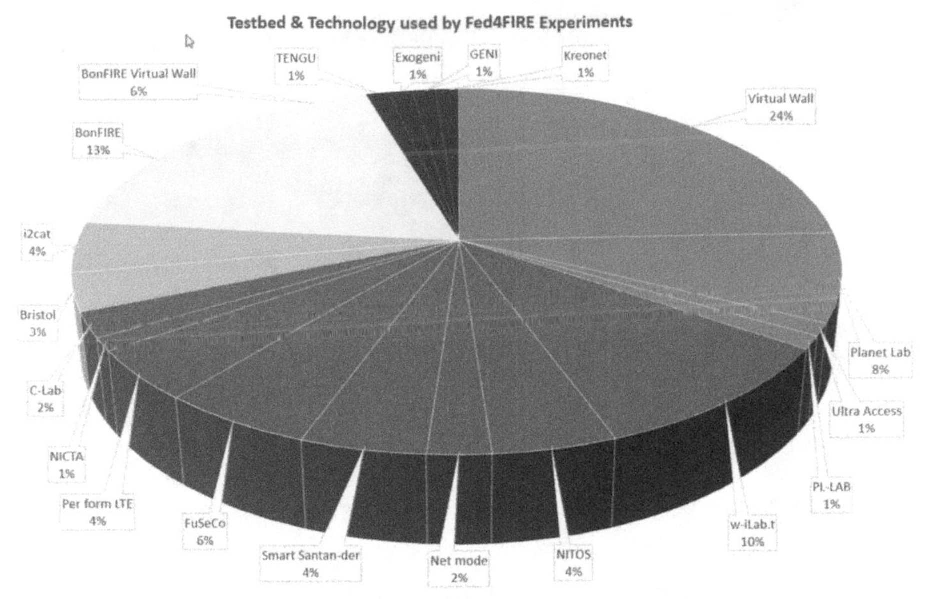

Figure 3.6 Utilization of Fd4FIRE testbeds by experiments.

Figure 3.7 already illustrates the need for a federated facility as more than 70% of the experiments make use of more than 1 testbed. What is even more clearly demonstrating the value of Fed4FIRE is the fact that if one uses the categories of technologies as defined above (wired/wireless/cloud/open flow/ other), more than half of the experiments use testbeds which are positioned

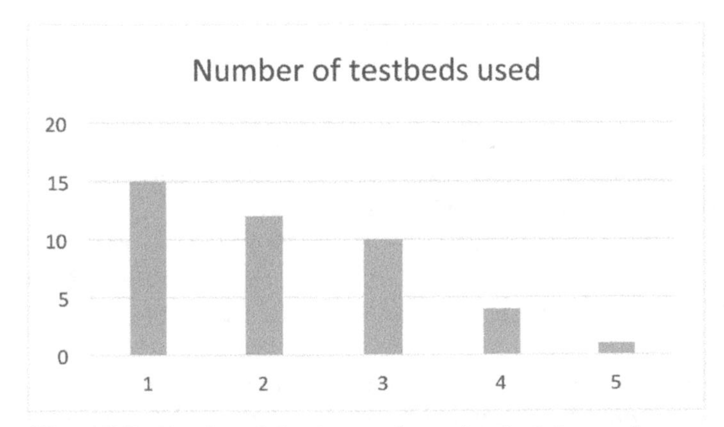

Figure 3.7 Number of simultaneously used testbeds in experiments.

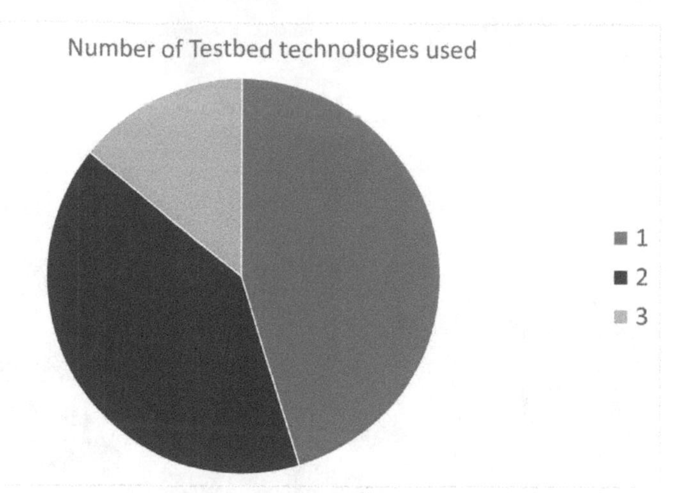

Figure 3.8 Number of simultaneously used test bed technologies in experiments.

in different technology areas (Figure 3.8). This clearly demonstrates the added value of a federated facility like Fed4FIRE covering different technologies.

3.4.3 Value to the Experimenter

Nearly all of the experimenters have chosen to submit an experiment to Fed4FIRE:

- To test and evaluate their products in a real environment which is by some companies used as sales argument and proof of the performance or reliability of their product to potential customers **"To test in a real testbed scenario some of the algorithms devised on paper"**
- To prepare their products for the market. "Fed4FIRE learned us that we are market-ready for large business"
- To test and evaluate scalability of their products or to carry out stress-tests on their products. Fed4FIRE clearly has the size to carry out these tests **"To identify problems with scalability"**
- Because of the uniqueness of the Fed4FIRE testbeds offering technologies which are not available in commercial testbeds: **"To access infrastructures that otherwise would not be reachable"**
- Because of the financial support received, an argument which is repeated by nearly all SMEs which ran an experiment on Fed4FIRE **"We would have spent thousands of euros to create an infrastructure for testing"**

From this feedback, which is collected from all experiments, it is clear that all experimenters indicate a significant to extreme impact on their business from the experiment. This impact slightly differs over the calls, but it is clear that the impact for SME's is more significant than for the standard Open Call experiments in which larger research groups or industrial partners participate.

3.4.4 Support Provided by the Federation to SMEs

Through its Open Calls for SMEs, Fed4FIRE has the objective to make the federated infrastructure easier and more directly available for execution of innovative experiments by experimenters at SMEs. The experiments envisaged were of a short duration (maximum 4 months) and examples included but were not limited to testing of new protocols or algorithms, performance measurements, service experiments.

Specific benefits for SMEs were identified as:

- Possibility to perform experiments that break the boundaries of different FIRE testbeds or domains (wireless, wired, OpenFlow, cloud computing, smart cities, services, etc.)
- Easily access all the required resources with a single account.
- Focus on your core task of experimentation, instead of on practical aspects such as learning to work with different tools for each testbed, requesting accounts on each testbed separately, etc.
- A simplified application process with a dedicated review process by external judges

An extra benefit which is offered towards SMEs is the dedicated support from specific Fed4FIRE members. Each SME, preparing a proposal was appointed a supporting Fed4FIRE consortium partner (the "Patron") which was in charge of dedicated (advanced) support of the experiment. This Patron received additional funding to provide this support in setting up, running and analysing the results of the experiment.

This support was provided in 2 layers:

A. Basic support

- Guaranteeing that the facility is up and running (e.g. answering/solving "could it be that server X is down?")
- Providing pointers to documentation on how the facility can be used (e.g. "how to use the virtual wall testbed" => answer: check out our tutorial online at page x")

- Providing pointers to technical questions as far as relevant (e.g. answering “do you know how I could change the WiFi channel” => answer: yes, it is described on following page: y”; irrelevant questions are for example “how to copy a directory under Linux”)

B. Dedicated (advanced) support includes all of the following supporting activities by the patron:

- Deeper study of the problem of the SME: invest effort to fully understand what their goals are, suggest (alternative) ways to reach their goals. To put it more concretely (again using the example of the Virtual Wall testbed), these SMEs do not need to know the details on the Virtual Wall or how it should be used, they will be told what is relevant to them and can focus on their problem, not on how to solve it.
- Help with setting up the experiments (e.g. “how to use the virtual wall” => answer: the tutorial is there, but let me show you how what is relevant for you, let me sit together with you while going through this example and let us then also make (together) an experiment description that matches what you are trying to do.
- (Joint) solving of practical technical problems (e.g. “do you know how I could change the WiFi channel” => yes, it is described on page y, in your case you could implement this as following: . . ., perhaps we should quickly make a script that helps you to do it more easily, . . .).
- Custom modifications if needed: e.g. adding third-party hardware and preparing an API for this.
- Technical consultancy during/after the experiments (e.g. “I do get result x but would have expected y, what could be the problem?”).

All of the SMEs, submitting a proposal to run an experiment sought this support already while preparing their proposal.

3.4.5 Added Value of the Federation

The following quotes are taken form some of the reports of the experiments that ran on Fed4FIRE. They clearly illustrate why experimenters come to Fed4FIRE

- We wouldn’t be in this position now if we hadn’t had access to Fed4FIRE facilities
- There is no alternative to Fed4FIRE as a platform hosting different technologies

- Fed4FIRE is independent of any other infrastructure, ... for companies is very important to avoid vendor lock-in, ...
- Running the experiment at a commercially available testbed infrastructure would have been unlikely mainly because of the novelty of some implemented solutions.
- The federation's main contribution is making individual facilities visible and usable through a homogenous set of standards and tools.
- Diversity and quantity of the nodes ... different technologies, types -outdoor/indoor-, different locations, possibility to combine infrastructures and resources.
- To develop projects that can provide services at European level, with millions of potential users at the same time, it is necessary to have a test infrastructure with sufficient technical resources.
- An experiment in Fed4FIRE is so close to reality that any development carried out in the environment can be migrated to a commercial platform.
- Thanks to the Fed4FIRE federation we had the chance to test our platform in a production – like environment. If there were no federation, our tests would have been less effective for our business objectives.

3.5 Operating the Federation

3.5.1 Federation Model, Structure and Roles

The operational model follows a service oriented approach that crucially provides services to both experimenters and testbeds, as both experimenters and testbeds are needed in adequate quantities and varieties for a successful federation.

Towards experimenters, the Federator offers identity management through single sign-on, a portal with basic information about the federation, at least one stand-alone tool for resource management, comprehensive documentation, First Level Support, advice and brokering, and reporting on KPIs (testbed availability, usage, performance of federation services, etc). Towards testbed providers, the Federator facilitates technical interoperation, provides compliant tools and portal, promotes the federation, and acts as a broker between experimenters and testbeds and reports on KPIs. The Federator also promotes the usage of tools that are developed externally to the federation and can provide added value. Towards the European Commission, the Federator reports on KPIs about the federation's operation.

Through these tools and the "one-stop shop" approach (Figure 3.9), Fed4FIRE natively supports the "Experimentation as a Service" concept, where the resources needed for an experiment can be acquired and accessed as one package by the experimenter. Fed4FIRE follows the FitSM management approach for its federation services. FitSM[8] is a free and lightweight standards family aimed at facilitating service management in IT service provision, including federated scenarios.

3.5.2 Financial Approach of the Federation

In the financial model, funding and revenues are coming from national, regional and local sources, the European Commission and private/industry sources (note that the latter will typically be limited). The costs are made by the federator, the facility providers and the experimenters (Figure 3.10).

The federation will organize Open Calls for experimentation, with a budget per experiment ranging from 5K to 100K euro, including financial support for testbed providers to provide technical support and consultancy services where required.

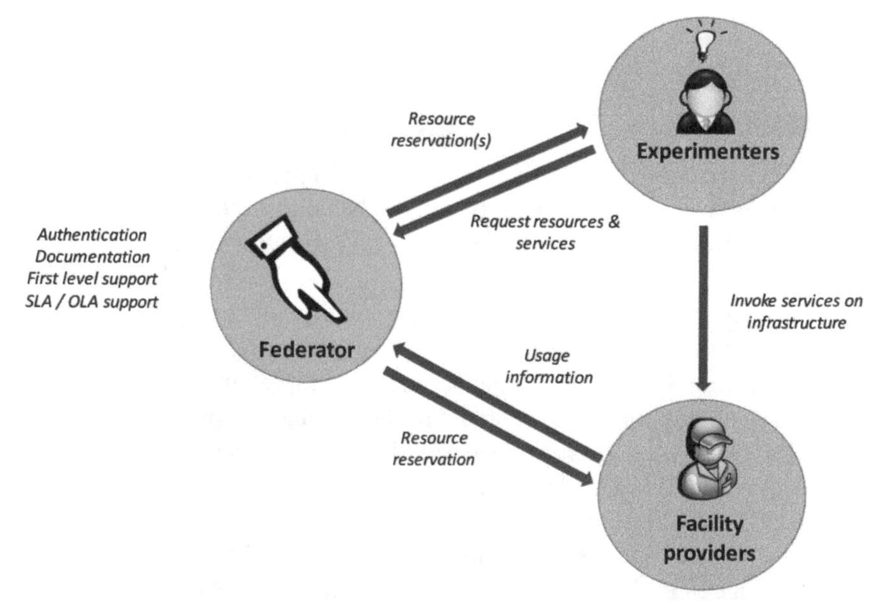

Figure 3.9 One-stop shop approach in Fed4FIRE federation.

[8]http://www.fitsm.temo.org

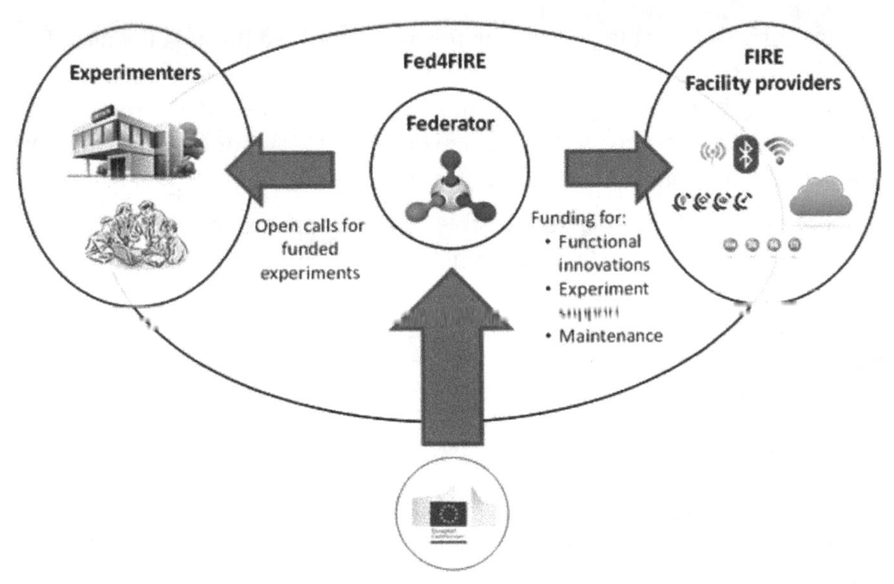

Figure 3.10 Financial flow within federation of testbeds.

3.5.3 Organization of the Federation

The primary stakeholders in the federation, the experimenters and the testbed providers, delegate the management of the federation to the Federator and the control of the federation to the Federation Board, the policy-making body.

The federation's governance model is based on three layers, related to governance (how the Federator and Federation Board are managed), operational issues (how the Federator operates) and financial aspects (costs and revenue/funding). The federation deals with policies in the following areas:

- Testbed and Experimenter Commitments and Eligibility Requirements: the key policy is to be as open and accommodating as possible, because a major success factor is to expand the federation membership.
- Resource Management: although the federator will allow the reservation of the resources on the testbeds, it is the final responsibility of the testbeds to manage the usage of their resources, as long as they fulfil the agreed Service Level Agreements (e.g. provide a minimum amount of resources, guaranty a certain up-time).
- Stakeholder Engagement (Communications and Marketing): the key objectives of these policies are to recruit experimenters and testbeds to expand the federation.

- Future Direction for the Federation: this is determined through the use of four key metrics: Fairness, Cost efficiency, Robustness and Versatility.
- Contractual Relationships and Terms and Conditions: the terms and conditions (T+C) for the federation cover a set of T+C for experimenters and another compatible set of T+C for testbed facilities.

Furthermore, the federator is responsible for the operation of and support for the federation services, for community building through Summer Schools (for experimenters) and Engineering Conferences (to drive technical developments) and for international collaboration with US, Brazil, China, South-Korea, Japan and others.

3.6 Summary

The Future Internet experimentation require a broad availability of facilities offering testing resources which apply the latest developed networking solutions and computing technologies, including testbeds established by the most relevant actual and recent research activities across Europe and world-wide. The Fed4FIRE project has established a European Federation of Testbeds and developed necessary technical and operational federation framework enabling the federation operation. With its 23 tesbeds, the Fed4FIRE represents the largest federation of testbeds in Europe which allows remote testing in different areas of interests; wireless, wireline, open flow, cloud, etc.

The Fed4FIRE architecture has been built by taking requirements from various stakeholders into account, including testbed providers and experimenters, with sustainability in mind and aiming to support as many actions from the experiment lifecycle as possible. Various user friendly tools established by the Fed4FIRE project enable remotely usage of the federated testbeds by experimenters who can combine different federation resources, independently on their location, and configure it as it is needed to perform the experiment.

The Fed4FIRE Federation offers its testbeds for use and experimentation to a wide community and to all interested parties, which can use the federation facilities through the mechanism of Open Calls for Experiments, partially funded by EC, or by using Open Access to the federation facilities. Since start of Fed4FIRE operation, more than 50 experiments have been completed and more than 150 experimentation proposals have been received from SMEs, other industry stakeholders, as well as academic and research institutions.

In respect to the federation operation, by using its powerful federation tools Fede4FIRE is applying so-called "one-stop shop" approach, natively supporting the "Experimentation as a Service" concept, where the resources needed for an experiment can be acquired and accessed by the experimenter through one single contact point of the federation – its Federator. Finally, Fed4FIRE elaborated a number of possible organization and funding models for the federation, which are planned to be exploited in the near future, aiming at establishment of a sustainable European Federation of Testbeds.

4

A Platform for 4G/5G Wireless Networking Research, Targeting the Experimentally-Driven Research Approach – FLEX –

Nikos Makris[1], Thanasis Korakis[1], Vasilis Maglogiannis[2], Dries Naudts[2], Navid Nikaein[3], George Lyberopoulos[4], Elina Theodoropoulou[4], Ivan Seskar[5], Cesar A. Garcia Perez[6], Pedro Merino Gomez[6], Milorad Tosic[7], Nenad Milosevic[7] and Spiros Spirou[8]

[1]University of Thessaly
[2]Ghent University – iMinds
[3]EURECOM
[4]COSMOTE
[5]WINLAB, Rutgers University
[6]Universidad de Malaga
[7]University of Nis
[8]INTRACOM SA Telecom Solutions

4.1 Introduction

The proliferation of smart mobile devices and data hungry mobile applications are driving the demand for faster mobile networks. Long Term Evolution (LTE), the 4th Generation of mobile network technology standardized by the 3rd Generation Partnership Project (3GPP) [1], aims at satisfying this demand by offering faster connection speeds at both the downlink and the uplink, increased network capacity and better coverage. The rapid penetration of LTE in different countries creates a vast field for innovation in terms of mobile broadband services. At the same time, research for the next generation mobile networks has already begun with the examination and evaluation of candidate technologies and architectures. Given the practical requirement for backward compatibility between successive technologies, it is rational to assume that these technologies, often referred to as Beyond 4th and towards the 5th Generation (B4G and 5G), will naturally evolve from the extension of LTE with new advanced features.

Evaluation of the performance of innovative broadband services over LTE and of candidate post-LTE technologies requires rigorous testing and validation. While network simulation software has evolved significantly over the years, it cannot still capture the complex real world environment, and field tests are still considered essential at the late stages of development. To that end, the existence of network testbed facilities plays a significant role in understanding the complexities associated with real use and therefore in building better solutions.

In Europe, since its establishment in 2008, the Future Internet Research and Experimentation (FIRE) initiative [2] has contributed in bridging the gap between visionary research and large-scale experimentation on new networking and service architectures and paradigms. Through the successful organization of several waves of research projects, an extensive and multidisciplinary open network testbed facility has been developed. Despite the diversity in the FIRE facilities in terms of available infrastructure and access technologies, a lack of truly open and operational LTE testbeds had been identified (and cellular testbeds in general). By "open" we mean that the facilities are available to external experimenters and that the latter can configure the testbed to some extent, according to their needs. By "operational" we refer to flexibility in accessing the core, gateways, access points and user equipment of the testbeds, and the capability to run full end-to-end services.

This lack was certainly not due to reduced interest from the community. On the contrary, there is a steadily increasing demand from the research community, including the industry, to have access to LTE and beyond experimentation facilities in different countries. However, the constraints typically posed by operators and large vendors, typically due to commercial considerations, restrict the configuration capabilities to an extent, which usually discourages testbed operators from deploying such infrastructure.

FLEX (FIRE LTE testbeds for Open Experimentation) [3] aims to remove these constraints through the development of a truly open and operational LTE experimental facility. Based on a combination of **truly configurable commercial equipment, truly configurable core network software, fully open source components, and on top of those, sophisticated emulation and mobility functionalities**, this facility allows researchers from academia and industry to test services and applications over real LTE and beyond infrastructure, or experiment with alternative algorithms and architectures of the core and access networks.

4.2 Problem Statement

Several EU funded projects have paved the way for the federation of isolated testbed islands across Europe. Excellent examples of them are the OpenLab [4] and the Fed4FIRE [5] projects, which have addressed both the control and experimental plane federation of heterogeneous FIRE resources. With the control plane we mean the way that the resources are discovered, represented and reserved inside federations, whereas with the experimental the option to include resources from heterogeneous testbeds, decoupled from their geographical location, and bundle them in one single large scale experiment. Yet, the focus on these federations lies only on the support of generic nodes, meaning just an abstract representation of any testbed resource, with a limited number of parameters being defined by the experimenters.

FLEX is addressing this lack of experimentation services for LTE and beyond resources, by integrating all the LTE hardware extensions to the state-of-the-art control and management services of the testbeds. Three core FIRE testbeds have been extended with LTE support initially, and two more have been added to the consortium after the completion of an infrastructure upgrade Open Call process. All of the FLEX testbeds, have been federated over the GÉANT network [6], thus enabling dedicated guaranteed end-to-end connections from one testbed to another able to bear the traffic, and the setup of novel experiments for decentralized architectures.

Moreover, FLEX is offering two setups; 1) a commercial equipment based testbed, for the development of novel services and 2) an open-source setup for the development and evaluation of new protocols, leveraging the LTE protocol stack. The commercial equipment is fully programmable, provided by the partners of the project, and through the definition of high level APIs, experimenters can take access over them. As for the open source solution, the project is using the open source solution of OpenAirInterface (OAI) [7], that allows the execution of a full stack LTE eNodeB or User Equipment (UE) over commodity hardware with a compatible RF front-end.

The testbeds that are available within FLEX are publicly available 24/7, remotely accessible and provided free-of-charge. The five experimental facilities, along with their capabilities, are detailed in the following subsection.

4.2.1 FLEX Testbeds

The five experimental facilities, that are comprising the FLEX testbed, are resources rich in heterogeneous equipment, each of them allowing the configuration of several parameters along with the LTE configurations, and enabling the experimentation at a very large scale. Following, we list the capabilities of the five different FLEX islands (see Figure 4.1).

4.2.1.1 NITOS testbed

NITOS testbed [8], is a heterogeneous testbed, located in the premises of University of Thessaly (UTH), in Greece. The testbed facilitates access to open source and highly configurable equipment, allowing for innovations through the experimental evaluation of protocols and ideas in a real world environment. The experimental ecosystem is consisting of several wireless and wired networking components, coupled with powerful nodes and a cloud computing infrastructure. The key equipment components in NITOS are the following: 1) Over 120 nodes equipped with IEEE 802.11 a/b/g/n/ac compatible equipment, and using open source drivers. The nodes are compatible also with the IEEE 802.11s protocol for the creation of wireless mesh networks, 2) Commercial off-the-shelf (COTS) LTE testbed,

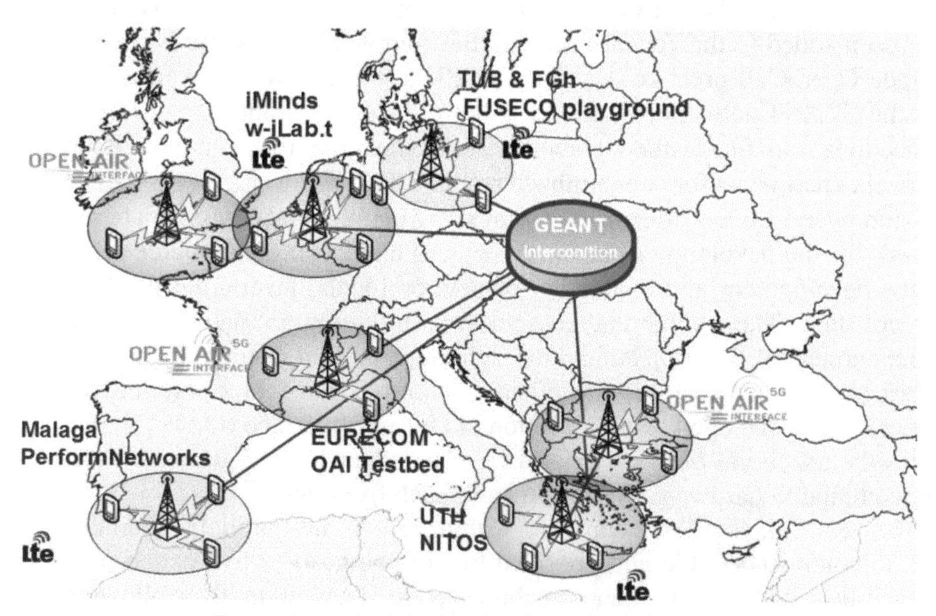

Figure 4.1 The FLEX testbed federation in Europe.

consisting of a highly programmable LTE macrocell (Airspan Air4GS), two femtocells (ip.access LTE 245F), an experimenter configurable EPC network (SiRRAN LTEnet) and multiple User Equipment (UE), such as USB dongles and Android Smartphones, 3) Open Source LTE equipment, running over commodity Software Defined Radio (SDR) equipment, by the adoption of the OpenAirInterface [7] platform. OpenAirInterface can be set to operate as either a femtocell or UE, whereas its accompanying open source network is provided (OpenAirCN), 4) COTS WiMAX testbed, based on a highly programmable WiMAX base station in standalone mode, along with several open source WiMAX clients (USB dongles and Smartphones), 5) A Software Defined Radio (SDR) testbed, consisting of 10 USRPs N210, 8 USRPs B210, 2 USRPs X310 and 4 ExMIMO2 FPGA boards. MAC and PHY algorithms are able to be executed over the SDR platforms, with very high accuracy, 6) The nodes are interconnected with each other via 5 OpenFlow hardware switches, sliced using the FlowVisor framework, and allowing multiple experimenters control the traffic generated from their experiments using any OpenFlow controller, 7) a Cloud Computing testbed, consisting of 96 Cores, 286 GB RAM and 10 TBs of hardware storage. For the provisioning of the cloud, OpenStack is used.

The equipment is distributed across three different testbed locations, and can be combined with each other for creating a very rich experimentation environment. The nodes are running any of the major UNIX based distributions.

4.2.1.2 w-iLab.t testbed

The w-iLab.t [9] is an experimental, generic, heterogeneous wireless testbed and provides a permanent testbed for development and testing of wireless applications. w-iLab.t hosts different types of wireless nodes: sensor nodes, Wi-Fi based nodes, sensing platforms, and cognitive radio platforms. Each of the devices can be fully configured by the experimenters. The wireless nodes are connected over a wired interface for management purposes. This interface can also be used as a wired interface. Hence, heterogeneous wireless/wired experiments are possible. Furthermore, iMinds hosts the Virtual Wall, which consists of 2 testbeds:

- Virtual Wall 1 containing 206 nodes
- Virtual Wall 2 containing 159 nodes

The Virtual Wall offers network impairment (delay, packet loss, bandwidth limitation) on links between nodes and is implemented with software impairment. Additionally, some of the nodes are connected to an OpenFlow switch to be able to do OpenFlow experiments in a combination of servers, software OpenFlow switches and real OpenFlow switches. Moreover, the following equipment has been installed in order to enable LTE experimentation in the testbed: 1) 2 ip.access LTE femtocells, 2) SiRRAN LTEnet EPC solution with 9 licenses, 3) 22 LTE UEs as USB dongles, 4) 2 Emulated Mobility Frameworks consisting of 4 (big) and 3 (mini) shielded boxes respectively. The boxes are interconnected with each other via COAX cables. The attenuation of the RF components that are placed in the boxes is controlled by programmable attenuators, 5) 2 additional ip.access femtocells accompanied by 2 LTE dongles that are part of the (big) Emulated Mobility Framework, 6) 2 ExMIMO2 FPGA boards and 3 USRPs B210 equipped with RF front-ends compatible with OpenAirInterface. The testbed is also using 20 programmable moving robots, that can be used for real mobility experiments [10]. The users are able to draw interactively a trajectory that each robot will follow during their experiment. Each of the robots is equipped with a Nexus 6P smartphone to enable LTE experimentation. The control of the LTE experimentation can be done using Signal and Spectrum Analyzers or a USRP N210 equipped with an LTE compatible RF front-end.

4.2.1.3 OpenAirInterface testbed

Facilities at EURECOM that are available to the project include an 8-node testbed, equipped with the OAI compatible RF front-ends, UEs and VMs acting as core networks. The OAI testbed [11] nodes include: 1) 4 machines that can be used for running OAI as eNodeB, equipped with the appropriate SDR platforms (2 of them using USRPs B210 and 2 of them ExMIMO2), 2) Dedicated services are executed on top of them, for the orchestration of the experiments, such as OpenStack [12] and JuJu [13], 3) 4 nodes that are equipped with COTS UEs, that can be used for running the OpenAirCN platform (OAI EPC), 4) 2 more UEs as Android Smartphones.

4.2.1.4 PerformNetworks testbed

PerformNetworks [14], formerly PerformLTE, provides multiple scenarios to enable experimentation with different levels of realism [15]. The testbed has been extended in the project with interoperability tools that have been used to perform interoperability testing with equipment available in other FLEX testbeds. Currently, the federated part of the testbed is composed by: 1) T2010 conformance testing units by Keysight Technologies, that can be used to provide LTE end to end connectivity to commercial UEs in any standardized FDD or TDD band. These units have been extended during project to support communication with standard core networks. 2) LTE release 8 small cells (Pixies) by Athena Wireless working on band 7. 3) Polaris Core Network Emulator (EPC), providing multiples instances in SGW, PGW, MME, HSS and PCRF (more details in [16]). This EPC has been successfully integrated with macro and pico-cells from Alcatel Lucent and with small cells from Athena Wireless and Sirran Technologies, 4) Several LTE UEs, working on different bands, successfully integrated with the T2010 units and the small cells, 5) ExpressMIMO2 and USRP SDR cards, 6) SIM cards from a Spanish LTE operator to be used on commercial deployments.

4.2.1.5 FUSECO playground

FUSECO Playground [17] allows FLEX experimenters to execute even larger scale experimentation with more LTE resources, in handover with 2G, 3G, Wi-Fi, and in collaboration with cloud services. FUSECO integration with the existing FLEX infrastructure adds values by supporting 5G research activities with NFV, SDN, etc. The hardware resources that FUSECO playground is offering to FLEX are summarized in the following: 1) ip.access LTE 245F eNodeB, supporting LTE FDD bands 7 and 13, 2) OpenEPC 3GPP Evolved Packet Core, 3) Virtualized LTE Network Functions (e.g. PDN-GW, SGW, MME) over SDNs, 4) 3 LTE dongles UEs and 3 Android Smartphones, 5) ip.access Nano3G E16 (model 239A) UMTS IMT 2100 (supporting LTE FDD bands 1, 2/5 and 4), 6) 3 Wi-Fi APs Cisco Aironet 3602e (supporting 802.11 a/b/g/n/ac), 7) Radio Signal Attenuation System with a frequency range from 700 MHz to 3 GHz, allowing the configuration of attenuation of 1–127 dB in 1dB steps, 8) OpenIMS Core (IMS Call Session Control Functions (CSCFs) and a lightweight Home Yes (ssh & OMF/FRCP Subscriber Server (HSS), which together form the core elements of all IMS/NGN access) architectures as specified today within 3GPP, 3GPP2, ETSI TISPAN and the Packet Cable initiative. The four components are all based upon Open Source software (e.g. the SIP Express Router (SER)).

4.3 Background and State-of-the-Art on Control and Management of Testbeds

In this section we provide some information on the state-of-the-art tools for testbed management and control, as well as federation setup, that existed prior to FLEX, along with some insights on how these have been extended in order to serve the goals paved by the project. These tools include control tools for the management of the testbeds and federations, experimental plane tools, for conducting experiments over the testbed, as well as monitoring methodologies, for collecting measurements over the distributed testbed resources.

4.3.1 Slice-based Federation Architecture (SFA)

Slice-based Federation Architecture (SFA) [18] is used in order to facilitate testbed federations, via providing a standardized interface. It provides a minimal interface, which enables testbeds of different technologies and/or belonging to different administrative domains to federate without losing control of their resources.

SFA provides a secure, distributed and scalable narrow waist of functionality for federating heterogeneous testbeds. However, there are barriers to entry to using SFA: a testbed owner would normally need to implement the certificate-based authentication and authorization mechanisms used by SFA, as well as coders and parsers for files that describe the resources on their testbed.

Some examples of well-known tools that take advantage of the SFA architecture are jFed [19], mySlice [20], OMNI [21], used to graphically represent an experiment including resources from multiple sites.

4.3.2 cOntrol and Management Framework (OMF)

The management of several heterogeneous resources is a significant issue for a testbed operator. The testbeds, which are participating in FLEX have adopted the cOntrol and Management Framework (OMF) [22] for the administration and experiment orchestration with the underlying resources. OMF was initially developed in ORBIT by Winlab and currently its development is being led by NICTA along with the contributions of other institutions like Winlab and UTH. FLEX has adopted the "cOntrol and Management Framework (OMF)" for providing experimentation services on top of the FLEX testbeds. The framework allows for the transparent configuration of the underlying resources, via the submission of a simple experiment description in a high level language. The experimenter is able to submit this kind of description to the testbed, and the different OMF components communicate with each other and set up the experiment topology.

Currently, two different releases of the OMF framework are supported: OMF5.4 and OMF6. OMF version 6 has introduced radical changes in the architecture and philosophy of the framework. The main concept of the new architecture is that everything is being treated as a resource and for every

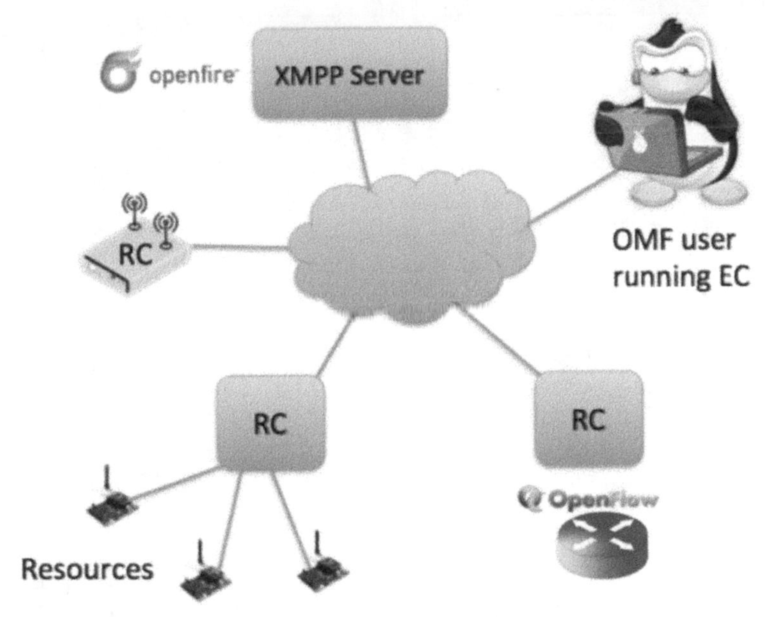

Figure 4.2 The OMF6 architecture.

resource there is a dedicated resource controller (RC) responsible for controlling it. OMF 6 moves towards to an architecture, which incorporates loosely connected entities, that communicate with a "publish-subscribe" mechanism by exchanging messages that have been standardized (Figure 4.2).

In overall, OMF 6 aims to define the communication protocol between all the entities rather than their specific implementation.

The messages of this communication protocol that are being exchanged are defined in the federated resource control protocol (FRCP [23]). This novel protocol defines the syntax of the messages, but not the semantics that are subject to the different implementations concerning the various kinds of resources (see Figure 4.2).

On the other hand, version 5.4 of the OMF framework is the most mature of the frameworks released under the 5th release. It supports interoperability with legacy OMF components. Although the exchange of messages is not standardized like in the 6th version, the testbed administrator is able to define a sequence of messages along the components and handle them appropriately. The different building blocks of OMF are the following, as shown in Figure 4.3:

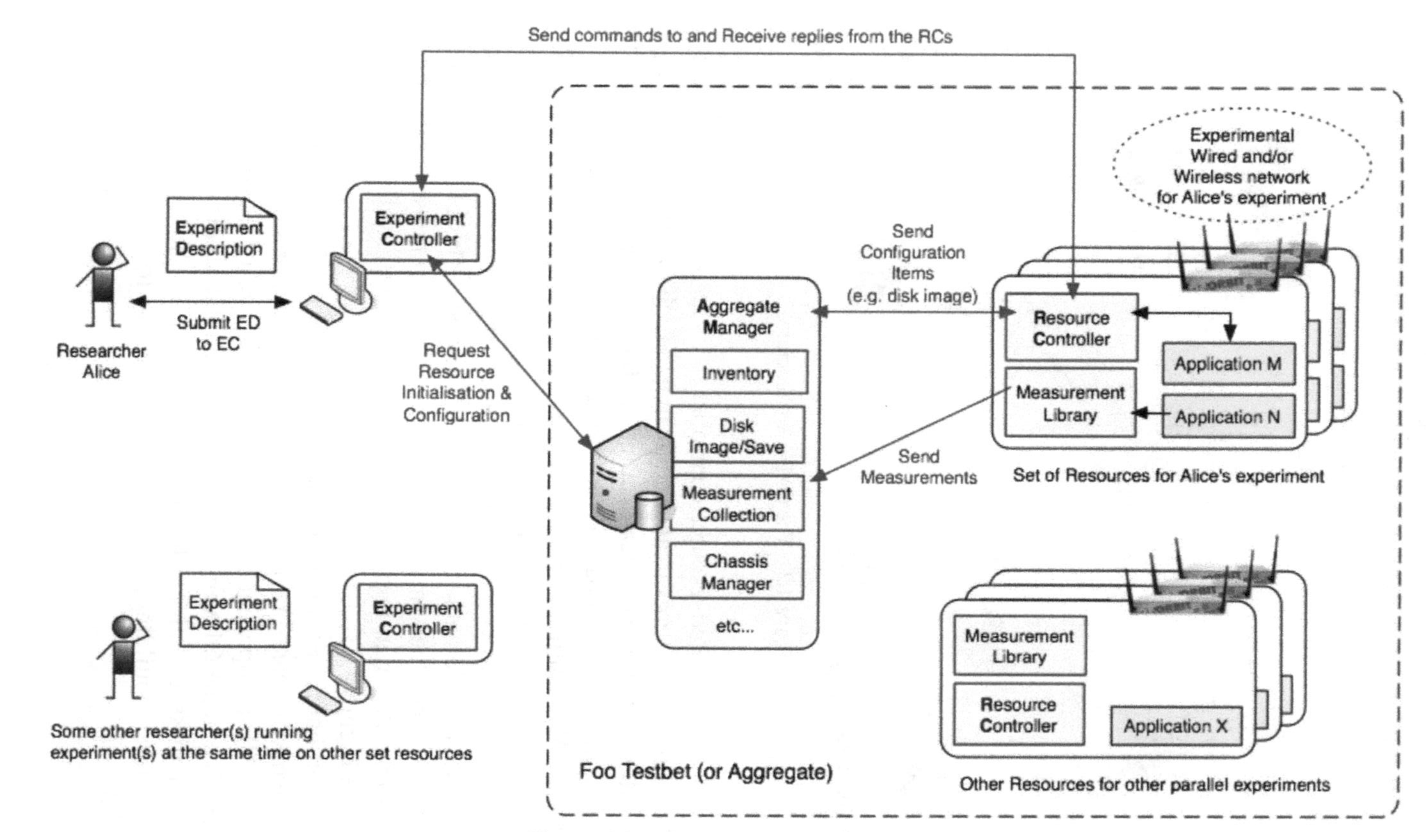

Figure 4.3 The OMF-5.4 architecture.

1. The OMF Experiment Controller (EC): The EC is in charge of receiving the experiment description in a high level language named OMF Experiment Description Language (OEDL) and generating the appropriate OMF messages sent to the Resource Controller.
2. The OMF Resource Controller (RC): The RC is in charge of parsing the OMF messages created by the EC and translating them in the appropriate commands for configuring the resources, installing/starting specific applications etc. The RC is generating OMF messages for monitoring the experiment process.
3. The OMF Aggregate Manager (AM): The AM is providing administration services for the testbed, like for example loading/saving an image on a node, turning a node on/off, etc.

4.3.3 OML

OMF Measurement Library (OML) [24] is acting complementary to the OMF framework and can be used for collecting distributed measurements from new or existing applications (Figure 4.4). Although initially it was developed to support the OMF framework, currently it can be used as a stand-alone library. OML is now a generic software framework for measurement collection.

OML is quite flexible and can be used to collect data from any source, such as statistics about network traffic flows, CPU and memory usage, input from sensors such as temperature sensors, or GPS location measurement devices. It is a generic framework that can be adapted to many different uses.

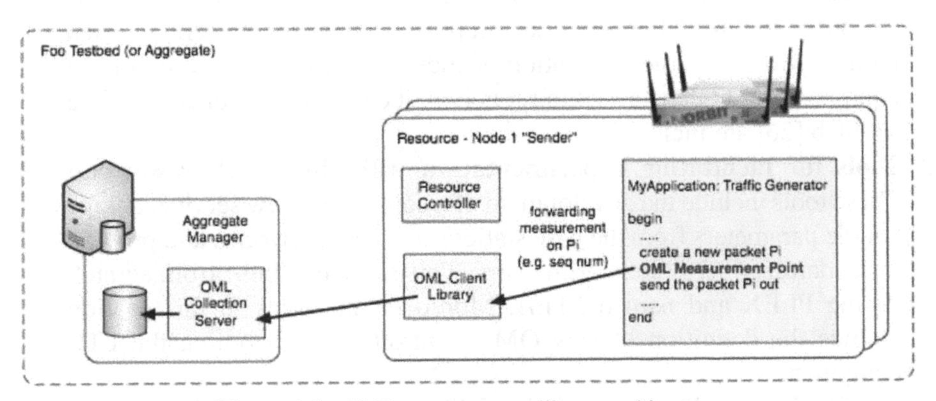

Figure 4.4 OML measurement library architecture.

Networking researchers who use testbed networks to run experiments would be particularly interested in OML as a way to collect data from their experiments.

OML consists of two main components:

- OML client library: the OML client library provides a C API for applications to collect measurements that they produce. The library includes a dynamically configurable filtering mechanism that can perform some processing on each measurement stream before it is forwarded to the OML Server. The C library, as well as the native implementations for Python (OML4Py) and Ruby (OML4R) are maintained.
- OML Server: the OML server component is responsible for collecting and storing measurements inside a database. Currently, SQLite3 and PostgreSQL are supported as database back-ends.

4.4 Approach

In order to enable the experimentation potential of the distributed FLEX platform, the resources offered by the consortium needed to be fully aligned with the testbed tools and frameworks. To this aim, FLEX has built extensions based on the aforementioned frameworks, as well as new platforms completely from scratch, in order to facilitate the experimenter access and usage of the LTE resources. The extensions and tools that FLEX has built are summarized in the following principles:

1. **Extensions for handling the LTE resources and SFA based federation**: These include the definition of new Resource Specifications (RSpecs) for the LTE network components that are present in each facility. Moreover, the integration of these RSpecs and handling of the equipment by higher layer tools, such as jFed, NITOS broker [25] and Emulab [26] are included.
2. **Tools for facilitating experimentation with the FLEX resources**: These tools include the development of a completely new service, able to handle parameters from the base stations and core networks, and provide a standardized API to experimenters. This service is built from scratch during FLEX and named LTErf. Moreover, the tools in this section include the definition of new OMF controllers for handling the LTE equipment.
3. **Monitoring applications of the LTE network status**: Monitoring applications have been developed by COSMOTE, the largest mobile

operator in Greece, along with UTH. The applications are aiming at both the depiction of network related information (e.g. Cell-Id, RSSI/RSRP, LAC/TAC) and the identification of possible network issues (e.g. poor/no coverage, unsuccessful handover). The tools are designed so as to fulfill the commercial requirements both in terms of presentation and functionalities. The tools developed are utilized in the context of FLEX project during the project time course by the project partners as well as by COSMOTE's engineering staff, mainly.

4. **A toolkit for enabling handover experimentation over FLEX**: As handover experimentation is of major importance for next generation and 5G technologies, FLEX members have developed a rich toolkit for enabling user-friendly experimentation and definition of handover experiments. The handover experiments that are currently supported include S1- and X2-based for LTE, as well as an SDN based handover scheme for cross-technology based handovers (e.g. LTE to Wi-Fi/WiMAX/Ethernet).
5. **Mobility emulation and real-mobility framework**: FLEX is providing sites offering real mobility, through either predefined trajectory control (iMinds) or fully uncontrolled mobility (UTH) inside the coverage area of a macrocell setup. Using the information collected through these real-world setups, including the signal fading for the different wireless channels, etc., FLEX is able to provision an emulation mobility platform using the programmable attenuation platforms for the LTE network. Through this framework, mobility patterns are used as predefined patterns, which can be programmed in the emulators by the experimenters.
6. **Functional federation of the testbeds**: This principle includes the operational engagement of the extensions for the control and experimental plane tools, as well as the physical interconnection of the testbeds over the GÉANT network in Europe. Using the extensions for the federation, resources from different testbeds inside FLEX are able to be bundled in one single experiment description, including scenarios of cross-platform interoperability (e.g. OAI femtocells and commercial macrocells from NITOS in Greece, controlled by an EPC network setup in Eurecom testbed in France).

The following section is describing in detail the extensions that FLEX has built in order to provision truly open LTE and beyond resources to the research community.

4.5 Technical Work

4.5.1 Control Plane Tools

The control plane tools that FLEX has focused are the ones that existed in the FIRE community before FLEX. The extensions to these tools are summarized in the following list:

- Extensions to the NITOS Scheduler – Portal platform
- Extensions to jFed
- Extensions to the NITOS Brokering tool

4.5.1.1 NITOS Scheduler

The NITOS Scheduler [27] is a framework developed by UTH, dedicated to the control and provisioning of testbed resources. It is developed in the spirit of serving as many users as possible without any complicated procedures. Its functionality relies on the OMF architecture. NITOS resources, namely nodes and wireless channels, are associated with the corresponding slice during the reserved time slots, in order to enable the user of the slice to execute an experiment. UTH has enabled Wi-Fi spectrum slicing support in NITOS, meaning that various users may use the testbed at the same time, without interfering with each other, since each one of them is using different spectrum blocks. The service can be adopted with very minor changes from any NITOS like testbed. It is worth to mention that already the Eurecom FLEX site is operating by adopting the NITOS Scheduler platform. It consists of a web frontend and a database backend for selecting and applying the appropriate firewall rules (for accessing the resources) and the spectrum restrictions (for not colliding with other experiments). In order to incorporate the FLEX resources, NITOS Scheduler has been extended in order to be able to parse the RSpecs regarding the LTE resources. Moreover, the web-frontend has been extended allowing the advanced filtering of the testbed resources, based on their type and frequency of operation.

4.5.1.2 jFed

jFed [19] is a framework that allows a user to design an experiment using resources of any of the Fed4FIRE's resource pool. It makes it possible to learn the SFA architecture and related APIs, and also to easily develop java based client tools for testbed federation. jFed is built around a low level library that implements the client side of all the supported APIs. A high level library manages and keeps track of the lifecycle of an experiment. On top of these two libraries various components have been built with different useful functionalities. The most important are:

- jFed Experimenter GUI (Graphical User Interface) and CLI (Command Line Interface) that allow experimenters to provision and manage their experiments.
- jFed Probe GUI and CLI that assist testbed developers to test their API implementations.
- jFed Automated tester GUI and CLI that perform extensive automated tests of the different testbed APIs.

The jFed framework that is used in FLEX has been extended to support LTE experimentation. Hence an experimenter can design his/her experiment and use the available LTE equipment. The equipment includes resources that are filtered through their defined RSpecs, regarding either base stations, EPCs or UEs. Moreover, the experimenter can alter the parameters that are used for setting up their experiment (e.g. transmission power, IP address of MME and PGW, etc.).

4.5.1.3 NITOS brokering

Fed4FIRE [5] project has been working towards federating experimental facilities using one unified framework. The Broker entity, which is designed by the Fed4FIRE project and implemented by the two partners who are also participating in the FLEX project (UTH and NICTA) is offering the means for resource discovery, reservation and provisioning of federated infrastructure to the testbed users. Broker's responsibilities contain the advertisement of testbed's resources to the interested users, but also the reservation and provision of them. It is a way to easily federate OMF testbeds under the scope of SFA [18]. However, it is not limited serving the SFA specification with the XML-RPC interface. Broker should be seen as the main way for experimenters to interact with an experimental facility. It offers additional interfaces beyond XML-RPC, like RESTful and XMPP which leverages the new OMF Messaging System. The main functions of the Broker are communication (through the Broker's available interfaces), Authentication/Authorization, Scheduling and AM Liaison.

The brokering service adopted by NITOS-like testbeds has been developed over the OMF6 framework and support the following configurations towards allowing the efficient provisioning of the project's testbeds:

- Discovery of the available LTE equipment in each testbed (base stations, EPCs and UEs).
- Configuration of this equipment tailored to each experimenter's needs (e.g. using a NITOS base station with a 3rd party EPC network using only the Internet connection).

- Intercommunication among the the testbeds for the resource reservation.
- Setting up the proper user accounts for accessing the LTE components.
- Configuring the appropriate access rules on each testbed for isolating concurrent experiments among different users.

The broker entity is interfacing the scheduler of each testbed and based on the resources creates the appropriate RSpecs for advertising the testbed components. It is also featuring multiple APIs for interfacing the SFA API that it provides. The supported APIs are three; 1) an SFA client based, using for example applications like SFI [28], 2) a REST based and 3) an FRCP [23] based.

4.5.2 Experimental Plane Tools

The extensions that are described in this section regard the following:

- The definition of the LTErf [29] service, for handling all the FLEX component parameters and easing the testbed federation, by allocating end-to-end isolated paths.
- The extensions to the core OMF framework for supporting experimentation with the LTE resources.

4.5.2.1 The FLEX LTErf service

One of the main challenges in provisioning an Open LTE testbed is the provided API for the configuration and setup of the involved LTE components. The LTE components we refer to are the base stations, EPC network, monitoring and datapath functions. In the following sections we refer to the **"LTErf"** [29] service that has been developed through the FLEX project, aiming for providing open and configurable APIs to the experimenters that take advantage of the FLEX testbeds.

The service is built on top of the OMF AM entity and provides a REST based interface for interacting with it. It is configured to reply with either an XML format or plain text, depending on the query and the representation that is requested by the end users. The APIs that are provided to the users are abstractly divided to four classes:

- *Base stations*: The wireless parameters, as well as the configuration of the base stations regarding their EPC interconnection should be the same among different vendors of hardware. Examples of such common parameters are the channel bandwidth, transmission power, etc.
- *EPC networks*: Similar to the base station approach, different EPC networks should provide similar functionality and thus provide the same

API for configuring them. Examples of such configurations are the different network configurations (IP addresses and ports for the S1-MME, S11, S6, S1-AP, etc. interfaces), Access Point Names (APNs) that will be used, etc.

- *Datapath configurations*: Setting a datapath, meaning the way that traffic will be routed beyond the EPC network, through a common API, regardless of the datapath chosen (eg. Internet/GÉANT). For the cases of the GÉANT network, the experimenter can set a VLAN tag for the traffic that will be exchanged, thus creating an end-to-end isolated slice on the wired network.
- *Monitoring functions*: As the equipment is already providing an API for the collection of network performance measurements, the service appropriately handles them and visualizes them to the end user.

The service has a modular architecture as shown in Figure 4.5. The different northbound interfaces for the subservices are mapped to resource specific drivers for controlling and configuring the diverse components. These drivers

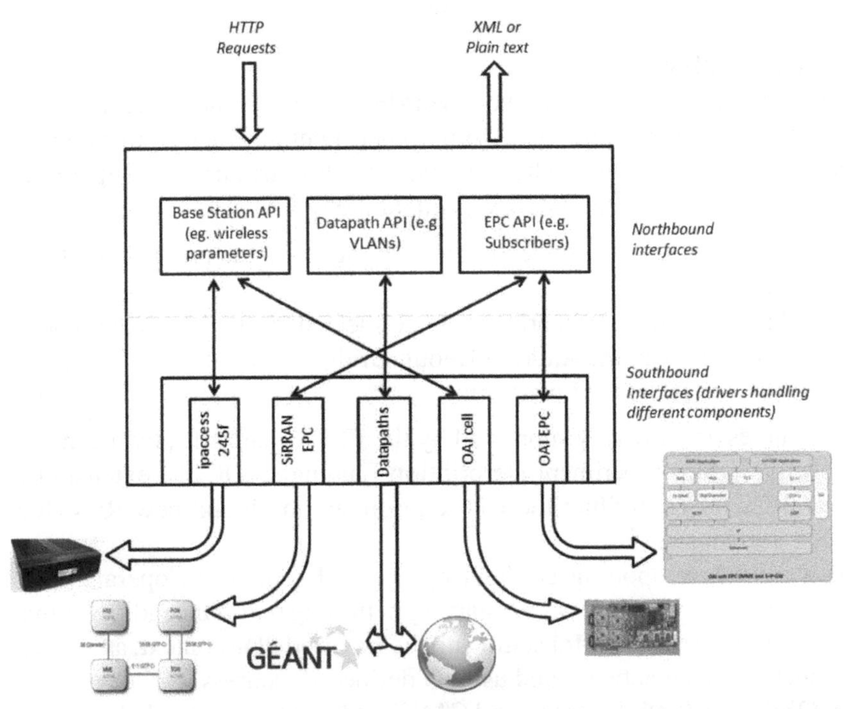

Figure 4.5 The LTErf service architecture; single northbound interfaces are mapped to several southbound depending on the type of the equipment.

consist the southbound interface, written in the Ruby language, able to handle the different methods of accessing the resources (e.g. SNMP/SSH access for the components). Upon service startup, a configuration phase is employed where the available resources (specified in a configuration file) are given to the service.

Different modules on the southbound interface are used to configure the different components are discovered and identified. During this phase, these drivers are initialized and set-up. From now on, the user is interacting with the web interface of the service, by addressing each resource using an identifier, like for example node1/node2 for the different base stations involved. The service parses any requests and delivers them to the appropriate driver for setting the respective resource.

The existing cellular solutions that are currently supported by the LTErf service are the following: 1) ip.access femtocells, 2) OpenAirInterface cells, 3) Airspan Air4GS LTE macrocells, 4) OpenBTS components, for configuring 2G/3G circuit-switched networks along with the 4G and beyond ones, 5) the Keysight T2010 conformance testing" units, 6) The SiRRAN EPC instances, 7) OpenEPC instances and 8) OpenAirInterface EPCs.

4.5.2.2 OMF extensions

As OMF has been widely deployed worldwide, FLEX has extended the available OEDL language for specifying experimental resources in order to include LTE resources as well. The LTE resources that are currently supported by incorporating them in an OMF experiment are:

1. LTE USB dongles, for connecting testbed nodes to the provisioned LTE networks,
2. LTE Android enabled Smartphones, connected to the FLEX networks and controlled over the Android Debug Bridge (ADB),
3. UE instances of the OAI platform.

These resources are currently supported by the FLEX platforms, by means of the respective OMF Experiment Description Language (OEDL) extensions, extended EC's for controlling the LTE equipment and brand new RCs (for both OMF versions).

The syntax is supporting configuring the LTE dongle to operate as a modem/USB mass storage device, restarting it, turning on/off the radio, setting an APN that will be activated for setting up the required PDP context, attaching and connecting to the network and using a defined IP address.

The OMF ECs (both for OM6 and OMF5.4) have been extended in order to support the updated experiment syntax and the generation of the OMF messages that are sent to the respective OMF RCs. For the case of the OAI UE, the same API is used as in the case of the LTE dongles, yet the vast configurability of the platform is allowing for the further extension of it in order to support more configuration parameters.

The RCs are responsible for receiving and decoding the OMF messages (FRCP for the case of OMF6) and translating them to the appropriate commands. For the case of the LTE dongles, the diversity of the available dongles inside the FLEX federation is posing several barriers that have to be overcome by the RCs. To this aim, the RCs are using the standardized protocol of AT commands [30] for interacting with the LTE dongles. The RCs for the smartphone components have been developed in the same spirit the respective ones for the LTE dongles.

Regarding the smartphone control, two RCs have been developed; an OMF5.4 RC for controlling the smartphone over the Android Debug Bridge (ADB) and an OMF6 RC for controlling it over the Wi-Fi interface. For the case of the ADB, the smartphones are connected in the NITOS testbed to the lightweight Raspberry-Pi based nodes that UTH has developed, or to standard NITOS nodes, via the USB connection.

4.5.3 Monitoring Applications

COSMOTE and UTH have developed over the FLEX platform three mobility/performance-related tools (Figure 4.6). The tools are decomposed to:

(a) Client applications running on Android devices, in "on-demand" mode, "on-event" mode or "periodically".
(b) Server-side infrastructure utilized to collect, store and process the related mobility/performance measurements.
(c) A graphical environment (WebGUIs) with advanced filtering and presentation capabilities, through which the measurements will be depicted.

4.5.3.1 FLEX QoE tool

The purpose of this tool is to present 2G/3G/HSPA/HSPA+/4G network related information (including BSs locations/capabilities/name, cell reselections/locations info, handover locations/info, etc.) in real time, over Google Maps. It is also able to measure and depict QoE related measurements in real time, such as signal strength (RSSI, RSRP, RSSNR, RSRQ, etc.), latency, maximum download bitrate, maximum upload bitrate and upload the QoE related measurements to a dedicated server for storage, post processing. The collected measurements are depicted via a user friendly web interface.

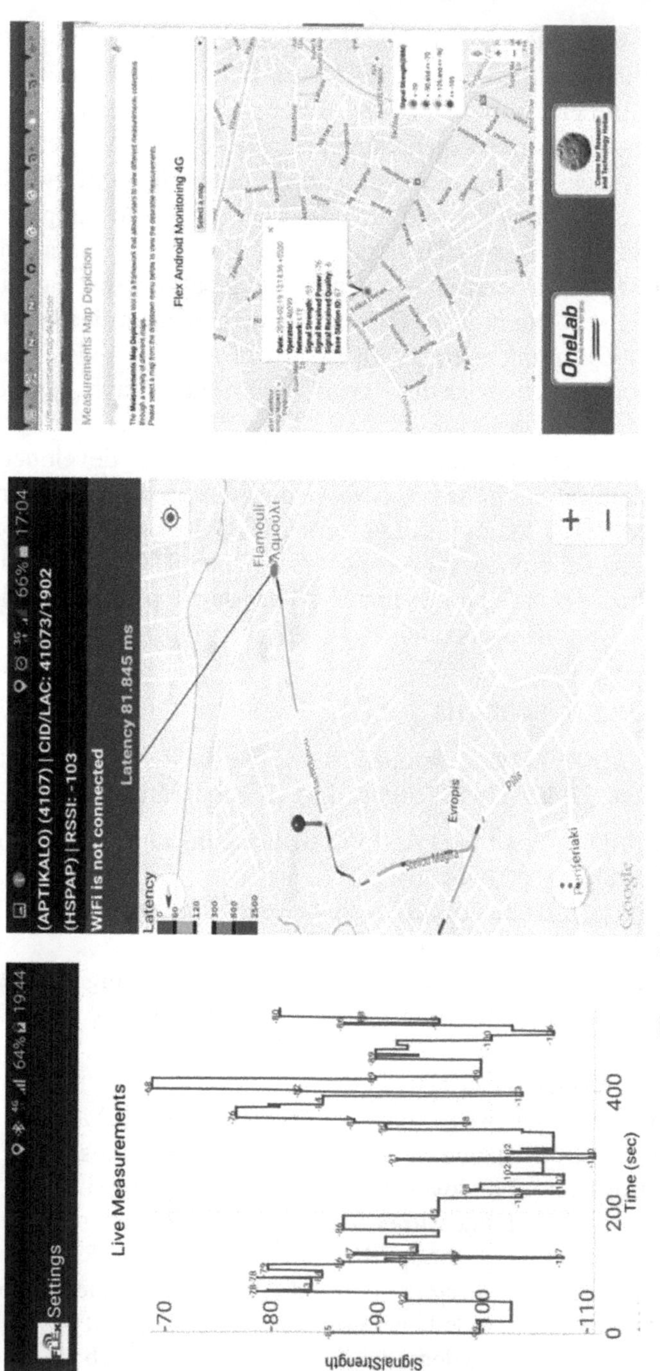

Figure 4.6 Mobile operator android tools for monitoring in FLEX.

4.5.3.2 FLEX_problems

The aim of the FLEX_problems tool, is to notify the MNO, in real-time, on network issues/problems (e.g., areas exhibiting huge number of cell reselections, poor coverage, no coverage, high number of handover failures). The client application runs on Android devices and could start either at power on, or manually. The application could be: (1) utilized by MNO staff (mobile UI is required in this case) and/or (2) offered by a MNO as a commercial application (running in background – no mobile UI required). In either case a graphical environment (WebGUI) shall be made available to the MNO so as to be informed on those network events. More specifically, the basic features of the FLEX_problems client App are the following: 1) Presents at terminal screen 2G/3G/4G network-related info (BS name, BS-id, RAT, cell-id, LAT/TAC, RSSI/RSRP, RSRQ, etc.), 2) "Listens" to the environment (2G/3G/4G) continuously and the terminal status (offhook, busy), 3) In case of an event (cell change on idle, handover, low-RSSI) it uploads, in real-time, to a dedicated server, the relevant measurements. 4) If the network is not available (handover failure, no coverage), it queues the "measurements" and uploads them (automatically) upon "network recovery", 5) Presents at terminal screen info, in real-time, regarding the number of cell reselections, handovers, poor coverage location identified, along with the number of queued messages (if any).

4.5.3.3 FLEX_netchanges

The aim of this application is to (automatically) measure the network performance in terms of signal strength (RSSI, RSRP, RSSNR, RSRQ, CQI), latency, maximum download bitrate, maximum upload bitrate) periodically (e.g., every X minutes). The application could be: (1) deployed by an MNO, on its own terminals distributed at specific locations – terminal operation could be remotely controlled and/or (2) offered by the MNO as a commercial app (running in background – no mobile UI required in this case). This application can serve as "real-time" network probes, in order the MNO to be notified on network performance e.g. in cases of Self-Organized Networks, network changes, etc.

4.5.4 Handover Toolkit

The handover toolkit available across the FLEX testbeds is an open framework that allows the configuration of the handover parameters for facilitating this type of experimentation. The following setups are supported:

- S1-based handover, using the commercial FLEX equipment.
- X2-based handovers, using the OpenAirInterface equipment.
- Cross-technology handover frameworks, using SDN and any types of LTE equipment.

4.5.4.1 S1-based handovers

In accordance with the FLEX project requirements to support experimentation of handover scenarios, SiRRAN and ip.access have extended the capabilities of their equipment (femtocells and EPC) to include S1 based handovers, between eNodeBs, connected to a single MME. Although S1 handover is normally utilised to facilitate transfer between eNodeBs that are connected to different MMEs, the NITOS and w-ilab.t testbed installations of the SiRRAN EPC use only a single MME component, so the functionality was designed in the EPC with this in mind. Initial development and testing was performed in SiRRAN's labs, using ip.access LTE245 and E40 radios.

4.5.4.2 X2-based handovers

Regarding the setup of the X2-handovers using the OpenAirInterface platform [31], within FLEX the extensions to support this type of handover procedure has been developed. X2 handover has several advantages compared to the conventional S1/MME handover used by other FLEX testbeds. The main key-features are described below:

1. The whole procedure is performed directly by the eNBs (without EPC). There is a direct tunnel formed between source and target eNBs for downlink data forwarding in handover execution time.
2. MME is involved only when the handover procedure is completed in order to setup the new network path.
3. The UE release context at the source eNB side is triggered directly by target eNB.

Thus, X2 handover minimizes the latency of the EPS network. A handover experiment in OAI can be performed using a different set of parameters that are managed via configuration/command-line (User CLI) inputs. User CLI provides certain commands for runtime control and monitoring of the OAI X2 handover. The parameters that can be adjusted are time to trigger, hysteresis parameter for this event, the frequency specific offset of the frequency of the neighbour cell, the cell specific offset of the neighbour cell, the frequency specific offset of the serving frequency, the cell specific offset of the serving cell, the offset parameter for this event, coefficient RSRP/RSRQ, parameter for exponential moving average (EMA) filter for smoothing any abrupt measurements variations. The developments take place over the OAI networking stack, thus enabling for the further extension and development of new policies for handover (e.g. [32]).

4.5.4.3 Cross-technology Inter-RAT SDN based handovers

Regarding the cross-technology inter-Radio Access Technology (RAT) handover framework, it is based on the OpenFlow technology [33], able to perform seamless handovers among different technologies (e.g. Wi-Fi to LTE, LTE to Ethernet, Wi-Fi to Bluetooth, LTE to WiMAX, etc.). The architecture adopted for the realization of the framework in NITOS is depicted in the Figure 4.7.

The framework is called **OpenFlow Handoff Control (OHC)** [34] and is consisting of two different entities; the mobile clients and the destination servers. During a handoff, network address changes take place at the mobile host, which break the established connections if no proper management is applied. These changes are induced by the different gateway used by each RAN, or by the NAT process that is always present before the traffic is routed to the Internet. With the OHC scheme the changes are handled at two points; on the client that performs the handoff and just before the traffic reaching the destination server. By using the OpenFlow technology, we are able to establish custom flows on a network switch, by mangling the exchanged traffic accordingly so as the connections are not dropped.

The key for applying our scheme relies on creating virtual OpenFlow enabled switches. To this aim, on the mobile node we employ the architecture illustrated in Figure 4.7; we place all the available networking interfaces in a single switch. By relying on the Open vSwitch framework [35] for the creation of our switches, the switches residing on the mobile node are OpenFlow enabled. The Operating System on the mobile node communicates only with the bridge device as a network interface and uses it as the default interface for any outgoing/incoming traffic from the mobile node. The controller that is establishing the flows on this virtual switch is in charge of selecting the appropriate southbound interface (e.g. Wi-Fi, LTE) for sending out the traffic.

The respective changes for adopting our framework have to take place before the traffic is delivered to the destination application. As we described, in the case that the bridge on the mobile node has an IP address of the 10.0.0.0/24 subnet while the Wi-Fi interface bears an IP address of the 192.168.0.0/24 subnet, the flow on the switch will change the source IP and MAC address

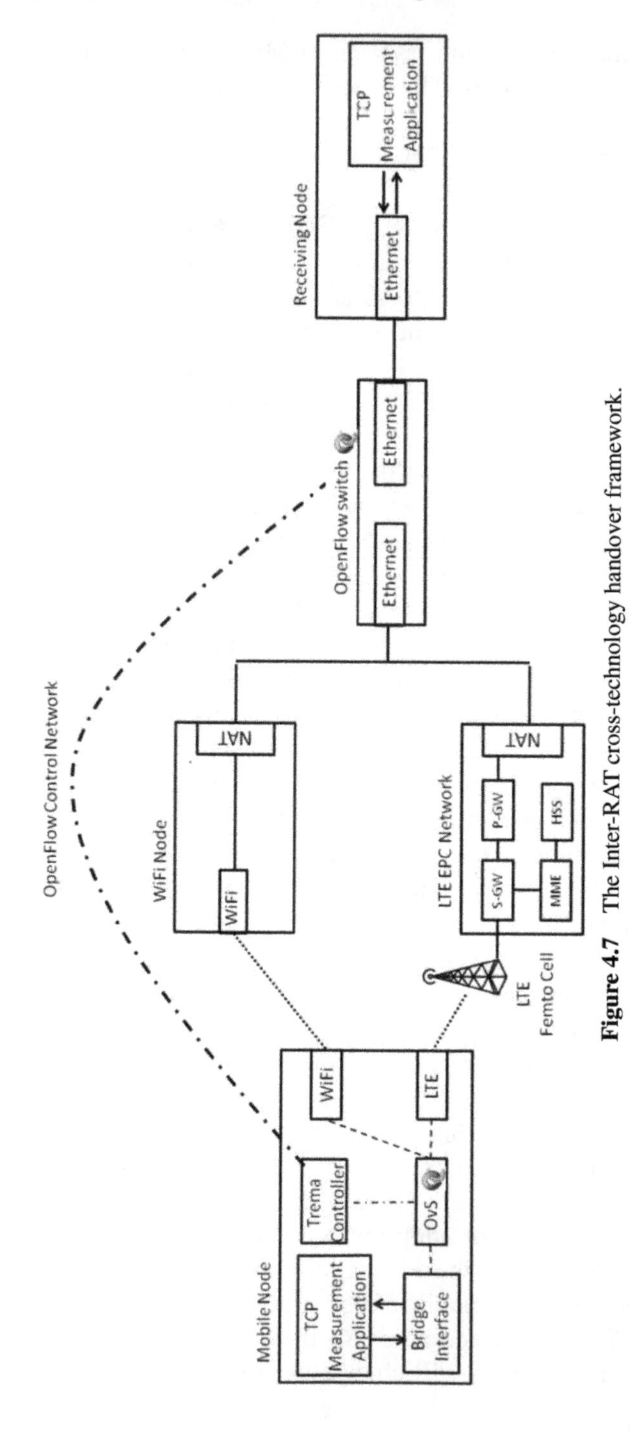

Figure 4.7 The Inter-RAT cross-technology handover framework.

of each outgoing packet to match the address of the Wi-Fi interface, and the respective destination MAC address to match the one of the target Wi-Fi Access Point. For the incoming packets, the opposite procedure has to take place.

The testbed application of our framework is the following. On our mobile node we use Open vSwitch (OvS) for our bridging solution, and enable its control from an OpenFlow controller residing on the same machine. We employed the Trema framework [36] as our solution for implementing our OpenFlow controller. Finally, we unified both the operation of our aforementioned algorithms (server side and mobile node side) in one controller instance, which is able to control multiple datapaths (mobile node and NITOS OpenFlow switch).

A comparison of the FLEX inter-RAT framework for LTE to Wi-Fi handovers against other state-of-the-art solutions for cross-technology handovers or higher-layer solutions is shown in Figure 4.8. As it is illustrated, both achieved throughput and delay through this scheme are better, compared to other technologies, and as if the interfaces were acting as standalone connections to their network.

4.5.5 Mobility Emulation Platforms

Data captured from the real network setup are used in order to feed the mobility emulation platforms. The data that is used for generating the patterns is collected from monitoring applications, residing at the FLEX testbed nodes, and after their anonymization (removing all the user sensitive information, such as the phone's IMEI, the card's IMSI, etc.) are fed to the emulation

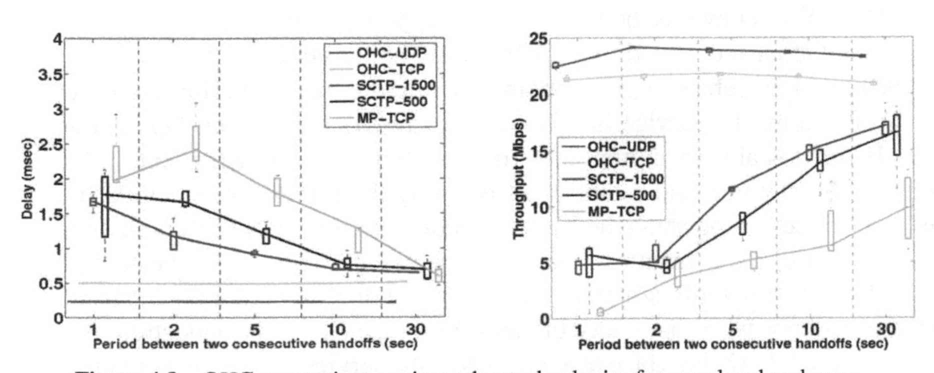

Figure 4.8 OHC comparison against other technologies for seamless handovers.

platform. The selection of a tool like the Qosmotec platform (by iMinds) is crucial, as it provides the experimenters with the potential to fully replicate a real world mobility experiment with the emulation platform.

Path loss models can be used to calculate the reduction in power density of the signal between two radio devices. The results of path loss model calculations can be used to feed the emulation mobility platforms (attenuators, LTE cells and UEs) and emulate signal attenuation. The simplest path loss model is the free-space path loss (FSPL) model that presents the loss in signal strength on a line-of-sight path without any obstacles [37]. The calculations are straightforward but do not model real conditions. For cellular networks, the Walfisch-Ikegami (COST231 project) [38] and Erceg model [39] are frequently used. The ITU-R P.1238 model [40] is developed for indoor conditions. Most of the models are used for lower frequencies (<2 GHz), but by adding a certain correction factor, they can still be used for higher frequencies.

4.5.6 Functional Federation

In order to enable the functional federation across the FLEX islands, dedicated end-to-end slices have to be reserved from one testbed to another, utilizing the GÉANT network. The tools that enable such access are the the LTErf service and jFed. LTErf has been developed in a manner that allows user defined datapath control. However, the incorporation of LTE resources in the testbed network creates several issues that are not present when dealing with other resources than the LTE ones. Since no ARP protocol is used on the LTE access network, and until data reaches the EPC, the EPC service is endowed with the process of handling the ARP messages for the data incoming to the EPC for the PDN-GW and towards the UE. As the address with which the EPC replies to any ARP request destined to the UE is always the same, we had to create a book-keeping mechanism for mapping the appropriate traffic flows to each UE. To this aim, the service is able to generate dynamically an OpenFlow controller that is able to appropriately map each request to each client based on the APN they use, and establish accordingly the traffic flows. Similar to this, the service is supporting the VLAN creation through an HTTP command, and adding it to the datapath so that the experimenter can create end-to-end isolated slices of the infrastructure, incorporating different components from different testbeds with guaranteed bit rates. Since the GÉANT connections are delivered as a VLAN interface at the testbeds, the service enables the creation of dedicated QinQ VLANs inside them, per each user request.

The jFed provisioning of end-to-end slices is based on VLANs which are provisioned and then stitched together at points where they meet. The workflow in the jFed tool is as follows:

- The experimenter draws in an experimenter tool a link between two nodes on different testbeds (which is translated in an RSpec).
- When the tool starts provisioning, it first calls the Stitching Computation Service (SCS) which calculates a route between the two testbeds based on the layer 2 paths it knows. The SCS augments the RSpec with this information.
- The tool then knows also intermediate hops in the path (e.g., GÉANT, Internet2) and can call them to set up the path.
- In the end, all the parts of the links and nodes become ready, and the experiment is ready.

For this fully automatic stitching, the VLAN numbers are dynamically chosen based on free VLAN overviews, tries and retries.

4.6 Results and/or Achievements

The experimentation potential that the FLEX platform is fulfilling is mirrored in the different number of use cases and scenarios that can be executed over the testbed. Indicatively, we present some experiments that have been successfully executed over the FLEX testbed, along with some experimental results. We focus on the following scenarios:

1. Spectrum coordination schemes for LTE in unlicensed bands, using semantics.
2. The development of an offloading framework using the commercial equipment.

4.6.1 Semantic Based Coordination for LTE in Unlicensed Bands

One of the types of different experiments that can be executed over FLEX testbeds deal with Dynamic Spectrum Access (DSA) for heterogeneous technologies, along with their spectrum coordination algorithms. To this aim, several works have been executed demonstrating the coordination of spectrum for different technologies, using either the commercial LTE equipment [41] or the OAI setup [42].

In this subsection, we focus on the LTE and Wi-Fi coexistence in an unlicensed band environment. Wi-Fi and LTE are different RATs designed for specific purposes at different frequencies. In the cases when they are required to coexist in the same frequency (e.g. LTE in Unlicensed bands) time and space, increased interference is caused to each other along with an overall system degradation because of a lack of inter-technology compatibility.

For LTE-U (LTE in Unlicensed bands) operation, several challenges have to be tackled for the efficient coexistence of LTE and Wi-Fi technologies. The key differences among the two technologies lie in the medium access method; Wi-Fi uses CSMA/CA, a "listen before talk" method in order to access the medium. In case of an unsuccessful transmission, the Wi-Fi device executes an exponential backoff algorithm before accessing the medium again. Contrary to that, and since LTE is designed for use under a licensed band environment, LTE is using OFDMA (Orthogonal Frequency Division Multiple Access). The coexistence of the two different technologies within the same band, can seriously affect the performance of Wi-Fi. Therefore, efficient spectrum management and power control should be employed for accommodating both of these technologies within the same band. In this use case, we focus on the spectrum coordination solution called CoordSS [42], which is using semantics for the coordination between Wi-Fi and LTE.

Figure 4.9 presents a conceptual overview of the CoordSS networking architecture. Three verticals and three horizontals can be identified in the architecture. The following verticals represent different views on top of the same set of foundational concepts:

- Network Environment – represents the "real" world. This includes hardware devices as well as physical phenomena (such as frequencies) along with their properties.

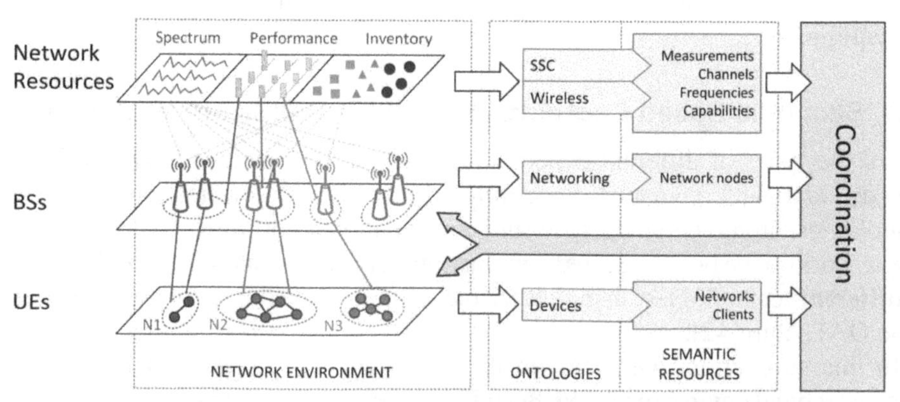

Figure 4.9 CoordSS Network model for semantic based coordination.

- Ontologies – are used to formalize domain specific knowledge that is independent of the context. They contain semantic definitions related to the meaning and purpose of the network environment. Ontologies are created by the domain experts and can be viewed, understand and managed by the humans as well as by the machines.
- Semantic resources – are the results of a semantic annotation of the network environment by mapping between the environment and ontologies. More precisely, if there is a physical resource that can be understood using the given set of ontologies it becomes the semantic resource.

Horizontals represent the main concepts in our network model. In the coordination algorithm, they play the roles of sources and/or destinations.

- Network resources – constitute the state and capabilities of the environment where BSs and UEs are working. They are the primary source of data for reasoning during the coordination. On the networking environment level, we are using spectrum sensing devices (such as Wiser [43]), connection bandwidth monitoring applications (such as *iperf*) and the inventory repository (Note that FLEX testbeds regularly provide such a service). The ontologies level consists of the Spectrum Sensing Capability (SSC) ontology (for describing spectrum sensing) and the Wireless ontology (for describing frequencies, channels and radio bands). And at last, semantic resources level contains data for FFT analysis of frequencies, connection speed, device parameters and their changes over time.
- BSs – nodes that provides access points for UE. They are a backbone for network communication. The OAI [7] ontology is used to describe such devices. The coordination protocol uses a semantic representation of BSs to decide which parameters can be changed to improve networking. Such parameters include their power signals, position (if applicable) and communication channel.
- UEs – client nodes that form networks so they can send and receive data among them. We can have multiple networks, and one UE can belong to any number of networks (but we view it as a separate UE for each network). Therefore, each device is identified by a network name to which it wishes to belong to. Semantic resources for UEs contain client demands for communication.

Coordination is centralized on one machine that is running the CoordSS Coordination server (CCS). The CCS is responsible for running the coordination algorithm, providing client/server communication with the network resources, mapping network resources to semantic resources, maintaining a semantic store that holds ontologies and semantic resources and executing SPARQL queries. The coordination algorithm is invoked in case the network environment changes, namely when a new BS or UE is introduced or when network resources fluctuate (e.g. changes are observed regarding the performance or spectrum). Clients send their spectrum, performance and node description to the server. This data is in a native format. CCS maps such data to semantic resources and stores them in the semantic store. The semantic store is used for storing and retrieving triplets, basic building blocks of ontologies and semantic resources. SPARQL queries are the standard way for retrieving semantic data, and are used by the coordination algorithm for all reasoning as well.

The main objective of the CoordSS coordination algorithm is to assign radio channels to the networks that are under its control. Any network that participates in our algorithm must have all of its nodes (UEs and BSs) registered to the CCS. Registered nodes send data to the CCS and also receive control messages from it. In our case, only channel allocation control commands are sent, but more elaborated control is also possible. When the algorithm decides to assign a channel to a network, commands are sent to all the nodes belonging to that network to switch to the new channel configuration.

There are two possible scenarios that we consider:

1. (S1) The network is part of the network environment and all of its nodes are aware of the CCS. This network does not have a channel assigned to it, but the coordination algorithm is responsible to provide one.
2. (S2) An uncoordinated network appears in the network environment (LTE or Wi-Fi network). This network uses its own algorithm for channel assignment. This network can interfere with existing coordinated networks. Our algorithm detects such a situation and resolves any interference by re-assign channels of the coordinated networks.

For the experimental evaluation of the proposed algorithm, we employ the NITOS testbed of the FLEX federation. The rich environment that NITOS is offering is utilized in order to configure the suitable environment for the experimental evaluation in real world settings of the CoordSS framework. To this aim, we employ the following testbed components:

- A pair of USRP B210 models, that will serve as the RF front-end of the deployed LTE network.
- Several Wi-Fi enabled nodes, that will be used as the contending traffic in the unlicensed under study bands.

- The OpenAirInterface (OAI) platform, that provides the execution of a 3GPP EUTRAN over commodity hardware, with the appropriate RF front-end. The OAI platform has been extended in order to allow its operation in the unlicensed bands.

The experiment topology is shown in Figure 4.10. The following methodology was used during the experiment. At first, only Wi-Fi stations were involved. Each Wi-Fi network would randomly choose a channel, and the resulting throughput was measured. This procedure was repeated 100 times and the average throughput was calculated. After that, wireless node 1 randomly chose a channel, wireless node 2 received a channel from CoordSS server, and the throughput was measured. The results are shown in Table 4.1. The second part of the experiment, besides the coordinated Wi-Fi networks, involved the LTE eNB, with and without coordination. A similar procedure, was applied. The results are shown in Table 4.2.

The results show the importance of the coordinated spectrum usage. Due to a relatively low number of the involved nodes, the average throughput is not very much improved by the coordination. However, the coordinated network has more stable throughput than the uncoordinated one, i.e. the difference between the lowest and the highest throughput is rather large in uncoordinated network. We should also have in mind that the output power of the USRP B210 is relatively low (10 dBm). Therefore, the influence of the dedicated LTE eNB on Wi-Fi would be much higher.

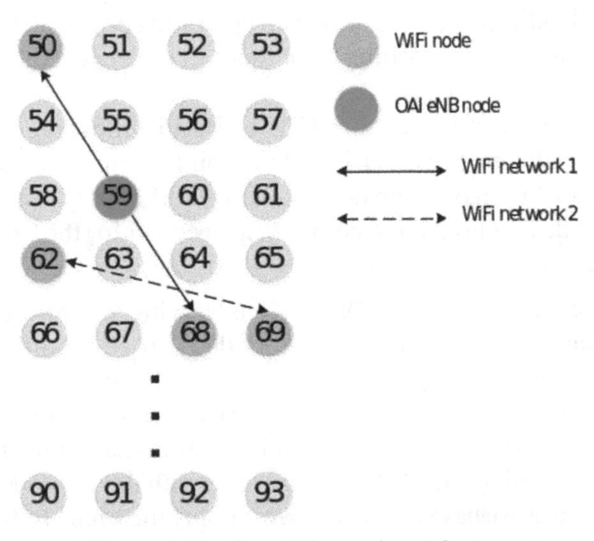

Figure 4.10 CoordSS experimental setup.

Table 4.1 Coordinated and uncoordinated shared spectrum access with Wi-Fi stations

	Wi-Fi Throughput (Mb/s)		
	Min	Average	Max
Uncoordinated	11.5	19.6	22.8
Coordinated	22.8	22.8	22.8

Table 4.2 Shared spectrum access with coordinated Wi-Fi networks and (un)coordinated LTE eNodeBs

	Wi-Fi Throughput (Mb/s)		
	Min	Average	Max
Uncoordinated	10.6	16.7	22.8
Coordinated	22.8	22.8	22.8

4.6.2 FLOW LTE to Wi-Fi Offloading Experiments

As the explosion of Internet and mobile data traffic has placed significant pressure on cellular networks, data offloading to complementary networks (e.g. Wi-Fi) seems to be the most viable solution. For the operators, in contrast to network planning strategies for upgrading, expanding and building up new infrastructure, which means extra capital and operational costs (CAPEX and OPEX), offloading can offer a sufficient and low cost solution for cellular load decongestion. Mobile Data Offloading is also significantly important for the mobile users, who can further benefit from short-range links so as to achieve better performance and experience better quality of communication by shifting to complementary networks. FLOW architecture aspires to address the challenges that offloading brings and create an open and applicable framework for implementing advanced offloading techniques in heterogeneous networks (LTE & Wi-Fi).

The FLOW experiment is realizing LTE to Wi-Fi offloading techniques over the FLEX testbeds (Figure 4.11). The components that have been developed during FLOW have been described in detail in [44]. Nevertheless, we provide a brief description of the components needed for the execution of the offloading framework:

1. **Wi-Fi Access Gateway (WAG)**: WAG is serving the role of the the actual gateway of the Wi-Fi mesh network that is used for offloading the LTE clients. Although the implementation of such a device would seem straightforward, in the FLOW framework we differentiate the traffic that is exchanged from the offloaded clients in order to meet some minimum requirements paved by the SLA that they have with the network provider. To this aim, and as we have described, we employ the Linux traffic queues for traffic shaping services.

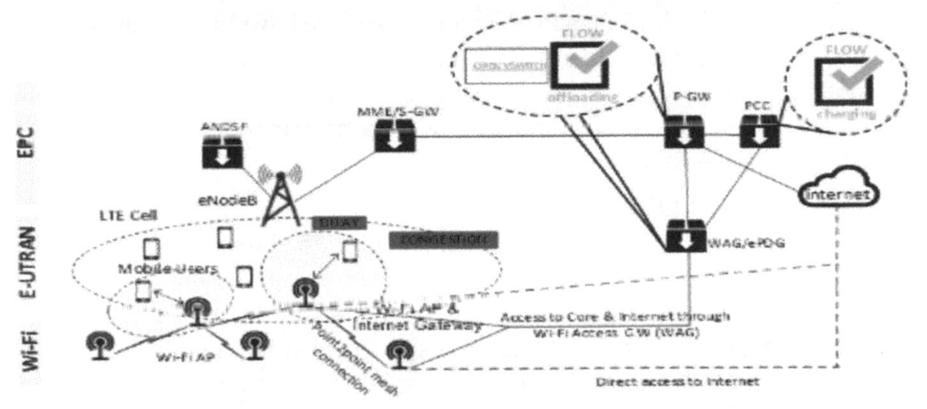

Figure 4.11 FLOW offloading framework.

2. **PDN Gateway (PGW)**: The LTE PGW interface is in-charge of terminating the SGi interface towards the PDN. In the case of multiple PDNs, more than one PGW will be available for the UE of the network, depending on the Access Point Names (APNs) used for the network. With FLOW we extend the functionality of the PGW in order to enable the operation of our offloading scheme. We implement an Open-vSwitch [35] bridge that enables the dynamic bridging of two different entities, Wi-Fi mesh network and the LTE network, and attaches the FLOW framework to take care of the low level network functions that have to be employed for the proper operation and routing of packets to the Internet.
3. **FLOW offloading framework**: The FLOW offloading framework has been designed in order to coordinate the interaction among the WAG and PGW elements. By employing a Software Defined Networking manner, we bridge the heterogeneous RANs and through a controller service we are able to select the respective RAN from the network provider's perspective. The policies that we implement for the offloading process are based on the load that each femtocell can provide and some predefined SLAs that each client has contracted with the provider. Moreover, based on the QCI parameters per UE in LTE, we allocate each of the offloaded clients to the respective traffic queue, upon which we schedule the transmissions of the respective data to the WAG and then the rest of the Wi-Fi mesh network.
4. **PCC (Policy Control & Charging)**: The PCC unit is in charge of applying the proper control of the policies and charging of the clients per subscriber basis, and based on the QoS class that they belong. As FLEX

components do not include a PCRF component, we have implemented it over the FLOW network to allow monitoring of each client. We are able to both monitor the data that a UE exchanges over the LTE or the Wi-Fi network, and is relying and interacting with the aforementioned schemes (FLOW, PGW, WAG).

For the setup of the FLOW offloading experiment, we employ the IEEE 802.11s extensions that are available in the NITOS testbed images. They are used for forming a multi-hop Wi-Fi mesh network for offloading the LTE clients. As in the NITOS indoor testbed all the nodes are able to "see" each other, we isolate the access nodes by adding specific next hop neighbours in order for the traffic that we send to use at least 2-hop paths before reaching the WAG gateway. The WAG component is also located in the NITOS testbed and is connected via a 1 Gbps connection to the EPC server that we use.

Regarding the WAG configuration, we use a tap-based tunnel for the communication of the EPC and the WAG components (Figure 4.12). We choose this type of connection as the PDN-GW is also represented a tap interface. On the node that is playing the WAG role, we use Open-vSwitch on the node in order to bridge the two interfaces (tap and Wi-Fi mesh). Based on a predefined set of IP addresses that we use for the Wi-Fi clients, sharing the same IP range with the LTE ones, we allocate them to a different traffic queue inside Open-vSwitch. Using external applications, such as the "tc" [45] traffic control service, we are able to throttle appropriately the traffic that is delivered to each client, based on the delivery IP address of each client. For the application of the different QoS profiles that each UE is using, we utilize the functionality that SiRRAN's LTEnet is offering, allowing us to setup different subscriber groups with multiple subscribers. Based on this configuration and groups, the EPC is able to throttle the traffic either on the DL or on the UL

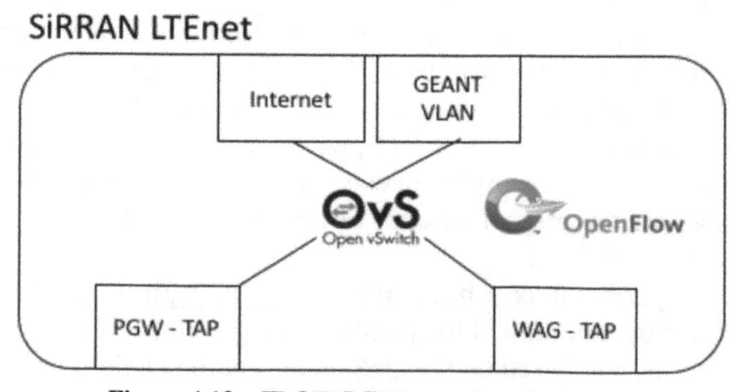

Figure 4.12 FLOW PGW extensions for FLEX.

that they exchange over the EPC network. This already supported functionality alleviates the employment of similar traffic throttling solutions for the LTE network, contrary to what happens for the Wi-Fi mesh network.

As within LTEnet the traffic that is delivered to the PGW interface is represented as an Ethernet tap interface, we used an altered version of the default "GÉANT" datapath that is available in NITOS as our starting point. This "GÉANT" datapath [29] is enabling the bridging of the PGW interface (that is reflecting an APN inside the network) and the GÉANT VLAN termination point in NITOS. The architecture that we have employed is depicted in Figure 4.13.

The cornerstone of the FLOW offloading management framework relies on the operation of the controller managing and establishing flows on the Open-vSwitch bridge on the LTEnet installation. For our initial tests and the experimental evaluation of the offloading frameworks, we developed a framework based on some predefined SLAs for all the involved clients.

The FLOW controller is in charge of the following actions:

1. Based on the first packet that it receives from the LTE client, checks whether the client's SLA can be met from the current capacity and bearer allocation at the LTE network.
2. Decides whether to offload the client or not.
3. In case that the client will not be offloaded, the controller establishes the proper flows that allow the communication of a client from the PGW interface to the Internet or the GÉANT network.
4. If the client will get offloaded the following actions are triggered:
    a. The controller triggers an assisting FLOW application running at the EPC which communicates the offloading message via the testbed control network to the UE. Another similar application that is installed on the testbed node, handles the message and instructs the wireless network interface to connect to the Wi-Fi mesh network. From now on, the offloaded UE will use the Wi-Fi network as the default gateway for sending traffic.
    b. The controller Is communicating a similar message to the WAG component. The WAG, based on the SLA for network capacity, allocates the node on the proper HTB queue of the system, thus scheduling appropriately and shaping the DL traffic that the client will receive over the Wi-Fi network. Finally, the controller establishes the appropriate flows on the Open-vSwitch bridge of the EPC network to use the WAG-tap interface as the default interface for the specific UE.

5. Continues monitoring the environment conditions, through similar messages received from the Wi-Fi mesh. In case that a client has left the LTE network, and the SLA of an offloaded client can be met from the LTE network, it reinstructs the client to connect to the LTE network, following a similar procedure like the one described in step 2.
6. Monitors the traffic load that each client has sent over the WAG/PGW interface in order to apply the pricing and charging functions.

The overall architecture that we adopted for an initial setup at the NITOS testbed is depicted in Figure 4.13. The setup at this point has been mapped over the NITOS testbed.

For the evaluation of the FLOW experiment, we performed offloading based on some pre-defined SLAs for the LTE network. The SLAs that we used for each LTE node are summarized in Table 4.3.

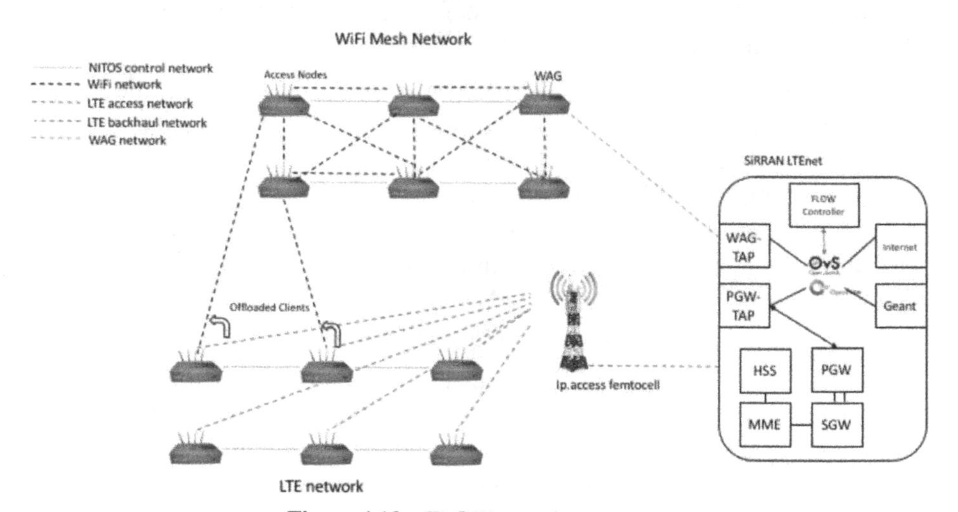

Figure 4.13 FLOW experiment setup.

Table 4.3 SLA setup for the FLOW offloading experiment

NITOS LTE Node	SLA for DL Bandwidth
Node054	15 Mbps
Node058	20 Mbps
Node074	10 Mbps
Node076	30 Mbps
Node077	7.5 Mbps
Node083	5 Mbps

The total capacity of the LTE network that the NITOS testbed is serving per femtocell is approx. 70 Mbps for the DL channel. Similarly, the total throughput (meaning the measured throughput from a client application) that the Wi-Fi mesh is achieving when using 2 hops is approx. 18 Mbps. Based on these given facts, and on the qualitative results that we expect to get from the theoretical framework that we have applied, always the client that is has the highest demand on DL bandwidth will be offloaded to the Wi-Fi mesh network. If his/her demand cannot be met by the Wi-Fi network, the second highest in demand client will be selected to be offloaded, or else the third, etc.

Below we present some first experimental results and how the clients have been reallocated to use the Wi-Fi network, for the given SLAs. As we can see, the experimental analysis (Figure 4.14) matches the theoretical framework expectations. It is worth to mention, that for the validity of our results we used Wi-Fi bands in the 5 GHz band, so that there is no external noise or no overlapping with the rest of the 802.11 frequencies.

This experiment is depicting an example run from the FLOW offloading over the NITOS testbed. The clients are admitted to the LTE network every 10 seconds, and the FLOW framework is handling these requests for offloading them to the Wi-Fi network. For this experiment run, node054 and node058 are using the LTE channel for the first 20 seconds.

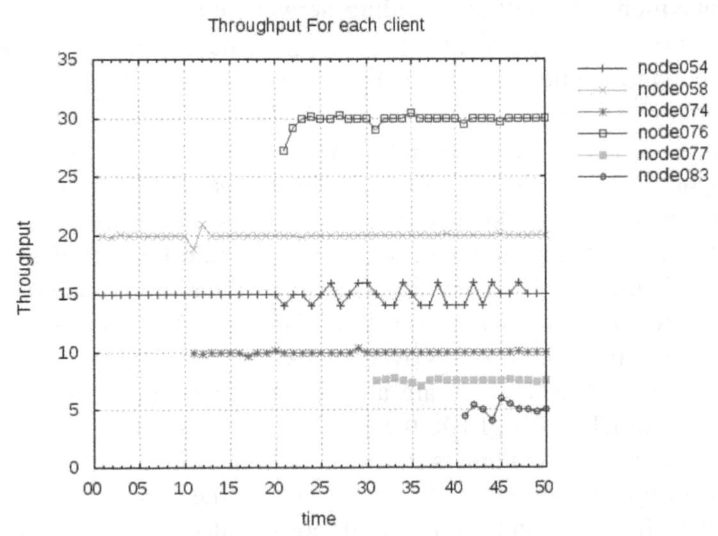

Figure 4.14 Throughput per each (offloaded) client.

When node074 is admitted to the network, the framework checks whether the node can be served by the Wi-Fi mesh network. However, the request for 30 Mbps DL traffic cannot be met by the Wi-Fi network and therefore the framework selects the second highest demanding client, which is node054. Similarly, when the rest of the clients are admitted to the LTE network, their total demand does not exceed the total LTE channel capacity. When the last client (node083) is admitted to the LTE network, the requested capacity will exceed the one that can be provided by the LTE channel. Therefore, the framework selects the most demanding client that is already served by the LTE network to offload to the Wi-Fi mesh. Nevertheless, the SLAs must be met at the Wi-Fi mesh network as well. Therefore, the choice that will make the best utilization of the network is the node083 itself, as it will be able to both get the remaining capacity of the mesh network and meet its demand for bandwidth.

Discussion

The potential of the FLEX federation of 4G and beyond testbeds has been demonstrated through the execution of some example experiments over the infrastructure. Yet, these are only a small portion of the experimentation capabilities of the platform, as several more have been proposed and are currently under execution. These include aspects regarding contemporary 4G network deployments, for either providing network measurements under a completely controlled environment, or developing new products designed for 4G and beyond applications, as well as aspects that will be addressed by the upcoming LTE releases and ultimately the 5G protocols, like for example narrow-band LTE development, Device-to-Device communications, NFV/VNF applications for the EPC, software defined backhauling for cellular networks, and even the development of software-defined base stations.

The platforms that are built through FLEX include high configurable equipment that is used for both development and evaluation of technologies for contemporary mobile networks, as well as for setting the cornerstone for the development of the first 5G pilots over the testbeds, using the open source software. Examples of such cases are also the experimental evaluation of functional splits for LTE over FLEX, the development of duplex schemes for wireless communications and others.

The high programmability of the platform and the vast potential that it has provides the community with the unprecedented chance to experimentally evaluate aspects for 5G networks using the existing infrastructure.

Moreover, the measurements that are provided by the testbeds, are given through open access to the community, thus enabling the implementation of algorithms regarding Big Data analysis, data mining techniques, etc.

Conclusions

FLEX is providing the infrastructure and platforms for the experimentally driven evaluation of scenarios including mobile broadband and potentially 5G networks. FLEX is filling a crucial gap in the existing infrastructures for the development of the Future Internet platforms, as it is the first pilot project that enhances FIRE's resource pool with cellular technologies.

In this chapter, we have presented briefly the FLEX platforms, and described the tools that have been developed in order to enable meaningful experiments to be executed over FLEX. These include tools for conducting federated experiments across the FLEX testbeds, always in line with the existing Fed4FIRE efforts in Europe, as well as for the user-friendly experimentation with the underlying equipment. Finally, some indicative use cases that take advantage of the infrastructure and platforms have been presented, as a means to demonstrate the potential of the platform. These include some crucial issues that are considered by the research community, such as the Wi-Fi and LTE coexistence in an unlicensed environment, as well as the Wi-Fi to LTE offloading process.

References

[1] "3GPP global initiative for mobile broadband standards", [online] Available: http://3gpp.org

[2] "Future internet research & experimentation", [online] Available: http://ict-fire.eu

[3] "FIRE LTE Testbeds for Open Experimentation", [online] Available: http://flex-project.eu

[4] Fdida, Serge, Timur Friedman, and Thierry Parmentelat. "OneLab: An open federated facility for experimentally driven future internet research." In *New Network Architectures*, pp. 141–152. Springer Berlin Heidelberg, 2010.

[5] "Federation for FIRE testbeds – Fed4Fire", [online] Available: http://fed4fire.eu

[6] "GEANT Pan-European network", [online] Available: http://geant.org

[7] Nikaein, Navid, Mahesh K. Marina, Saravana Manickam, Alex Dawson, Raymond Knopp, and Christian Bonnet. "OpenAirInterface: A flexible platform for 5G research." *ACM SIGCOMM Computer Communication Review* 44, no. 5 (2014): 33–38.

[8] "NITOS – Network Implementation Testbed using Open Source platforms", [online] Available: http://nitlab.inf.uth.gr

[9] Bouckaert, Stefan, Wim Vandenberghe, Bart Jooris, Ingrid Moerman, and Piet Demeester. "The w-iLab. t testbed." In *International Conference on Testbeds and Research Infrastructures*, pp. 145–154. Springer: Berlin, Heidelberg, 2010.

[10] Maglogiannis, Vasilis, Dries Naudts, Ingrid Moerman, Nikos Makris, and Thanasis Korakis. "Demo: Real LTE Experimentation in a Controlled Environment." In *Proceedings of the 16th ACM International Symposium on Mobile Ad Hoc Networking and Computing*, pp. 413–414. ACM, 2015.

[11] "OpenAirInterface Testbed", [online] Available: https://oailab.eurecom.fr/oai-testbed

[12] Sefraoui, Omar, Mohammed Aissaoui, and Mohsine Eleuldj. "OpenStack: toward an open-source solution for cloud computing." *International Journal of Computer Applications* 55, no. 3 (2012).

[13] "JuJu Service Modeling", [online] Available: http://www.ubuntu.com/cloud/juju

[14] "PerformNetworks testbed", [online] Available: http://morse.uma.es/performnetworks

[15] Merino-Gómez, Pedro. "PerformLTE: A Testbed for LTE Testing in the Future Internet." In *Wired/Wireless Internet Communications: 13th International Conference, WWIC 2015*, Malaga, Spain, May 25–27, 2015, Revised Selected Papers, vol. 9071, p. 46. Springer, 2015.

[16] "Polaris Networks, EPC Emulators Test Tools", [online] Available: http://www.polarisnetworks.net/epc-emulators.html

[17] "FUSECO Playground", [online] Available: https://www.fokus.fraunhofer.de/go/en/fokus_testbeds/fuseco_playground

[18] Peterson, L., R. Ricci, A. Falk, and J. Chase. "Slice-based federation architecture (SFA)." *Working draft, version* 2 (2010).

[19] "jFed tool." [Online]. Available: http://jfed.iminds.be/

[20] Augé, Jordan, Loïc Baron, Timur Friedman, and Serge Fdida. "Supporting the experiment lifecycle with MySlice." In *Invited talk@ GENI Engineering Conference, GEC15-Oct*, pp. 23–25. 2012.

[21] Mitchell, Tom. "GENI aggregate manager API." In *Geni Engineering Conference 2010 (8th GEC)*, pp. 1998–2001. 2010.

[22] Rakotoarivelo, Thierry, Maximilian Ott, Guillaume Jourjon, and Ivan Seskar. "OMF: a control and management framework for networking testbeds." *ACM SIGOPS Operating Systems Review* 43, no. 4 (2010): 54–59.

[23] Vandenberghe, Wim, Brecht Vermeulen, Piet Demeester, Alexander Willner, Symeon Papavassiliou, Anastasius Gavras, Michael Sioutis et al. "Architecture for the heterogeneous federation of future internet experimentation facilities." In *Future Network and Mobile Summit (FutureNetworkSummit), 2013*, pp. 1–11. IEEE, 2013.

[24] Singh, Manpreet, Maximilian Ott, Ivan Seskar, and Pandurang Kamat. "ORBIT Measurements framework and library (OML): motivations, implementation and features." In *First International Conference on Testbeds and Research Infrastructures for the DEvelopment of NeTworks and COMmunities*, pp. 146–152. IEEE, 2005.

[25] Stavropoulos, Donatos, Aris Dadoukis, Thierry Rakotoarivelo, Max Ott, Thanasis Korakis, and Leandros Tassiulas. "Design, architecture and implementation of a resource discovery, reservation and provisioning framework for testbeds." In *Modeling and Optimization in Mobile, Ad Hoc, and Wireless Networks (WiOpt), 2015 13th International Symposium on*, pp. 48–53. IEEE, 2015.

[26] Hibler, Mike, Robert Ricci, Leigh Stoller, Jonathon Duerig, Shashi Guruprasad, Tim Stack, Kirk Webb, and Jay Lepreau. "Large-scale Virtualization in the Emulab Network Testbed." In *USENIX Annual Technical Conference*, pp. 113–128. 2008.

[27] "NITOS Scheduler for NITOS-like testbed", [Online]. Available: https://github.com/NitLab/NITOS-Scheduler

[28] "SFI command line SFA tool", [Online], Available: http://www.fed4fire.eu/sfi/

[29] Makris, Nikos, Christos Zarafetas, Spyros Kechagias, Thanasis Korakis, Ivan Seskar, and Leandros Tassiulas. "Enabling open access to LTE network components; the NITOS testbed paradigm." In *Network Softwarization (NetSoft), 2015 1st IEEE Conference on*, pp. 1–6. IEEE, 2015.

[30] "ETSI TS 127 007 v 10.3.0, Digital cellular telecommunications system (Phase 2+); Universal Mobile Telecommunications System (UMTS); LTE; AT command set for User Equipment (UE) (3GPP TS 27.007 version 10.3.0 Release 10)".

[31] Alexandris, Konstantinos, Navid Nikaein, Raymond Knopp, and Christian Bonnet. "Analyzing X2 Handover in LTE/LTE-A." In *WINMEE 2016, Wireless Networks: Measurements and Experimentation*, May 9, 2016, Arizona State University, Tempe, Arizona, USA.

[32] Alexandris, Konstantinos, Nikolaos Sapountzis, Navid Nikaein, and Thrasyvoulos Spyropoulos. "Load-aware Handover Decision Algorithm in Next-generation HetNets." In *WCNC 2016, IEEE Wireless Communications and Networking Conference*, 3–6 April 2016, Doha, Qatar. 2016.

[33] McKeown, Nick, Tom Anderson, Hari Balakrishnan, Guru Parulkar, Larry Peterson, Jennifer Rexford, Scott Shenker, and Jonathan Turner. "OpenFlow: enabling innovation in campus networks." *ACM SIGCOMM Computer Communication Review* 38, no. 2 (2008): 69–74.

[34] Makris, Nikos, Kostas Choumas, Christos Zarafetas, Thanasis Korakis, and Leandros Tassiulas. "Forging Client Mobility with OpenFlow: an experimental study." In *WCNC 2016, IEEE Wireless Communications and Networking Conference*, 3–6 April 2016, Doha, Qatar. 2016.

[35] "Open vSwitch, An Open Virtual Switch", [Online]. Available: http://openvswitch.org/

[36] Thomas Dietz. "Trema Tutorial." (2012).

[37] Rappaport, Theodore S. *Wireless communications: principles and practice*. Vol. 2. New Jersey: Prentice Hall PTR, 1996.

[38] Cichon, Dieter J., and Thomas Kurner. "Propagation prediction models." *COST 231 final report* (1999): 134.

[39] Erceg, Vinko, K. V. S. Hari, M. S. Smith, Daniel S. Baum, K. P. Sheikh, C. Tappenden, J. M. Costa et al. "Channel models for fixed wireless applications." (2001).

[40] Shellhammer, Steve. "Overview of ITU-R P. 1238–1, Propagation Data and Prediction Methods for Planning of Indoor Radiocommunication Systems and Radio LAN in the Frequency Band 900 MHz to 100 GHz." *doc.: IEEE802* (2000): 15–00.

[41] Passas, Virgilios, Nikos Makris, Stratos Keranidis, Thanasis Korakis, and Leandros Tassiulas. "Towards the efficient performance of LTE-A systems: Implementing a cell planning framework based on cognitive sensing." In *1st International Workshop on Cognitive Cellular Systems (CCS)*, pp. 1–5. IEEE, 2014.

[42] Milorad Tosic, Valentina Nejkovic, Filip Jelenkovic, Nenad Milosevic, Zorica Nikolic, Nikos Makris, and Thanasis Korakis. "Semantic Coordination Protocol for LTE and Wi-Fi Coexistence". *In European*

Conference on Networks and Communications 2016: Testbeds and Experimental Research, 26 June 2016, Athens. 2016.

[43] Milošević, Nenad, Zorica Nikolić, Filip Jelenković, Valentina Nejkovič, Milorad Tošić, and Ivan Šeškar. "Spectrum sensing for the unlicensed band cognitive radio." In *Telecommunications Forum Telfor (TELFOR), 2015 23rd*, pp. 250–252. IEEE, 2015.

[44] "FLEX D5.13 FLOW: Experiment Description and Requirements", [Online]. Available: http://flex-project.eu

[45] Hierarchy Token Bucket (HTB), [Online]. Available: http://linux.die.net/man/8/tc-htb

5

MONROE: Measuring Mobile Broadband Networks in Europe

Özgü Alay[1], Andra Lutu[1], Rafael García[2], Miguel Peón-Quirós[2], Vincenzo Mancuso[2], Thomas Hirsch[3], Tobias Dely[3], Jonas Werme[3], Kristian Evensen[3], Audun Hansen[3], Stefan Alfredsson[4], Jonas Karlsson[4], Anna Brunstrom[4], Ali Safari Khatouni[5], Marco Mellia[5], Marco Ajmone Marsan[2,5], Roberto Monno[6] and Hakon Lonsethagen[7]

[1]Simula Research Laboratory
[2]IMDEA Networks
[3]Celerway Communications
[4]Karlstad University
[5]Politecnico di Torino
[6]Nextworks
[7]Telenor Research

Abstract

Mobile broadband (MBB) networks (e.g., 3G/4G) underpin numerous vital operations of the society and are arguably becoming the most important piece of the communications infrastructure. Given the importance of MBB networks, there is a strong need for objective information about their performance, particularly, the quality experienced by the end user. Such information is valuable to operators, regulators and policy makers, consumers and society at large, businesses whose services depend on MBB networks, researchers and innovators. In this chapter, we introduce the MONROE[1] measurement platform:

[1]MONROE is funded by the European Union's Horizon 2020 research and innovation programme under grant agreement No. 644399. For more information, please visit https://www.monroe-project.eu/

An open access, European-scale, and flexible hardware-based platform for measurements and custom experimentation on operational MBB networks with WiFi connectivity. The platform consists of mobile and stationary nodes that are flexible and powerful enough to run most measurement and experiments tasks, including demanding applications like adaptive video streaming. Access to such a platform enables accurate, realistic and meaningful assessment of the performance of MBB networks by continuously monitoring these networks via active testing (e.g., delay test, web performance test, download speed test) and context metadata collection (e.g., connection mode, signal strength parameters). The multihoming feature of MONROE allows for the comparison of different networks under similar conditions as well as the exploration of new ways of aggregating providers to increase performance and robustness. In this chapter, we showcase the monitoring capabilities of the platform by analyzing preliminary performance measurement results. Considering that MONROE is *open* to external users, we further discuss a representative set of measurements and experiments to highlight the potential use cases of the platform. We argue that mobile measurements over operational networks, hence platforms such as MONROE, are crucial not only for characterizing and improving the user experience for services that are running on the current 3G/4G infrastructure, but also for providing feedback on the design of upcoming 5G technologies.

5.1 Introduction

Wireless and mobile access to the Internet have revolutionized the way people interact and access information. Mobile broadband (MBB) networks have become the key infrastructure for people to stay connected everywhere they go and while on the move. According to Cisco's Global Mobile Data Traffic Forecast [1], in 2015 the number of mobile devices grew to a total of 7.9 billion, exceeding the world's population. Also, fourth generation (4G) traffic exceeded third generation (3G) traffic for the first time in 2015 [1].

The society's increased reliance on MBB networks has made provisioning ubiquitous coverage the highest priority target for mobile network operators, as well as focusing on performance and user quality of experience (QoE). MBB coverage and performance experienced by the end-users are of great importance to many stakeholders including mobile subscribers, regulators, governments and businesses whose services depend on MBB networks. This also motivates researchers and engineers to further enhance the capabilities of mobile networks, by designing new technologies to cater for plethora of new applications and services, growth in traffic volume and a wide variety of user devices. In this dynamic ecosystem, there is a strong need for both open objective data about the performance and reliability of different MBB operators, as well as open platforms for experimentation with operational MBB providers. On the one hand, objective performance data is essential for regulators to ensure transparency and the general quality level of the basic Internet access service [2], especially in light of an evolution of service offerings beyond the best-effort traffic mode, including a balanced approach to net neutrality. On the other hand, custom experimental approaches are key to forwarding our understanding and driving innovation in MBB networks.

Characterizing the performance of home and mobile broadband networks requires systematic end to end measurements. Several regulators have translated this need into ongoing nationwide efforts, for example, the FCC's Measuring Broadband America initiative [3] in the USA. Operators and independent agencies sometimes perform drive-by tests to identify coverage holes or performance problems. These tests are, however, expensive and do not scale well [4]. Another approach is to rely on end users to run performance tests by visiting a website (e.g., [5]) or running a special measurement application (e.g., [6]). The main advantage of this approach is scalability: it can collect millions of measurements from different regions, networks and user equipment. However, with such an approach, repeatability is hard and one can only collect measurement data at users' own will, with no possibility to either monitor or control the measurement process. Furthermore, mostly due to privacy reasons, these measurements do not provide rich context information and metadata, e.g., location, type of user equipment, type of subscription, and connection mode (2G/3G/4G); however, metadata is critical when analyzing the results. Also, such a setup does not provide active measurements that can reveal important information on stability and availability of a network, since this requires long and uninterrupted measurement sessions. Finally, this approach limits the possibility of testing novel applications and services since this might require configuration changes (e.g., customized kernels).

MONROE is the first European platform for open, independent, multihomed, large-scale monitoring and assessment of performance of mobile broadband networks in heterogeneous environments. Access to such a platform allows for the deployment of extensive measurement campaigns to collect data from operational MBB networks. The availability of this vast amount of data allows us to advance our understanding of the fundamental characteristics of MBB networks and their relationship with the performance parameters of popular applications. This is crucial not only for improving the user experience for services that are running on the current 3G/4G infrastructure, but also for providing feedback on the design of upcoming 5G technologies.

In the remainder of this chapter, we summarize the current state of the art in Section 5.2. We then expand on the MONROE vision in Section 5.3, where we provide an overview of the MONROE goals and the key features of the measurement platform. In Section 5.4, we describe the current architecture design of the MONROE platform. We discuss in Section 5.5 how the MONROE user access and scheduling system is designed and how users can deploy their experiments. In Section 5.6, we present initial results from basic measurements running on operational MONROE nodes active in Norway, Sweden and Spain. We show that the MONROE system enables efficient MBB performance monitoring, operator benchmarking and complex network analytics. Finally, we conclude the chapter in Section 5.8.

5.2 Background and State of the Art

During the past years, we have seen increased interest in the networking community from different parties (e.g., researchers, operators, regulators, policy makers) in measuring the performance of mobile broadband networks. In this section, we aim to provide a condensed but comprehensive review of some of the most relevant approaches that strive to shed light on the mobile broadband ecosystem.

Large scale research measurement platforms such as RIPE Atlas [7], BISmark [8] or PlanetLab [9] share many common goals with MONROE. However, these platforms do not operate in the mobile environment. In order to cater to the need of open large-scale MBB measurements and to address the scarcity of available measurement platforms, several crowd-sourcing approaches emerged over the past years, either from the research environment, e.g., Netalyzr [6], NetPiculet [10], or commercial-oriented, e.g., OpenSignal [11], RootMetrics [12] or MobiPerf [13]. These approaches leverage the wide adoption of mobile devices in the world and depend on the willingness of end-users to run the proposed tests. We note that the common vision of these tools is to identify and monitor a set of significant metrics which can accurately describe mobile broadband performance to the interested parties. For example, commercial-oriented OpenSignal proposes a complete approach for building MBB coverage maps by retrieving the connectivity-related metadata from user devices and characterizing multiple radio access technologies in the same ares. They introduce the notion of "time coverage" which provide s statistics for the time a device has been using a certain radio access technology in order to provide the end-user the possibility to make informed decisions in terms of the preferred MBB provider in a certain area. Similarly, RootMetrics defines a set of key performance metrics which allows for network benchmarking, with the intent of rating different providers available in a certain geographical area. Additionally, tools such as NetPiculet or Netalyzr aim to shed light on the infrastructure and the performance of broadband providers with the purpose of informing protocol and application design.

There are several research projects [6, 14–17] that use custom-designed apps to crowdsource and measure the performance of MBB providers and popular Internet applications, with a main focus on web browsing [18] and video streaming [19]. For example, MobiPerf [13] enables mobile network performance analysis [14]. The app builds on top of the Mobilyzer open library [20] and tracks a series of network performance metrics, such as HTTP benchmark downloading latency and bandwidth, traceroute with latency to different hops, ping latency, DNS lookup latency, TCP uplink and downlink throughput or RRC states metrics. Other similar relevant measurement efforts from the research community include [21–23].

With the increasing popularity of web and video-related services over MBB networks [24], there is a magnitude of research studies that focus on understanding the correlation between the network quality of service (QoS) metrics and the quality of experience (QoE) of the end-users [24–26]. In particular, this is appealing to operators, who continuously strive to provide the best service to their subscribers in order to increase their customer base. At the same time, the end-users themselves are looking for relevant metrics that can objectively assess the performance of popular applications over different MBB providers. In addition to the application performance, another important concern for the users is the energy efficiency of bandwidth intensive applications [27, 28].

Even more, alongside the attention coming from end-users, businesses or operators, there is rising interest from regulators for defining and monitoring a representative and unitary set of metrics that accurately captures the performance of today's broadband services in practice. In this sense, several of them (e.g., FCC, Ofcom and Anatel) have translated these efforts into national projects in collaboration with commercial partners such as SamKnows [29], which specializes in home and mobile broadband performance evaluation.

However, in order to allow for an open an unitary approach as well as the comparability of measurements, a common open framework is needed. This has been hard to achieve due to the proprietary nature of the measurement efforts, as is the case of [11, 12, 29], making it difficult for regulators to view measurement results from a harmonized and macroscopic scale. In this sense, several open measurement methodologies [30, 31] have been proposed with the goal of supporting the creation of inter-operable large-scale testbeds and advance a common approach on network performance characterization. The Internet Engineering Task Force (IETF) Large-Scale Measurement of Broadband Performance (LMAP) is currently working towards standardizing an overall framework for large-scale measurement platforms.

The MONROE platform complements the existing experimental platforms by providing unique features in the field of network-controlled mobile measurements. Three key aspects of MONROE that makes the platform unique are: repeatability and controllability of measurements for precise and scientifically verifiable results (even for the mobile scenarios), support for demanding applications such as web and video services and support for protocol and service innovation. These aspects sets up MONROE in an excellent position to advance the state-of-the-art measurement tools and platforms.

5.3 MONROE Approach and Key Features

MONROE's goal is to build a dedicated infrastructure for measuring and experimenting in MBB and WiFi (IEEE 802.11) networks, comprising both fixed and mobile hardware measurement nodes. The platform integrates 450 nodes scattered over four European countries (Italy, Norway, Sweden and Spain) and a backend system that collects the measurement results, offering tools for real-time traffic flow analysis as well as powerful visualization tools. We designed the MONROE nodes to be flexible and powerful enough to run most measurement and experiment tasks, including demanding applications like adaptive video streaming. The current MONROE node is an Accelerated Processing Unit (APU) with AMD 1 GHz dual core 64 bit processor and 4 GB DRAM. Each MONROE node connects simultaneously to three MBB networks through three MiFis using commercial grade mobile subscriptions. The nodes also provide WiFi connectivity[2] through a built-in dual band AC WiFi card. MONROE nodes have built-in support for collecting metadata such as cell ID, signal strength and connection mode. The nodes are equipped with GPS for tracking their location.

[2]The access points for WiFi will be provided when applicable for stationary nodes.

The MONROE platform allows external users to test their novel applications and services that run over MBB networks with WiFi connectivity. Through a user-friendly web client, external experimenters can schedule and deploy their own experiments on the MONROE nodes. Experimenters can use the MONROE platform to run measurements of different MBB providers at regular intervals over long time periods and under similar conditions.

The MONROE platform complements the existing experimental platforms such as RIPE Atlas [7] by providing unique features in the field of network-controlled mobile measurements. MONROE builds on the existing NorNet Edge (NNE)[3] [32] and extends its functionality, scale and coverage. The main features of MONROE are:

1) **Large-scale and wide geographical coverage**: MONROE is composed of 450 nodes that are widely distributed across Norway, Sweden, Italy and Spain, as we illustrate in Figure 5.1. MONROE is able to collect measurements under diverse conditions, from major cities to remote islands (including one node in Svalbard, in the Arctic). There is a dense deployment of nodes in a few main cities (e.g. Oslo, Stockholm, Madrid, Torino, etc.), giving a more detailed view of network conditions in urban areas.
2) **Mobility**: 150 MONROE nodes are deployed on trains and buses in order to cover both rural and urban areas. These nodes are instrumental to provide insights on the mobility characteristics of MBB.
3) **Multihomed**: Each MONROE node is connected simultaneously to three mobile broadband networks, which makes it possible to conduct a wide range of measurements and experiments that compare the performance of each network, or explore novel ways of combining resources from each network. Along with MBB networks, MONROE also provides WiFi connectivity to allow experimenting on different access technologies and explore methods such as traffic offloading.
4) **Flexible and powerful MONROE nodes**: The MONROE nodes are designed such that they are flexible and powerful enough to run most measurement and experiment tasks, including demanding applications like adaptive video streaming. Furthermore, MONROE enables experimenting novel services and applications on MBB networks by allowing configuration changes such as kernel modifications.

[3]NNE is currently in an operational state, with a functioning system for node management, deployment of experiments, handling of data etc. as well as real-time visualization of measurements (demo available at http://demo.robustenett.no).

Figure 5.1 Geographical distribution of MONROE Nodes. MONROE builds on the existing NorNet Edge (NNE) infrastructure, consisting of 200 dedicated operational nodes spread across Norway.

5) **Rich context information**: In addition to information about network, time and location for experiments, MONROE nodes have built-in support for collecting metadata from the externally connected modems, including cell ID, signal strength and connection mode.
6) **Open access**: MONROE is open to external users and makes it easy to access the system and deploy experiments on all or a selected subset of the nodes.
7) **Visualization and Open Data**: The MONROE platform has a measurement system that collects basic experiment results and then stores them in a database. Interested parties can then consume the measurement results through a real time visualization system. Furthermore, the results are provided as *Open Data* in regular intervals.

5.4 MONROE System Design

We designed the MONROE platform to make it easy for external experimenters to run their customized measurements. In this section, we expand on the MONROE system design and review the main building blocks and their functions. We illustrate the MONROE framework in Figure 5.2. Notably, MONROE not only allows to monitor and analyze the behavior of MBB network connections in real-time, but also to store measurement data jointly with metadata in the form of open data for offline analysis. The MONROE system comprises:

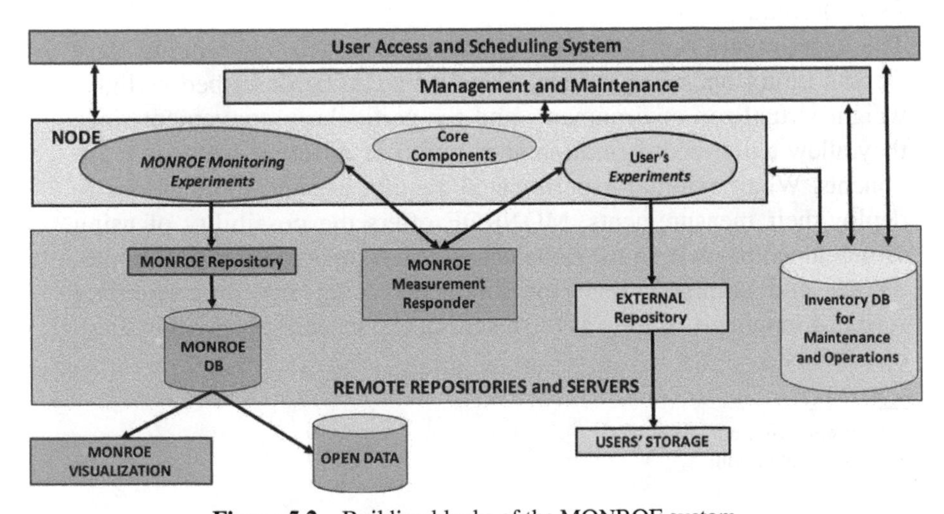

Figure 5.2 Building blocks of the MONROE system.

1. *User access and scheduling system*: The scheduling system handles the MONROE measurements through a user-friendly interface consisting of an AngularJS-based web portal. As part of the MONROE federation with the Fed4FIRE initiative of the European Commission[4], the user access follows the Fed4FIRE specifications in terms of authentication and provisioning of resources. The portal allows to access the MONROE scheduler, which is in charge of setting up the experiments without requiring the users to directly interact with the nodes (i.e., no login access to the node environment).
2. *Management and maintenance system*: The operations team uses this system to manage and maintain the MONROE testbed. It involves an *Inventory* that keeps all the information (e.g., the status of each node, status of different connections, location of the nodes, etc.) required for operations and maintenance. It also involves a *Monitoring Agent* that monitors and reports the health of the system (e.g., logging, performance monitoring, self checks for services etc.).
3. *Node modules*: The software on the measurement nodes includes the core management components and the set of experiments. The core components consist of the main software (watchdog, routing, network monitor, etc.) running on the node and make sure that the node is operational. An important core component is the Metadata Multicast, which is responsible for collecting and multicasting the metadata such as node status, connection technology and GPS. We provide a messaging API in order to relay real-time metadata to experimenters through ZeroMQ in JSON format.
   The experiments run in Docker[5] containers, which are running on a Debian Linux operating system. Containers can be described as lightweight virtualized environments and are particularly convenient since they allow agile reconfiguration and control of different software components. When external experimenters require kernel modifications to deploy their measurements, MONROE offers the possibility of using virtual machines within the node ecosystem. Experimenters can implemented and configure their measurements using any programming/ scripting language, as long as the resulting experiment runs within these constraints.

[4]http://www.fed4fire.eu/
[5]http://www.docker.com

In order to monitor and assess the performance of MBB networks, MONROE continuously runs a basic set of experiments (*MONROE Monitoring Experiments* in Figure 5.2). Current deployed basic experiments include: continuous background measurements (e.g., ping to predefined servers), periodic bandwidth-intensive measurements, and a traffic analyzer developed in the mPlane project (`Tstat`). In Section 5.6.1, we expand on these measurements and analyze preliminary results. Apart from this, MONROE enables many other experiments for its external users (*User's Experiments* in Figure 5.2), which we further exemplify in Section 5.7.

4. *Repositories and Database*: The MONROE system supports external repositories to collect experimental data. Data transfer from nodes to the repositories is based on a set of agents that follow a publisher/subscriber model. We collect the results of the *MONROE Monitoring Experiments* in the MONROE repository and we subsequently import them to a centralized database for offline analysis. The database is based on a non-relational technology, oriented to time series analysis, and highly scalable to manage large volumes of data. We designed the database schema around the concept of experiments instead of physical nodes, with a clear distinction between experimental measurements and metadata. Several measurement responders we host in the MONROE backend act as measurement servers for certain experiments.
5. *Visualization and Open Data*: A near real-time visualization and monitoring tool enables stakeholders to access a graphical representation of the MONROE platform status in terms of deployment of the nodes, status of each device, as well as results of *MONROE Monitoring Experiments*. The results of selected measurements are provided as *Open Data* in regular[6].

5.5 Experiment Deployment

MONROE is an open platform for external users to experiment with MBB networks through active measurements. In this section, we detail the process an external user needs to follow in order to access the MONROE platform and we detail the MONROE components each experimenter interacts with. The work flow involves three main phases, as illustrated in Figure 5.3: Experiment Design, Testing and Experimentation.

[6]https://zenodo.org/collection/user-h2020_monroe

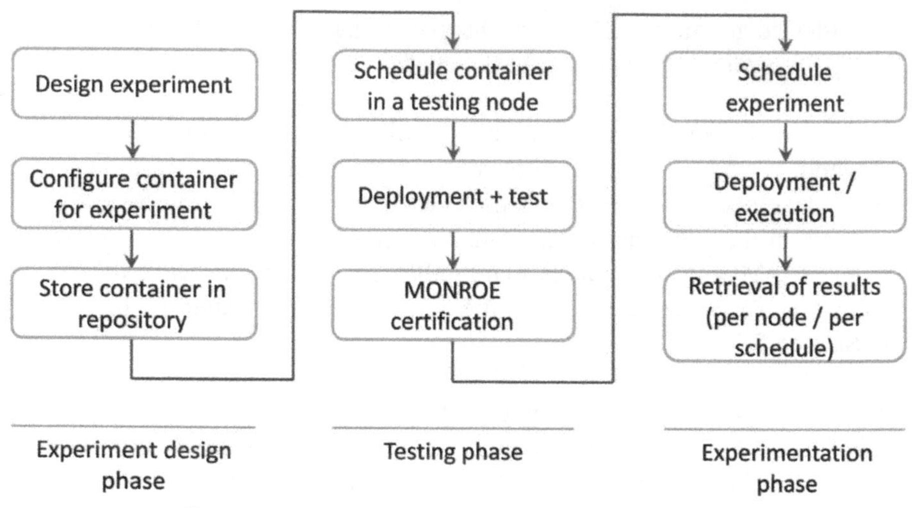

Figure 5.3 Experiment creation and deployment phases.

Experimenters have to define the measurements they want to obtain and decide how to implement them. Experiments run inside Docker containers, so they can consist of virtually any piece of software. During the testing phase, a MONROE administrator checks that the behavior of the container adheres to a set of minimum safety and stability rules; approved images are cryptographically signed and moved to our repository. Finally, the experimenter uses a web-based interface to schedule the experiment, selecting the number and types of nodes and suitable time-slots. Once the experiment is deployed and run, the results of experiments are automatically collected and transferred to a repository maintained by MONROE. Alternatively, experimenters can also choose to transfer/stream the results to their preferred location using their own independent solution.

Experiments can collect active and passive traffic measurements from multiple MBB networks. For active measurements the platform provides both standard/well-known tools (e.g., ping, paris-traceroute) and project-crafted ones. For passive measurements, it embeds tools such as `Tstat` [33] to analyze the traffic generated. Moreover, each node passively generates a metadata stream with modem and connectivity status, and the measurements of several embedded HW sensors (GPS, CPU usage, temperature, etc.). Experimenters can either subscribe their experiments to the stream in real-time or consult the database afterwards. Considering that experimenters can deploy any additional measurement tools, the set of possible measurements is flexible and open.

We provide User Access to the experimental platform via a web-based MONROE Experimenters Portal that enables users to schedule and run new experiments. The portal allows to access the MONROE scheduler, which is in charge of setting up the experiments without requiring the users to access the nodes. Since we federated MONROE within Fed4FIRE in order to build a large-scale, distributed and heterogeneous platform – authentication and provisioning of resources follows the Fed4FIRE specifications. In the following sections, we provide details on MONROE's federation with FED4FIRE, user authentication, experimenters portal and scheduler.

5.5.1 MONROE as a Fed4FIRE Federated Project

The Fed4FIRE Portal is a common and well-known tool where registered users can select and access an available testbed (e.g., the MONROE platform). The Fed4FIRE Portal is powered by MySlice software[7] and offers a directory of all FIRE testbeds, tools and links to project websites. In other words, the portal acts as an experimentation bridge to resources and their corresponding control tools.

To be able to join MONROE and run their experiments, the external users must first become familiar with the terminology and the tools of the Fed4FIRE federation and, in particular, with the MONROE project documentation. The available documentation of Fed4FIRE describes the federation of testbeds as a generic environment.

The user must apply for a Fed4FIRE account and download the corresponding required certificates, which should be associated with an existing MONROE experimentation project. The Fed4FIRE introductory documentation explains how to go through these particular steps. We note that the user must specify an already existing MONROE project, or alternatively, create a new one. In Section 5.5.2 we expand on how to complete the user authentication phase.

Once granted access to the platform, the user is recommended to follow and execute a MONROE tutorial, which describes those elements that are specific to the MONROE testbed, including the AngularJS client developed in the project for user access and experiment scheduling. Those users that plan to run measurement experiments in MONROE testbed should be familiar with the contents of the MONROE tutorial. To reserve the

[7]MySlice: http://myslice.info

resources for a specific experiment, the experimenter has to use the MONROE scheduler (Section 5.5.4), which can be accessed through the MONROE User Access client (Section 5.5.3). With the above, the experimenter can reserve the resources up to the limit granted to him/her by the MONROE consortium.

5.5.2 User Authentication

In this section, we describe the Fed4FIRE AAA policies and procedures, and how we adapt them to the MONROE project.

A federation is a collection of testbeds (or *"islands"*) that share and trust the same certification authorities and user certificates. Fed4FIRE realizes a federation of a large number of wired, wireless and OpenFlow-based testbeds principally located in Europe. Each island manages its resources using dedicated tools and can decide which kind of certificates (and from which authorities) it wants to accept. In this context, Fed4FIRE works with X.509 certificates to authenticate and authorize experimenters (users) on its testbeds. The authority which provides valid certificates in the Fed4FIRE federation is located at the iMinds infrastructure. The certification authority has the concept of Projects which bundle multiple users. Any user can requests for the creation of a new project, but it must be authorized by the Fed4FIRE administrators. Subsequently, the project responsible can approve new experimenters for that particular project, without prior approval from Fed4FIRE administrators.

MONROE shares and trusts the certificates generated by the iMinds authority, and therefore, is a member of the Fed4FIRE federation. We note that all the project functions and operations in MONROE depend on the user certificates, including resource reservation, measurements deployment and downloading experiment data. MONROE does not support other certification authorities or other federations (e.g., GENI).

Each partner in the MONROE consortium manages its own private project inside Fed4FIRE. Similarly, external institutions could have their own private projects upon request and approval by the MONROE Project Board. Individual researchers cannot join the MONROE testbed, as all the users must belong to at least one project (which corresponds to an institution that is managing it). However, each institution can easily invite new users and grant access to their respective projects offering the available resources which the MONROE administrators manage.

5.5.3 The Experimenters Portal (MONROE User Access Client)

Through the Experimenters Portal, verified external users can obtain access to the MONROE platform and deploy their measurements. After providing the necessary credentials to authenticate with the MONROE User Access client, the user can visualize a historic of all its experiments and check their current status (Figure 5.5). Clicking on any row of the table shows the details of the experiment selected.

Before scheduling new experiments, users can verify the current state of the MONROE resources. The *"Resources"* tab (Figure 5.4) allows experimenter to query all the existing resources in the MONROE platform and their time availability, using multiple filters if required.

In the *"New experiment"* tab, the user can create a new experiment and input the required parameters. The basic experiment details include the identifying name and the docker script to run the experiment. In the Experiment Size group, the user specifies the number of nodes required to run the experiment, and the desired characteristics of those nodes using filters that allows to select, e.g., the location of the nodes to use in the experiments, their hardware/software version, static or mobile nodes, testing nodes for preliminary/debugging tests, etc. Furthermore, the user can select the operator of interest and then define the maximum amount of data to be transferred per experiment over that interface/operator. This data limit is enforced during the experiments in order to avoid exceeding the mobile data quotas. In the Experiment Duration the user specifies the duration of the experiment by providing a starting and stopping date-time, or by clicking the "as-soon-as-possible" check box.

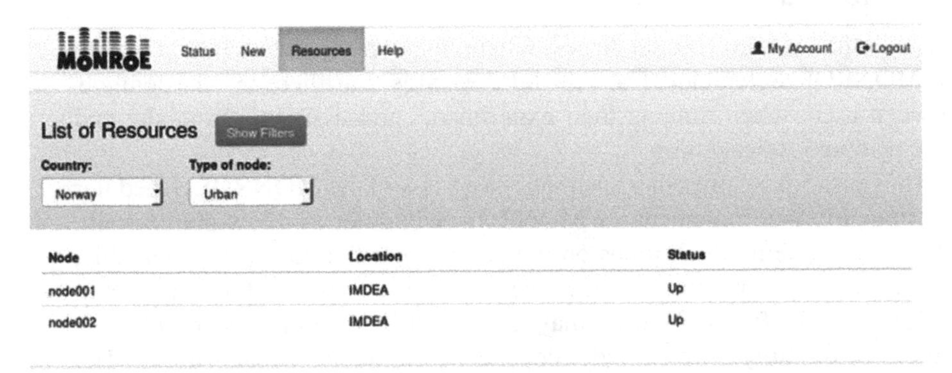

Figure 5.4 Resources availability in MONROE.

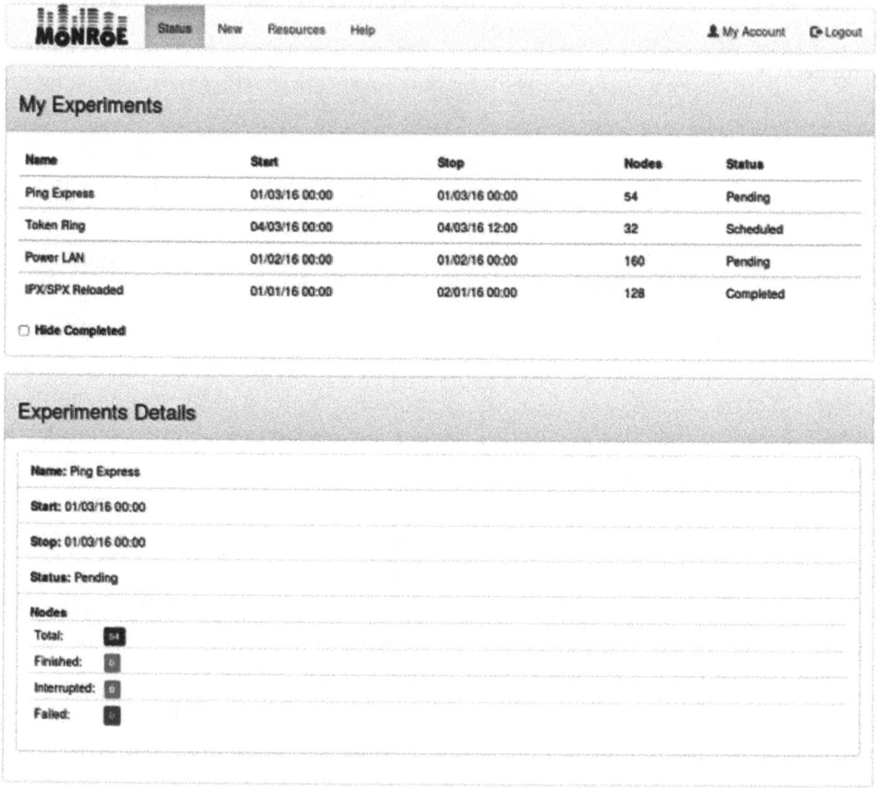

MONROE Status New Resources Help My Account Logout

My Experiments

Name	Start	Stop	Nodes	Status
Ping Express	01/03/16 00:00	01/03/16 00:00	54	Pending
Token Ring	04/03/16 00:00	04/03/16 12:00	32	Scheduled
Power LAN	01/02/16 00:00	01/02/16 00:00	160	Pending
IPX/SPX Reloaded	01/01/16 00:00	02/01/16 00:00	128	Completed

Hide Completed

Experiments Details

Name: Ping Express

Start: 01/03/16 00:00

Stop: 01/03/16 00:00

Status: Pending

Nodes

Total:

Finished:

Interrupted:

Failed:

Figure 5.5 MONROE experiment status.

5.5.4 MONROE Scheduler

Through the MONROE User Access Client, the experimenters interacts with the MONROE Scheduler. The scheduler ensures that there are no conflicts between users when running their experiments and assigns a time-slot and node resources to each user.

In Figure 5.6 we present a schematic overview of the MONROE Scheduler functionality. We implement the MONROE Scheduler as a low-connectivity scheduling system which relies on the assumption that nodes are available, independent of short-time loss of connectivity. Due to the multihoming setup of the MONROE nodes, they may contact the scheduler from different addresses, possibly with provider-dependent modifications and filters. The Scheduler consists of two components – the *scheduling server* running in a central, well-known location and the *scheduling client* running on the nodes (Figure 5.6).

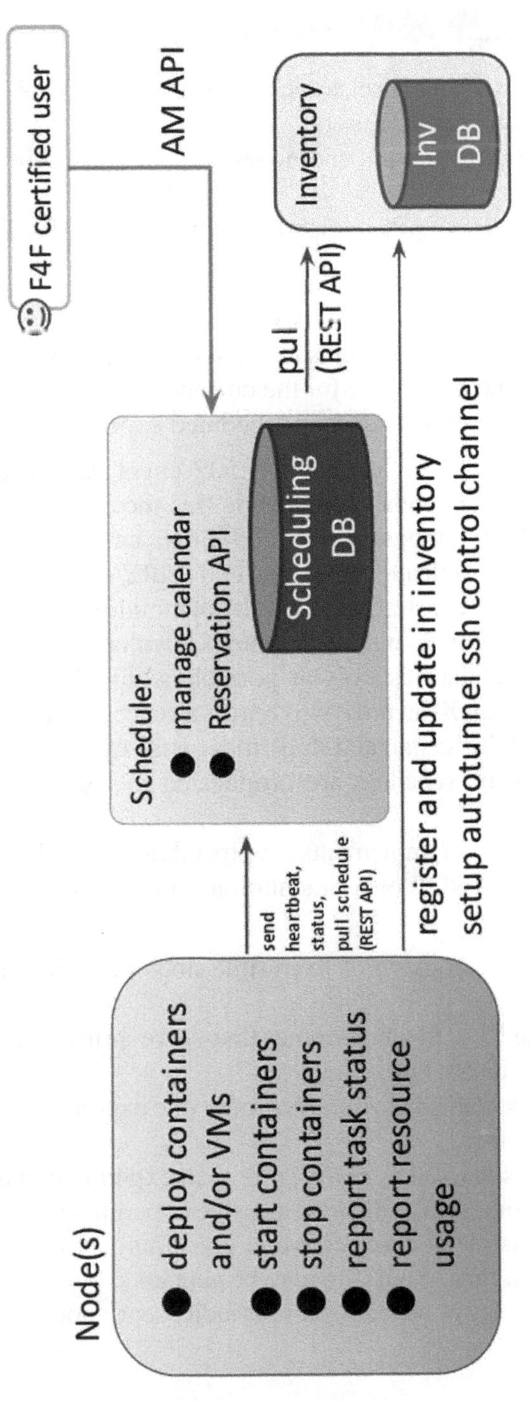

Figure 5.6 Scheduling system.

The scheduling server:

- takes care of the experiment schedule and resolves conflicts
- assigns roles to authenticated users
- provides a REST API to users and nodes to query and edit scheduling status
- provides an XML-RPC API compatible with the Fed4FIRE AM API definition

The scheduling client:

- sends a regular heartbeat and status to the scheduling server
- fetches the experiment schedule for the current node
- downloads, deploys, starts and stops scheduled experiments

Authentication to the server is based on X.509 client certificates. Users, administrators and nodes all authenticate using this mechanism and use the same scheduling API. By importing the Fed4FIRE certification authority certificate, users may authenticate using their Fed4FIRE credentials.

Due to the connectivity constraints especially of mobile nodes, deployment of experiments on the node is not immediate. Download and deployment of experiments will take place as early as possible within the constraints of available space on the node. The node will report a successful deployment to the scheduler and schedule the start and stop times for the experiment container internally. Changes in the schedule are propagated to the node whenever possible.

The MONROE Scheduler implements the procedures and policies we have defined to guide the MONROE experimentation. These include, but are not limited to:

- The scheduler allows booking of fixed time slots for each measurement experiment.
- Priority is defined by the first-come first-serve principle, while the consortium will monitor fairness.
- If an experiment is marked as *exclusive*, only one experiment may run at a given time on a node.
- If an experiment is marked as *active*, one such experiment may run at a given time on a node, while allowing passive experiments.
- If an experiment is marked as *passive*, a given number of such experiments may run at a time. No traffic may be generated by the experiment.
- User experiments may be scheduled as periodic, continuous, or one-time.

- Only experiments for which a time slot has been booked in advance may be run.
- Nodes may be of different types (static, mobile, urban, rural, certain country, etc. . .) defined by the MONROE project. Booking requests can select to use or reject these filters.
- A booking over several nodes or several time periods is treated as atomic (i.e., if one of the booking periods or nodes is unavailable, the entire booking is rejected). Several bookings over different nodes or time periods may be linked to an atomic unit

In order to determine the resource requirements, each user needs to schedule its experiment to first run on the testing nodes (Testing Phase in Figure 5.3). This step allows us to monitor the resource usage of each experiment. If the usage is within defined constraints, the MONROE administrators move on to approve the user experiments by means of a cryptographic signature. Only then, the experiment image is cleared to be scheduled on regular nodes.

The scheduling process on the node (Deployment Phase in Figure 5.3) defines three actions: (i) deployment, (ii) start and (iii) stop of the experiment. The deployment step may take place at any time before the scheduled start time, and should finish before the experiment starts. In this step, the scheduler reserves the requested resources and loads the experiment image onto the nodes. During the start process, the scheduler sets the resource quotas and starts using the experiment image a container system where experiments will run. The stop action notifies the experiment of its impeding shutdown, then removes the container after a short grace period. Measurement results may be stored on disk, and will be transferred during and after the termination of the experiment as connectivity allows.

5.6 Network Measurements and Analytics with MONROE

The MONROE platform continuously runs a set of basic measurements with the purpose of characterizing the state of the MBB providers in Europe. Interested parties can consume the data through the MONROE visualization GUI, thus making MONROE a solution for near real-time network performance monitoring. In Figure 5.7, we show a snapshot from the MONROE monitoring interface tracking a node in terms of both RTT and signal strength. Alternatively, we provide the measurement results as open data which external users can access and use for running network analytics.

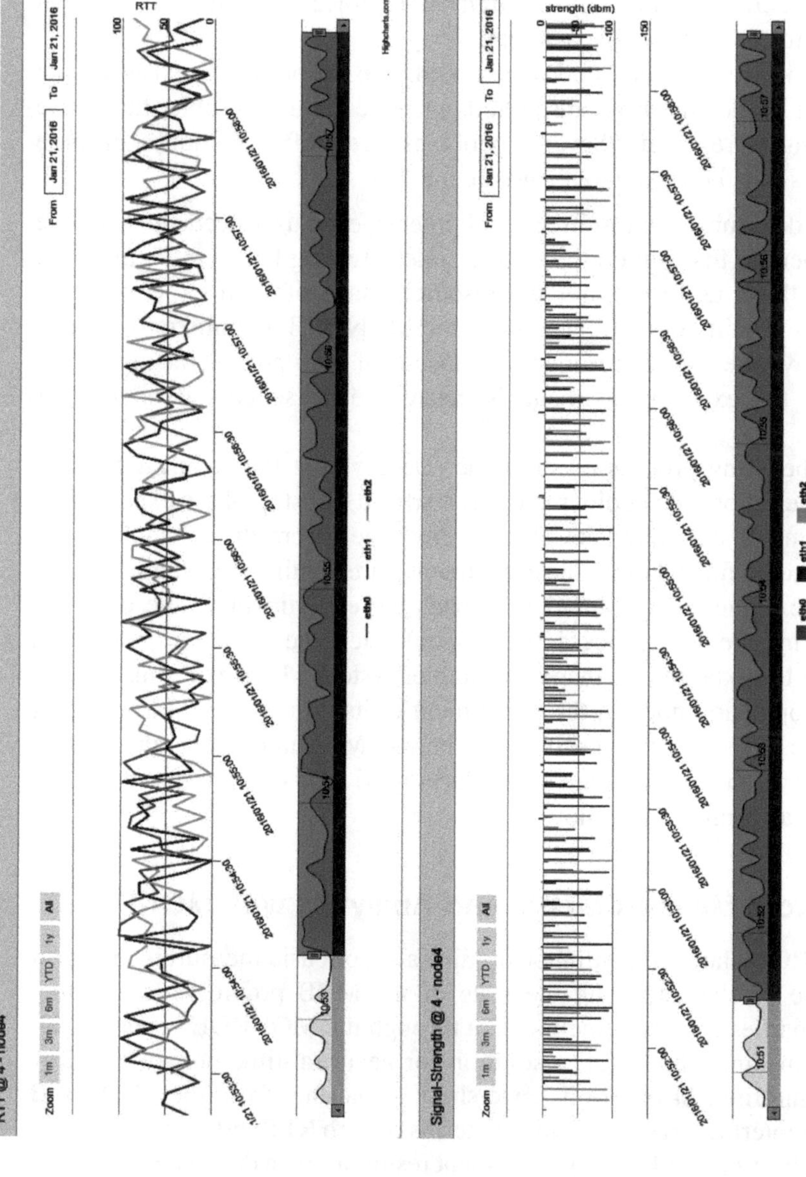

Figure 5.7 MONROE visualization GUI snapshot for RTT and signal strength monitoring in near real time.

5.6.1 MONROE Monitoring Experiments

The MONROE Monitoring Experiments currently include (but are not limited to) i) continuous ping measurements towards a fixed target in Sweden, ii) a simple bulk data download, and iii) web browsing performance measurements. The MONROE nodes also continuously run Tstat [33], a passive monitoring tool developed within the mPlane project [34]. Tstat extracts information from the flow of packets being transmitted and received by each node. This facilitates the use of the MONROE platform as an analytic tool for troubleshooting and root-cause analysis. In this section, we report preliminary measurement results illustrating the capabilities of the platform towards performance monitoring and network analytics.

a) RTT Measurements: Each MONROE node runs a ping measurement every second on each active interface against the same target measurement server we host in the MONROE backend in Sweden. Figure 5.8 shows the violin plot for the RTT samples we collected during one week (from the 8th of July until the 15th of July 2016) from 30 stationary nodes connected in total to 7 different operators in 3 countries. Each "violin" shows the probability density of the RTT at different values, the higher the area, the higher

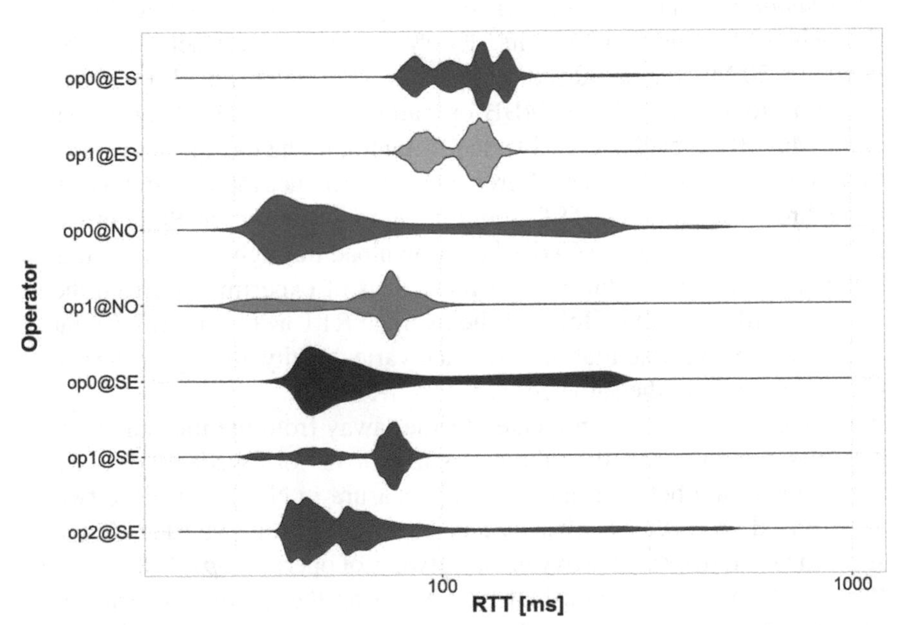

Figure 5.8 Violin plots of the RTT measurements for different operators in Spain (ES), Norway (NO) and Sweden (SE).

the probability of observing a measurement in that range. We observe that the RTT measurements exhibit typically a multimodal distribution, corresponding to different access delays faced by different radio access technologies (e.g., 3G/4G).

The results are intuitively expected: nodes in Norway and Sweden that are closer to the target measurement server (which we host in the MONROE backend in Sweden) exhibit lower delay than the nodes in Spain. However, the variance of the measurements is much higher than in fixed networks, showing that MBB introduces complexity even for basic tests, such as RTT monitoring. Given that the ping experiment is running continuously, some of this variation can be due to interactions with other experiments running on the MONROE nodes. The repetitive measurements allow us to track this key parameter in time and capture the experience of customers using mobile subscriptions similar to those active on the MONROE node. By analyzing the RTT time series, we plan to further identify delay trends and correlate them with the time of the day, the geolocation of the measurement node and the rich context information we collect from the devices (e.g., RAT changes and variations in the signal strength). This uniquely enables us to work towards understanding congestion patterns in the networks.

b) Download Capacity Measurements: In Figure 5.9, we illustrate downlink throughput measurement results. Every two hours, we schedule the download of a 50 MB file on 30 stationary MONROE nodes on all interfaces corresponding to seven different MBB operators from an HTTP server we host in the MONROE backend in Sweden. Running in the background, Tstat analyzes this traffic and generates different key performance metrics, including download throughput and the RTT from the client to the server. Plots in the top row of Figure 5.9 show the CDF of the download throughput, while plots on the bottom show the evolution over three days of experiments (from the 22nd of July until the 24th of July) of the average RTT as observed by Tstat during the transfer. We note that performance varies wildly among countries, among operators within the same country and over time.

As expected, nodes in Spain located further away from the measurement server display a higher RTT than the nodes in Norway or Sweden. Also, we see a clear separation between the RTT we measure in Norway for the two operators. Based on further analysis we perform with Tstat, we identify the presence of a non-transparent proxy in the network of operator *op1*. We further note the impact of the web proxy when monitoring the goodput metric for both operators in Norway: *op1* benefits from the proxy and displays a higher goodput than *op0*.

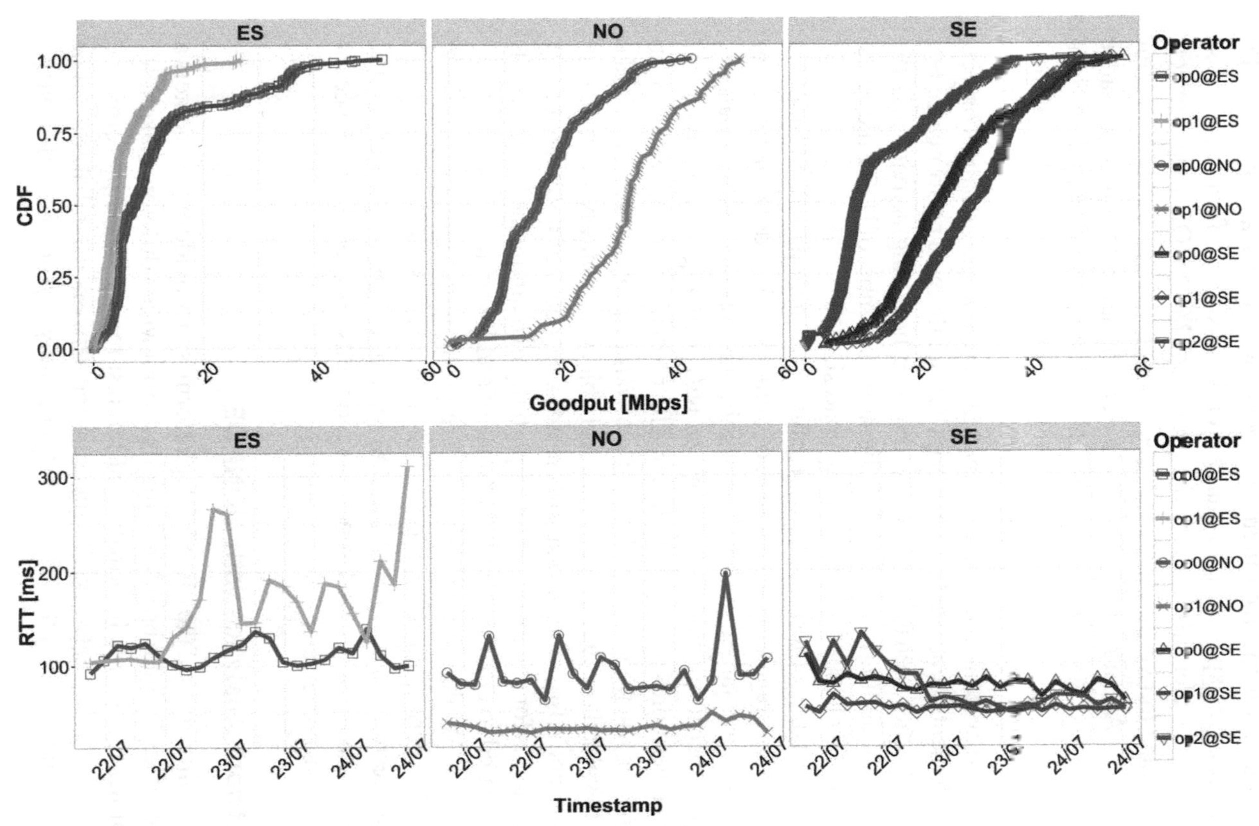

Figure 5.9 CDF of the download goodput (*top*) and average RTT (*bottom*) for different operators in different countries.

c) Web Browsing Performance: Aside from the basic measurements that run continuously on the measurement nodes, we design and periodically schedule a specific experiment to gauge web browsing performance across multiple MBB providers in different countries. Each MONROE node connects on each interface to two different websites[8], which we chose based on their popularity in the Alexa ranking, but also based on their different appearance and rendering style. As part of the experiment design, the web performance test breaks down the times used for different phases in a web transaction at each interface of the MONROE node: time to resolve the DNS name, time to connect to web server and time to download the web content and all its objects (including elements generated by javascript). Also, the web performance test tracks several other metrics to describe the web browsing activity and the target website, including number of DNS iterations, number of HTTP redirects, number of HTTP elements or HTTP download size.

In Figure 5.10, we illustrate the CDF of the complete page load time and the CDF of the average time to first byte of content broken down per country and per website we target. We observe significant variance in both metrics. This happens because some pages (e.g., *en.wikipedia.org*) consist of fewer objects, and therefore can complete faster. The median object counts per web page are 69 for *www.bbc.com* and 14 for *en.wikipedia.org*. Other pages take longer to download because they have several objects that may be fetched from multiple servers. Also, for the Spanish operators, we detected multiple number of DNS iterations for *www.bbc.com*, thus partially explaining the higher TTFB metric compared to other operators in Norway and Sweden.

Discussion: While these experiments are preliminary, they clearly show the need of experimental investigation to understand 3G/4G network and application performance. The MONROE platform offers researchers the unique opportunity to run and repeat experiments to provide evidence of complicated phenomena.

5.6.2 Network Analytics with MONROE

One of the main targets of the MONROE platform is to provide experimenters a rich dataset of key mobile broadband metrics, from which different stakeholders can further extract the information of interest regarding the performance and reliability of MBB networks. To measure the network in a reliable and fair way, it is crucial to identify the metrics that accurately capture the performance

[8]The two websites we target are *"www.bbc.com"* and *"en.wikipedia.org"*.

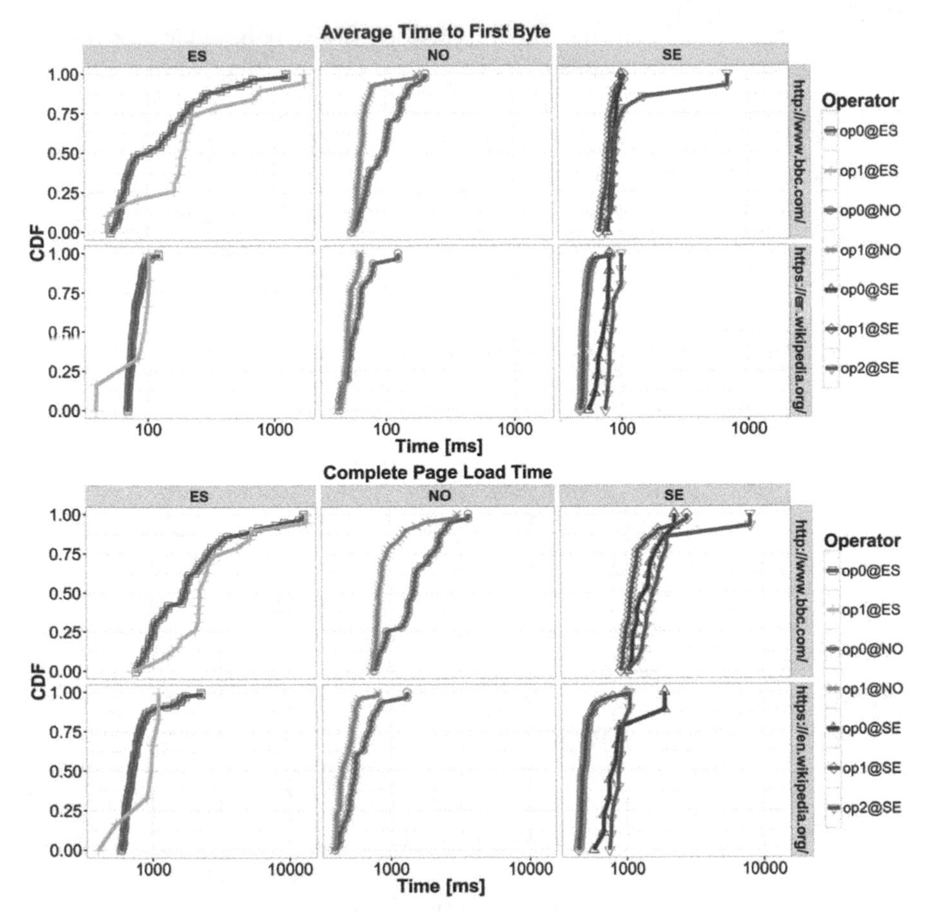

Figure 5.10 Web performance results: the Average Time to First Byte and the Complete Page Load Time for operators in Spain (ES), Norway (NO) and Sweden (SE) for two target websites www.bbc.com and en.wikipedia.org.

and the conditions under which we evaluate these metrics. Different stakeholders have different requirements on the metrics supported by the MONROE platform. For example, on the one hand, regulators need connectivity, coverage and speed information collected from a third-party, independent platform to monitor whether operators meet their advertised services, and as a baseline for designing regulatory policies. On the other hand, operators are interested in time series reporting of operational connectivity data to identify instability and anomalies. Furthermore, application developers need to cross-check QoS parameters against the behavior of the underlying network to design robust services and protocols. From the above considerations, it is clear that the collection of data cannot be limited to transmission and packet-level statistics, but there is an obvious need for rich metadata to be associated with the performance and reliability measurements.

The network metadata enables MONROE to capture the network context under which we measure the key performance metrics. The parameters we report include but not limited to provider name, radio access technology (RAT) type, RAT-specific parameters (e.g., RSRP, RSRQ, RSSI) and network connectivity status. Network metadata is crucial not only for coverage information but also during the analysis of the measurements in order to understand the underlying factors that affect the performance.

a) Mimicking Drive Tests for Mobile Coverage: One essential aspect when monitoring MBB providers is characterizing the coverage offered to unveil complex patterns of different radio access technologies (RATs) in an area. Network operators regularly test different network parameters of their deployed infrastructure for network benchmarking, optimization, troubleshooting and service quality monitoring. This is usually done via drive-testing where measurements are either collected by a vehicle with an embedded GPS device and other measurement equipments e.g. a laptop or by using mobile phone with an engineer roaming around the streets and roads of a region so that to have an end-user experience. However, there are major drawback to this approach, mainly the high cost it entails in terms of time and labor, and also that it does not cover most of the region where there are customers. The mobile MONROE nodes (placed on public transport vehicles) enable mimicking the drive tests measurements resulting in a dataset similar to the ones operators work with. Piggy-backing network measurements onto public transportation vehicles via MONROE offers additional benefits, including ensuring repeatability of drive runs on the same route, in similar busy-hour conditions, since the MONROE node is active in the times when the trains or buses carry passengers to their destinations. This approach emerges as a cost-effective alternative to the drive test performed by operators, with the added perk of allowing other parties, including public transport companies, to assess and compare the MBB coverage along their infrastructure at a zero added cost.

In Figure 5.11, we illustrate the measurement location from the mobile nodes active aboard trains inside Oslo are in Norway. We color-code the data points to show the radio access technology we read from the modem connected to one of the operators we measure. We observe that majority of time the node has 3G coverage and intermittent 4G coverage.

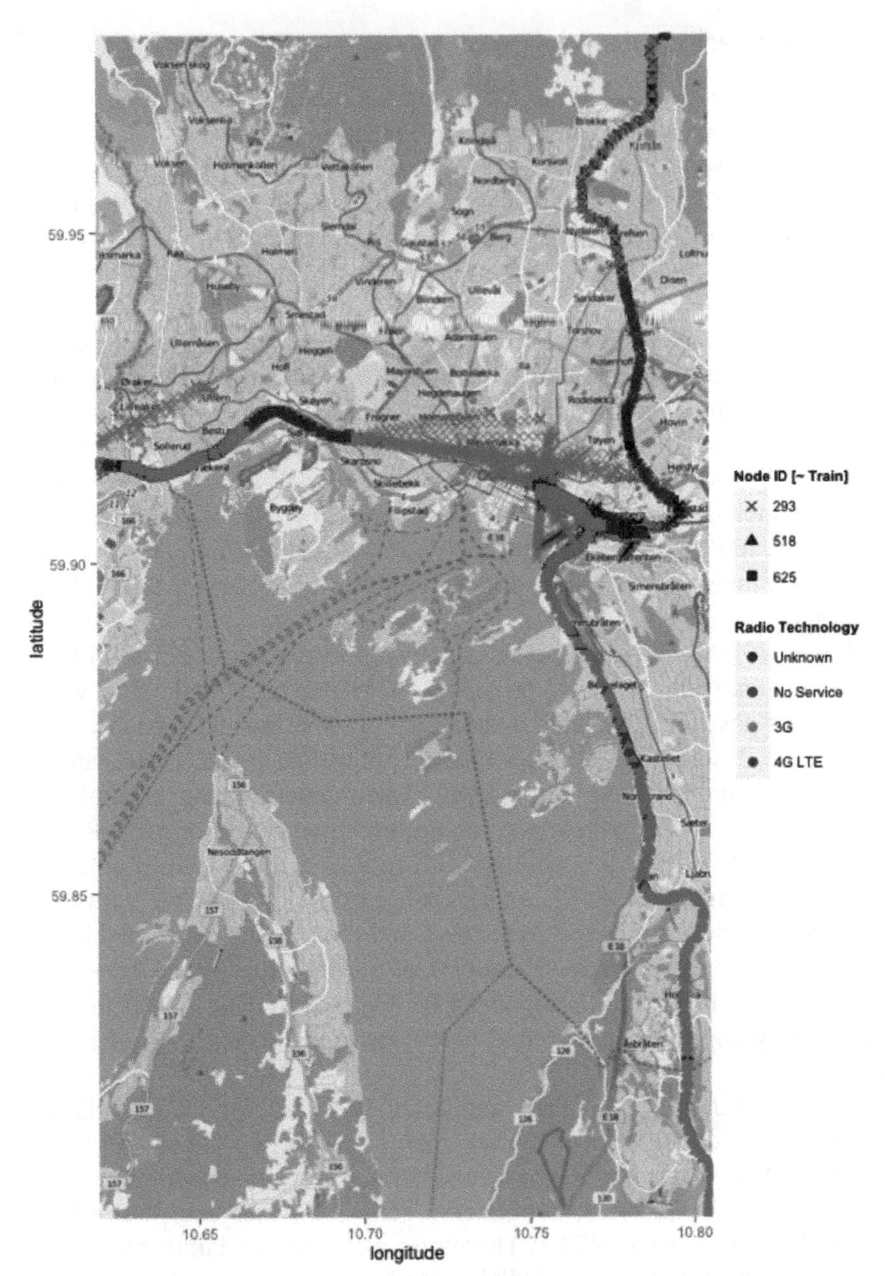

Figure 5.11 Coverage reading from MONROE nodes operating aboard trains in Oslo, NO.

5.7 User Experiments

Along with being a near real-time monitoring and benchmarking platform, MONROE is an open platform for experimentation with MBB networks. Below, we list a set of representative examples that MONROE users are currently curating. This serves to further illustrate the value of the MONROE platform and the variety of experiments it can accommodate.

a) Service Oriented Quality of Experience: A first dimension to explore comes from the great interest in how users perceive individual services and applications over different terminals (e.g., mobile phones, tablets, and computers). The recent proliferation of user-centric measurement tools such as Netalyzr [6] to complement available network centric measurements validate the increasing interest in integrating the end user layer in network performance optimization. MONROE enables experimentation with essential services and applications, including video streaming, web browsing, real-time voice and video, and file transfer services. The service oriented measurements give a good bases for investigating the mapping from Quality of Service to Quality of Experience.

b) Protocol Assessment: A second dimension to explore consists in the assessment of existing and new protocols in MBBs on a scale that was previously not possible. The large availability of experimental resources in MONROE is well suited to assess networked applications under a wide range of network conditions, while still giving experimenters strong control of the testing environment. Furthermore, the multihoming aspect of MONROE nodes makes it ideal for experimenting with protocols that exploit multiple connections opportunistically, e.g., in parallel or by picking the one with the best available service to increase robustness and performance, or to achieve the best cost-performance ratio. Examples of such protocols and services include, but are not limited to, Multipath TCP, Device-to-Device for offloading or public safety applications, portable video streaming services or e-health services.

c) Middlebox Impact: Another significant use case for MONROE is related to the use of middleboxes. These can range from address and port translators (NATs) to security devices to performance enhancing TCP proxies. Middleboxes are known to introduce a series of issues and hinder the evolution of protocols such as TCP. Therefore, measuring and understanding their behavior is essential. Since middleboxes of different types are ubiquitous in MBB networks, a platform such as MONROE offers an excellent vantage point from which to observe and characterize middlebox operation in real world deployments.

d) Knowledge Discovery and Network Analytics: Beyond mere service and protocol assessment, MONROE offers the possibility to develop mechanisms to augment network performance by learning from measurements. This use case involves post processing of data, to deepen the understanding of network behaviors. The goal is to identify causalities and correlation of different parameters that can individually or collectively affect the performance and reliability of the network. In order to identify unexpected data patterns that deserve attention, one should go beyond data-mining and correlation approaches, and rather use knowledge description techniques, such as the Kolmogorov complexity method [35] or the minimum description length theory [36]. Such approaches are beneficial for different stakeholders including operators, vendors, developers and service providers. Therefore, we envision MONROE to have a significant impact on different sectors of industry through these knowledge discovery approaches, while helping to improve the performance of their products leading to a better user experience for the end users.

5.8 Conclusions

In this chapter, we introduce the MONROE platform: an open and industry-grade platform for MBB measurements and experiments. The MONROE platform enables accurate, realistic and meaningful assessment of the performance of MBB networks by continuously monitoring these networks via active testing (e.g., delay test, web performance test, download speed test) and context metadata collection (e.g., connection mode, signal strength parameters). Furthermore, MONROE provides the perfect setting to test novel services and protocols thanks to its flexible and powerful nodes with multihoming support. In this chapter, we showcase the monitoring capabilities of the platform by analyzing preliminary performance measurement results. We further describe various examples of experiments that are supported by the platform in order to illustrate the unique features of the MONROE platform.

We argue that mobile measurements over operational networks are essential to understand the fundamental characteristics of mobile ecosystem as well as to establish the quality of end user's experience for different services. Such information is valuable to many different stakeholders including operators, regulators, policy makers, consumers, society at large, businesses whose services depend on MBB networks, researchers and innovators. For example, MONROE measurement results provide insights that can enable operators with more accurate radio resource and infrastructure planning, more cost-efficient investments, and better network utilization. Operators can also explore differentiated and specialized services, as well as their requirements and impact on applications. Application developers for mobile devices can use the platform to test various applications and services over MBB. With better knowledge about MBB and the ability to test services, MONROE will contribute to service providers innovating more and realizing innovative services. Internet of Things and smart city services will lead in this direction as more vertical specific applications and services will be developed along with the evolution towards 5G. Due to multihomed support, innovations regarding network selection, handover and aggregation can be developed to make applications more robust with better adaptability and increased quality; for this, multipath TCP and Device-to-Device communications are instrumental. These are a few examples of the opportunities in the MBB field that requires extensive research efforts from both industry and academia, and the MONROE platform with its unique features is the key enabler to achieve them.

References

[1] *Cisco Visual Networking Index: Global Mobile Data Traffic Forecast Update, 2015–2020 White Paper*, http://www.cisco.com/c/en/us/solutions/collateral/service-provider/visual-networking-index-vni/mobile-white-paper-c11-520862.html, Cisco Systems, Inc., February 2016, accessed: 2016-08-01.

[2] networld2020, "Service level awareness and open multi-service internetworking – principles and potentials of an evolved internet ecosystem," 2016.

[3] FCC, "2013 Measuring Broadband America February Report," FCC's Office of Engineering and Technology and Consumer and Governmental Affairs Bureau, Tech. Rep., 2013.

[4] Tektronix, "Reduce Drive Test Costs and Increase Effectiveness of 3G Network Optimization," Tektronix Communications, Tech. Rep., 2009.

[5] OOKLA, "http://www.speedtest.net/."

[6] C. Kreibich, N. Weaver, B. Nechaev, and V. Paxson, "Netalyzr: illuminating the edge network," in *Proceedings of the 10th ACM SIGCOMM conference on Internet measurement*. ACM, 2010, pp. 246–259.

[7] R. Atlas, "https://atlas.ripe.net/."

[8] S. Sundaresan, S. Burnett, N. Feamster, and W. De Donato, "Bismark: a testbed for deploying measurements and applications in broadband access networks," in *2014 USENIX Conference on USENIX Annual Technical Conference (USENIX ATC 14)*, 2014, pp. 383–394.

[9] PlanetLab, "https://www.planet-lab.org/"

[10] Z. Wang, Z. Qian, Q. Xu, Z. Mao, and M. Zhang, "An untold story of middleboxes in cellular networks," in *Proc. of SIGCOMM*, 2011.

[11] "OpenSignal™," http://opensignal.com

[12] "RootMetrics™," http://www.rootmetrics.com/

[13] "MobiPerf," http://www.mobiperf.com

[14] A. Nikravesh, D. R. Choffnes, E. Katz-Bassett, Z. M. Mao, and M. Welsh, "Mobile Network Performance from User Devices: A Longitudinal, Multidimensional Analysis," in *Procs. of PAM*, 2014.

[15] M. Molinari, M.-R. Fida, M. K. Marina, and A. Pescape, "Spatial interpolation based cellular coverage prediction with crowdsourced measurements," in *Proceedings of the 2015 ACM SIGCOMM Workshop on Crowdsourcing and Crowdsharing of Big (Internet) Data*. ACM, 2015, pp. 33–38.

[16] M. K. Marina, V. Radu, and K. Balampekos, "Impact of indoor-outdoor context on crowdsourcing based mobile coverage analysis," in *Proceedings of the 5th Workshop on All Things Cellular: Operations, Applications and Challenges*. ACM, 2015, pp. 45–50.

[17] A. Le, J. Varmarken, S. Langhoff, A. Shuba, M. Gjoka, and A. Markopoulou, "Antmonitor: A system for monitoring from mobile devices," in *Proceedings of the 2015 ACM SIGCOMM Workshop on Crowdsourcing and Crowdsharing of Big (Internet) Data*. ACM, 2015, pp. 15–20.

[18] Z. Wang, F. X. Lin, L. Zhong, and M. Chishtie, "How far can client-only solutions go for mobile browser speed?" in *Proceedings of the 21st International Conference on World Wide Web*. ACM, 2012, pp. 31–40.

[19] F. Wamser, M. Seufert, P. Casas, R. Irmer, P. Tran-Gia, and R. Schatz, "Yomoapp: A tool for analyzing qoe of youtube http adaptive streaming in mobile networks," in *Networks and Communications (EuCNC), 2015 European Conference on*. IEEE, 2015, pp. 239–243.

[20] A. Nikravesh, H. Yao, S. Xu, D. Choffnes, and Z. M. Mao, "Mobilyzer: An open platform for controllable mobile network measurements," in *Proceedings of the 13th Annual International Conference on Mobile Systems, Applications, and Services*. ACM, 2015, pp. 389–404.

[21] J. Sommers and P. Barford, "Cell vs. WiFi: On the Performance of Metro Area Mobile Connections," in *Proc. of IMC*, 2012.

[22] W. L. Tan, F. Lam, and W. C. Lau, "An empirical study on 3G network capacity and performance," in *INFOCOM 2007. 26th IEEE International Conference on Computer Communications. IEEE*. IEEE, 2007, pp. 1514–1522.

[23] M. Z. Shafiq, L. Ji, A. X. Liu, J. Pang, and J. Wang, "Characterizing geospatial dynamics of application usage in a 3G cellular data network," in *INFOCOM, 2012 Proceedings IEEE*. IEEE, 2012, pp. 1341–1349.

[24] J. Huang, F. Qian, Y. Guo, Y. Zhou, Q. Xu, Z. M. Mao, S. Sen, and O. Spatscheck, "An in-depth study of LTE: Effect of network protocol and application behavior on performance," in *Proceedings of the ACM SIGCOMM 2013 Conference on SIGCOMM*, ser. SIGCOMM '13. New York, NY, USA: ACM, 2013, pp. 363–374. [Online]. Available: http://doi.acm.org/10.1145/2486001.2486006

[25] P. Casas, M. Seufert, and R. Schatz, "YOUQMON: A System for On-line Monitoring of YouTube QoE in Operational 3G Networks," *SIGMETRICS Perform. Eval. Rev.*, Vol. 41, No. 2, pp. 44–46, Aug. 2013. [Online]. Available: http://doi.acm.org/10.1145/2518025.2518033

[26] F. Fund, C. Wang, Y. Liu, T. Korakis, M. Zink, and S. Panwar, "Performance of dash and webrtc video services for mobile users," in *Packet Video Workshop (PV), 2013 20th International*, Dec 2013, pp. 1–8.

[27] J. Huang, F. Qian, A. Gerber, Z. M. Mao, S. Sen, and O. Spatscheck, "A close examination of performance and power characteristics of 4G LTE networks," in *Proceedings of the 10th international conference on Mobile systems, applications, and services*. ACM, 2012, pp. 225–238.

[28] N. Thiagarajan, G. Aggarwal, A. Nicoara, D. Boneh, and J. P. Singh, "Who killed my battery?: analyzing mobile browser energy consumption," in *Proceedings of the 21st International Conference on World Wide Web*. ACM, 2012, pp. 41–50.

[29] *Methodology and technical information relating to the SamKnows™ testing platform – SQ301-002-EN*, SamKnows™, 2012.

[30] M. Bagnulo, P. Eardley, T. Burbridge, B. Trammell, and R. Winter, "Standardizing Large-scale Measurement Platforms," *SIGCOMM Comput. Commun. Rev.*, Vol. 43, 2013.

[31] S. Sen, J. Yoon, J. Hare, J. Ormont, and S. Banerjee, "Can they hear me now?: A case for a client-assisted approach to monitoring wide-area wireless networks," in *Proc. of IMC*, 2011.

[32] A. Kvalbein, D. Baltrūnas, J. Xiang, K. R. Evensen, A. Elmokashfi, and S. Ferlin-Oliveira, "The Nornet Edge platform for mobile broadband measurements," *Elsevier Computer Networks special issue on Future Internet Testbeds*, 2014.

[33] A. Finamore, M. Mellia, M. Meo, M. M. Munafo, P. D. Torino, and D. Rossi, "Experiences of internet traffic monitoring with tstat," *IEEE Network*, Vol. 25, No. 3, pp. 8–14, May 2011.

[34] P. Casas, P. Fiadino, S. Wassermann, S. Traverso, A. D'Alconzo, E. Tego, F. Matera, and M. Mellia, "Unveiling network and service performance degradation in the wild with mplane," *IEEE Communications Magazine*, Vol. 54, No. 3, pp. 71–79, March 2016.

[35] M. Li and P. M. Vitnyi, *An Introduction to Kolmogorov Complexity and Its Applications*, 3rd ed. Springer Publishing Company, Incorporated, 2008.

[36] P. D. Grünwald, *The Minimum Description Length Principle (Adaptive Computation and Machine Learning)*. The MIT Press, 2007.

6

PerformNetworks: A Testbed for Exhaustive Interoperability and Performance Analysis for Mobile Networks

Almudena Diaz, Cesar A. Garcia-Perez, Alvaro Martin, Pedro Merino and Alvaro Rios

Universidad de Málaga, Andalucia Tech, Spain

Abstract

PerformNetworks (formerly PerformLTE) is a FIRE facility located at University of Málaga devoted to LTE and 5G technologies experimentation. This testbed is one of the first to provide mobile technologies in FIRE, featuring a unique combination of commercial-off-the-shelf technology with conformance and research equipment. This chapter will provide the details about the testbed which provides mobile connectivity through different experimentation scenarios, moving between emulation and real-world environments. The configurations offered cover a broad spectrum of experiments, from applications and services to innovative network solutions. The chapter will also describe the experiences in the context of FIRE including: the federation with Fed4FIRE technologies; the use of experimentation technologies like those in the FLEX project, the support for several experiments (MobileTrain, SAFE and LTEUAV) from SMEs coming from different sectors; the exploitation as the core testbed in two new H2020 FIRE+ Innovation Actions and the evolution of the testbed to overcome future challenges in mobile networks research and innovation.

Keywords: LTE, 5G, Mobile Communications, QoS, QoE.

6.1 Introduction

PerformNetworks[1] is a testbed which is building and maintaining an experimentation eco-system, that will provide access to experimenters, to state-of-the-art mobile technology. Its primary objective is to provide an advanced and realistic experimentation environment for researchers, developers, manufactures, SMEs and mobile operators.

The testbed is intended to address the main trends of current mobile deployments, providing tools to characterize the behavior of networks under different conditions, providing insights into how the protocols and the services can be optimized. It is therefore very important to give developers, mobile operators and manufactures a very accurate view of any component of the network behavior in order to implement the right policies regarding resources management.

PerformNetworks supports developments and the improvement of deployments around mobile technologies through:

- Delivery of a full testing platforms that properly support the configuration of full stack mobile technologies including radio access network, core network and performance measurements.
- Delivery of measurement tools, for discovering the precise impact of radio and core configurations on devices and applications. This is critical for device manufacturers and operators to ensure that applications and devices can take full advantage of the potential offered by upcoming 5G mobile technologies.
- Delivery of advanced results based on the correlation of data collected at different points of the network and at different levels of the protocol stack, to obtain a complete characterization of mobile applications under different radio and core configurations.

PerformNetworks can play many roles in the field of mobile experimentation, as a first approach it can be used as a platform to track-and-trace network configurations and the QoS delivered at the users level. Power consumption is also a major issue in the design of mobile devices and in mobile applications, and also greatly affects the quality of the subscriber experience. Therefore accurate measurements of power consumption in mobile devices are provided.

The testbed also aims to become a reference interoperability platform where manufacturers and researchers can check the interoperability of

[1]http://performnetworks.morse.uma.es/

commercial and/or experimental solutions. Finally mobile devices are one of the first point of contacts with the new mobile technologies and the testbed supports the interconnection of commercial and experimental devices as well as the installation of external applications.

PerformNetworks has been successfully used in several experiments as part of Fed4Fire project, and is used in several FIRE projects: FLEX, TRIANGLE and Q4HEALTH.

6.2 Problem Statement

The objectives of PerformNetworks are in line with the future technological requirement of 5G networks, which directly relate to User Experience, Device, System Performance, Business Model, Enhanced Service, Management and Operation, as stated in [1].

Now that the mobile Internet has come of age, the main stakeholders and also other small actors need access to realistic and extensive experimentation to ensure the success of their solutions. Simulations and theoretical solutions are not enough to test the performance of their solutions. All too often it is difficult to correlate data from simulations with the real world, this is why our testbed comprises real hardware, such as commercial mobile devices and eNodeB emulators, which include real signal processing, base stations and an EPC (Evolved Packet Core).

Moreover this equipment is very expensive and so unaffordable for researchers, developers or SMEs. The PerformNetwork testbed provides all these stakeholders access to an environment where they can deploy and test their solutions.

Another important fact is that the vast majority of testbed's users have only a limited knowledge of mobile technology. In most cases they are looking to test their solutions in a very realistic mobile scenario, however they do not know how to configure the testbed to meet their testing requirements. This is why consultancy is also an important part of our testbed, we translate their testing requirements into test plans which reproduce the network conditions that are relevant for them.

Finally, even in the case experimenters have the resources and the knowledge to deploy their own testing network it is difficult to deploy real pilots due to spectrum regulations. Besides the technical issues, researchers have to reach agreements with operators who are the owners of the spectrum and might be sceptical about leasing it. This is why we offer three different scenarios. As a first option, the experimenter can use the most controlled and configurable platform, a complete proprietary LTE network, built on top of a eNodeB emulator, employed by certified laboratories, where radio conditions can be fully configured and mobility scenarios can be reproduced. Once the configurations have been evaluated in this scenario, the same experiment or new experiments can be validated with real eNodeBs deployed in a proprietary LTE network integrated by commercial eNodeBs and an EPC. In the last scenario, PerformNetworks enables the remote evaluation of the experiments by providing access to on-the-shelf devices connected to LTE commercial networks deployed in Málaga.

6.3 Background and State of the Art

This section provides an overview of the different mobile networks tools and platforms for research and experimentation currently available for experimenters, depicting their capabilities. Firstly, the available tools, an then the commercial solutions, are described. A brief overview of some of the most important European testbeds, devoted to wireless communication, is also given.

6.3.1 Research Tools for Wireless Communications

There are some tools that can be used for experimentation (besides commercial equipment). The most widely used in research papers are the simulators, mainly ns-3 and Riverbed Modeler (formerly OPNET). Simulators can provide inexpensive, systematic results but the reliability of these results can vary depending on the problem and the tool used.

One of the most common simulation tools is ns-3[2] which includes some functionality for LTE. The support is provided by the LTE-EPC Network Simulator (LENA) [2], an open source module that was designed to evaluate some aspects of LTE systems such as Radio Resource Management, QoS-aware Packet Scheduling, Inter-cell Interference Coordination, Dynamic Spectrum Access as well as simulate End-to-End IP connectivity. The ns-3 framework can be used as an emulator, although the performance results can be limited [3].

Riverbed Modeler (formerly OPNET) is a commercial solution that provides an LTE simulation platform designed according to 3GPP

[2]https://www.nsnam.org/

Rel. 8 specifications. OPNET implements most of LTE's basic features and also includes powerful statistical evaluation tools.

Open source implementations for Software Defined Radio (SDR) are becoming very popular, the price of the hardware has lowered and the availability and quality of the solutions is better. These types of solutions can provide a realistic environment with total control of the stack, the major drawback to them is the coverage specifications.

Open Air Interface (OAI)[3] wireless technology platform offers an open-source software-based implementation of LTE UE, E-UTRAN and EPC, compatible with many different SDR solutions such as ExpressMIMO2, USRPs, BladeRF and SodeRa. The solution was created by Eurecom[4] and is now managed via the OAI Software Foundation (OSA). OAI includes tools to configure, debug and analyze several aspects of LTE layers and channels and can interact with commercial equipment [4].

Another solution gaining popularity is the LTE libraries (srsLTE and srsUE)[5] designed by Software Radio Systems (SRS)[6] compatible with SDR applications and covering compliant with the 3GPP Release 8. The srsLTE library provides common functionality for LTE UE and eNB with support, when available, of the VOLK acceleration libraries. srsUE is based on srsLTE and provides the basic functionality of an LTE UE.

The emulator equipment can provide very realistic results operating with commercial devices whilst maintaining a high level of reproducibility in the results. This type of equipment normally provides end to end functionality and sometimes can also include the effects of the channel. The major drawback is the price of the solutions which is very high and the focus on the radio access which limits interoperability with the EPC network. These emulators are traditionally provided, to be used in design verification, conformance testing and/or signaling protocol testing.

For instance the E7515A UXM by Keysight Technologies[7] is conformance testing equipment for Release 10 LTE devices. UXM allows users to validate the functional and RF performance of their UEs, providing end-to-end LTE-Advanced connectivity as well as a highly configurable network and radio access parameters. The unit is capable of providing data rates of up to 1 Gbps

---

[3]http://www.openairinterface.org/

[4]http://www.eurecom.fr/en

[5]https://github.com/srsLTE/srsUE

[6]http://www.softwareradiosystems.com/

[7]http://www.keysight.com

in downlink, multiple cells, carrier aggregation, MIMO and fading emulation all provided in a single box.

An example of a signaling protocol tester is CMW500 by Rohde & Schwarz[8], which provides developers of wireless devices access to a radio access network emulation, including a network operability test. This equipment offers MIMO 2X2, multi-cell and data rates up to 150 Mbps in the downlink and is able to support other technologies such as 2G, 3G and Wi-Fi.

6.3.2 Wireless Testbed Platforms

There are many different platforms available for mobile experimentation, like [5], where European 5G platforms are described. In the context of FIRE, there are three main testbeds: Fuseco, NITOS and w-iLab.t. The role of ORBIT is also very important as it provides one of the most common experimentation frameworks, OMF. The ORBIT testbed[9] is a wireless network emulator for experimentation and realistic evaluation of protocols and applications. ORBIT provides a configurable mix of both cellular RATs (WiMAX and LTE) and Wi-Fi, together with Bluetooth, ZigBee and SDR platforms.

Fuseco Playground[10], by Fraunhofer, is an open testbed for R&D of mobile broadband communication and service platforms. Fuseco integrates several RATs (DSL/WLAN/2G/3G/LTE/LTE-A) together with M2M, IoT, sensor networks and SDN/NFV. This testbed can be used directly at the Fraunhofer premises in Berlin, and in many cases, remotely.

The NITOS Future Internet Facility [6], is a testbed which provides support for research into wired and wireless networks. NITOS provides a heterogeneous experiment environment, including Wi-Fi, WiMAX, LTE and Bluetooth wireless technologies, SDR, SDN and sensor networks. The NITOS testbed can be used remotely.

w-iLab.t[11], by iMinds, is a wireless testbed for the development and testing of wireless applications. w-iLab.t offers two different LTE networks for testing, including both ip.access femtocells and SiRRAN EPCs. Furthermore, this testbed provides Wi-Fi, sensor node and cognitive networking experiment platforms. w-iLab.t testbed is accessible remotely.

[8]https://www.rohde-schwarz.com

[9]http://www.orbit-lab.org/

[10]https://www.fokus.fraunhofer.de/go/en/fokus_testbeds/fuseco_playground

[11]http://ilabt.iminds.be/iminds-wilabt-overview

6.4 Approach

Since its inception the focus of the PerformNetworks testbed has been to provide access to the researchers with the wide array of tools present in a commercial scenario without losing the control and configuration options available in the academic world. With this approach the testbed's users gain two advantages over the deployment of the simulators usually used in the scientific realm. On the one hand, thanks to the use of commercial equipment the researchers have the same level of access as the network operator and can perform realistic tests without the simplifications and assumptions that are part of using simulators in the experiments. On the other hand the researcher maintains all the flexibility of access to every layer of the network without fear of disrupting the normal operation of a commercial operation setup.

This flexibility usually imposes an additional burden on the researcher tasked to interconnect and configure all the nodes of the test network. So, having experienced these problems first hand, the testbed team has created a set of sensible defaults and ready-to-use configurations so researchers, while still being able to change the network as needed, only have to focus on the important parts of their tests.

The way to achieve this relies on the interchangeability of most of the components available to researchers and is based on the experience and knowledge accumulated by the testbed operators during the initial setup and the many experiments running over it. Based on the feedback from researchers, the testbed team can suggest architectures and configurations that best adapt to a specific experiment, run while maintaining a low-level complexity for the parameters that falls outside the scope of the experiment. This in turn guarantees an optimal performance.

Figure 6.1 outlines the architectural components a researcher may choose to use in his experiments:

- A commercial *Evolved Packet Core* (EPC) from Polaris Networks[12] with all the entities upgraded to the 3GPP standard Release 11. The experimenter has direct access to the *Mobility Management Engine* (MME), *Serving Gateway* (SGW), *PDN Gateway* (PGW), *Home subscriber server* (HSS), the *Policy and Charging Rules Function* (PCRF), and the new entities *Access Network Discovery and Selection Function* (ANDSF) and the *Evolved Packet Data Gateway* (ePDG).

[12]http://www.polarisnetworks.net/

- A virtualization server where the experimenter can deploy several virtual machines and interconnect them in an arbitrary way to increase the complexity of the setup. One typical use of this component is to install network software, for example *Open vSwitch*, to route the traffic to and from the EPC with different priorities.
- Commercial small cells to be used as *eNodeB*.
- Several *Software Defined Radio* (SDR) cards that can be used as *eNodeB* with the appropriate software, like *OpenAirInterface*, in case the researcher requires access to the code running the nodes.
- The *T2010 Conformance Tester* from Keysight Technologies which is used by mobile manufacturers worldwide to precisely measure the radio performance of new devices. The ones present of the testbed have been modified, as described in the following sections, to provide standard connectivity to commercial core networks, offering full end to end experimentation scenarios.
- An array of attenuators, RF switches, channel emulators and equipment to measure power consumption in the device under test.
- Various COTS UE with Android and Linux operating systems where researchers can install their own apps and programs.

Using this equipment the testbed offer to experimenters an iterative approach. The experimenter can go to the fine tune of the components of the network

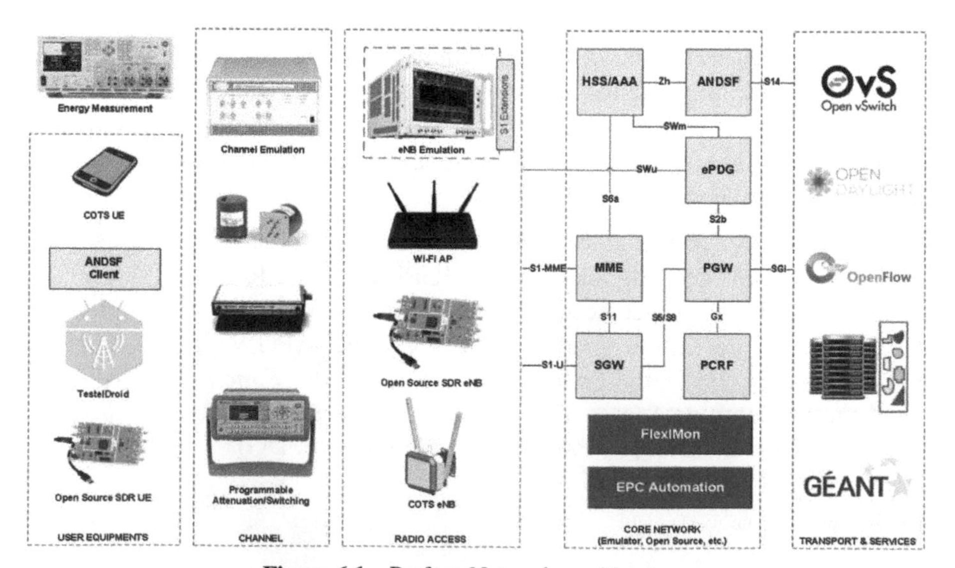

Figure 6.1 PerformNetworks architecture.

including the effects of the channel on a controlled environment using conformance testing equipment. In these scenarios they can obtain reproducible and systematic results maintaining a realistic environment where they can use commercial UEs and EPCs. In a more advanced stage of their research experimenters can validate their approach employing an indoor deployment, using commercial base stations to have an idea on how their solutions could perform on the operator networks but also using eNB based on SDR if they need modifications on the radio access stack. Finally researchers can use commercial deployments to measure the performance on real networks but also obtaining information on several KPIs of the radio access.

6.5 Technical Work

An important part of the technical work on the PerformNetwork testbed has consisted in the interconnection of the different equipment, the use of heterogeneous equipment hardens the interoperability. Several tools have been also designed for the use of the testbed. In this section a modification to the T2010 conformance testing equipment to support standard S1 interface is described. Fleximon is an interoperability tool designed to provide remote monitorization of communication interfaces. There have also been some developments to provide support for some federation technologies such as OMF resource controllers or aggregate managers.

6.5.1 T2010 Standard S1 Interface Extension

The *T2010 Conformance Tester* by Keysight Technologies allows manufacturers to test LTE end-to-end connections in a highly configurable way. Its primary function is to ensure new UE models adhere to the *3GPP* standard, but it can also be used to test non-ordinary conditions such as different power profiles, fading scenarios or exotic resource assignment. The *T2010* measuring capabilities are concentrated in the lower level of the stack, but it also implements most of the eNodeB protocols as well as a basic EPC emulation, so the UE being tested acts as if it is connected to a real LTE network.

The objective of the *T2010 Standard S1 Interface Extension*, developed at the University of Malaga within the framework of the *FlexFormLTE* project, was to extend the functionality of the T2010 with a standard *S1* interface so the user would be able to choose between the limited emulated *EPC* or connect to a fully functional external one. A complete *S1-MME* module was created, with hooks to the existing interfaces of the T2010 so control and user planes are created in the upper levels while maintaining the radio connection to the UE controlled by the equipment.

Thanks to the extension developed within this project the testbed now has a powerful new tool, combining the feature-rich T2010 physical and radio configuration with a realistic connection to a commercial *EPC*.

6.5.2 Fleximon

One of the main challenges a mobile network researcher faces during a experimentation campaign is how to extract signaling information from the components without disrupting the normal operation of the system. The control software usually only reports an aggregate of the events that have been recently fired in the network without detailed information about the data passed between the entities involved. One way to obtain this information is to capture all the traffic in a specific interface of the EPC, but the operator is usually reluctant to give the researcher access to their internal network and the amount of data captured this way can be overwhelming. *FlexiMon* is a tool within the scope of the FLEX project designed with this scenario in mind. The objective is to provide the network operator with a tool that opens a data path to the experimenter without modifying the network workflow, and gives the researcher a powerful platform where he/she can develop monitoring and statistic analysis software for that data.

It comprises two independent modules, written in C++ to lower the penalty hit in performance and with portability between different systems as a requirement. The first module, aptly called *FlexiCapture*, runs in any device with access to the network interface between one *eNodeB* and its corresponding MME. From there it identifies the traffic of protocols configured by the researcher (currently *SCTP*, *GTP* and/or *S1AP*) without altering the flow of data between the two entities. A copy of any matching packet is then relayed to the other module called *FlexiView*, which is running in the researcher's desktop, to be processed. *FlexiView* can save the traffic it receives in *pcap* format for future analysis with any standard tool like, for example *Wireshark*[13], but its main feature is an API which can be used to implement any real time processing in the traffic. With it a researcher can easily extend the monitoring capabilities of the application as if it were running inside the operator network. Also, to fully integrate this tool within the framework used in several *FIRE* projects, there is also the possibility to send the measurement results to an *OML* server for storage in a database.

[13]https://www.wireshark.org/

The following is a not intended to be an exhaustive list of modules which have already been implemented using this API:

- Identification and monitoring of an specific user as soon as it connects to the network.
- Amount of data and throughput of each user being monitored.
- Basic sanity checks in the *GTP* and *SCTP* protocols using the periodic *echo request* and *responses* used in them
- Several performance figures in the *S1* interface such as attach procedure duration, dedicated bearer creation success rate, etc.

6.5.3 TestelDroid

UE devices in PerformNetworks run Testeldroid [7], our custom tool for monitoring device performance parameters and data traffic, to collect experiment data. We have modified Testeldroid so that it sends that information as an OMF stream to an instance of an OML database which we are also running in our testbed.

TestelDroid is a passive monitoring software tool for Android devices. This tool collects not only simple metrics such as throughput, but also radio parameters such as received signal strength, radio access technology in use, the actual IP traffic and more to obtain a fully detailed picture to help characterize the traffic performance of mobile applications.

6.5.4 FIRE Technology

Currently, the experiment control is done through an OMF experiment controller (EC) deployed on one of our nodes. This controller can be accessed via SSH. We have also deployed a web frontend to the experiment controller called LabWiki [8], created by NICTA[14].

As described the PerformNetworks testbed has a moderate number of specialized pieces of equipment. Most of this equipment offers an interface based on SCPI (Standard Commands for Programmable Instruments), which is used to control its operation through a Resource Controller (RC) which triggers the configuration commands to the instruments. An specific Resource Controller for SCPI instruments have been developed, which is based on an XML definition that provides a mapping between high level functionality of the instruments and SCPI commands. However to support our latest equipment the Resource Controller available through the official distribution of the OMF

[14]https://www.nicta.com.au/

framework has been used, this will simplify the integration of future versions of the OMF framework. For the pieces of equipment that do not support this interface specific configuration scripts, that are also issued via a standard RC, have been designed.

PerformNetworks has deployed GCF, an implementation of SFA AM (Aggregated Manager) created by GENI[15], as its solution for resource discovery and provisioning. Due to the nature of the testbed, consisting mainly in specialized hardware, reservation is manual and exclusive, i.e. only a single experiment can be run on top of it at any one time.

The AM provides a federated SSH access to the Experiment Controller (EC) of the testbed. Resource description is done via RSpecs. The current RSpec definition of the testbed provides a monolithic specification of the EC of the testbed. This definition can be used to gain SSH access to the EC of PerformNetworks using, for example, the jFed Experimenter GUI tools.

The EC also contains the reference experiments described in OEDL. The experimenters can modify and launch their customized experiments using OMF EC procedures available in the EC. Figure 6.2 provides a general picture of the orchestration framework deployed in the PerformNetworks testbed.

6.6 Results and Achievements

The PerformNetworks testbed has been integrated in several of the EU FIRE initiatives. The first integration of the platform was performed in Fed4Fire where the PerformLTE testbed was federated to be exposed to third parties as well as remotely operated. This federation was initially based on an SCPI-enabled resource controller for industrial equipment [9] developed by the MORSE group and the provision of an aggregate manager and several experiment controllers to enable remote ssh access to external users.

6.6.1 SME Experiments

In the context of Fed4FIRE several SMEs have run their experiments on PerformNetworks, gaining access to highly complex and expensive equipment which they have used to improve their businesses. Some of these experiments are MobileTrain, SAFE and LTEUAV.

MobileTrain was an experiment executed by Naudit[16] and consisted in several test campaigns to improve their QoS tools using packet-train [10]

[15]https://portal.geni.net/

[16]http://www.naudit.es/

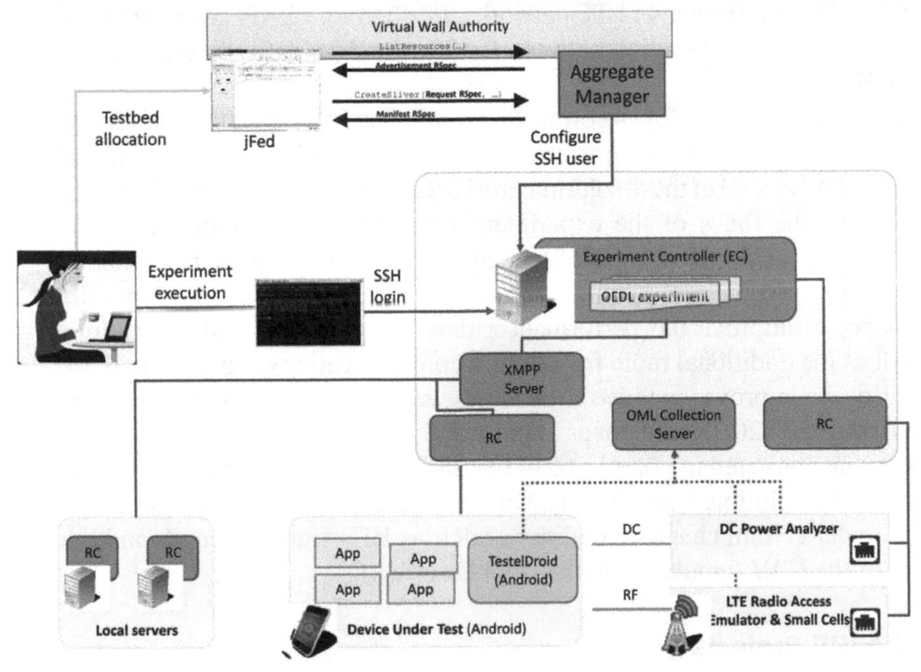

Figure 6.2 PerformNetworks orchestration architecture.

measurement techniques. The setup was based on the T2010 emulator that provided the LTE connectivity with a dedicated stratum 2 Precision Time Protocol (PTP) which was deployed for this experiment to obtain more accurate one way delay measurements.

SAFE, an experiment run by RedZinc[17], was motivated by the need to study the performance of LTE-transmitted video in emergency situations. The video was streamed live via a wearable platform designed by the company. They also had developed an engine to produce QoS enforcement in the network which was integrated, during the experiment, with the IMS interface of the testbed's core network. This experiment used an Alcatel Lucent *pico-cell* prototype that was employed in scenarios that required dedicated bearer establishment (generated via the Rx interface of the EPC) and the T2010 conformance testing equipment to emulate scenarios with mobility. As a result of the experiment, RedZinc were able to develop their own Rx driver to communicate with standard core networks and they obtained a first optimization of

[17]http://www2.redzinc.net/

their BlueEye platform over LTE networks. Furthermore SAFE was the origin of a subsequent collaboration between RedZinc and MORSE that resulted in the Q4Health project[18].

LTEUAV was run by Aeorum[19], a company that develops solutions based on computer vision, unmanned vehicles management and artificial intelligence. Several of these solutions are based on Unmanned Aerial Vehicles (UAV) and the focus of the experiment was precisely to optimize of the communications for these types of scenarios. The main problem the company encountered was the optimization of the video streaming captured from the UAVs so, to improve the performance, they used mobile communications as well as the traditional radio frequencies communications available in these scenarios. To improve the video streaming several test campaigns were undertaken using the T2010 emulator as a controlled reproducible environment. The experiment consisted in optimizing of the different video parameters, such as video feed resolution, frame rate and encoding, based on the response of the video under certain channel conditions such as LTE signal strength and the speed of the UAV (emulated by the Doppler effect).

6.6.2 FIRE Projects

PerformNetworks is used in several FIRE projects. It is integrated into the FLEX project, which was specifically oriented towards LTE and 5G experimentation. The testbed was also federated with the FLEX technology with the development of an LTE.rf controller for the T2010, which was extended to support standard S1 communications, and the EPC. With these new capabilities, the PerformNetworks testbed was used to perform interoperability testing with the different pieces of equipment present in other testbeds of the project so as to identify any potential problems. This resulted in an interoperability report which presented all the results and suggested guidelines to improve the definition of experiments involving different FLEX platforms.

In 2016 two new innovation actions using PerformNetworks have been accepted, the results of which will be a testbed improvement to accelerate time to market of products from companies in different sectors. One of these actions is the *Triangle* [11] project (described with more detail in chapter REFERENCE_TO_TRIANGLE_CHAPTER). In this project PerformNetworks is going to be evolved to support different experimenter profiles, trying

[18]http://www.q4health.eu/

[19]http://aeorum.com/

to provide them with useful tools in a language they can understand. The project focuses on 5G certification so several extensions to the testbed are foreseen, such as the introduction of LTE Release 12 equipment, supporting very high throughput, the interconnection of certification equipment with commercial core networks via Software Defined Network enabled switches and the exploration of dual connectivity to support heterogeneous wireless communications. The other action is the *Q4Heatlh* [12] project (described in more detail in chapter REFERECE_TO_Q4HEALTH_CHAPTER) which is the natural continuation of the Fed4FIRE experiment SAFE. In Q4Health PerformNetworks will be extended to support ultra low latency services by combining NFV and SDN techniques, the EPC has been upgraded to support Release 12/13 features such as seamless handover with non 3GPP technologies or MME relays, and a new optimized version of the RedZinc BlueEye platform is expected to be ready by the end of 2018.

6.6.3 Research Activities

PerformNetworks is also used by the MORSE group with research an academic purposes. The research activities are developed for many different reasons, gathering requirements for future releases of the testbed, improving the experimental interfaces and optimization and characterization of mobile networks.

The exploration of new functionality for the testbed has resulted in different research contributions. In order to improve the support for mission critical communications in the testbed by means of Commercial-Off-The-Shelf (COTS) technology, the use of the standards were analyzed on [13], driven by the particular use case of LTE communications for railway signaling. In this paper the requirements for railway communications which include traffic prioritization, broadcast services, location dependent addressing, etc. were analyzed, providing standard alternatives when available, and providing a quantitative analysis of the fulfillment of these requirements. In [14] the future standard architecture for IoT applications is analyzed, covering aspects such as the addressing, energy consumption, and congestion avoidance.

In [15] a framework for VoIP measurement analysis, including MOS estimation based on Perceptual Evaluation of Speech Quality (PESQ), RTP processing and more, was developed and used to extract voice measurements from test campaigns involving the public Spanish high speed railway. This tool has been also used in cooperation with Spanish operators wishing to have a characterization of their network and its basic extraction engine, named

TestelDroid, is available on the Google Play Store[20]. A more detailed analysis was performed in [16], that was not only limited to voice calls (around 400 calls were performed) but also included FTP and ping measurements providing a comparison between two Spanish mobile operators.

From the results of these measurement campaigns the limitation of using third party networks became clear. Operators were not willing to open up their networks so more complex measurements could be taken, and even less inclined to setup their equipment to optimize certain services. A campaign of measurements to see the performance of prioritized railway signaling traffic over live networks giving coverage for high speed trains was studied and finally in the context of the Tecrail[21] project a setup to perform such measurements was designed. An agreement with an Spanish operator was reached and Alcatel Lucent provided LTE base stations which were deployed along the railway tracks. These base stations were connected to the PerformNetworks EPC, giving access to all the measurements on the network and also enabling the configuration of a service level agreement for the different services under test by means of the establishment of dedicated radio bearers on the network. Additionally an emulated European Train Control System (ETCS) service, designed by AT4 Wireless[22], was used on top of this infrastructure. The combination of emulated and commercial equipment in a realistic environment provided support to a unique experiment and became one of the distinguishing features of the PerformNetworks testbed that started to evolve in this direction.

This VoIP toolset was then used with the T2010 to provide and test end-to-end connectivity under different channel conditions. In this setup, measurements from the LTE network stack (e.g.: MAC BLER/Throughput, CQI, etc.) could be extracted and were correlated with the measurements from the application level, providing insights into how certain network conditions translate into QoE performance indicators. Energy consumption has also been explored with the tools of PerformNetworks, for instance in [17] a runtime verification system was developed based on the measurements extracted from commercial devices, that were stimulated with execution traces. In [18] the use of the T2010 and a power analyzer offered results on the power consumption of mobile phones when performing voice calls over an LTE network under different network conditions and also with different network configurations.

[20]https://play.google.com/store/apps/details?id=com.ad.testel
[21]http://www.tecrail.lcc.uma.es
[22]https://www.at4wireless.com/

The provision of experimentation interfaces has also been explored in several research papers. For instance in [9] we described the approach taken to support OMF/OML on the testbed that consisted in the abstraction of the functionality of the instrument in an XML definition. This XML provides the mapping between the high level functions of the equipment and the low level configuration which is done by means of SCPI commands. This is used by a resource controller which interprets high level commands to trigger the appropriate configurations and a transformation tool, that generates the OEDL interface based on the XML file. The use of this approach considerably simplified the integration of other experimentation interfaces such as LTE.rf, which could be done by implementing a different transformation. In [19] the modifications done to TestelDroid in order to support SCPI commands and the OML library are described. With the new modifications, the tool was integrated with the rest of the SCPI compliant equipment already present in the testbed and is now able to generate real-time measurements in an OML database.

PerformNetworks can offer many different types of results and is now being evolved to attract more users, especially those with little background in mobile communications, and to support future 5G mobile communications acting as a testbed enabler.

6.7 Discussion

PerformNetworks has evolved considerably over the last few years. The main focus has been to provide highly realistic experimentation environments while maintaining a high level of customization and flexibility. This trend is still very much present in the PerformNetworks testbed roadmap but more requirements have been identified.

PerformNetwork should offer consultancy services. Many of the external experimenters using the testbed are not experts in wireless communications, they come from different domains and their solutions make use of the wireless connectivity. The testbed interfaces were designed with the figure of research experimenters in mind, an expert on mobile communications who wished to set up all the components of the network. However most of the experimenters are from many different domains, normally vertical sectors, and lack the knowledge and time to learn how to setup the full network. From the second quarter of 2016 onwards PerformNetworks has offered its consultancy services via the University of Málaga branded as the UMA Mobile Network Laboratory.

Another important aspect of the experiments for future mobile communication is the scale. One of the targets of 5G technologies is to increase user capacity by 1000, and the role of IoT in future technologies is clear and comes with capacity requirements. To enable these experiments PerformNetworks is following two main research lines. On the one hand going live by broadcasting on commercial frequencies is considered a key aspect to facilitate these experiments. To do so it is mandatory to engage operators which are the owner of spectrum licenses. Obtaining their permission to broadcast can be difficult, normally (the regulation is different in every country) they have legal responsibility on the signals broadcasted in their frequencies. A possible idea could be to share their frequencies and/or equipment via RAN sharing technologies. The other enabler could be in the form of massive UEs emulators, that could be implemented with SDR technologies.

Open equipment is very important to enable future mobile communications. The testbed is trying to provide as many modifiable components as possible, like for example OpenAirInterface (PerformNetworks is part of the OpenAirInterface Software Alliance), which provides source code for UE, eNB and EPC; or srsUE[23], which centres on the UE. The PerformNetworks tools which are not protected by intellectual properties agreements with third parties will also become open source.

In addition, MORSE will also cover new research projects that will be part of future releases of the testbed, some of these topics are:

- SDN Applications validation and verification. The use of formal methods and runtime verification is currently being explored.
- NFV functionality, especially the CloudRAN features. There are ongoing efforts to implement new network functions to enable optimized network procedures and low latency communications.
- Mission critical communications are still on the testbed's radar, especially those involving high speed scenarios, such as railways, high availability or ultra low-latency services.
- Advanced network probes. In the last few years probes for the core network and Android phones have been developed, so the testbed will be extended with new tools to provide even more information from the stacks, making them deployable on commercial mobile networks.

6.8 Conclusion

This chapter has provided an overview of the PerformNetworks testbed from its origin to its future evolution. The testbed has been used by many different companies as well as by the MORSE group both for research and

[23]https://github.com/srsLTE/srsUE

innovation activities. We have described some of the challenges present in mobile experimental platforms and have provided an overview of the different tools which might be useful to the testbed's users, as well as the status of the most relevant FIRE testbeds in this field.

We have also depicted some of the implementations and integrations that have been done in the context of the testbed. This includes the extension of a conformance testing equipment to boost the number of available scenarios with channel emulation, the implementation of an interoperability tool capable of monitoring a communication interface remotely, providing information and statistics of the status of the different processes in the network. An Android application to perform drive tests of QoS and QoE has also been provided, together with some details on the implementation of the different experimentation and federation interfaces.

Some of the external experiments executed on the platform have been described with details about their requirements and their achievements. We have also outlined the research activities of the group, covering the analysis of different services on both live and emulated networks, the execution of pilots to enhance the realism of the deployments, the correlation of the information from different levels of the stack and the efforts to provide of remote access interfaces. Finally we have discussed future research activities for the testbed including some details on its possible roadmap.

We expect that PerformNetworks will become a reference platform for future 5G technologies and will attract more experimenters, by offering simplified interfaces as well as consultancy services to improve their products or research.

References

[1] NGMN Alliance. NGMN 5G White Paper. Technical report, Next Generation Mobile Networks, February 2015.

[2] Nicola Baldo, Marco Miozzo, Manuel Requena-Esteso, and Jaume Nin-Guerrero. An open source product-oriented lte network simulator based on ns-3. In *Proceedings of the 14th ACM International Conference on Modeling, Analysis and Simulation of Wireless and Mobile Systems*, MSWiM '11, pages 293–298, New York, NY, USA, 2011. ACM.

[3] T. Molloy, Zhenhui Yuan, and G. M. Muntean. Real time emulation of an lte network using ns-3. In *Irish Signals Systems Conference 2014 and 2014 China-Ireland International Conference on Information and Communications Technologies (ISSC 2014/CIICT 2014). 25th IET*, pages 251–257, June 2014.

[4] Navid Nikaein, Mahesh K. Marina, Saravana Manickam, Alex Dawson, Raymond Knopp, and Christian Bonnet. Openairinterface: A flexible platform for 5g research. *SIGCOMM Comput. Commun. Rev.*, 44(5): 33–38, October 2014.

[5] Networld 2020 ETP. 5G Experimental Facilities in Europe. Technical report, Networld 2020 European Technology Platform, March 2016.

[6] Dimitris Giatsios, Apostolos Apostolaras, Thanasis Korakis, and Leandros Tassiulas. *Methodology and Tools for Measurements on Wireless Testbeds: The NITOS Approach*, pages 61–80. Springer Berlin Heidelberg, Berlin, Heidelberg, 2013.

[7] A. Alvarez, A. Diaz, P. Merino, and F. J. Rivas. Field measurements of mobile services with android smartphones. In *Proc. IEEE Consumer Communications and Networking Conf. (CCNC)*, pages 105–109, 2012.

[8] Thierry Rakotoarivelo, Guillaume Jourjon, Olivier Mehani, Max Ott, and Michael Zink. Repeatable Experiments with LabWiki. Technical report, National ICT Australia, October 2014.

[9] C. A. García-Pérez, M. Recio-Pérez, Á. Ríos-Gómez, A. Díaz-Zayas, and P. Merino. Extensive and repeatable experimentation in mobile communications with programmable instruments. In *2016 13th International Conference on Remote Engineering and Virtual Instrumentation (REV)*, pages 30–36, Feb 2016.

[10] M. Ruiz, J. Ramos, G. Sutter, J. E. Lopez de Vergara, S. Lopez-Buedo, and J. Aracil. Accurate and affordable packet-train testing systems for multi-gigabit-per-second networks. *IEEE Communications Magazine*, 54(3):80–87, March 2016.

[11] Andrea F. Cattoni; German Corrales Madueno; Michael Dieudonn; Pedro Merino; Almudena Diaz Zayas; Alberto Salmeron; Frederik Carlier; Bart Saint Germain; Donal Morris; Ricardo Figueiredo; Jeanne Caffrey; Janie Baos; Carlos Cardenas; Niall Roche and Alastair Moore. And end-to-end testing ecosystem for 5g. In *Networks and Communications (EuCNC), 2016 European Conference on*, June 2016.

[12] Cesar A. Garcia-Perez; Alvaro Rios; Pedro Merino; Kostas Katsalis; Navid Nikaein; Ricardo Figueiredo; Donal Morris and Terry O'Callaghan. Q4health: Quality of service and prioritisation for emergency services in the lte ran stack. In *Networks and Communications (EuCNC), 2016 European Conference on*, June 2016.

[13] A. Díaz Zayas, C.A. García Pérez, and P. Merino Gomez. Third-generation partnership project standards for delivery of critical

communications for railways. *Vehicular Technology Magazine, IEEE*, 9(2):58–68, June 2014.

[14] A. Díaz-Zayas, C. A. García-Pérez, M. Recio-Pérez, and P. Merino. 3gpp standards to deliver lte connectivity for iot. In *2016 IEEE First International Conference on Internet-of-Things Design and Implementation (IoTDI)*, pages 283–288, April 2016.

[15] F. Javier Rivas Tocado, A. D. Zayas, and P. M. Gómez. Performance study of internet traffic on high speed railways. In *World of Wireless, Mobile and Multimedia Networks (WoWMoM), 2013 IEEE 14th International Symposium and Workshops on a*, pages 1–9, June 2013.

[16] F. J. Rivas Tocado, A. Díaz Zayas, and P. Merino. Characterizing traffic performance in cellular networks. *IEEE Internet Computing*, 18(1): 12–19, Jan 2014.

[17] Ana Rosario Espada, María del Mar Gallardo, Alberto Salmerón, and Pedro Merino. *Runtime Verification of Expected Energy Consumption in Smartphones*, pages 132–149. Springer International Publishing, Cham, 2015.

[18] Á. M. Recio-Pérez, A. Díaz-Zayas, and P. Merino. Characterizing radio and networking power consumption in lte networks. *Mobile Information Systems*, 2016.

[19] A. Díaz-Zayas, M. Recio-Pérez, C. A. García-Pérez, and P. Merino. Remote control and instrumentation of android devices. In *2016 13th International Conference on Remote Engineering and Virtual Instrumentation (REV)*, pages 190–195, Feb 2016.

7

Large Scale Testbed for Intercontinental Smart City Experiments and Pilots – Results and Experiences

Louis Coetzee[1], Marisa Catalan[2], Josep Paradells[2], Anastasius Gavras[3] and Maria Joao Barros[3]

[1]CSIR Meraka Institute, Pretoria, South Africa
[2]i2CAT Foundation, Barcelona, Spain
[3]Eurescom GmbH, Heidelberg, Germany

Abstract

The challenges that cities face today are diverse and dependent on the region they are located. Inherently cities are complex structures. To improve service delivery in these complex environments the cities are being augmented by "Internet of Things" (IoT) and "Machine to Machine" (M2M) type of technologies that lead to the emergence of extremely complex Cyber-Physical Systems (CPS), often referred to as "Smart Cities". To support choices for technology deployments in Smart Cities, one has to gain knowledge about the effects and impact of those technologies through testing and experimentation. Hence experimentation environments are required that support the piloting and evaluation of service concepts, technologies and system solutions to the point where the risks associated with introducing these as part of the cities' infrastructures will be minimised.

With this rational, the TRESCIMO (**T**estbeds for **R**eliable **S**mart **C**ity **M**achine to **M**achine **C**ommunication) project deployed a large scale federated experimental testbed across European and South African regions, allowing for experimentation over standardised platforms and with different configurations. Among others, the main requirement for the testbed federation was to cater for the different contextual dimensions for Smart Cities in Europe and South Africa. The testbed is composed of a standards-based M2M platform (openMTC), using standard FIRE SFA-based management tools (FITeagle) and including a variety of sensors and actuators (both virtual and physical). Furthermore, a Smart City Platform attached to openMTC hosts applications for a variety of stakeholders (i.e. experimenters or typical end-users). A series of experiments were conducted with the TRESCIMO testbed to validate the plug-and-play approach and Smart City Platform-as-a-Service architecture. This architecture is positioned to provide smart services using heterogeneous devices in different geographical regions incorporating multiple application domains. This chapter elaborates on, and validates the TRESCIMO testbed by presenting the experimental results and experiences from two trials executed in South Africa and Spain.

7.1 Introduction

Urbanization is a universal phenomenon with cities experiencing a significant growth in population. This in turn is increasingly stressing services provided in cities. Aspects related to the economic, societal and environmental challenges need to be effectively addressed to ensure quality of life of citizens as well as economic and environmental sustainability. Example challenges include finding means to address unstable power supply in cities in developing countries (i.e. South Africa) or ensuring a cleaner and greener environment for both developed (i.e. Spain) and developing countries.

Smart Cities have been touted as a possible solution in addressing challenges in cities. A Smart City is associated with an environment containing sensors and actuators able to observe and influence, and appropriate communications mechanisms into back-end platforms hosting applications. Using the data acquired from the environment, applications can make smarter decisions to the benefit of the city and its inhabitants.

The concept of interfacing with the physical world and linking the data with digital services is referred to as Machine-to-Machine (M2M) and Internet of Things (IoT). In realising a Smart City through M2M and IoT the technological challenges are ranging from developing cost-effective sensors, supporting and maintaining these sensors, creating or using appropriate network connectivity means, utilising fit-for-purpose platforms as well as developing domain appropriate applications need to be resolved. Other aspects related to scale, heterogeneity, interoperability, and adherence to evolving standards complicate the context even more.

Introducing technology just for technology's sake is not appropriate, especially in an environment with financial constraints and with gaps in available resources (people as well as technological infrastructure). In a situation where technology is introduced, care should be taken to ensure the required societal and environmental impact as well. To minimize risk when introducing smart services in a city, especially when moving from a lab to a real world context suitable experiments need to be conducted first. With these experiments a better understanding of the challenges and potential for impact and innovation become possible.

Testbeds for **R**eliable **S**mart **C**ity **M**achine to **M**achine **C**ommunication (TRESCIMO) is a project aimed at understanding the complete context (both technology as well as society) when smart city solutions are created and rolled out in a city. The context also refers to instances where services and solutions might be geospatially far apart and if a service and architecture developed for one area can be utilised effectively in another area. TRESCIMO created an intercontinental research facility using state of the art standards and technologies for experiments associated with the real world.

Section 7.2 presents the TRESCIMO architecture and describes the trials executed in Spain and South Africa. Furthermore the section elaborates on the components used for the trials. Section 7.3 presents the trial experiments and results, while Section 7.4 presents views on the results. Section 7.5 concludes.

7.2 TRESCIMO Architecture

TRESCIMO created experimental facilities in the context of Smart Cities dealing with mass urbanization in both developed and developing worlds. These facilities aimed to identify and implement appropriate architectures for Smart Cities. The facilities also serve as means to investigate the utility and impact of services related to smart and green technological social innovation (e.g. the societal impact in energy management or greener environments).

Four dimensions were considered in TRESCIMO: a federated research testbed, a Platform-as-a-Service Proof-of-Concept, and for validation a Smart Energy trial and an Environmental Monitoring trial. Figure 7.1 depicts the reference architecture for TRESCIMO. Software components were developed that integrate and federate in a plug-and-play manner to experiment with, and address a variety of requirements [4–7] .

Figure 7.2 depicts the architecture and software components used to realise the reference architecture presented in Figure 7.1.

The software components in TRESCIMO utilises state of the art standards (e.g. oneM2M, CoAP, Core-Link, and OMA LWM2M device management) or innovates by leveraging prior art where no clear standards have yet emerged. Based on the needs of a particular set of use-cases the components can be

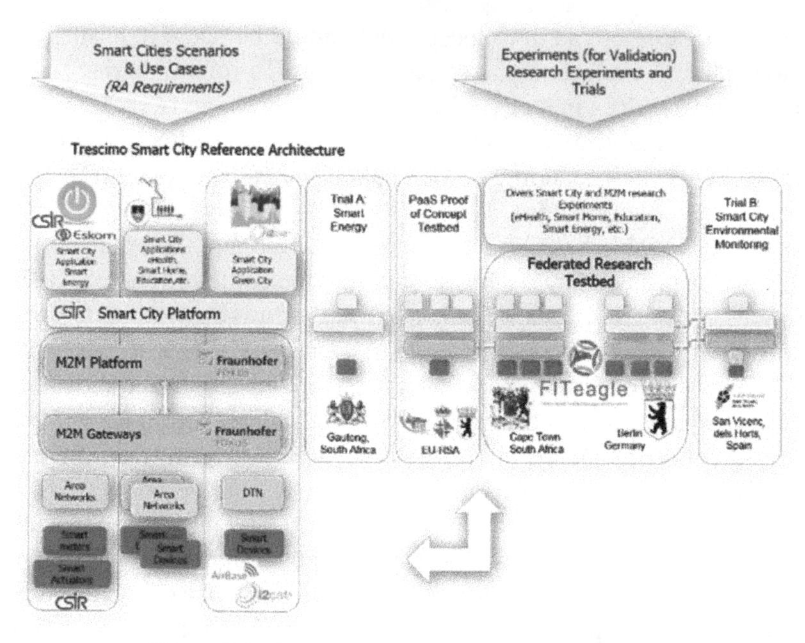

Figure 7.1 Reference architecture and experiments.

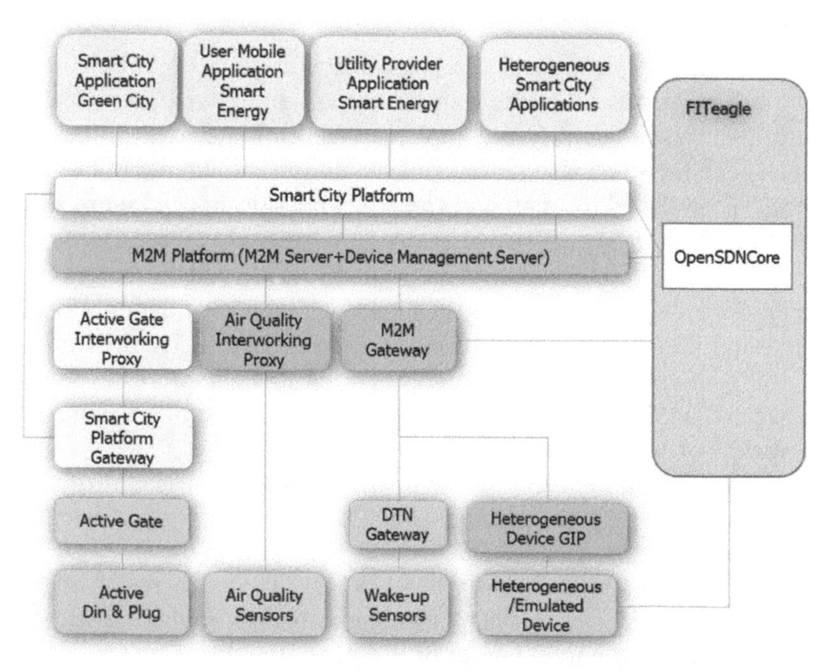

Figure 7.2 Integrated prototype architecture.

integrated by deploying an appropriate combination of software components. To validate the architecture and associated concepts a Smart Energy trial and a Smart Environmental Monitoring trial were conducted. The Smart Environmental Monitoring trial was executed in Vicenç dels Horts, Barcelona, Spain, while the Smart Energy trial was ran in Sandton, Fourways, Sunninghill and Randfontein in Johannesburg, South Africa.

7.2.1 Smart Environmental Monitoring Trial

The Smart Environmental Monitoring trial utilises components as depicted in Figure 7.3. The trial uses smart sensors (wake-up devices and air quality sensors), gateways with delay-tolerant features to activate the wake-up sensors, an openMTC gateway and platform (oneM2M compliant), the Smart City Platform (SCP) and a visualisation application (Green City application). The aim of the trial was to deploy a solution that monitors non-critical environmental and pollution parameters in a city without the need for deploying or relying on purpose built infrastructure.

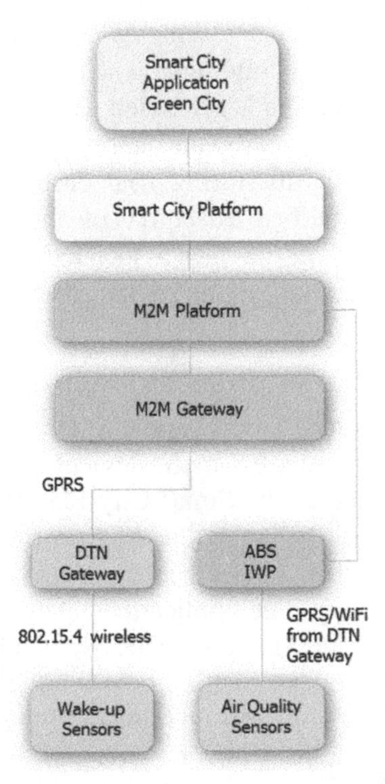

Figure 7.3 Smart Environmental Monitoring use-case architecture.

The system is based on a Delay Tolerant Network (DTN) concept, where a gateway, installed in a public transportation bus, is used as the sole element to collect the data from sensors installed in the city close to the route followed by the bus. To prevent battery powered sensors (installed in light posts, bus stops and other street furniture) from battery starvation while continuously waiting for the next gateway to collect their information, an energy-efficient radio wake-up mechanism has been implemented. This mechanism uses two separate radio interfaces in the low-power sensor nodes and the sensor: an 868 MHz interface that consumes less than 3 μW in listening state; and an IEEE 802.15.4 radio interface that is only active to transmit or receive data. Sensors are mostly in a "sleeping" state (only the low-power radio is active) and isolated (no network is present). When a collector device (the gateway installed on the bus) comes close to the sensors, the communications interface in the sensor is enabled, triggered by the low-power radio, and observations are captured and communicated to a gateway from where they are transferred via the M2M platform, through the SCP and finally to the environmental visualisation dashboard. The radio wake-up mechanism has been designed with enhanced features allowing device addressing and an extended range of tens of meters. In addition, air quality sensors, equipped with a WLAN or GPRS interface, were installed in buildings owned by the municipality since they require continuous power. The DTN-based gateway provides a WLAN interface to collect the data from nearby air quality sensors.

The Smart Environmental Monitoring trial dashboard is presented in Figure 7.4. It provides functionality to a user to view observation readings over time for a specific resource (either the ones associated with the delay-tolerant network or those connected directly to the backend).

7.2.2 Smart Energy Trial

Figure 7.5 depicts the components used for the Smart Energy trial. The trial used Internet enabled energy measurement devices (referred to as Active devices) and a gateway linked to the Smart City Platform via a Smart City Platform Gateway application. The Smart City Platform hosts a web dashboard application for the energy utility as well as a mobile enabling application which is linked to a mobile app. The applications are capable of visualising the consumption and actuate individual devices by switching them on or off based on user demand. The communication between the Active devices and the gateway uses a 6LoWPAN network, while communication to the Smart City Platform uses the 3G cellular network.

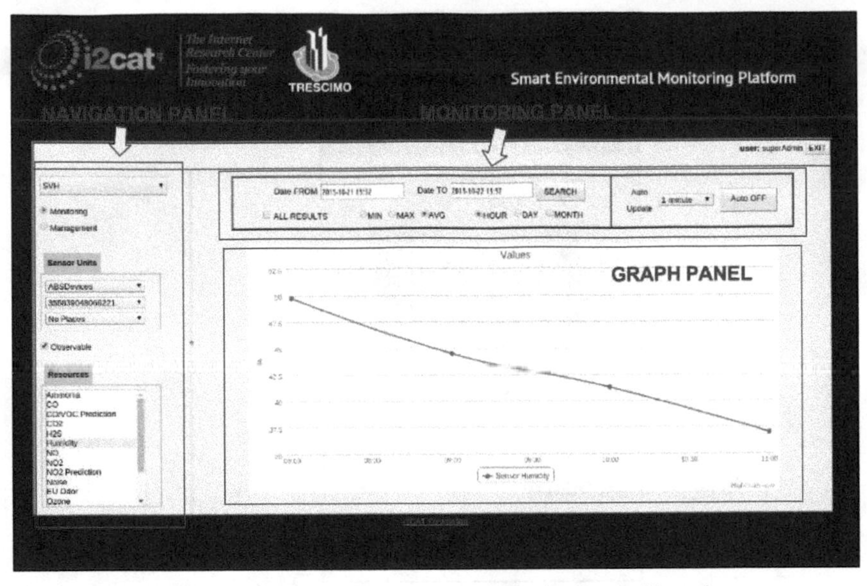

Figure 7.4 Smart Environmental Monitoring dashboard.

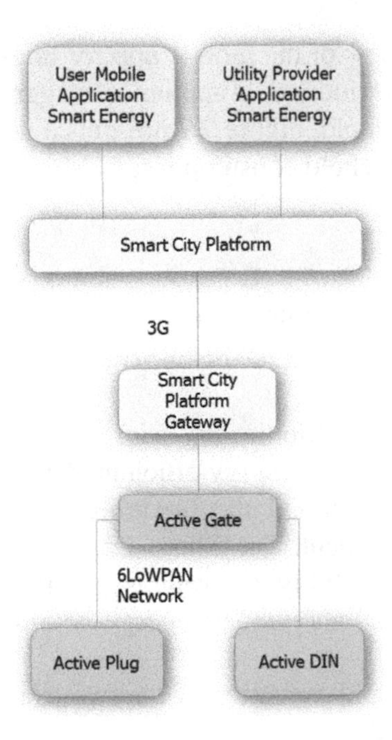

Figure 7.5 Smart Energy use-case architecture.

Figure 7.6 Energy mobile application.

Figure 7.6 presents a view of the mobile app for household owners. The app presents consumption per individual appliance or aggregated consumption for all appliances in a household. Figure 7.7 presents a web dashboard for an alternative view on the household consumption.

7.3 Trial Results

In addition to verifying the TRESCIMO plug-and-play methodology and Smart City Platform-as-a-Service concept, two trials with different aims were conducted.

The Smart Environmental Monitoring trial verified the feasibility of deploying infrastructure-less and energy-efficient data acquisition systems for Smart Cities and demonstrated the functionality of the TRESCIMO architecture in a real deployment. The Smart Energy trial focused on verifying the technological feasibility as well as gaining deeper understanding of customer behaviour when smart energy solutions are installed in households.

Through the validation and execution of the two trials numerous experimental results were obtained.

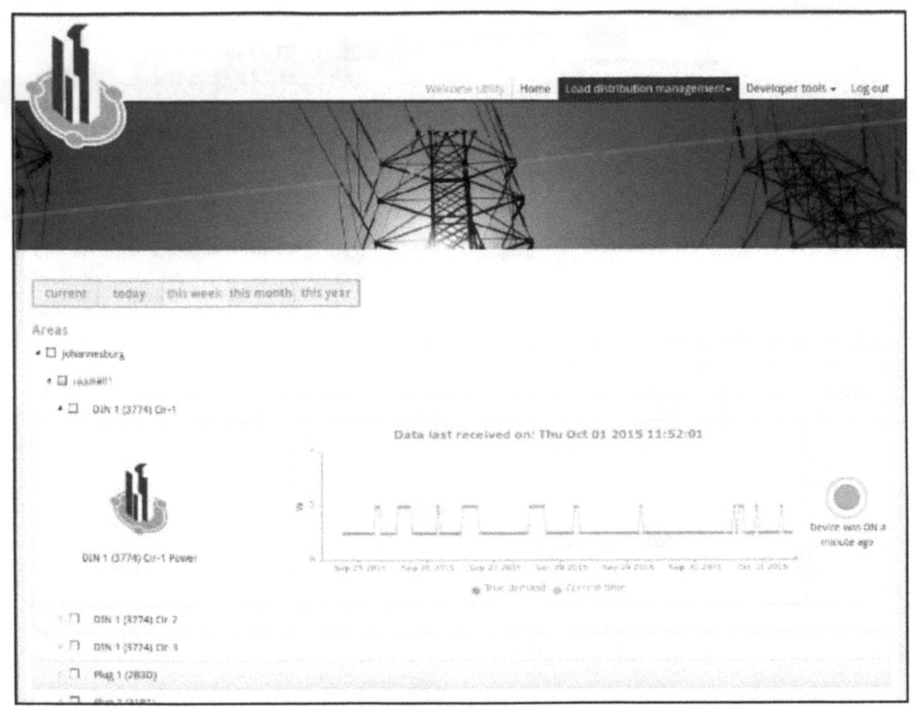

Figure 7.7 Energy web dashboard.

7.3.1 Smart Environmental Monitoring Trial

Figure 7.8 depicts the various components chosen from the TRESCIMO technology stack for the Smart Environmental Monitoring trial.

7.3.1.1 Scenario and experiments

The trial was deployed in Sant Vicenç dels Horts, a Spanish city of about 28000 inhabitants close to a cement factory. Due to this last aspect, the municipality has a special interest in solutions to monitor environmental parameters and pollution in the urban area. The following devices were installed in the city:

- Five devices (provided by Airbase) dedicated to air quality and pollution monitoring;
- Thirty four low-power wake-up devices equipped with batteries and various environmental sensors (light, barometric pressure, temperature and humidity);

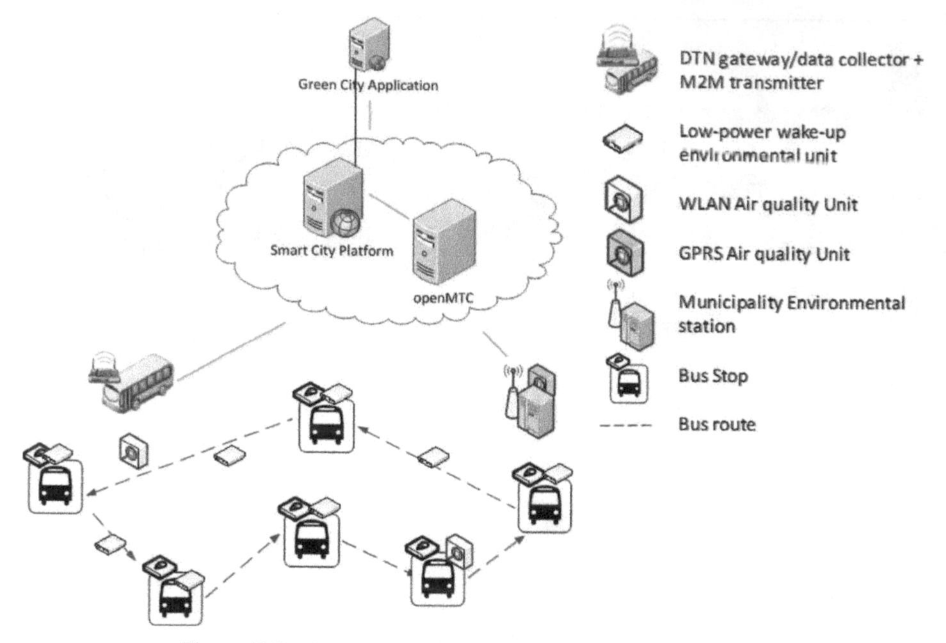

Figure 7.8 Smart environmental monitoring trial use case.

- One gateway device installed in a public transportation bus and two devices installed in additional vehicles to support the evaluation.

Figure 7.9 shows the placement of the sensor devices (in green the wake-up sensors and in yellow the air quality units) and the routes followed by the bus (data collector). The trial began in October 2015 and has been kept running after the finalisation of the project in December 2015.

Figures 7.10 and 7.11 depict several devices as installed in the city on light poles, bus stops and buildings.

Two types of experiments were conducted for the trial:

- **Acquisition of data from the sensors distributed in the city**. The objectives of this experiment are: 1) to prove that data can be collected in a delay-tolerant manner; 2) to study the performance of the DTN and wake-up based system in a real scenario and 3) to provide environmental data, that is useful to the municipality as a potential end-user of the solution, for surveillance or informational purposes. This information will serve also as input for future experimenters (e.g. to test or validate algorithms against real data).

Figure 7.9 Routes followed by the bus and location of the sensor units: low-power wake-up (green) and Airbase air quality (yellow) devices.

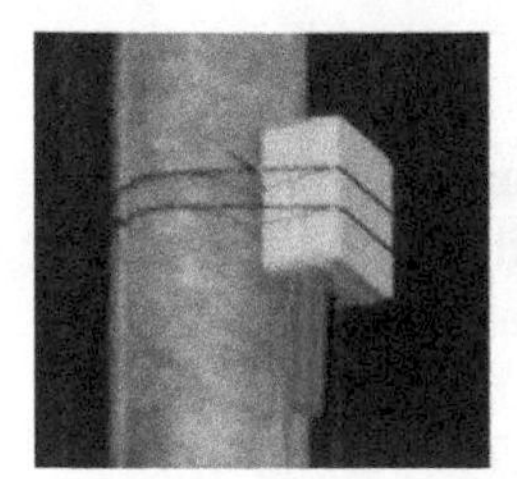

Figure 7.10 Sensor devices. Barometric wake-up device (*left*). Temperature, humidity and light wake-up device (*center*). Airbase WLAN air quality device (*right*).

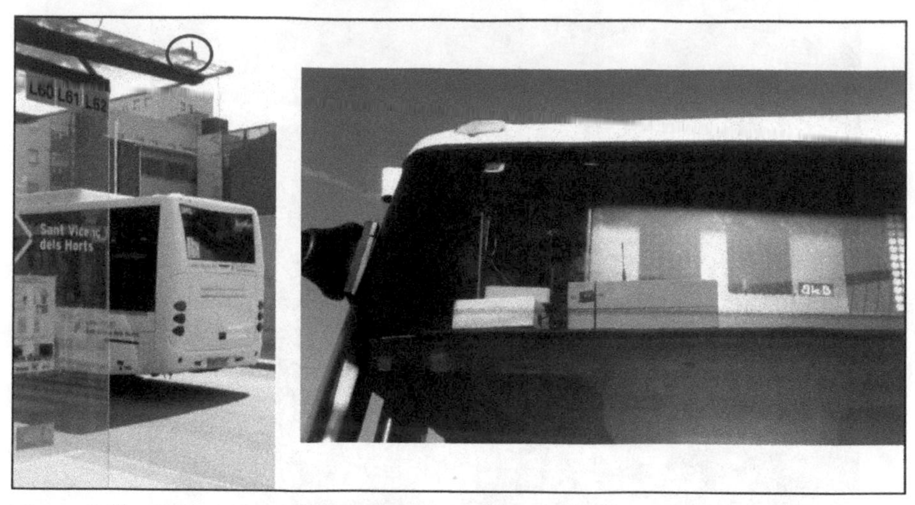

Figure 7.11 Delay Tolerant Network devices. Bus passing close to a low-power wake-up sensor device installed in a bus stop. Detail of the equipment (gateway) installed in the bus.

- **Communication from the collector to the low power devices**. The aim is to validate the bidirectional communication between the collector and the wake-up sensor devices. Bidirectional communication allows the collector to gather data and to interact with the devices (e.g. for reconfiguration, performing firmware updates over the air or polling) in a delay-tolerant manner.

For the Smart Environmental Monitoring trial, the following Key Performance Indicators (KPI) have been identified:

- **Acquisition of data from the Wake-up and Airbase sensors and functionality validation of the full stack**. The data monitored from the sensor devices should be collected and forwarded through the TRESCIMO architecture. It should be possible to view the information using the client web interface (Figure 7.4).
- **Device energy consumption (for the wake-up sensors)**. Wake-up devices should provide proof of low consumption and maximizing of their battery lifetime and, thus, minimize the cost of maintenance of the installed devices.
- **Communication range (for the wake-up sensors)**. This parameter is directly linked to the scalability and flexibility of the solution. The range must be large enough to confirm that a moving vehicle can collect the information without the need for stopping or reducing its speed.

Furthermore, the range of the solution determines the size of the area where sensors can be installed and, thus, the amount of devices that can be supported by each route. Before installing the wake-up units in Sant Vicenç dels Horts, individual tests were performed in a controlled scenario with the transmitter and the receiver in Line of Sight (LoS) conditions to get an idea of the optimal performance (best case) in terms of range that can be expected. The minimum range observed in these experiments was about 36 meters and almost 50% of the devices responded to wake-up signals at a distance of 50 meters or greater. The expected performance in the real scenario should be close to these values.

- **Communication time window and amount of data that can be transmitted or received during the wake-up process**. These figures can help to determine how much information can be sent from the sensor devices to the collector in the bus (data gathering) and in the opposite direction. These figures establish the capabilities of the system to support device configuration or firmware updates over the air.

For the evaluation of the trial, the following tools and inputs for the analysis are used:

- Tracking of the data monitored by the sensor devices; namely, environmental parameters, battery consumption, and timestamp when the collector module in the DTN-gateway acquires the data. This information is stored during the trial and can be retrieved from the Smart City Platform (SCP). It can be visualised by a user through the web visualisation dashboard interface (Figure 7.4).
- Tracking of the GPS location data on the buses. Location and timestamp observations are sent each time a wake-up process is triggered so that it can be correlated with the wake-up process and with the reception of the sensor data. This information is stored during the trial and can be retrieved from the Smart City Platform (SCP).
- Tracking of the functionality of the wake-up mechanism. The following parameters are recorded for each wake-up process: timestamp when the wake-up node responds to the triggered radio signal, number of attempts performed until a successful wake-up is received, distance to the sensor node, and unsuccessful and unexpected wake-ups. The distance to the sensor node when a data message is received is an indicator of the effective communication range. Unsuccessful wake-ups are determined when the transmission of the wake-up signal exceeds a given number

of retries, which is a configurable parameter in the DTN-based gateway. This information added to the statistics about the number of attempts performed for the nodes in the trial provide insight into the performance of the wake-up mechanism. Unexpected wake-ups indicate the reception of data from a node that has not been prompted; this can help to detect interferences from external sources that might affect the performance of the overall system.

7.3.1.2 Evaluation results

Key results obtained from the evaluation of the Smart Environmental Monitoring trial taking into account the aforementioned Key Performance Indicators are presented in the following subsections.

7.3.1.2.1 *Visualisation and monitoring of the data transmitted by the sensor devices*

A subset of data monitored during the trial is shown to illustrate the end-to-end performance of the system. The monitored samples were obtained by the wake-up low power sensors and the Airbase air quality devices. Note that data is sent by the devices, collected by the gateway, forwarded by the openMTC platform and stored in the Smart City Platform (SCP); thus, the full TRESCIMO architecture can be validated. Further results have been reported in the project deliverable which is publicly available [3].

Figure 7.12 illustrates the visualisation of data monitored by the wake-up sensor devices during the period from November to January. Sensors were programmed to capture instantaneous data samples only when the wake-up is performed. Thus, connectivity gaps at night and on Sundays are visible. The operation of the system during the trial months was also affected by the unavailability of the bus due to mechanical problems and maintenance operations. This prevented the gateway from collecting data from the sensors for hours or even days at a time. The information gap observed in the web application from the 20th to the 28th of November 2015 is a result of this.

Figure 7.12 displays the changes of the temperature in Device_16. The device is installed on a light pole that has direct solar exposure. The first week of November has been especially warm in Barcelona and its surroundings. This explains the high values (above 30°C) monitored by the temperature probe. It is noticeable how the maximum temperature dropped during December and January, as would be expected for the winter season.

Figure 7.13 displays the NO_2 hourly average measurements captured by one of the Airbase air quality stations. The Airbase devices allow data sampling

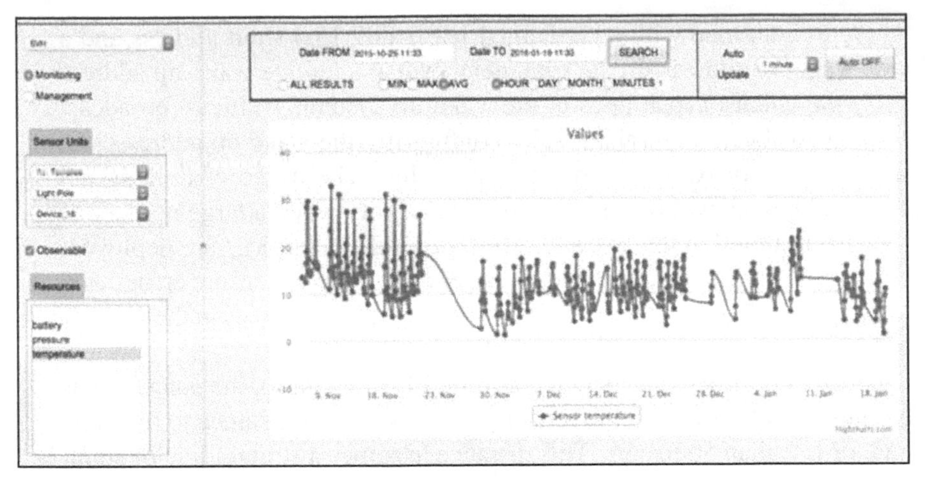

Figure 7.12 Temperature measurements captured by a low power wake-up device in Sant Vicenç dels Horts.

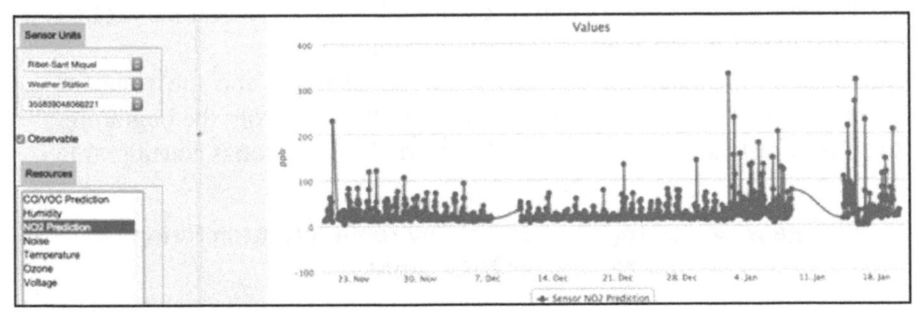

Figure 7.13 NO_2 measurements captured by the Airbase air quality device.

and storage while no network is available. Devices were configured to obtain a measurement every 10 minutes. Spikes whose values are slightly over the recommended healthy limit are noticeable. According to the EPA Air Quality Level [2], values above 101 ppb over one hour period are considered unhealthy for sensitive groups. Though spikes appear in a spurious manner, a continuous surveillance of the air quality will be useful to the municipality to control their repeatability and analyse their possible causes.

7.3.1.2.2 *Performance of the DTN and wake-up system*

As commented previously, one of the enhanced features of the deployed wake-up system is the support for device addressability. Each wake-up sensor device has been programmed with a predefined IEEE 802.15.4 short address and a 2-bytes wake-up address. In the trial, devices use a unique wake-up address to verify the unicast capabilities of the wake-up system. Multicast/broadcast addressing has been also validated by configuring the wake-up addresses of devices in close proximity with the same value. The usage of unicast and multicast addresses will be an interesting capability when a large number of sensors are installed in the city and different kinds of services are deployed. In this way, it is possible to wake up a sensor or a group of sensors on demand (for example, for configuration needs), while the rest of devices in the vicinity remain in low power mode.

To obtain empirical results in a controlled LoS scenario, the gateway has been configured to wake-up the sensor devices when its distance to the units is equal or less than 50 meters. This distance assumes a straight line of sight; however, in a real deployment the distribution of streets, driving directions and objects (buildings, other vehicles, and traffic signals) act as obstacles in the communication between the gateway in a moving bus and the wake-up sensor. To improve the success rate of communication in such uncontrolled scenario, the gateway can execute several wake-up attempts.

Figure 7.14 illustrates the average wake-up distance and the standard deviation (in meters) for the sensors involved in the trial from the beginning of November until the end of April. As observed, the deviation is considerable

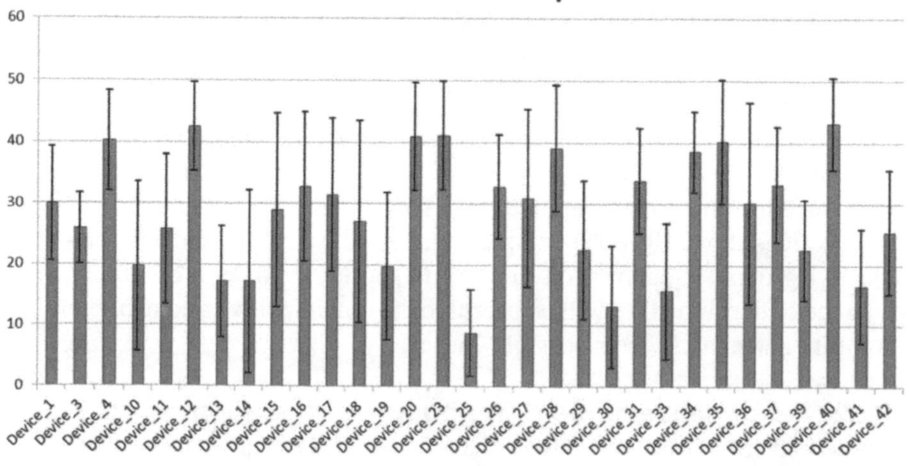

Figure 7.14 Average wake-up range of the DTN and wake-up based solution (in meters).

in all the cases; however, this is an expected result in a real mobile scenario where the performance of the communication can be affected by a multitude of external and variable factors. Most of the sensor devices show effective wake-up distances greater than 20 meters. A small number of devices show a poorer result. This can be explained because of their location on street edges, turnarounds or behind traffic signals. Collectively, the mean range observed for all the devices over the full trial period is greater than 28 meters. This confirms that wake-up technologies are a feasible option to retrieve information from the city. Wake-up nodes installed behind a traffic light or a street crossing sign experience a lower performance in terms of effective range and higher percentage of unsuccessful wakeups. Unsuccessful wake-ups can occur due to two reasons: (a) the maximum number of wake-up attempts is reached or (b) the bus goes out of the wake-up range of the sensor. The first cause can be explained by the bus turning a corner without direct visibility to the sensor device, especially if the bus comes from a non-preference road or there is a traffic light that forces a stop for a long duration. In the second case, it should be noted that the amount of time the vehicle is in the range of the sensor and, thus, the possibility to wake the device up and establish communication will decrease with higher speeds. On average, the percentage of unsuccessful wake-ups is below 8%. This can be considered a good performance in a real deployment and under non-ideal and variable conditions. Finally, in almost all the cases a maximum wake-up range exceeding 48 meters was observed. The significant wake-up range validates the promising capabilities of the wake-up mechanism implemented and deployed in the trial. These results serve as input to determine what the best locations for the sensor devices are. The results provide insight into the optimal settings to maximize the performance of the wake-up system and to infer some recommendations that can be useful for future deployments.

7.3.1.2.3 *Consumption of the wake-up sensor devices*

In the trial, the battery consumption of the low-power devices is reported as a parameter in every data message. A trend over time can be visualised and monitored. The energy usage of the device sensors over the long-term can thus be monitored. Figure 7.15 provides a screenshot of the Smart Environmental dashboard interface showing the average daily battery consumptions from November to January for a sensor device (Device_41). The fact that no relevant battery drops are observed in this period confirms that the device energy consumption is performing as expected and that the devices are in a low-power mode status most of the time. Note that the nominal value of the battery used for the wake-up sensor devices is 3.6 Volts.

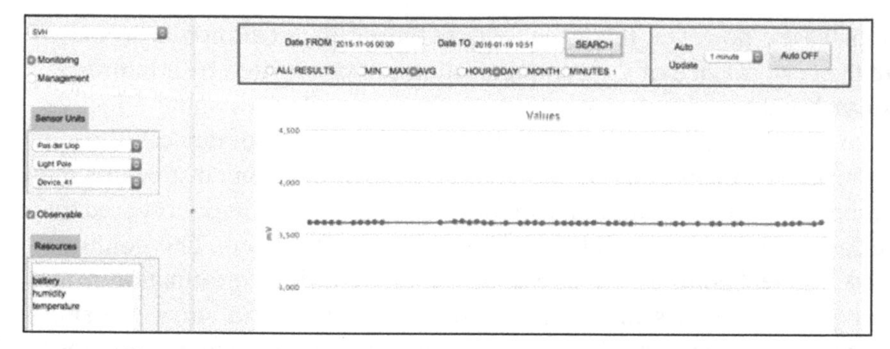

Figure 7.15 Battery evolution of the low power wake-up devices.

7.3.1.2.4 *Performance of the data collection process and device update capabilities*

Tests were performed to determine the communication time window and the amount of data that can be transmitted in the uplink direction (from the sensor device to the DTN-based gateway). To conduct this test the sensor device was configured into a mode where packets are sent in a continuous manner to the coordinator in the gateway. Tests were performed at several speeds to simulate different scenarios. To allow for repeatability of the test and provide more flexibility to control the speed of the mobile gateway, a particular vehicle was used for this evaluation. The experiments were performed using one of the sensors of the deployment installed on a lamp post and in the middle of a straight street (to maximize the visibility between the gateway and the sensor device). From the results obtained it can be concluded that the infrastructure-less system implemented in the trial allows the devices to store and, at a later time, send a considerable amount of data (between 30 and 40 kB) at a speed of 30 km/h between two consecutive bus journeys. This is interesting for a real world deployment as the frequency of public transportation might be notably low; for example, as in the case of the trial, some buses do not drive over the weekend.

To validate the bidirectional functionality, it was confirmed that the gateway is capable of changing the sampling rate and the wake-up address of the sensor units in a delay-tolerant manner. Furthermore, it was possible to send a message to the unit to reboot it and to query its current firmware version and configuration settings. By default, the wake-up sensor device operates in low-power mode; thus, once the wake-up is performed, a data request to the coordinator in the gateway is performed requesting data. At that moment, the configuration message is sent to the sensor unit. Once received, the wake-up device needs to confirm the instruction with an acknowledgement (either positive or negative) that the operation has been completed and the setting has been updated or discarded. When several consecutive packets need to be transmitted to the environmental equipment (e.g. to perform an over-the-air firmware update), the DTN-gateway would send a message to the gateway to indicate that it must switch to active mode (always listening) so that data is transmitted faster. As the IEEE 802.15.4 link is peer-to-peer and symmetric, the amount of data that can be transmitted during a wake-up process is equivalent to the results obtained in the bulk data tests performed from the sensor unit to the gateway.

7.3.2 Smart Energy Trial

In the Smart Energy trial (Figure 7.16), 30 Eskom households were equipped with the Active devices for monitoring the energy consumption (one Active-Gate using 3G backhaul to the Smart City Platform, two ActivePlugs for appliances and an ActiveDIN used for higher current appliances such as a geyser or pool pump).

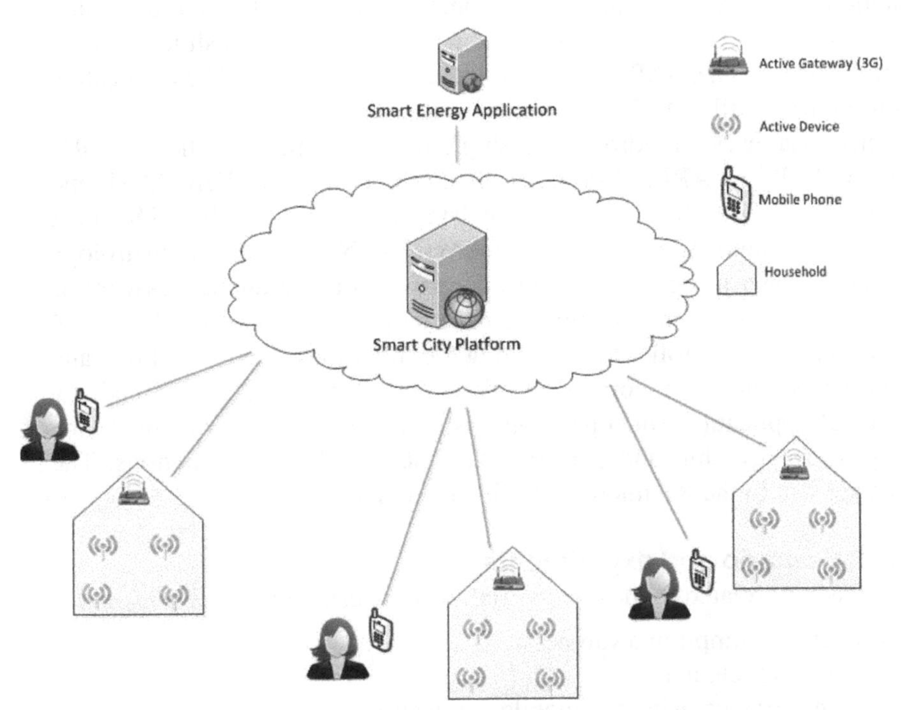

Figure 7.16 Smart Energy trial use case.

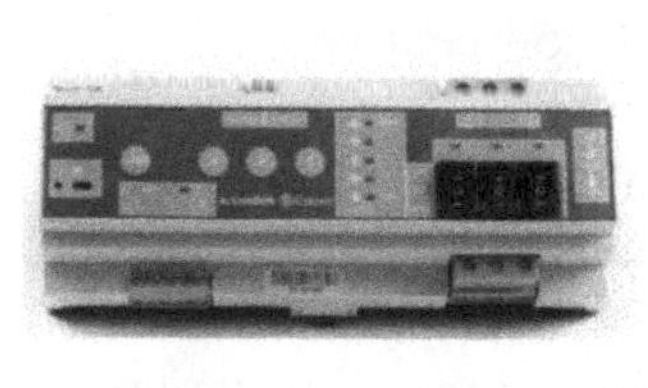

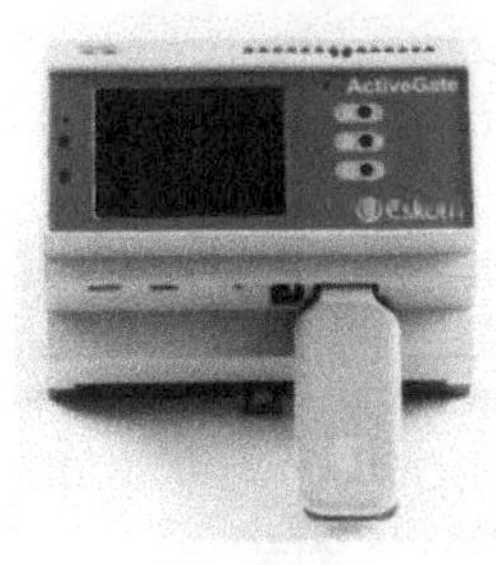

Figure 7.17 Active devices (ActiveDIN on the *left*, ActivePlug *center* and ActiveGate on the *right*).

ActiveGate is a processing and routing platform, while the ActivePlug and ActiveDIN are energy management devices. Figure 7.17 depicts the ActiveDevices. In addition, a household owner had the Smart Energy mobile app installed on his smart mobile phone.

The three devices (ActivePlug, ActiveDIN and ActiveGate) communicate using a 2.4 GHz 802.15.4 radio module based on the STM32W108 System-on-Chip (SoC) from STMicroelectronics. The RF microcontroller (the STM32W108 SoC) performs the low power wireless mesh networking function and hosts a CoAP server with the device resources. The application is built in the Contiki-OS framework.

ActiveGate uses an Odroid-U3+ single board computer with a 1.7 GHz Exynos4412 Prime ARM Cortex-A9 quad-core processor, 2GB RAM, and various external interfaces. The ActiveGate runs Ubuntu 14.04 LTS Linux as operating system. The ActivePlug and ActiveDIN use STPM01 metrology circuitry for measuring voltage, current, power, line frequency as well as active, reactive, apparent, and fundamental energy consumption and an ARM Cortex-M4 microcontroller for managing the metrology, load switching, and interface functions. The Cortex-M4 microcontroller from Atmel contains a bare-metal application (no operating system) that continuously reads the energy metrology chip and performs the energy related calculations. The results are sent to the RF microcontroller at a rate of 2 Hz.

7.3.2.1 Scenario and experiments

Four aspects as related to the energy trial were investigated:

- Energy consumption awareness;
- Behavioural change;
- User experience using the mobile application, and
- Technology performance metrics.

To gain understanding into the homeowner, questionnaires were utilised (one during installation and another during decommissioning). The questionnaires also served as platform for the trial participants to voice their opinions regarding the particular technology solution and similar systems in general. Technology performance metrics were obtained through experiments and measurements through the stack using the various physical installations.

7.3.2.2 Evaluation results

7.3.2.2.1 *Energy consumption awareness*

Table 7.1 presents results as extracted from the pre-trial questionnaire in relation to awareness. It should be noted that all the participants were from a high "Living Standards Measure" category and also had pre-existing smart meters installed.

Trial participants responded as follows in the post-trial questionnaire (Table 7.2):

An important aspect highlighted is the participants' energy consciousness. In the context of the energy constraints during the trial this is insightful as it implies that through this technology people can become even more cognisant of energy limitations.

7.3.2.2.2 *Behavioural change*

As the trial participants already had smart meters installed, comparisons over the course of the trial with readings from the year prior to the trial were possible. Results indicate that no clear and consistent change in consumption was visible. The consumption was varied and ranged from significantly increased consumption, significantly decreased consumption, and very small changes. This indicates that users in general did not utilise (or were not able to utilise) the smart mobile app to control their load. However, load control

Table 7.1 Pre-trial questionnaire summary

	Yes	No
Awareness of energy consumption: *Do you track your consumption?*	62%	38%
Response to behaviour change request: *Do you respond to TV and radio power alert requests to switch off appliances when requested?*	80%	20%
Willingness to change behaviour: *Would you change your consumption patterns for reduced rates or rebates?*	85%	15%
Device control: *Do you have timers for control of devices installed?*	71%	29%
Control preference: *Do you prefer to switch your non-essential loads yourself?*	86%	14%

Table 7.2 Post-trial questionnaire summary

	Yes	No
Energy Consciousness: *Are you more energy conscious than before the trial?*	69%	31%
Change in consumption: *Did you notice any changes in your consumption?*	Reduction: 54% Increase: 0%	No change: 46%
Motive for change: *What will potential motive for change be in response to reduced rates or rebates?*	Financial: 31% Security of Supply: 46% Social: 8% Security of supply and financial: 46%	
Control preference: *Do you prefer to switch your non-essential loads yourself?*	85%	15%
Communication Medium: *Would you prefer to receive messages via your cell phone or rather alerts via TV or radio?*	100%	0%

was possible and utilised by some participants as illustrated in the following two figures. Figure 7.18 depicts consumption readings on a geyser (hot water boiler) where the household occupant did not control its appliance. This is in contrast to Figure 7.19 where the occupant did choose to intervene and control when the geyser should be switched on.

7.3.2.2.3 *Mobile app*

Trial participants only had access to information from connected appliances via the mobile app as depicted in Figure 7.6. The web interface as presented in Figure 7.7 was used by the project partners to verify operation of the trial components. Results indicate low utilisation of the mobile app. This can be attributable to challenges experienced with the mobile app itself. For example it was reported that quite often login via the app was problematic. Furthermore a low general interest was observed in gaining access to the current state of consumption. User utilisation varied considerably. Results indicate that four trial participants made use of the app (two significantly more than the other two), while most trial participants did not.

Participant 7 logged in 196 times with 45 "on" and "48" off commands. Participant 12 logged in 79 times with 42 "on" commands and 40 "off" commands. Participant 18 logged in 208 times, with 95 "on" and 79 "off" commands. The fourth participant logged in on 51 occasions and executed 14 "14" on and 16 "off" commands. Viewed in conjunction with Figure 7.20, participant 7 and 18 experienced good uptimes of the complete system.

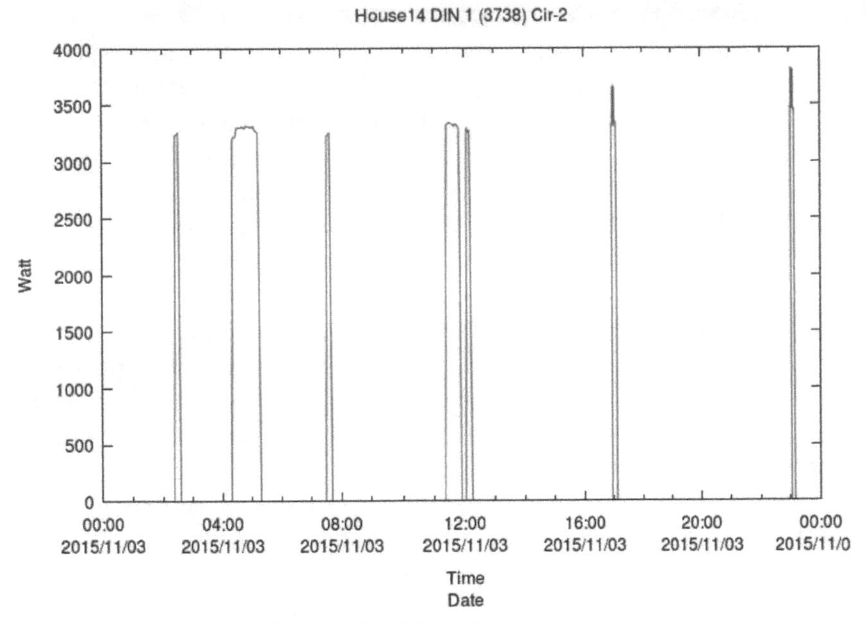

Figure 7.18 No appliance control.

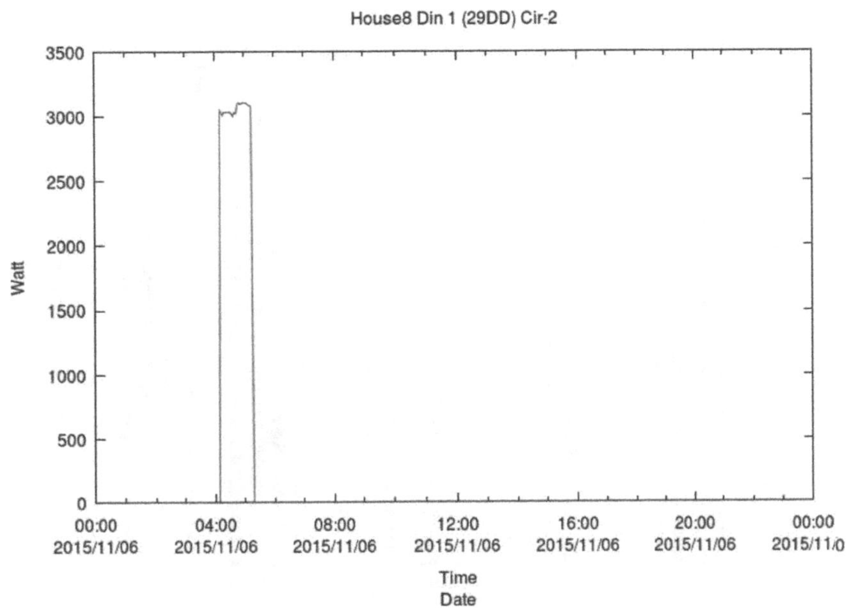

Figure 7.19 Controlled appliance.

7.3.2.2.4 Technology performance metrics

The technology performance metrics reveal a number of interesting aspects. These are attributable to the stability of the technology and communication effectiveness, as well as constraints related to trial participant access during the trial. Figure 7.20 depicts the measured uptime per household during the duration of the trial. The uptimes vary considerably within households. The uptimes are calculated based on the number of observation data points captured in the database (i.e. data flow throughout the complete stack from sensor to application). This measurement is a good indication of the overall performance of the technology stack. However, no conclusions can be made as to which component impacted on the performance when challenges were experienced. For instance, what in the stack prevented data flow (i.e. was it a failure in backhaul connectivity, a device that has gone down or unavailability of other components in the stack)?

Throughout the duration of the trial, updates of software on the accessible ActiveGates were done. This included monitoring and control software able to detect if a software component has failed and, thus, needs to be restarted. However, this functionality and new software releases could only be installed on those devices having adequate communication. Uptimes in

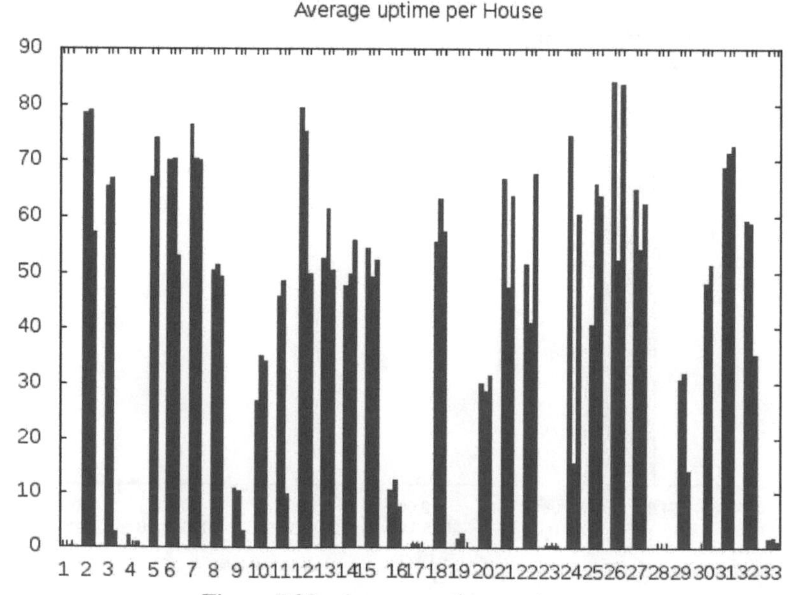

Figure 7.20 Average uptime per house.

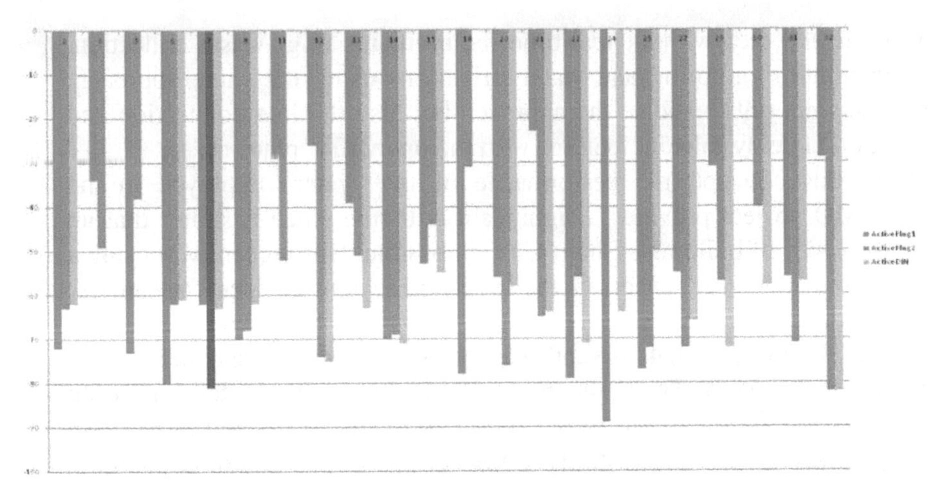

Figure 7.21 6LoWPAN.

general improved for installations with good communication, while those with poor communication (who would have benefited more from the updates) were limited to the initial configurations and releases. The Active devices made use of 6LoWPAN connectivity within a household. A significant variation in signal strength is visible between devices in a household as well as between households. Figure 7.21 depicts RSSI measurements in some households. Within households the signal strength varied significantly.

7.4 Discussion

7.4.1 Smart Environmental Monitoring Trial Observations

In relation to the system performance, the results of the Smart Environmental Monitoring trial obtained so far are promising and confirm the expectations. It demonstrated that solutions based on radio wake-up systems and DTNs allow for information collection while minimizing the number of devices that need to be deployed and maintained. Sensor devices have been designed to ensure energy-efficiency and maximize the battery lifetime and as a consequence reduce the operating expense (OPEX). Furthermore, the improvements achieved by the project with the enhanced wake-up system led to communication ranges of more than 40 meters (28 meters on average). Experiments confirm that the range is sufficient to retrieve data from a moving vehicle. Finally, addressing techniques permit to univocally determine the sensor device to be woken up; this opens the possibility to deploy differentiated services in the city without real-time requirements (e.g. waste collection, environmental monitoring and water irrigation) using the same approach. The results not only validate the approach but also the interconnection and integration of delay tolerant features with the openMTC platform.

With this concept, the performance of the system deployed in the TRESCIMO project provides a significant outcome since it shows that an alternative way of building a Smart City is possible. Until recently, sensing a city required deploying sensors on the street and a set of devices (forwarders) that can collect data from sensors and transfer the data to a collecting point (gateway). From there, data is sent to a central element where data is stored and can be processed. The deployment of forwarders and gateways in the city is costly since they need to be connected to the mains and in the case of the gateway to have connectivity to the network. The solution used in the Smart Environmental Monitoring trial solves some of the difficulties listed. It suppresses the use of forwarders and instead uses gateways installed on vehicles (public transportation buses in this case) that move along the city. The installation and maintenance of a gateway is much simpler since it can be done when the bus is in the garage where sufficient power is readily available. The main limitation of the solution is the lack of real-time reporting. This is the reason why this solution is described as being delay tolerant. However, there exist many smart city services without real-time requirements (e.g. environmental monitoring, garbage collection, street furniture maintenance, water irrigation, and smart meters) to which this solution is applicable. The approach can have a further impact since the bus can be equipped with sensors that measure relevant parameters while the bus is moving. This offers an enhanced paradigm for data acquisition; often sensors capture data at a fixed location while through the instrumented bus sensing becomes possible along a variety of routes.

Another outcome from the project is the availability of an experimental network in the city. The equipment deployed in the city is and will remain available to any experimenter. In fact, sensors can be accessed quite easily to retrieve data from them directly since very simple mechanisms are used. Also, the gateways on the vehicles are integrated with the openMTC platform; so their resources could be accessible by a third party through the M2M platform.

A relevant outcome is the municipality recognition. The city is close to a cement factory and citizens are concerned about air pollution. This is an issue in the municipality and proof of this is the fact that the city has two fixed environmental stations, one from the autonomous government and one from the cement factory. This is very rare, since most of the municipalities in Spain have no monitoring station at all. The usage of the TRESCIMO technology provides detailed environmental monitoring. This allows citizens to be more aware of pollution in the environment. The municipality of Sant Vicenç dels Horts are pleased with the experience gained in the trial and is convinced that the model supports the building of "more cost effective Smart Cities" based on delay tolerant networks. Using a delay tolerant approach is more suitable for a medium size city than deploying and maintaining a purpose built infrastructure.

7.4.2 Smart Energy Trial Observations

Important aspects and learning were gained in the process of running the South African Smart Energy trial. Actual residential Eskom customers were included in the trial. This necessitated approval from a number of business divisions within Eskom. It also implied that intrusion into the participant's home and daily lives be kept to a minimum and that mechanisms were in place to provide training to the customer, provide continuous support (in the form of a call centre), minimize any possible risk to the participant's property and ensure that the household was restored to the same state upon decommissioning. In minimizing intrusion into the participant's home (a total of only three in-person engagements were done per participant), support and maintenance of devices and gateways could only be done online. This in itself created a problem when a device was offline as no means were available to reset a particular device. It also became clear during the duration of the trial the inherent tension in providing a near perfect operational environment where all risks were removed against a research and development context where failures and downtimes are expected (in hardware, communication, as well as services).

A number of challenges were experienced during the trial. Most significantly backhaul connectivity from the household to the Smart City Platform proved to be a challenge. The trial used cellular communication hosting a VPN connection. Cellular coverage in South Africa varies significantly. In the trial, bandwidth throughput to gateways varied from 1.3 Mb/s to only about 20 Kb/s. In some cases, no connectivity from household to backend was possible. Naturally, the low bandwidth was problematic as connectivity was intermittent, over-the-air updates were difficult and interaction through the gateways at times almost impossible. This however is a valuable observation and result from the trial. The assumption has been made that cellular connectivity would be sufficient, but it is not the case, thus requiring other connectivity solutions in addition to the cellular network. A DTN solution as for the Environmental Monitoring could have been useful, but it was not planned and thus not deployed in South Africa.

The 6LoWPAN signal strength in a household varied significantly and was in some cases very poor (depending on where the devices were installed relative to the gateway). This affected uptimes and data flow. The current gateway made use of an internal low gain antenna. The signal strengths indicate that this is not adequate. In lab setups and testing, gateway external antennas were used. With the external antennas, the stability and uptimes were excellent, in some instances six weeks went by without any communication failures. This implies that in future experiments the gateway will have to be fitted with an external 6LoWPAN antenna, in addition to an improved backhaul connection mechanism.

The trial was impacted by hardware failures, in particular Channel 1 in a number of ActiveDINs failed when under high load. This required an electrician to replace the ActiveDIN, or rewire Channel 1 to an unused channel.

The mobile app served as a means for the participant to access his own energy consumption. In minimizing possible disruption to the participant, a choice was made to use a trial specific email address for user authentication. This in retrospect was problematic as the user often defaulted into using his personal email with the result that he was not able to log in. Results from the trial were further skewed due to the downtimes experienced in connectivity. It can be noted that participants made use of the app where reliable connectivity was available. However, a broader set of results would have been possible if enhanced uptimes were obtained throughout the trial.

User experience from the trial was predominantly positive. Feedback indicated that opportunities exist to enhance the system (in hardware and software service reliability, connectivity, look and feel, and ergonomics of the devices), but also that the utilisation of next generation smart devices using the latest standards such as 6LoWPAN, CoAP can form the basis of smart demand side management solutions.

Through the trial, insight into the participant behaviour was obtained. Awareness of energy consumption was raised. Feedback from the participants also indicated that they would prefer to control their own environments and not have the utility do so remotely. Given this it is interesting to note that this was not a function often used by the participants.

7.4.3 General Observation

A key observation from the results presented is the utility and functionality of the TRESCIMO testbed. The aim was to come up with a plug-and-play approach supporting reconfiguration based on needs of a specific context. In the execution of the two trials, different components were used to experiment with and gather the results. The results obtained and the ability to execute the two very different trials supports the TRESCIMO approach and usability which were key requirements in the TRESCIMO vision.

7.5 Conclusion

The masses of people moving to cities are straining services provided by those cities. Smart City concepts are required to enhance the efficiencies of existing services, or to create new services. The impact and value of services are not always well understood. Similarly, the technologies and architectures required to actually implement those services are still evolving. To address these issues (i.e. to experiment with appropriate architectures in cities with applications introducing value) real world experimentation is required. TRESCIMO has created an international, intercontinental research testbed aimed at creating such an environment and to also validate the technologies, services and better understand the societal value introduced through these services.

The TRESCIMO architecture is based on standardized protocols and technologies where they exist, or by creating new innovative solutions for the technology stack where no standards exist or technology gaps are present. The architecture resulted from efforts to define and implement a reference M2M solution, which could be adapted and applied to very diverse use cases and scenarios and to different contexts. To prove the validity and flexibility of the solution, two trials were conducted, each with different aims.

In Spain, a Smart Environmental Monitoring trial was deployed that focused on the usage of an infrastructure-less system based on delay-tolerant networks to supervise environmental parameters and air pollution in Sant Vicenç dels Horts. The results of the trial led to promising results and have raised the interest of the municipality. The city is surrounded by several factories and, thus, pollution is a critical issue for its citizens. Furthermore, the proposed solution does not require a big investment in infrastructure and would be applicable to multiple services in the city that do not rely on real-time requirements. This aspect is very interesting for small and middle-sized cities that usually have limited resources. Finally, the results in Spain could in future open doors for new technology possibilities in South Africa; for example, the same approach used for infrastructure-less sensing can rapidly be deployed.

The South African trial focused on Smart Demand Side Energy Management. It explored the technical feasibility of components from the TRESCIMO technology stack for monitoring and managing instrumented appliances in a household. Furthermore, the trial interfaced with household occupants to better understand their needs and the perceived value of having access to Smart Energy management systems. The technology components used for the trial were validated and showed that such systems can be utilised in a developing world context. It further showed that these types of solutions have value to the occupant, given that the reliability of the technology is at an acceptable level.

Acknowledgments

The TRESCIMO project has received funding from the European Union's Seventh Programme for research, technological development and demonstration under grant agreement no. 611745, as well as from the South African Department of Science and Technology under financial assistance agreement DST/CON 0247/2013. The authors thank all the TRESCIMO partners for their support in this work.

References

[1] A. Corici, R. Steinke, T. Magedanz, L. Coetzee et al. "Towards Programmable and Scalable IoT Infrastructures for Smart Cities", *13th International Workshop on Managing Communications and Services*, March 14–March 18 2016, Sydney, Australia.

[2] Nitrogen Oxides (NOx), Why and How They are Controlled, US Environmental Protection Agency Office of Air Quality Planning and Standards, EPA-456/F-99-006R, November 1999.

[3] D.4.3. TRESCIMO Experiment results. Available at: http://trescimo.eu/wp-content/uploads/2015/11/D4.3-TRESCIMO_final.pdf

[4] D.2.3. Final Requirements and scenarios. Available at: http://trescimo.eu/wp-content/uploads/2015/01/TRESCIMO_D2-3_v1.0.pdf

[5] D.2.1. TRESCIMO Scenario Specification. Available at: http://trescimo.eu/wp-content/uploads/2015/01/TRESCIMO_D2.1_v1.0.pdf

[6] D.2.2. TRESCIMO User and Technical Requirements. Available at: http://trescimo.eu/wp-content/uploads/2015/01/TRESCIMO_Deliverable2_2_Ver1.1.pdf

[7] L. Coetzee, A. Escobar, A. Corici et al. "TRESCIMO: European Union and South African Smart City Contextual Dimensions", 2015 IEEE 2nd World Forum on Internet of Things (IEEE WF-IoT) – Enabling Internet Evolution, December 2015, Milan, Italy.

8

BonFIRE: A Multi-Cloud Experimentation-as-a-Service Ecosystem

Michael Boniface[1], Vegard Engen[1], Josep Matrat[2], David Garcia[2], Kostas Kavoussanakis[3], Ally Hume[3] and David Margery[4]

[1]University of Southampton IT Innovation Centre
[2]Atos
[3]University of Edinburgh
[4]INRIA

8.1 Introduction

The demand for ways to explore and understand how applications and services behave in a shared software defined infrastructures is increasing. Completely new applications are emerging, alongside "Big Data" and the convergence of services with mobile networks and the Internet of Things (IoT) all exploiting Cloud scalability and flexibility along with integration with software defined networks. These innovative technologies are creating opportunities for industry that requires a new collaborative approach to product and services that combines, commercial and funded research, early-stage and close-to-market applications, but always at the cutting edge of ideas.

The range of application sectors places significant challenges for cloud infrastructure and application providers. How to manage infrastructure resources considering the new types of demand? How will applications behave on a shared virtualised resource? This is not a new problem and some of the issues are now being addressed by Platform-as-a-Service providers, but the landscape is changing again as the convergence of cloud computing and dynamic software-defined networks picks up pace. The merging of industries and technology requires a collaborative approach to product and service innovation that allows technical and businesses exploration across the traditional boundaries of telecommunications and cloud infrastructures.

In this chapter we summarise six years of cloud and services experimentation at the BonFIRE facility which ran its last experiment on 30 May 2016. We show how BonFIRE delivered impact and broke new ground for technically advanced and sustainable Experimentation-as-a-Service (EaaS) platforms supporting cloud and service innovation with cross-cutting networking affects.

8.2 A Cloud and Services Experimentation Service

BonFIRE was a multi-site experimentation service for research and development of novel cloud and networking products and services. BonFIRE allowed customers to outsource testbed infrastructure on-demand by offering the four key capabilities necessary for experimentation: control, observability, usability and advanced cloud/network features (e.g. cross site elasticity, bandwidth on-demand). These features lead to reduced barriers to entry for providers of innovative Cloud offerings.

BonFIRE provided infrastructure capacity to support medium scale cloud experiments through a permanent infrastructure providing a hub that was used as the foundation for growth to larger scale experiments through additional on-request resources and relationships with 3rd party suppliers. BonFIRE operated a multi-cloud broker that brought together pan-European providers of cloud and network infrastructure. Uniquely, BonFIRE offered capabilities to control cloud computing and network infrastructure using a single interface, in this way experimenters could explore cross-cutting effects of applications, clouds and networks, in scenarios with increasing levels of realism. Software technologies could be deployed on demand either on a single site with highly controllable emulated networking or on multiple sites with controlled wide-area networking. No other public cloud or network provider offered this capability at the time. With a prioritisation on ensuring accuracy and confidence in results, BonFIRE allowed experimenters to control and observe the behaviour of physical and virtualised infrastructure in ways that was not offered by existing public cloud providers (e.g. Amazon, Rackspace, or Flexiant). BonFIRE achieved the differentiation by targeting Research Technology and Development (RTD) phases of the technology lifecycle rather than downstream production deployments of customer technology. BonFIRE capabilities were designed for testing and experimentation, rather than production runs where business drivers require operational decisions that prioritise service level guarantees and scale rather than controllability and observability.

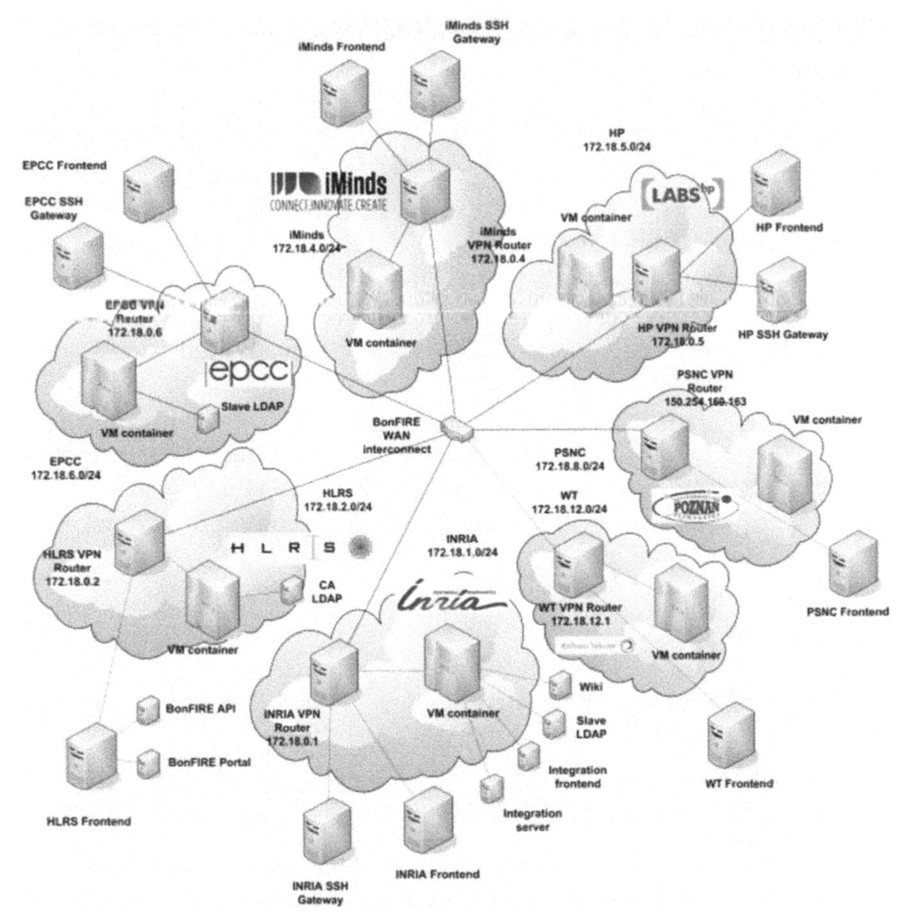

Figure 8.1 The BonFIRE infrastructure.

BonFIRE's targeted experimenters where those with insufficient capital or requirement for long-term investment in dedicated testbed facilities themselves. This includes Small and Medium Sized Enterprises (SMEs), academic researchers, and research collaborations (e.g. EC Projects). BonFIRE was not a "mass" market service, but at the same time, most users are largely self-supporting and the service was not tailored for each customer. Supporting experimenters in the development of service strategies was a key part of EaaS along with tools to transition technology from service design to service operation in production environments. BonFIRE recognised that transitioning new services from an experimental facility to production environments efficiently was essential to reduce the time to market by interoperating with production cloud providers to ensure technology could be transferred to mainstream deployment easily.

BonFIRE offered a multi-site, geographically distributed set of federated testbeds. At its peak, BonFIRE included seven sites across Europe, which offer 660 dedicated cores, with 1.5 TB of RAM and 34 TB of storage (See Figure 8.1). An additional 2,300 multi-core nodes could be added to BonFIRE on user-request using additional capacity at testbed sites, each heterogeneous in terms of Cloud managers, with OpenNebula[1], HP Cells[2] and VMWare employed; the hypervisors and the types of hardware employed are also very varied. In addition to Cloud resources, BonFIRE allowed access to the Virtual Wall emulated network facility with proxy access to Amazon EC2 resources, access to FEDERICA[3] and the AutoBAHN Bandwidth on Demand[4] service of GÉANT. More recently BonFIRE was integrated within the European Federation of future internet testbeds FED4FIRE[5] enabling many new experiments wanting to explore clouds in the context of Internet of Things and mobile networking.

8.3 Technical Approach

Design Principles and Architecture

BonFIRE offered services based on unique design principles that were not easily obtained in public clouds but are important for cloud-based testing on novel future internet applications. These principles included:

- Controllability: allow experimenters to control the infrastructure at multiple levels by specification their resourcing requirement not only on virtualisation level, but also on the underlying physical level (e.g. deploy two VMs on the same physical host).

[1]New applications emerge exploiting Cloud scalability and flexibility along with integration with software defined networks.

[2]HP Labs cloud-computing test bed projects –Cells as a Service, http://www.hpl.hp.com/open_innovation/cloud_collaboration/projects.html

[3]Peter Szegedi et al., "Enabling future internet research: the FEDERICA case", IEEE Communications Magazine, Vol. 49, No. 7, pp. 54–61, July 2011.

[4]GÉANT Services – AutoBAHN, http://geant3.archive.geant.net/service/autobahn/Pages/home.aspx

[5]Vandenberghe, W., Vermeulen, B., Demeester, P., Willner, A., Papavassiliou, S., Gavras, A., ... & Schreiner, F. (2013, July). Architecture for the heterogeneous federation of future internet experimentation facilities. In *Future Network and Mobile Summit (FutureNetworkSummit), 2013* (pp. 1–11). IEEE.

- Scalability: allow experimenters to construct high-scalable infrastructure for running their experiment by adjusting the size of the infrastructure at runtime.
- Federation: provide seamless integration and unique access to cloud services under different domains of control through standard protocols.
- Heterogeneity: support provisioning of different infrastructure consisting of various VM types and networking resources from geographically distributed cloud constituents.
- Networking: provide highly networked resources allowing experimenter to emulate complex and dynamic internetworking environments for their experiments.
- Observability: allow experimenter to define and gather infrastructure-level, both virtual and physical level, and application-level metrics to evaluate and analyse experimental results.

BonFIRE was designed and operated to support testing of cloud applications based on the notion of deploying software defined infrastructure resources in ways that allows testing to monitor what's going on inside the cloud allowing understanding of the performance and behaviour of the system under test, the causes of their degradation and the opportunities to improve them. BonFIRE was not a site for production running or for routine application development. BonFIRE was for experimentation through empirical investigation, which can be in a wide variety of research areas including but not limited to elasticity, cloud reliability, networking, heterogeneous clouds and federation. Different levels of access were offered including basic cloud infrastructure, impact of cloud on an existing application, investigation of new scenarios such as next generation mobile networks.

BonFIRE provided an experimentation platform which is not only highly controllable at all levels, but also offered tools to enable experimenters to investigate in-depth. Designed for usability and versatility experimenters could quickly get down to the details of their work, often under strict time-constraints. On top of this, BonFIRE offered unique testbeds for cross-cutting research in network effects, bandwidth on demand, and heterogeneous servers, and advanced tools such as the ability to emulate contention effects. All features were offered through the BonFIRE Resource Manager (RM), facilitating access to the disparate and geographically distributed resources, and in the management plane, and perhaps above all in the choice of well-defined interfaces which enable researchers to define, control, run and re-run their experiments according to their needs.

A high level view of the BonFIRE architecture is shown in Figure 8.2. Users can interact with BonFIRE using a web Portal, an Application Programming Interface using the Experiment Manager (EM) using a declarative, multi-resource, deployment descriptors or using the BonFIRE RM that provided a RESTful, Open Cloud Computing Interface (OCCI) [REF] interface to create and manage resources one at a time. Interactions with the BonFIRE API were programmed or scripted using a variety of tools. BonFIRE used a centralized broker-wrapper architecture for federation implemented in the RM. The RM service maps user requests to the appropriate infrastructure site and used an implementation of the wrapper pattern to translate these requests to the appropriate format for each site.

Components

In this section we describe in more the components within the BonFIRE architecture.

Portal

The Portal offers the experimenter a graphical interface to the BonFIRE capabilities. It has a view of the experimenter's data, the running experiments, and the available platform capabilities. The Portal accesses the functionalities

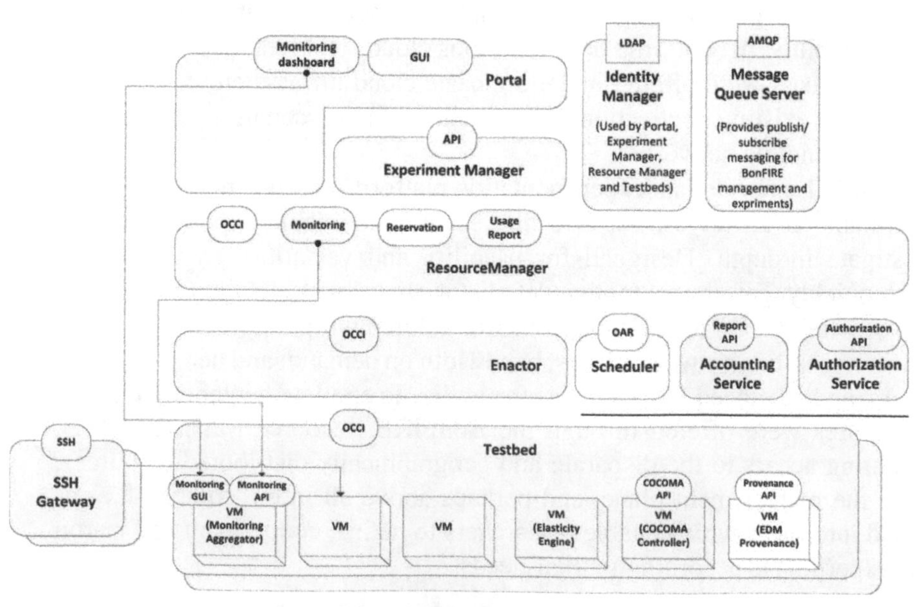

Figure 8.2 The BonFIRE architecture.

exposed by the BonFIRE Application Programming Interface (API). Every function performed through the Portal could be performed by the experimenter without using the Portal by issuing the respective HTTP requests directly to the API. The task of the Portal, however, is to make this process much more convenient and provide a concise overview of the resources and options available to the experimenter. The Portal furthermore provides additional documentation and guidance to the user. The Portal is implemented as a web application written in the python programming language and implemented as a set of plugins to the content management system and web application framework Django[6].

Experiment Manager (EM)

The Experiment Manager (EM) provides a simple RESTful HTTP interface to allow users to create a managed experiment by uploading an experiment descriptor file. The experiment description is parsed and validated immediately, and the user is notified of the success or failure of this stage. The experiment will be deployed in the background by making successive calls to the RM, and the user can check the status by doing a HTTP GET on the managed experiment resource. Through the use of GET, the user can also download the experiment log file, which lists messages on the progress of the experiment. The EM keeps track of a 'managed experiment' resource, which has a status and a link to the URL of the experiment on the RM. The managed experiment can also be deleted from the EM; this will also delete the experiment on the RM.

BonFIRE's investment to ease of use was the inception of a domain-specific, declarative experiment descriptor. The JSON-formatted BonFIRE Experiment Descriptor covers all BonFIRE features that are invoked at deployment time. Unlike the transactional OCCI interface, the user submits a single document to the EM interface. The EM identifies dependencies between resources and decides on order of execution. Consider for example an experiment that has a monitoring Aggregator using a separate storage at Cloud Site A; one compute at Cloud Site A and another one at Cloud Site A B. The EM will first create the storage; then creates the Aggregator and take its site-supplied IP; and finally create the VMs and pass that Aggregator IP to them as part of their context. The Experiment Descriptor is the cornerstone of usability for BonFIRE, the vehicle for Experimentation-as-a-Service. In the context of Cloud testing, what the users want to do is deploy large scale experiments, on

[6]https://www.djangoproject.com/

various facilities. What they then want to do is run the same experiment, under controlled conditions, to build the statistical confidence that their findings are correct and collect the data that prove it. What they may also want to do is to change the deployment to different target systems, to observe the effect.

Resource Manager (RM) and APIs

The RM is the component that provides the resource-level API through which users, and higher layers such as the Portal and EM, interact with BonFIRE. The RM is the entry point for programmatic resource level interactions with BonFIRE. The RM API is an open interface based on the Open Cloud Computing Interface (OCCI)[7] that allows experimenters to build their own clients or use direct Command Line Interface (CLI) calls to the API, which can be embedded in scripts. Through the API, BonFIRE allows experimenters to select the site on which to deploy their VM. A motivation might be a particular application topology the user is interested in studying, in which specific components of the application can be placed at specific sites. One step up from observing, the BonFIRE user can specify themselves on exactly which host to place their VM. This feature could be used to deploy their VM on the specific kind of hardware that they prefer, and BonFIRE's sites have different hardware both between them and inside them.

The Portal is an example GUI client of the RM API. Others include a client toolkit called Restfully and the BonFIRE Command Line Interface (CLI). Restfully[8] is a Ruby library that utilizes the RESTful BonFIRE API to allow deployment and control of the experiment. The experimenter can develop the logic that they need on scripts and add very complex, runtime functionality, as allowed by Ruby and its powerful libraries. The Command Line Tools are a powerful way of scripting deployment and control. They are a Python-based toolkit that encapsulates the OCCI and exposes an intuitive interface that covers all aspects of the BonFIRE functionality.

Enactor

The Enactor shields the technical details of how to communicate with each specific testbed from the higher level RM. Once the RM has decided to perform an action on a testbed, the Enactor is in charge of transforming that request onto suitable format for the appropriate testbeds through a collection of adaptors. Adaptors where classified into four different categories: OCCI adaptors (that

[7]OGF Open Cloud Computing Interface Working Group, http://www.occi-wg.org/

[8]Restfully, https://github.com/crohr/restfully/blob/master/README.md

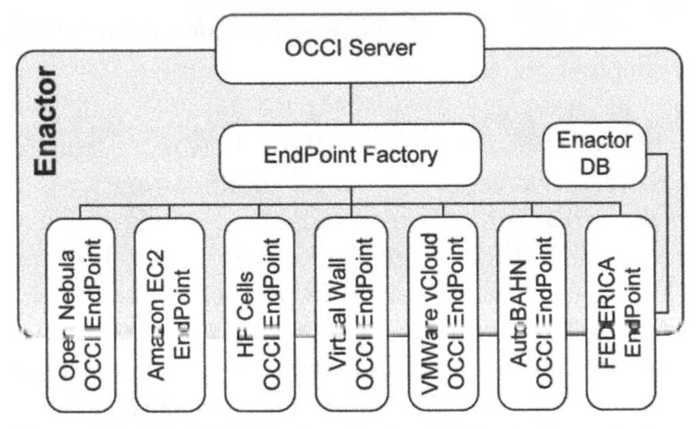

Figure 8.3 BonFIRE cloud and network infrastructure adapters.

are subdivided into five different types: OpenNebula, HP Cells, VirtualWall, and VMWare vCloud), Amazon EC2, AutoBAHN, and FEDERICA. It is possible to add other kind of adaptors outside those categories, making BonFIRE easily extendable.

The Enactor was not responsible for the security of the incoming call – but in counterpart it must enforce secure communication with the testbeds. The Enactor authenticates itself against testbed APIs (for example, by presenting a valid certificate, while user attributes are passed as HTTP headers – the testbed APIs can log/use them as they wish for auditing/accounting purposes). The Enactor supports multiple, concurrent, possibly time-consuming requests. It is a non-blocking service, capable of serving other requests while asynchronously waiting for a response from one of the testbed APIs.

Monitoring

BonFIRE provides its users with experiment monitoring facilities that support three types of metrics: VM metrics, application metrics and infrastructure metrics. BonFIRE provided this functionality through the use of the Zabbix open source monitoring software[9]. The Zabbix system adopts a client/server approach where the monitoring aggregator plays the role of the server and monitoring agents are the clients. Experimenters are free to deploy aggregators and agents in what in whatever way they wish but BonFIRE provides explicit support for the pattern where a single monitoring aggregator is deployed for each experiment. This aggregator collects data from several monitoring agents

[9]www.zabbix.com

deployed throughout the experiment and possibly also from infrastructure level aggregators deployed at each testbed.

The aggregator has been made available in the form of a dedicated virtual machine image containing an installation of the Zabbix monitoring software. This image is deployed like any other virtual machine image – no further configuration by the experimenter is required. The only requirement for the VM running the aggregator is that it must have an IP address that is reachable from the other VMs in the experiment and by the Resource Manager and Portal. This is necessary to enable the monitoring agents deployed on the individual machines to contact the aggregator and to enable the Resource Manager and Portal to expose the Zabbix API and web interface respectively.

A monitoring agent software is also included preinstalled within the images provided by BonFIRE. It needs to be configured with the IP address of the monitoring aggregator. This configuration is realized through the contextualization mechanisms of OCCI. After startup, the agent will register itself with the monitoring aggregator, from which point on the agent machine is fully integrated within the experiments monitoring system. The experimenter has the ability to further configure the agent by defining personalized metrics which should be evaluated and sent to the aggregator. This can be done through the standard mechanisms of the Zabbix software or via the contextualization section of a BonFIRE OCCI request.

The experimenter has multiple options on where to store the monitoring data of an experiment. The monitoring data can be stored either inside or outside the aggregator image. In the second option, the database of the aggregator is stored in an external, permanent storage that is mounted as an additional disk to the aggregator VM. This option enables more flexibility, the experimenter can set, on-demand, the storage size for the monitoring data, and this data is also available after the experiment's expiration or deletion. As a third option, the experimenter can use an external storage resource that was already in previously experiment. All these options are available through the BonFIRE Portal. By default the aggregator is created with an external, permanent storage with 1 GB size. As well as monitoring at the VM level, BonFIRE also supports monitoring at the infrastructure level. Those testbeds that support infrastructure monitoring have an infrastructure monitoring aggregator that gathers information regarding the whole testbed. An experiment aggregator fetches monitoring data of predefined, privilege metrics relating to those physical machines that host its virtual machines. The experiment aggregator fetches this data through the monitoring API.

Elasticity Engine

The Elasticity Engine supports three possible approaches for elasticity in BonFIRE: manual, programmed and managed. The BonFIRE components support manual elasticity by providing a Portal that allows various monitoring metric to be observed and the RM's OCCI API through which resources may be created or deleted. Additionally the architecture supports programmed elasticity via the Resource Manager's monitoring and OCCI APIs. This is done by the elasticity engine (EE) a stand-alone component able to manage the experiment based on some Key Performance Indicators (KPIs). It is basically a rules engine which can be configured via OCCI. It can be deployed inside a compute resource used by the experiment. In this way it is possible to create an elastic experiment using the portal, the experiment manager, or directly sending requests to the resource manager.

The basic functionality of the elasticity engine is to automatically increase or decrease compute resources in a running experiment. The experimenter has to pre-configure his own image using the SAVE_AS functionality. Once the image is ready he has to communicate this information to the elasticity engine which will deploy or remove compute resources automatically based on some rules expressed by the experimenter.

In order to distribute the load between different compute resources, the elasticity engine deploys a load balancer which is included in the BonFIRE standard images. The load balancer is part of the standard pool of images. It provides internally two different kinds of load balancer: HTTP and SIP. The first one is based on the open source HAProxy, with an additional HTTP interface for being managed remotely by the EE. The second one is based on Kamailio, an open source SIP proxy which offers also some functionalities of dispatching messages. Figure 16 shows an example of architecture of an elastic experiment.

CoCoMa: Controlled Contentious and Malicious Patterns

One of the main common characteristics of cloud computing is resource sharing amongst multiple users, through which providers can optimise utilization and efficiency of their system. However, at the same time this raises some concerns for performance predictability, reliability and security:

- Resource (i.e. CPU, storage and network) sharing inevitably creates contention, which affects applications' performance and reliability.
- Workloads and applications of different users residing on the same physical machine, storage and network are more vulnerable to malicious attacks.

Studying the effect of resource contention and maliciousness in a cloud environment can be of interest for different stakeholders. Experimenters may want to evaluate the performance and security mechanisms of their system under test (SuT). On the other hand cloud providers may want to assess their mechanisms to enforce performance isolation and security.

The Controlled Contentious and Malicious patterns (COCOMA) components provides experimenters the ability to create specific contentious and malicious payloads and workloads in a controlled fashion. The experimenter is able to use pre-defined common distributions or specify new payloads and workloads. VM images can be created that allow the injection of CPU, memory and disk I/O contention patterns to the physical host. COCOMA allows these types of contention to be combined and also allows variation of the intensity of contention across time. Still, all this control is not enough and affects other users on a multi-tenant physical host. To combat this, BonFIRE grant users exclusive access to physical hosts. This eliminates contention on the local disk, the memory and the CPU of the physical host, and combined with COCOMA gives BonFIRE users unique control across the whole range of zero to maximum isolation.

Networking

BonFIRE's multi-Cloud services has extensive support for controlled networking experiments. BonFIRE includes the Emulab-based [REF] Virtual Wall facility, which allows users to construct not only compute and storage resources, but also networks with user configurable bandwidth, latency and packet-loss characteristics. The user can modify these metrics at runtime, using BonFIRE's API or Portal. The Virtual Wall also allows users to inject background traffic to their networks and change the network buffering strategies. BonFIRE is also an early adopter of the GÉANT AutoBAHN pilot service of bandwidth on demand provision. AutoBAHN allows users to set up a point-to-point link with predefined bandwidth between two sites in its deployment. With the help of GÉANT, Janet and PIONIER, BonFIRE exposes this functionality to end-users that deploy their VMs on the EPCC and PSNC testbeds. Although it only allows control of bandwidth, AutoBAHN is more realistic than the Virtual Wall in that it involves real, rather than emulated network devices. In our experience, the key benefit of AutoBAHN for testers is not so much guaranteeing the quality of service, which is GÉANT's intended use, but rather policing it to within the limits of the user specification, so as to allow users to evaluate their system under known network conditions.

BonFIRE was committed to bridge the gap between advanced networking functionalities and the target cloud user community. To this end we enriched our interface to ease adoption of the network features. For example, AutoBAHN requires routing set-up on the newly created compute resources. BonFIRE exposes routing at the familiar, OCCI level, and provides simple directives as well as guidelines to declare routing on VM instantiation. This allows our users an easy, error-free way to specify routing without accessing the resource after it has been instantiated. Importantly, they get the network service without needing to go down to its level.

Experiment Data Provenance

An Experiment Data Manager (EDM) for Provenance (Prov) is used to describe the provenance of an Experiment, resources (compute, storage and network) within the experiment(s), any software/services running on the resources, any particular components as part of software/services, any users interacting with entities in an experiment. The EDM Prov will build upon the W3C PROV Data Model (PROV-DM)[10], which is a recent specification that stems from work on the Open Provenance Model (OPM)[11] with many existing vocabularies, applications and libraries/services. The PROV-DM core model allows extensions, such as subtyping (software agents running software). Other extensions for BonFIRE will be identified and made available to experimenters. PROV-DM model is very flexible, allowing experimenters to capture provenance of anything within their experiments. The model also supports bundles and collections of entities, allowing provenance of provenance. PROV-DM therefore offers a very powerful framework for experimenters to use in BonFIRE. The EDM Prov will comprise several components and will be made available in a VM image that experimenters can deploy as an optional service in BonFIRE. Other components in BonFIRE, like COCOMA, or services deployed by the experimenters on different compute resources may also generate provenance events, which need to be sent to the EDM Prov. To achieve this, the contextualisation functionality in BonFIRE can be used to provide those components with the IP of the EDM Prov, in the same way it is currently used for passing the Zabbix Aggregator IP to VMs with Zabbix Agents for monitoring.

[10]https://www.w3.org/TR/prov-dm/

[11]http://openprovenance.org/

Authentication, Authorization and Accounting

The authentication solution adopted by BonFIRE is based on existing state-of-the-art components such as Apache modules and Lightweight Directory Access Protocol (LDAP). To secure the connections between the components of the BonFIRE architecture server certificates are needed. These certificates are issued by the BonFIRE Certificate Authority (CA). The components behind the Resource Manager validate HTTP requests by using the BonFIRE Asserted ID Header field. These components trust the request from an authenticated user, because of the existing X-BonFIRE-Asserted-ID header field. The LDAP server and the BonFIRE CA are deployed on a VM with private IP address at HLRS. For security reasons access to that server is restricted. The BonFIRE CA is based on OpenSSL and the LDAP server for storing centralized information based on OpenLDAP.

The Authorization Service is used by the Resource Manager to control access to certain resource types and sites on a per-group basis. For example, the authorization service may restrict users in a group so that they can only use two named BonFIRE sites. Additionally, the Authorization Service also monitors current usage on a per-group basis and can be used to control the maximum amount of resources used by a group at any given time. The Authorization Service was added to support the degree of capability management that is required for BonFIRE open access phase.

The Accounting Service records all the usage of BonFIRE and can produce usage reports. These usage reports are essential to understand usage of BonFIRE with a view to informing sustainability decisions. The accounting reports were also envisaged as a precuser to any future billing system.

8.4 Federation of Heterogeneous Cloud and Networking Testbeds

BonFIRE offered a federated, multi-site cloud testbed to support large-scale testing of applications, services and systems. This is achieved by federating geographically distributed, heterogeneous clouds testbeds where each exposes unique configuration and/or features while giving to the experimenters (users) a homogeneous way to interact with the facility. BonFIRE supported five different types of Cloud testbed:

- OpenNebula: The currently operated OpenNebula version 3.6 includes an implementation of an OCCI server based on the OCCI draft 0.8. In order to provide valuable cloud functionality, additional fields of use were added by the BonFIRE developers in order to improve and extend the whole OCCI software stack of OpenNebula.

- HP Cells: The OCCI at HP Cells is completely stateless, so there is nothing that can get out of sync with the BonFIRE cental services or with the Cells state. BonFIRE-specific information such as groups, users, etc. are not stored, so the information retrieved on each request from the Enactor is filtered according to the permissions of the requesting user. This OCCI server was implemented specifically to support the BonFIRE project.
- Virtual Wall: The Virtual Wall emulation testbed is not a typical cloud environment, as it lacks the ability to dynamically add computes to an already running experiment. However, its functionality offers a first step to bridge the gap between network and cloud experimentation. The Virtual Wall offers the same OCCI resources as the other testbeds in BonFIRE, but their implementation is very different due to its underlying framework, Emulab. For instance, the Virtual Wall maps Compute resources to physical nodes, which prevents virtualisation, but allows the experimenter to take full control of the hardware. In response to the need of experimenters to share larger amounts of storage between different Compute resources, the Virtual Wall implements a notion of shared storage based on the Network File System (NFS).
- VMWare vCloud: vCloud does not offer by default an OCCI API. Similar to the case of HP Cells, an OCCI server was developed inside the BonFIRE project that interacts with the VMWare vCloud Director API to support VMWare Cloud facilities. The OCCI server is stateless, all the requests coming from the Enactor are translated and mapped to the proprietary API.
- Amazon EC2: The Amazon EC2 endpoint at the Enactor makes use of the API that Amazon provides to connect remotely to their Cloud services. The endpoint only allows to manage two kind of resources: storages and computes that are mapped to their Amazon equivalents, volumes or images and instances. In order to deal with the large volume of information returned, BonFIRE caches some OCCI queries in the Enactor, like listings of EC2's numerous storage resources.

BonFIRE supports experimentation and testing of new scenarios from the services research community, focused on the convergence of services and networks. In order to support network experimentation, BonFIRE is federated with the iMinds Virtual Wall testbed; and is interconnected with two network facilities: FEDERICA and AutoBAHN. The most distinctive features of the iMinds Virtual Wall are related to its networking capabilities. Whereas the other BonFIRE testbeds only provide a best-effort variant of the Network resource, the Virtual Wall implements three different types of Network resources: Default Networks that provide basic connectivity between two or more Computes; Managed Networks that provide controllable QoS (parameters that can be adjusted are bandwidth, packet loss rate and delay) over the network links; and Active Networks, that, on top of the functionality of Managed Networks, also provide the possibility to control the background traffic (UDP and TCP connections with dynamically adjustable packet size and throughput) on a network link. These networks provided by the Virtual Wall are emulated, using the Emulab software. FEDERICA is an infrastructure composed of computers, switches and routers connected by Gigabit Ethernet circuits. Through the Slide-based Federation Architecture (SFA) paradigm, FEDERICA offers to BonFIRE experimenters iso-lated network slices by means of virtualizing routers. This interconnection is aimed to help experimenters to investigate application performance through better control of the underlying network. The following changes were carried out in.

BonFIRE to incorporate these new network resources: the router resource was added to the BonFIRE OCCI and the network resource was enhanced with two new attributes: network link and vlan. Finally, since FEDERICA offers an SFA interface as federation API, it was necessary to implement an SFA endpoint at Enactor level. The FEDERICA SFA interface expects a unique XML request, where all the slice resources and their configuration are specified. This differs from the BonFIRE architecture, where each resource is requested in a single OCCI call. The main function of the BonFIRE SFA endpoint is to transform BonFIRE's OCCI information model to the SFA information model.

The federation between BonFIRE and the AutoBAHN beta-functionality offered by the GEANT facility allows the experimenters to request QoS guaranteed network connectivity services between VMs deployed on EPCC and PSNC testbeds. Overcoming the Best Effort limitation of the public Internet, dedicated network services can be established on demand for each experiment, with guarantees in terms of bandwidth, reduced jitter and service reliability. This option is fundamental to offer a controlled connectivity between VMs, so that the experimenters can evaluate the performance of their applications in environments able to emulate a variety of network conditions. In BonFIRE, a BoD service is represented by a new type of OCCI resource: the site link. Once the resource is created, it can be used to connect two networks created in the

BonFIRE sites at the edge of the site link: the traffic between the VMs attached to these networks is routed through the dedicated service. The processing of the OCCI requests for site link resources is managed at the enactor through a dedicated AutoBAHN end-point that is in charge of translating the OCCI specification into the AutoBAHN BoD service format. The Enactor endpoint acts as an AutoBAHN client.

8.5 Federation within the Broader FIRE Ecosystem

BonFIRE's infrastructure resources are only part of a highly complex and diverse Future Internet ecosystem consisting of infrastructure, services and applications. Through the EC FP7 FED4FIRE project[12], BonFIRE became part of a wider Experimentation-of-a-Service ecosystem offering access to heterogeneous Future Internet resources for experimentation such as cloud computing, wired and wireless networks, sensor networks and robotics deployed in laboratory and real world environments. The goal of FED4FIRE was to bring together European testbeds so that their resources may be used in a uniform manner by experimenters using their resources.

FED4FIRE has adopted a standardised protocol for resource reservation. The FED4FIRE federation performed a survey of its initial set of testbeds[13] and found that the most commonly used protocol for resource reservation and provisioning is the Slice-based Federation Architecture (SFA)[14]. Given that many of the federation's testbeds already supported SFA, plus the added advantage of compatibility with GENI testbeds, the SFA was adopted as the common protocol for the FED4FIRE federation, and tooling and guidance has been developed within the FED4FIRE project to support the SFA protocol, which testbeds can use to help them support the SFA protocol, thus reducing the cost of entry to the FED4FIRE federation for testbeds.

The "Slice" in the SFA is a client-side construct that is used as an identifiable container to collect resources from different provider in. The user may make a request to an SFA-compliant testbed, quoting their slice ID, and request that resources from the testbed be placed within the slice.

[12]http://www.fed4fire.eu/

[13]Vandenberghe, W., Vermeulen, B., Demeester, P., Willner, A., Papavassiliou, S., Gavras, A., Sioutis, M., Quereilhac, A., Al-Hazmi, Y., Lobillo, F. and Schreiner, F., 2013, July. Architecture for the heterogeneous federation of future internet experimentation facilities. In Future Network and Mobile Summit (FutureNetworkSummit), 2013 (pp. 1–11). IEEE.

[14]Peterson, L., Ricci, R., Falk, A. and Chase, J., 2010. Slice-based federation architecture (SFA). Working draft, version 2.

The FED4FIRE federation's choice of the SFA brings with it an access token format, the GENI Credential[15]. This enables users to use re-sources reserved in their slices, and owners of slices to grant access for other users to resources within the slice. In its basic form, the Slice Credential is a signed XML document containing the ID of the slice, certificate of the slice's owner and the ID of the slice. The Slice Credential also contains the rights the owner has on the slice, and whether the owner can delegate rights to others. There is another form of Slice Credential, the Delegated Slice Credential, and this enables the owner of a slice to grant permissions to other users on the slice.

BonFIRE had its own mechanisms for resource allocation and used different access tokens. Hence, a mapping had to be established between the BonFIRE resource allocation protocol and the FED4FIRE's chosen standard of SFA. Figure 8.4 shows the different concepts the SFA-compliant testbed and BonFIRE use. The slice is a container held by the user and is used to group resources from different testbeds together. In an SFA testbed, the user presents the slice and asks the testbed provider to allocate resources to it. In BonFIRE, the existing approach is to create an experiment at a testbed, which resources are allocated to (this is indicated by the dashed arrow in Figure 8.4). To enable holders of SFA slices to use BonFIRE a mapping between the slice identifier and a BonFIRE experiment was needed.

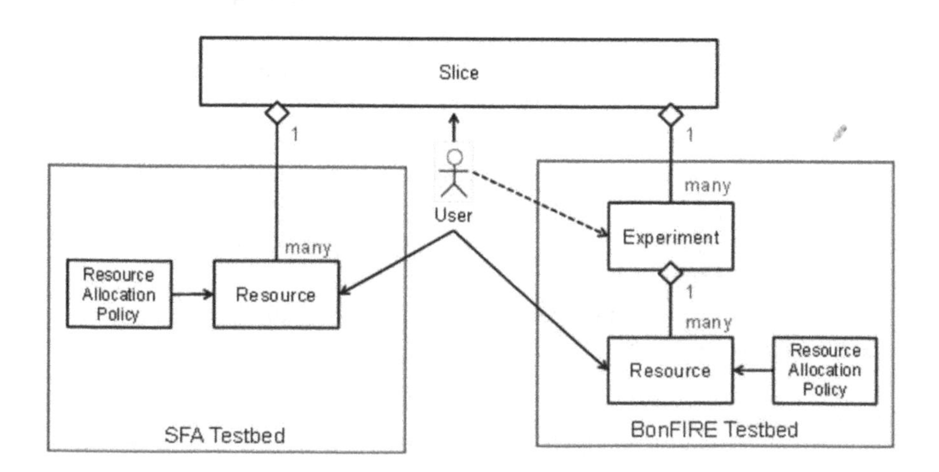

Figure 8.4 SFA-BonFIRE mapping.

[15]Available from http://groups.geni.net/geni/wiki/GeniApiCredentials

A Thin Aggregate Manager was developed that maps the BonFIRE experiment to a slice presented by the user. The existing components (specifically the Resource Manager) can continue to use the existing BonFIRE experiment ID. In use, the user also requests resources and presents their slice credential. The Thin Aggregate Manager requests an experiment be created by the Resource Manager, and the Resource Manager creates the experiment and allocates resources to it. The experiment and resource IDs are returned to the Thin Aggregate Manager.

8.6 Pioneering Open Access Experimentation and Sustainability

BonFIRE pioneered open access and sustainability of European experimentation services within the FIRE Ecosystem. In February 2013, BonFIRE launched the 1st Open Access initiative providing free access to both commercial organisations, academic institutions and other European projects outside of the BonFIRE consortium. Open access was developed as part of BonFIRE's sustainability activity as it transitioned through distinct operational phases on its route to a service offering beyond the lifetime of the project. Each phase had a distinct financial model that influences the governance and decision making of the experimentation services, and importantly the relationship with experimenters (Facility Users) as shown in Figure 8.5. The effect was that BonFIRE was no longer driven by the needs of a funded research project but by the features demanded by the experimenters external to the consortium. This was an important step towards an operational experimentation facility concerned with efficiency, accountability and customer satisfaction.

The lifecycle phases in BonFIRE's strategy are described below:

- **Pre-project conceptualisation**: concerned with defining the concept of a social and network media facility and getting buying from all stakeholders. This includes primarily supplies of services, technologies and other assets such as venue operations, technology providers and initial investors. The result of this phase is a public funded Project to implement the facility.
- **Project driving experiments**: concerned with implementing the facility in terms of technical and operational aspects. There are no Facility Users but Driving Experiments that define requirements and testcases to validate the facility offerings. The result of this phase is the 1st operational facility available for Facility users.

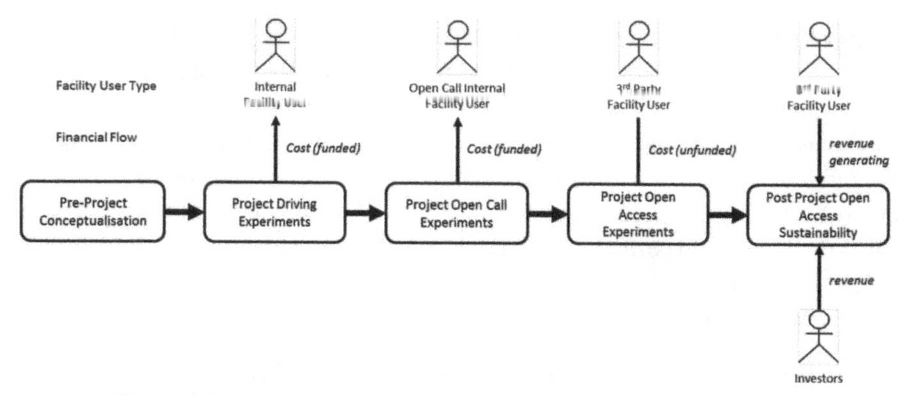

Figure 8.5 Transitions in governance and experimenter relationships.

- **Project open call experiments**: concerned with selecting and executing a set of experiments funded by the facility. The Facility Users are paid to run experiments and are acceded to the project contract with the facility providers. In return for payment Facility Users help facility providers understand how to improve the service offering by testing software and operational policies. The result of this phase is an enhanced facility that has been tailored to meet the needs of users.
- **Project open access experiments**: concerned with selecting and executing a set of experiments that are not funded by the facility. The Facility Users must pay their own costs and are not acceded to the project contract. Facility Users are therefore 3rd parties and access to IPR where needed must be governed by an appropriate license. Further legal agreements may be necessary to attribute rights, responsibilities and legal liability. Allowing 3rd party access allows the project to understand the legal and operational requirements required for post project facility use. The phase does not cover the mechanisms for revenue generation but unfunded experiments do provide test cases for simulating future business models including costs and revenues.
- **Post-project sustainability**: concerned primarily with continuing facility services. Exploitation agreements between partners were established to define how BonFIRE foreground can be used and post project governance structures implemented. Project partners must align themselves with operational roles and commit appropriate levels of resources to sustain activities.

BonFIRE successfully managed the transition between the different phases was a key factor in the success of the project. Each has placed demands on governance in terms of technical and operational requirements for the facility. For example, transitioning from open call to unfunded experiments requires the project to deal with access to the facility by third parties. During the final year the project has been concerned with the transitions from experiments funded by "Open Call" to experiments using BonFIRE through "Open Access" agreements. Finally in December 2013 the BonFIRE Foundation was established to operate the BonFIRE multi-site Cloud testing facility beyond the lifetime of the project, which continued operations until May 2016 some 18 months after the initial funded research project. The BonFIRE Foundation comprised members from world-leading industrial and academic partners, dedicated to continue to deliver services that enable developers to research new, faster, cheaper, or more flexible ways of running applications with new business models.

The BonFIRE Foundation was highly successful hosting over 50 experiments addressing a range of cloud computing challenges and through participation in the Fed4FIRE Federation BonFIRE has supported a further 11 experiments. Table 8.1 describes a few highlights from open access experiments.

The 11 Fed4FIRE experiments have used BonFIRE and finished their work successfully. Highlights included IPCS4FIRE focusing on the orchestration of cloud and user resources for efficient and scalable provisioning and operations

Table 8.1 Example open access experiments

Experiment	Description
MODA Clouds Alladin (Atos)	Atos Research and Innovation, Slovakia, are investigating a multi-Cloud application in BonFIRE that delivers telemedicine health care for patients at home. The application provides an integrated online clinical, educational and social support network for mild to moderate dementia sufferers and their caregivers. The aim of the experiment is to analyse the application behaviour in a multi-Cloud environment and improving its robustness and flexibility for peak load usage.
Sensor Cloud (Deri)	Digital Enterprise Research Institute (DERI) at the National University of Ireland, Galway, came to BonFIRE for testing scalability and stability of a stream middleware platform called Linked Stream Middleware (LSM, developed for the EC-FP7 OpenIoT and Vital projects). The experiment in BonFIRE utilises multiple sites with sensors generating up to 100,000 streaming items per second consumed by up to 100,000 clients. The data processing modules such as data acquisition and stream processing engines are run on the BonFIRE cloud infrastructure.

(Continued)

Table 8.1 Continued

Experiment	Description
SWAN (SCC)	This is an experiment conducted by SSC Services to analyse how one of their software solutions, SWAN, can handle large amounts of data transferred between business partners under different networking conditions. SSC Services have utilised the iMinds Virtual Wall site to achieve fine-grained control of the networking conditions in order to identify critical Quality of Service (QoS) thresholds for their application when varying latency and bandwidth. Moreover, investigating possible actions and optimisations to the SWAN components to deal with worsening conditions, to be able to deliver the expected QoS to the business partners.
ERNET	ERNET India are developing software for moving e-learning services into the Cloud and are using BonFIRE to analyse the benefits of Cloud delivery models, including multi-site deployment. In particular, they investigate fault tolerance.
JUNIPER	BonFIRE also facilitates other research projects, giving access to multiple partners to perform an experiment. One of these projects is the EC-FP7 project JUNIPER (Java Platform for High-Performance and Real-Time Large Scale Data), which deals with efficient and real-time exploitation of large streaming data from unstructured data sources. The JUNIPER platform helps Big Data analytic applications meet requirements of performance, guarantees, and scalability by enabling access to large scale computing infrastructures, such as Cloud Computing and HPC. In JUNIPER, the BonFIRE Cloud premises are used to initially port pilot applications to a production-like Cloud infrastructure. The JUNIPER experiment benefits from the availability of geographically distributed, heterogeneous, sites and the availability of fine grained monitoring information (at the infrastructure level) to test and benchmark the developed software stack. Another important advantage of BonFIRE to JUNIPER is that some of the sites owning HPC facilities, e.g., HLRS (Stuttgart), provide a transparent access (bridge) from Cloud to HPC, which is of a great importance for JUNIPER experiments.

of security services. As a result of their experiment, IPCS4FIRE were able to explore best-practices and share the optimal design with users to automatically provision and protect virtual machines without manual intervention, while minimising the time required to achieve this protection. SCS4FIRE performed experiments on the validation of Secure Cloud Storage system for multi-cloud deployments. SCS4FIRE optimized their methodology to automate the transfer of virtual machines and encrypted data volumes between multiple cloud sites, while maintaining continuous access for end users.

Finally SSC researched big data analysis components on Smart City data using cloud resources. SSC were able to validate that their Super Stream Collider middleware can achieve high scalability, continuous accessibility and high performance, for more than 100.000 clients.

8.7 Conclusions and Outlook

From Sept 2010 until May 2016, FIRE experimentation ecosystem has incorporated the BonFIRE multi-Cloud experimentation facility alongside testbeds in the networking, sensors and smart cities. The BonFIRE facility was unique in supporting services and network experimentation across multi-cloud sites focusing on a blueprint for experimentation and incorporating methodology and techniques to support repeatability and reproducibility. BonFIRE took these notions further, to deliver a facility based on four pillars: observability, control, advanced features and ease of use for experimentation. The end result was a facility that differed substantially from public Cloud offerings. Public Cloud providers will never offer the internal tracelogs and parameters of the clusters since it is highly sensitive data for their business, whereas this information is essential in research by experimentation to understand the behaviour of the Cloud applications. Also, public Clouds did not offer detailed level of control over physical and virtual resources, since their objective is to hide the complexity and operation from the users and reduce costs. Advanced features, such as user-specified bandwidth on demand and controlled networks were greatly received by the services experimenters, but are not in line with public Cloud offerings, while domain-specific tooling for experimentation is naturally not a concern. BonFIRE was funded between 2010 and 2013 and continued to be operated by the BonFIRE Foundation.

There are many emerging opportunities and requirements for Cloud-based experimentation facilities in the future driven by the needs of applications and services communities, and the ongoing convergence of software defined infrastructures. We see two major areas of expansion: embracing Big Data and enabling Mobile scenario testing. Researchers are exploring how to deal with the characteristics and demands of data within services, infrastructures, sensor networks and mobile devices, while the uptake of smartphones motivates the combination of mobile networks and Cloud computing. It is necessary to cover the full data lifecycle across multiple experimentation platforms facilities providing the necessary data interface, format, optimized transfer mechanisms, data analytics and management toolset to extract value from experimental data.

On the other hand, as the data traffic demand from mobile phones and tablet applications is exponentially growing (e.g. video, VoIP, Gaming and P2P) networks are developing to offer more capacity , higher throughput and better QoS. Future 5G networks and concepts dominate the research arena. Many telecom operators and network equipment manufacturers are embracing Network Functions Virtualization (NFV) techniques since it is envisaged that this will change the telecom industry landscape. Industry in ETSI is doing a great effort with the first sets of specifications and the "traditional" Cloud community has a lot to offer to the "virtualisation and softwarisation" of networks. Notably, this is a central research topic in the 5G PPP initiative where large-scale validation of these network virtualisation techniques are expected and experimentation platforms can play a role. A key lesson learnt from BonFIRE is that there is great value to be had from offering high-level interfaces for experimentation. Experimentation as a Service is a fact, not an endeavour, and the only way forward is to offer a truly PaaS tooling environment for experimenters on top of the IaaS layer, no matter what this infrastructure is.

Six years after the project kick-off, BonFIRE concluded its successful journey on 30 May 2016. In this period BonFIRE delivered impact consistently, breaking new ground in experimentation platforms and service delivery models across both technical and sustainability fronts. Open Access was highly successful with new and returning users, like EC FL7 RADICAL project renewing its Open Access for a third year and BonFIRE supporting the project right up until RADICAL's final review. Utilisation was high, with EPCC and Inria at times completely full and oversubscribed. The stability of the infrastructure has been remarkable, with two short, unplanned outages, both down to external factors. The services have now been decommissioned and no further access will be possible but the legacy of the BonFIRE initiative has provided a pioneering blueprint for current and future experimentation-as-a-service platforms exploring Next Generation Internet technologies.

Acknowledgements

The authors would like to thank all partners in the BonFIRE consortium.

The BonFIRE project has received funding from the European Commission as part of the FP7 work programme.

9

EXPERIMEDIA – A Multi-Venue Experimentation Service Supporting Technology Innovation through New Forms of Social Interaction and User Experience

Michael Boniface[1], Stefano Modafferi[1], Athanasios Voulodimos[2] David Salama Osborne[3] and Sandra Murg[4]

[1]IT Innovation Centre
[2]The Institute of Communications and Computer Systems, National Technical University of Athens
[3]Atos
[4]Joanneum Research

9.1 Introduction

New media applications and services are revolutionising social interaction and user experience in both society and in wide ranging industry sectors. The rapid emergence of pervasive human and environment sensing technologies, novel immersive presentation devices and high performance, globally connected network and cloud infrastructures is generating huge opportunities for application providers, service provider and content providers.

These new applications are driving convergence across devices, clouds, networks and services, and the merging of industries, technology and society. Yet the developers of such systems face many challenges in understanding how to optimise their solutions (Quality of Service – QoS) to enhance user experience (Quality of Experience – QoE) and how their disruptive innovations can be introduced into the market with appropriate business models.

In this report, we present the results of a new multi-disciplinary collaborative approach to product and service innovation that brings together users, technology and live events in a series of experiments conducted in real world settings. Through experimentation we have explored a broad range of technical, societal and economic challenges faced by technology providers each aiming to create and exploit new multimedia value chains in markets such as leisure and tourism, cultural and heritage, and sports science and training.

The experiments highlight the features of multimedia systems and the future opportunities for companies, as the Internet continues to transition towards the increasingly connected world of Internet of Things and Big Data. We know that putting user values at the heart of design decisions and evaluation is the key to success, and that long term benefits to providers of technology, services and content must derive from enhanced user experience. Engaging users in real-world settings to co-design and assess how technology can be used is now more important than testing how technology will be operated.

We have only scratched the surface of possibility in novel networked multimedia systems yet we believe that the individual and collective results in the report are significant as they are grounded in real-world evidence. A new way of conducting research and innovation has been created that maximises the potential for commercial exploitation and societal impact. We think this is extremely important and when adopted will lead to greater benefits for all.

9.2 Networked Multimedia Systems

Multimedia is the combination of multiple forms of content and is a fundamental element of applications in areas such as communication, entertainment, education, research and engineering. The convergence of technologies for distributed multi-stakeholder systems, data analytics and user experience is dramatically changing the way multimedia systems need to produce, deliver and consume content.

Providers of multimedia systems are now looking to create value by linking people to each other and to locations (both real and virtual) in such a way as to capture the popular imagination, and exploit the desires of consumers to share their experiences, thus creating new channels for revenue creation and advertising.

To create such experiences requires innovative applications that focus on: enhanced personalisation, non-linear story-telling; interactive immersive experiences; creation of social communities which allow people to use 3D environments to communicate and interact with each other; the capture and reproduction of the real world in 3D; and the creation of perceptual congruity between real and virtual worlds.

Of course, these innovative applications will place significant demands on network and content management infrastructures as providers attempt to deliver guaranteed Quality of Service and enhanced Quality of Experience to communities that dynamically organise themselves around socially distributed, fixed and mobile content. These additional demands will require investment in infrastructure but the expectation is that by linking multimedia and enhanced real-world experiences, consumers will be prepared to make long lasting commitments.

9.3 A Multi-Venue Media Experimentation Service

EXPERIMEDIA is a multi-venue experimentation service for research and development of novel Internet products and services aiming to deliver new forms of social interaction and user experience. EXPERIMEDIA was developed as part of a European research project of the same name within the Future Internet Research and Experimentation initiative (FIRE) [1] (Figure 9.1).

The EXPERIMEDIA project set out to develop and operate a unique facility offering researchers and companies what they need to gain insight into how Future Internet technologies can be used and enhanced to deliver added value media experiences to consumers. The approach aimed to deliver, reusable, cost-effective testing and experimentation facilities, platforms, tools and services for social and networked media systems. The EXPERIMEDIA project developed four foundation elements necessary for experimentation of multimedia systems conducted in real world environments:

- Smart venues: attractive locations where people go to experience events and where experiments can be conducted using smart networks and online devices;
- Smart communities: online and real-world communities of people who are connected over the Internet and available for participation in experiments;
- Live events: exciting real-world events that provide the incentives for individuals and smart communities to visit the smart venues and to become participants in experiments;

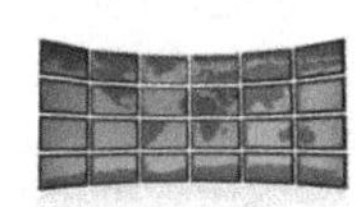

Figure 9.1 Four foundation elements of a multi-venue media experimentation service.

- Service Platform: state-of-the-art Future Internet testbed infrastructure for social and networked media experiments supporting large-scale experimentation of user-generated content, 3D internet, augmented reality, integration of online communities and full experiment lifecycle management.

The combination of live events, venues, user communities and an advanced technology platform accelerates product and service innovation by allowing companies to co-create solutions in real contexts with end-users. EXPERIMEDIA characterises live events as "any cooperative human activity that can be enhanced through access to real-time information delivered by the Internet". Examples live events include:

- 1000 spectators attending a two day ski championship at a ski resort.
- An athlete participating in a one hour sports training session with a coach and sports scientist.
- A group 50 students attending a one hour interactive virtual reality presentation about ancient Greece.
- A small group of hikers on a day trip on a mountain, a round of golf or a trail run.

There are many socio-technical and economic benefits to experimenters of using live events as the basis of trials and experimental studies. Each live event captures a distinct user experience to be enhanced along with providing temporal and spatial constraints associated the activity such as location, technical constraints associated with available infrastructure and socio-cultural constraints associated with the user communities. Dealing with contextual factors is a major challenge for experimenters aiming to develop generic solutions for Internet deployments and to understand how to address barriers to adoption of technology. In addition the ability of media technologies to connect people in real-time across distant locations can create new opportunities for interaction with live events. From an economic perspective, live events provide technology providers with access to an entry point to a potential market. This entry point can lead to significant direct and indirect sales (Table 9.1).

The EXPERIMEDIA Service Platform consists of a set of media services that have been instrumented for deep levels of observability for use within experimentation and technology trials. Each service has a corresponding service model with QoS metrics that are reported and available to the customer during experimentation. Such detailed metrics are necessary for customers to explore the relationship between QoS and QoE. These types of metrics are typically not available from equivalent commercial services. In addition, a provenance model is offered that allows user-centric activities and interactions

Table 9.1 Benefits and opportunities for experimenters

Socio-Technical Benefits for Experimenters: Testing Opportunities	Economic Benefits for Experimenters: Exploitation Opportunities
observation of individual and community behaviours	access to a potential market, direct sales
experience of scaling for large-scale short-lived communities	working with a customer's customers
adaptation to the environment, considering physical, social and ethical constraints	creation of high impact showcases, indirect sales
adaptation of content according to individual and/or group preferences	engagement and collaboration with stakeholders, potential partners/ suppliers
real-time orchestration allowing for adaptive narratives	
sensors and devices for detection and tracking of feature points	
device capabilities both remote and at a venue	
cooperative or collaborative frameworks including dealing with selfish or malicious users	

to be tracked and linked to the detailed metrics reported by the other entities involved. This capability is important to allow experimenters to track users in open studies and to explore correlations between QoE, system interaction and system performance (Figure 9.2).

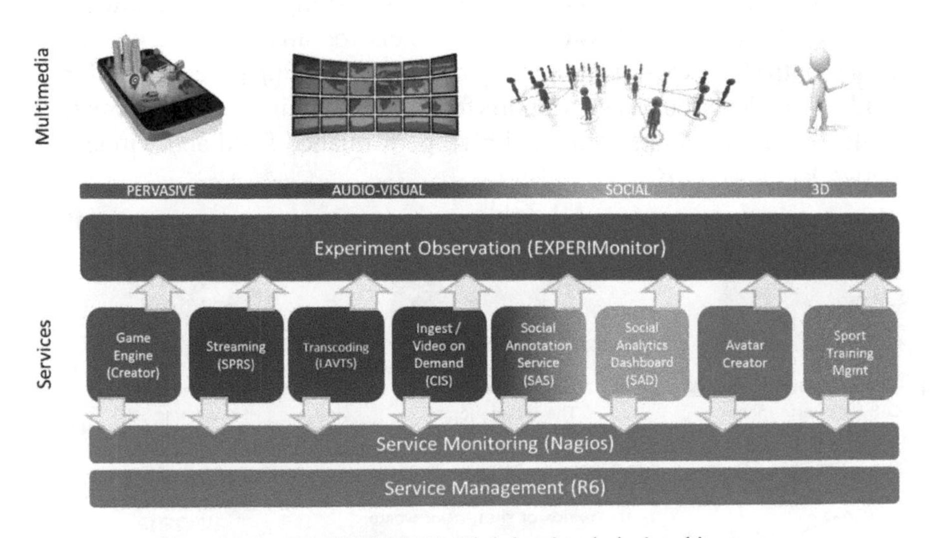

Figure 9.2 EXPERIMEDIA High-level technical architecture.

From the platform point of view the reusability across experiments is a key point enabling multi-domain applications. The media services are technology enablers whose capability allows users to achieve added value through use, either by design (i.e. the purpose is known in advance) or more frequently by openness (i.e. the purpose is opportunistically established by the user). Technology enablers are a key part of future innovation in programmes such as FIRE and the Future Internet Public Private Partnership. Networked Multimedia technology enablers must address the needs novel applications and services allowing them to exploit a range of social, audio/visual, pervasive content and 3D content. The platform offers services to support different types of content considering the distinct characteristics and lifecycles (authoring, management and delivery).

9.4 Smart Venues and Experiments

Smart Venues are real world locations that offer live events, communities, infrastructure and relevant data assets to experiments. Smart venues have distinct characteristics and provide context for experimentation. EXPERIMEDIA has three smart venues covering important application sectors for multimedia systems including outdoors and leisure, cultural learning and sports training and science (Figure 9.3).

- Centre d'alt Rendiment (CAR), Spain, is a high performance sports training centre which gives support to athletes competing at an international level. CAR offers a professional environment for small scale (5 participant) controlled experiments aiming to improve training programmes for students, athletes, coaches and sports federations within a dedicated smart building with a private cloud and high performance fixed and wireless network connectivity.

Figure 9.3 EXPERIMEDIA smart venues.

- The Foundation of the Hellenic World (FHW), Greece, is a cultural centre that offers real and virtual exhibitions, congresses and performing arts events aiming to educate people about the Hellenic World. FHW offers a public environment for medium scale (30 participant) experiments aiming to improve visitor experience and the quality of learning through multimedia exhibitions, virtual and immersive reconstruction, and serious games. FHW offers a 3D, dome shaped virtual reality theatre, exhibition places, and cave systems.
- Schladming, Austria, is one of the leading international ski resorts in Austria and part of the Ski Amadé network covering 28 ski areas and towns that make up the largest ski area in Europe. Schladming offers a public environment for medium scale (50 participant) open trials of technology aiming to improve visitor experience within the region. The ecosystem is complex and potential activities are broad but most relevant are winter and summer outdoor sports such as skiing, hiking, mountain biking.

We funded a series of 16 experiments through two open calls. The experiments were conducted by researchers and SMEs at three Smart Venues throughout Europe covering a broad range and complimentary multimedia topics.

- Schladming Smart Venue
  - DigitalSchladming: hyper local social content syndication and filtering
  - MediaConnect: ubiquitous interactive and personalised media
  - PinPoint Schladming: augmented reality mobile applications
  - iCaCoT: interactive UHD camera-based coaching and training
  - Smart Ski Goggles: real-time information delivered to wearable data goggles
- CAR Smart Venue
  - Live Synchro: accurate analysis of choreographed team sports
  - 3D Media in Sports: non-invasive reconstruction of biomechanics
  - CONFetti: interactive 3D video conferencing for collaborative sports training
  - 3D Acrobatics: wireless sensor motion capture and 3D visualisation
  - 3DRSBA: remote 3D sports biomechanics analysis
  - CARVIREN: multi-factor athlete tracking using real-time video and sensor information
  - Augmented Table Tennis: automatic notation analysis system based on vibration sensors and on table surface projection

- FHW Smart Venue
  - NextGen Digital Domes: learning, interaction and participation using social and augmented content.
  - REENACT: serious games and immersive media.
  - BLUE: personalised museum experiences using cognitive profiling.
  - PLAYHIST: serious games with real-time 3D reconstruction of moving humans.

A significant dilemma is balancing research versus innovation activities. Geoff Nicolson of 3M once said "Research is turning money into knowledge, whereas Innovation is turning knowledge into money". Very few organisations complete the full lifecycle in the scope of an experiment. In many cases, impact is achieved much later either in-house by other groups (e.g. industry organisation) or by others exploiting knowledge published research institutions. In fact for research institutions the link between knowledge generation and exploitation in innovative services is significantly weaker. However, by creating multi-disciplinary teams including domain experts, social scientists, legal experts and technologists working with end users it is possible to overcome barriers and accelerate adoption in target markets.

Smart venues are concerned with offering innovative services that deliver enhanced user experience. Knowledge is only a route to that goal. The first open call experiments had an emphasis on knowledge creation rather than innovation due to the characteristics of the partners performing the work. As a consequence, the impact of those experiments was far less and the project strategy was changed to create experiments driven by SMEs for the second open call. Overall six experiments were executed by SMEs, nine by research institutions and one by industry. 18 technology outcomes where identified from the experiments with impact classified as follows:

- Commercialisation (5 of 18): benefit is exploitable in revenue generating products and services.
- Further Trials (4 of 18): promising outcomes justifying further investment in trials to scale up to produce quantitative results or to explore qualitatively in a new application domain.
- Further Research (8 of 18): benefit looks feasible but could not be sustained without significant research and development.
- Barrier (1 of 18): benefit could not be delivered.

Significant commercial opportunities have been delivered to experimenters highlighting the innovation potential of EXPERIMEDIA. Smart Ski Goggles will launch a commercial service in the Ski Amade region for the 2014/2015 ski season and there are ongoing negotiations for the commercialisation of the associated lift waiting time service. CARVIREN, 3D Acrobat Sports and 3DRSBA resulted in commercial contracts with the CAR Smart Venue. DigitalSchladming MyMeedia service remains operational 12 months after the experiment and is part of IN2's "staging" strategy and business model. iCaCoT is in negotiation with Schladming Ski School for use of interactive UHD video and annotation system as part of their skier training offering. 3D Media in Sports has received significant commercial interest from weightlifting and cycling communities following a large scale trial with the Movistar cycling team. Augmented Table Tennis has created significant commercial interest from TV broadcasters and the International Olympic Committee.

9.5 Users at the Heart of the System

User centricity is a critical element in the design and development of multimedia systems aiming to enhance user experience. Understanding the needs, wants and limitations of end users must be given extensive attention throughout the design process. We have adopted two main principles in our user centric design processes:

- users are the primary beneficiaries, and other benefits to providers of services and technology will follow from user benefits.
- users who participate in observations are also those same users that realise the primary benefits.

These principles reflect the shift towards the democratisation of Internet services where users play a greater role in generating information and the need to recognise explicitly the cost and benefit of participation. In general terms, designers must consider a multi-stakeholder data value chain where observations are acquired, data are processed by multimedia capabilities and data are transformed into benefits presented to users.

Observation is the process of closely watching and monitoring users and their context. User observations are processed as an inherent part of content delivery (e.g. location and activity tracking in geo-location services) or are used to understand the experience itself (e.g. a user satisfaction survey). From a user's perspective, observations have a cost either directly in terms of time and attention during an experience, or indirectly in terms of loss of right to self-determination (i.e. privacy). Context observations are processed to give additional meaning to Quality of Experience (e.g. a user had a good time in a group of 15 close friends) and importantly to optimise the Quality of Service delivered by service providers. As context plays a significant influential role in Quality of Experience it is typically the case that service providers have to manage context, including both real-world (e.g. how many people participating) and multimedia context (e.g. how much infrastructure resource, quality of virtual presentations, etc.).

Analysing the experiments we can define six categories of user observations from a total of 95 different user observations:

- Satisfaction (32 of 95): feedback about relative satisfaction with their experience covering aspects such as utility, emotional, subjective, economic, usability and usefulness.
- Online Activities (32 of 95): direct interaction with an application (e.g. interaction logs, web site statistics) that complements the real word activities, and is strongly related with the nature of the experiments.
- Real-World Activities (16 of 95): activity recognition, for example, biomechanics representing the position of body components (e.g. the angle formed by bones in an athlete while performing), higher level human activities (e.g. weightlifting, skiing).
- Collaboration (7 of 95): the relationship to a group, in terms of interpersonal relationships, social interaction, group dynamics (e.g. questions in a group presentation), group enhancement.
- Location (6 of 95): the absolute or relative position of a user where relative means with respect to external elements (e.g. a ski-run).
- Cognitive (1 of 95): the capacity to process information and apply knowledge (e.g. psychometric profile).

The absolute value of observations related to a category is not a measure of importance. A single type of observation can be the most important in a given experiment as it is the most significant factor in delivering the benefit to a user. "Collaboration" highlights that multimedia features aim to benefit users by supporting interaction. The "satisfaction" group is typical of any experimental environment and it is propaedeutic to evolve from experiment to exploitation (Figure 9.4).

Context is more complex as by definition it is anything not related to a user that can influence Quality of Experience. Analysing the experiments we can establish two main high level context categories from the 56 context observations:

- **Real-World Context**: observations related to people and environment conditions associated with real-world activities.

- **Online Context**: observations related to the performance characteristics of the system under test covering aspects such as content quality and infrastructure utilisation.

The significant number of context observations acquired means that the surrounding environment plays a significant role in multimedia systems. In fact, very often the benefit delivered to the user is the combination of context and personal information. Real-world context is highly dependent on the Real-World Activity. Within EXPERIMEDIA this is defined by the nature of the live events being studied at Smart Venues. Real world context is difficult to observe automatically and in a general way considering the specific nature of live events. EXPERIMEDIA has focused on observing users with some cases of capturing Real-World Context where this is an essential part of the experience and the cost is not prohibitive. In controlled experiences such as those at the CAR where Real-World Activities are well-defined and constrained the Real-World Context is known and can be captured out of band. In more dynamic and open situations at Schladming and FHW it is necessary to observe Real-World Context either directly (e.g. definition of Points of Interest within a geographic region, queue waiting times, etc.) or indirectly (e.g. inferences about group dynamics from temporal/spatial analysis or online interaction).

Making inferences about Real-World Context and Activities from Online Context and Activities is an essential part of multimedia systems and experimentation especially in situations where the cost of direct observation is

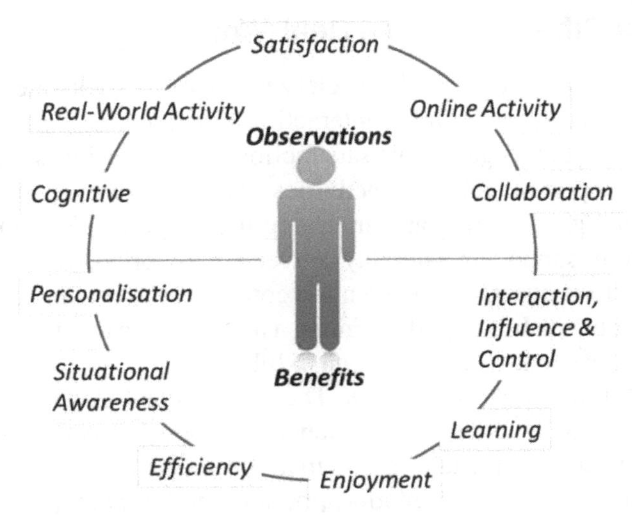

Figure 9.4 User centric observation and benefits model.

prohibitive either through software or feedback from users. EXPERIMEDIA's hybrid metric and provenance model offers a foundation for such analytics. The hybrid approach provides the ability to collect large quantities of measurement data (e.g. service response times, network latency, user satisfaction, etc) whilst allowing for exploration of causation between observations within such data (e.g. user satisfaction in relation to service response time). Also, it is recognised that Internet of Things domain has made significant progress in acquiring real-world context across a broad range of dynamic situations. There is an opportunity to deliver increased benefits by strengthening the relationship between User and Real-World Context observations.

Online Context is of significant interest to service providers who use this information to manage resources and optimise the delivery of multimedia services, including adaption of the quality of content. As such Online Context is an important facet of experiments that focus on the relationship between QoE and QoS. Of course this depends on the nature of the study but the advantage of the EXPERIMEDIA Platform is that it is already instrumented for Observation of Online Context to ensure that important technical information was available to experimenters. Typically experiments have identified the significant Online Context observations related to delivery of a desired Quality of Experience. These include the quality of context (e.g. accuracy of biomechanics data, video quality), network performance (e.g. delay, bandwidth) and cloud performance (e.g. CPU utilisation).

9.6 Making a Difference in the Real-World

Digital technologies are most useful to society when used to deliver enhanced real-world impact and benefits. Online interaction alone, such as digital games, can bring enjoyment but longer lasting satisfaction is achieved by using digital technologies in support of real-world activities. We focus our experiments on this area by defining, measuring and analysing user experience (UX) where multimedia systems support the interplay between real-world Live Events and online activities. Live events create the main context for user experience. We have explored events such as a sports training, a night out in a town, attendance at large scale sports events, and visiting an exhibition.

Studying UX is a complex endeavour. The International Standard Organisation (ISO 9241-210) defines User Experience as "a person's perceptions and responses that result from the use or anticipated use of a product, system or service". UX includes all users' emotions, beliefs, preferences, perceptions, physical and psychological responses, behaviours and accomplishments that occur before, during and after the use of product, system or service. The experiments themselves focused on distinct UX aspects enhanced through multimedia features. Exploring the experiments we identify seven high level user benefit categories from 61 measurable benefits:

- Learning (22 of 61): acquisition or improvement of a skill/ability, a key goal of the CAR and FHW.
- Efficiency (11 of 61): support for increasing the productivity processes in terms of time, effort or cost to complete the intended task or purpose. Efficiency is a common quantifiable measure for all activities associated with live events.
- Interaction, Influence & Control (10 of 61): interacting with the surrounding context for influence and control. (e.g. remote access to training sessions, or incorporation of a remote expert in an education session).
- Situational Awareness (10 of 61): understanding of when/where/why something is happening, so as to maximize the active participation of the user in the experience. This benefit pertains to the delivery of the right thing (information/support/other) exactly when it is needed.
- Enjoyment (5 of 61): the enjoyment a user has in the performed activities, a primary goal of Schladming Venue as a tourist destination.
- Personalization (3 of 61): tailoring the information to maximize user satisfaction including expressing themselves in social networks.

The majority of benefits are produced through processes that enhance raw data collected from multiple information sources. "Learning" is a primary benefit in all CAR and FHW experiments due to learning being a key objective of the venues. NextGen Digital Domes focused on how augmented reality can prime student knowledge prior to virtual reality presentations whereas REENACT introduced a role playing game that allowed participants to enact and discuss historical events. "Situation awareness" is another common user benefit demonstrating how through sensors and analytics users are provided with better knowledge of surrounding context. Geo-spatial and temporal data were essential elements of Smart Ski Goggles and Pinpoint Schladming. "Influence and control" demonstrates the increased possibility of controlling and influencing real-world situations through remote interaction with multimedia (using or being part of the content). PlayHist, CONFetti and 3DRSBA all use networked collaborative working to enable remote users to interact and influence training and learning sessions whilst measuring the efficiency or setting up 3D capture equipment. Greater "enjoyment" is an important benefit across all venues but was not expressed significantly at professional environments such as CAR where objective performance gains were a priority.

9.7 Real-Time Interactive and Immersive Media

The games industry has a significant impact on business and innovation models of the digital era. In many ways, the games industry are forerunners of innovative content, services and business models of a growing digital economy. Consequently the games industry is preparing the way for the other sectors where the digital revolution has not started yet. An industry-changing dynamic is the transformation of multiplayer gaming, built on vast networks of players interlinked by broadband across continents and growing further still by leveraging social networks. With capabilities strengthened further by the generational leaps in 3D graphics, gameplay mechanics, and collaborative platforms, gaming is partnering with and spurring growth in other media segments.

Gaming technologies have been a source of inspiration in EXPERIMEDIA through the adoption of game engines, 3D sensors and advanced presentation technologies across a range of applications. Novel algorithms have been developed in 3D Media in Sport using data from the low cost Kinect sensor, built for the Xbox console. Using 3D information, the algorithms provide athletes and coaches with real-time performance insights in both weightlifting (i.e. speed and trajectory) and cycling (i.e. aerodynamics) applications. Serious games were adopted by REENACT and PlayHist as a way of increasing quality of learning for students visiting the FHW Smart Venue and presented in the immersive Tholos Dome and on mobile devices. A set of abstract game design patterns were defined as part of the second methodology to provide constructs for creating effective gameplay independent of specific game types and technology implementations.

The multi-domain coverage of the EXPERIMEDIA Platform has created opportunities for transfer of multimedia technologies developed within the lifetime of the project across sectors. Technical advances in one sector can be rapidly transferred to other sectors via the platform, accelerating the opportunity for innovation. For example, real-time 3D reconstruction of moving humans from Kinect is a core capability of the EXPERIMEDIA platform. Initially the capability was developed for high performance sports training the generic capability of 3D acquisition from visual and depth sensors was identified to have potential for collaboration between remote users in different situations to be placed into virtual environments. This led to use of the technology at the FHW Smart Venue for including expert actors into serious games within PlayHist.

What is clear is that novel real-time interactive media delivery mechanisms are transforming social interaction models and immersive experiences. People increasingly connect to each other for work and leisure using augmented and realistic 3D reconstructions of the real world delivered over heterogeneous networks in real-time to indoor and outdoor locations. These capabilities are driving infrastructure requirements. A 3D reconstruction of a moving human from a Kinect sensor produces 100 MB of data a frame (future HD sensors will have much higher data volumes), and with transmission rates of 8 fps with compression of 1:30 a bandwidth of at least 8 Mb/sec is required. Quality of experience requirements in tele-immersive applications requires synchronisation precision of less than 100 ms with a fixed end-to-end latency. Data demands are driving the need for experiments exploring QoS and UX techniques such as end-to-end QoS over fixed and wireless networks, context-based content/infrastructure adaptation and synchronised stream and event processing.

We know that live events are a major driving force for mass audiences. Through digital production, broadcasters can now deliver content more efficiently, flexibly and with greater scalability. However, audiences are demanding enhanced real-time participation in live events and this goes beyond what is possible with current models of media creation and consumption. The next logical step in media production will be the creation of more meaningful relationships between the players at live events, the spectators and the massive online communities at home or on the move. Currently, broadcasters are only skimming the surface of social interactions: posting of viewers opinions such as tweets or blogs alongside programme summaries, capturing an essence of audience engagement through "likes", encouraging personalised media production through user submitted photos or videos, etc. Broadcasters, Games Providers, Event Managers, and to some extent the online communities themselves, must work together closely to offer more engaging and immersive user experiences which can encompass all of the different actors across the various zones of participation.

9.8 Economic and Social Viability of Data Value Chains

Data value chains are at the core of the future digital economy, bringing opportunities for digital developments that build on the increasing availability and processing of all types of data. Today, data value chains focus on intelligent use of data to enable the creation of new products, the optimisation of the production or delivery processes, the improvement of the market, new organisation and management approaches, and the reinforcement of research and development cost reduction of operations, increase of efficiency and better and more personalised services for citizens [2]. However, it's clear that although big data and value chains are driving the new industrial revolution, without design and engagement of the creative industries, such information at worst is meaningless and at best sub-optimal [3].

We have designed and explored many data value chains associated with outdoor leisure activities and sports performance. Smart Ski Goggles is delivering a commercial service to be launched in 2014/2015 ski season. The service enhances visitor experience while skiing on a mountain by delivering real-time information and navigation system using state-of-the-art data goggles incorporating a heads up display. Information about lifts, slopes, weather, hospitality, social media and navigation are integrated into a single application allowing users to explore the region according to their interests. Mixed data were considered including a combination of open, closed, free and personal data. Data and service providers within the local region were engaged to explore cost, revenue and price points for business models supporting long term viability of the service.

What is clear from engaging in a regional ecosystems is that dealing with closed data is fundamental to economic viability. Many business models of the web are built on advertising where data assets can attract large scale online populations. This is not the case for regional data assets that are highly localised. For example, Pinpoint Schladming delivered augmented geospatial open data but the limited user base in Schladming and the availability of information through other channels reduced the potential value of geo-location data application.

Another challenge with data value chains building on open data is as soon as a data asset attains value, owners will have a tendency to close data to protect value rather than contribute it back to the open data pool. Also value is often realised due to scarcity resulting from production costs including costs (e.g. privacy, time, etc) for users involved in observations. For example, lift waiting time was considered high value for skier navigation but the camera installation and video analytics costs were high. Viable solutions required commercial agreements between lift operators, technology providers and the mobile application provider. Price points in business models must consider that the benefit to users must be greater than the cost of data production.

Data value chains were central to improving training programmes and athlete performance at the CAR Smart Venue making extensive use of wearable and non-invasive techniques to capture biomechanics and physiological information. High performance training is a complex endeavour requiring continuous support from specialists responsible for analysis of multi-factor data. Coaches and doctors need accurate measurements in order to offer the correct feedback for performance improvements and the avoidance of injuries. Feedback must be timely and often instantaneous to increase the efficiency of training sessions. 3D Media in Sports used 3D information from Kinect cameras for real-time calculation of cyclists' aerodynamic performance and optimal weightlifting speed and trajectory. 3D Acrobatics used wearable inertia sensors to calculate detailed biomechanics data whereas CARVIREN used wearable device (WIMU) to collect a wide range of athlete data.

The success of solutions in CAR's environment were not driven by economics but the cost of participation by athletes in terms of ergonomics, inconvenience or time. Training sessions are carefully scheduled and choreographed. Wearable technologies that inhibit movement or take significant time to put on or calibrate are deemed unacceptable unless the information captured has significant benefits (e.g. injury avoidance). As a consequence, current techniques have been lab-based and not part of everyday training routines. Experiments conducted in EXPERIMEDIA demonstrated the possibility of moving advanced measurement techniques from the lab to the field without introducing significant costs to the athletes. What we see at CAR in terms of multi-factor measurements will be representative of wider society in future as communities realise visions for quantifying self through wearable technologies.

9.9 Innovation whilst Respecting Privacy

Multimedia systems are developed with human participants and in particular require an increasing understanding of human behaviour and experience to provide meaningful collective experiences to individuals and society. Acquisition, processing and protection of personal data is an essential system feature which must be provided in the context of privacy legislation. Of course, the privacy debate has raged in recent years as US social network providers experiment with society's appetite for disclosing personal information. In many ways, European service providers are not operating on a level playing field but if we believe in preserving and promoting European values, legislation that incorporates such values must be respected.

We have successfully delivered European product and service innovation in the context of EU privacy directives such as Directive 95/46/EC; ii) Directive 2000/31/EC; iii) Directive 2010/13/EU. Although compliance with the correct ethical oversight directives is often perceived as a barrier to progress, performing experimentation in their frame can in fact prepare solutions for

European markets. We use a Privacy Impact Assessment (PIA) methodology to uncover potential privacy risks with multimedia systems and at the same time propose mitigation strategies. Early analysis of the PIA allows for sufficient time to implement the necessary amendments and safeguards to ensure that privacy is taken into account by design, rather than being added at the end of the project development. With the appropriate safeguards, systems were able to collect personal data, profile users and track users indoors and outdoors. Some of the features included the use of secure data storage, encrypted transfer, controlled and auditable access for different classes of data distributed over the same channel and obscuring/removing user identities at source (e.g. in the user's own smartphone or home network, depending on application) to prevent direct user tracing.

BLUE used personal data to correlate cognitive profiles with movements and personal preferences, to see if this knowledge can enhance user experiences in their visit of museums. The cognitive profiles where calculated using a Facebook game and are sensitive personal data. BLUE analysed privacy consequences by exploring questions such as whether the profile would be published on or at least known by Facebook? What if an employer sees it? What if the cognitive style is identified wrongly? An analysis of Facebook's Platform Policy highlighted there is no obligation to send back to Facebook the interpretation or observations on cognitive profiles of the user derived from information extracted from Facebook APIs. If however, the user chooses to publish these results on their profile, then they will be available to their friends, as well as to Facebook.

This example highlights a significant challenge for multimedia systems building on popular social networking sites. PinPoint Schladming, Digital Schladming, MEDIAConnect, BLUE, REENACT and CARVIREN all built on the Facebook Application Programming Interfaces (APIs). Developers are required to use the API in accordance with rules on leveraging content from the underlying social networks as defined in developers' Terms & Conditions ("T&C"). What's clear is that compliance with the Social Networks' T&C can significantly influence system architecture considering rules for publishing content and the increasingly stringent rules for extracting content. Platform providers monitor closely the application ecosystem and demand that the developers cooperate with them, especially in case the application requires a large amount of API calls. Through Terms and conditions Social Network providers maintain their position of power within multimedia systems that rely on social media content.

9.10 Conclusions

Multimedia systems are characterised by those that acquire, process and deliver multiple forms of content in services and applications where user experience is a significant factor for their success. The features of multimedia systems are extremely broad covering all aspects of content lifecycles such as low level signal and image processing, data fusion, transcoding, compression and decompression, network transmission, and rendering. Multimedia systems evolve and are intrinsically linked to content forms that they support.

In recent years, the forms of content available and way content is produced and consumed has changed significantly. Mobile devices, wearable technologies, sensors, cameras and online services are acquiring an increasing array of pervasive, social, audio-visual and 3D content about real world environments and how individual and communities behave. In addition, novel immersive environments, augmented reality devices and high definition displays are transforming user experiences.

Through a multi-domain approach we have identified and explored a cross-section of challenges that are associated with multimedia features and their application. We have presented the features of and opportunities for networked multimedia systems building on the results of experiments conducted at the EXPERIMEDIA facility. We have demonstrated the benefits of the EXPERIMEDIA approach for delivering innovative products and services to specific markets as represented by Smart Venues by conducting experiments at Live Events. Significant commercial opportunities have been delivered by experiments highlighting the innovation potential of EXPERIMEDIA experimentation services realised by ensuring users who participate in observations must also be the same users that realise the primary benefits.

Risks in implementing multimedia solutions in a live context where lots of people are involved are various. For example, defining technology solutions without a business cases or not being able to properly address privacy issues. Both of these can be mitigated, if not completely removed, using EXPERIMEDIA methodologies demonstrating that concerns, for example, regarding privacy and ethical oversight are not a barrier to innovation in experimentally driven research.

CAR's high performance training plans across multiple sports have been radically changed through multi-factor sensing, high definition video and video conferencing technologies. New knowledge has been generated that shows how Quality of Learning can be improved through serious games, personalisation and interactive media technologies at FHW. Real-time geo-spatial information and social recommendation have enhanced visitor experience at Schladming.

What is clear is that networked multimedia systems have huge potential for socio-economic impact and will be transformed through the continuing convergence of infrastructure technologies and the increasing availability of data from IoT platforms and Big Data analytics. However, to realise the benefits of this digital revolution users and user benefit must be at the centre of design processes, and creative experience designers will have a major role to ensure that the explosion of data can be turned into enhanced experiences and sustainable data value chains.

Acknowledgements

The authors would like to thank the following experts for advice and review during the production of this chapter. Gillian Youngs (Faculty of Arts, University of Brighton), Theodore Zahariadis (Synelixis), Tom Gross (Human-Computer Interaction Group, University of Bamberg), Federico Alvarez (Universidad Politécnica de Madrid), Freek W. Bomhof (TNO), David Geerts (K U Leuven).

The EXPERIMEDIA project has received funding from the European Commission as part of the FP7 work programme.

References

[1] European Future Internet Research and Experimentation.
[2] OECD (2012) report on 'Exploring data-driven innovation as a new source of growth', DSTI/ICCP(2012)9/REV1, p. 13.
[3] Designing the Digital Economy, Embedding Growth Through Design, Innovation and Technology, http://www.policyconnect.org.uk/apdig/sites/site_apdig/files/report/463/fieldreportdownload/designcommissionreport-designingthedigitaleconomy.pdf

10

Cross-Domain Interoperability Using Federated Interoperable Semantic IoT/Cloud Testbeds and Applications: The FIESTA-IoT Approach

Martin Serrano[1], Amelie Gyrard[1], Michael Boniface[2], Paul Grace[2], Nikolaos Georgantas[3], Rachit Agarwal[3], Payam Barnagui[4], Francois Carrez[4], Bruno Almeida[5], Tiago Teixeira[5], Philippe Cousin[6], Frank Le Gall[6], Martin Bauer[7], Ernoe Kovacs[7], Luis Munoz[8], Luis Sanchez[8], John Soldatos[9], Nikos Kefalakis[9], Ignacio Abaitua Fernández-Escárzaga[10], Juan Echevarria Cuenca[11], Ronald Steinke[12], Manfred Hauswirth[12], Jaeho Kim[13] and Jaeseok Yun[13]

[1]Insight Centre for Data Analytics, Ireland
[2]IT Innovation, UK
[3]INRIA, France
[4]UNIS, United Kingdom
[5]Unparallel Innovation, Lda (UNPARALLEL, Portugal)
[6]Easy Global Market, France
[7]NEC Europe Ltd. UK
[8]University of Cantabria, Spain
[9]Athens Information Technology, Greece
[10]Sociedad para el desarrollo de Cantabria, Spain
[11]Ayuntamiento de Santander, Spain
[12]Fraunhofer FOKUS, Germany
[13]Korea Electronics Technology Institute, Korea

10.1 Introduction

The Internet-of-Things (IoT) [61] has been identified as one of the main pillars of the world's economies and the technology enabler for the evolution of the societies and for the future developments and improvement of the Internet

[4]. A large number of research activities in Europe have been working in this direction i.e. FP7 projects in the context of Future Internet Research and Experimentation (FIRE) initiative. FIRE projects have already demonstrated the potential of IoT technologies and deployments in a number of different application areas including transport, energy, safety and healthcare. FIRE deployments and project results have also demonstrated the advantages of implementing Smart Cities testbeds (national and EU scale) both have been extensively reported in [5]. Smart City testbeds are the key places for large demonstration of IoT concepts and technology. Smart cities testbeds are prone to be large scale, highly heterogeneous and target a diverse set of application domains.

In Smart cities despite the growing number of IoT deployments, multiple installations and related testbeds, the majority of deployed IoT applications tend to be self-contained, thereby forming application silos [50]. Recent research efforts have been focused on demonstrate the capacity of IoT systems to be part of an overall arch-systems called federation (e.g., FP7 Fed4FIRE), in a federated environment it is possible the co-existence and co-operation of multiple infrastructures (including IoT testbeds). The Federation is the first step to the integration of these silos, since they provide a wide range of indispensible low-level capabilities such as resource reservation and negotiation. Nevertheless, these efforts tend to be heavyweight and do not adequately deal with the need to access diverse IoT datasets in a flexible and seamless way. In a federation one of the mayor challenges is the data centric integration and the combination of data silos that is identified as a under investigation area for IoT [4], and with a very rich potential both in terms of novel experimentation (e.g., in the scope of living labs and IoT testbeds) [49] and in terms of added-value enterprise applications. Related to data, the ability to combine and synthesize data streams and services from diverse IoT platforms and testbeds remains a challenge and multiple researches follows the promise to broaden the scope of potential data interoperability applications in size, scope and targeted business context. In the Internet of tings area the ability to repurpose and reuse IoT data streams across multiple experimental applications can positively impact the Return-on-Investment (ROI) associated with the usually costly investments in IoT infrastructures and testbeds. The integration, combination and interoperability of IoT silos is fully in-line with the overall FIRE vision that makes part of the Horizon 2020 program, which aspires to allow European experimenters/researchers to investigate/develop leading-edge, ubiquitous and reliable computing services, as well as seamless and open access to global data resources.

The futuristic vision of integrating IoT platforms, testbeds and associated silo applications is related with several scientific challenges, such as the need to aggregate and ensure the interoperability of data streams stemming from different IoT platforms or testbeds, as well as the need to provide tools and techniques for building applications that horizontally integrate silo platforms and applications. The convergence of IoT with cloud computing is a key enabler for this integration and interoperability, since it allows the aggregation of multiple IoT data streams towards the development and deployment of scalable, elastic and reliable applications that are delivered on-demand according to a pay-as-you-go model. During the last 4–5 years we have witnessed several efforts towards IoT/cloud integration (e.g., [29, 39]), including open source implementations of middleware frameworks for IoT/cloud integration [23, 52] and a wide range of commercial systems (e.g., Xively (xively.com), ThingsWorx (thingsworx.com), ThingsSpeak (thingspeak.com), Sensor-Cloud (www.sensor-cloud.com)). While these cloud infrastructures provide means for aggregating data streams and services from multiple IoT platforms, they are not fully sufficient for alleviating IoT fragmentation of facilities and testbeds. This is because they emphasize on the syntactic interoperability (i.e. homogenizing data sources and formats) rather on the semantic interoperability of diverse IoT platforms, services and data streams.

Recently several IoT projects [33] have started to work on the semantic interoperability of diverse IoT platforms, services and data streams. To this end, they leverage IoT semantic models (such as the W3C Semantic Sensor Networks (SSN) ontology [16, 58]) as a means of achieving interoperable modeling and semantics of the various IoT platforms. A prominent example is the FP7 OpenIoT project, a (BlackDuck) award winner open source project in 2013, which has been developed and released as an open source blueprint infrastructure [51] addressing the need for semantic interoperability of diverse sensor networks at a large scale (see also https://github.com/OpenIotOrg/openiot). The semantic interoperability of diverse sensor clusters and IoT networks is based on the virtualization of sensors in the cloud. At the heart of these virtualization mechanisms is the modeling of heterogeneous sensors and sensor networks according to a common ontology, which serves as harmonization mechanism of their semantics, but also as a mechanism for linking related data streams as part of the linked sensor data vision. This virtualization can accordingly enable the dynamic discovery of resources and their data across different/diverse IoT platforms, thereby enabling the dynamic on-demand formulation of cloud-based IoT services (such as Sensing-as-a-Service services). Relevant semantic interoperability techniques are studied in depth as part of the fourth activity chain of the IERC cluster (IERC-AC4) (see for example [17]). Similar techniques could serve as a basis for unifying and integrating/linking geographically and administratively dispersed IoT testbeds, including those that have been established as part of FIRE projects. Such integration holds the promise of adding significant value to all of the existing IoT testbeds, through enabling the specification and conduction of large-scale on-demand experiments that involve multiple heterogeneous sensors, Internet Connected Objects (ICOs) and data sources stemming from different IoT testbeds.

Based on the above-mentioned Sensing-as-a-Service paradigm, dynamic virtualized discovery capabilities for IoT resources could give rise to a more general class of Experiment-as-a-Service (EaaS) applications for the IoT domain. EaaS services are executed over converged IoT/cloud platforms, that are developed on the basis of the technologies outlined above. EaaS services are not confined to combinations of sensor queries (such as Sensing-as-a-Service), but they would rather enable the execution of fully-fledged experimental workflows comprising actuating and configuration actions over the diverse IoT devices and testbeds. The benefits resulting from the establishment and implementation of an EaaS paradigm for the IoT domain include:

- The expansion of the scope of the potential applications/experiments that are designed and executed. Specifically, the integration of diverse testbeds for offering to the European experimenters/researchers with the possibility of executing IoT experiments that are nowadays not possible.
- The ability to repurpose IoT infrastructures, devices and data streams in order to support multiple (rather than a single) applications. This increases the ROI associated with the investment in the testbeds infrastructure and software.
- Possibility for sharing IoT data (stemming from one or more heterogeneous IoT testbeds) across multiple researchers. This can be a valuable asset for setting up and conducting added-value IoT experiments, since it enables researchers to access data in a testbed agnostic way i.e. similar to accessing a conventional large scale IoT database.
- The emergence of opportunities for innovative IoT applications, notably large scale applications that transcend multiple application platforms and domains and which are not nowadays possible.
- The avoidance of vendor lock-in, when it comes to executing IoT services over a provider's infrastructure, given that an EaaS model could boost data and applications portability across diverse testbeds.

Beyond the interconnection and interoperability of IoT and smart cities testbeds, semantic interoperability tools and techniques could also enable the wider interoperability of IoT platforms, which is a significant step towards a global IoT ecosystem.

10.2 Federated IoT Testbeds and Deployment of Experimental Facilities

Addressing the need of IoT federated infrastructures and following the interoperability need and the use of semantics IoT/cloud Testbeds and applications the FIESTA project aim to be a globally unique infrastructure for integrated IoT experimentation based on the federation of multiple interoperable IoT testbeds. FIESTA targets the main objective for defining and implementing a Blueprint IoT Experimental Infrastructure that can offer services and tools for external applications and mainly for enabling the concept Experimentation as a Service "EaaS". FIESTA look at researching and establishing a novel blueprint infrastructure for IoT platforms/testbeds interoperability and EaaS (Experimentation-as-a-Service), which enables researchers, engineers and enterprises (including SMEs) to design and implement integrated IoT experiments/applications across diverse IoT platforms and testbeds, through a single entry point and based on a single set of credentials. The EaaS infrastructure facilitates experimenters/researchers to conduct large scale experiments that leverage data, information and services from multiple heterogeneous IoT testbeds, thereby enabling a whole new range of innovative applications and experiments.

FIESTA has implemented the testbed agnostic access to IoT datasets, providing tools and techniques enabling researchers to share and access IoT-related datasets in a seamless testbed agnostic manner i.e. similar to accessing a large scale distributed database. This also has involved the use of linking diverse IoT datasets, based mainly on the linked sensor data concept. FIESTA has implemented tools and techniques for IoT Testbeds Interoperability and Portability by providing tools and techniques (semantic models, directory services, open middleware, tools) for virtualizing and federating geographically and administratively dispersed IoT platforms and testbeds. Special emphasis was done in the specification and implementation of common standardized APIs for accessing the underlying testbeds, thereby boosting the portability of IoT experiments. FIESTA has also research and implement the meta-cloud infrastructure along with accompanying tools (i.e. portal, development, workflow management, monitoring) facilitating the use of the EaaS infrastructure for the design, implementation, submission, monitoring and evaluation of IoT/cloud related experiments and related integrated applications.

FIESTA developed a global market confidence programme (as a Sustainability Vehicle) for enabling IoT platform/testbed providers and IoT solutions providers to test, validate and ensure the interoperability of their platforms/solutions against FIESTA standards and techniques. The programme includes a certification suite for compliance testing. As part of pursuing this objective, FIESTA ensures the development and realization of a clear sustainability path for the project's results. Furthermore, it defined ways for collaboration with other bodies and working groups, which are currently working (at EU level) towards the establishment of similar initiatives, such as the IoT forum. FIESTA is implemented in the way to be a blueprint experimental infrastructure for EaaS on the basis of the federation and virtualization of real-life IoT testbeds, but also on the basis of real-life experiments that have be designed, executed and evaluated over them. These span the areas of pollution monitoring, crisis management, crowdsensing as well as enterprise/commercial activities and emphasize portability and testbed agnostic access.

FIESTA implemented a stakeholders engagement program to guarantee the expansion in terms of experiments and testbeds by meaning of the involvement of third parties towards a global IoT experimentation ecosystem). The FIESTA ecosystem is to attract and engage stakeholders beyond the project consortium as third parties through managing an open calls process, but also through the mobilization of (third-party) research communities with a strong interest in IoT. FIESTA permanently works towards the identification and generation of reference activities to elicit and document a range of best practices facilitating IoT platform providers and testbed owners/administrators to integrate their platform/testbed within FIESTA, along with best practices addressed to researchers, engineers and organizations wishing to use the FIESTA meta-cloud EaaS infrastructure for conducting innovative applications and experimentation.

In order to validate the global and federated character of the FIESTA infrastructure, FIESTA has already established collaborations and liaisons with IoT partners in Asia (Korea) and USA. In particular, the consortium includes a Korean partner (KETI), that has also a established IoT collaborations with US organizations (thanks to the Inria's collaboration with the Silicon Valley as part of the Inria@Silicon Valley programme). Note that KETI's participation in the consortium has allowed the integration/federation of a testbed located in

Asia (i.e. KETI's testbed) to the FIESTA EaaS infrastructure. At the same time, the above-listed collaborations ensures the global dissemination and outreach of the project's results, while also broadening the scope of participation in the third-party selection processes of the project (i.e. open calls) on the basis of participants from Asia and USA.

FIESTA has allocated a significant share (31%) of its foreseen budget to the introduction of third-parties (through the open calls process), notably third-parties that have started the undertaken and the conduction of new experiments and/or the blending/integration of new testbeds within the FIESTA infrastructure. Note that the stakeholders' community of the project also serves as a basis for validating the global market confidence programme of the project. The active engagement of the stakeholders in the project, but also in the third-parties selection process are boosted by FIESTA partners already animating ecosystems of researchers and enterprises (i.e. SODERCAN, Com4innov), as well as from participants from non-EU countries (i.e. KETI from Korea). Links to participants from Asia and USA are also sought (through KETI and the Inria@Silicon Valley programme). The ultimate vision of FIESTA is to provide the basis of a global IoT experimentation ecosystem.

10.3 Cross-Domain Interoperability

FIESTA project has opened new horizons in the development and deployment of IoT applications and experiments not only at a EU but also global scale, based on the interconnection and interoperability of diverse IoT platforms and testbeds FIESTA has created an ecosystem of IoT experimentation. To this end, FIESTA provides a blueprint experimental infrastructure, tools, techniques, processes and best practices enabling IoT testbed/platforms operators to interconnect their facilities in an interoperable way, while at the same time facilitating researchers and solution providers in designing and deploying large scale integrated applications (experiments) that transcend the (silo) boundaries of individual IoT platforms or testbeds. FIESTA enables researchers and experimenters to share and reuse data from diverse IoT testbeds in a seamless and flexible way that has open up new opportunities in the development and deployment of experiments and for exploiting data and capabilities from multiple testbeds. The blueprint experimental infrastructure provided by FIESTA includes a middleware for semantic interoperability, tools for developing/deploying and managing interoperable applications, processes for ensuring the operation of interoperable applications, as well as best practices for adapting existing IoT facilities to the FIESTA interoperability infrastructure.

The FIESTA infrastructure empowers the Experimentation-as-a-Service (EaaS) paradigm for IoT experiments, while also enables experimenters to use a single EaaS API (i.e. the FIESTA EaaS API) for executing experiments over multiple IoT federated testbeds in a testbed agnostic way i.e. like accessing a single large scale virtualized testbed. Experimenters are therefore able to learn easily how to connect with the EaaS API and accordingly use it to access data and resources from any of the underlying testbeds. To this end, the underlying interconnected testbed provides common standardized semantics and interfaces (i.e. FIESTA Testbed Interfaces) enables the FIESTA EaaS infrastructure to access their data, resources and other low-level capabilities (Figure 10.1). Note that the FIESTA EaaS infrastructure is accessible through a cloud computing infrastructure (conveniently called FIESTA meta-cloud), on the basis of a cloud-based on-demand paradigm.

FIESTA also includes a directory service (conveniently called FIESTA meta-directory), where sensors and IoT resources from multiple testbeds are registered. This directory enables the dynamic discovery and use of IoT resources (e.g., sensors, services) from all the interconnected testbeds. Overall, the project's experimental infrastructure provides to the European

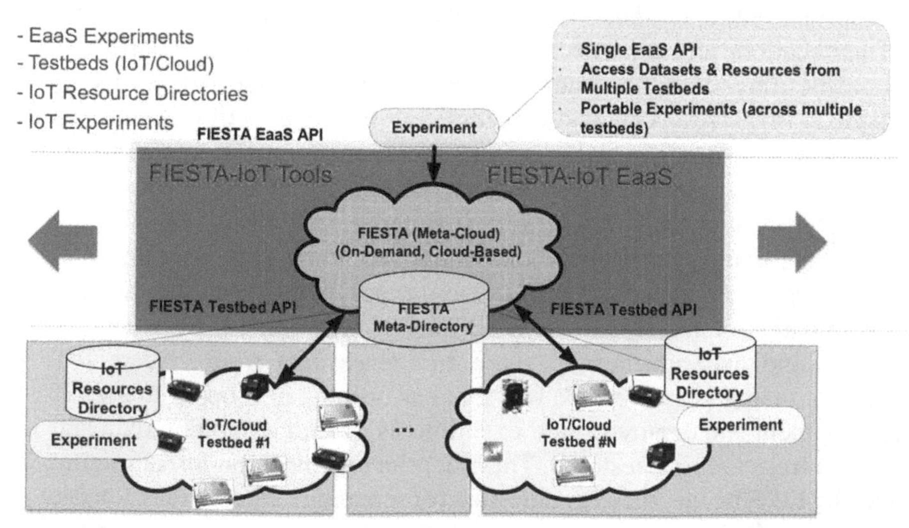

Figure 10.1 FIESTA interoperability model for heterogeneous IoT testbed experimentation.

experimenters in the IoT domain with the following unique capabilities (Figure 10.1):

- Access to and sharing of IoT datasets in a testbed-agnostic way. FIESTA provides researchers with tools for accessing IoT data resources (including Linked sensor data sets) independently of their source IoT platform/testbed.
- Execution of experiments across multiple IoT testbeds, based on a single API for submitting the experiment and a single set of credentials for the researcher.
- Portability of IoT experiments across different testbeds, through the provision of interoperable standards-based IoT/cloud interfaces over diverse IoT experimental facilities.

FIESTA technology leverages recent results on IoT semantic interoperability, notably results produced as part of the AC4 activity chain of the IERC, as well as within relevant projects in the IoT (e.g., FP7 OpenIoT) and FIRE (e.g., Fed4FIRE) areas. In particular, IoT projects offers the foundations of semantic interoperability at the IoT data and resources levels, while FIRE projects contribute readily available results in the area of reserving and managing resources across multiple testbeds. On the basis of these results, FIESTA research, design and deliver an open middleware infrastructure (i.e. semantics and APIs) for the virtualization and federation of IoT testbeds that enable sharing and access to a wide range of IoT-related datasets. FIESTA's infrastructure comprise semantic models enabling the virtualization, as well as middleware libraries facilitating the streaming and semantic annotation of IoT from the various testbeds in a single unified cloud infrastructure (FIESTA cloud). The FIESTA cloud therefore aggregates, manages and linked data from the various testbeds, while at the same time providing methods and tools that enables researchers to access them in a flexible and testbed-agnostic way. Therefore, the FIESTA cloud act as a meta-testbed, which integrates, linked and uses information sources from a variety of IoT/cloud testbeds.

FIESTA cloud enables European experimenters/researchers to design, implement, execute and evaluate IoT experiments based on data from various IoT testbeds all over Europe. To this end, FIESTA also offers a wide range of tools facilitating experimenters in the above tasks. These include: a) A portal infrastructure serving as a single entry point for setting up and submission of IoT experiments and the monitoring of their progress, b) Tools for designing and enacting experiments in terms of IoT/cloud services and workflows, c) Tools for sharing, linking and accessing datasets in a testbed agnostic way,

d) Tools and techniques for monitoring and managing the FIESTA cloud, including monitoring of all the necessary aspects of the underlying testbeds and e) Tools and techniques for monitoring the status of experiments and collecting data for evaluating the experiments. These tools are an integral element of the project's Experiment-as-a-Service paradigm for the IoT domain.

FIESTA establishes, implement and support a global market confidence programme, on the basis of its blueprint infrastructures and processes, that encourages and facilitate stakeholders to comply with the FIESTA interoperability guidelines and accordingly to deploy large scale innovative interoperable IoT applications. The FIESTA global market confidence programme includes a certification/compliance suite enabling platform providers and solution providers to test and ensure the level of interoperability of their platforms and services. This programme is a main vehicle for the sustainability of the project's results, as well as for impact creation at a global scale. Note that the programme is used as a vehicle for the sustainability of the project's results. During its lifetime FIESTA boost and ensure the engagement and participation of multiple platforms providers within Europe (including both consortium members and third-parties) in the FIESTA global confidence programme. Based on this engagement, FIESTA ensures the proper design, implementation, validation and fine-tuning of the programme.

FIESTA integrates diverse IoT testbeds (three in EU and one in Korea), towards providing experimenters with the possibility of designing, implementing, executing and evaluating sophisticated IoT (EaaS based) experiments that are not possible nowadays. To this end, the project leverages recent advances and results associated with semantic interoperability for IoT applications towards federating multiple IoT testbeds. FIESTA specify the scope of the IoT platforms and testbeds integration, federation and interoperability in terms of the functionalities that should be supported, the business/research actors that have access to specific functionalities of the testbeds, their EaaS model, as well as type of experiments that are enabled. FIESTA attempts to cover all aspects of IoT testbeds integration, including technology aspects (i.e. the technologies needed), business aspects (including how to run the confidence programme and ensure the longer term sustainability of the FIESTA model), organization (e.g., the processes needed to deploy/operate interoperable platforms and applications), as well as innovation aspects.

FIESTA has been validated on the basis of the federation of four existing real-life diverse IoT testbeds (provided by partners UNICAN/SDR, UNIS, Com4Innov and KETI), which include prominent European FIRE testbeds (such as SmartSantander), as well as Korean testbed (accessible through partner KETI). FIESTA first federate these testbeds and accordingly with testbed specifications validate the federated/virtualized infrastructure on the basis of a range of EaaS experiments covering both e-science and e-business purposes. The project's experiments (which are detailed in following paragraphs) unveil the unique capabilities of the FIESTA infrastructure in terms of testbed-agnostic data sharing, execution of experiments across multiple testbeds, as well as ensuring the portability of IoT experiments across different testbeds.

In order to accomplish its goals, the project issue, manage and exploit a range of open calls towards involving third-parties in the project. The objective of the involvement of third-parties is two-fold:

- To ensure the design and integration (within FIESTA) of more innovative experiments, through the involvement of additional partners in the project (including SMEs). The additional experiments focuses on demonstrating the added-value functionalities of the FIESTA experimental infrastructure.
- To expand the FIESTA experimental infrastructure on the basis of additional testbeds. In this case the new partners undertake to contribute additional testbeds and to demonstrate their blending and interoperability with other testbeds (already adapted to FIESTA). As part of this blending, the owners of these testbeds also engage with the project's global market confidence programme, which provide them with the means to auditing the interoperability and openness of their platforms.

The involvement of third-parties therefore play an instrumental role for the large scale validation of the FIESTA experimental infrastructure, but also for the take-up of the project's global market confidence programme on IoT interoperability. It is also a critical step to the gradual evaluation of FIESTA towards an infrastructure/ecosystem for global IoT experimentation, as shown in Figure 10.2.

Beyond the validation of the FIESTA infrastructure on the basis of practical experiments and the integration of additional IoT testbeds, the project specify concrete best practices for the federation of testbeds (addressed to testbed owners/administrators) wishing to become part of the virtualized meta-cloud infrastructure of the project. Similar best practices are also produced for European researchers and enterprises (including SMEs) wishing to design and execute experiments over the FIESTA EaaS infrastructure. These best practices have been disseminated as widely as possible, as part of the project's efforts to achieve EU-wide/global outreach. The attraction and engagement of researchers and enterprises in the use of the FIESTA EaaS infrastructure is another vehicle for the sustainability and wider use of the project's results,

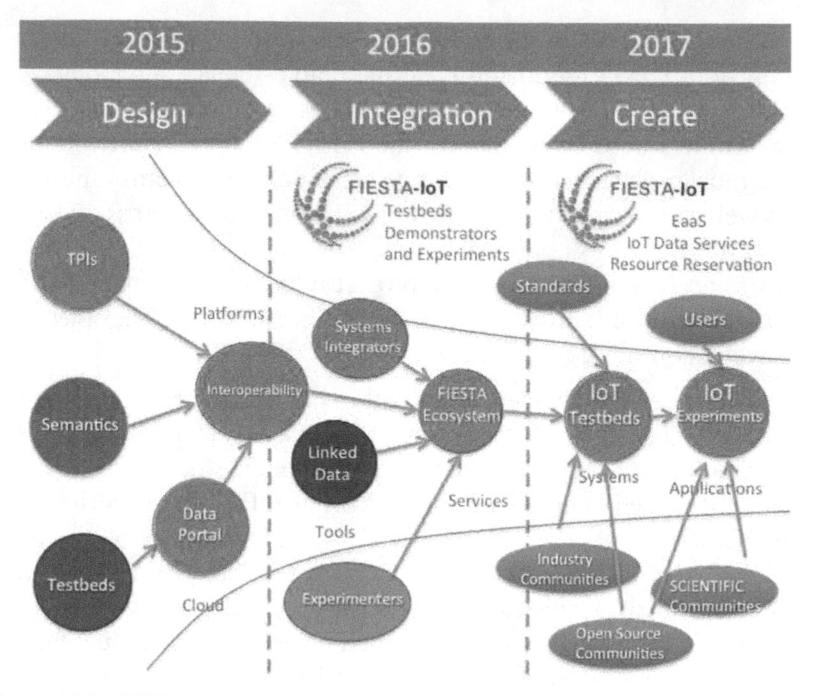

Figure 10.2 FIESTA evolution towards an ecosystem for global IoT experimentation.

which complement the global market confidence programme outlined above. This is overall in-line with the vision of establishing a global ecosystem for IoT experimentation (as already shown in Figure 10.2)

10.4 Experimentation as a Service

The FIESTA overall approach comprises a range of research activities that aims at setting up and validating the FIESTA EaaS model and associated blueprint experimental infrastructure, as well as a range of exploitation and sustainability activities that deals with the design and activation of the project's global market confidence project on IoT interoperability. A set of demonstration activities have been carried out in order to showcase the capabilities of the FIESTA infrastructure on the basis of the design and execution of novel experiments.

The FIESTA project's methodology towards researching and providing the FIESTA Experimentation as a Service (EaaS) paradigm, involves the following groups of activities, and the details are further analysed in following paragraphs:

- Analysing requirements for EaaS experimentation in the IoT domain, and specifying the detailed technical architecture of the FIESTA experimental (meta-cloud) infrastructure, including its (meta) directory of IoT resources.
- Research towards virtualizing access to the individual testbeds and their resources. This includes the provision of common standards-based interfaces and APIs (i.e. FIESTA Testbed APIs) for accessing datasets and resources in the various testbeds, according to common semantic models (ontologies).
- Research towards creating the FIESTA meta-cloud EaaS infrastructure, which enables experimenters to access data and resources from any of the underlying testbeds in a testbed agnostic way i.e. similar to accessing a single large scale virtualized testbed.

FIESTA Engineering Requirements: The FIESTA engineering requirements activities have produced the requirements associated with testbed-agnostic experimentation, as well as with the EaaS model to designing and conducting IoT experiments. They were planned early in the project's work plan and have produced the interoperability requirements and more, based on a variety of modalities for collecting and analysing requirements, including analysis of state-of-the-art, contact with stakeholders (including researchers and experimenters), analysis of the various IoT testbeds etc.

FIESTA Architecture and Technical Specifications: The FIESTA requirements have been taken into account towards producing detailed technical specifications for the EaaS model. Furthermore, a technical architecture have been established, specifying the FIESTA (meta-cloud) EaaS infrastructure, its tools, the meta-directory of IoT resources, as well as the interfaces of the above-listed components to individual FIESTA platforms and testbeds. The architecture drives the organization and integration of research tasks associated with the individual components of the FIESTA solution.

FIESTA Research on semantic interoperability for IoT (data and resources): The project's methodology includes a dedicated set of activities that aim at realizing IoT platforms/testbeds semantic interoperability at both the data and resources levels. To this end, FIESTA selects and extends the ontologies that provide the common semantics of the FIESTA interoperable infrastructure, while also working on the federation and linking of the heterogeneous data streams. As a result of the research, a set of blueprint middleware libraries enabling each testbed to adapt its data and resources to the common produced semantics.

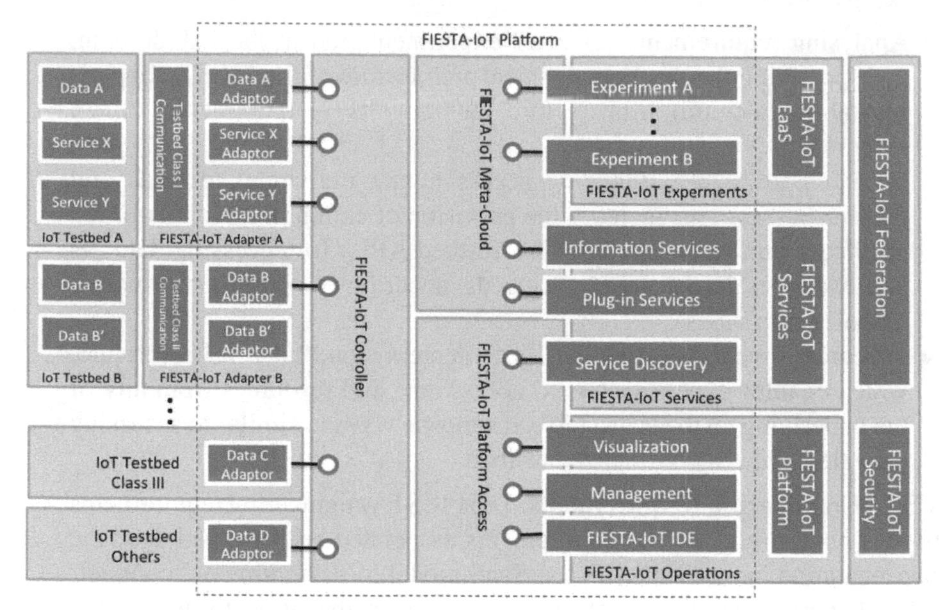

Figure 10.3 FIESTA EaaS experimental infrastructure overview.

FIESTA Research on virtualized access to IoT/cloud infrastructures: In addition to developing the models that ensures the common semantics of resources and data across various testbeds, FIESTA project have provided a set of standards-based portable interfaces for accessing the various IoT/cloud infrastructures. The interfaces ensure that the FIESTA infrastructure can be seamlessly expanded on the basis of additional platforms/testbeds that support the specified standards-based interfaces.

FIESTA is in-line with the directions identified and prioritized as part of recent FIRE roadmaps in the areas of IoT and its convergence with cloud computing and smart city applications. FIESTA project addresses the challenges identified in recent support actions (e.g., the AmpliFIRE Support Action) for the FIRE domain. Figure 10.3 illustrates the main elements of the FIESTA EaaS infrastructure, which are further analysed in later paragraphs.

10.5 IoT Data Marketplace

FIESTA tools and techniques for accessing data in a testbed agnostic way defines a number of tools enabling submission of experiments, testbed agnostic access to (shared) data, as well as authentication and authorization of the users are implemented and make available over the FIESTA meta-cloud infrastructure. FIESTA meta-cloud infrastructure has provided a meta-cloud infrastructure enabling access to data and resources from a wide range of underlying testbeds. This infrastructure leverages the semantics and interfaces that make FIESTA meta-cloud to serve as single entry point of the EaaS infrastructure. It also includes a (meta) directory service, which enables dynamic discovery and dynamic access to resources from any of the underlying virtualized testbeds.

The project's demonstration activities are focused on validating and demonstrating the FIESTA IoT Data Marketplace on the basis of three experiments that are designed and executed by project partners, but also based on several experiments that are executed by third-parties to be selected based on open calls processes. FIESTA demonstration of IoT Data Marketplace in a way of innovative experiments on Testbed agnostic data access and by sharing that data as a means of validating the FIESTA infrastructure is generated by using a number of innovative experiments over the FIESTA infrastructure that is being developed and demonstrated by the end of the project duration.

The focus on the IoT Data Marketplace is in three fold: a) Access to data and services from multiple IoT testbeds, b) Experiments portability across testbeds (i.e. provided that testbeds provide the sensors and/or resources needed to execute the experiment and c) Dynamic discovery of sensors and resources across multiple testbeds. A great deal of the demonstration activities is also based on new experiments to be introduced as part of the Open Calls processes of the project.

10.6 FIESTA Platform Services and Tools

FIESTA intends to become a first of a kind experimental infrastructure, which provides researchers with the capabilities of accessing data and services from multiple IoT testbeds in a seamless and testbed agnostic way. This enables researchers to design and enact more sophisticated and more innovative experiments, as part of their projects and product development processes. The realization of the FIESTA vision requires significant scientific and technological advancements in the areas of semantic interoperability of IoT testbeds, the linking of related IoT data streams, the development of IoT architectures suitable for federating multiple (cloud-based) testbeds, the provision of standards-based interfaces for accessing the various IoT/cloud testbeds, as well as the development of an on-demand EaaS model to executing experiments. The scientific and technological objectives of the project are ground breaking since this allow researchers to experiment with data sets that stem for administratively and geographically dispersed testbeds, while at the same time ensuring the portability of the experiments across testbeds with similar/analogous capabilities. These advancements represent the scientific and technological ambition of the project. At the same time, the project has ambitious objectives associated with the sustainability and market take-up of the project's results, based on the establishment of the global market confidence programme for IoT interoperability. These ambitious targets are presented in the following paragraphs.

10.6.1 FIESTA Approach on Global Market Confidence Programme on Interoperability Service

A global market confidence programme on IoT interoperability has been designed as a vehicle for the sustainability of the project's results, but also as a means of offering these results in a structured way to many experimenters (i.e. individuals researchers and enterprises (including SMEs)). FIESTA operate the global market confidence programme on IoT interoperability, towards boosting the sustainability of the project's results, as well as towards using semantic interoperability as a vehicle for alleviating vendor lock-in and the related fragmentation of the IoT market.

The programme is designed to be validated on the basis of the auditing and certification of several IoT platforms for their interoperability against FIESTA standards and guidelines. IoT platforms/testbeds are contributed by project partners (based also on their background projects), but also by new participants joining the project following open calls processes.

The methodology of the project includes activities that aim at attracting stakeholders in the adoption and use of the project's results, based on the global market confidence programme of stakeholders. FIESTA caters for the support of these stakeholders, through providing focused training and consulting, relevant to the project's interoperability programme.

In addition to opportunities derived from the global market confidence programme on IoT interoperability, the FIESTA project conducts a wide range of dissemination and communication activities aiming at supporting the exploitation strategy and goals of the project. Likewise, all partners that are involved in exploiting the project in line with their business and research strategies, also, a set of created business plans in relation to the FIESTA exploitable products and services.

10.6.2 FIESTA Approach on Linking and Reasoning over IoT Data Streams Services

FIESTA's work on semantic interoperability of data streams is to ensure the accessibility of heterogeneous input streams in a uniform format, as well as the ability to support/implement a uniform access paradigm to these data. In addition to alleviating the complexity of the data access process, this interoperability also empower large-scale reasoning over the multiple diverse data streams, towards linking related data streams and enabling large scale experiments, as well as experiments with richer functionality.

The most promising approach towards linking data streams is the use of Linked Open Data (LOD) standards [30] along with semantic annotations and uniform access with RESTful services (REST: REpresentational State Transfer) down to the physical sensor level. Linked Data ensures a uniform data model based on an underlying graph-based/network model (vs. a traditional relational model) capable of representing arbitrary information models in an intuitive and straightforward way. Linked Data models are used already in many domains, such as the Web, enterprise information systems, e-government (e.g., http://data.gov.uk), social networks (e.g., W3C Semantic Interoperability of Online Communities (SIOC) standard), sensors data (W3C Semantic Sensor Networks Incubator Group (SSN-XG), http://www.w3.org/2005/Incubator/ssn/), etc. with a trajectory of massive further growth. Uniform access in a RESTful way using Linked Data originated from Web-based information systems and has become the standard on Web-based systems and for accessing social media, e.g., Twitter REST API (https://dev.twitter.com/docs/api), as well as for many enterprise service solutions. Recently, also the IoT world has committed to RESTful access through the on-going standardization of the COnstrained Application Protocol (COAP, https://datatracker.ietf.org/doc/draft-ietf-core-coap/) and Constrained RESTful Environments (CORE, http://datatracker.ietf.org/wg/core/charter/) by the IETF. A complete stack for Linked Data based on these abstractions has bee developed by the FP7 project SPITFIRE (Semantic Service Provisioning for the Internet of Things, http://www.spitfire-project.org/).

Dynamic cost models and support for scalable and efficient processing are missing [60] as are query approximation and relaxation techniques for "close matches" [32]. Stream query processors for Linked Streams can already provide reasoning support up the level of expressivity of SPARQL (http://www.w3.org/TR/rdf-sparql-query/). The most relevant systems are CQELS, C-SPARQL [10], and EP-SPARQL [3] among a number of research prototypes (e.g., Sparkwave, which, however, does not have comprehensive performance evaluation results available, thus not making it comparable to the above 3 systems). These systems all share the same approach of utilizing SPARQL-like specification of continuously processed queries for streaming RDF data. If more complex reasoning is required, other approaches like nonmonotonic logic programming are required. Stream processing engines which augment stream reasoning through this kind of approach are still limited, but include those such as the use of Prova [36, 59] and Streaming IRIS [37]. Although based on logic programming, these approaches do not gain the inherent benefits of Answer Set Programming (ASP) syntax and semantics in terms of expressivity. In terms of performance, Prova is more concerned about how much background (static) knowledge can be pushed into the system, while Streaming IRIS does not test complex reasoning tasks. To the best of the consortium's knowledge, the work by Do [21] is probably the only other current stream reasoning approach for the Semantic Web that utilizes ASP. Although the work is quite recent, their approach is still much more prototypical. More importantly, this approach does not pertain to continuous and window-based reasoning over stream data.

10.6.3 FIESTA Approach on Federating IoT Stream Data Management Services

As we are heading towards a world of billions of things [26], IoT devices are expected to generate enormous amount of (dynamically distributed) data streams, which can no longer be processed in real-time by the traditional centralized solutions. Thus, IoT needs a distributed data management infrastructure to deal with heterogeneous data stream sources which autonomously generates data at high rates [9]. An early system designed to envision a world wide sensor web [11] is IrisNet, which supports distributed XML processing over a worldwide collection of multimedia sensor nodes, and addresses a number of fault-tolerance and resource-sharing issues. A long the same line, HiFi [24] also supports integrated push-based and pull-based queries over a hierarchy where the leaves are the sensor feeds and the internal nodes are arbitrary fusion, aggregator, or cleaning operators. A series of complementary database approaches aimed to provide low-latency continuous processing of data streams on a distributed infrastructure. The Aurora/Medusa [13], Borealis [1], and TelegraphCQ [12], StreamGlobe [53], StreamCloud [27] are well-known examples of this kind. These engines provide sophisticated fault-tolerance, load-management, revision-processing, and federated-operation features for distributed data streams. A significant portion of the stream processing research merit of these systems has already made its way from university prototypes into industry products such as TIBCO StreamBase, IBM Stream InfoSphere, Microsoft Streamlight. However, such commercial products are out of reach of most IoT stream applications and there have not been any comprehensive evaluation in terms of cost effectiveness, performance and scalability. Due to this reason, there have emerged open source stream processing platforms from Apache Storm [54], S4 [55] and Spark [56] which were primarily built for some ad-hoc applications: Twitter, Yahoo!. While these platforms aim to support elasticity and fault-tolerance, they only offer simple generic stream processing primitives that require significant effort to build scalable stream-based applications.

The above systems provide steps in the right direction for managing IoT data streams in distributed settings. However, they have several federation restrictions in terms of systems of systems and system data organization. For system organization, most of distributed stream processing engines are extended from a centralized stream-processing engine to distributed system architectures. Thus, in order to enable the federation among stream processing sites, they have to follow strictly predefined configurations. However, in IoT settings, heterogeneous data stream sources are provided by autonomous infrastructures operated on different independent entities, which usually do not have any prior knowledge about federation requirements. In particular, a useful continuous federated query might need to compare or combine data from many heterogeneous data stream sources maintained by independent entities. For example, a tourist guide application might need to combine different data stream relevant to the GPS location of users, e.g., weather, bus, train location, flight updates, tourist events. Also, they might then correlate these streams with similar information from other users who have social relationships with the user via social networks such as Twitter, Facebook and also with back ground information like OpenStreetMap, Wikipedia. In such examples, stream data providers did not only agree how their systems are used to process those federated queries but also they did not agree on data schema/format to make data able to be queried for the federated query processing engine. Note however that the need of having uniform and predefined data schema and formats poses various difficulties for query federation on IoT applications using heterogeneous stream data sources.

In FIESTA the lack of standards has been studied as the major difficulty leading to restrictions, and the wide (and changing) variety of application requirements. Existing IoT Stream processing engines vary widely in data and query models, APIs, functionality, and optimization capabilities. This has led to some federated queries that can be executed on several IoT stream providers based on their application needs. Semantic Web addresses many of the technical challenges of enabling interoperability among data from different sources. Likewise, Linked Stream Data enables information exchange among stream processing entities, i.e., stream providers, stream-processing engines, stream consumer with computer-processable meaning (semantics) of IoT stream data. There have been a lot of efforts towards building stand-alone stream processing engine for Linked Stream Data such as C-SPARQL 10], SPARQLstream [10], EP-SPARQL, 6]. The data and query-processing model of Linked Stream data has been standard by W3C [46]. However, there are only few on-going efforts of building scalable Linked Stream Data processing engines for the cloud like Storm and S4 respectively, i.e., CQELS Cloud [31]. None of them supports federation among different/autonomous stream data providers.

10.6.4 FIESTA Approach on Semantic Interoperability for IoT/Cloud Data Streams Tools

The FIESTA EaaS approach to IoT experimentation is based on the semantic interoperability of diverse platforms. To this end the project exploits and extends recent developments in the area of semantic interoperability of IoT data streams. In the general area of data stream management for IoT, the landscape is divided between two major approaches for data stream processing [7, 48]: (i) in-network processing, which is close in essence to the Wireless Sensors and Actuators Networks (WSANs) work (peer-to-peer communication), and (ii) cloud-based processing, related to big data approaches (centralized client-server communication, where the cloud can be considered as an elastic server). With regard to (i), Data Stream Management Systems (DSMS) for WSANs may be classified into three broad families as follows:

- **Relational DSMSs** [2, 40] extend the relational model by adding concepts necessary to handle data streams and persistent queries, together with the stream-oriented version of the relational operators (selection, projection, union, etc.). State of the art DSMSs primarily differ with respect to: the expressiveness of the query language, the associated algebra, and the assumptions made about the underlying networking architecture. More specialized proposals [20] deal with issues as diverse as blocking and non-blocking operators, windows, stream approximation, and optimizations.

- **Macroprogramming-based DSMS** [42] are oriented toward the development of applications over WSANs, as opposed to the expression of data queries over the network. The macroprograms are typically specified using a domain-specific language, and are compiled into microprograms to be run on the networked nodes, hence easing the developer's tasks who has no longer to bother with the decomposition and further distribution of the macroprograms.
- **Service-oriented DSMSs** [38] aim at integrating with classical service-oriented architectures, thereby allowing to exploit the functionalities of the infrastructure (interaction and discovery protocols, registries, service composition based on orchestration or choreography, security infrastructure, etc.).

Cloud-based approaches, on the other hand, rely on the cloud infrastructure to collect, process and store the data acquired from the environment. In contrast to DSMSs for WSANs, cloud computing offers a simple way to perform easily a wide range of heavy computations and to deal with ultra-large streams at ultra-large scales [41, 52]. These characteristics make the cloud an interesting solution for the IoT, given the expected scale. The convergence between the cloud and the IoT, referred as "Cloud of Things", is relatively recent [52] and is pioneered by emerging approaches such as Sensor Clouds [63], IoT platforms [35] and Sensing-as-a-Service [64]. Basically, all approaches share common features and follow the same global process: sensor providers (users, cities, companies, etc.) join the Cloud of Things (CoT) by registering their sensors or sensor networks. Users can send requests to the CoT, which then collects data from a set, or a representative subset, of sensors that match the requirements of the requests. These data are processed by the CoT according to the computation expressed by the request, and the results are sent back to the users.

When combining IoT data streams originating from different sources, one can leverage semantic technologies for achieving interoperability. Most of the existing (semantic interoperability) efforts to provide uniform representations for entities in the Internet of Things (IoT), i.e., Things, sensors/actuators they host, and services they provide, adopt the semantic approach and exploit ontologies. A considerable portion of ontologies exploited in the IoT domain is inherited from efforts in the Wireless Sensor Networks domain. In the latter, the main focus is directed towards modeling sensor, actuators and their data (e.g., [15, 16, 47]). A commonly exploited ontology, to reason about sensors is SSN [16], provided by the W3C Semantic Sensor Network Incubator Group. SSN models sensing specific information from different perspectives: a) Sensor perspective to model sensors, what they sense and how they sense; b) System perspective to model systems of sensors and their deployment; c) Feature perspective to model what senses a specific property and how it is sensed; d) Observation perspective to model observation values and their metadata. Other sensor ontologies are also surveyed in [18]. Many of the ontologies surveyed therein provide a solid basis for the representation of sensors, actuators, and their data. However, those entities are only a portion of the IoT.

More efforts have been made recently to extend the ontologies with IoT-specific semantics, including Things, their functionalities, or their deployment spaces. For instance, Sense2Web [19] provides an ontology that models the following Thing-related concepts: the Entity (equivalent to a feature on interest); the Device, which is the hardware component (equivalent to a Thing); the Resource, which is a software component representing the entity; and the Service through which a resource is accessed. A resource can be a sensor, actuator, RFID tag, processor or a storage resource. Christophe et al. [14] focus more on the deployment spaces of Things rather than Things themselves, especially indoor locations. The ontologies provided by the authors provides a vocabulary to describe Objects, which are physical Things, their location, their capability, and virtual objects, which are higher level abstractions of the Things combining the above information together. Another example is the work in [62] where authors present an ontology that models services provided by Things; deployment information; Observations; Entities, which are real-world features to measure/act on, and finally Things.

10.6.5 FIESTA Approach on Semantic Interoperability for IoT/Cloud Resources Tools

FIESTA's work on semantic interoperability for IoT and Cloud resources that focuses on developing common annotation models for describing the resources and IoT data and providing validation and testing tools for semantic interoperability evaluation. The core models are constructed by investigating the existing semantic and ontology models including the IoT-A information models (i.e. resources, service and entity models developed in the FP7 EU IoT-A project, http://epubs.surrey.ac.uk/127271/), W3C Semantic Sensor Network Ontology (SSN Ontology) (http://www.w3.org/2005/Incubator/ssn/XGR-ssn-20110628/), EF7 IoT.est models). FIESTA uses the existing concepts, namespaces and semantic models and develops a set of core models to describe IoT resources (e.g. sensor devices, gateways, actuators) and their capabilities and features and also provide semantic models to describe Cloud services and Cloud based components. The existing semantic models such as W3 SSN, IoT-A models are usually developed for specific purposes and in the domain of the projects.

10.6.6 FIESTA Approach on Testbeds Integration and Federation Tools

Federation in FIESTA is understood to be: "an organization within which smaller divisions have some internal autonomy" [43]. In terms of testbeds this considers that each testbed operates both individually and part of a larger federation in order to gain value (larger user base, potential combinations with other testbeds to support richer experimentation, etc.). Typical testbed federation functions include: resource discovery (finding the required resources for an experiment); resource provisioning (management or resources such that they are available when required); resource monitoring (monitor operation in order to collect experimental results); and finally security (ensuring authorized users can access resources, and the federation provides a trusted base to keep experiment information secure). Different federation models can then be applied to implement the federation; the FedSM project defines a number of models including lightweight federation where there is little if any central control of these functions (by the federation) through to a fully integrated model where a central federation authority implements and provides the functions.

The FIRE programme has a long standing history in developing cutting edge testbed federations. In the field of networking research: Openlab provides access to tools and testbeds including PlanetLab Europe, the NITOS wireless testbed, and other federated testbeds to support networking experimentation across heterogeneous facilities. OFELIA is an OpenFlow switching testbed in Europe federating a number of OpenFlow islands supporting research in the Software Defined Networking field. CONFINE co-ordinates unified access to a set of real-world community IP networks (wired, wireless, ad-hoc, etc.) to openly allow research into service, protocols and applications across these edge networks. CREW federates five wireless testbeds to support experimentation with advanced spectrum sensing and cognitive radio. Finally, FLEX is a new FIRE project that works towards providing testbeds for LTE experimentation. In the field of software services, the Bonfire project created a federation of cloud facilities to support experimentation with new cloud technologies. Importantly, in terms of Internet of Things testbeds, SmartSantander provides a set of Smart City facilities through large-scale deployments of sensor networks atop which applications and services can be developed. Also, Sunrise is a federation of sensor network testbeds providing monitoring and exploration of the marine environments and in particular supporting experimentation in terms of the underwater Internet of Things. While each project typically performs federation within its own domain, the Fed4FIRE project is an initiative to bring together heterogeneous facilities across Europe so as to target experimentation across the whole Future Internet field i.e., networks, software and services, and IoT.

Many of the projects (crucially Fed4FIRE) employ OMF [45] and SFA [8] federation technologies. OMF is a control, measurement and management framework for testbeds. From an experimenter's point of view, OMF provides a set of tools to describe and instrument an experiment, execute it and collect its results. From a testbed operator's point of view, OMF provides a set of services to efficiently manage and operate the testbed resources (e.g. resetting nodes, retrieving their status information, installing new OS image). The OMF architecture is based upon Experiment Controllers that steer experiments defined in OEDL (OMF experiment Description Language), which is a declarative domain-specific language describing required resources and how they should be configured and connected. It also defines the orchestration of the experiment itself.

Outside FIRE, there have been a number of federation initiatives to support the wider Future Internet community. Two relevant ones are Helix Nebula and XIFI. XIFI is a federation of data centres connected to resources such as wireless testbeds and sensor networks; its goal is to support large-scale Future Internet trials before transfer to market. XIFI employs a federation architecture based around web technologies (e.g. OAUTH, OCCI, and open Web APIs). On the other hand, Helix Nebula – the Science Cloud is an initiative to build federated cloud services across Europe in order to underpin IT-intense scientific research while also allowing the inclusion of other stakeholders' needs (governments, businesses and citizens).

10.7 FIESTA-IoT Architecture

FIESTA deals with the federation, virtualization and interoperability of diverse IoT testbeds, notably testbeds that comply with different IoT architectures, including architectures developed by standardization bodies (e.g., OGC [44] and GS1/EPCGlobal [22]), as well as FP7 projects (such as SmartSantander [25]). These architectures serve application specific purposes and are characterized by increased penetration in specific industries. In addressing this heterogeneity, FIESTA attempt to map and describe IoT platforms complying with these architectures to a general-purpose meta-level architecture, which serves as a basis for the FIESTA virtualized architecture layer Figure 10.4. The foundation for developing this meta-architecture is the Architecture Reference Model [34], developed by the FP7 IoT-A project and the IERC cluster. The current status reached by IoT-A at the end of the project (November 2013), as far as the Architectural Reference Model (ARM) is concerned includes a set of Models, Views & Perspectives in addition to a comprehensive set of guidelines that explains how to use Model, Views and Perspectives in order

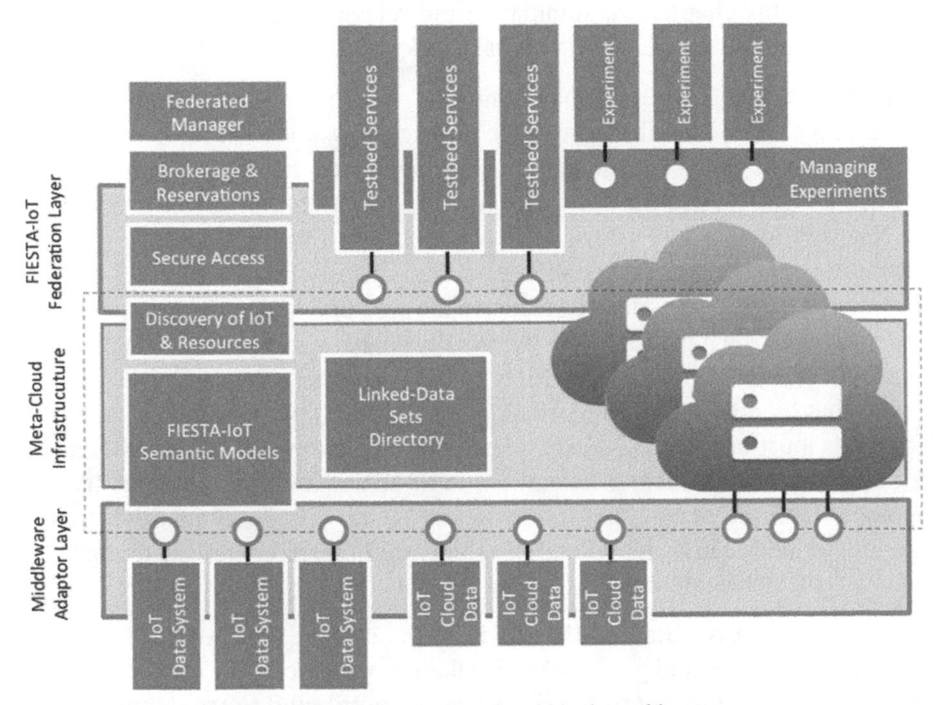

Figure 10.4 FIESTA functional blocks architecture.

to derive a concrete architecture. Part of the Guidelines is a large set of design choices that are linked to the perspectives, i.e. linked to some qualities that the system is expected to meet; part of those properties is system interoperability.

While the ARM provides some recommendations and tactics in order to achieve system interoperability, this essential quality is not guaranteed when applying the ARM to a concrete system as choices are left in the architect's hands. In order to boost the adoption of the ARM and make its usage easier the IoT Forum considers that the next step is to develop specific profiles that implement flavours of the ARM focusing on specific qualities of the system (e.g., ability to handle specific functional or non-functional requirements such as reliability, resilience, QoS awareness).

10.8 Conclusions

Fiesta has advance the state of the art in different directions, relevant activities are on-going work but most of the progress related with design, architecting and implementation have been completed and reported, in the various public documents, in this chapter we summarize the relevant contributions in the different relevant areas where FIESTA has work so far.

FIESTA Federation: Currently, there is no easy way to carry out experiments across a range of IoT facilities without having deep knowledge in sensor networks technology, communication technologies and platform configuration. FIESTA has opened up this space to provide a richer experimentation space that appeal to a wider range of experimenters (both in industry and research). Existing federation technologies are typically heavyweight in the effort required to add and control testbeds; in the case of OMF an experimental controller needs to be integrated with the facility so that standard conformance is achieved. While standards solve integration problems they often do so in a way that hinders long term sustainability (detracts new joiners)—new IoT testbeds must be able to quickly come and go as new technology trends emerge. FIESTA's approach to federation, built up semantic interoperability technologies and the meta-cloud infrastructure to provide novel methods to ensure that testbeds can be integrated in a lightweight manner and ensure that sustainability is not hindered.

FIESTA Architecture: Since 2014 the IoT Architecture Reference Model (IoT-ARM) sustenance and profile work is taken care by the WG "Technology and Openness" of the IoT Forum. FIESTA have contributed to the definition of the "Semantic Interoperability" profile based on the FIESTA achievements in that matter and envisages getting ARM/profile certification for some of the "Semantic Interoperability" enablers implemented in the course of the project. Overall, FIESTA maps several concrete architectures to the IoT-ARM, as a means to studying and ensuring their interoperability. The testbeds to be interconnected and virtualized in the project are the starting point for these mappings, while additional mappings can be realized as part of open calls that is well know ask for the federation of additional testbeds.

FIESTA Semantic Interoperability: A common limitation to all surveyed ontologies is that they still mostly lack a very important requirement: modeling the physics and mathematics, which are at the core of any sensing/actuation task, as first class entities. In more detail, it is important to relate various quantifiable and measurable (real-world) features in order to define, in a user understandable and machine-readable manner the processes behind single or combined sensing/actuation tasks. This correlation enables the system exploiting the ontologies to have a better understanding of the sensing/actuating task at hand and consequently better analyse its outcomes or substitute it more efficiently if need be, i.e., if required sensors/actuators are not available, or if the functionalities they provide do not fully satisfy the task at hand. FIESTA deals with these interoperability issues to allow researchers to design and submit interoperable experiments that are able to understand the semantics of sensing and actuating tasks and accordingly to select sensors/actuators that are suitable for executing the specified experiments. As a starting point the project leverages the W3C SSN ontology, along with mathematical formulas introduced in [28] in order to represent sensing/acting processes in a universally accepted language (i.e. algebra). FIESTA deploy middleware implementing such algebra over the federated testbeds, as well as appropriate mapping techniques for streaming tasks, in order to allow researchers to specify experiments based on combinations/compositions of sensing and actuating processes. From an implementation perspective, FIESTA deploys middleware (residing at the individual testbeds) endowing the testbeds with interoperability capabilities, along with middleware (residing at the FIESTA meta-cloud infrastructure) empowering discovery of IoT resources and compositions of sensing and actuating processes from multiple testbeds.

FIESTA has progress the state-of-the-art by introducing re-usable and common core models to describe the IoT and Cloud resources. The built models are based on the existing and common IoT models to maximize the interoperability among different providers and test-beds. At the design level, FIESTA provide semantic interoperability check and validation services using a common portal and web services to allow service developers and test-bed providers check and evaluate the interoperability of their meta-data descriptions based on the FIESTA core models and also other existing common models. The results of semantic interoperability check and evaluation gives feedback to the semantic model designers and test-bed providers on the level of interoperability between their resource descriptions and the commonly used resource description frameworks. At the deployment level, FIESTA provides wrappers and matching services to enable translation of the resource descriptions from the test-beds to the FIESTA's core models and/or other existing common models. At the run-time level, FIESTA enables test-bed providers and Cloud service developers to publish, query and access large set of semantically annotated resource descriptions according to different semantic description models and representation frameworks (i.e. using different semantic models and also different representation formats). This enables the test-bed and Cloud service providers and developers to test and evaluate efficiency of different solutions and also to measure the level of interoperability between different schemes and also to enable the resource providers to adapt common models or use wrapper to enhance the semantic interoperability between their resource descriptions and other resources that are described within the FIESTA framework that are distributed over different test-beds.

FIESTA Linking and Reasoning: FIESTA have improved the state of the art in this area by providing highly efficient approaches for efficient processing of linked data streams typical for applications in the IoT and smart cities areas. FIESTA's work is based on CQELS. Based on this basic reasoning functionality, the project provides layered reasoning formalisms at different levels of complexity (uncertainty, nonmonotonicity, recursion) for adaptive trade-offs between scalability and expressivity as required by experimental applications in the areas addressed by the FIESTA testbeds.

FIESTA Federating IoT Data Streams: FIESTA has advanced the state-of-the-art in federated processing for IoT data through enabling semantic-based interoperability among stream processing engines using Linked Stream Data. FIESTA enables semantically-self-described stream data items to automatically travel from its point of origin (e.g., sensors) downstream to applications, through autonomously passing through many stream engines. Each of the stream engines might provide potential stream data for the targeted stream-based computation that can be expressed in a standardized continuous query language, i.e, an extension of SPARQL [57]. FIESTA also support automatic discovery of stream data at run-time based on context represented as semantic links in stream data. This enables the federation of schema-free and semantic-based data aggregation without prior knowledge about stream data format, data schema and origins of the input stream data. FIESTA also has targeted the provisioning for a standardized RDF-based stream protocol to facilitate the semantic-based interoperability among the federation setting.

Acknowledgements

This work is funded by the European Commission under the EU-H2020 Project Grant "Federated Interoperable Semantic IoT/cloud Testbeds and Applications (FIESTA)" with the Grant Agreement No. CNECT-ICT-643943.

References

[1] Daniel J. Abadi, Yanif Ahmad, Magdalena Balazinska, Ugur Cetintemel, Mitch Cherniack, Jeong-Hyon Hwang, Wolfgang Lindner, Anurag S. Maskey, Alexander Rasin, Esther Ryvkina, Nesime Tatbul, Ying Xing, and Stan Zdonik. "The Design of the Borealis Stream Processing Engine," Proc. 2nd Biennial Conf. Innovative Data Systems Research (CIDR 05), 2005; http://www-db. cs.wisc.edu/cidr.

[2] Giuseppe Amato, Stefano Chessa, and Claudio Vairo. Mad-wise: a distributed stream management system for wireless sensor networks. Software: Practice and Experience, 40(5), 2010.

[3] D. Anicic, P. Fodor, S. Rudolph, and N. Stojanovic. EP-SPARQL: a unified language for event processing and stream reasoning. In Proceedings of the 20th International World-Wide Web Conference, pages 635–644. ACM, 2011.

[4] Scott Kirkpatrick (lead), Michael Boniface et al., «FUTURE INTERNET RESEARCH AND EXPERIMENTATION: VISION AND SCENARIOS 2020», AMPLIFIRE Support Action (Grant Agreement: 318550) Deliverable D1.1, June 2013.

[5] Sathya Rao, Martin Potts, et al., «FIRE PORTFOLIO CAPABILITY ANALYSIS», AMPLIFIRE Support Action (Grant Agreement: 318550) Deliverable D2.1, October 2013.

[6] Darko Anicic, Paul Fodor, Sebastian Rudolph, and Nenad Stojanovic. 2011. EP-SPARQL: a unified language for event processing and stream reasoning. In Proceedings of the 20th international conference on World wide web (WWW '11). ACM, New York, NY, USA, 635–644.

DOI=10.1145/1963405.1963495 http://doi.acm.org/10.1145/1963405.1963495

[7] Luigi Atzori, Antonio Iera, and Giacomo Morabito. The internet of things: A survey. Computer Networks, 54(15), 2010.

[8] Augé, J. and Friedman, T (2012) The Open Slice-based Facility Architecture available at http://opensfa.info/doc/opensfa.html

[9] Magdalena Balazinska, Amol Deshpande, Michael J. Franklin, Phillip B. Gibbons, Jim Gray, Mark Hansen, Michael Liebhold, Suman Nath, Alexander Szalay, and Vincent Tao. 2007. Data Management in the Worldwide Sensor Web. IEEE Pervasive Computing 6, 2 (April 2007), 30–40. DOI=10.1109/MPRV.2007.27 http://dx.doi.org/10.1109/MPRV.2007.27

[10] Davide Francesco Barbieri, Daniele Braga, Stefano Ceri, and Michael Grossniklaus. 2010. An execution environment for C-SPARQL queries. In Proceedings of the 13th International Conference on Extending Database Technology (EDBT '10), Ioana Manolescu, Stefano Spaccapietra, Jens Teubner, Masaru Kitsuregawa, Alain Leger, Felix Naumann, Anastasia Ailamaki, and Fatma Ozcan (Eds.). ACM, New York, NY, USA, 441–452. DOI=10.1145/1739041.1739095 http://doi.acm.org/10.1145/1739041.1739095

[11] Jason Campbell, Phillip B. Gibbons, Suman Nath, Padmanabhan Pillai, Srinivasan Seshan, and Rahul Sukthankar. 2005. IrisNet: an internet-scale architecture for multimedia sensors. In Proceedings of the 13th annual ACM international conference on Multimedia (MULTIMEDIA '05). ACM, New York, NY, USA, 81–88. DOI=10.1145/1101149.1101162, http://doi.acm.org/10.1145/1101149.1101162

[12] Sirish Chandrasekaran, Owen Cooper, Amol Deshpande, Michael J. Franklin, Joseph M. Hellerstein, Wei Hong, Sailesh Krishnamurthy, Samuel R. Madden, Fred Reiss, and Mehul A. Shah. 2003. TelegraphCQ: continuous dataflow processing. In Proceedings of the 2003 ACM SIGMOD international conference on Management of data (SIGMOD '03). ACM, New York, NY, USA, 668–668. DOI=10.1145/872757.872857 http://doi.acm.org/10.1145/872757.872857

[13] Mitch Cherniack,, Hari Balakrishnan, Magdalena Balazinska, Don Carney, Ugur Cetintemel, Ying Xing, and Stan Zdonik ., "Scalable Distributed Stream Processing," Proc. 1st Biennial Conf. Innovative Data Systems Research (CIDR 03), 2003, pp. 257–268; http://www-db.cs.wisc.edu/cidr.

[14] Benoit Christophe. Managing massive data of the Internet of Things through cooperative semantic nodes. In IEEE Sixth International Conference on Semantic Computing, (ICSC), pages 93–100. IEEE, 2012.

[15] Michael Compton, Holger Neuhaus, Kerry Taylor, and Khoi-Nguyen Tran. Reasoning about sensors and compositions. In SSN, pages 33–48, 2009.

[16] M. Compton, P. Barnaghi, L. Bermudez, R. G. Castro, O. Corcho, S. Cox, et al.: "The SSN Ontology of the Semantic Sensor Networks Incubator Group", Journal of Web Semantics: Science, Services and Agents on the World Wide Web, ISSN 1570–8268, Elsevier, 2012.

[17] Philippe Cousin, Martin Serrano, John Soldatos, «Internet of Things Research on Semantic Interoperability to address Manufacturing Challenges», In the Proc. Of the 7th International Conference on Interoperability for Enterprise Systems and Applications, I-ESA 2014, Albi, France, March, 24–28, 2014.

[18] Mathieu d'Aquin and Natalya F Noy. Where to publish and find ontologies? A survey of ontology libraries. Web Semantics: Science, Services and Agents on the World Wide Web, 11:96–111, 2012.

[19] Suparna De, Tarek Elsaleh, Payam Barnaghi, and Stefan Meissner. An Internet of Things platform for real-world and digital objects. Scalable Computing: Practice and Experience, 13(1), 2012.

[20] Nihal Dindar, Nesime Tatbul, Reneèe J. Miller, Laura M. Haas, and Irina Botan. Modeling the execution semantics of stream processing engines with secret. The VLDB Journal, 2012.

[21] T. Do, S. Loke, and F. Liu. Answer set programming for stream reasoning. Advances in Artificial Intelligence, pages 104–109, 2011.

[22] EPCglobal: The EPCglobal Architecture Framework, EPCglobal Final Version 1.2 Approved 10 September 2007.

[23] Geoffrey C. Fox, Supun Kamburugamuve, Ryan Hartman Architecture and Measured Characteristics of a Cloud Based Internet of Things API Workshop 13-IoT Internet of Things, Machine to Machine and Smart Services Applications (IoT 2012) at The 2012 International Conference on Collaboration Technologies and Systems (CTS 2012) May 21–25, 2012 The Westin Westminster Hotel Denver, Colorado, USA.

[24] Michael J. Franklin, Shawn R. Jeffery, Sailesh Krishnamurthy, Frederick Reiss, Shariq Rizvi, Eugene Wu,Owen Cooper, Anil Edakkunni, Wei Hong: Design Considerations for High Fan-In Systems: The HiFi Approach. CIDR 2005: 290–304

[25] J.A. Galache, V. Gutiérrez, J.R. Santana, L. Sánchez, P. Sotres, J. Casanueva, L. Muñoz, "SmartSantander: A Joint service provision facility and experimentation-oriented tesbed, within a smart city enviroment", Future Network & Mobile Summit, Lisbon (Portugal), July 2013.
[26] Forecast: The Internet of Things, Worldwide, 2013. http://www.gartner.com/document/2625419? sthkw=G00259115
[27] Vincenzo Gulisano, Ricardo Jimenez-Peris, Marta Patino-Martinez, Claudio Soriente, and Patrick Valduriez. 2012. StreamCloud: An Elastic and Scalable Data Streaming System. IEEE Trans. Parallel Distrib. Syst. 23, 12 (December 2012), 2351–2365. DOI=10.1109/TPDS.2012.24 http://dx.doi.org/10. 1109/TPDS.2012.24
[28] Sara Hachem, Thiago Teixeira, and Valérie Issarny. 2011. Ontologies for the internet of things. In Proceedings of the 8th Middleware Doctoral Symposium (MDS '11). ACM, New York, NY, USA, Article 3, 6 pages.
[29] Mohammad Mehedi Hassan, Biao Song, Eui-nam Huh: A framework of sensor-cloud integration opportunities and challenges. ICUIMC 2009: 618–626
[30] Tom Heath and Christian Bizer (2011) Linked Data: Evolving the Web into a Global Data Space (1st edition). Synthesis Lectures on the Semantic Web: Theory and Technology, 1:1, 1–136. Morgan & Clay-pool.
[31] Jesper Hoeksema and Spyros Kotoulas. High-performance Distributed Stream Reasoning using S4,Ordring workshop of ISWC2011. 2011.
[32] Hogan A., Mellotte M., Powell G., and Stampouli D. (2012) Towards fuzzy query-relaxation for RDF. In Proceedings of the 9th international conference on The Semantic Web: research and applications (ESWC'12), Elena Simperl, Philipp Cimiano, Axel Polleres, Oscar Corcho, and Valentina Presutti (Eds.). Springer-Verlag, Berlin, Heidelberg, 687–702.
[33] Martin Bauer (IOT-A), Paul Chartier (CEN TC225), Klauss Moessner (IOT.est), Nechifor, Cosmin-Septimiu (iCore), Claudio Pastrone (ebbits), Josiane Xavier Parreira (GAMBAS), Richard Rees (CEN TC225), Domenico Rotondi (IoT@Work), Antonio Skarmeta (IoT6), Francesco Sottile (BUTLER), John Soldatos (OpenIoT), Harald Sundmaeker (SmartAgriFood), «Catalogue of IoT Naming, Addressing and Discovery Schemes in IERC Projects», electronically available at: http://www.the internetofthings.eu http://www.theinternetofthings.eu/sites/default/files/%5Buser-name%5D/IERC-AC2-D1-v1.7.pdf)

[34] Andreas Nettsträter, Martin Bauer, et al. Internet-of-Things Architecture Project (IoT-A), Deliverable D1.3, «Updated reference model for IoT v1.5», July 2012.

[35] M. Kovatsch, M. Lanter, and S. Duquennoy. Actinium: A RESTful runtime container for scriptable Internet of Things applications. In Proc. of the 3rd International Conference on the Internet of Things, IOT '12, 2012.

[36] A. Kozlenkov and M. Schroeder. Prova: Rule-based java-scripting for a bioinformatics semantic web. In Data Integration in the Life Sciences, pages 17 30, Springer, 2004.

[37] N. Lanzanasto, S. Komazec, and I. Toma. Reasoning over real time data streams. http://www.envision-project.eu/wpcontent/uploads/2012/11/D 4-8v1-0.pdf, 2012.

[38] Le-Phuoc, D., Dao-Tran, M., Xavier Parreira, J., & Hauswirth, M. (2011). A native and adaptive approach for unified processing of linked streams and linked data. The Semantic Web–ISWC 2011, 370–388.

[39] Kevin Lee, «Extending Sensor Networks into the Cloud using Amazon Web Services», IEEE International Conference on Networked Embedded Systems for Enterprise Applications 2010, 25th November, 2010.

[40] Samuel R. Madden, Michael J. Franklin, Joseph M. Hellerstein, and Wei Hong. Tinydb: an acquisitional query processing system for sensor networks. ACM Trans. Database Syst., 30(1), March 2005.

[41] Samuel Madden. From databases to big data. IEEE Internet Computing, 16, 2012.

[42] Ryan Newton, Greg Morrisett, and Matt Welsh. The regiment macroprogramming system. In Proceedings of the 6th international conference on Information processing in sensor networks, IPSN '07, pages 489–498, New York, NY, USA, 2007. ACM.

[43] "federation, n.". OED Online. March 2014. Oxford University Press. http:// www.oed.com/view/Entry/ 68930?redirectedFrom=federation (accessed March 28, 2014).

[44] Open Geospatial Consortium, OpenGIS Sensor Observation Service version 1.0.0, Arthur Na, Mark Priest, 2008-02-13.

[45] Thierry Rakotoarivelo, Max Ott, Guillaume Jourjon, Ivan Seskar, "OMF: a control and management framework for networking testbeds", in ACM SIGOPS Operating Systems Review 43 (4), 54–59, Jan. 2010.

[46] RDF Stream Processing Community Group, http://www.w3.org/commun ity/rsp/

[47] David J Russomanno, Cartik R Kothari, and Omoju A Thomas. Building a sensor ontology: A practical approach leveraging ISO and OGC models. In IC AI, pages 637–643, 2005.

[48] S. Sagiroglu and D. Sinanc. Big data: A review. In Collaboration Technologies and Systems (CTS), 2013 International Conference on, 2013.

[49] H. Schaffers, A. Sällström, M. Pallot, J. Hernandez-Munoz, R. Santoro, B. Trousse, "Integrating Living Labs with Future Internet Experimental Platforms for Co-creating Services within Smart Cities". Proceedings of the 17th International Conference on Concurrent Enterprising, ICE'2011, Aachen, Germany, June 2011.

[50] Gregor Schiele, John Soldatos, Paul Lefrere, Nathalie Mitton, Kahina Hamadache, Manfred Hauswirth 'Towards Interoperable IoT Deployments in Smart Cities', submitted for publication to the European Conference on Networks and Communications EuCNC, Bologna, Italy, June 23–26, 2014 (EuCNC 2014).

[51] Martin Serrano, Manfred Hauswirth, John Soldatos, "Design Principles for Utility-Driven Services and Cloud-Based Computing Modelling for the Internet of Things", International Journal of Web and Grid Services (to appear), 2014.

[52] J. Soldatos, M. Serrano and M. Hauswirth. "Convergence of Utility Computing with the Internet-of-Things", International Workshop on Extending Seamlessly to the Internet of Things (esIoT), collocated at the IMIS-2012 International Conference, 4th6th July, 2012, Palermo, Italy.

[53] Bernhard Stegmaier, Richard Kuntschke, and Alfons Kemper. 2004. StreamGlobe: adaptive query processing and optimization in streaming P2P environments. In Proceeedings of the 1st international workshop on Data management for sensor networks: in conjunction with VLDB 2004 (DMSN '04). ACM, New York, NY, USA, 88–97. DOI=10.1145/1052199.1052214 http: //doi.acm.org/10.1145/1052199.1052214

[54] Distributed and fault-tolerant realtime computation. http://storm.incubator.apache.org/

[55] Distributed Stream Computing Platform. http:// incubator.apache.org/s4/

[56] Lightning-fast cluster computing. http://spark.apach e.org/

[57] SPARQL 1.1 Query Language. http://www.w3.org/ TR/sparql11-query/

[58] Kerry Taylor, «Semantic Sensor Networks: The W3C SSN-XG Ontology and How to Semantically Enable Real Time Sensor Feeds», 2011 Semantic Technology Conference, June 5–9, San Francisco CA, USA.

[59] K. Teymourian, M. Rohde, and A. Paschke. Fusion of background knowledge and streams of events. In Proceedings of the 6th ACM International Conference on Distributed Event-Based Systems, DEBS '12, pages 302–313, New York, NY, USA, 2012, ACM.

[60] Umbrich J., Karnstedt M., Hogan A., Parreira J. (2012) Hybrid SPARQL queries: fresh vs. fast results. In Proceedings of the 10th International Semantic Web Conference, 2012.

[61] Ovidiu Vermesan & Peter Friess, «Internet of Things - Global Technological and Societal Trends», The River Publishers Series in Communications, May 2011, ISBN: 9788792329738.

[62] Wei Wang, Suparna De, Ralf Toenjes, Eike Reetz, and Klaus Moessner. A comprehensive ontology for knowledge representation in the Internet of Things. In Trust, Security and Privacy in Computing and Communications (TrustCom), 2012 IEEE 11th International Conference on, pages 1793–1798. IEEE, 2012.

[63] M. Yuriyama and T. Kushida. Sensor-cloud infrastructure – physical sensor management with virtualized sensors on cloud computing. In Network- Based Information Systems (NBiS), 2010 13th International Conference on, 2010.

[64] Jia Zhang, Bob Iannucci, Mark Hennessy, Kaushik Gopal, Sean Xiao, Sumeet Kumar, David Pfeffer, Basmah Aljedia, Yuan Ren, Martin Griss, et al. Sensor data as a service – a federated platform for mobile data-centric service development and sharing. In Services Computing (SCC), 2013 IEEE International Conference on, 2013.

11

Combining Internet of Things and Crowdsourcing for Pervasive Research and End-user Centric Experimental Infrastructures (IoT Lab)

Sébastien Ziegler[1], Cedric Crettaz[1], Michael Hazan[1], Panagiotis Alexandrou[2,3], Gabriel Filios[2,3], Sotiris Nikoletseas[2,3], Theofanis P. Raptis[2,3], Xenia Ziouvelou[4], Frank McGroarty[4], Aleksandra Rankov[5], Srdjan Krčo[5], Constantinos Marios Angelopoulos[6], Orestis Evangelatos[6], Marios Karagiannis[6], Jose Rolim[6] and Nikolaos Loumis[7]

[1]Mandat International, Switzerland
[2]University of Patras, Greece
[3]Computer Technology Institute and Press Diophantus, Greece
[4]University of Southampton, United Kingdom
[5]DunavNET, Novi Sad, Serbia
[6]University of Geneva
[7]University of Surrey, United Kingdom

11.1 Introduction

The Internet of Things will be massive and pervasive. It will impact many and diverse application domains such as environmental monitoring, transportation, energy and water management, security and safety, assisted living, smart homes and eHealth, etc. Developing and testing technologies in conventional research labs appears to be insufficient to really grasp, fine tune and validate new IoT technologies. Moreover, an approach purely focused on technical requirements may lead to a missed target if the end-user perspective is not properly taken into account. End-user acceptance is probably as much important as technical performance, and better understanding their acceptance and satisfaction is critical.

IoT Lab (www.iotlab.eu) is a European research project [1], which has developed a hybrid research infrastructure combining Internet of Things (IoT) testbeds together with crowdsourcing and crowd-sensing capabilities. It enables researchers to use IoT testbeds, including in public spaces, while collecting inputs from end-users through crowdsourcing and crowd-sensing. It enables researchers to exploit the potential of crowdsourcing and Internet of Things testbeds for multidisciplinary research with more end-user interactions. IoT Lab approach puts the end-users at the centre of the research and innovation process. The crowd is at the core of the research cycle with an active role in research from its inception to the results' evaluation. It enables a better alignment of the research with the society and end-users needs and requirements. On the other side, IoT Lab aims at enhancing existing IoT testbeds, by integrating them together into a testbed as a service and by extending the platform with crowdsourcing and crowd-sensing capacities.

11.2 Approach

In order to achieve such aims, IoT Lab has researched complementary set of technologies and approaches, including:

- Crowdsourcing and crowd-sensing mechanisms and tools, by developing a smart phone application enabling researchers to collect real time feedbacks from research participants. It also enables participants to share data from their smart phone embedded sensors.
- Integration of heterogeneous testbeds together, by federating together several European IoT testbeds located in different parts of Europe.
- Virtualization of IoT testbeds and crowdsourcing resources into a fully integrated Testbed as a Service;

The IoT lab framework has been designed and developed bearing in mind two key objectives:

- Enabling and supporting multidisciplinary researches;
- Ensuring privacy by design and a full compliance with European personal data protection obligations, including the newly adopted General data Protection Regulation.

In order to validate the designed model, several research and experiments have been tested, including "Crowd-driven research".

We will now present with more details some key technological developments.

11.3 Architecture

IoT Lab platform architecture design addressed a double challenge: on one hand, it had to integrate diverse IoT-related testbeds (static, portable, mobile) located in different regions of Europe; on the other hand, it had to integrate smart phones with existing FIRE testbed infrastructures, thus representing a novel approach with respect to existing crowdsourcing solutions. An architecture generation process started with the analysis of technical and end user related requirements derived from selected representative use cases in order to identify key platform components, their functionalities, interaction patterns, interfaces and communication links and enable fully supported experimentation through both crowd and IoT interactions.

At the top level key components are:

- **IoT Lab Accounts Manager** for the profile management of all users' accounts, including the access control and support for incentives and reputation
- **IoT Resources Management Interface** based on Fed4FIRE enablers enabling interactions with IoT components from testbeds and smart phones and access to collected IoT data
- **Crowd Interaction Management Interface** completely independent from Fed4FIRE, that handles interaction with participants, including editors to set up a survey, and enables access the collected crowd knowledge data.

The architecture derivation process followed an IoT-A methodology [2] to support interoperability and scalability and to enable use of a wide range of heterogeneous devices and testbeds from different application domains thus satisfying a high number of requirements. Privacy by design concept is followed to ensure participants are requested minimal information and, that for each research and its belonging experiments a clear description of the required user and device data is presented.

IoT Lab architecture illustrating its federation strategy and modularity is presented at Figure 11.1. Each individual static testbed facility uses a SFAWrap via which the testbed resources are exposed through the Aggregate Manager.

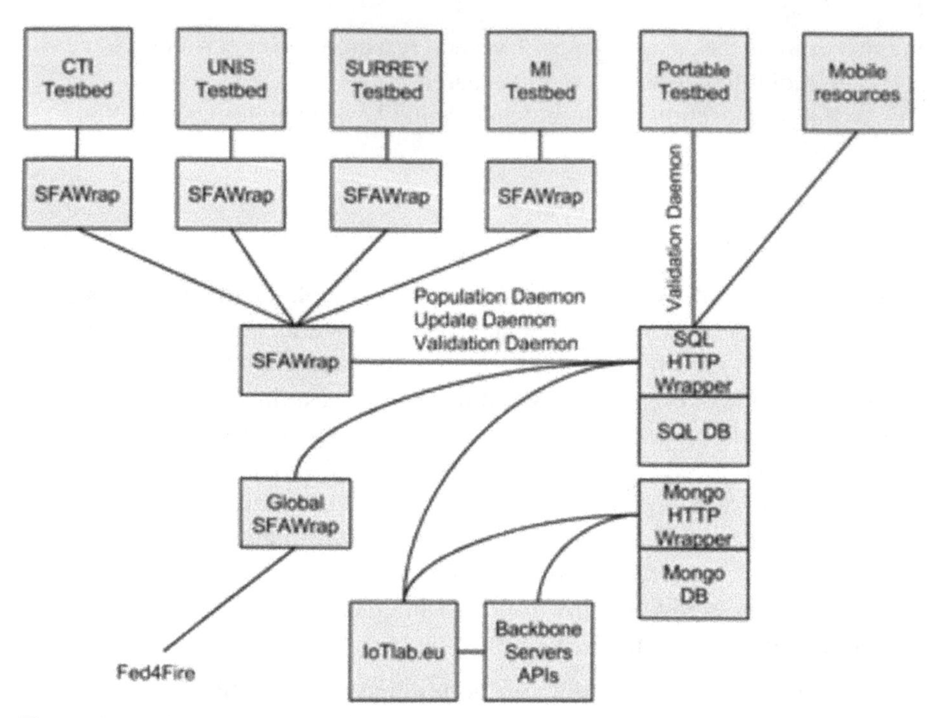

Figure 11.1 Overview of the IoT Lab architecture defining the federation strategy and showing the modular architecture.

All information regarding the type of the resources, their availability and the way of accessing and interacting with them is stored in an SQL database acting as a Resource Directory. Access to this Resource Directory is provided via a HTTP API. Resources that are provided in an ad-hoc manner, such as those of portable testbeds or the crowdsourced resources via the IoT Lab smartphone application, are registered to the system by directly accessing the Resource Directory. This registration process is regulated by a validation daemon. Although these resources do not utilize the SFAWrap (the wrapper is not designed to address the ephemeral nature of such resources), they do use the same resource description schemes and tools (e.g. RSpec documents). All resources stored in the Resource Directory (individual and portable testbeds and crowdsourced resources) are exposed to third party entities via a global SFA Wrapper that wraps around the database. This way, all registered resources are virtualized and exposed via the common interfaces of Fed4FIRE enabling other facilities to discover them. At the end-user application layer of the IoT Lab platform, a researcher conducting the experimenter can access all available resources via the IoT Lab Web page. After having been identified, a researcher can create a new research project, review and select required resources, define the experiment and dispatch it for execution at the back-end of the platform. During the execution of the experiment, all collected measurements are stored in a second database, the Measurements Database. Measurements Database is developed using MongoDB to better address the nature of the stored information as well as their expected big volume.

A view of deployed IoT Lab architecture is presented in Figure 11.2 [3] illustrating all the modules and their belonging components: Account and Profile Manager, Resource Manager, Experiment Manager, User Interface (Web and Mobile app), Testbeds (static, portable/mobile and smartphone) as devices, Communication components and Security and Privacy.

IoT Lab architecture represents a service based architecture for IoT testbeds exposing all the testbed operations as services (Testbed as a Service), enabling federation of diverse resources in a scalable and standardised way and enabling smooth and seamless integration of crowdsourced resources. Researchers performing experiments via Testbed as a Service can via a common interface (Web UI) access a diverse set of resources and conduct experiments.

The IoT Lab network architecture with all components (application, testbeds and server) is shown in Figure 11.3. The current platform is scalable to a considerable number of mobile and testbed resources [4]. For the average scenario with the IoT Lab server working at 50% of its capacity, we can have 2.8 M devices connected to the platform, whereas for the testbeds 24 M devices can be connected. Even if in a very remote use case the number of resources reaches or exceeds the limit, the server capacity can be increased in order to support all connections and data.

11.4 Heterogeneous Tesbeds Integration

IoT Lab brought together several pre-existing IoT testbeds from UK, Switzerland, Greece, Serbia and Sweden, including:

- University of Surrey smart campus testbed;
- Mandat International Smart HEPIA and Smart Office testbeds;
- University of Geneva IoT testbed;
- Dunavnet EkoNet testbed of mobile environmental sensors;
- CTI in Patras IoT testbed;

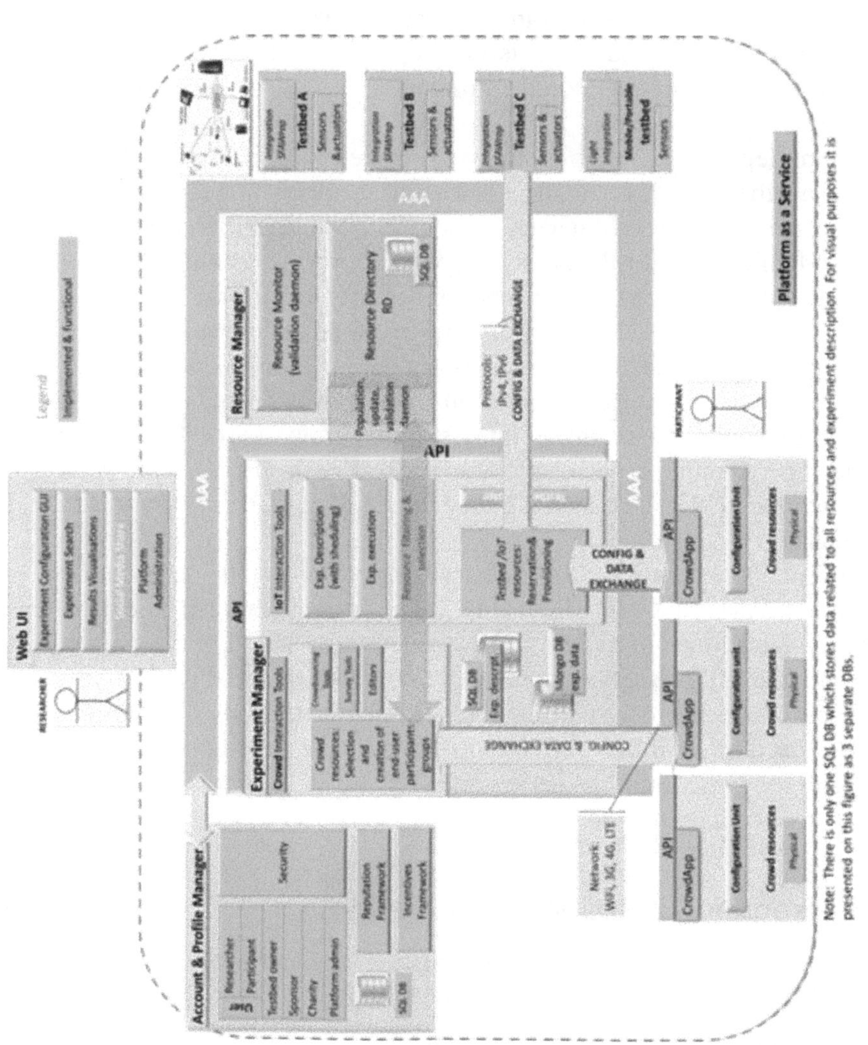

Figure 11.2 IoT Lab platform deployment.

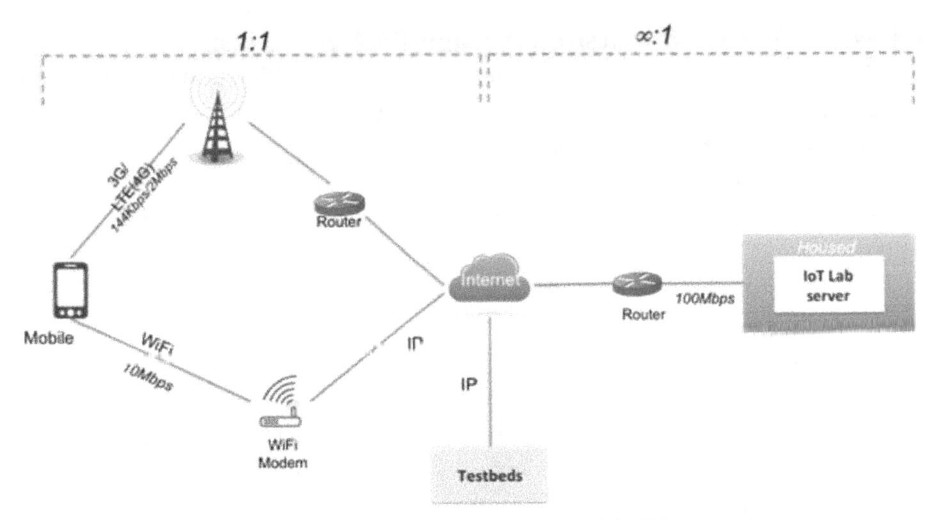

Figure 11.3 IoT Lab – network architecture with all its components.

The various testbeds were developed with distinct technologies and architectures. In order to enable a proper integration of these various and heterogeneous resources together and to enable end-to-end interconnection, the consortium opted to leverage on IPv6 as a network integrator. It leveraged on IoT6 European research project results [5] and initial attempts to enable IPv6-based testbeds integration between Europe and China [6].

An important challenge was related to the diversity of compliance levels with IPv6. Being distributed cross various countries, the corresponding ISP services offer was uneven too. We ended up with four distinct testbed profiles in terms of network configurations and connectivity,- all to be integrated together. The Figure 11.1 details the various cases:

- **Case A – Local IPv6 integration, including with non-IP IoT devices:**
  In this case, the ISP constraints were avoided through a direct integration. However, the testbed included both IPv6 and non-IP IoT devices, using communication protocols such as KNX, ZigBee, EnOcean, BACnet and others. In order to integrate these heterogeneous devices, a UDG proxy has been used to generate consistent and scalable IPv6 addresses to the legacy devices.
- **Case B – Remote full end-to-end IPv6 compliance:**
  In this case (TB-B), the testbed integration was achieved through end-to-end IPv6 integration, including 6LoWPAN end nodes directly parsed into IPv6 addresses.

- **Case C – Remote IPv6 testbed through IPv4 ISP access:**
  In this case (TB-C), in order to overcome the lack of IPv6 connection at the ISP level, the testbed integration has been performed through v6 in v4 end-to-end tunnelling, with a very limited latency impact.
- **Case D – Remote IPv4 testbed:**
  Finally, one of the testbed was fully and exclusively IPv4 based (TB-D). In this context, we decided to use a UDG proxy on the server side to map IPv6 addresses on top of the local IPv4 addresses.

The address definitions across the testbeds were maintained consistent by clearly separating the management of the Host ID on one side (IoT address) from the Network ID (Testbed address). This simple approach resulted in a consistent and highly scalable model, enabling the Testbed as a Service (TBaaS) to use a fully integrated and homogenized addressing scheme, including with mobile devices.

Another challenge was related to the heterogeneity of communication protocols used in some of the testbeds. In order to overcome this challenge, IoT Lab leveraged on the Universal Device Gateway (UDG) [7], a multi-protocol control and monitoring system developed by a research project initiated in Switzerland. It aimed at integrating heterogeneous communication protocols into IPv6. The UDG control and monitoring system enables cross protocol interoperability. It demonstrated the potential of IPv6 to support the integration among various communication protocols and devices, such as KNX, X10, ZigBee, GSM/GPRS, Bluetooth, and RFID tags. It provides connected device with a unique IPv6 address that serves as unique identifier for that object, regardless its native communication protocol. It has been used in several research projects, including by IoT6, where it has been used as an IPv6 and CoAP proxy for all kinds of devices.

In IoT Lab, the UDG platform has been used as a locally deployed proxy in the local testbed (TB-A in the Figure 11.4) and as a cloud-based proxy in some other cases (TB-C and TB-D in the Figure 11.4). However, for communication protocols which are non-compliant with the Internet Protocol, a local deployment was required.

11.5 IoT Lab Smart Phone Application

IoT Lab intends to put the end users in the centre of research and innovation. It required the development and introduction of a tool that offers ubiquitous and seamless interaction capabilities with the crowd participants. A specific

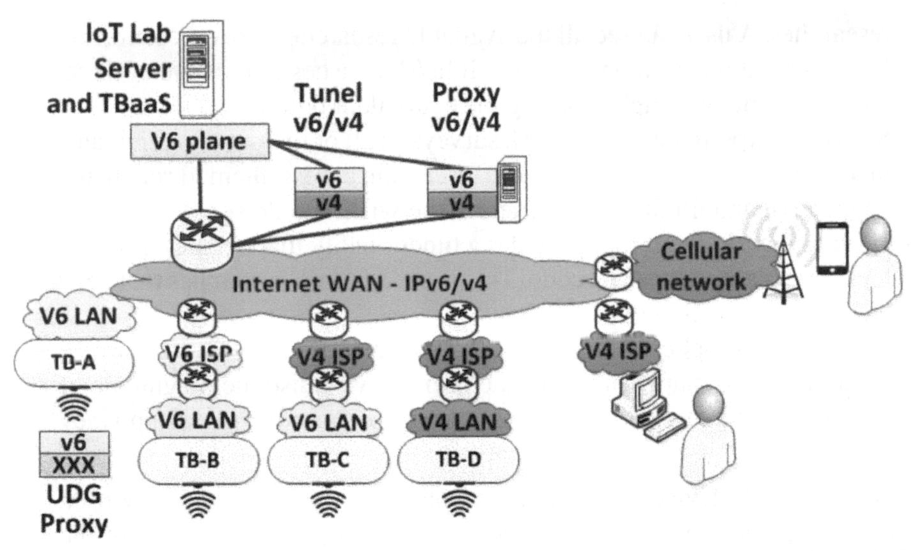

Figure 11.4 IoT Lab IPv6-based network integration representing the four main testbed profiles.

IoT Lab smart phone application was developed, which can be installed to all the devices that run Android OS 4.1.1, or later.

Ensuring a user-friendly interface with a state of the art user experience led to focus on the user experience and feedbacks with iterative adaptations and fine tuning during the project. Moreover, frequent updates made sure that all the found bugs were solved, as well as, the provided functionalities were optimised constantly.

Smart Phone Application Functionalities

The mobile smart phone application provides a set of functionalities that are described below:

- **Add idea**: through a limited number of steps, any user can be part of the platform and express a new idea. The predefined options help make this process faster and users more keen to use it.
- **Rank idea**: every user can see and rank the aforementioned proposed ideas. By selecting one out of the list of all the available, the user can see more information about it and rank it using the provided tools.
- **Available researches**: IoT Lab application is, among others, a tool for crowdsourcing and crowdsensing. Hence, it can be used during ongoing

researches. A user can see all the available researches, browse them, and learn more about each one of them. If he/she wishes to join one of them, it can be done by simply clicking the equivalent button.
- **Surveys**: Experimenters can push surveys to all or to a set of participants in the context of a research. The user can access them through the application and fill them whenever he/she wishes to do so.
- **Update location**: our tool provides a functionality that updates mobiles' location by scanning a QR code. This is used during researches that need fine grained location updates.
- **Map**: IoT lab application can display all the resources of our platform on an anonymised map. This helps the users to visualise the magnitude of our project and feel part of the community, without having any privacy issue.

Additionally, each device that runs our application can be potentially used as a multi-purpose sensing mote. In order to do that, the user/owner of the device can to explicitly allow the application to make the embedded sensors of the device available to future researches, or to manually join a research. Moreover, since IoT Lab was designed with respect to users' privacy, each time one's device is about to be exploited in a crowdsensing scenario, multiple notifications are sent to the device informing about the ongoing background tasks. More about the IoT interactions and experiment composition will be presented in the next section of this chapter.

Crowdsensing Using IoT Lab Application

Crowdsensing takes place as a part of an ongoing research. As described in the previous section, a device can be manually or automatically assigned to one research according to user's settings and configuration. The background mechanism that sets crowdsensing to work is Google Cloud Messaging (GCM).

Protocol Selection

Before digging more into the steps that need to be taken during a crowdsensing experiment, it is important to present the reasons that led us to the selection of the used IoT communication protocol. The deciding factors were

- computational requirements,
- bandwidth usage,
- scalability,
- robustness,
- support from the community.

MQTT is a lightweight-by-design IoT protocol that is widely aligned with our system requirements, and was the other candidate except GCM. IBM claims [8] that in real life scenarios we can preserve 4.1% energy per day, just by switching from HTTPS to MQTT. Additionally, there is a plethora of free brokers that can numerous active connections at all time. Finally, the Eclipse community hosts and supports the MQTT project over the past years.

On the other hand, GCM is a service created and provided by the IT giant, Google. With a dedicated community ready to answer and tackle all the emerged problems, GCM was a great candidate. Moreover, GCM is not limiting the number of active devices.

Both protocols were selected after reflecting the type of communication needed between the devices and the back end. Due to their nature, smartphones do not have a static IP. Hence, we were troubled by the need to be able to access specific devices from the back end. MQTT and GCM offer mechanisms that handle message delivery.

We chose GCM other approaches, as it is open source, scalable, free for a big amount of users, and is optimised in terms of energy consumption during idle states. Additionally, all the back end support is handled by Google itself and we do not need to do any more provisioning.

Mobile Crowdsensing

Google Cloud Message carries JSON messages that can be easily modified and are used in order to send sensing triggers to a specific, or a set of mobile devices. An example of such a message is displayed in Figure 11.5.

As presented if Figure 11.6 bellow, the steps that take place during a crowdsensing experiments are the following:

- Back end sends a notification to all the devices that are about to participate in the crowdsensing experiment. The notification is delivered using the GCM.
- After a period of time, the crowdsensing loop starts.
- A sensing trigger is pushed to the mobile phone using GCM.
- When the trigger is received, the OS is responsible to "wake up" the IoT Lab application.
- IoT Lab application analyses the sensing request and samples the desired measurement.
- The measurement is stored to the IoT Lab database using the appropriate API.

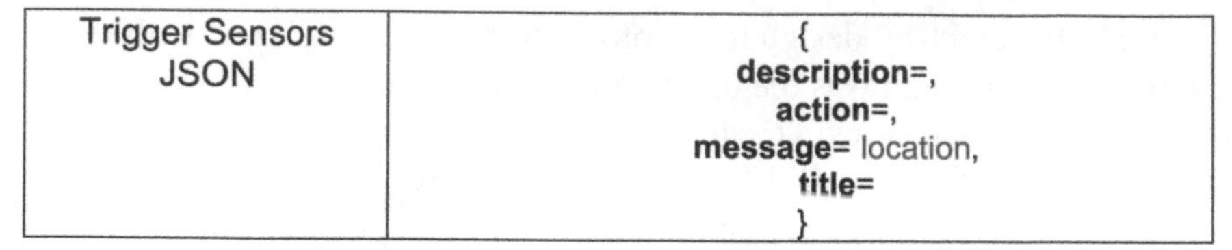

Trigger Sensors JSON	{ **description=**, **action=**, **message=** location, **title=** }

Figure 11.5 A sensing trigger message.

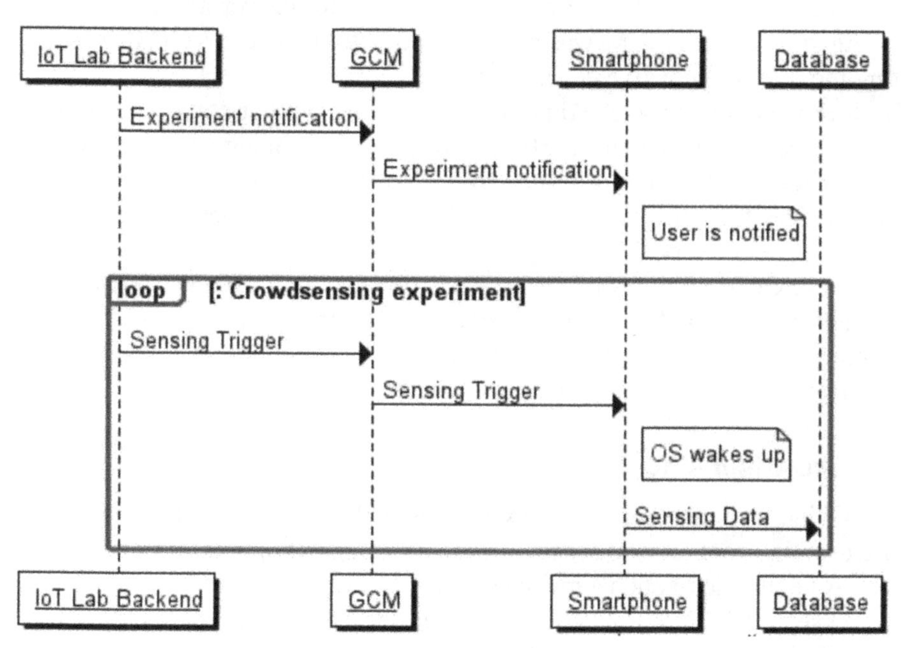

Figure 11.6 Sequence diagram of the Crowdsensing steps.

11.6 Testbed as a Service

The IoT Lab platform federates a variety of resources, ranging from static IoT devices to mobile phones. The role of these mobile devices is twofold: they can function as multi-purpose sensing motes (i.e., using accelerometer, GPS, luminosity) or as a source of interaction with their owners. From the above we can distinguish the two kinds of experiments: The first one with IoT devices either mobile or static and the second one involving the owners of mobile phones. These experiments are realised through IoT and Crowd Interactions functionalities.

IoT Interactions

In IoT interactions the experimenter is provided with a list of available resources that he can view and reserve for their experiment. After the experimenter chooses and reserves the desired resources, he/she is prompted to the experiment composition module. In the background, an XML schema called RSpec is used to transfer the information regarding the resources reserved between the reservation module and the experiment composition module along with some meta-information on the experiment itself; e.g. duration and period of execution, human readable description of the experiment, etc. This information is incorporated in the RSpec document via tags such as the <research id> tag, that provides the id of the parent research of the experiment to be composed, the <experiment title> tag which provides the title the experimenter has given to the experiment to be composed and the <experiment desc> which provides a short description of the experiment.

Experiment Composition

The experiment composition module receives this information and provides a simple but powerful mechanism with which the experimenter can define the details of how resources will be used in the context of "If This Then That" (IFTTT) scenarios. The final experiment consists of a set of these scenarios.

The experiment composition module allows the experimenter to set the following actions:

- **Get a value from specified resources**. The frequency of the reading request is set in minutes or hours and includes one or more resources. The resources must be of type "sensor" and must be included in the experiment before the experimenter enters the main composition module. This action is called "reading". As an example, a reading can be "Get a value from sensor 1 every 5 minutes between these 2 dates and times".
- **Set a condition**. A condition can be the average, absolute, minimum or maximum value of one or more resources being greater, equal or lesser than a set value. In the case of multiple resources a logical operator can be set. An example of a condition can be "The maximum value of sensor 1 OR the maximum value of sensor 2 to be greater than 5".

- **Set an outcome**. An outcome is an action that can be taken. This action is either to take more measurements from sensors or to actuate an actuator. Outcomes also include a logical operator in case there is more than one conditions. An example of an outcome could be "Actuate actuator 1, if all conditions are met (with logical AND)".
- **Define an action**. Actions are combinations of conditions and outcomes. Actions are set in an "IF-THEN" form in order to clarify their meaning. An example of an action can be "IF condition1 AND condition2 are true THEN perform outcome 1". The logical operator AND is actually defined in the outcome and not in the conditions, as specified above.

After the experiment scenario has been defined, it is dispatched to the execution module. The scenario is described in an XML schema called Experiment Description XML schema (ED XML). The Experiment Description XML defines a parent tag <experiment> </experiment> that encloses all other elements. The <measurements> tag defines the measurements database server information along with the <ip> and <port> sub-tags inside it. The next tag is a random identifier tag. This is generated during the ED creation randomly and is used to uniquely identify the experiment description. The tag that provides this identifier is the <identifier> tag.

Readings are included in the <reading> tag. Inside this tag, a <frequency> tag with a "unit" property defines the frequency of the reading while <start> and <end> tags define the start and end of the readings period for the specified reading. The <resources> tag then defines which resources have to be probed for a reading every time it's needed. These are defined using <id> tags that include properties "component", "resource id", "port", "ip", "protocol" and "path". The combination of these properties allows the execution engine to identify and reach the resources directly.

Actions are defined using the <action> tag. These include <conditions> and <outcome> tags. The <conditions> tag include the aggregation and logical operations as a tag and property respectively (e.g. <average logic="and">). Inside this tag, the resources are defined using an <id> tag and also the threshold is defined using a <threshold> tag. The <outcome> tag includes a property for the logical operator and inside the tag, resources are defined (either sensors or actuators) using <id> tags as above. An example of an ED XML is shown in Listing 1.1 in the Appendix.

Experiment Execution

When an experimenter finalizes the definition of an experiment at the Experiment Composition module, an Experiment Description XML document is created which is transferred to the Experiment Execution module which proceeds in parsing it and finding all necessary information in order to start running the Experiment.

At first, the research ID, the experiment title and the experiment description are identified and posted as a new 'research' entity in the Resource Directory database. As already described, the Experiment Description XML document contains a number of readings and action tags. Each of these tags will spawn a new celery job to handle their tasks.

Each reading tag has several resources with their contact information and a frequency with which they are to be read. Every one of those readings, spawns a celery task tasked with obtaining the measurements from the resources in the time and with the frequency specified by the experimenter. When the time to obtain measurements comes the task creates a new task responsible for the next set of measurements. When the measurement comes, a new task responsible for the action tag will be called. Inside the actions tag there are a number of tied conditions and outcomes. Their information is parsed and summarized in two lists: one for the conditions called conditionsList and one for the outcomes called outcomeList. A task for the function called conditionChecker(), with the two aforementioned lists as parameters is called after the resolution of each reading tag. This task, will evaluate the logic of conditionsList as specified in the Experiment Description XML. If it is evaluated to 'True', then the outcomes from outcomeList will be executed.

Crowd Interactions

In Crowd interactions the experimenters ask for inputs from the smartphone users through surveys and questionnaires (Figure 11.7). The surveys are constructed using LimeSurvey which is integrated within our platform. The process of filtering and selecting the user in order to engage him/her in the specific research includes the following mechanisms available through the architecture: survey queries, survey lists, geofencing and project code.

Survey Queries: A query is a mechanism that allows the experimenter to filter crowd users in a meaningful way in order to select the users needed for the post of a mobile query. The filtering function is based on the socio-economic profile of the user which they voluntarily include during anonymous registration through the mobile app. The query is defined and then saved in the experimenter's profile so that it can be easily reused in the future, which makes it a very powerful tool as the crowd users constantly change in number

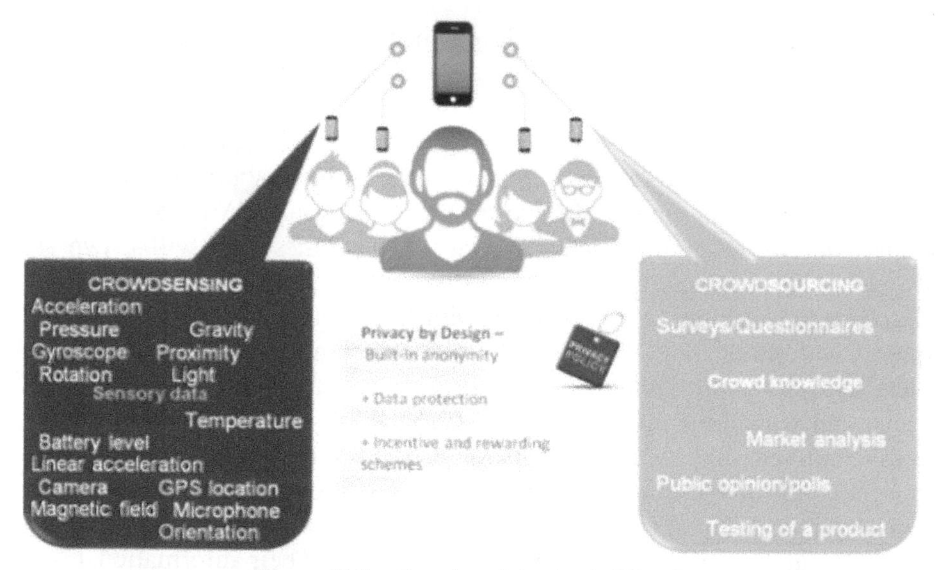

Figure 11.7 Crowd participation in TBaaS.

throughout the architecture's lifetime. Queries, although static themselves, provide dynamic results in the form of sets of users that fit the set criteria.

Survey Lists: Every time a query is used, an up-to-date list of crowd users that meet the query's criteria is presented. The experimenter then has the opportunity to select individual recipients to form a survey list. A survey list is a static list of survey recipients that is used to send a survey to the mobile devices recipients. The content of the user list is anonymous and only social and economic data are associated with each entry. When the final survey list is compiled, it is saved under the experimenter's profile and can be used as the destination list in which to post a survey. A special case of a survey recipient list is the "all users" static list which includes all available users of the architecture.

Geofencing: Geofencing refers to the experimentation activity in which it is possible to setup a virtual perimeter on a real world geographic area and utilize this perimeter for determining if a mobile resource enters the area defined by the perimeter, exits such an area or is located inside or outside this area. This could be achieved, for example through the use of the GPS sensors, which are usually available on modern smartphones.

Project Code: A project code is a mechanism that allows the experimenter to advertise an experiment (e.g. through social media) and select all the users that responded to his call. It allows both the filtering of crowd resources as well as creating a survey list.

11.7 Virtual & Modelled Testbeds

In order for research in networks and systems to be conducted in a systematic way, there is a need for environments that will provide the necessary control and tools for designing and conducting experiments with the aforementioned attributes. Such integrated environments are called testbeds; i.e., facilities particularly designed for conducting scientifically correct experiments in order to test analytic results, computational tools, architectures and technologies.

Typically, testbeds are developed with a focus on a particular class of applications (e.g. wired networks, IoT systems, etc.). Apart from the development of the system under study per se, auxiliary components are also developed in parallel that help define the parameters of the experiment and monitor the operation of the system. Typical examples include tools for automatic reconfiguration of the system architecture (e.g. selecting a specific sub-set of the resources), automatic definition of parameters such as generated volume of data, on-line monitoring of the operation of the system and data collection for post-experiment processing. Such toolsets standardize the experimenting process and alleviate a great burden from the researchers thus helping them focus on the actual research.

However, despite their great advantages, testbeds also pose limitations on experimental research. The way a testbed is designed and developed designates (sometimes to a significant extent) the way experiments can be conducted and therefore may greatly affect research. The hardware that is being used, the size and the architecture of the testbed are indicative factors which have great effect on experiment design. For instance, a facility may be focusing on IoT applications (e.g. use case scenarios for smart rooms) and can be equipped with specialized hardware for monitoring energy consumption. On the other hand, it may provide limited support for developing and evaluating low power routing algorithms.

In this context, software-based facilities can be used in order to alleviate such restrictions. An existing physical testbed can be qualitatively and quantitatively extended with the aid of software-based facilities. IoT Lab has identified and investigated two different classes of such facilities. On one hand virtual testbeds, which quantitatively augment a physical testbed via emulated nodes, and on the other hand modelled testbeds which qualitatively extend a testbed via specialized simulation software.

Virtual Networks as Testbeds

Sometimes, the need arises for a physical testbed to be augmented quantitatively, but the physical resources are limited and cannot be easily extended on demand. In order to address such cases and provide the facility providers with a higher degree of agility, IoT Lab has proposed a method for augmenting an existing physical testbed with virtual nodes. Of course, the proposed method is not generic and does not apply to any kind of testbed facility. Following the thematic scope of IoT Lab, the proposed method addresses IoT experimenting facilities with a focus on studying use case scenarios (e.g. instead of evaluating networking algorithms and protocols)

In this method, the Cooja network simulator is used, which is an actual compiled and executing Contiki system available also in its latest release of 3.0. The advantage of this system is that Contiki is compiled for the native platform as a shared library which is then loaded into Java using Java Native Interfaces making the system fully compatible with physical resources running the same Contiki code. Apart from the fact that the simulated resources do not actually sense the environment and are not physically placed in the same space as their physical counterparts, the resulting resources are identical to the physical ones, running the same firmware and interfaces.

The simulated nodes form a virtual network which communicates with the provider's gateway. The gateway then exposes the virtual network to the rest of the IoT Lab platform using the same methods and interfaces as the physical nodes. This allows for the experimenter to discover, reserve and utilize them using the standard IoT Lab interfaces and processes, thus augmenting the testbed and extending the availability of resources as needed.

When advertised to the IoT Lab platform, the simulated resources are marked as virtual so as to be identifiable from the experimenters, who will choose whether they want/need to use them along with the actual physical resources. The pool of simulated resources is predefined for each testbed and each resource is utilized only when needed. This choice is made in order to mitigate any potential issues regarding the stability of the provider's gateway and the quality of service provided to the experimenters. The size of the pool of the simulated resources depends on the capabilities of the gateway and is to be decided by the provider.

The simulated resources report sensor values by either taking into account only other simulated resources in the system (isolated simulation environment) or by being interlaced with physical resources of the same provider. These resources will be interlaced with the physical resources of the testbed in the sense that the sensor values reported by the simulated nodes will be extrapolated from the values measured by the physical motes. The extrapolation will be based on their relational (virtual) position in the space of the modelled testbed.

Modelled Testbeds

Some of the restrictions posed by testbed facilities come from the limited number of available resources (e.g. sensor motes) as well as the usually fixed positions of the resources in the area of deployment. In a typical physical IoT testbed adding more resources or changing their topology may not be so easy (either due to lack of hardware or due to configurations needed). In an effort to mitigate such issues IoT Lab studies modelled testbeds.

In this study, modelled testbeds although operating and heavily relying on software, are tightly connected with existing physical testbeds; both in terms of semantics and in terms of operation. This way, the benefits coming from both solutions are combined. On one hand, physical testbeds provide the desired level of realism – an issue that commonly emerges in simulation studies – and on the other hand modelled testbeds provide the desired agility and ease of deployment. The modelled testbed contains data on which physical resources are taken into consideration as well as which virtual resources were created in its context (virtual resources created in the context of a modelled testbed are not shared or used along with resources of other modelled testbeds). In terms of semantics, a modelled testbed is connected to the physical testbed it models. So, it also maintains data on the physical space of the modelled testbed in the form of building topology data.

Regarding physical resources, these are described in the Resource Specification XML (aka RSpec) along with the paths needed for them to be accessed and serve measurement queries. A similar mechanism is provided for the virtual resources of a modelled testbed in the form of a Virtual Measurements Interface. This interface provides paths to be used by the experiment execution module of the IoT Lab platform for each virtual resource that is contained in a modelled testbed. Behind the scenes, it also calculates the measurement values that the virtual resources return as a response to measurement queries. These responses are based on the real measurements obtained by the physical resources as well as their relative placement in the 3D space.

As an indicative example, consider a modelled testbed modelling a given IoT testbed, which is equipped with several environment sensors (ambient luminance, temperature, relative humidity, etc.). An experimenter spawns a new virtual temperature sensor, in the context of this modelled testbed, and places it in between two other temperature sensors which correspond to physical sensor motes in the physical testbed. When queried, the sensor values that the virtual sensor will report, will be a function of the real values measured by the two physical sensors. For instance, this function can be defined as the weighted average of the two real measurements with respect to the distance among the sensors. The actual form of this computing function can vary and therefore, can be defined by the testbed owner. We also investigate the possibility of each modelled testbed to support several such functions and give the ability to the experimenter to freely choose among them. The specific forms of the function could include several types of average (in terms of central tendency) depending on the relative distances and number of neighbouring physical resources of the same sensory type and could be weighted depending on several other topology data, such as walls blocking direct line-of-sight between physical and virtual resources.

11.8 Privacy by Design

IoT Lab is deeply committed to respect and embed privacy and personal data protection. The whole platform is designed and developed with a *"privacy by design"* approach. The privacy and personal data protections are part of the project's requirements and are impacted the platform architecture, as well as the technologies used. Any data collection is based on the prior informed consent of the users and the potential use of personal data will be fully in line with the European directives and regulations.

The European Personal Data Protection Norms

Personal data protection is a fundamental requirement and objective of IoT Lab. The project committed to align and fully abide to the European personal data protection norms. It voluntarily decided to align with the newly adopted General Data Protection Regulation (GDPR).

According to the GDPR, article 4, "*"personal data" means any information relating to an identified or identifiable natural person ('data subject'); an identifiable natural person is one who can be identified, directly or indirectly, in particular by reference to an identifier such as a name, an identification number, location data, an online identifier or to one or more factors specific to the physical, physiological, genetic, mental, economic, cultural or social identity of that natural person;"*

In its recital 26, the GDPR states that: *"The principles of data protection should apply to any information concerning an identified or identifiable natural person. Personal data which have undergone pseudonymisation, which could be attributed to a natural person by the use of additional information, should be considered to be information on an identifiable natural person. To determine whether a natural person is identifiable, account should be taken of all the means reasonably likely to be used, such as singling out, either by the controller or by another person to identify the natural person directly or indirectly. To ascertain whether means are reasonably likely to be used to identify the natural person, account should be taken of all objective factors, such as the costs of and the amount of time required for identification, taking into consideration the available technology at the time of the processing and technological developments."*

The same recital highlights that: *"The principles of data protection should therefore not apply to anonymous information, namely information which does not relate to an identified or identifiable natural person or to personal data rendered anonymous in such a manner that the data subject is not or no longer identifiable. This Regulation does not therefore concern the processing of such anonymous information, including for statistical or research purposes."*

The Dilemma and the IoT Lab Approach

The main dilemma in IoT Lab Privacy policy is between complete end-user controlled process and the scope of the platform to serve and support researches. On the one hand, the project intends to maximize personal data protection. However, if users can modify/delete the provided data, it will impact and change a posteriori the results of the research, which is a real problem for the researchers that are using the platform. This can be considered as a trade-off between a complete end-user controlled process and the purpose of the platform to serve and support researches. IoT Lab, being a research oriented platform, is assigning the priority to the researcher. The adopted policy will be based on clear prior informed consent mechanisms. Participants will explicitly accept to give away experiments data, provided that they are fully anonymized.

IoT Lab main purpose is to support the research community by providing a tool enabling researchers to perform experiments, collect data and publish their results, without any risk that their results may be compromised by later modifications or manipulations. The capacity of IoT Lab to anonymize the collected data is hence of upmost importance. By failing to do so, the platform should enable the participants to access, modify and delete their data at any time. This would translate in modifying research results at posteriori. It would be a major problem for researchers, as their published results could be later changed by the participants' posterior interaction.

In order to address this complex situation, IoT Lab has adopted a dual strategy:

- IoT Lab has researched and aimed at ensuring systematic, complete and effective anonymity of participants and anonymization of data collected from the participants in line with Recital 26 of the GDPR. The IoT Lab platform voluntarily intends not to know who are the natural persons taking part in its experiments.
- In parallel, IoT Lab has developed mechanisms that enable, in case of technology or jurisprudence evolution, to access and delete specific data sets provided by the participants.

IoT Lab is committed to fully respect the European personal data protection norms, and is treating other specific data sets, such as information related to the researchers, as personal data, by enabling the non-anonymized data subjects to access, modify, and delete their personal data, as well as to benefit from the right to be forgotten. Moreover, the platform has adopted a very clear and explicit prior informed consent mechanism, as well as the possibility for the participants to control and modify at any time the data they share and provide to experiments.

Our Strategy and Technical Measures

Based on the considerations in the previous subsections, our consortium has taken full measures to implement applicable EU policies and good practices in order to ensure the privacy of the data subjects who participate in experiments with the IoT Lab platform. Our consortium has decided to adopt a privacy protection strategy based on the following anchor points:

- *Full compliance with European personal data protection norms.* We have followed the guidelines given by the EU privacy protection legislation (e.g. EU Data Protection Supervisors, Opinion 05/2014 on Anonymization Techniques – ARTICLE 29 DATA PROTECTION WORKING PARTY etc.) so as to be fully compliant with existing EU legislation with regard to protecting the privacy of the IoT platform participants.

- *Leverage on effective participants data anonymity* as specified in Recital 26 of the GDPR.
- *Principle of proportionality.* The IoT platform will never ask from a participant any information not directly linked to an experiment or research conducted through the platform. This precludes the collection of any personal information leading to the identification of the participant as it is not directly linked with the types of experiments allowed by the platform.
- *Clear Prior informed consent mechanism.* We have implemented a user consent mechanism which is ubiquitous throughout the interactions of a participant with the platform. At any step of the interactions where any kind of information is send by the participant to the platform (e.g. sensor data), a specially designed interface informs the participant about this and asks for his/her explicit consent to perform the sending operation.
- *Sliced informed user consent*: We have implemented a sliced (granular) user consent mechanism whereby it is ensured that the crowdsourcing tool users are timely informed about the policies of the IoT Lab for privacy, anonymity and security when a given data processing is going to take place.
- *The right to be forgotten.* Even if their data are fully anonymized, the participants can at any time easily access their profile, modify it and delete it. The modification or deletion of profile is immediate, and is applied to any new data collection. Modification of deletion of profile is not impacting previously collected data as long as these data are deemed fully anonymized. As an additional protection and safeguard, a complementary mechanism enables the administrator to manually give access, modify and delete data sets according to the anonymized user ID.
- *Role-based access control*: an identity management scheme is implemented with a role-based authentication and authorisation policy. In this scheme, individual identifiers are assigned to all the types of users of the platform that are used for their authentication, authorization and management of privileges across the platform. The access rights differ from user to user, depending on the role of the user (administrator, researcher, participant, sponsor, charity, etc.).
- *Actively ensuring that collected data from the participants are effectively anonymized and cannot be linked to an individual.* This measure enables to treat the collected data as non-personal data from the start. However, in order to give full flexibility and generality to the platform, we have

developed complementary mechanisms that will enable the participants to delete their data on request (or automatically if they wished so).

- *Decreasing raw data granularity.* Raw data are not personal data per se; however, when combined with other pieces of information they may enable the data controller to infer some detailed information on users. Therefore, limiting raw data granularity when not necessary is a way to prevent potential unnecessary combination of the latter with other information relating to an individual.

11.9 Incentive Mechanisms and Model

While ensuring that end-user privacy is protected, as presented above, it is equally important to motivate and engage with the crowd not only to participate (initial use) but also to sustain its engagement at all times (continued use). Thus, keeping participants motivated and engaged across time, while accounting for their individual evolution within the system is of critical importance for the success of any crowd-driven ecosystem whose participatory value creation processes are driven by users (Ziouvelou et al., 2016). Existing research in the area indicates that enjoyment, career concerns, satisfying intellectual interest, increase of status, community support, feeling affiliated and creating social contacts are a few of the most important motives for crowdsourcing and crowdsensing systems (Brabham (2010), Kaufmann et al., (2011), Nov (2007)); which vary with the type of crowd-driven initiative.

The IoT Incentive Model

In the context of IoT Lab we have placed special emphasis on the motivation and engagement of the crowd-participants as well as on the rest of the ecosystem stakeholders via the design of an incentive mechanism that triggers motivation and engages user participation while accounting for the evolutional parameter of the user within the system.

Based on our analysis of a variety of different incentive models, the most appropriate model for IoT Lab has been found to be a "*hybrid gamified incentive model*" that combines two key types of incentives, namely: (i) intrinsic and (ii) extrinsic incentives, while it also includes innovative approaches that aim to enhance both the extrinsic, intrinsic and social motives such as the "*gamification approach*" (Figure 11.8). Such an amalgamation will not only motivate users' participation during their initial usage decision but also play

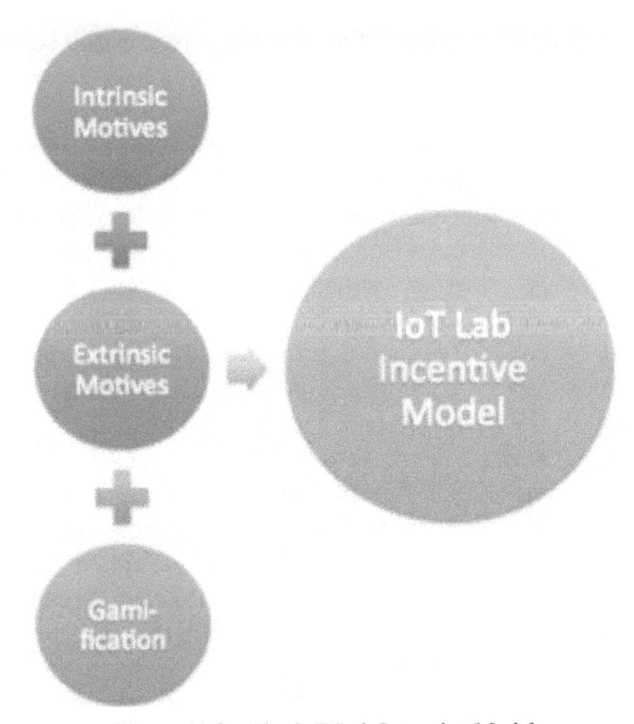

Figure 11.8 The IoT Lab Incentive Model.

a critical role during the subsequent user decisions facilitating a continued and engaged use of the IoT Lab system. In addition this model accounts for the dynamic evolution of the ecosystem as well as its users via the integration of a gamification practices that will act as an important incentivisation scheme that will enhance user experience and will sustain their on-going engagement.

Gamification by Design

The IoT Lab hybrid-gamified incentive model, integrates a number of key gamification [9] techniques such as points, badges and leaderboards (Morschheuser et al., 2016). A *point-based reward system* has been designed taking into account the specificities of the IoT Lab experimentation process for the crowd participants and the researchers, awarding points/credits for the different actions of the users inside the IoT Lab platform.

Having adopted a "*social good business model*" IoT Lab will allow its community members to allocate the points/credits collected by participating in the experiment to a charity of their choice, out of a list that will be provided by the platform. This approach is based on the assumption that a research sponsor provides a budget for an experiment, out of which a small amount of the budget ("social revenue distribution") will be used for the platform maintenance and the rest will be allocated to the users so that they can in turn re-allocate them to the charities proportionally to their point/credit distribution. This will enhance further the intrinsic motives of the crowd participants, as they will be contributing to a greater cause that goes beyond contributing to emerging research.

Furthermore users will also be able to earn badges for different activities (resulting in different levels), providing a sense of accomplishment for the different types of user-effort (simple/complex crowdsourcing and crowdsensing tasks) and signify user status and progress within the IoT Lab ecosystem. In addition, users will be able to track their performance over time and subjective to anonymous other users of the ecosystem via leaderboards.

Finally a novel incentivisation scheme has been designed for the purposes of the project. The ***"Reputation Scoring" (R-Score)*** (Figure 11.9) is a dynamic scoring mechanism that aims to enhance the user engagement within the platform while considering the user behaviour in a qualitative

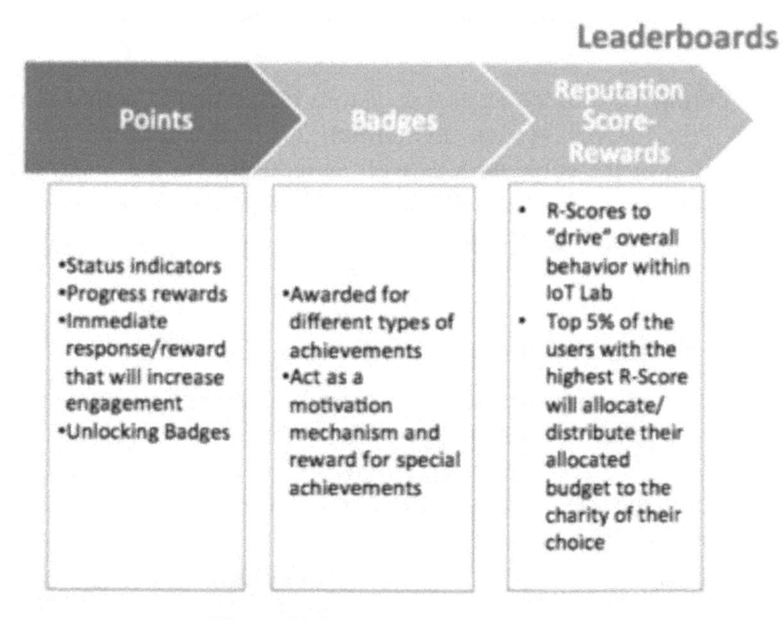

Figure 11.9 IoT Lab Leaderboards.

and quantitative manner. The R-Score, accounts for the users' overall activity (crowd-driven research value-added) from different perspectives and associated KPIs namely: (a) *Incentive-based KPI* (i.e., account for points and badges gathered by the user, among others); (b) *Crowd-driven research KPI* (i.e., proportion of proposed ideas, rate of proposed ideas, evolution of ideas into experiment, among others) and (c) *Behaviour KPI* (i.e., usage history, experiment contribution score, market assessment contribution score, among others). As such the R-score facilitates a different rewarding that encourages users on-going contribution. The R-score based rewards will be provided to the top 5% of the users with the highest R-Score: (i) *Social rewards:* Top Contributor Reward & Badge and (ii) *"Good-cause" reward:* Distinct badge and ability to do select the charity of their choice to receive part of the IoT Lab donations that will be allocated to the user.

11.10 Examples of IoT Lab Based Researches

Energy Efficiency

An energy saving scenario is being run in the University of Patras. The end goals are to monitor the energy consumption, to automate the lighting and climate and to save energy. The scenario uses static and crowd lent IoT devices together with surveys, as a way to learn the crowd's opinion. The first step is to monitor the energy consumption. Then a group of crowd users using project code is created and a message is sent, informing them about the experiment and their role in it. The research requires passive light measurements from their sensors as well as opportunistic ones for their location within the building. These values determine whether or not the lights and air-condition will be turned on or off. Follow up questionnaires determine the user's satisfaction and the need to read just the parameters of the experiment. Key challenges are the need to engage the crowd with a strong suit of incentives and to optimize the environmental parameters (i.e., light and cooling) of the space while trying to maximise energy saving.

Smart HEPIA

A smart building testbed infrastructure has been deployed in the HEPIA building of Geneva, a branch of the University on Applied Sciences Western Switzerland. The testbed enables to monitor and interact with two floors of the building. It includes temperature, light, humidity and presence monitoring, energy metering, as well as actuation on heaters, blinds and lighting system.

The testbed has been integrated to the IoT Lab platform and is used with a dual purpose: to support education of ICT engineers and to support research activities.

The Smart HEPIA deployment is used to research new solutions for improving energy efficiency of the building. Students are using the IoT Lab testbed as a service to experiment and measure the impact of various algorithmic solutions. The project is closely followed by the local authorities, which have designed the building as a reference one for future energy optimization strategies in all publicly owned constructions.

Brewery

In cooperation with the Brewery of Heineken Group at the industrial area of Patras (Greece), a use-case scenario that uses the IoT Lab platform runs at the department of New Cellars of the factory (Figure 11.10). The end goal of this use-case is to show the ability of the IoT Lab platform to serve as a useful tool for the industrial community to implement IoT technologies in their Factories and use their equipment as a service. Via this use-case it is able to achieve energy saving in satisfactory levels for the energy managers of the factory

Figure 11.10 Heineken factory in Patras, Greece.

and at the same time to provide the optimal conditions for the employees, the production and the equipment.

In this application, there are sensors to monitor the ambient conditions in this department (light level, temperature and humidity) in accordance with the use of it by the employees by taking in account the human presence (PIR sensors). Also actuators are connected to the electrical panel of the lighting system which can control the lights of this department. Moreover, the energy consumption from the lighting system of this department is measured from energy meters that are connected to the IoT Lab platform and their measurements are provided to the platform as resources.

All these sensors and actuators are provided as resources over the IoT Lab platform to the key operator of the department. Then the energy manager of this department, composes via the IoT Lab platform the appropriate scenario for the lighting system to be adapted automatically and provide the light level that is needed at any time with no energy wastage.

Also depending on the readings from the sensors, the energy meters and the actuators, it is possible for the platform to send a notification to the key operator as an alarm (in case of conditions out of limits) or a report (with aggregated data).

The key challenges of this use case are

- to develop the wireless sensor network in an industrial environment with many restrictions because of the hard nature of this environment
- to assure that the platform is robust enough to guarantee stable operation of the system to provide safety, good quality of service and ease of use for the employees
- to achieve a good level of energy saving that makes the use of Iot Lab platform a sustainable solution in real applications for energy saving.

EkoNet Novi Sad

Measurement of the air quality represents an important aspect of quality of life in the cities, as well as for running responsible operations in different industries.

ekoNET portable testbeds [10] composed of low cost sensor based monitoring devices (EB800/RPi800) enable a real time monitoring of the air quality (gas and particle sensors, sensors for air pressure, humidity, temperature and noise measurements) in urban and rural areas and they can be deployed either indoor or outdoor. Advantages include high mobility and portability, easy installation, cheaper sensor technologies and a better utilisation of data.

Each device includes a GPS module for location and GPRS mobile network interface for data transfer.

The ekoNET solution with portable testbeds is integrated within the IoT Lab platform providing a description of all resources to IoT Lab database and enabling the access to measurements from EkoNET sensors via web service. ekoNET devices are deployed at several locations in Serbia including Novi Sad city buses (MobiWallet Serbian Pilot [11]), several schools in Belgrade (CitiSense project [12]), and an open pit mine in Serbia as well as at test sites in Australia and Canada.

The IoT Lab platform with an integrated ekoNET solution represents a valuable tool for setting up and deploying the use cases to address the air pollution in smart cities enabling collection of the people's perspectives and subjective feeling about the air quality as well as allowing the crowdsourcing of opinions to tackle the problem and propose solutions for reduction of air pollution.

The use case, set outdoor, in the city of Novi Sad, combines geo-localised environmental data collected by the bus mounted ekoNET devices with geo-localised inputs from the crowd on perception of the air quality and their happiness level collected through a simple survey all via IoT Lab platform. It explores the correlation between the crowd happiness level and environmental conditions taking also into account the crowd socio-economic profile. Results obtained through this use case will benefit the local administration to reduce the air pollution in the city. As part of incentive scheme each completed survey will contribute towards a small donation to local charity thus making a step forward towards the happier city.

Similar use case is planned for schools to explore relation between air quality in schools and satisfaction, performance and behavior of pupils.

11.11 Conclusions

IoT Lab has been successful in developing, testing an using a new experimental infrastructure combining IoT and crowdsourcing. It is supporting a triple paradigm shift:

Extending IoT Research to End-users

Traditional IoT-related experiments are usually focused on the technical features and dimensions of IoT deployment. However, due to its ubiquitous and pervasive dimension, the IoT will require more and more end-user perspective to be taken into account. IoT Lab enables researchers to extend their experiments to this fundamental dimension: how are solutions accepted by end-users, where and what value they perceive in a given deployment, etc.

Enabling More Pervasive Experiments

IoT Lab enables the researchers to perform experiments in all sorts of environments, including among others smart buildings and smart cities. A set of initial experiment has started to assess the potential of IoT and crowdsourcing to assess the level of smartness and sustainability of any city. This work is a direct contribution to the ITU Focus Group on Smart Sustainable Cities [23]. In other words, IoT Lab enables research to leak outside of traditional labs by exploring IoT deployments in real environment with real end-users providing real time feedbacks.

Crowd-driven Research Model

Finally, IoT Lab is enabling and testing a new model of crowd-driven experiments. The key concept is to enable anonymous participants (the crowd) to suggest research topics and to rank them. According to the results, the favorite ideas will be proposed to researchers for selecting and implementing some of them. The results are expected to be shared with the participants (the crowd) in order to get their inputs and their assessment of the generated results. The idea is to explore the potential of a bottom-up research model on the IoT based on crowdsourcing and closer interactions between the researchers and potential end-users as illustrated in Figure 11.11.

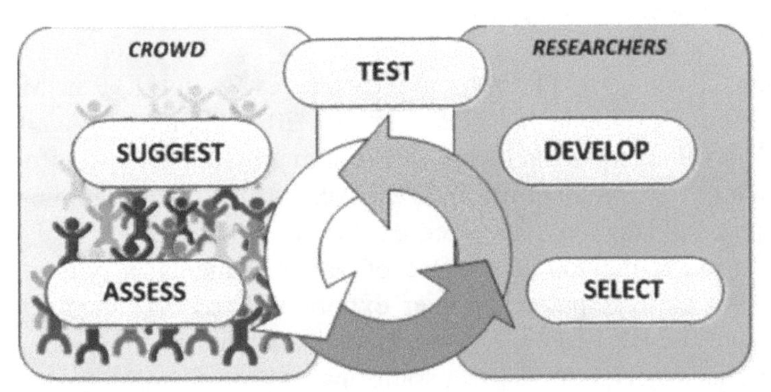

Figure 11.11 Crowd-driven Research Model enabling anonymous end-users to trigger and drive experimentation process in cooperation with researchers.

A non-for-profit association has been established to jointly maintain the IoT Lab platform and make it available to the research community. The platform is also supporting new research projects, such as F-Interop, which is developing online testing tools for the IoT.

References

[1] IoT Lab is a European research project from the FP7 research programme: http://www.iotlab.eu
[2] Internet-of-Things Architecture, http://www.iot-a.eu
[3] IoT Lab deliverables D1.2, D1.3 and D1.4 on IoT Lab architecture and component specification, and Jokic, S. et al., (2015), IoT Lab Crowdsourced Experimental Platform Architecture, 5th International Conference on Information Society and Technology (ICIST)
[4] Panagiotis Alexandrou et al., (2016), A Service Based Architecture for Multidisciplinary IoT Experiments with Crowdsourced Resources, Ad-hoc, Mobile, and Wireless Networks, Volume 9724 of the series Lecture Notes in Computer Science pp. 187–201.
[5] IoT6 European research project, http://www.iot6.eu
[6] S. Ziegler, M. Hazan, H. Xiaohong, L. Ladid, "IPv6-based test beds integration across Europe and China", in Testbeds and Research Infrastructure: Development of Networks and Communities, Springer, and Trident Conference 2014 proceedings.
[7] UDG is an IPv6-based multi-protocol control and monitoring system using IPv6 as a common identifier for devices using legacy protocols. It was developed by a Swiss research project and used by IoT6 for research purpose. More information on UDG ongoing developments at: www.devicegateway.com
[8] K. Holm, "Using MQTT Protocol Advantages Over HTTP in Mobile Application Development," IBM, 18 10 2012. [Online]. Available: https://www.ibm.com/developerworks/community/blogs/sowhatfordevs/entry/using_mqtt_protocol_advantages_over_http_in_mobile_application_development5?lang=en. [Accessed 02 2016].
[9] Gamification is defined as the use of game elements in non-gaming systems so as to improve the user experience and user engagement, loyalty and fun (Deterding, Khaled, Nacke, and Dixon, 2011).
[10] eKonet bus project, http://ekonet.solutions/
[11] Mobiwalet project, http://www.mobiwallet-project.eu/
[12] Citi Sense project, http://co.citi-sense.eu/

12

Describing the Essential Ingredients for an Open, General-Purpose, Shared and Both Large-Scale and Sustainable Experimental Facility (OpenLab)

Harris Niavis[1], Thanasis Korakis[1], Serge Fdida[2] and Loic Baron[2]

[1]CERTH, University of Thessaly, Greece
[2]UPMC Sorbonne University, France

12.1 Introduction

OpenLab was an instrumental project, delivering OneLab, the first heterogeneous federation of testbeds open for services on August 2015 and having served hundreds of users since.

OpenLab brought together the essential ingredients for an open, general purpose, and sustainable large-scale shared Future Internet Research and Experimentation (FIRE) Facility, by advancing early prototypes of this Facility. Its main goal was to advance the community by pushing the envelope of a more mature facility, targeting user interface, control and experimentation planes for highly heterogeneous testbeds as well as monitoring and first line support tools. It brought together the most experienced experts and teams in a 3 years project.

The early prototypes, coming from former FIRE initiatives OneLab and Panlab, as well as other valuable sources, included a set of demonstrably successful testbeds: PlanetLab Europe, with its 150 partner/user institutions across Europe; the NITOS and w-iLab.t wireless testbeds; two IMS telco testbeds for exploring merged media distribution; the GSN green networking testbed; the ETOMIC high precision network measurement testbed; and the HEN emulation testbed. Associated with these testbeds were similarly successful control- and experimental-plane software. OpenLab advanced these prototypes with key enhancements in the areas of mobility, wireless, monitoring, domain interconnections, and the integration of new technologies such as OpenFlow. These enhancements are transparent to existing users of each testbed, while opening up a diversity of new experiments that users can perform, from wired and wireless media distribution to distributed and autonomous management of new social interactions and localized services, going far beyond what can be tested on the current Internet. OpenLab's interoperability work brought FIRE closer to the goal of a unified Facility and provided models that were promoted to the Future Internet PPP. Finally, the project, through two open calls, supported users in industry and academia, notably those in FP7 Future Internet projects, who proposed innovative experiments using the OpenLab technologies and testbeds. Besides supporting users with a single portal and authentication mechanism, providing a direct access to their preferred testbeds, It opened an avenue for radically new needs covering the so-called verticals (smart cities, industrial Internet, transportation, environement, e-Health where several heterogeneous technologies have to be combined in a single experiment. This is a unique feature that is brought to the experimenters by OpenLab/OneLab.

12.2 Problem Statement

Experimentally-driven research is key to success in exploring the possible futures of the Internet. An open, general-purpose, shared experimental facility, both large-scale and sustainable, is essential for European industry and academia to innovate today and assess the performance of their solutions.

These were exciting times for those involved in creating new computing and communications applications, exploiting new technologies, and in seeing the world we live in change with the results. OpenLab aimed to play a key role in making these changes happen, and making Europe the hub for these changes.

Since computer applications now reach the home, the automobile, and the street, they go beyond making business and government services more efficient, and now form part of our social fabric. New ideas that start at the edge of the Internet, or of the telecommunications network, do not wait to be carefully deployed, or "rolled out" by industry, but are instead pulled out by users from App Stores, to be tried at modest or sometimes zero expense. However, their ultimate success or failure often depends on how well the infrastructure supports their requirements for interactivity and the responsiveness with which the network as a whole delivers the data and rich media that the users expect and now require. To a greater extent than ever before, the usability and naturalness of an application controls its fate. That was a time in which the portfolio of testbeds developed through the Future Internet Research and Experimentation Initiative (FIRE) could have a major impact.

A second area in which the creation of new computer applications and the businesses that they support was changing rapidly was the Cloud. New services were developed and then deployed to businesses only as they were needed. The results were a promising source of growth for business of all sizes.

At the same time as we are seeing a wealth of services being deployed in data centres, we are seeing an ever-wider distribution of data generation and storage. Computing devices are getting smaller, giving impetus to an increasing use of data input by cameras, sensors, and crowd-sourcing from cell phones and vehicles. Local data storage is cheap, whereas communications are expensive and still slow, so we expect an increasing fraction of the world's information to remain close to where it is first captured and reduced to analyzable form. However, once made analyzable, these local, managed pools of data can be accessed and used from around the globe, wherever the ultimate users are found.

Innovations today come from all parts of the world. Skype is one example with its origins in Europe. Scandinavian firms have led the definitions of 3G and LTE, or 4G, mobile technologies. CERN has contributed the Grid approach to high performance, cost-effective computing. The German automobile companies play a leading role in automating personal transportation, and Europe is a world leader in rapid intercity trains. The USB memory stick was first productized in Israel. And in Japan last year, half of the best-selling popular novels were composed on smart phones or Blackberries during commute time. Note that these innovations tended to combine novel elements on more than one scale – for example, centralized information and smart phone applications with localized inputs, or more widely distributed data integrated by applications smart enough to exploit available resources that change as the user's location changes. But all these new ideas need better infrastructure to support them, need tuning based on understanding the properties of this new internet. FIRE's purpose is to provide the environment in which we can make the Future Internet happen on a small scale today, allowing innovators to tune while exploring these new ideas. OpenLab therefore put a special emphasis on linking testbeds of different types to prototype these fast growing applications.

12.3 Background and State of the Art

In this section, we provide a state-of-the-art for the topics addressed in OpenLab, at the time of the beginning of the project.

12.3.1 Federation in the Control and the Experimental Plane

Being able to stitch together several, potentially very different testbeds was at that time an area of very active research and development. The SFA architecture and OMF were two approaches for federation developed in the OneLab project. GENI had been pursuing this goal as well, advancing a set of four **control frameworks**, each with its own approach to federation [1]. Two of these clusters, namely PlanetLab (cluster B) and ProtoGeni (cluster C) were based on the SFA architecture. Although their initial implementations were not interoperable, an ongoing effort was carried out in this direction. Built on top of a secure layer for issuing API calls, SFA defined basically three kinds of services, namely a Registry that provides an index to the resources known to the system, an Aggregate Manager that manages the target testbed, and a Slice Manager that performs routing and aggregation of data in the meshed federation. SFA is highly decentralized, in that ideally a given user, although registered at any one authority in the federation, is able to browse and allocate resources from the entire mesh. In order to be able to cope with any sort of testbed, including the ones that have not yet been designed or invented, SFA makes no assumption as to the actual resource description languages that are left to the underlying testbeds and simply forwarded through the control plane. The ORBIT (cluster E) paradigm had a rather different approach, as security concerns were traditionally less crucial in this environment, where access control can be safely implemented at a single entry point for each testbed. OMF, the software behind ORBIT and many more testbeds worldwide, had been embracing a more holistic approach which combines the control, experimental, and measurement plane through a common set of design principles underpinning its suite of tools and services. One distinguishing feature in the context of the control plane is OMF's focus on efficient use of resources, as wireless testbeds tend to be owned by single entities that have strong economic incentives to have their users be as efficient as possible when running experiments. As an outcome of the Panlab/PII project, Teagle [2] built upon a resource federation framework to control distributed, highly heterogeneous resources. The model was more centralized compared to SFA, in the sense that a common information model allowed the detailed description of resources. Centralized services such as the orchestration and provisioning of federated resources make extensive use of the common resource descriptions to allow for different granularities of resource abstraction. The aim was to support the user in working with the federated resources, exposing the necessary level of detail without overwhelming him with complex configuration requirements. Teagle as a trusted entity controls resources like physical and virtual machines, devices, and software across pan-European testbeds.

Concerning **experiment control**, it turns out that various user tools could exist that supported methodologies of defining experiments and best practices to conduct experiments. Such description of an experiment contained: i) resource requirements: what kind of resources are needed for the experiment, ii) resource configuration: user defined parameters applied on a resource, iii) resource relationships: resources might publish to or consume data from other resources, iv) workflow information: describe the needed provisioning tasks to be performed in order to create the experiment.

Regarding data storage and federation of data, the nmVO2 [3] represented a significant leap ahead in the state of the art infrastructures (DatCat, Perf-SONAR, MOME, etc.) that was storing meta-information of the monitoring and measurement data and was returning a pointer to a zipped file hosted by the owner of the data.

12.3.2 Wireless Testbeds

At the time of OpenLab many wireless testbeds for evaluating algorithms and protocols and validating communication techniques had been deployed. The most widely known were ORBIT [4] (WiMAX, Wi-Fi, Bluetooth, Zigbee and cognitive radio), MIT's Roofnet [5] and WARP [6] at Rice University, which focused on software-defined radio. These testbeds were open to the research community; however, they were limited by design drawbacks (such as the focus on static configurations, the very diverse resource descriptions, the very specific use policies) that prevented users from exploiting the many interesting features in an efficient way and that hinders tested facility owners to provision their facilities with better utilization of their resources. Other testbeds such as DieselNet at UMass [7] and the EU N4C testbeds [8] were focusing on more disruptive technologies such as delay-tolerant and opportunistic networking. Those testbeds were however closed to external experimenters. The OneLab community aimed to extend those testbed's initiatives by supporting a better framework for management and scheduling. NICTA [9], WINLAB [10], UTH [11] and other institutions had collaborated to develop and adopt a more efficient scheme for testbed management and control, based on their needs.

As a result OMF, a framework for unified testbed management, and the NITOS scheduler, a resource reservation scheduler providing slicing features, were developed, giving the opportunity to wireless network researchers to experience a more efficient and user-friendly experimental environment. Several institutions worldwide had adopted OMF for their testbeds.

12.3.3 Wired and Emulation Testbeds

Several EU funded projects promoted OpenFlow at the beginning of OpenLab, like OFELIA [12] which created an experimental facility based on OpenFlow. Five interconnected islands based on OpenFlow infrastructure had been created to allow experimentation on multi-layer and multi-technology networks. EU FP7 CHANGE explored the capabilities of OpenFlow to develop architecture for flow processing platforms within the network and individual processing of different flows. Additionally it tried to develop on-path and off-path flow processing and be the basis for flexible deployment of innovative services. The FP7 project SPARC [13] aimed at implementing a new split in the architecture of Internet components. The project would investigate splitting the traditionally monolithic router/switch architecture into separate forwarding and control elements. This functional split supports network design and operation in large-scale networks with multi-million customers who require a high degree of automation and reliability.

In the domain of the OpenFlow technology, there was clear orientation towards modelling the OpenFlow protocol functionality so that it could be offered to the wired platform users as a collection of network resources. As the wired platform would be a collection of physical and virtual network infrastructures, OpenFlow resources might correspond to a physical network infrastructure, a virtual one or a mixture of both.

Another important field addressed with new testbed enhancements is the domain of media streaming applications. As of today, experimentation on media streaming has been restricted in the field of research for coding algorithms with the scope to lighten the traffic volume of multimedia content. The VITAL++ FIRE project1 has demonstrated the immense interest of the wider ICT community for experimentation on P2P multimedia content routing with optimum use of network resources. In the frame of the VITAL++ project, research has focused on demonstrating the feasibility of accommodating some particular P2P routing algorithms in an IMS testbed. The main challenge set in this project was the design of a generic mechanism for P2P algorithms incubation across testbed networks.

12.4 Approach

OpenLab's proposition was to bring together the essential ingredients for an open, general-purpose, shared experimental facility, both large-scale and sustainable. OpenLab objective was to build and open the OneLab facility. We wanted to extend early prototypes of testbeds, middleware, and measurement tools so as to provide more efficient and flexible support for a diverse set of experimental applications and protocols. The prototypes include a set of demonstrably successful testbeds: PlanetLab Europe, with its 153 partner/user institutions across Europe; the NITOS and wiLab.t wireless testbeds, two IMS (IP Multimedia Subsystem) telco testbeds for exploring merged media distribution; a green networking testbed; the ETOMIC high precision network measurement testbed; and the HEN emulation testbed. Associated with these testbeds were similarly successful control- and experimental-plane software. OpenLab wished to advance these prototypes with key enhancements in the areas of mobility, wireless, monitoring, domain interconnections, and the integration of new technologies such as OpenFlow. These enhancements were planned to be transparent to existing users of each testbed, while opening up a diversity of new experiments that users could perform, extending from wired and wireless media distribution to distributed and autonomous management of new social interactions and localized services, going far beyond what could be tested on the current Internet. OpenLab results will advance the goal of a unified Future Internet Research and Experimentation (FIRE) facility. Finally, OpenLab issued open calls to users in industry and academia to submit proposals for innovative experiments using the OpenLab's technologies and testbeds, and devoted one million euros to funding the best of these proposals.

OneLab came from a vision originated in 2005, built on several issues related to experimentally driven research. The networking community was facing a few successes in its ability to build testing tools (like PlanetLab or Emulab) but many more failures due to well-identified causes. In addition, a challenge that is still open for our community is to develop reproducible research, meaning that one should be able to reproduce the results that are published and supports a discovery.

This vision considered that an experimenter, namely, the one that was using the facility, should had access to an ecosystem or a "market" of various resources managed by different authorities. For this purpose, the experimenter would register to one such authority that would act as a mediator towards its peers.

The beauty of this model was grounded on the observation that there existed plenty of valuable resources out there that one could benefit if an open access was provided. Some of these resources might be unique, or the sum, or combination of them might be valuable. As it might not be the role of the resource owner to manage the users, this was delegated to an authority according to some constraints and obligations. In addition, it became quickly evident there is not a single testbed that fit all needs and that, solely, a federated model would succeed to embrace the vision (Figure 12.1).

OneLab project was one of the pillar of the European FIRE1 initiative and the initiator of the federation concept.

Enabling this vision required to define an architecture that supported the underlying concept of federation that was originally introduced in OneLab. Federation empowers to run services and tests using resources provided by autonomous organizations. Three main technology accelerators were identified:

- Virtualization,
- Open Source,
- Open Data.

Virtualization allows synthetic polymorphism (diversity of technologies) from one platform. In addition, it can create policy and security boundaries

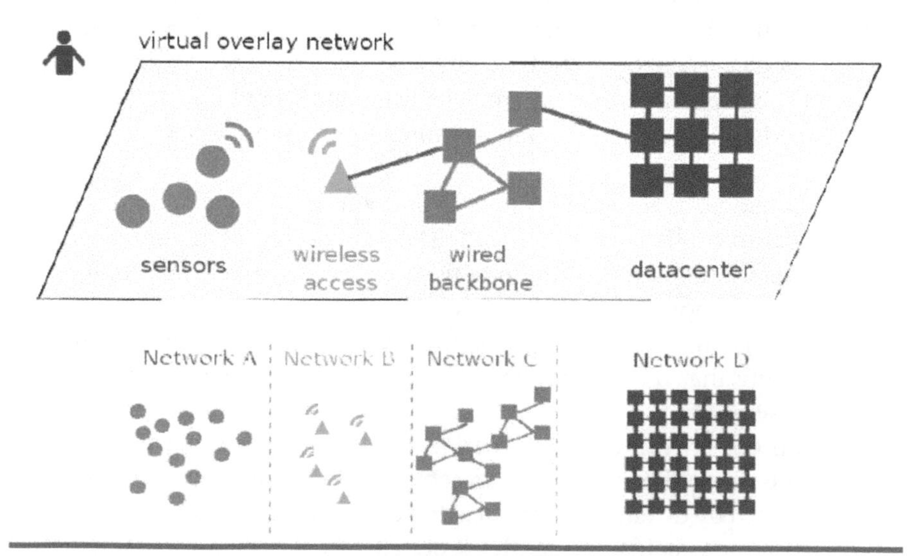

Figure 12.1 The federation of heterogeneous resources (provided by testbeds).

that are not the same as physical boundaries. **Open Source** was criticized at the beginning of OneLab as it did not support a given business model to pay for the usage of resources. Nevertheless, this approach has been proved credible in order to build a community of contributors to support the architecture. This does not even preclude any business model for the future use of the facilities. Finally, **Open Data** has not been given high-priority but will become dominant in the future as the first class citizen in any experiment is the data that have been produced and our ability use, document, and eventually share them.

Therefore, it became instrumental to address the following questions:

- What is the right level of abstraction, the minimum set of functionalities to be adopted to share resources owned by various authorities?
- What is the governance model that best supports subsidiarity?
- And finally, is there a business model or how can we enforce sustainability?

OpenLab's work in all these areas was assessed at two interoperability testing events (also known as "bakeoffs" or "plugfests"), that were planned to be the occasions to see the extent to which the tools worked across the testbeds.

12.5 OpenLab Prototypes

This section describes the prototypes, a set of existing tools and testbeds, as of January 2011, illustrating the maturity of many of these technologies and their readiness to be integrated into a larger facility.

The prototypes described here emerged mostly from earlier FIRE efforts, notably the OneLab2 and PII projects and the work of IBBT. In addition, OpenLab brought in a number of excellent contributions from elsewhere, such as UCL's HEN testbed, which was developed with UK national funding.

We organised our work on tools into two categories: control plane tools and experimental plane tools. Control plane tools largely work behind the scenes to support basic testbed operations and federation, whereas experimental plane tools were visible to the user and depend upon the control plane in order to function. The distinction between these two planes was similar to the notions of kernel and user space in the operating systems arena.

The tools that OpenLab started with were each typically specific to a single testbed environment. Our embrace of multiple tools with similar capacities was intentional: each had a particular way of doing things that would be attractive to a particular type of user with particular needs. Indeed, some tools such as the OMF [9] Experiment Controller and PLE's MySlice [14] interface came with existing communities of hundreds of users who were already comfortable working with them. OpenLab extended these tools' coverage. By requiring a heterogeneous set of tools to function across multiple testbeds we put our emerging interoperability standards to the test.

Other prototypes were individual testbeds, or were tools that were specific to a testbed or type of testbed. We grouped these prototypes into two broad categories, wireless and wired, corresponding to two broad research networking communities. The groupings were not strict: some technologies described in one category could also be applied in another.

12.5.1 Wireless Prototypes

12.5.1.1 NITOS (Network Implementation Testbed using Open Source code)

NITOS [15] is an OMF-based wireless testbed in a campus building at UTH in Volos, Greece. It consisted of 45 nodes equipped with a mixture of Wi-Fi and GNU-radios, as well as cameras and temperature and humidity sensors. Two programmable robots provided mobility. This publicly available testbed supported experiments across all networking layers. In addition to OMF, the testbed employed locally developed tools: the NITOS scheduler, a resource reservation application, and TLQAP, a topology and connectivity monitoring tool.

12.5.1.2 w-iLab.t

The w-iLab.t testbed [16] is a wireless mesh and sensor network infrastructure deployed across three floors of the IBBT office building in Ghent, Belgium. It contained 200 locations, each equipped to receive multiple wireless sensor nodes and two IEEE 802.11a/b/g WLAN interfaces. Wi-Fi and sensor networks operated simultaneously, allowing complex and realistic experiments with heterogeneous nodes and multiple wireless technologies. In addition, shielded boxes used to accommodate nodes that can be connected over coax cables to RF splitters, RF combiners and computer controlled variable attenuators, thus allowing fully reproducible wireless experiments with emulated dynamically changing propagation scenarios. With an in-house designed hardware control device, unique features of the testbed included the triggering of repeatable digital or analogue I/O events at the sensor nodes, real-time monitoring of the power consumption, and battery capacity emulation.

12.5.1.3 DOTSEL

The DOTSEL testbed at ETH Zürich is focused on delay-tolerant opportunistic protocols and applications. It was composed of 15 Wi-Fi equipped Android Nexus One devices that were carried by staff members, and five Wi-Fi a/b/g ad-hoc gateways.

12.5.2 Wired Prototypes

12.5.2.1 PLE (PlanetLab Europe)

PLE [17] is the European arm of the global PlanetLab system, the world's largest research networking testbed, which gives users access to Internet-connected Linux virtual machines on over 1000 networked servers located in the United States, Europe, Asia, and elsewhere. Nearly 1000 scientific articles mention the PlanetLab system each year, including papers in such prestigious networking and distributed systems conferences as ACM SIGCOMM, ACM CoNEXT, IEEE INFOCOM, ACM HotNets, USENIX/ACM NSDI, ACM SIGMETRICS, and ACM SIGCOMM IMC. Researchers use PLE for experiments on overlays, distributed systems, peer-to-peer systems, content distribution networks, network security, and network measurements, among many other topics.

Established in 2006 and developed by the OneLab initiative, PLE is today overseen by four OpenLab partners: UPMC, INRIA, HUJI, and UniPi. UPMC handles testbed operations and INRIA co-leads, along with Princeton University, development of MyPLC, the free, open-source software that powers PlanetLab. The PlanetLab Europe Consortium has 150 signed member institutions: mostly universities and industrial research laboratories, each of which hosts two servers that it makes available to the global system. These institutions are home to 937 users. On a typical recent day, 244 were connected to on-going experiments.

OpenLab extends both the PlanetLab software and the PlanetLab Europe Consortium.

12.5.2.2 HEN (Heterogeneous Experimental Network)

HEN [18], built between 2005 and 2010 by UCL, provides 100 server-class machines with between 6 and 14 NICs each, interconnected by a Force10 E1200 switch with 550 Gigabit ports and 24 10-Gigabit ports. This infrastructure allows emulation of rich topologies in a controlled fashion over switched VLANs that connect multiple virtual machines running on each host. The precise control of topology and choice of end-host operating system possible on HEN are particularly valuable facilities to networking and distributed systems researchers.

Many dozens of researchers actively use HEN: at Stanford University, the University of Lancaster, NYU, the Nokia Research Centre, and NEC Labs Europe, to name a few. UK- and EU-funded projects, including the EPSRC-funded Virtual Routers project, EPSRC-funded ESLEA project, EU FP7-funded Trilogy project, and EU FP7-funded CHANGE project, have all generated the bulk of their experimental results on HEN. Results have been published in prestigious networking and distributed system venues including ACM SIGCOMM, ACM HotNets, USENIX/ACM NSDI, USENIX Security, ACM CCR, ACM CoNEXT, Presto, FDNA, PMECT, ICDCSW, and LSAD.

12.5.2.3 The WIT IMS testbed

The TSSG/WIT NGN IMS testbed [19] is an Irish nationally-funded initiative serving telecom firms seeking to develop or test NGN services. It provides them with advanced multimedia services, such as conference calling and handling of presence information. The testbed is a carrier grade NGN platform based on the Ericsson IMS Communications System (ICS). The SIP based horizontal network architecture includes an Ericsson IMS core and the components for managing sessions, addressing, subscriptions and IMS interworking components with the relevant gateways for connectivity to other networks. The testbed has recently been upgraded with pico/femto cells to allow secure remote access to the test facility. The network also includes support systems for handling provisioning, charging, device configuration and operation and maintenance.

Clients include IP centrex companies, a location based service provider, and developers of pico/femto cell technology. International customers have conducted testing in the area of IMS security and testbed interconnection using the GSMA Pathfinder service operated by Neustar.

12.5.2.4 The University of Patras IMS testbed

The University of Patras IMS testbed supports PSTN testing scenarios: calls between a PSTN network and any PSTN number (including international and mobile numbers); and calls between IP phones (either soft phone or hard phone) and any PSTN number (including international and mobile numbers). The testbed has been used in numerous interoperability experiments with the carrier grade network of Telecom Austria, and the NGN testbeds of Siemens AG in Munich and Telefónica TID in Madrid. It is currently hosts experiments from the FP7 VITAL++ project. Integration of the testbed into the Teagle framework was carried out under the PII project. In OpenLab, the testbed will be enhanced to incubate P2P/NGN QoS reservation algorithms and establish experimentation paths taking advantage of the OpenFlow protocol.

12.6 Technical Work

12.6.1 Federation in the Control and the Experimental Plane

Research in the networking area has fostered the emergence of a wide variety of experimental testbeds. The vision, that OpenLab had been promoting from its very beginning, had it that disruptive innovations in the networking area would emerge from giving researchers easy access to all these resources in an open and consistent way, thus creating opportunities to conjugate all the new capabilities at a large scale.

OpenLab's activities had been instrumental in bringing this vision to reality, by tackling the general issue of testbed federation. Our achievements in this field were very substantial, both at the design and implementation levels. On the design front, OpenLab had been an active contributor to the architecture and specification of SFA, that offers a common way for testbeds and tools to expose, discover and provision resources – what we call "Control Plane" – in an homogeneous way, and that was defined inside an international community that, despite being informal, has representatives from virtually any kind of networking testbed in the developed countries. Still on the design side, OpenLab had proposed **FRCP**, a testbed-neutral layer for managing live resources – what we call "Experimental Plane" – to serve the same kind of purpose as SFA but during experimentation and not only in the preparation phases. FRCP has likewise reached a very wide consensus over the community and is starting to be widely available.

Like always when general adoption is at stake, proposing specifications and architecture is not enough if it does not come with at least one reference implementation. This is why OpenLab developed **SfaWrap** [20] and **OMF6** [9] that provided such a reference implementation of SFA and FRCP respectively, see also Figures 12.2 and 12.3.

Building on top of this architectural foundation, OpenLab had created a legal framework for operating a wide and heterogeneous federation of testbeds that spanned beyond OpenLab per se, and that we had named **the OneLab Consortium** (more details in Section 12.7.2). Starting with the OneLab Portal (see Section 12.7.2.2), researchers can enjoy all the benefits of testbeds federation on for example PlanetLab Europe that operates old-school

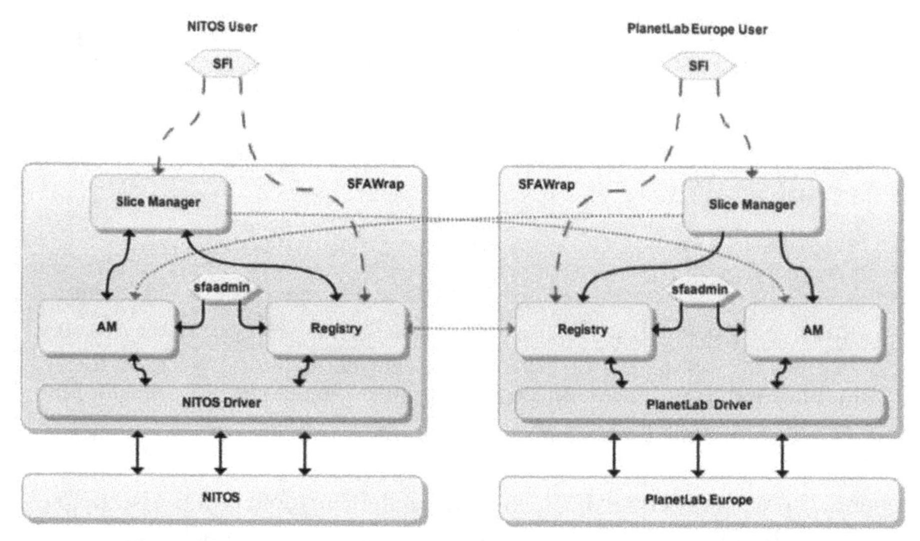

Figure 12.2 NITOS and PlanetLab Europe federation via SFAWrap.

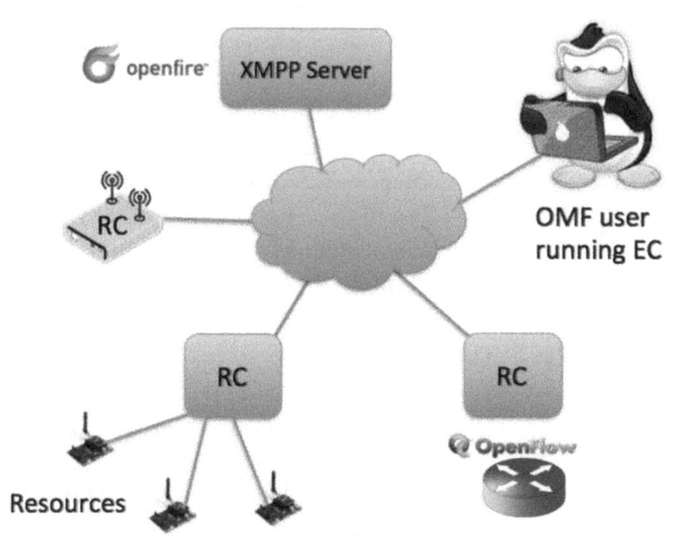

Figure 12.3 OMF 6 architecture.

wired servers, IoTLab that offers sensor nodes, NITOS that features WiFi nodes; several other European testbeds are in the process of joining.

This portal features higher level, more experimenter-oriented tools, that were developed within OpenLab; in this category let us quote **MySlice** [14], alongside with its companion **manifold**, that runs under the hood in the OneLab Portal (see Section 12.7.2.2), and that offers a set of web-based tools for dealing with heterogeneity, in terms of both resources and measurements, but in a uniform manner, as depicted in Figure 12.4.

However, using the portal was not the only option, and third-party tools had also been implemented, that directly took advantage of SFA and FRCP to offer alternative all-in-one tools for researchers, like for example **NEPI** [21, 22].

To summarize, OpenLab has created a complete paradigm for deploying a federation of testbeds, and is now operating the OneLab Portal as a first production-grade such federation. We are hoping to provide valuable help to the research community thanks to this new tool, and are confident that the conceptual assets of OpenLab will be further enhanced by on-going and future projects, like Fed4Fire.

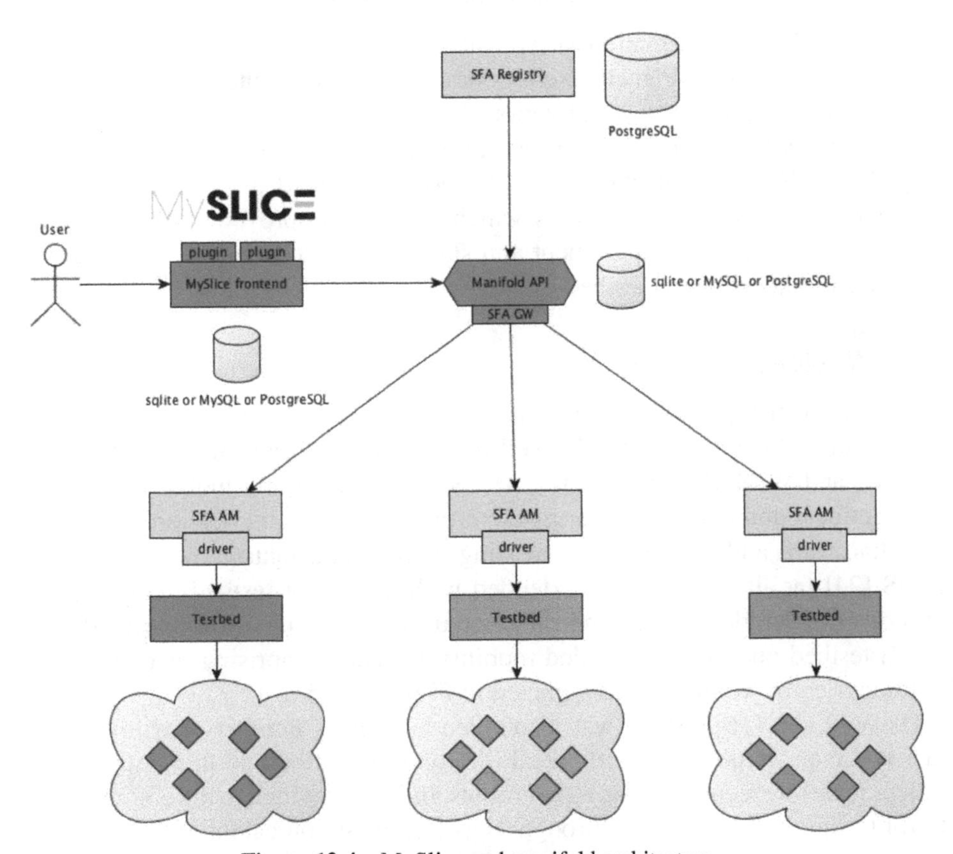

Figure 12.4 MySlice and manifold architecture.

Concerning experiment control, the most visible outcome is all the set of enhancements and added features brought to NEPI for supporting generically SFA and FRCP; those capabilities obviously extended the tool's scope drastically, as a very welcome addition to the already existing testbed-specific methods for accessing resources (like e.g. raw ssh). Concerning interoperability of measurement data, the MOI ontology was concluded with participation of TUB, iMinds and UAM. A data federation tool called GrayWulf for SQL data sources was developed that enables users to access various databases through a unified SQL-based querying interface. Both approaches integrated TopHat, EtomicDB and nmVO databases operated by different OpenLab partners. Concerning usage control, a detailed usage accounting activity was performed, for each of the testbeds participating in OpenLab.

Finally, two experiments were conducted through the two Open Calls of the project using and validating the activities of this section:

- SNIFFER Experimentation: a replicable base for long-running service using OpenLab and PlanetLab environment in order to better observe and track the long-term growth of various Storage Networks (Grids, Clouds, Content Delivery Networks, Information-Centric Networks) [23].
- ECLECTIC Experimentation: a new tool for testbed management for Peer-to-Peer (P2P) applications which includes improved support for resource allocation, deployment and state-of the-art monitoring over a range of experimental testbeds.

12.6.2 Wireless Testbeds

In the context of OpenLab, three existing FIRE wireless testbeds (NITOS, w-iLab.t and DOTSEL) were enhanced in terms of hardware with features like: LTE and WiMAX technologies, new wireless fixed and mobile nodes and directional antennas and in terms of software with the integration of the above hardware additions into the existing control and management tools. NITOS [24] facility was greatly extended with an indoor testbed featuring new powerful wireless nodes and directional antennas, a WiMAX testbed, an LTE testbed and an uncontrolled mobility testbed comprising of mobile phones carried by volunteers, as depicted in Figures 12.5 to 12.7.

The w-iLab.t [16] testbed was also extended with a number of wireless nodes and more importantly with a real life mobility testbed. In this testbed, wireless nodes were mounted on top of robots and are provided to the community for testing mobility issues, through user-friendly graphical interfaces and tools. The experimenter is able to use a web graphical interface, which enables

Figure 12.5 Extensions of the NITOS testbed: Icarus nodes on the left and directional antennas on the right.

Figure 12.6 WiMAX/LTE Base Station in NITOS testbed.

him/her to draw the desired path of the robots, simulate the scenario prior its execution or auto-detect collisions between the mobile nodes. Moreover, the testbed's administrator is able to monitor vital metrics like the exact status of each robot (docked, idle, active etc.), its remaining power and the access point that is connected, through the aforementioned GUI.

Figure 12.7 Mobile robots in w-ilab.t testbed.

Regarding the software developments of the wireless testbeds during the project, some of the main achievements are the extensions built for OMF, which is the framework that OpenLab and most of the wireless testbeds in the world use to orchestrate their testbed's resources. All the hardware extensions mentioned above, are now OMF compatible, by developing Resource Controllers (RC) for the WiMAX testbed, for the mobile phones of the uncontrolled mobility testbed and for the mobile nodes of the controlled mobility testbed enabling the user to control those resources through OMF. This way, the experimenter is able to design, develop and deploy an experiment using for example the WiMAX Base Station, some wireless nodes and some mobile phones, where he/she is able to test an Android application developed by him/her, or provided by the testbed. The most important aspect of this is that all these heterogeneous resources can be handled through one single OMF script, regardless of their physical location, as depicted in Figure 12.8.

Finally, the experimenter is able to design and deploy complex experiments on top of more than one wireless testbeds (for example NITOS and w-iLab.t) and take advantage of the diverse federation aspects enabled through the OMF framework. This way all the hardware extensions happened in the context of the project were integrated into OMF, providing an SFA interface, namely the necessary hook for the connection and the communication with other federation frameworks.

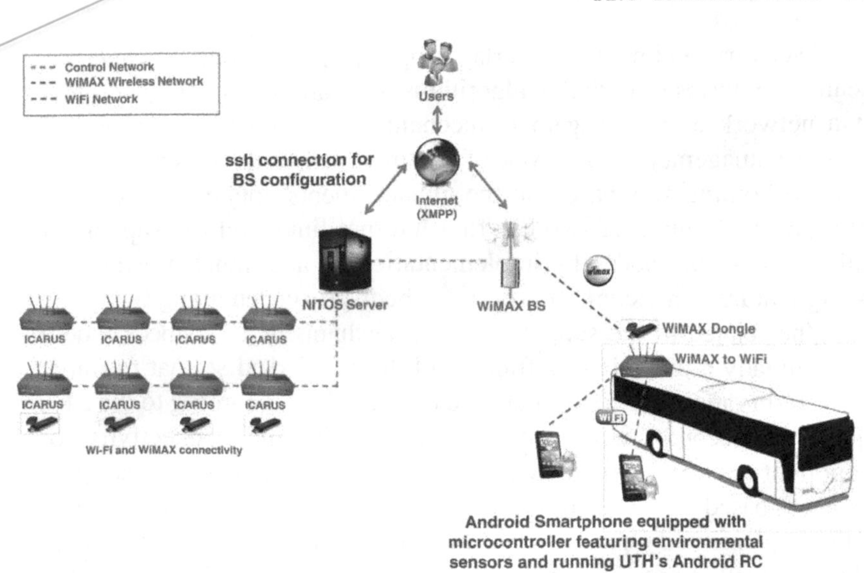

Figure 12.8 Demonstration of a complex experiment controlling heterogeneous resources with a single script.

Finally, two experiments were integrated in the context of this section's activities through the two Open Calls:

- SAVINE experiment: a Social-aware Virtual Network embedding framework for a wireless content delivery scenario [25].
- EXPRESS experiment: an innovative resilient SDN system able to withstand attacks, failures, mistakes, natural disasters and able to keep operating also in fragmented and intermittently connected networks [26].

12.6.3 Wired Testbeds

Regarding wired testbeds, OpenLab focused on the evolution of the four wired testbeds (PLE, OSIMS, WIT and HEN) participating in the project towards the support of a number of new features including: Software Defined Networking (SDN) capabilities, implementation of QoS support mechanisms, multi-homing functionality and finally support of testbed interconnectivity both from control and networking perspective to allow seamless usage of testbeds. SDN features were implemented, as planned, on the basis of support of OpenFlow. The OpenVSwitch was ported and installed in the testbeds and also integrated with each testbed's control framework and procedures for the provision of additional functionality on top of it (e.g policy enforcement, resource discovery, and network overlay support). QoS support mechanisms were realised in terms of both P2P algorithms incubation and policy enforcement via network flow management mechanisms integrated in the overall control and management framework of the involved testbeds (OSIMS and WIT). Multi-homing was based on the enhancements applied to the HEN testbed with the addition of two external Internet links and the support of external access to the nodes by implementation of additional firewall and addressing features. Interconnectivity of testbeds proceeded along two axes. The first one related to the support of SFA mechanisms in testbeds where there was already a provisioning framework to be adapted so that resource discovery, reservation and configuration can be applied according to the SFA principles. The second one related to the networking interconnectivity for the support of experiments involving more than one testbed. The adopted approach was based on the use of Layer 2 Overlays that can be instantiated by use of the implemented OpenFlow enhancements. Control and management procedures as well as provisioning mechanisms in each testbed were updated and enhanced to support the network interconnectivity approach.

Finally, four experiments were integrated in the context of this section's activities through the two Open Calls:

- ALLEGRA: deployment and "proof-of-concept" testing of a lightweight greedy geographical routing algorithm.
- ANA4IoT: extension of the OpenLab/FIRE testbeds by mixing the available infrastructure (physical or virtual) to build a new scenario for testing different approaches and evaluate the capabilities to cope with IoT requirements.
- PSP-SEC: evaluation of a PSP SDN application running on top of OPENER, with the aim of delivering security in a way that BGP is not even aware that it is being secure.
- WONDER: experimentation with and evaluation of WebRTC service delivery mechanisms namely IMS and Web service delivery approaches.

The experiments have helped both experimenters and testbed owners to collect valuable feedback and focus better on issues relating either to their experimentation aspects or to their testbed mechanisms respectively.

12.7 Results and/or Achievements

A major achievement of the OpenLab project is the opening of the OneLab facility, the first open federation of heterogeneous testbeds, making OneLab sustainable and independent from the OpenLab project ending in September

2014. OneLab materialized the efforts of the OpenLab consortium during the course of the project. It also clearly stressed our ambition to deliver a Service as the main contribution of the project and not to focus on the underlying technology, although very important.

In section 12.7.1 we describe the particular outputs of the project and in section 12.7.2, we present the major outcome of the project, namely the OneLab Experimental Facility.

12.7.1 OpenLab Main Outputs

We benefited from the experience in architecting the Internet to design our architecture model. It is grounded on two principles:

- The "Hourglass" model of the Internet that identifies the IP protocol as the convergence layer. We'll define one for the Federation of testbed resources;
- The peering model of the Internet that relies Customer sand Providers and define peering agreements in a way that there is not a single point of control. Here, we will clearly identify Experimenters, Testbed owners or providers and the Facility itself that rule them all.

We therefore have defined the following abstractions:

- Resource: Testbed ensures proper management of nodes, links, switches.
- User/Experimenter: Testbed guarantees the identity of its users.
- Slice: A distributed container in which resources are shared (sharing with VMs, in time, frequency, within flowspace, etc.). It is also the base for accountability.
- Authority: An entity responsible for a subset of services (resources, users, slices, etc.).

SFA (Slice-based Federation Architecture) was designed as an international effort, originated by the NSF GENI framework, to provide a secure common API with the minimum possible functionality to enable a global testbed federation.

The fundamental components for testbed federation were built incrementally, as the understanding about the requirements became better understood. The first international realization of federation arose in 2007, as a mutual investment from PlanetLab Central, managed by Princeton, and Planet Europe, established by UPMC and INRIA in Europe. It was then, enlarged to both private and public instances of PlanetLab, allowing a user registered under one of these authorities to benefit from resources own by any other authority.

Nevertheless, resources were homogeneous and the usability was tight to PlanetLab users. It then became of utmost importance to enlarge and extend the federation principle to other type of resources, a more scalable model of federation and an increased ease of use. In parallel, started the important effort to complement and populate the architecture with components mandatory for the entire experiment life cycle.

The experiment lifecycle comprises the following steps:

1. User account & slice creation
2. Authentication
3. Resource discovery
4. Resource reservation & scheduling
5. Configuration/instrumentation
6. Execution
7. Repatriation of results
8. Resource release

Step 1 is handled by the Home Authority of the User, the one the user has registered with. Steps 2 to 4 and 8 concern all involved authorities. Steps 5 to 6 are not in SFA but other components such as OMF have been developed for this purpose. OMF is a control, measurement and management framework that was originally developed for the ORBIT wireless testbed at "Winlab, Rutgers University". Since 2008, OMF has been extended and maintained by NICTA and UTH as an international effort.

SFA provides a secure API that allows authenticated and authorized users to browse all the available resources and allocate those required to perform a specific experiment, according to the agreed federation policies. Therefore, SFA is used to federate the heterogeneous resources belonging to different administrative domains (authorities) to be federated. This will allow experimenters registered with these authorities to combine all available resources of these testbeds and run advanced networking experiments, involving wired and wireless technologies.

Another component of the SFA layer is the Aggregate Manager (AM), which is required in each SFA-compliant testbed. The AM is responsible for exposing an interface that allows the experimenters to browse and reserve resources of a testbed. The SFA AM exports a slice interface that researchers interact with to set up, control, and tear down their slices. When the Control and Management Framework (CMF) of a testbed is not SFA-compliant, a so-called SFA driver is required to translate SFA originated queries into queries for the testbed. This driver wraps the CMF and exposes a standard interface to the AM. SFA Wrap [20], was then designed to ease the adaption of SFA by testbed owners so that they only have to develop the part related to the specificities of their own testbeds. This is a shared development model that has been widely adopted by the testbed community.

The SFA layer is composed of the SFA Registry, the SFA AMs and drivers. The SFA Registry is responsible to store the users and their slices with the corresponding credentials.

MySlice [14] was introduced as a mean to provide a graphical user interface that allows users to authenticate, browse all the testbeds resources, and manage their slices. This work was important to provide a unified and simplified view of many hidden components to the experimenter. At the same time, it provides an open environment for the community to enrich the portal through various plugins specific to each testbed or environment. The basic configuration of MySlice consists on the creation of an admin user and a user to whom all MySlice users could delegate their credentials for accessing the testbed resources. In order to enable MySlice to interact with heterogeneous testbeds, MySlice has to be able to generate and parse different types of RSpecs (Resource Description of the testbeds); this task is performed by plugins. MySlice has been widely adopted by the community and is currently an international effort. As of today MySlice has been adopted by the following testbeds (or Projects): FIT (France), F-Lab (France), FanTaaStic (EU), Fed4Fire (EU), OpenLab (EU), FIBRE (Brazil), FORGE (EU), CENI (China), SmartFire (Korea) and III (Taiwan).

Finally, **Instrumentation and Measurement** has always been considered as a major component since the development of the Internet of testbeds concept. Indeed, core activities involved in experimentation are related to identifying, assessing and providing a set of tools and methodologies to create an empirical evidence base by measuring and adequately representing Internet data (Metrology) and, in a subsequent process, information (Mediametry) thanks to experimental investigation. The role of Instrumentation and Measurement is therefore strategic as it aims at providing the experimental validation and assessment of the scientific principles supported in the experiment. It is also involved in measuring unknown quantities, ranging from low-level parameters, such as packet loss, to high-level as individual user utilities and concerns. Today, every testbed has integrated more or less mature tools to support the user's experiment. Federating heterogeneous testbeds creates the necessity to organize measurement tools and methodologies, as well as consider a relevant architecture for this purpose. An important effort has also been dedicated by the community but has taken more time to materialize although many tools are already available such as the Manifold framework in operation in the OneLab facility.

12.7.2 The OneLab Experimental Facility

A major achievement in OpenLab exploitation is the birth of a sustainable OneLab Experimental Facility [27], with its defined governance and management structure independent and beyond any projects' lifetime. The Consortium Agreement for the OneLab Experimental facility was signed in March 2014 by five OpenLab partners: UPMC, UTH, INRIA, iMinds and TUB. The OneLab Experimental Facility Parties are partners both from the OpenLab project and from the FIT project.

- Among the Parties are joint participants in the OpenLab project, which is funded under the Seventh EU Framework Programme for Research and Technological Development, and which includes as part of its work programme the establishment of a consortium to facilitate the use of testbeds for networked computer communications;
- Among the Parties are joint participants in the FIT project, which is funded under the French national Equipements d'Excellence programme, and which is currently building a group of testbeds for networked computer communications, and which has opted to promote the use of the FIT testbeds in the context of a larger consortium that includes other testbeds;
- The Parties having collaborated in these and other projects, and having also worked severally, in recent years on testbeds for networked computer communications, this work now having reached a state of maturity in which the testbeds can be presented collectively, or in a "federation", to users.

The Parties have agreed to create a consortium called OneLab, that, for the first time in this domain, facilitates the use of multiple testbeds for networked computer communications, exists independently of any individual project, and exists beyond the lifetime of individual projects;

1. Through OneLab, to create and manage the OneLab Experimental Facility, an overarching management structure with the technical systems necessary to support a group of testbeds, referred to as Affiliated Testbeds, which will have signed agreements with OneLab. This support is to be guided by a principle of "subsidiarity", whereby only those functions that are best performed by OneLab are allocated to the Consortium, all other

functions being reserved to the legal entities that manage each Affiliated Testbed;

2. Through OneLab, to facilitate the use of the Affiliated Testbeds by individuals who are connected to legal entities that have entered into signed Membership Agreements with OneLab, the legal entities thereby becoming Members. These individuals include, but are not limited to, researchers who are salaried by a Member, students who are enrolled at a Member that is an institution of higher education, and others who have a visitor's status with a Member; The Parties wish by means of this Agreement to set the terms and conditions for implementing the OneLab Consortium,

The OneLab Experimental Facility materialises in practical terms in an established governance structure of the OneLab Consortium and in OneLab Portal.

12.7.2.1 OneLab Consortium

OneLab Consortium Agreement defines the structure, the functions, the rights and responsibilities of Parties for governing and managing OneLab Experimental facility. The Consortium defines three bodies, which are: the Governing Board, which is the sole decision-making body of the Consortium; the Board of Affiliates, an advisory body that deliberates on issues relating to the Affiliated Testbeds; the General Assembly, an advisory body that deliberates regarding OneLab's plan of activities. These are, collectively, the Consortium Bodies. In the context of these bodies, three categories of participants in the Consortium are defined:

- Governors, which are the Parties to this Agreement;
- Affiliates, which are, principally, those legal entities responsible for Affiliated Testbeds;
- Members, which are, principally, those legal entities that bring users to the Affiliated Testbeds.

The role of the Coordinator is also defined, the coordinator being the legal entity that represents and manages OneLab. Furthermore, the roles of two officers appointed by the Coordinator are defined:

- the President, who carries out the tasks related to representing OneLab;
- the Executive Director, who carries out the tasks related to managing OneLab.

OneLab Experimental Facility welcomes Affiliates and Members to benefit from and to enlarge even more the impact of shared, federated, networked computer communication resources.

12.7.2.2 OneLab Portal

OneLab value proposition is the ease of use and support in experimentation. OneLab Portal [28] offers easy access to testbeds; through OneLab, the experimenter, (a researcher, a developer, and innovator) can easily test the software system in any one, or any combination of the following networked communication environments:

- Internet-overlaid testbeds
- Wireless, sensing, and mobility testbeds
- Broadband access and core testbeds
- Network emulation environments

OneLab aims to attract users beyond its own testbed providers and its immediate stakeholders, to offer resources also for e.g. educators, learners (FIRE FP7 project FORGE, EIT ICT Labs Master School) and SMEs. There is much to understand in networked communication testbeds: each platform's hardware capabilities, how the available software environments be configured and loaded onto a platform, the many features of the experiment control tools, etc. At OneLab, we offer a skilled team that is happy to assist throughout the experimentation cycle.

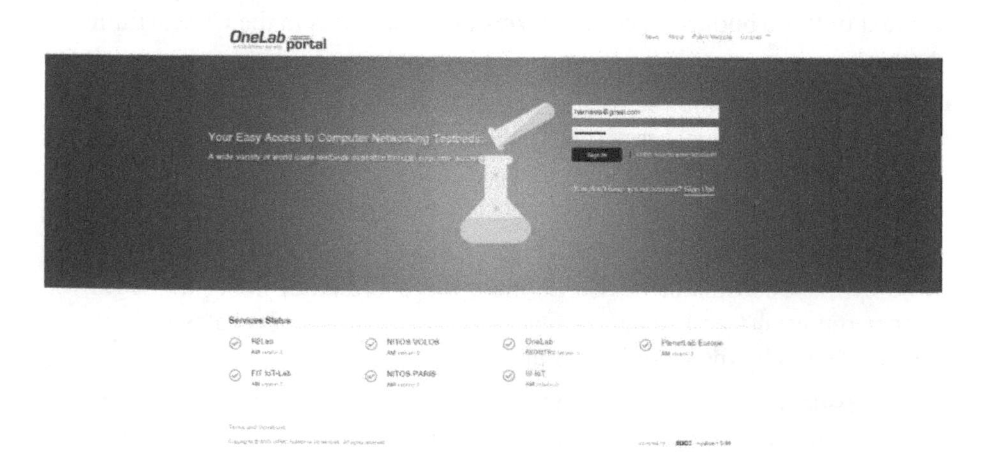

Figure 12.9 The OneLab Portal.

- For all users – academic and industrial: OneLab provides with online tutorials, documentation, and invitations to hands-on workshops and community events at which experimenters provide feedback and platform and tool developer describe their plans.
- For SMEs & other industrial users: OneLab team can accompany through the entire process of designing and running your experiment and interpreting its results.

OneLab provides tools that make it easy to use its testing environments:

- Through OneLab Portal one can access any of the testing resources. If the testing is repeated in more than one environment, there is no new account to open, no new system to learn.

Through OneLab Portal one can select highly capable experiment control tools. These are free open-source tools which are evolving to meet the needs of an ever-growing community of experimenters, and can be tailored, if needed, for particular requirements.

12.8 Conclusions

OpenLab has made a major contribution by deepening the capabilities of its various testbeds, inherited from FIRE's former OneLab and Panlab initiatives as well as other valuable sources. OpenLab advanced early FIRE prototypes that had proven their worth; this integrated infrastructure project associated and extended them, enhancing the value of the FIRE portfolio of facilities.

Combining these advanced infrastructures provided many examples of opportunity, with media distribution and localized services delivered to wireless clients perhaps capturing the most attention. As a result, our second major effort in OpenLab was to provide interoperability of these different services, allowing access and authorization to one to permit access to others, within the policies for usage and security required by each. Such common access methods form a control plane for the FIRE testbeds.

OpenLab wanted to strengthen the current offering by constructing a standard set of experiment deployment procedures, or a federated experimental plane, through which resources could be described, found, and reserved or allocated immediately. This also required implementing standards for the description of experimental configurations, for a real or simulated workload, and for the forms in which the resulting data will be logged, aggregated, analysed, and archived. Monitoring tools were critical for understanding usability issues that affect new applications and would also give us a means of seeing how testbed utilization develops over time. For testbeds to be sustainable, they should evolve to meet new interests. Only sustained access to testbeds will permit the creation of long-running experimental services so that we can understand their strengths, their weaknesses and the degree to which they meet users' expectations. This sustainability has been one of the strengths of PlanetLab and of PlanetLab Europe in recent years.

The generic control and experimental planes introduced above also needed to be instantiated and extended in each of the wireless and wired testbeds included in this project. The chief tool we employed for wireless is OMF (cOntrol and Management Framework), a framework that is used to control around 20 testbeds worldwide, including several in Europe. The tools have been expanded to support both controlled and uncontrolled experimental conditions. The wired testbeds in OpenLab allow innovation within both the telco and data network paradigms (IMS and IP protocols). The availability of OpenFlow protocols further enriches the mix of activities that can be supported and the depth into the networking stack to which experiments can probe or prototype. The Heterogeneous Experimental Network (HEN) in operation at UCL brings the two types of environments closer together. SFA (Slice-based Federation Architecture) was deployed and extended as the envelope to federate the various technology-specific control and experiment planes.

OpenLab was instrumental to push the envelope of knowledge and tools in FIRE such that the OneLab Facility was successfully launched as an independent facility on August 2015, supporting a broad and diverse set of experiments.

References

[1] "GENI Spiral 2," [Online]. Available: groups.geni.net/geni/wiki/Spiral Two
[2] "Teagle Portal," [Online]. Available: http://www.fire-teagle.org/
[3] "nmVO," [Online]. Available: http://nm.vo.elte.hu/
[4] "ORBIT Testbed Portal," [Online]. Available: http://www.orbit-lab.org
[5] "MIT Roofnet," [Online]. Available: https://en.wikipedia.org/wiki/Roofnet
[6] "WARP: Wireless Open Access Research Platform," [Online]. Available: http://warp.rice.edu/
[7] "UMass DieaselNet," [Online]. Available: https://dome.cs.umass.edu/umassdieselnet

[8] "N4C: Networking for Communications Challenged Communities," [Online]. Available: http://www.n4c.eu/
[9] "cOntrol and Management Framework," [Online]. Available: https://mytestbed.net/
[10] "WINLAB," [Online]. Available: http://www.winlab.rutgers.edu/
[11] "NITlab: Network Implementation Testbed Laboratory," [Online]. Available: http://www.winlab.rutgers.edu/
[12] "OFELIA experimental facilities," [Online]. Available: http://www.fp7-ofelia.eu/
[13] "SPARC FP7 project," [Online]. Available: http://www.fp7-sparc.eu/
[14] "MySlice," [Online]. Available: https://www.myslice.info/
[15] "NITOS Future Internet Facility," [Online]. Available: http://nitlab.inf.uth.gr/NITlab/index.php/nitos.html
[16] "w-iLab.t testbed," [Online]. Available: http://wilab2.ilabt.iminds.be/
[17] "PlanetLab Europe," [Online]. Available: www.planet-lab.eu
[18] "HEN, Heterogeneous Experimental Network," [Online]. Available: mediatools.cs.ucl.ac.uk/nets/hen
[19] "TSSG/WIT NGN IMS testbed," [Online]. Available: ngntestcentre.com
[20] "SFA Wrap," [Online]. Available: http://www.sfawrap.info
[21] A. Quereilhac, M. Lacage, C. Freire, T. Turletti and W. Dabbous, "NEPI: An integration framework for network experimentation," in *19th International Conference on Software, Telecommunications and Computer Networks (SoftCOM)*, 2011.
[22] "NEPI," [Online]. Available: http://nepi.inria.fr/
[23] A. Bak, P. Gajowniczek, M. Pilarski and M. Borkowski, "Automated Discovery of Worldwide Content Servers Infrastructure – the SNIFFER Project," in *FedCSIS*, Warsaw, Poland, 2014.
[24] "NITOS Future Internet Facility," [Online]. Available: http://nitlab.inf.uth.gr/NITlab/index.php/nitos.html
[25] A. Leivadeas, C. Papagianni and S. Papavasileiou, "Demonstration of a social aware wireless content delivery paradigm," in *IEEE Infocom 2014*, Toronto, Canada, 2014.
[26] S. Fdida, T. Korakis, H. Niavis, S. Salsano and G. Siracusano, "The EXPRESS SDN Experiment in the OpenLab Large Scale Shared Experimental Facility," in *1st International Science and Technology Conference on Modern Networking Technologies: SDN & NFV*, Moscow, Russia, 2014.
[27] "OneLab," [Online]. Available: https://onelab.eu/
[28] "OneLab Portal," [Online]. Available: http://portal.onelab.eu

13

Wireless Software and Hardware Platforms for Flexible and Unified Radio and Network Control (WiSHFUL)

Nicholas Kaminski[1], Spilios Giannoulis[2], Ilenia Tinnirello[3], Peter Ruckebusch[2], Piotr Gawlowicz[4], Domenico Garlisi[3], Jan Bauwens[2], Anatolij Zubow[4], Robin Leblon[5], Pierluigi Gallo[3], Ivan Seskar[6], Luiz, A. DaSilva[1], Sunghyun Choi[7], Jose de Rezende[8] and Ingrid Moerman[2]

[1]Trinity College Dublin
[2]iMinds-Ghent University
[3]CNIT
[4]Technische Universtät Berlin
[5]nCentric Europe
[6]Rutgers University
[7]Seoul National University
[8]Universidade Federal do Rio Janerio

13.1 Introduction

In the last years, we have assisted to an impressive evolution of wireless technologies for short distance communication (like IEEE 802.11, IEEE 802.15.4. Bluetooth Low Energy, etc.) due to the need of coping with the heterogeneous requirements of emerging applications, such as Internet of things, the Industry 4.0, the Tactile Internet, the ambient assistant living, and so on. Indeed, for optimizing the technology performance in these scenarios, it is often required to support some forms of *protocol adaptation*, by allowing the dynamic reconfiguration of protocol parameters and the dynamic activation of optional mechanisms, or some targeted *protocol extensions*. In both cases, prototyping, testing and experimentally validating potential solutions is a complex task, which generally requires significant time and resource investment. On one side, off-the-shelf wireless interfaces are based on radio chips which implement only the obligatory parts of the standards and arbitrarily selected optional parts, with only partially documented interfaces and with drivers being either closed or limited in functionality. On the other side, many powerful Software Defined Radio (SDR) platforms, while offering excellent flexibility at the physical layer, typically have limited performance and lack high-level specifications and programming tools as well as standard APIs for developing protocols.

Consequently, testing of new solutions often proves problematic, as experimenters can only rely on the limited optimization space enabled by the drivers, or on *open* software architectures where many functionalities have to be written from scratch and are tightly dependent on the specific hardware platform. In many cases, different experimentation platforms have to be considered for working on specific optimizations, because each platform supports a different level of complexity and controllability. This heterogeneity further slows down the innovation process, because experimenters have to be familiar with platform-specific architectures and programming tools before prototyping their solutions.

To overcome the aforementioned shortcomings and reduce the threshold for experimentation, we propose a novel approach within the European project WiSHFUL [1]. The project main goal is the design and development of a software architecture enabling a flexible radio and network control of heterogeneous experimentation platforms, based on standardized wireless technologies and SDRs, through unified programming interfaces. More specifically, the architecture is devised to allow:

- *Maximal exploitation of radio functionalities* available in current radio chips, as opposed to today's radio drivers that restrict radio functionality. For example today's radio drivers for IEEE 802.11 do not support TDMA (Time Division Multiple Access) operation, while the hardware perfectly supports it.
- *Clean separation between radio control and protocol logic*, as opposed to today's monolithic implementations, which do not allow to work separately on the logic for enabling specific protocol features and the definition of these features.

To frame this effort, several driving scenarios were identified to capture the challenges associated with the *increasing density* and *heterogeneity* of wireless devices in a concrete and tangible manner. These scenarios directly present a set of relevant and significant requirements for developing the functionalities required by the WiSHFUL control framework in order to investigate the challenges of future wireless systems experimentation. Each showcase focuses on a different source for inter-device and inter-technology interference and displays a scenario, which requires novel experimentation functionalities.

Following the definition of this set of motivating scenarios, an architecture is presented to support future wireless experimentation. This architecture is constructed to address the requirements of the tangible scenarios, capturing key challenges of future systems while allowing for extensions to support investigation of as yet unforeseen challenges.

13.2 Background

The need for fine-grained control of communication networks is well demonstrated by the interest of the scientific community in solutions that enable *software defined networking*, (SDN). *OpenFlow* [2], for instance, is a good example of an SDN-enabler because it allows researchers to control routers, without knowing the internals of vendor-specific implementations. OpenFlow focuses on controlling the forwarding rules between devices (e.g. switches, routers and wireless access points) connected by means of pre-installed links (usually wired). However, it does not explicitly deals with wireless links, where conditions change over time and strongly depend on interference and propagation conditions. Indeed, for wireless links the use of forwarding functionalities, which have inspired the match/action abstraction used for wired link, cannot be adequate for capturing the inter-link and inter-network dependencies, despite the fact that some extensions have been proposed, e.g. OpenRadio, for classifying the traffic flows on the basis of PHY-related fields and configuring the transmission power of the links. Actually, a closer look reveals that the wireless community has arguably anticipated, if not even inspired, the wired networking shift towards centralized controllers, for example with the CAPWAP protocol (Control And Provisioning of Wireless Access Points) [3] for the remote control of wireless access points. However, the CAPWAP control model was based on parametric control of technology-specific configuration parameters. WiSHFUL goal is more forward-looking, and resides in i) devising a generic programming model for wireless devices and wireless links, based on technology-independent programming abstractions and ii) showing that they can be handled with a network control framework which include global and local controllers.

To accomplish this goal, WiSHFUL pushes towards the identification of viable abstractions for radio behavior, by integrating four different platforms exposing high-level programming models for heterogeneous wireless technologies while taking into account the emerging solutions and standardization work concerning *reconfigurable radio systems* (ETSI-RRS) [4]. The four supported platforms are: *Wireless MAC Processor* (WMP) for IEEE-802.11 radios [5], *Time-Annotated Instruction Set Computer* (TAISC) for IEEE-802.15.4 radios [6], the *Implementing Radio in Software* (IRIS) for SDRs [7] and finally the popular Atheros chip based cheap-of-the-self wireless cards running the ATH9k driver [8]. Moreover, the WiSHFUL control framework complements OpenFlow, by enabling the coexistence of local and global controllers devised to react to the network events at different time scales. In the next phase WiSHFUL also plans to extend to support cross-layer control from the network layer and above as well, providing SDN like characteristics regarding the management and fine-tuning of control knobs ranging from routing protocols parameters and realization of flow control to transport layer parameters like TCP window for example. GITAR [9] supports the cross layer parameter control, especially in the context of WSNs, but can be used in all platforms that are supported within WiSHFUL as a cross layer parameter management component.

13.3 Motivating Scenarios

The emerging wireless ecosystem is characterized by a heterogeneous mix of technologies, operators, and service providers attempting to coexist in a single environment, and featuring a high-density deployment of wireless devices. High heterogeneity in device capabilities (in terms of spectral bands, coverage, management functionalities, networking models, etc.) combined with limited open, vendor-independent configuration interfaces complicate achieving the often conflicting goals of independent providers and integration of technologies to provide coherent service. Indeed, wireless devices often employ multiple radio interfaces, spanning over several standards (such as LTE, Wi-Fi, ZigBee and Bluetooth) or offering more esoteric capabilities in the form of programmable interfaces, based on software defined radio (SDR) techniques.

Experimental-driven research is essential for analyzing the performance of this eco-system, because of the difficulty in simulating or modelling the interactions between heterogeneous technologies, protocol configurations, environments and network operators. We consider some exemplary scenarios in order to identify the *functional requirements* and *control models* required for testing optimization and coexisting strategies dealing with the complexity of the wireless eco-system. From the analysis of these scenarios, we identified two main groups of functional requirements: i) configuring the radio of each wireless node, in terms of set-up of physical transmission parameters, bandwidth allocation, medium access schemes and prioritization mechanisms for different transmission queues, ii) configuring the network-wide policies for dealing with different traffic flows, by defining logical links and paths between nodes, mapping of traffic flows into transmission queues, performing flow control among multiple links and interfaces, etc. Moreover, it is required to introduce monitoring functionalities at different levels for collecting statistics about the radio performance and the local channel views.

13.3.1 Interference Management among Overlapping Cells

In dense wireless networks, co-channel interference is a fundamental problem, especially in the case of WiFi technologies working on the unlicensed ISM bands characterized by the availability of a few orthogonal channels and by the coexistence of multiple independent networks. Ultimately, this scenario examines questions related to the dynamic control of multiple Access Points in a coordinated manner. A possible solution for controlling co-channel interference is working on the adaptation of contention parameters and transmission opportunities used by co-located APs. Some research work has suggested the use of airtime as a metric to quantify the channel resources that are granted to each AP. The airtime is the sum of the channel holding times used by a given cell during a reference time interval. To enforce any decision about the network configuration, it is also required to represent a network global view, by considering the interference relationships among the APs, which depend on the specific location of the stations. In particular, it is required to detect hidden nodes, which may experience severe collision rates.

Consider the example network given in Figure 13.1. This scenario assumes four active flows in the following QoS classes – the first three are best effort (BE) while the last one is voice. Each flow is assigned to one of the two APs. Furthermore, let us assume that AP1 and AP2, are operating on the same radio channel. In such a case a cell-edge user like node STA2 may suffer from interference due to hidden node, i.e. the downlink traffic from AP1 to STA2 will collide with traffic originated from AP2. By solving the hidden node problem, the performance of all nodes in neighboring wireless networks can be improved.

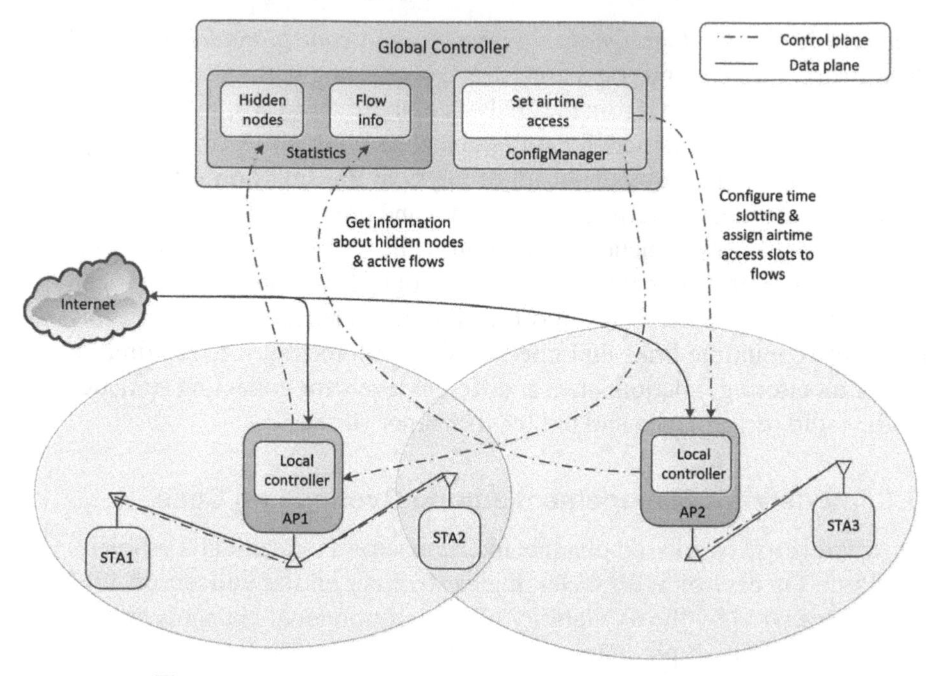

Figure 13.1 Traffic-aware 802.11 airtime management scenario.

The challenges of this scenario may be addressed by monitoring the performance of each AP. Such monitoring would make degradation associated with inefficient management clear, thus allowing rescheduling of flows to avoid interference. To accomplish this goal, global monitoring of network performance would be required. Specifically, some control entity would need the ability to monitor the active flows for detecting hidden nodes and to define appropriate channel access patterns and assign airtimes for solving the hidden node problem by dividing the competing flows in the time plane. Furthermore, tight time synchronization between APs is required for time-slotting airtime. This may be achieved by usage of PTP running over backbone interfaces.

13.3.2 Co-existence of Heterogeneous Technologies

In dense wireless networks, the co-existence of heterogeneous technologies using the same wireless resources is challenging. Indeed, although technologies working on ISM bands intrinsically deal with mechanisms for managing interference, such as carrier sense, adaptive modulations, spreading solutions, etc., it has been demonstrated that they can experience severe throughput degradation in case of coexistence of heterogeneous links, because of asymmetries in recognizing other technologies and reacting to their presence. A central controller could overcome these problems, by supporting a harmonized spectrum allocation across separate wireless technologies. This will enhance the performance in both networks and make the quality of service (QoS) characteristics (such as throughput, latency and reliability) more predictable.

As a reference example, we consider the coexistence between IEEE-802.11 (Wi-Fi in 2.4 GHz band) and IEEE-802.15.4e (time-slotted channel hoping, TSCH) illustrated in Figure 13.2. The simultaneous operation of both networks in close proximity will inevitably lead to performance degradation due to cross-technology interference. This is because of contention-free explicit scheduling of radio resources in TSCH and the unreliability of carrier-sensing (listen-before-talk) mechanism used in Wi-Fi as far as detecting IEEE 802.15.4 transmissions is concerned, rendering Wi-Fi unable to sense any wireless transmission of the other technology. The QoS in both networks can be increased by making them aware of each other.

One can imagine multiple co-existence schemes for Wi-Fi and TSCH. Some basic schemes can be implemented by only modifying the sensor network. More advanced, and also promising, schemes require cooperation between the networks. This scenario examines a traffic-aware interference

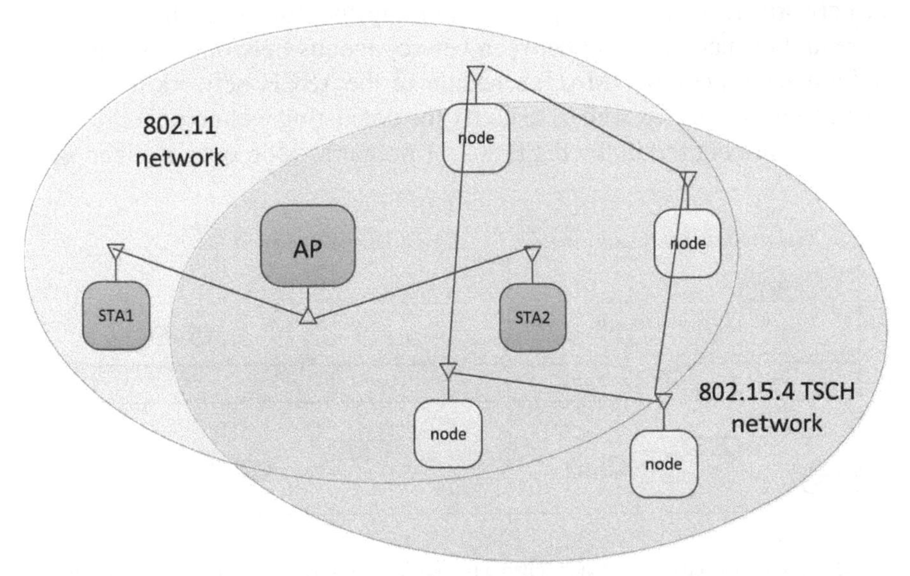

Figure 13.2 Example illustrating two co-located wireless networks of different technology.

avoidance scheme where, depending on the network load in both networks, other decisions are made. For such a scheme two possible cases, illustrated by Figure 13.3, must be considered. In the first case the sensor network is highly loaded. Here it is more efficient to perform any interference avoidance in the Wi-Fi network, thus reducing the overhead on the more loaded sensor network. To accomplish this, the sensor network would need to provide the scheduling information to allow the Wi-Fi network to delay transmissions to points in time where a collision will not occur. In the second case the network load in the Wi-Fi is high, suggesting that excluding the spectrum used by Wi-Fi from the hopping scheme of the sensor network is a more promising approach to co-existence.

More advanced approach is to use a cross-technology TDMA protocol to coordinate the transmission between both types of nodes and reduce interference to a minimum. The system runs a TDMA radio program on the Wi-Fi nodes, adapts time slots to traffic requirements, keeps free some slots that are implicitly reserved to TSCH, and uses the remainder for transmission, in order to minimize cross interferences.

To support the experimental investigation of this scenario, a great deal of functionality is required. A mechanism for the discovery of co-located wireless networks within interference range is certainly necessary to identify whether a problem exists. Furthermore, a range of mechanisms to support mutual network awareness is required, including the ability to share information regarding network load between heterogeneous networks, to expose the medium access control (MAC) schedule of the TSCH network to other coexisting technologies, as well as to notify the coexisting technologies about the wireless channel used in the IEEE 802.11 network. Moreover, mitigation

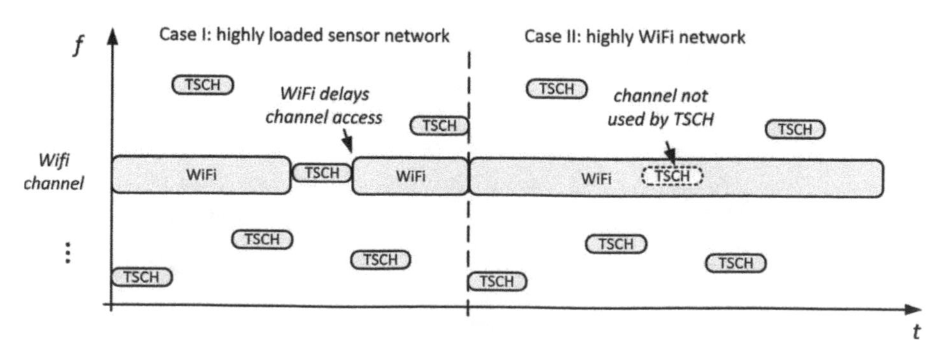

Figure 13.3 The proposed co-existence scheme for avoiding interference between Wi-Fi and TSCH.

functionality must be available, potentially including the configuration of spectrum access in the Wi-Fi network, configuration of channel exclusions within the TSCH network, time synchronization between both networks, and the tuning of MAC parameters according to frames size and slot allocation.

13.3.3 Load and Interference Aware MAC Adaptation

It is well known that contention-based access protocols work better than scheduled-based protocols in case of intermittent and unpredictable traffic flows [10]. Moreover, the contention parameters can be optimized as a function of the time-varying number of nodes which have traffic to transmit. However, for most wireless technologies, the choice of contention-based or scheduled-based access protocols, as well as the configuration of the contention parameters or schedule periods can only be configured statically, and cannot be adapted to the varying network conditions.

In order to experimentally validate the possibility to perform MAC layer adaptations or to switch from one MAC protocol to another as a function of an estimate of the network topology and contention level, it is required that nodes can infer about the number of neighbors, network congestion and node visibility by monitoring elementary channel events (busy intervals, hello messages, collisions). Moreover, pre-defined MAC protocols such as CSMA and TDMA can be abstracted from the physical layer and available for different technologies by exposing the same list of configurable parameters, including the contention windows for CSMA and the frame size for TDMA protocols. Under these assumptions, different adaptation logics can be developed for maximizing the network throughput, minimizing the delay jitters or the packet losses, regardless of the specific node technology (Figure 13.4).

As an example, we initially consider only a few active wireless nodes using a CSMA base MAC.

13.3.4 In-Situ Testing

Wireless testbeds are imperative for testing innovative technologies such as protocols, hardware, and several other modules of any wireless solution. Many of these technologies will serve in dynamic wireless environments and under challenging conditions. For the sake of maintainability and experiment repeatability, however, testbed infrastructure is often fixed. Relocating nodes is difficult since their power supply and/or network connections

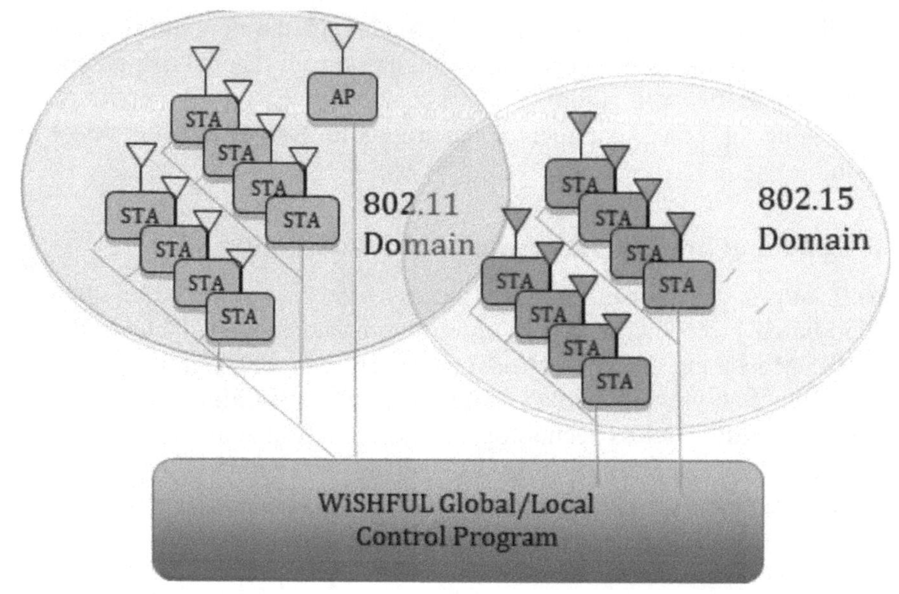

Figure 13.4 Deployment of a single local program across several platforms.

are mounted on wall sockets. The testbed environment is thus less dynamic and the conditions are more stable, thus making the evaluation of experimental wireless solutions in testbeds less realistic.

A portable testbed that can be easily deployable on remote, real-world locations is clearly necessary. Such a testbed would need to be straightforward to deploy where needed, include rugged equipment and self-contained power. Furthermore, a wireless mesh backbone to ensure connectivity between the nodes would be required to allow operation in a variety of environments. This backbone would need to employ the sort of interference management suggested by previously discussed scenarios. Finally, the portable testbed must operate in a transparent manner to allow users to examine the phenomena of interest.

Taking the successful Fed4FIRE approach [9] as a model for the use of testbed, the following steps, illustrated in Figure 13.5, would be required on the portable testbed during experiment life-cycle:

1. When the experimenter arrives at the location, the flight case is plugged into the power grid and the servers and switches boot. Optionally, the experimenter can connect the switch uplink to the Internet.
2. As the servers boot, the backbone also configures itself automatically. It creates a wireless mesh among the nodes.

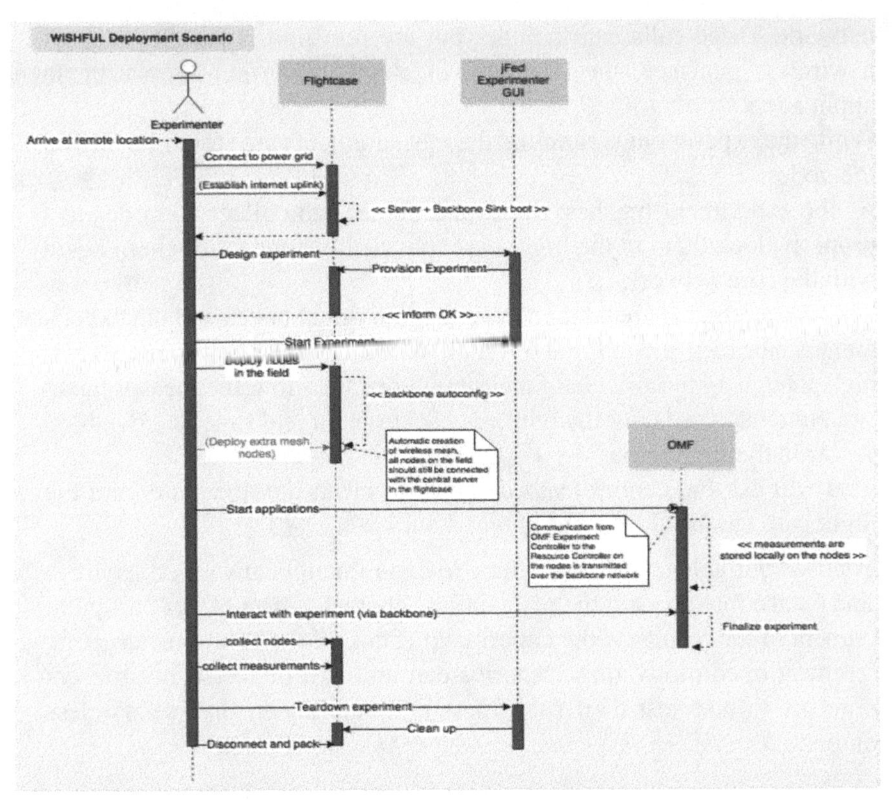

Figure 13.5 Sequence diagram for the deployment of the portable testbed.

3. When everything is up and running, the experimenter launches the jFed tool from a laptop that is either inside the flight case or connected to the central switch. The experiment is designed or loaded from a previous run.
4. jFed will perform the needed actions via the testbed management server and the nodes will be provisioned with the desired software.
5. After this process, the user is informed and the actual experiment is started.
6. The user will deploy all nodes in the field; they remain connected and accessible via the wireless backbone.
7. If there should be a bad wireless link between one or several nodes, an extra backbone node can be added to optimize the mesh network.
8. Via Orbit management framework (OMF), the experimenter starts his experiment. OMF will make the calls to the nodes over the backbone

network. These calls can include (but are not limited to) the setup of a wireless interface, the changing of channels or the starting of an application.

9. While the experiment is running, the measurements are stored locally on the nodes.
10. As the experiment finishes, the experimenter can collect all nodes and properly dock them in the flight case, physically connecting them again with the core network.
11. The measurements are fetched from the individual nodes and the experiment can be torn down. If the throughput of the wireless mesh network is high enough, or the amount of measurement data is low, the measurement can be transported over the wireless backbone in real time to a database server in the flight case.
12. jFed will ask the central testbed server to clean the nodes up, and the flight case can be closed and plugged out.

The proposed portable testbed will allow to experiment in any given environment and to take into account the real wireless characteristics of this particular environment in the results of the experiment. Thus all the solutions targeting the aforementioned motivating scenarios can and will be tested in different environments to also test their robustness and stability in diverse wireless environments.

13.4 WiSHFUL Software Architecture

Experimentation is certainly a vital tool in the development of future wireless solutions. Furthermore, as illustrated by the above discussion of scenarios for future wireless networks, a large variety of functionality must be supported to investigate the challenges most relevant in the advancement of wireless communications. Moreover, the increasing diversity of wireless solutions and competing radio technologies, along with the ever more stringent requirements on the reliability of test results, has caused wireless test facilities to evolve to be exceedingly complicated imposing steep learning curves for new experimenters. Therefore, as the need for investigating a broad range of scenarios grows, so does the difficulty in doing so.

For these reasons, the WiSHFUL project directly targets lowering the experimentation threshold by developing flexible, scalable, open software architectures and programming interfaces to prototype novel wireless solutions. Specifically, WiSHFUL develops mechanisms for unified radio control to provide developers with deep control of physical and medium access components without requiring deep knowledge of the radio hardware platform and unified network control to allow the rapid creation, modification, and prototyping of protocols across the entire stack. These mechanisms chiefly take the form of UPIs that operate across a range of hardware platforms. In this way WiSHFUL empowers experimentation facilities with the capability to experiment with emerging wireless technologies.

13.4.1 Major Entities

The WiSHFUL architecture, illustrated in Figure 13.6, contains several entities designed to support the investigation of future networks. First and foremost within this architecture is the collection of UPIs, with each UPI providing specific functionality to experimenters. The radio interface (UPI_R) consists of a set of functions that ensure uniform control of the radio hardware and lower MAC behavior across heterogeneous devices. The functions provided herein take a generic form in order to provide experimenters with consistent operation over hardware specific implementations. The network interface (UPI_N) parallels the UPI_R with a set of functions that provides uniform control over the upper MAC and network layer protocol behavior across various devices. Again, the UPI_N consists of generic functions to provide a consistent and straightforward experimentation experience across heterogeneous platforms. The global interface (UPI_G) extends the reach of the control provided by both the UPI_R and the UPI_N across several devices in a coordinated and generic manner. The generic functions of UPI_R, UPI_N, and UPI_G are supported by monitoring and configuration engines (MCEs) that contain and manage the platform specific implementations of UPIs within WiSHFUL empowered facilities. Naturally, the UPI_R and UPI_N are supported by a local MCE, while the UPI_G employs a global MCE. Finally, the hierarchical control interface (UPI_HC) enables hierarchical communication between CPs structured in a standard manner. Note that this interface does not directly interact with hardware, but rather provides experimenters with the means to explore hierarchical control by offering a convenient method of inter-control program communication.

The separation between radio and network functionality occurs within the MAC layer of the OSI stack. In particular, WiSHFUL considers the Upper MAC and higher layers as network control functionality, relegating the Lower MAC and lower layers to radio control functionality. The Upper MAC is responsible for inter-packet states that are not time critical, including framing and management functions, where some form of negotiation between

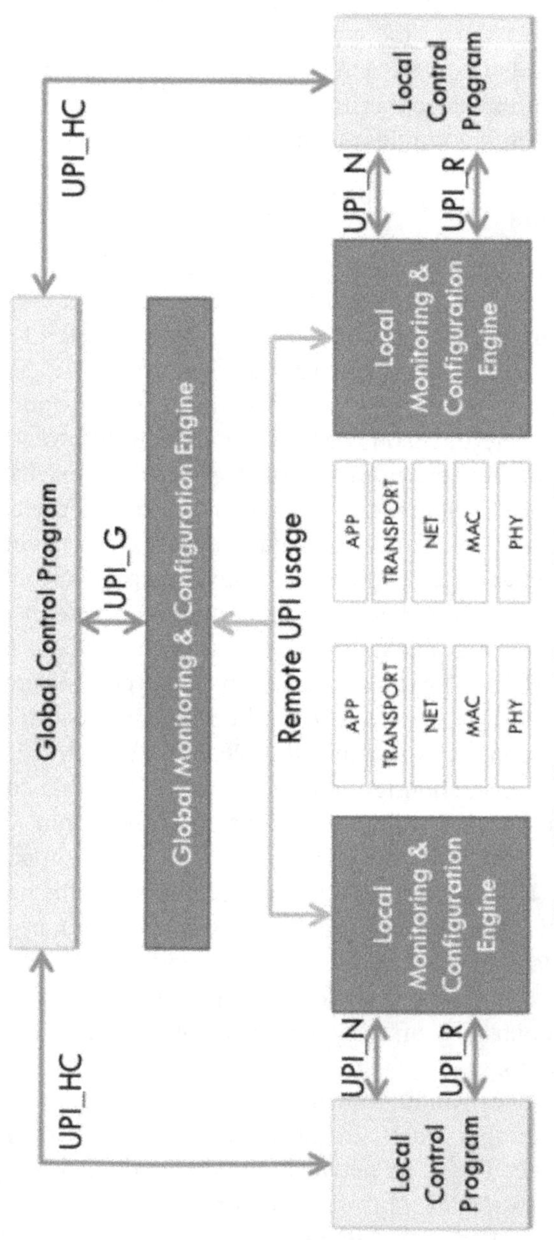

Figure 13.6 Conceptual diagram of WiSHFUL architecture.

nodes is required. The Lower MAC, on the other hand, directly interacts with the physical layer (PHY) transmission and reception operations, where minimization of processing latency is certainly critical. Typical Lower MAC functions include sending and receiving data, back-off, inter-frame spacing, and slot synchronization. As such, this distinction reflects the focus on inter-device coordination within the network control and more direct hardware operations within the radio control.

13.4.2 User Control

The interfaces of the WiSHFUL architecture are designed to support the user in controlling wireless hardware and the accompanying protocol stacks. WiSHFUL views user control as being embodied in CPs, which are either local or global in nature. In general, CPs are user defined software that implement the controlling logic for a wireless experiment and makes use of the UPI_R and/or UPI_N for hardware/protocol control. Local Control Programs (LCPs) are those that use the local information and abilities of a single device, while Global Control Programs (GCPs) interact with a group of devices.

The WiSHFUL architecture supports a two-tier control hierarchy. These two tiers work in a coordinated manner, being orchestrated at the global level. Indeed, GCPs can instantiate LCPs on wireless nodes, performing a sort of control by delegation, or can act directly on the wireless nodes in a coordinated manner. Control by delegation is needed when the reconfiguration decisions or the parameters to be monitored have strict time constraints, which cannot be guaranteed by the control network. In fact, the physical channel used for conveying control messages to/from the GCP can be unreliable and introduce some latencies. Since radio performance depends on highly variable network conditions (e.g. channel propagation, fading, interference, access timings, etc.), control by delegation is particularly important for radio control. The architecture also supports hybrid approaches, in which some control operations are managed at the global level, while some others are delegated to wireless nodes. The coordination between global and LCPs is achieved by employing the UPI_HC. Currently, the WiSHFUL framework follows a proactive approach. A CP has to trigger the execution of UPI functions on the wireless node under control. This polling-based approach might be not sufficient for every CP's implementation requirements. Therefore, it is planned to offer support also for a reactive approach in the near future. Here the user will be able to define a trigger, i.e., when a certain condition is fulfilled, a registered callback function is executed to handle the event.

13.4.3 Hardware Interfacing

Figure 13.7 illustrates how the WiSHFUL radio control works on three different platforms, namely the Iris SDR framework [6], TAISC [5] and WMP [4]. The global MCE runs remotely on a Linux machine and allows implementing node configuration that depends on network-level decisions and can be executed in a time-coordinated fashion among multiple nodes. Each of the WiSHFUL enabled nodes runs a local MCE that offers the same local services and the same UPI functions on different platforms by means of a specific connector module (CM). This unified approach unloads the experiment from the burden of dealing with a multiplicity of configuration and utility tools, such as iw, iwconfig, iptables, iwlist, iperf, b43fwdump, etc. These tools, indicated in Figure 13.7 as local control services, are heterogeneous upon platforms/operating systems and depend on the hardware and software configuration of the device under test.

The CM operates in conjunction with local MCEs to expose the uniform UPI functions on different hardware and software radio platforms.

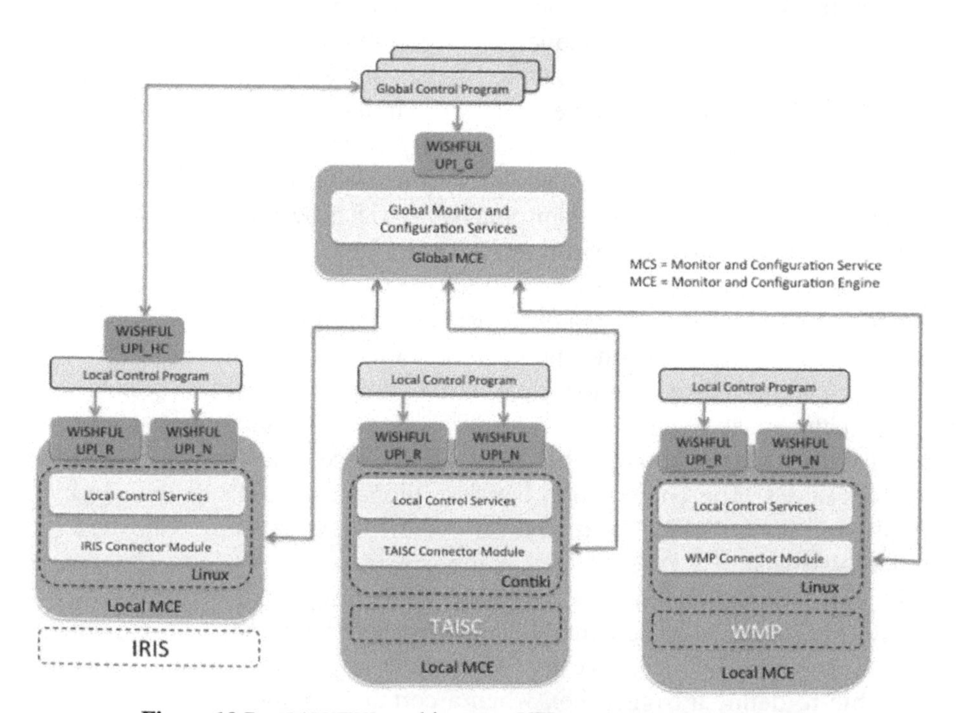

Figure 13.7 WiSHFUL architecture, UPIs, and supported platforms.

The module achieves two main goals: i) diverting platform-independent UPI calls to platform-dependent implementations and ii) providing a unified way to deal with a plethora of tools provided by heterogeneous operating systems (e.g. iw, iwconfig, iptable) or platforms (e.g. bytecode-manager for the WMP). Note that certain UPI functionality may or may not be supported by every platform, depending on the capabilities of the platform and the implementation status of the CM.

Figure 13.8 illustrates the interaction from MCE to the CM and subsequently the radio platform. The local MCE delegates each UPI call to the appropriate CM that executes the call using platform-specific sub modules. Currently, all local MCEs and CMs are implemented in Python, except from sensor nodes that, in addition to the Python implementation, also have a native implementation using GITAR [6]. The native implementation is used when the sensor nodes are decoupled. In case they have a Linux host PC (e.g. in testbeds) the Python implementation can be used. This allows to easily prototype wireless solutions for sensor networks that can also work in real deployments, when their host PCs are not available.

13.4.4 Basic Services and Capabilities

Alongside, the UPIs themselves, the WiSHFUL framework offers a number of basic services that are summarized here.

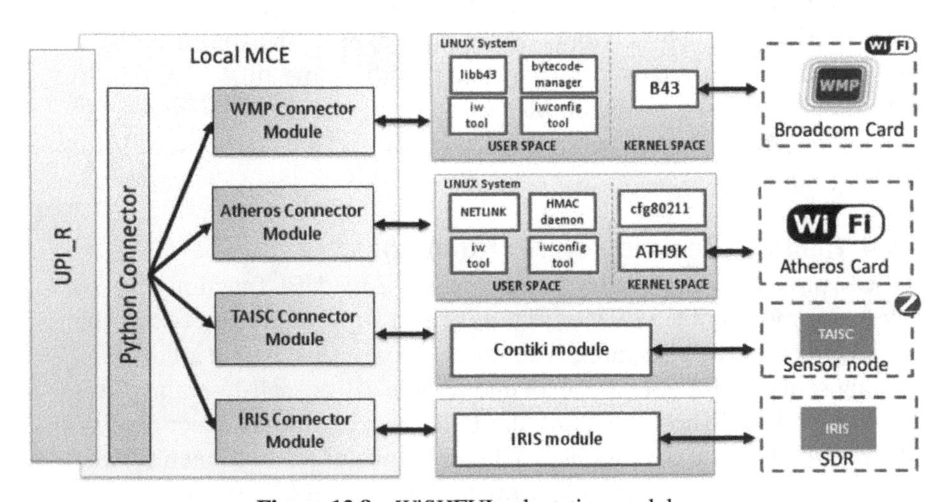

Figure 13.8 WiSHFUL adaptation modules.

13.4.4.1 Node discovery

A GCP often requires functionality for automatic node discovery. WiSHFUL provides the protocol developer an easy way to define the set of nodes he wants to control. Any wireless node belonging to the same experiment group can be controlled by a GCP using the WiSHFUL UPIs. From that set of nodes the user can either select all of them or just a sub-set.

13.4.4.2 Execution semantics

The WiSHFUL MCE (local and global) supports two execution semantics. The first is a synchronous blocking UPI call where the caller, i.e. the CP, is blocked until the callee, i.e. any UPI function, returns. The second option is an asynchronous non-blocking UPI function call. Here any UPI call returns immediately. The caller has the option to register a callback function so that he can receive the return value of the UPI call at a later point in time.

13.4.4.3 Time-scheduled execution of UPI functions

Besides the possibility of immediate execution of UPI functions either using a blocking or non-blocking scheme, the WiSHFUL MCEs also provide the possibility for time-scheduled execution of UPI functions at a particular point in time. This is important if nodes need to coordinate their actions in time, e.g. a set of nodes must perform a time-aligned switching to a new channel. The possibility for time-scheduled execution of UPI functions is especially important for GCPs if a non-real-time backbone networking system like Ethernet is used. In such networks we cannot expect that the WiSHFUL control commands are received by all nodes at the same time, e.g. due to CSMA non-deterministic behavior. Moreover, network congestion and delay are also reasons for providing hierarchical control over UPI_HC between local and GCPs.

13.4.4.4 Remote execution of UPI functions

WiSHFUL provides full location transparency. Any UPI function can be executed either locally by a LCP or remotely by a GCP. In the latter case, the WiSHFUL global MCE transparently serializes all input and output arguments. The calling semantic for both the local and remote calls is call-by-value. This has to be considered when extending the UPIs with additional functionality. Finally, as with the local execution also the execution of remote functions can be time-scheduled. This is especially important if a given UPI function needs to be executed at the same time on a set of wireless nodes.

13.4.4.5 Time synchronization

A wide range of WiSHFUL applications, like the centralized control of channel access, requires a tight time synchronization among wireless nodes. The way the wireless nodes are time synchronized is platform and architecture-dependent. Basically, we distinguish between systems based on whether a backbone network exists. Here in order not to harm the performance of the wireless network the nodes are time synchronized using the backbone (e.g. Ethernet) and some time-protocol like Precision Time Protocol (PTP). Wireless nodes without a backbone have to rely on other techniques for time synchronization (e.g. through global positioning system, GPS).

13.4.4.6 Packet forgery, sniffing and injection

WiSHFUL provides a wide range of functionality for packet forgery, sniffing and injection. A CP can use this to create and inject network packets into any layer of the network stack of a node or to receive copies of packets.

13.4.4.7 Deployment of new UPI functions

WiSHFUL provides an open and extensible architecture, which can be easily extended by new UPI functions. Any new introduced UPI function can be implemented in a different way for different platform and software architecture. Therefore, in WiSHFUL for each platform there is separate CM, as discussed above.

13.4.4.8 Global control

To enable remote usage of UPI functions using the UPI_G interface, a system supporting remote procedure calls is required. For this purpose the arguments of UPI functions have to be serialized and sent to proper node. The proposed framework provides a user-friendly interface that hides all complexity of serialization and transferring data between GCP and nodes.

13.4.4.9 Remote injection and execution of user code

To enable support for global management and control of the deployment of a WiSHFUL controlled experiment, the proposed framework supports "on-the-fly" injection of user defined functions (constituted of UPI) to be executed locally on a node directly from a GCP.

13.5 Implementation of Motivating Scenarios and Results

In order to clarify the potentialities of the WiSHFUL architecture and unified programming, in this section we present some examples of control logic and protocol adaptations developed for the motivating scenarios presented in Section 13.3. The goal is not designing a novel optimization logic for each scenario, but rather demonstrating the flexibility of the proposed approach by separating the logic for controlling the experimentation platforms from the specific transmission mechanisms running on the platform.

13.5.1 Interference Management Among Overlapping Cells

We decompose this scenario into two tasks: 1) hidden node detection and 2) hybrid TDMA-MAC management. For investigation of both tasks, we implemented a WiSHFUL enabled Wi-Fi network.

13.5.1.1 Hidden node detection

The first task to be solved for the efficient airtime management showcase is the detection of wireless links, which are suffering from performance degradation due to hidden terminals (Figure 13.9). Specifically, only flows using links, which are suffering from the hidden node problem, should be assigned to exclusive time slots. Hence, WiSHFUL provides functionality, which detects links which are suffering from the hidden node problem.

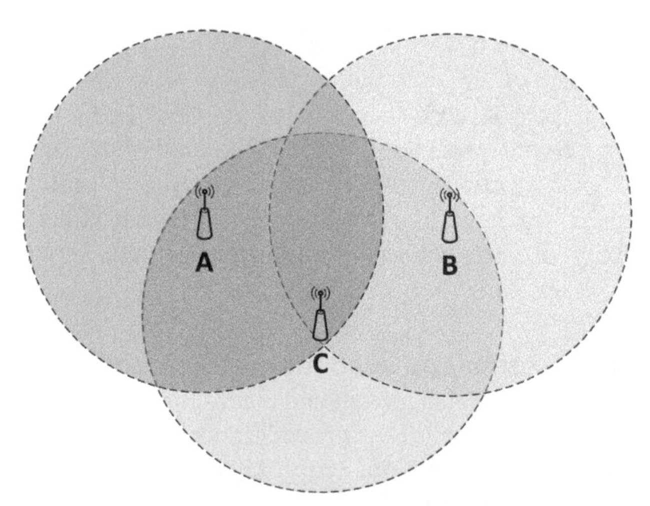

Figure 13.9 Example illustrating a hidden node scenario. As nodes A and B are outside their carrier sensing range the packet transmissions from A and B would collide at node C.

13.5.1.1.1 *Application of WiSHFUL framework*

For hidden node detection WiSHFUL provides the following UPI network functions which are used by GCPs.

Given a set of nodes using the specific wireless interface,radio channel (e.g. 6) and detection threshold (e.g. 0.9) this function returns a boolean matrix indicating which nodes are inside each node's carrier sensing range and which are outside:

```
def getNodesInCarrierSensingRange(self, nodes, wifi_intf, rfCh, detection_th)
```

Furthermore by using this function that returns a boolean matrix indicating which nodes are inside each node's communication (reception) range and which are outside it is possible to reach to a conclusion if there is a hidden node for any pair of nodes forming a link in the network:

```
def getNodesInCommunicationRange(self, nodes, Wi-Fi_intf, rfCh, detection_th)
```

These two functions are used to detect links hidden by some node. As an illustrative example, consider the case where nodes A and B are outside of carrier sensing range and C is inside the reception range of both A and B. In this case packet transmission from A to C and B to C must use exclusive time slots in order to prevent performance degradation due to packet collisions. The technical details of this functionality is further discussed in deliverable D4.2 [12].

13.5.1.1.2 *Results*

The used algorithm performs two steps. First, we use the UPI functionality to estimate which nodes are in carrier sensing range and which are outside. The algorithm uses the following approach:

- It first compares the measured isolated broadcast transmit rate of each node with the one achieved by transmitting concurrently with some other node in the network. If the latter is smaller we know that the two nodes are in carrier sensing range.
- Second, we use the UPI functionality to estimate which nodes are in communication range. The corresponding UPI function sets each wireless node in sniffing mode. Then, in each round a single transmitter is transmitting raw 802.11 broadcast frames while the other nodes are capturing the received frames.

With the information which nodes are in carrier and reception range we are able to estimate which links are suffering from hidden nodes and hence must be protected.

13.5.1.2 Hybrid TDMA MAC

Enterprise IEEE 802.11 networks need to provide high network performance to support a large number of diverse clients like laptops, smartphones and tablets as well as capacity hungry and delay sensitive novel applications like mobile HD video and cloud storage. Moreover, such devices and applications require much better mobility support and higher QoS and quality of experience (QoE).

IEEE 802.11 uses a random access scheme called distributed coordination function (DCF) to access and share the wireless medium. The advantage of DCF is its distributed and asynchronous nature making it suitable for unplanned ad-hoc networks which have no infrastructure. The main disadvantage is its inefficiency in congested networks. Moreover, it suffers from performance issues due to hidden and exposed node problem which is a severe problem in high density enterprise networks.

In contrast to DCF, in TDMA the channel access is scheduled in a synchronized and centralized manner, and hence is able to provide the required high QoS/QoE requirements of enterprise environments. WiSHFUL allows to build TDMA on top of today's off-the-shelf Wi-Fi hardware by providing a flexible and extensible software solution. Currently, the focus is being set on the downlink whereas in the future also the uplink will be considered for support from the TDMA scheme.

Following the software-defined networking (SDN) paradigm we separate the control plane from the data plane and provide an application programming interface (API) to allow local or global CPs to configure the channel access function. In particular it allows to configure the TDMA downlink channel access by defining the number and size of time slots in the TDMA super-frame. Moreover, for each time slot a medium access policy can be assigned which allows to restrict the medium access for particular stations (identified by their MAC address) and traffic identification (e.g. VoIP or video). The latter can be used to program flow-level medium access. Finally, for each time slot we can configure whether carrier-sensing is activated or not. The latter would results in the classical TDMA MAC while keeping carrier-sensing within each slot allows for transparent coexistence with legacy networks that are not aware of the TDMA scheme being used within the WiSHFUL enabled network. The data plane itself resides in each AP and is controlled by the WiSHFUL runtime system.

The control plane in our design is managed by either a global or local WiSHFUL CP which takes as input the channel access scheme specified by the applications. Any application is responsible to decide on how to map the per-flow QoS requirements on the channel access. An example would be to measure which wireless links are suffering from hidden node problem and to assign exclusive time slots for flows requiring high QoS. However, the provided centralized coordination for channel access requires a tight time synchronization among APs. In WiSHFUL time synchronization is performed using the wired backhaul network and hence is not harming the performance of the wireless network. The utilized Precise Time Protocol (PTP) gives an accuracy in microsecond level. The WiSHFUL MCE running on each AP locally is responsible for coordination of channel access as configured by the local or global CP.

13.5.1.2.1 *Application of WiSHFUL presentation of UPIs used*

The UPIs provided by WiSHFUL to set-up and control a hybrid TDMA MAC are as follows:

```
def installMacProcessor(self, node, interface, mac_profile)
def updateMacProcessor(self, node, interface, mac_profile)
def uninstallMacProcessor(self, node, interface, mac_profile)
```

The UPI functions allow the installation, reconfiguration at runtime and uninstallation of a hybrid TDMA MAC. The mac_profile is an object-oriented representation of the hybrid MAC configuration (Figure 13.10).

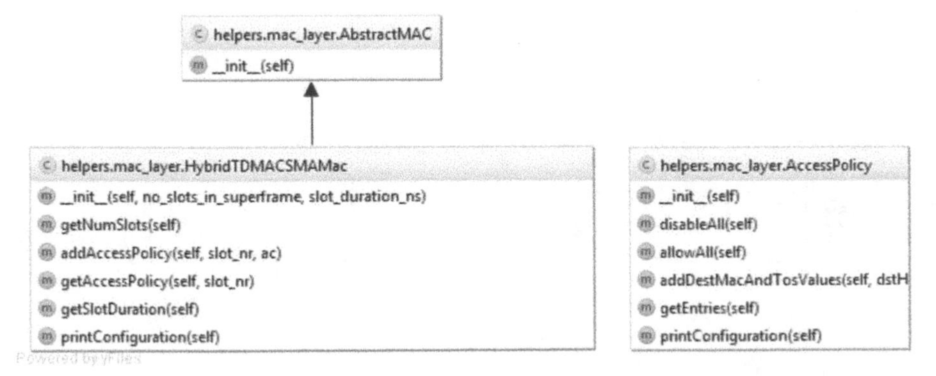

Figure 13.10 UML class diagram showing the hybrid MAC relevant configuration.

The following example shows how to set-up a new hybrid MAC instance:# *create new MAC for each node* HybridTDMACSMA-Mac(no_slots_in_superframe=7,slot_duration_ns=20e3)

```
# assign access policy to slot 0
acBE = AccessPolicy()
# MAC address of the link destination
dstHWAddr = '12:12:12:12:12:12'
# best effort
tosVal = 0
acBE.addDestMacAndTosValues(dstHWAddr, tosVal)
slot_nr = 0
mac.addAccessPolicy(slot_nr, acBE)
# assign time guard slot 1
acGuard = AccessPolicy()
acGuard.disableAll() # guard slot
slot_nr = 1
mac.addAccessPolicy(slot_nr, acGuard)
# UPI call
mac = radioHelper.installMacProcessor(node, iface, mac)
```

Finally, Figure 13.11 illustrates the hybrid MAC being configured to assign exclusive time slots to two wireless links which are hidden to each other. In order to account to time synchronization inaccuracy guard slots are added.

13.5.1.2.2 *Results*

Figure 13.12 depicts how the UPI functionality is implemented on a Linux system using an Atheros Wi-Fi chip and the Ath9k wireless driver. When the locally running WiSHFUL agent receives a command for the setup of a hybrid MAC TDMA from the GCP (*installMacProcessor()* command), it starts

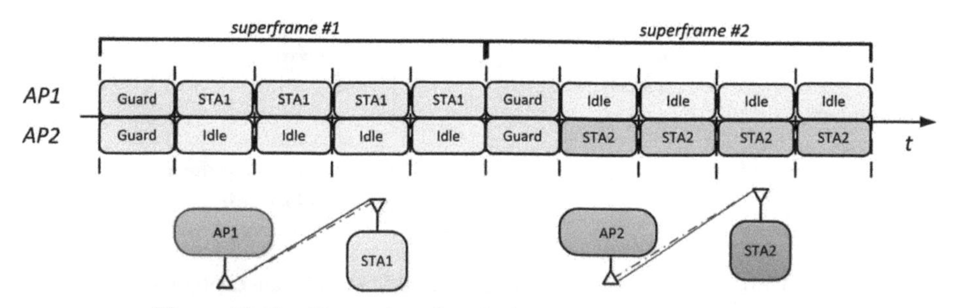

Figure 13.11 Illustration of exclusive slots allocation in TDMA.

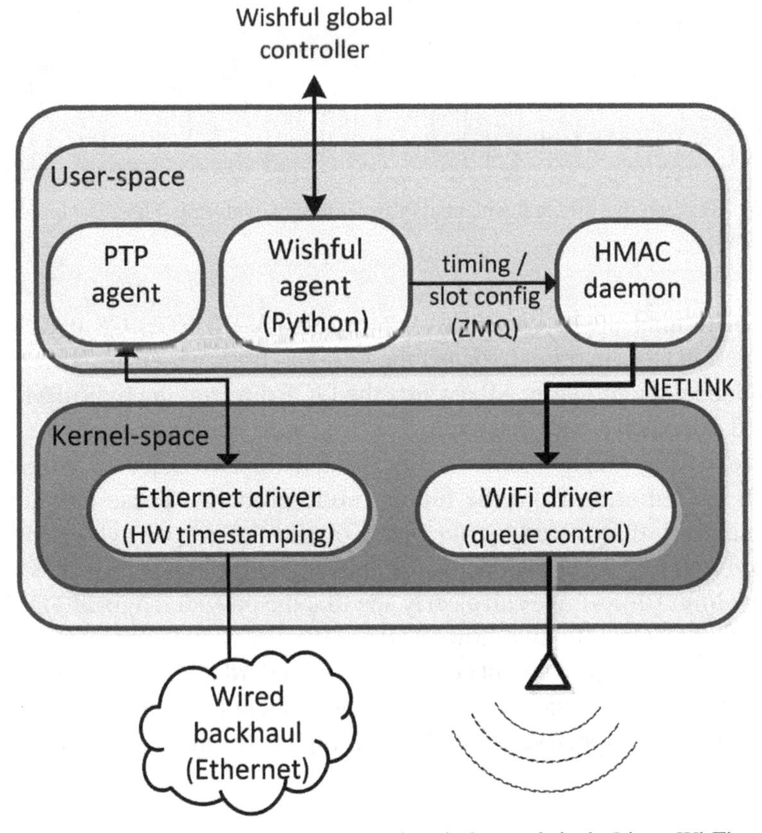

Figure 13.12 Overview of the components on the wireless node in the Linux-Wi-Fi prototype.

the HMAC daemon. The agent controls the (re)configuration of the HMAC daemon using a message passing system (ZMQ). The task of the daemon is to pass slots configuration information to the wireless network driver using the NETLINK protocol. Moreover, it is responsible to inform the wireless driver about the beginning of each time slot. The patched wireless driver uses the slot configuration information to control which network queues are active and which are frozen. Only packets from active queues are allowed to be sent while the others are buffered.

In order to evaluate the proposed efficient airtime management scheme, experiments were conducted in the TWIST 802.11 testbed. Ubuntu 14.04, Intel i5s with a wired Ethernet NIC from Intel supporting HW timestamping and an Atheros 802.11n wireless chip were used in order to setup the experimental network deployment.

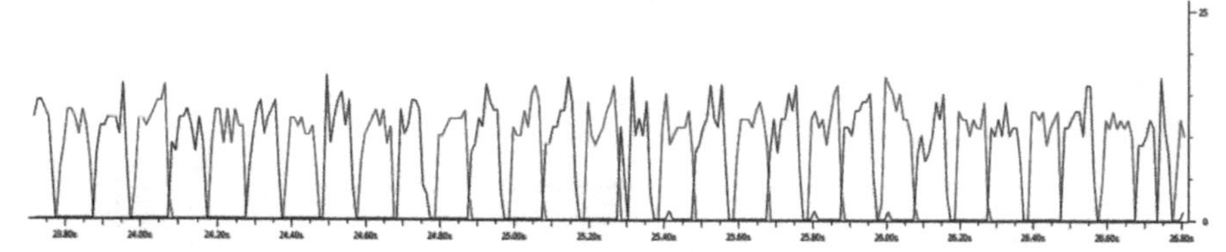

Figure 13.13 IO graph illustrating the number of packets sent over time. The color indicates a particular flow.

At the beginning of the experiment the global WiSHFUL CP used the network function UPIs in order to detect the wireless links which are suffering from the hidden node problem. Afterwards the GCP directed the hybrid MAC on these nodes in such a way that exclusive time slots were assigned.

In the following graphs results are presented for two selected wireless links which are suffering from the hidden node problem. These two links were automatically discovered by our protocol and the proper hybrid MAC was set-up. Figure 13.13 shows the IO graph where the color indicates the two different links (flows). We can clearly see that the provided hybrid TDMA scheme is able to isolate the two flows as desired.

The performance improvement compared to standard 802.11 DCF is show in Figure 13.14. On this particular link the throughput could be increased by a factor of 5.2 and 2.8 respectively which is an impressive increase in network capacity.

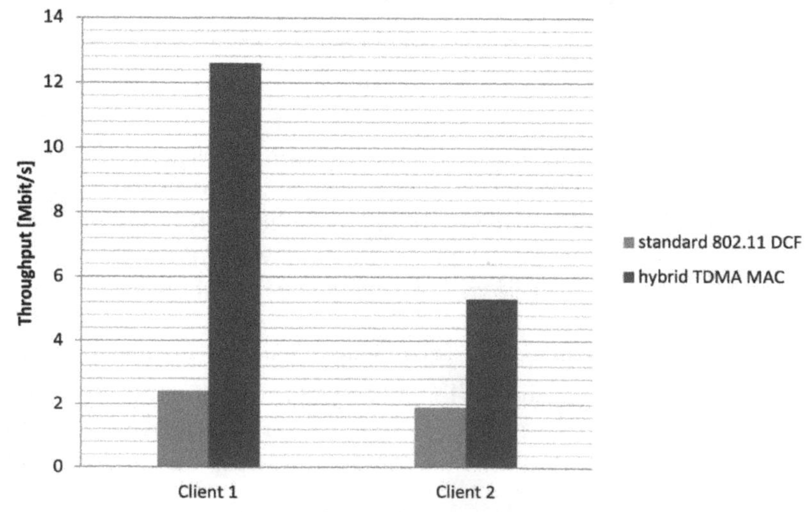

Figure 13.14 TCP/IP performance.

The described efficient airtime management scheme was fully implemented and the source code is available in the WiSHFUL project's Github public repository [13].

13.5.2 Co-existence of Heterogeneous Technologies

Up until this time the coexistence of different wireless technologies in the same domain has been inadequately supported and mostly is based on simply selecting different wireless channels to divide into the frequency plane the heterogeneous technologies that use the same spectral band in general like the ISM band at 2.4 GHz.

The WiSHFUL control framework aims to provide solutions also in the time plane based to inter-technology communication and synchronization. Table 13.1 lists the communication technologies that are currently supported and summarizes, for each technology, the available operating systems, hardware platforms and drivers.

The demonstration set-up presented in this scenario is deployed in the iMinds w.iLab.t testbed and comprises of 32 Contiki sensor nodes with an IEEE-802.15.4 radio and 14 Linux nodes with two IEEE-802.11 radios. Both showcases are executed simultaneously and can be demonstrated remotely. During execution, live measurements are taken and can be presented in two formats: 1) live graphs displaying performance statistics and b) real-time spectrum scanning plots using a universal software radio peripheral (USRP) device. The following configuration options for both showcases are possible:

13.5.2.1 Configuration options for the basic showcase

The experimenters can configure the Wi-Fi channel (2.4 GHz ISM band) and select the bandwidth (20/40 MHz) used for sending Wi-Fi frames. To mitigate interference, experimenters can choose the TSCH channels

Table 13.1 Supported platforms, OSs and drivers

	Supported Platforms, Operating Systems and Drivers		
Technology	Operating System	Platform	Driver
IEEE-802.11	Linux, Windows	Atheros, Broadcom	Ath9k, NDIS driver, WMP
IEEE-802.15.4	Contiki, TinyOS	MSP430, CC2x20, CC283x	Contiki/TinyOS drivers, TAISC
SDR	Linux, Windows	USRP, Xylink ZebBoard	Iris, LabView, GNU radio

that must be blacklisted. It is also possible to add an extra external Wi-Fi interference stream on a different channel and investigate the impact of uncontrolled cross technology interference.

13.5.2.2 Configuration options for the advanced showcase

The experimenters can also dynamically change the cross-technology TDMA schedule, e.g. allocation of slots between Wi-Fi and TSCH networks. Moreover, they can also specify a different synchronization pattern on-the-fly and add multiple concurrent streams in the TSCH network.

13.5.2.3 Results

An example of the live performance statistics monitored during execution of the first, basic showcase is given in Figure 13.15. The graph shows the overall average network throughput measured over time. From the results, it can be clearly seen that there is a substantial loss of throughput when there is Wi-Fi interference. After blacklisting the affected TSCH channels, the throughput rises up again close to its previous value. By changing the configuration parameters described in Section V. A, an experimenter can witness an immediate impact on the performance. Note that other statistics such as packet loss, jitter, TX throughput can be measured as well.

While executing the more advanced showcase it is also possible to monitor performance statistics in combination with real-time spectrum scanning using

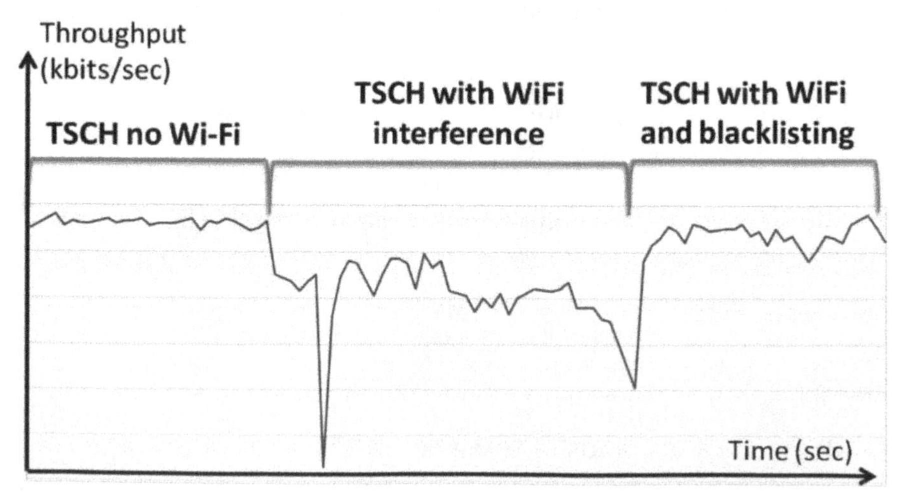

Figure 13.15 Live performance statistics showing the average network throughput (kbits/sec) over time.

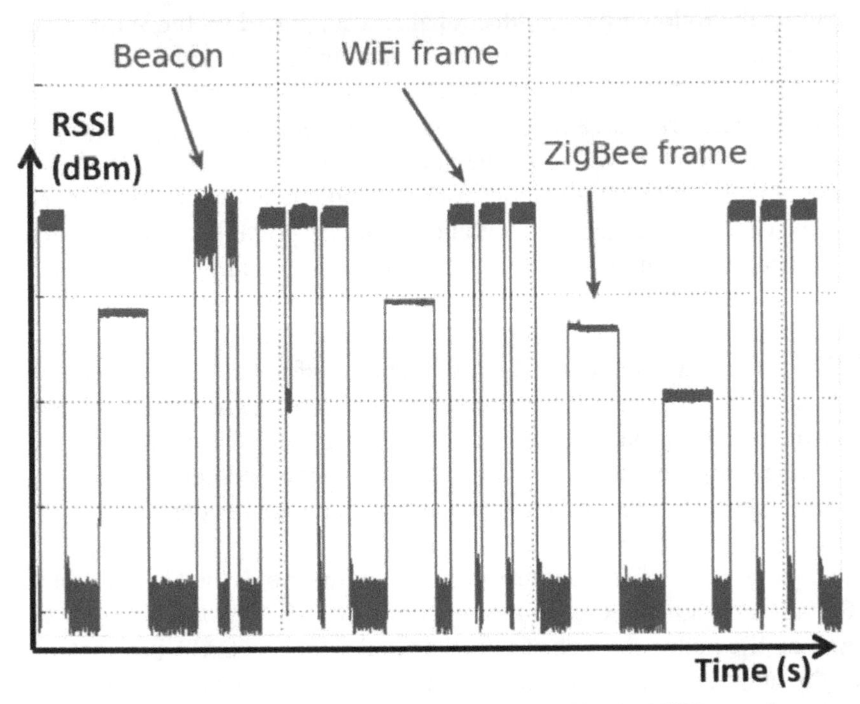

Figure 13.16 Live capture of RSSI (dBm) measured by the USRP over time.

USRP devices. Figure 13.16 illustrates the cross-technology synchronization beacon and TDMA schedule in real-time using an energy detection plot (the y-axes is RSSI in dBm). When configuring this showcase, experimenters will have an immediate feedback on the USRP plot.

The results from both showcases demonstrate the effectiveness of cross-technology interference mitigation and the ability to quickly set-up, investigate and fine-tune an interference scenario using the WiSHFUL control framework.

13.5.3 Load and Interference Aware MAC Adaptation

Here, the application of WiSHFUL in order to enable technology-independent MAC adaptation logic is presented. By employing the WiSHFUL framework it is possible to: i) dynamically tune the parameters of contention-based protocols based on load and interference conditions, and ii) switch between protocols. The logic can work on Wi-Fi or IEEE 802.15.4 nodes, regardless of the PHY layer capabilities and even on cognitive radio platforms, by exploiting the following main functionalities supported by the WiSHFUL UPI: sensing capabilities of wireless nodes, local tuning of CSMA contention windows, and global coordination of MAC protocol switching.

A wireless network with a time-varying number of active nodes under the same contention domain (where all the nodes are in radio visibility) is taken into account and a wired ethernet network is available as a control network between the GCP, the wireless stations and the access point. Each node runs a local optimization function that is loaded by the GCP for tuning the contention window of a CSMA protocol as a function of the network load.

In particular, a tuning function called Moderated EDCA backoff (MEDCA) is used, whose goal is the minimization of the delay jitters on the channel access times. Since these jitters depend on the exponential backoff mechanism, which introduces short-term throughput unfairness among the stations, the tuning function automatically finds a fixed contention window equal to the average contention window value experienced under exponential backoff.

When the number of stations crosses a given threshold, the GCP disables the LCP and coordinates the on-the-fly protocol switch from MEDCA to TDMA in all the nodes. As part of this, the GCP sets the TDMA parameters, such as the number of slots and the slot allocation to each station, based on the number of active flows.

13.5.3.1 Application of the WiSHFUL framework

Once calculated according to the MEDCA scheme, new contention window values are set through the UPI_R function responsible of configuring lower layer parameters as follows:

```
#update CW value
UPI_myargs = { 'interface' : 'wlan0', UPI_RN.CSMA_CW : cw, UPI_RN.CSMA_CW_MIN : cw, UPI_RN.CSMA_CW_MAX : cw}
upiRNImpl.setParameterLowerLayer(UPI_myargs)

#The GCP may activate TDMA by calling the UPI_R function setActive:
UPIargs = {'position' : position, 'radio_program_name' : 'TDMA', 'path' : radio_program_pointer_TDMA,
'interface' : 'wlan0' }
rvalue = global_mgr.runAt(node, UPI_RN.setActive, UPIargs, exec_time)

#Finally, the GCP configures the TDMA parameters of each station through the UPI_R utility function set_TDMA_parameters.

tdma_params={'TDMA_SUPER_FRAME_SIZE' : superframe_size_len, 'TDMA_NUMBER_OF_SYNC_SLOT' : len(mytestbed.Wi-
Finodes), 'TDMA_ALLOCATED_SLOT': node_index}

set_TDMA_parameters(node,log,mytestbed.global_mgr,tdma_params)
```

13.5.3.2 Results

First focusing on contention window tuning, we activated six wireless nodes running CSMA with exponential backoff contending under greedy traffic sources towards a common access point. Figure 13.17 shows the throughput performance achieved by each station. The short-term and long-term throughput variability exhibit here results from the exponential backoff mechanism (short-term) and the location-dependent interference conditions suffered by each station (long-term).

For three of the above nodes, we activated the MEDCA backoff scheme. Figure 13.18 shows that these stations achieve an average throughput comparable to that experienced with exponential backoff, but with smaller fluctuations.

Turning to MAC protocol switching, Figure 13.19 shows the measured packet loss and throughput before, during, and after the switch occurs. Radios operated in CSMA for 90 seconds then switched to TDMA. Here we examined 32 sensor nodes, which sent iPerf traffic to a single sensor node acting as a sink and used the output of the iPerf server to generate the graphs. All nodes were in the same collision domain.

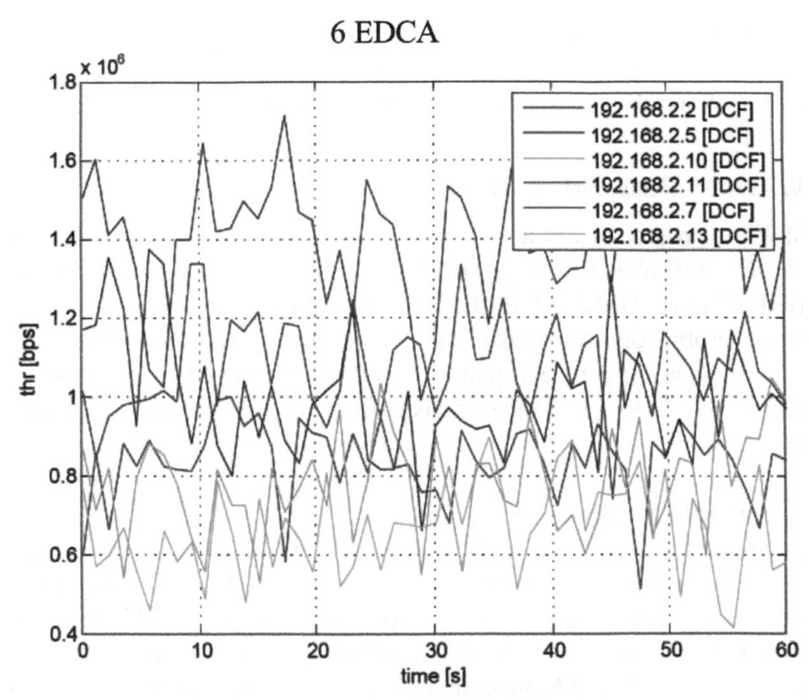

Figure 13.17 Throughput performance of 6 wireless nodes executing CSMA with exponential backoff.

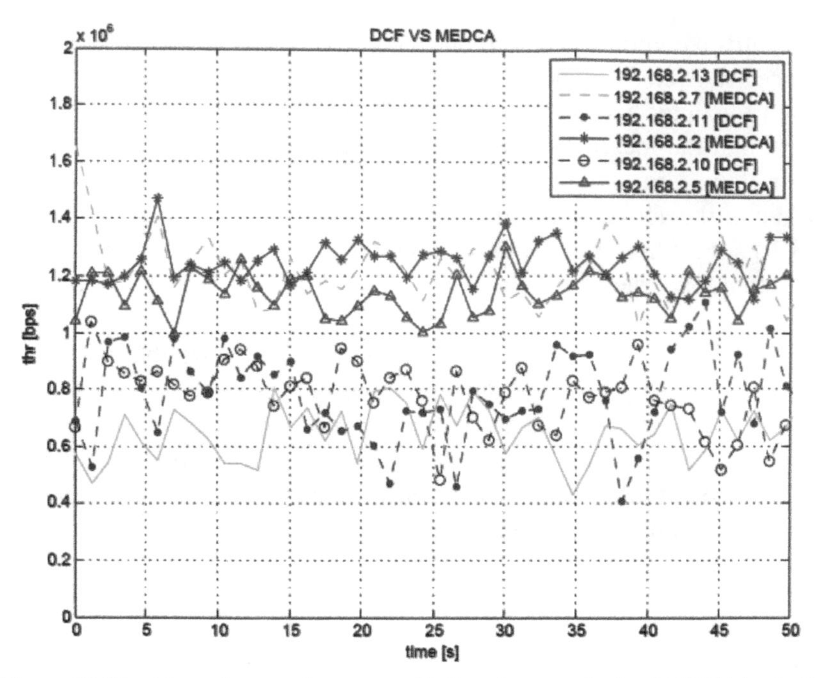

Figure 13.18 Throughput performance with 3 stations employing MEDCA backoff.

13.5.4 Wireless Portable Testbed

The WiSHFUL project offers access to several wireless testbeds, such as TWIST (TUB), w-iLab.t (IMINDS), IRIS (TCD), Orbit (Rutgers University) and a FIBRE Island at UFRJ. All of these testbeds are installed in either office environments or other dedicated testbed environments. Because some research requires doing measurement campaigns or actual testing in heterogeneous environments, the WiSHFUL project also offers a portable testbed to the community.

13.5.4.1 Portable testbed setup

The architecture of the portable testbed is presented in Figure 13.20. As can be seen there are two distinct wireless networks (blue and yellow) present in the testbed, namely BN (Backbone) network and DUT (Device Under Test, or Experiment Node) network. These two networks will be configured and controlled by the Experiment Management Servers. The blue arrows represent a highly reliable wireless backbone that allows the user to place the nodes anywhere in the field without having the practical disadvantages of using cables.

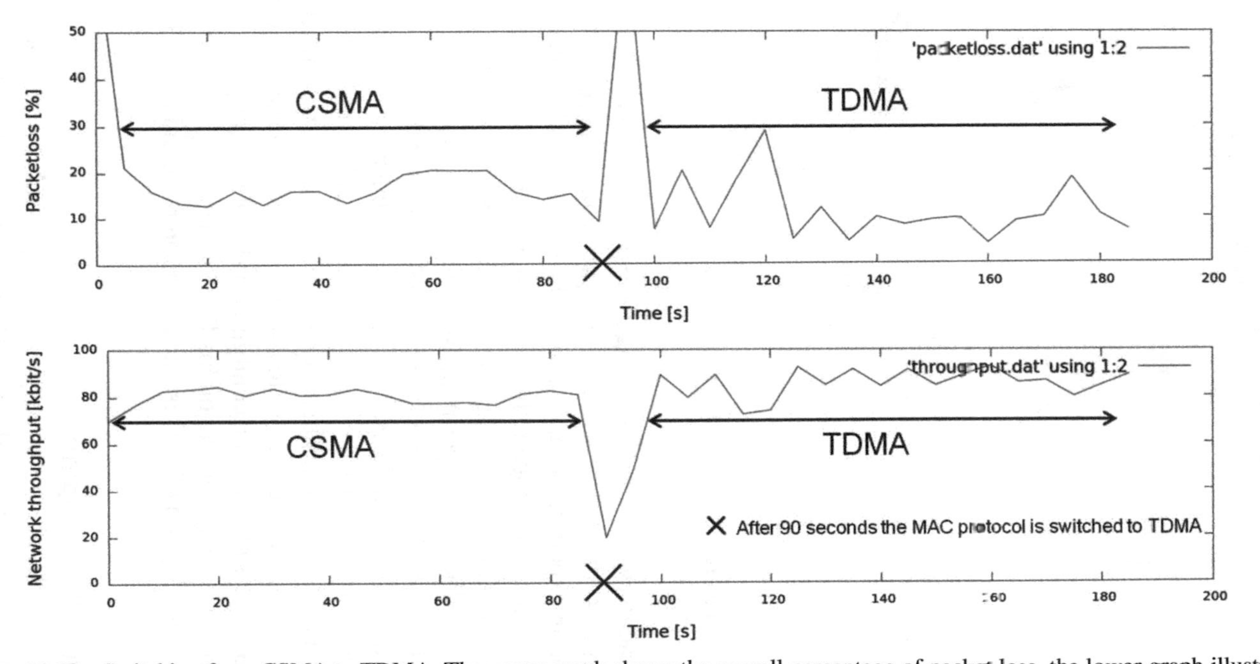

Figure 13.19 Switching from CSMA to TDMA. The upper graph shows the overall percentage of packet loss, the lower graph illustrates the overall throughput. The X marks the switch.

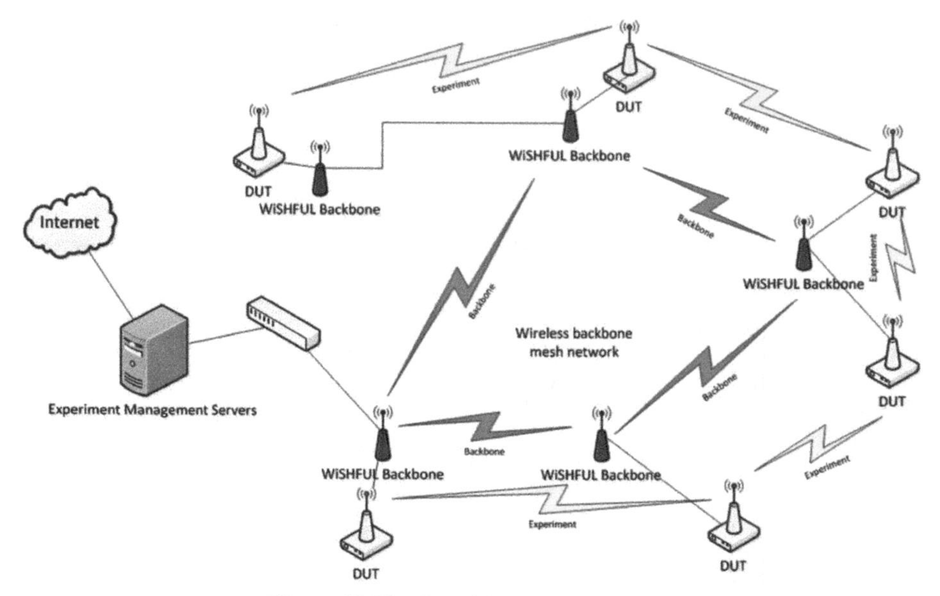

Figure 13.20 Portable testbed overview.

It also allows interaction with the nodes during the experiment. As the Portable Testbed introduces an additional network to an experiment, it is implemented in such a way that an experimenter is not overwhelmed with additional and complicated configuration procedures. In D6.1 [14], it is shown that Portable Testbed follows the "Plug and Play" approach and an experimenter should be able to use the same Testbed and Experiment Management tools as on the fixed testbeds. It has to be noted that an experimenter does not have the possibility to directly control the behavior of the Backbone network, but he is able change the channel that the Portable Testbed uses. Moreover, logical L2 networks are provided to interconnect DUT nodes in order to make them unaware whether they are connected to Portable Testbed or to a regular wired Ethernet network. This approach also reduces the required configuration because an experimenter does not have to configure any routing on his DUT nodes.

A more detailed description of the testbed setup can be found in D6.1 and D6.2 [15].

13.5.4.2 Hardware & packaging

In order to provide flexible means of transport for the portable testbed, an easy to carry, robust and spacious case is desired. It also needs protective material on the inside so the delicate electronics are not damaged during transport. Plywood flight cases are used to secure the hardware in transport.

A primary flight case hosts the central switch and experiment management servers. The EMS is a single, powerful embedded PC that hosts several VM's for each of the testbed core services. The DUT nodes are stacked in several secondary flight cases. These are made from aluminum and robust plastic and are slightly lighter than the primary case. To fix the nodes inside the case, foam is used: a base of hard foam is glued to the bottom of the case and is cut specifically to fit the DUTs. In the top of the briefcase, softer, more flexible polyurethane foam is used as its only function is to push down softly on the nodes so they stay in place while transporting the cases.

DUT devices are COTS Intel NUC (Next Unit of Computing) devices of model D54250WYKH. These are basically headless barebone PC's. They consist of an Intel Core i5 4250U processor, 4 GB of RAM, a gigabit Ethernet port, several USB ports, a 320 GB hard disk and two Wi-Fi cards: one 802.11n (WPEA-121N/W) and one 802.11ac card (WLE900VX 7AA). The nodes are by default equipped with an 802.15.4 sensor node and a Bluetooth USB dongle. The USB connections of the node can be used to attach extra hardware (e.g. LTE dongles or other USB compatible hardware). The DUT features a default embedded Linux operating system to which the experimenter can gain full (root) access. The experimenter has full control over the operating system and the software packages that are installed on the DUT. The DUT can also be used as a proxy to access all USB peripherals of the node, like sensor nodes. If the embedded PC provided by WiSHFUL does not satisfy the experimenter's needs, other hardware can be used as long as it can interface over Ethernet with the backbone nodes. A more detailed description of testbed hardware can be found in D6.2.

13.6 Conclusion

Advancing wireless communications requires overcoming several challenges. Herein, several such challenges have been examined in the form of motivating scenarios. These scenarios outline a number of requirements on experimentation platforms for investigating the future of wireless communications. The WiSHFUL project directly addresses these challenges and requirements by defining a software framework to support unified experimentation across several platforms beyond the today's standards. Examples have been provided through several case studies that apply the WiSHFUL framework to the motivating scenarios and results obtained are presented. It is evident that the use of the WiSHFUL framework provides the necessary functionality to enable advanced wireless experimentation while in parallel it lowers the learning curve for any experimenter across multiple heterogeneous wireless communication technologies.

Acknowledgment

The research leading to these results has received funding from the European Horizon 2020 Program under grant agreement n645274 (WiSHFUL project).

References

[1] Fortuna, C, et al. (2015). Wireless Software and Hardware platforms for Flexible and Unified radio and network controL. European Conference on Networks and Communications (EuCNC 2015), Workshop on 5G Testbeds and Hands-on Experimental Research, Paris, France, June 29, 2015.

[2] N. McKeown, T. Anderson, H. Balakrishnan, G. Parulkar, L. Peterson, J. Rexford, S. Shenker, and J. Turner. 2008. OpenFlow: enabling innovation in campus networks. SIGCOMM Comput. Commun.

[3] IETF Control And Provisioning of Wireless Access Points (CAPWAP) Protocol Specification, https://tools.ietf.org/html/rfc5415

[4] ETSI TR 102 682, Reconfigurable Radio Systems (RRS); Functional Architecture for Management and Control of Reconfigurable Radio Systems, 2009.

[5] I. Tinnirello, G. Bianchi, P. Gallo, D. Garlisi, F. Giuliano, F. Gringoli, "Wireless MAC Processors: Programming MAC Protocols on Commo dity Hardware" IEEE INFOCOM, March 2012.

[6] Bart Jooris; Eli De Poorter; Peter Ruckebusch; Peter De Valck; Christophe Van Praet; Ingrid Moerman; TAISC: a cross-platform MAC protocol compiler and execution engine; Under submission for Computer Networks.

[7] Sutton, Paul, et al. "Iris: an architecture for cognitive radio networking testbeds." Communications Magazine, IEEE 48.9 (2010): 114–122.

[8] A. Zubow and R. Sombrutzki, "A low-cost MIMO mesh testbed based on 802.11n," 2012 IEEE Wireless Communications and Networking Conference (WCNC), Shanghai, 2012, pp. 3171–3176.

[9] P. Ruckebusch, E. De Poorter, C. Fortuna, and I. Moerman, (2015). GITAR: Generic extension for Internet-of-Things Architectures enabling dynamic updates of network and application modules. Ad Hoc Networks, January 2016, Vol. 36, Part 1, Pages 127–151.

[10] Wauters, Tim, et al. "Federation of Internet experimentation facilities: architecture and implementation." European Conference on Networks and Communications (EuCNC 2014). 2014.
[11] A. C. V. Gummalla and J. O. Limb, "Wireless medium access control protocols," in IEEE Communications Surveys & Tutorials, Vol. 3, No. 2, pp. 2–15, Second Quarter 2000.
[12] Deliverable D.4.2, WiSHFUL project, H2020, 2016, http://www.WiSHFUL-project.eu/deliverables
[13] WiSHFUL project Github repository, https://github.com/WirelessTestbedsAcademy
[14] Deliverable D.6.1, WiSHFUL project, H2020, 2016, http://www.WiSHFUL-project.eu/deliverables
[15] Deliverable D.6.2, WiSHFUL project, H2020, 2016, http://www.WiSHFUL-project.eu/deliverables

PART III

Research PROJECTS and CASES Using Experimentation TESTBEDS

14

Estimating the Dimension of Your Wireless Infrastructure by Using FIRE Testbeds

Sergio Gonzalez-Miranda, Lorena Bourg, Viveca Jimenez and Alejandro Almunia

Planet Media Studios S.L.,
C/ Serrano Galvache, 56 (Parque Norte), Edificio “Abedul”, 4a. Planta, 28033, Madrid

14.1 Introduction

Nowadays, the audience for live events such as: concerts, theatre, sport matches, etcetera is demanding real-time information more than ever. Spectators require data about all possible details of who is taking part and what is happening on the event. This phenomenon takes special relevance for instance on sports events where spectators continuously consult handheld devices to view additional information such as: statistics, repeated plays, players’ information, comments from professionals, friends’ points of view and etcetera. This information is attractive to them because they are able to best judge what is going on. Summing up, providing additional premium content in real-time while a sports event is being broadcasted makes a huge difference to the spectators’ experience.

LIVEstats platform has been developed to reach this demand, providing live statistical information and 3D replays of sports events to the audience on their Smartphone or Tablet. Even though it was originally devised to serve as second screen application to be consumed at home premises, we realized that it had a huge potential if it were to be offered to the audience on site, at stadiums during football matches.

Nevertheless, as usual, in order to properly define both business and cost models, we had to perform the processes of testing the platform in real environments. Due to the special characteristics of the target venues considered, this process becomes unaffordable for an SME. This is how Fed4Fire tools became relevant to us and here is where the experiment titled ***LIVEstats On Fire*** came to life.

The experiment aimed to assess the performance of LIVEstats as an innovative cloud-based platform for the provision of enhanced 3D interactive content during a sports event in an outdoor scenario as if it was a real Stadium. We wanted to overcome the difficulties of testing the provision of this content using wireless technologies to serve hundreds of spectators through their devices at the same time.

In LIVEstats On Fire experiment we made use of the FIRE wireless facilities (including Wi-Fi, WiMAX and LTE) to assess the viability to provide LIVEstats service to the spectators in a real case scenario, with a massive number of users accessing the platform simultaneously while ensuring the delivery of a good quality of service. For that purpose we conducted a comprehensive evaluation of the required topology and configuration of the wireless network that should be deployed.

Results were analysed and incorporated to our business model with the objective to get aligned with the leading edge of technology for second screen applications in the sports market. Results also provided us with the appropriate knowledge in order to become able to characterize the physical infrastructure that is needed to provide content from our platform over wireless technologies to a large audience on real-time during a sports event on site on a sports stadium. With this information, we were able to enrich our business model for the platform by preparing an adjusted cost model in addition to the information and the required models for the deployment and commercialization of the system in real-case scenarios.

The experiment consisted of 3 set-ups to perform particular tests by using Wi-Fi, LTE and WiMAX nodes available at FIRE's NITOS testbed infrastructure.

Those tests were focused on:

1. Specifying the **relationship among the characteristics and number of nodes** of the network and the number of spectators that could be able to enjoy the service with a suitable Quality of Service.
2. Analysing the **factors and characteristics of the infrastructure** that are critical for the streaming experience, this is, which features make the difference about the number of spectators accepted by the system.
3. Defining the **relation among the topology of the network and the Quality of service** offered. Results will guide us to adapt the network distribution configuration to a specific stadium.

From results we have been able to answer many questions related to relevant and useful aspects and we are now in a much better position for understanding and modelling the physical infrastructure that may be needed in any real case scenario that we may face in the future. We obtained valuable information that will allow us to prepare a detailed cost model and prepare marketing actions in order to approach our key partners and potential customers.

We wouldn't be in this position now if we hadn't had access to Fed4FIRE facilities, because the particular scenario for our platform required human, time and physical resources that are too costly for us to address.

The experience has also been significantly enriching for our company in a variety of aspects, since we have been able to access in depth to a very interesting federation of testbeds, counting with dedicated support all along the process.

14.2 Problem Statement

The market for global sports rights has increased by an average of 5% over the past 5 years, reaching around €22 billion in 2014. This trend is expected to continue thanks to the continuous advances in technologies and premium sports services that allow providing more sophisticated and compelling contents to the audience. In Europe, the football industry is one of the most powerful and has a huge impact, attracting annually tens of millions of spectators to the stadiums and many hundreds of millions of viewers at their homes.

The factors that influence the decision making process of buying a ticket or to watch a football game at home are very diverse; but clubs have already faced the reality that watching a match on TV at home is most of the times much cheaper, easier and more comfortable for spectators. A recent Cisco study showed that 57% of fans prefer to watch the game at home. Therefore the latest trend is to redefine the concept of living the match on-site by offering a unique experience for the spectators, thus making it more appealing than ever before to attract fans to the stadium.

Adding connectivity, offering new ways of interacting with spectators and providing additional services and apps within the stadiums can vastly improve fan experience and keep people coming to games even when they have a 50-inch TV and comfortable chair at home. With the additional features over wireless technologies, spectators get an experience at the stadium that they cannot get anywhere else.

LIVEstats platform was born to enter the audio visual sports market and add value to the broadcasting of sports events by improving multiplatform viewer experience through an innovative concept of enhanced information access. It is a Cloud-based platform deployed currently using Amazon Web Services. It provides real-time on demand 3D content generation to enhance the interactivity between the viewer and a sporting event through a "second screen" approach (Tablet, Smart TV, and Smartphone). Using an innovative image recognition system that positions players in a specific area of the field on real-time, the platform creates an accurate 3D recreation of each play. These plays can be accessed and manipulated on demand by the viewer with their smart device through a web-based application built on HTML5, rotating field, changing angles and moving the timeline forward or backwards. Such information is supplemented by the statistical information provided by television operators during the sporting event.

Even if originally LIVEstats was thought to provide second screen capabilities in indoors scenarios, because of the current challenges that the live sports market is facing, in the business model of LIVEstats platform we are considering, as a way for commercialization, offering the service at stadiums during football games, as an optimized and more personalized way of interacting with spectators and bring them closer to the action. We strongly believe that the platform, providing a new level of interactivity and immersivity, has a great potential to lure the audience to the stadiums. With this idea in mind and in order to better define our value proposition and previous to setting the cost structure, we are facing right know a new challenge: we need to run specific tests and make validations that let us acquire detailed information about the specific infrastructure that would be required within a stadium to provide such a service. However, demonstrating the service and assessing its performance on a real environment would be too costly for us.

Here is where FED4FIRE comes into play, and, in particular, NITOS infrastructure. The NITOS outdoor testbed with multiple wireless interfaces will allow us to test the performance of the platform with heterogeneous (Wi-Fi, WiMAX, LTE) wireless technologies and check the viability and select the most suitable infrastructure to offer the service to a large audience through different devices on real time.

The main objective of the experiment was:

To characterize the physical infrastructure that is needed to provide LIVEstats platform premium content over wireless technologies to a large audience on real-time during a sports event on site on a sports stadium.

In order to achieve this main objective, our specific goals will be:

1. Defining and running **3 specific test scenarios** to assess the service performance with the different wireless technologies offered by NITOS: **Wi-Fi, WiMAX and LTE**. Results will later guide us to decide upon the most suitable configuration (including hybrid networks) for each specific scenario.
2. Specifying the relationship among the characteristics and number of nodes of the network and the number of spectators that could be able to enjoy the service with a suitable Quality of Service. This will be done by specifying the **maximum number of spectators** that would be able to simultaneously connect to the service through the wireless nodes provided by the NITOS infrastructure with a suitable Quality of Service in order to able to extrapolate the results for a wider audience (typically for stadiums with a capacity of 25.000 to 100.000 spectators) Our goal would be that at least 40% spectators in a stadium (users that are geographically very close from each other) would be able to simultaneously access to the 3D replays and get a good quality of experience.
3. Analysing the **factors and characteristics of the infrastructure** that are critical for the streaming experience, this is, which features make the difference about the number of spectators accepted by the system. From the extracted features, specifying on what basis they are relevant for the performance of the system and their specific values for each configuration in order to provide a good Quality of Service (starting from latencies at or below 200 msec.). This will let us know about the most cost-effective solution for each case.
4. Defining the **relation among the topology of the network and the Quality of service** offered. Results will guide us to adapt the network distribution configuration to a specific stadium.

14.3 Background and State-of-the-Art

14.3.1 Background

Planet Media is a leader company in the development of multichannel technological solutions (B2B, B2C & B2E) oriented towards Digital Transformation, Comprehensive Mobile and Web applications for Smart Cities, Media, Mobile marketing and loyalty strategies systems.

Planet Media leaded the creation of LIVEstats platform in the context of a research and innovation project with funding from the Spanish Ministry of

Economy. The platform was created to enter into the audio visual sports market and add value to the broadcasting of sports events by improving multiplatform viewer experience through an innovative concept of enhanced information access. It is a Cloud-based platform deployed currently using Amazon Web Services. It provides on demand interactive 3D replays of the most relevant plays during a sports event while the spectator is watching the TV. The replays are provided through a "second screen" approach (Tablet, Smart TV, and Smartphone) and are interactive, imitating cloud-gaming platforms, therefore enhancing the experience of the viewer during the sporting event. For that purpose, the platform uses an innovative image recognition system that positions players in a specific area of the field on real-time, and creates an accurate 3D recreation of each play. These plays can be accessed and manipulated on demand by the viewer with their smart device through a web-based application built on HTML5, rotating field, changing angles and moving the timeline forward or backwards. Such information is supplemented by the statistical information provided by television operators during the sporting event (Figure 14.1).

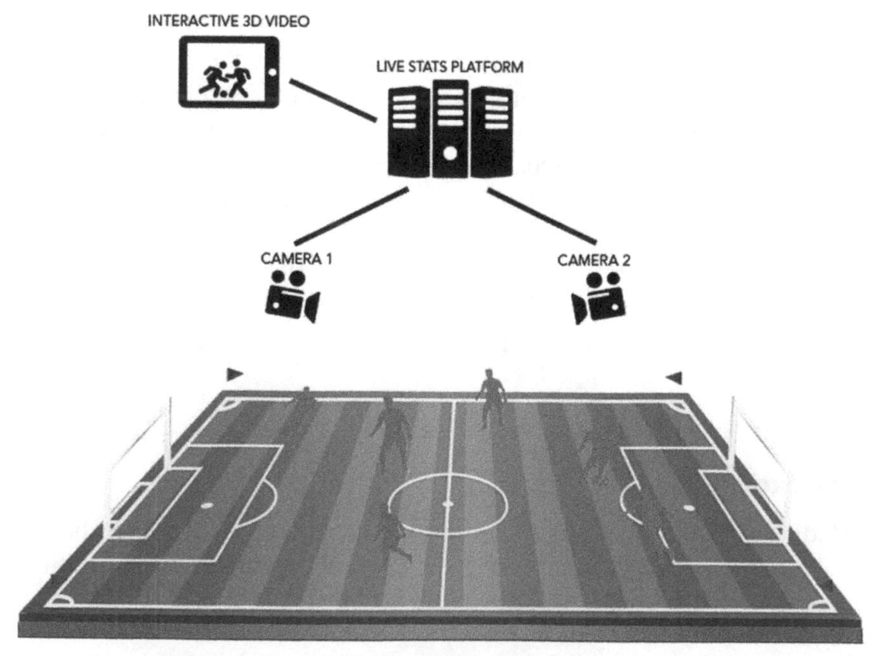

Figure 14.1 LIVEstats platform main use case scenario.

Originally LIVEstats was thought to provide second screen capabilities in indoors scenarios as we have described; but because of the current challenges that the live sports market is facing and the knowledge of the market that we have acquired thanks to our privilege position as service provider for some of the main broadcasters in Spain, such as RTVE, we have extended our vision. We have considered, for the business model of LIVEstats platform as a way for commercialization, offering the service at stadiums during football games, as an optimized and more personalized way of interacting with spectators and bring them closer to the action. We strongly believe that the platform, providing a new level of interactivity and immersivity, has a great potential to lure the audience to the stadiums. With this idea in mind and in order to better define our value proposition and previous to setting the cost structure, we were facing the challenge to run specific tests and make validations that let us acquire detailed information about the specific infrastructure that would be required within a stadium to provide such a service. However, demonstrating the service and assessing its performance on a real environment would be too costly for us.

Here is where FED4FIRE has come into play, and, in particular, NITOS infrastructure. By running an experiment using the NITOS outdoor testbed with multiple wireless interfaces we wanted to assess the performance of the platform with heterogeneous (Wi-Fi, WiMAX, LTE) wireless technologies and check the viability and select the most suitable infrastructure to offer the service to a large audience through different devices on real time.

Therefore, our main objective with this experiment has been:

To characterize the physical infrastructure that is needed to provide LIVEstats platform premium content over wireless technologies to a large audience on real-time during a sports event on site on a sports stadium.

The Figure 14.2 shows the concept of this infrastructure characterization. In a real scenario, the Stadium where the sports event takes place will have a certain number of Wireless nodes (which may be WiFI, LTE, WiMAX or a hybrid network), each of them with N instances of the service running. At each moment, there will be a number of users accessing the nearest node in their zone in order to request a 3D interactive replay from the match. The configuration and topology of the network that needs to be available will depend on the specific stadium infrastructure and the requirements of LIVEstats platform for providing an acceptable QoS. The modelling of these

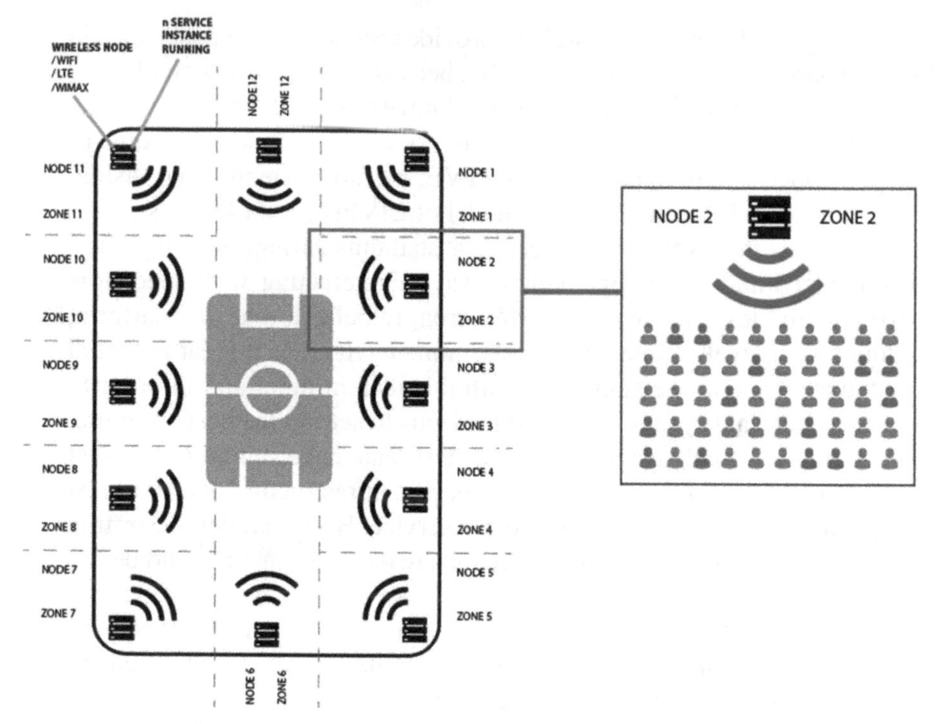

Figure 14.2 Infrastructure characterization in a Stadium.

requirements is the main objective of running the experiment with NITOS infrastructure.

Our specific goals included:

1. Defining and running **3 specific test scenarios** to assess the service performance with the different wireless technologies offered by NITOS: **Wi-Fi, WiMAX and LTE**. We want to use results to be able to select the most suitable configuration for each specific scenario that we may face in the future.
2. Specifying the relationship among the characteristics and number of nodes of the network and the number of spectators that could be able to enjoy the service with a suitable Quality of Service. This is done by specifying the **maximum number of spectators** that would be able to simultaneously connect to the service through the wireless nodes provided by the NITOS infrastructure with a suitable Quality of Service in order to able to extrapolate the results for a wider audience (typically for stadiums with a capacity of 25.000 to 100.000 spectators). Our goal would be that at least 40% spectators in a stadium (users that are geographically

very close from each other) would be able to simultaneously access to the 3D replays and get a good quality of experience.

3. Analysing the **factors and characteristics of the infrastructure** that are critical for the streaming experience, this is, which features make the difference about the number of spectators accepted by the system. From the extracted features, specifying on what basis they are relevant for the performance of the system and their specific values for each configuration in order to provide a good Quality of Service (starting from latencies at or below 200 msec.). This will let us know about the most cost-effective solution for each case.

Defining the **relation among the topology of the network and the Quality of service** offered. Results would guided us to adapt the network distribution configuration to a specific stadium.

LIVEstats on FIRE aimed at defining the specific infrastructures needed for suitable service provision and therefore the proper definition of its costs and scaling-up characteristics for a viable and beneficial exploitation plan. Carrying out the experiment described in this proposal will allow the identification and specification of the characteristic and requirements of the infrastructure capable of supporting the service provided.

LIVEstats solution is highly dependent on the infrastructures under which the data transmission is accomplished. Consequently, the main impact of this experiment, by means of NITOS support, was the definition of the technical requirements of the needed network. It will also allow the detection of emerging obstacles that might work as blocking issues regarding current technical features that may need to be further adapted. Additionally, the specification of the relationship between the technical infrastructure, the streaming experience and the number of spectators that can be served by it, allows the definition of a more precise cost model required to reach the market with adequate estimations about the required investments, pricing models and strategically partners needed to the efficient provision of the service.

This information also allows characterizing our suitable customer segments: by facilitating the description of the precise technical prerequisites that the stadium hosting the service must have and allowing the estimation of the cost that an itinerary infrastructure may have, the experiment will bring to light if smart stadium are necessary or if it is economically feasible having itinerant network infrastructures.

Moreover, the application of our solution in a real scenario provided us with valuable evidences for promotion and marketing activities towards our target customers. These are mainly the providers of audio visual content that have the rights for broadcasting the sports events and the sports clubs owning the stadiums where the events take place. With the results of this experiment, Planet Media is now in a better position to get its products closer to the market and reinforce their expertise on multi-device and multi-channel services related to the creation of content for the sports industry. Besides, this customer and market knowledge might evolve in relevant strategic alliances to be exploited with commercial purposes.

In summary, the experiment helped us to prepare an adjusted cost model for our business model in addition to the information and the required models for the deployment and commercialization of the system in real-case scenarios. Lastly, from a technical point of view, the technology assessed in this experiment has set the basis for further potential applications and platforms not only in the sports environment but others. The experiment was contextualized within the strategic plan of the company, which includes the following action areas:

- Progress on the international dimension of the company, by promoting the internationalization of the R&D activities and increasing participation in international reference projects strengthening our participation in FIRE and FIWARE communities.
- Increasing our competitiveness by adjusting our estimation of infrastructure costs in relation to new media and streaming products.

Boosting a favourable environment for the investment on sustainable R&D&I through active participation in national and European technological platforms.

14.3.2 State-of-the-Art

Professional sports leagues around the world are embracing the advances of technology by adding connectivity at the stadiums to enhance the fan experience. The Levi's stadium in Santa Clara, California (USA), which serves as the home of the San Francisco 49ers of the National Football League, is currently the most connected stadium in the world. Opened in August 2014, it has 1.500 access points and infrastructure to support Wi-Fi 30x faster than any other, allowing 60.000+ fans to simultaneously connect. It is followed by Barclays Center in Brooklyn, a multi-purpose indoor arena with seating capacity over 18.000 fans, which serves as the home for Brooklyn Nets basketball team. It supports large volumes of high-definition video and Wi-Fi traffic during the events through Cisco StadiumVision Mobile solution[1], which

[1]https://www.cisco.com/web/strategy/sports/stadiumvision_mobile.html

enables more effective use of scarce Wi-Fi spectrum by enabling reliable multicast (the same Wi-Fi transmissions can be shared by all mobile devices requesting the same content); high-quality and reliable video delivery to a massive number of mobile devices; and low delay delivery of in-venue content (including streaming video, audio, and data).

In Europe, Stadiums such as Real Madrid's Santiago Bernabéu, Manchester City's Etihad, Bayer Leverkusen's BayArena, Glasgow's Celtic Park and several stadiums in Scandinavia, have been converted into digital, connected, football venues. Through platforms such as Cisco® Connected Stadium Wi-Fi, all these stadiums aim to provide an all-encompassing multimedia fan experience through an average 10- to 20-Mbps connectivity, so that they for example may upgrade their seats when they walk into the stadium, get real-time video of the event, access to social media, order drink and food from their seats, get information about closest services and restrooms, etc.

There are specific solutions such as the freeD Arena System infrastructure[2] installed in the AT&T Stadium in Arlington, Texas (USA), which is composed of more than 24 cameras allowing offering a unique way for fans to view replays, creating a 3D effect and depth of field on a 2D plane.

In general, the replays that are offered currently at Stadiums are broadcasted through the stadiums HDTV's. The providers of such services are currently working to offer the streaming individually to users through their mobile phones and enable interactivity. With this experiment we would be aligning with the current market supply trend and get **on to the cutting edge for the provision of these services,** especially in Europe.

14.4 Approach

14.4.1 Methodology

The work plan for the experiment implementation was based on a three cycle methodology. Each cycle put the main focus on one of the wireless technologies supported by NITOS testbed (Wi-Fi, WiMAX and LTE) and was composed of 4 phases: 1) deployment of the LIVEstats platform over the specific configuration of NITOS testbed; 2) definition of the goals, conditions and indicators that need to be assessed during the experiment, 3) execution of the experiment and 4) analysis of results and extraction of conclusions. The first cycle focused on Wi-Fi technology and was the leading one, in the sense that the experience acquired during this test provideed relevant feedback

[2]http://replay-technologies.com/technology/

that served as guide for the performance of the subsequent tests. Phases 1, 2 and 3 were documented and further analysed in phase 4 in order to extract conclusions and provide valuable feedback for the Fed4Fire consortium.

14.4.2 Associated Work Plan

According to the described methodology, the work plan was divided in the following Work Packages (WPs) during the timeline of the experiment:

	Month 1	Month 2	Month 3	Month 4
WP1 Deployment	Wi-Fi		WiMAX	LTE
WP2 Pilot Definition	Wi-Fi		WiMAX	LTE
WP3 Pilot Execution		Wi-Fi	WiMAX	LTE
WP4 Analysis and Conclusions		Wi-Fi	WiMAX	LTE

WP1 Experiment deployment. This WP was in charge of the deployment of LIVEstats system under the infrastructures provided by NITOS testbed. An exhaustive previous analysis was done to adapt our system to the specific characteristics provided by Fed4Fire. A first phase of resources discovery, requirements, reservation and provision was also performed.

WP2 Pilot definition. Concurrently with the previous WP, the definition of the pilot to carry out the experiment was defined. Here we defined the test battery to be performed in the corresponding tests. In addition, we precisely defined the indicators to be evaluated. Pilot definition phases for tests 2 and 3 were based on the first one over Wi-Fi, taking into account the experience acquired and the lessons learnt in order to improve the test battery and obtain more specific-feature focused, precise and detailed results.

WP3 Pilot execution. This WP took the responsibility for the execution of the experiment by means of the accomplishment of the test defined within the previous WP. Three experimentation tests were done focusing on the three wireless technologies offered by NITOS: Wi-Fi, WiMAX and LTE. The tests to be performed were classified in two different collections; (1) Functional tests and (2) Performance tests.

WP4 Analysis and conclusions. The analysis of all the information extracted during the experiment timeline were done within this WP. Based on results we were able to specify the most relevant indicators and their level of priority for the deployment of a network in a real case. Moreover, relations among them and specific values required in order to provide a good quality of experience (e.g. relationships such as Bandwidth and number of spectators supported by the system; topology of the network and number of spectators) were described.

This WP also compiled all the feedback about the experience and results of running the experiment over the testbed facility. All this information was detailed in the **final report** of the experiment.

14.4.3 Experimentation Methodology

The tests performed were focused to assess the Quality of Service provided when the system is accessed according to the following diagram:

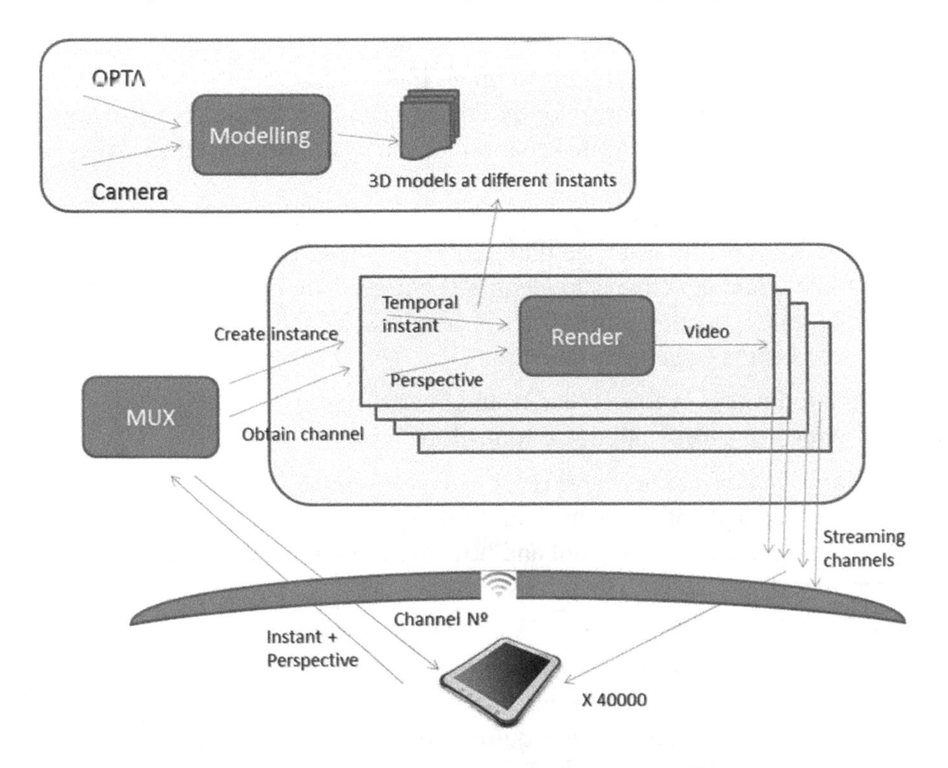

The spectator (using a tablet or smartphone) will request the visualization of a specific play at a specific instant in time from a specific perspective of a virtual camera.

1. The MUX (multiplexor module) verifies which of the current videos that are being rendered fits the most with those parameters. If there is no play fitting these parameters, a new render requested to the corresponding streaming channel. The MUX returns the selected channel to the user.
2. The render module takes the 3D model generated by the modelling server and generates the streaming from the model. (NOTE: a 3D play in

LIVEstats is generated using the video signal of a camera at the stadium and additional information coming from OPTA[3] information).
3. The user receives the streaming through the selected channel.

Specifically the tests to be performed were grouped into:

- Functional tests, aiming to verify that the system is working accordingly to the functionalities contemplated once it is implemented under NITOS infrastructure.
- Performance tests:
    - Stress test. This test aims to break the system under test by overwhelming its resources or by taking resources away from it. The main objective is to make sure that the system fails and recovers gracefully.
    - Load test. This test aims to put demand on a system and measuring its response so as to see until where it is able to work adequately. The main objective is to identify the maximum operating capacity of an application as well as any bottlenecks and determine which element is causing degradation.

Some of the indicators evaluated had been already identified, despite the fact that a more precise specification was to be done within WP2.

- *Network latency and packet loss*
- *Graphic quality*. Indicates how faithful is the quality of the streamed system screens on a thin client and how the graphic quality is degraded over imperfect network conditions.
- *Traffic characteristics*
    - Network delay. Time required transmitting a user's command to the server and a system screen back to the client.
    - Processing delay. Time required for the cloud system server to receive and process a user's command, and to encode and packetize the corresponding frame for the client.
    - System delay. Time required by the system software to process a user's command and render the next system frame that contains responses to the command.
    - Playout delay. Represents the time for the client to receive, decode, and display a *frame*.

[3]http://www.optasports.com/en/about/what-we-do/live-performance-data.aspx

- *Scalability of the render server*: We tested the most suitable configuration (regarding scalability) of the render server in order to offer the best experience with the minimum possible hardware cost.
- *Multiplexing capacity*: We tested the capacity of the multiplexing module that shares the signal among users.

14.5 Technical Work

14.5.1 Set-up of the Experiment

The tests performed aimed to assess the Quality of Service provided when the system is accessed were design according to the following Figure 14.3:

During the sports event, after a specific play (e.g. after a goal occurred during the football match) the spectator using a tablet or smartphone requests the replay on 3D on his/her device. This happens at a specific instant in time from a specific perspective of a camera.

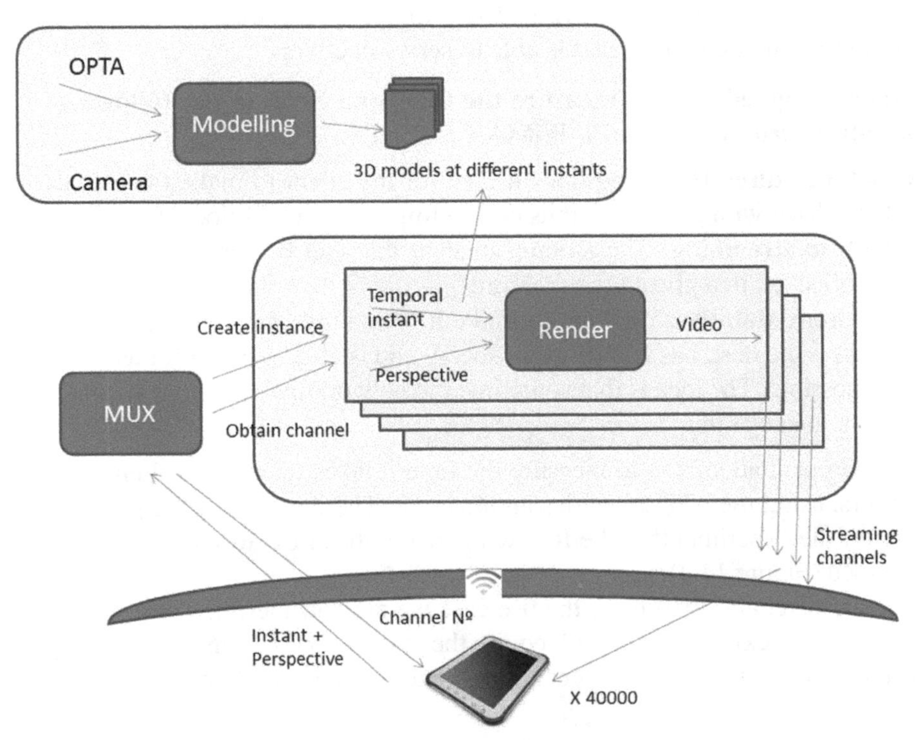

Figure 14.3 LIVEstats platform general architecture diagram.

1. The MUX (multiplexor module) verifies which of the current videos (3D replays) that are being rendered fits the most with those parameters (in terms of time and perspective). If there is no play fitting these parameters, a new render is requested to the corresponding streaming channel. The MUX returns the selected channel to the user.
2. The render module takes the 3D model generated by the modelling server and generates the streaming from the model. (NOTE: a 3D play in LIVEstats is generated using the video signal of a camera at the stadium and additional information coming from OPTA[4] information)
3. The user receives the streaming through the selected channel.

The Multiplexor module is needed because we are talking about providing an instant 3D replay to hundreds of users at the same (or very similar) time. This will probably happen during the most interesting plays or football players' actions: goals, offsides, faults, etc. The platform must be prepared to manage a high number of requests and deliver the 3D replay with an acceptable QoS within a delay that does not exceed the acceptable margins. The tests that we have performed in this experiment aim to verify that the system works accordingly (functional test) and is able to serve all clients.

We have designed tests to measure the following data, in the following available environments: WiFi, WiMAX and LTE.

- **Latency time**: How long does it take for the client to make the initial handshake with the server, thus connecting to it via WebSocket.
- **BPS in streaming**: The amount of data that can be sent through the WebSocket, in a given time measure.
- **Network stability**: We want to find out what kind of wireless environment is the best, has less drops in speed, and is able to be up for as long as possible. The idea is that sometime the network may fail, and we also want to gauge this.

The experiment attempted to measure the time it takes to serve N clients by X servers, using the NITLab nodes architecture. The diagram of the general set up for the experiments is the following, using the JFED tool provided by the testbed (Figure 14.4):

Given the characteristics of the use case we are considering, the optimal scenario for the experiment would require the reservation of as much nodes as possible: a high number of servers and a higher number of clients accessing

[4]http://www.optasports.com/en/about/what-we-do/live-performance-data.aspx

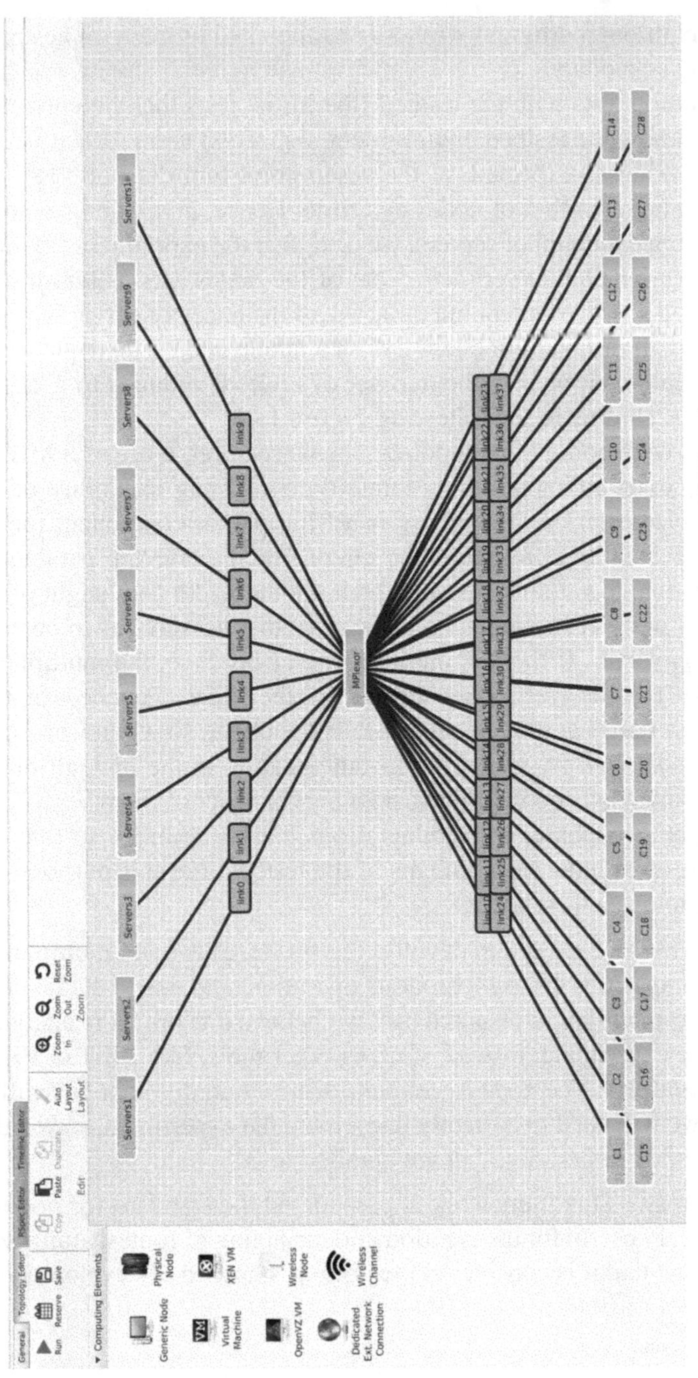

Figure 14.4 General setup of the experiments – JFED tool.

to those servers through different wireless channels; all of them orchestrated by the multiplexor module.

The preparation tests with the testbed (the initial tests that were made in order to get to know the testbed environment and tools) showed that not all nodes were working as expected, or the reservation tool did not allow the reservation of a high number of nodes at a time. Hence, given that we could not use an unlimited amount of servers, we designed the experiments in order to use two nodes for each experiment. One of the nodes acts as server hub, and the other nodes acts as client hub, that is, all instances of server live in a single node, and all instances of client live in a single node. The deployment fits the same initial concept, but the number of nodes is reduced to 2, server and client hub, according to the following Figure 14.5.

The servers were going to be sending, via WebSocket, a file of 8 MB for the client to receive, attempting to emulate the streaming capability of the WebSocket implementation. Also, there is a Multiplexor component for the test, whose functionality is to attend the client first, receive the parameters the client is sending to get an instance, which are the width and height of the client viewport, in order to emulate a number of distinct devices to connect to the streaming server. If there is one free server instance, the multiplexor component forwards the WebSocket address for the client to connect to it. If there is no free instance at the moment, it tells the client, so it can wait some time (2–3 seconds), and try to get the instance again. In the end, all of the clients must be attended and the file must be served to all of them.

In the preparation phase, some adaptations had to be made so that our platform complied with the requirements of the testbed. For this purpose, the following technologies were used:

- **Node.js**: This tool is used to develop the server, client and multiplexor. Due to its capacities for asynchronous operation, and due to the fact that node is non-intensive in resource use, we believe it is the more adequate tool for the experiment, instead of using other tools, like .NET or PHP.
- **Node.js modules**: The node.js standard distribution does not have all of the tools we will need to correctly implement the experiment, so we also use some node.js modules, that are detailed next:
    - **Express**[5]: This framework is considered the standard for node.js. It is very useful for the creation and managing of routes, along with the fact that it is considered the node.js standard of development.

[5]http://expressjs.com/

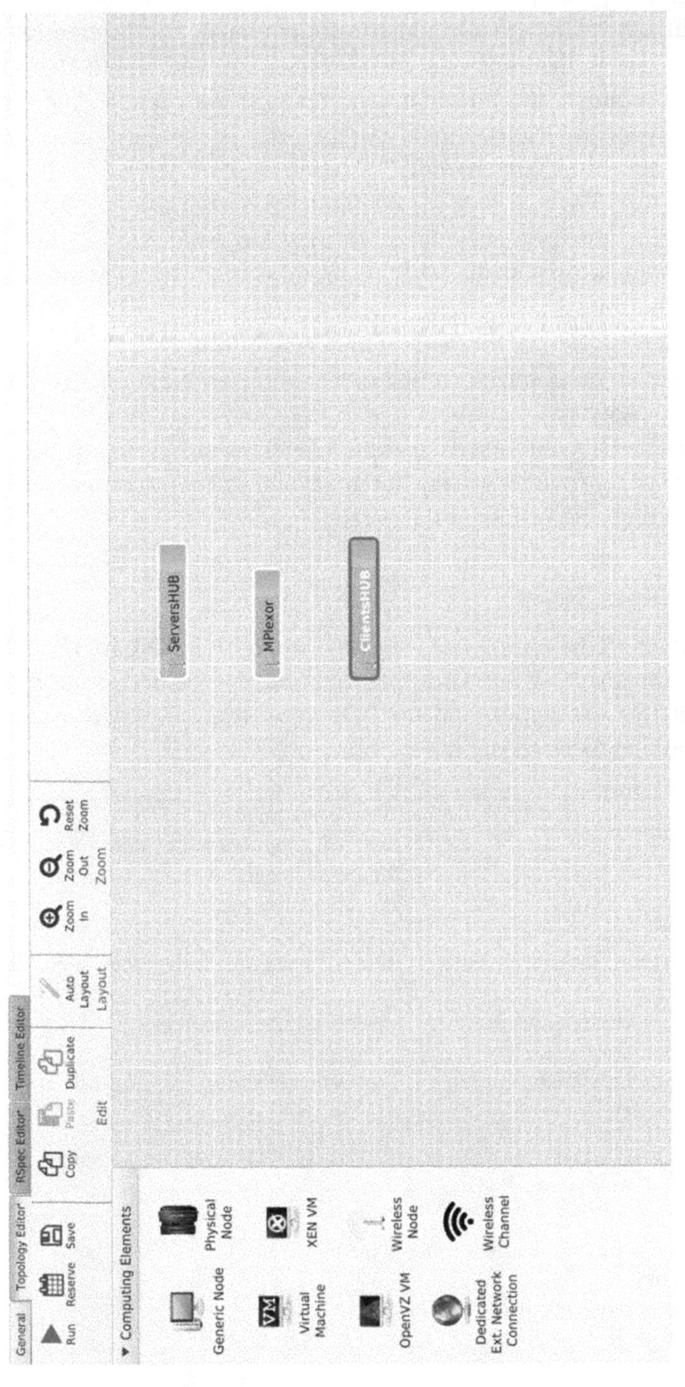

Figure 14.5 Setup of the experiments – JFED tool.

- **BinaryJS**[6]: This module contains functionality for the creation of a server that emits binary data via WebSocket, and clients that know how to connect to said server, and read the data sent by the server. There is no such functionality in the node.js standard distribution, therefore we use this module.
- **MySQL**[7]: We use a MySQL database to be able to store data regarding the instances, whether they are occupied or free. The communication with the MySQL engine is made thought the node.js module 'mysql'.
- **Chance**[8]: A library for random data generation. Due to the fact that we need to randomize certain aspects of the application, this library is very useful.
- **Chalk**[9]: This module is included in order to write to console, because it provides styling for text, and is therefore very useful for reading the results of the tests.

14.5.2 Preparatory Tests

This phase includes all the testing that was run in order to prepare for the core of the experiment: the tests to get to know the tools and the testbed environment: the possibilities for reservation and the best approaches for deploying our platform and tools in order to run the experiment:

Testbed Infrastructure Used	Tests Description
iMinds WiLab2	Reserving and accessing nodes (jFed, SSH) Mounting images on nodes Testing connectivity
NITOS lab:	
NODE1–040 (Nitos Outdoor Testbed: Grid, Orbit)	Reserving and accessing nodes (jFed, SSH)
NODE041–049 (Nitos Office Testbed: Icarus)	Mounting images on nodes
NODE050–085 (Nitos Indoor RF Isolated Testbed)	Testing connectivity
Results	Section B.2.1.1

[6]http://binaryjs.com/

[7]http://www.mysql.com/, http://github.com/felixge/node-mysql/

[8]http://chancejs.com/

[9]https://www.npmjs.com/package/chalk

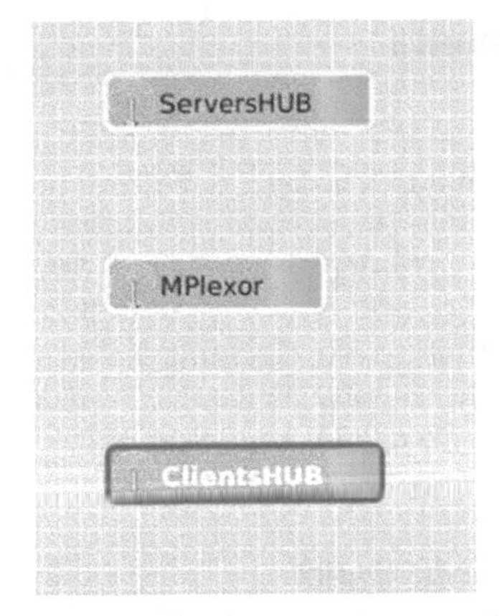

Figure 14.6 Components for the general set up of the experiments.

14.5.3 Laboratory Use Cases

The following sections describe the design of the experiments that were the core of our project. All of them follow the structure that was presented earlier in this document, using a server and a client hub, according to the set up (Figure 14.6).

14.5.3.1 Wi-Fi experiments

Identifier	WiFi 001	WiFi 002	WiFi 003
NITOS lab infrastructure used	WiFi nodes	WiFI nodes	WiFi nodes
No. of server instances	>30	>20	>40
Number of Clients	15–360	15–360	15–600
Repetitions	2	3	3
Objective	Time to serve all clients. Independent result and median value for the three of them together		
Results	Section B.2.1.2.1	Section B.2.1.2.2	Section B.2.1.2.3

14.5.3.2 LTE experiments

Identifier	LTE 001	LTE 002	LTE 003
NITOS lab infrastructure used	LTE nodes	LTE nodes	LTE nodes
No. of server instances	30	10	40
Number of Clients	15–360	15–360	15–600
Repetitions	3	3	3
Objective	Time to serve all clients. Independent result and median value for the three of them together		
Results	Section B.2.1.3.1	Section B.2.1.3.2	Section B.2.1.3.3

14.5.3.3 WiMAX experiments

Identifier	WiMAX 001	WiMAX 002	WiMAX 003
NITOS lab infrastructure used	WiMAX nodes	WiMAX nodes	WiMAX nodes
No. of server instances	30	30	30
Number of Clients	15–360	15–360	15–600
Repetitions	3	3	3
Objective	Time to serve all clients. Independent result and median value for the three of them together		
Results	Section B.2.1.4.1	Section B.2.1.4.2	Section B.2.1.4.3

14.5.4 Resources and Tools Used

Resources	
Virtual Wall (iMinds)	At first, we used iMinds because we had to test the experiment somewhere, and there was a certain confusion regarding the testbed we had to use. After speaking to Donatos Stravopoulos, we began using NITOS.
w-iLab.t (iMinds)	This testbed was used at first, when we were still in the learning process of how to interact with the platform via the jFED application. After a process of learning how to use the platform, we began using SSH to access it.
NITOS (UTH)	This is the testbed we used mainly. The nodes we used are mainly the following ones: Grid Nodes in the "Outdoor testbed" (node16 to node35). The Orbit nodes seemed to be working quite well (node02 to node09), we were advised not to fully rely on them, due to the fact that they are not very modern, and apparently, there was some errors associated with said nodes. In the "Indoor RF Isolated Testbed", we mainly used the LTE nodes (node054 to node058), that were AMAZING in response time and speed.
Tools	
Fed4FIRE portal	The reservation system works really well, allowing us to see beforehand what nodes are available, and specifying the kind of node in each case (this last bit was really useful).
JFed	We started using jFed at first, but the inability that it had to correctly interact with the scheduling functionality made it a bit cumbersome after a time, preferring, in time, to use SSH and other console commands to access the gateway and nodes.
OMF	OMF was used to create and mount images on the nodes. It was really useful, because once the node was reserved, and the image had been created, with all of the tools (and even source code) that we were going to use, it was a no-fuss kind of procedure. Really smooth, and

	very appreciated in order to maintain the homogeneity of the distinct environments.
JFed timeline	At first, we used jFed almost exclusively, so we consulted the availability of nodes via this tool. Later on, we developed a series of command-line aliases and tools that, together with the web portal reservation system, allowed us to be more efficient.
Other tools used	For the experiment, we have been using the following technologies mainly: NodeJS, WebSockets, MySQL and SSH, this last one being the main way to communicate between the client machine and the gateway, and then between the gateway and the node itself.

14.6 Results and/or Achievements

14.6.1 Technical Results Obtained

14.6.1.1 Preparatory tests

Initial testing gave many problems when trying to access resources. Reservation of nodes could not be completed, and, once they were finally reserved, attempts to mount images failed. Some of those attempts are described in the Table 14.1:

Table 14.1 Battery of preparatory tests with NITOS and IMinds WiLAB2 testbeds

Identifier	TESTBED	RESOURCE	EXPERIMENT	RESULT
Prep_001	NITOS	Nodes 005, 006, 007, 014, 015, 016, 021, 024, 046	Creating an image	Resource reservation failed
Prep_002	NITOS	Node 029, 033, 035, 005, 007	Creating an image	Resource reservation failed
Prep_003	NITOS	Node 035, 033, 014, 015, 021, 023	Creating an image	Resource reservation failed
Prep_004	NITOS	Node 005, 052, 085, 062,095	Creating an image	Resource reservation failed
Prep_005	NITOS	Node 033	Creating an image	Resource reservation failed
Prep_006	NITOS	Node 007	Creating an image	Reservation OK → SSH → Connection closed by remote host (KO)

(*Continued*)

Table 14.1 Continued

Identifier	TESTBED	RESOURCE	EXPERIMENT	RESULT
Prep_007	NITOS	Node 006	Creating an imagc	Reservation OK → SSH → Connection closed by remote host (KO)
Prep_008	IMinds WiLAB2	Internet	Creating an Image	Reservation of resources failed
Prep_009	IMinds WiLAB2	Airswitch	Creating an Image	Reservation of resources failed
Prep_010	IMinds WiLAB2	Coreswitch	Creating an Image	Reservation of resources failed
Prep_011	IMinds WiLAB2	Poeswitch	Creating an Image	Reservation of resources failed
Prep_012	NITOS	Node 005+ Channel 2 (wireless)	Creating an Image	Reservation OK → SSH → Connection closed by remote host (KO)

The very first errors were produced due to our lack of knowledge of the platform, we were not even aware about the reservation process that had to be followed. Once we got to know the reservation process, issues arised with the jFed tool, which didn't seem to work with the expected behaviour, and therefore returned fails with the reservations. We started then trying accessing the nodes via SSH, and here we got some errors with nodes about connection closed by remote host.

We need to remark that the feedback from the tools were limited in most cases. We were able to understand what was going on with the process only thanks to the support of NITOS technical team.

All these tests allowed us to gain a lot of knowledge about the testbeds. We concluded that for our experiment, we would be able to use only a couple of nodes at the same time.

We figured out that more than 200 clients would not be advisable because it would make the system clog itself.

Also, while running tests, we observed a curious behaviour: when servers were just started, the process was slower than when the servers had already made some executions. This could be due to several factors (cache, mostly), but it was interesting to see. After a "training"/warming up process they worked much better.

14.6.1.2 Wi-Fi experiments

14.6.1.2.1 *Wi-Fi 001*

Identifier	WiFi 001
NITOS lab infrastructure used	Servers hub: Node 068 Clients hub: Node 064
No. of server instances	31
Number of Clients	15–360
Repetitions	2
Objective	Time to serve all clients. Independent result and median value for the three of them together
Results	Completed

This experiment uses **WiFi** nodes ***node068*** as servers' hub and ***node064*** as clients hub. The amount of server instances used is 31. The experiment was repeated 2 times. The number of clients goes from 15 to 360, and the time results are shown in the graph below, showing the time it took each single experiment to run and the median time for the three of them.

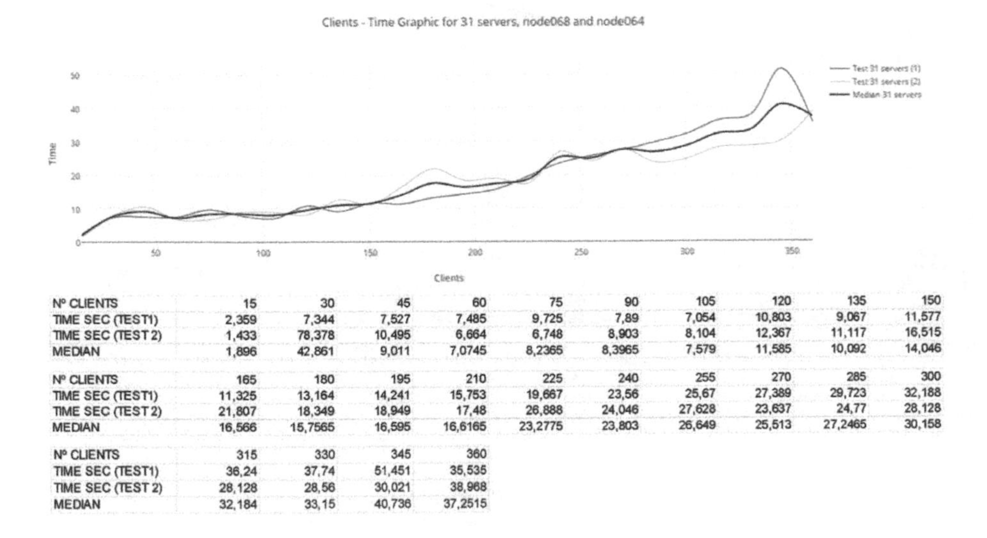

Nº CLIENTS	15	30	45	60	75	90	105	120	135	150
TIME SEC (TEST1)	2,359	7,344	7,527	7,485	9,725	7,89	7,054	10,803	9,067	11,577
TIME SEC (TEST 2)	1,433	78,378	10,495	6,664	6,748	8,903	8,104	12,367	11,117	16,515
MEDIAN	1,896	42,861	9,011	7,0745	8,2365	8,3965	7,579	11,585	10,092	14,046

Nº CLIENTS	165	180	195	210	225	240	255	270	285	300
TIME SEC (TEST1)	11,325	13,164	14,241	15,753	19,667	23,56	25,67	27,389	29,723	32,188
TIME SEC (TEST 2)	21,807	18,349	18,949	17,48	26,888	24,046	27,628	23,637	24,77	28,128
MEDIAN	16,566	15,7565	16,595	16,6165	23,2775	23,803	26,649	25,513	27,2465	30,158

Nº CLIENTS	315	330	345	360
TIME SEC (TEST1)	36,24	37,74	51,451	35,535
TIME SEC (TEST 2)	28,128	28,56	30,021	38,968
MEDIAN	32,184	33,15	40,736	37,2515

14.6.1.2.2 *Wi-Fi 002*

Identifier	WiFi 002
NITOS lab infrastructure used	Servers hub: Node 054 Clients hub: Node 058
No. of server instances	23
Number of Clients	15–360
Repetitions	3
Objective	Time to serve all clients. Independent result and median value for the three of them together
Status	Completed

This experiment uses **WiFi** nodes ***node054*** as servers hub and ***node058*** as clients hub. The amount of server instances used is 23. As usual, we repeated the experiment three times, and we show the time value for each experiment, along with the median time. The number of clients goes from 15 to 360, and the results are shown in the char below:

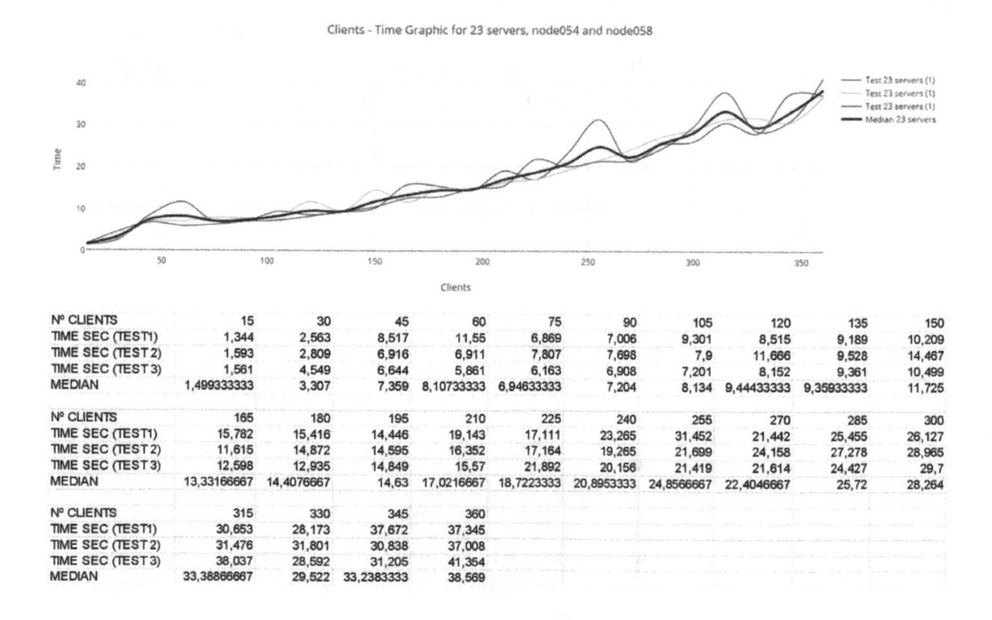

Nº CLIENTS	15	30	45	60	75	90	105	120	135	150
TIME SEC (TEST1)	1,344	2,563	8,517	11,55	6,869	7,006	9,301	8,515	9,189	10,209
TIME SEC (TEST 2)	1,593	2,809	6,916	6,911	7,807	7,698	7,9	11,666	9,528	14,467
TIME SEC (TEST 3)	1,561	4,549	6,644	5,861	6,163	6,908	7,201	8,152	9,361	10,499
MEDIAN	1,499333333	3,307	7,359	8,10733333	6,94633333	7,204	8,134	9,44433333	9,35933333	11,725

Nº CLIENTS	165	180	195	210	225	240	255	270	285	300
TIME SEC (TEST1)	15,782	15,416	14,446	19,143	17,111	23,265	31,452	21,442	25,455	26,127
TIME SEC (TEST 2)	11,615	14,872	14,595	16,352	17,164	19,265	21,699	24,158	27,278	28,965
TIME SEC (TEST 3)	12,598	12,935	14,849	15,57	21,892	20,156	21,419	21,614	24,427	29,7
MEDIAN	13,33166667	14,4076667	14,63	17,0216667	18,7223333	20,8953333	24,8566667	22,4046667	25,72	28,264

Nº CLIENTS	315	330	345	360
TIME SEC (TEST1)	30,653	28,173	37,672	37,345
TIME SEC (TEST 2)	31,476	31,801	30,838	37,008
TIME SEC (TEST 3)	38,037	28,592	31,205	41,354
MEDIAN	33,38866667	29,522	33,2383333	38,569

14.6.1.2.3 *Wi-Fi 003*

Identifier	WiFi 003
NITOS lab infrastructure used	Servers hub: Node 054 Clients hub: Node 058
No. of server instances	40
Number of Clients	15–600
Repetitions	3
Objective	Time to serve all clients. Independent result and median value for the three of them together
Status	Failed

This experiment collapsed the nodes. There were too many instance servers and clients defined for the experiment. The test failed to be completed.

14.6.1.3 LTE experiments

14.6.1.3.1 *LTE 001*

Identifier	LTE 001
NITOS lab infrastructure used	Servers hub: Node 050 Clients hub: Node 054
No. of server instances	30
Number of Clients	15–360
Repetitions	3
Objective	Time to serve all clients. Independent result and median value for the three of them together
Results	Completed

This experiment uses **LTE** nodes ***node050*** as servers hub and ***node054*** as clients hub. The amount of server instances is 30. As usual, we repeated the experiment three times, and in the following chart, the median time, along with each attempt time, is shown:

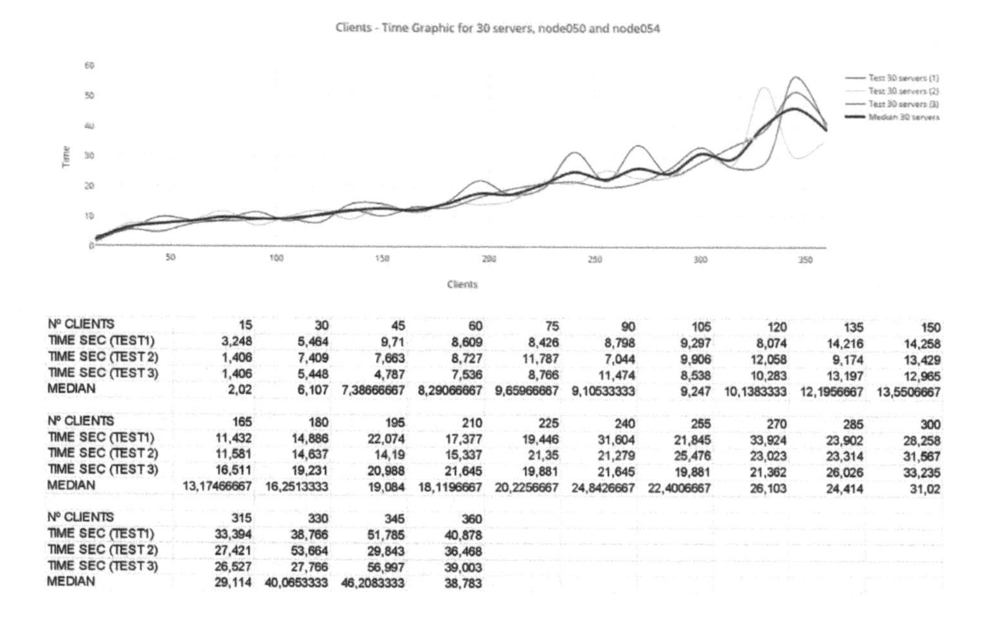

Nº CLIENTS	15	30	45	60	75	90	105	120	135	150
TIME SEC (TEST1)	3,248	5,464	9,71	8,609	8,426	8,798	9,297	8,074	14,216	14,258
TIME SEC (TEST 2)	1,406	7,409	7,663	8,727	11,787	7,044	9,906	12,058	9,174	13,429
TIME SEC (TEST 3)	1,406	5,448	4,787	7,536	8,766	11,474	8,538	10,283	13,197	12,965
MEDIAN	2,02	6,107	7,38666667	8,29066667	9,65966667	9,10533333	9,247	10,1383333	12,1956667	13,5506667

Nº CLIENTS	165	180	195	210	225	240	255	270	285	300
TIME SEC (TEST1)	11,432	14,886	22,074	17,377	19,446	31,604	21,845	33,924	23,902	28,258
TIME SEC (TEST 2)	11,581	14,637	14,19	15,337	21,35	21,279	25,476	23,023	23,314	31,567
TIME SEC (TEST 3)	16,511	19,231	20,988	21,645	19,881	21,645	19,881	21,362	26,026	33,235
MEDIAN	13,17466667	16,2513333	19,084	18,1196667	20,2256667	24,8426667	22,4006667	26,103	24,414	31,02

Nº CLIENTS	315	330	345	360
TIME SEC (TEST1)	33,394	38,766	51,785	40,878
TIME SEC (TEST 2)	27,421	53,664	29,843	36,468
TIME SEC (TEST 3)	26,527	27,766	56,997	39,003
MEDIAN	29,114	40,0653333	46,2083333	38,783

14.6.1.3.2 *LTE 002*

Identifier	LTE 002
NITOS lab infrastructure used	Servers hub: Node 050 Clients hub: Node 054
No. of server instances	10
Number of Clients	15–360
Repetitions	3
Objective	Time to serve all clients. Independent result and median value for the three of them together
Results	Completed

This experiment uses the same nodes and methodology as the previous one for LTE, the difference is the number of server instances, in this case they are 10. As seen, the time goes up quite a lot.

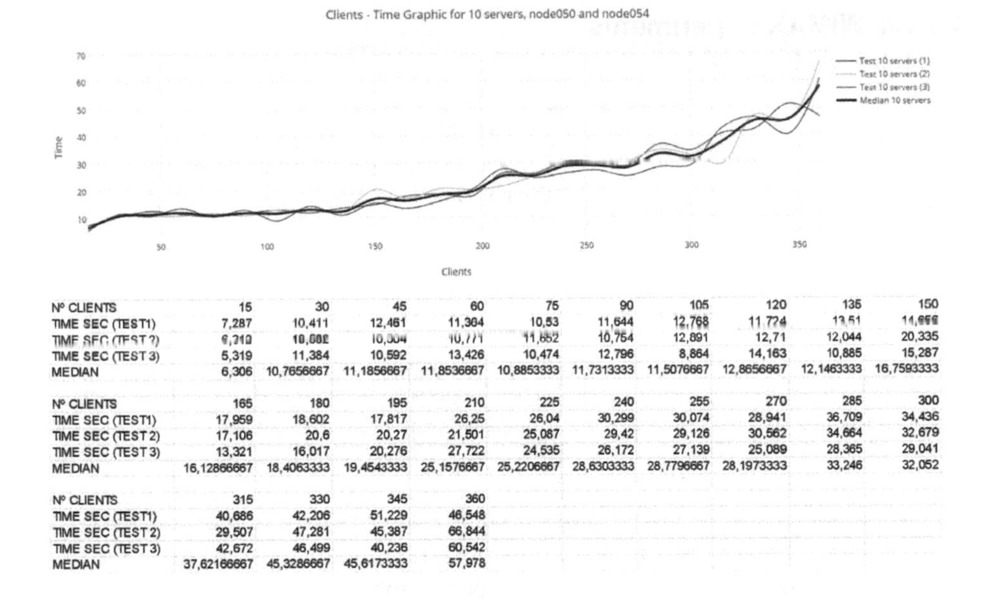

Nº CLIENTS	15	30	45	60	75	90	105	120	135	150
TIME SEC (TEST1)	7,287	10,411	12,461	11,364	10,53	11,644	12,768	11,724	13,51	[illegible]
TIME SEC (TEST 2)	[illegible]	[illegible]	[illegible]	10,771	11,652	10,754	12,891	12,71	12,044	20,335
TIME SEC (TEST 3)	5,319	11,384	10,592	13,426	10,474	12,796	8,864	14,163	10,885	15,287
MEDIAN	6,306	10,7656667	11,1856667	11,8536667	10,8853333	11,7313333	11,5076667	12,8656667	12,1463333	16,7593333

Nº CLIENTS	165	180	195	210	225	240	255	270	285	300
TIME SEC (TEST1)	17,959	18,602	17,817	26,25	26,04	30,299	30,074	28,941	36,709	34,436
TIME SEC (TEST 2)	17,106	20,6	20,27	21,501	25,087	29,42	29,126	30,562	34,664	32,679
TIME SEC (TEST 3)	13,321	16,017	20,276	27,722	24,535	26,172	27,139	25,089	28,365	29,041
MEDIAN	16,12866667	18,4063333	19,4543333	25,1576667	25,2206667	28,6303333	28,7796667	28,1973333	33,246	32,052

Nº CLIENTS	315	330	345	360
TIME SEC (TEST1)	40,686	42,206	51,229	46,548
TIME SEC (TEST 2)	29,507	47,281	45,387	66,844
TIME SEC (TEST 3)	42,672	46,499	40,236	60,542
MEDIAN	37,62166667	45,3286667	45,6173333	57,978

14.6.1.3.3 *LTE 003*

Identifier	LTE 003
NITOS lab infrastructure used	Servers hub: Node 050 Clients hub: Node 054
No. of server instances	40
Number of Clients	15–600
Repetitions	3
Objective	Time to serve all clients. Independent result and median value for the three of them together
Results	Failed

This experiment collapsed the nodes. There were too many instance servers and clients defined for the experiment. The test failed to be completed.

14.6.1.4 WiMAX experiments

14.6.1.4.1 *WiMAx 001*

Identifier	WiMAX 001
NITOS lab infrastructure used	Servers hub: Node 041 Clients hub: Node 044
No. of server instances	30
Number of Clients	15–360
Repetitions	3
Objective	Time to serve all clients. Independent result and median value for the three of them together
Results	Completed

This experiment uses **WiMAX** nodes ***node041*** as server hub and ***node044*** as clients hub. The number of instances is 30. The methodology to show the values in the graph are the same as in previous graphs:

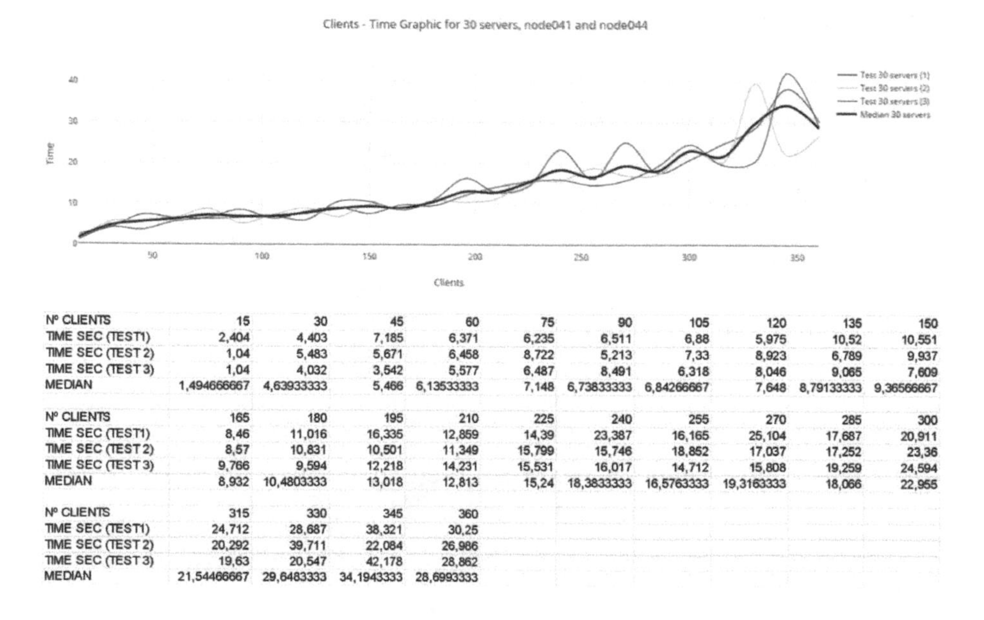

Nº CLIENTS	15	30	45	60	75	90	105	120	135	150
TIME SEC (TEST1)	2,404	4,403	7,185	6,371	6,235	6,511	6,88	5,975	10,52	10,551
TIME SEC (TEST 2)	1,04	5,483	5,671	6,458	8,722	5,213	7,33	8,923	6,789	9,937
TIME SEC (TEST 3)	1,04	4,032	3,542	5,577	6,487	8,491	6,318	8,046	9,065	7,609
MEDIAN	1,494666667	4,63933333	5,466	6,13533333	7,148	6,73833333	6,84266667	7,648	8,79133333	9,36566667

Nº CLIENTS	165	180	195	210	225	240	255	270	285	300
TIME SEC (TEST1)	8,46	11,016	16,335	12,859	14,39	23,387	16,165	25,104	17,687	20,911
TIME SEC (TEST 2)	8,57	10,831	10,501	11,349	15,799	15,746	18,852	17,037	17,252	23,36
TIME SEC (TEST 3)	9,766	9,594	12,218	14,231	15,531	16,017	14,712	15,808	19,259	24,594
MEDIAN	8,932	10,4803333	13,018	12,813	15,24	18,3833333	16,5763333	19,3163333	18,066	22,955

Nº CLIENTS	315	330	345	360
TIME SEC (TEST1)	24,712	28,687	38,321	30,25
TIME SEC (TEST 2)	20,292	39,711	22,084	26,986
TIME SEC (TEST 3)	19,63	20,547	42,178	28,862
MEDIAN	21,54466667	29,6483333	34,1943333	28,6993333

14.6.1.4.2 *WiMAx 002*

Identifier	WiMAX 002
NITOS lab infrastructure used	Servers hub: Node 047 Clients hub: Node 048
No. of server instances	30
Number of Clients	15–360
Repetitions	3
Objective	Time to serve all clients. Independent result and median value for the three of them together
Results	Completed

This experiment uses **WiMAX** nodes ***node047*** as server hub and ***node048*** as clients hub. The number of instances is 30. The methodology, value and graph wise, are the same as in past experiments:

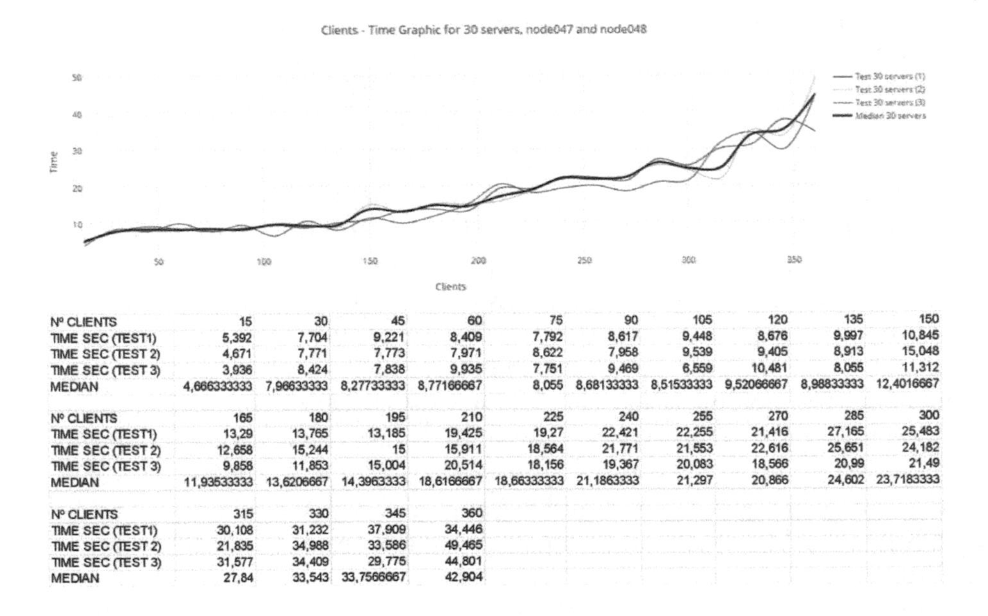

Nº CLIENTS	15	30	45	60	75	90	105	120	135	150
TIME SEC (TEST1)	5,392	7,704	9,221	8,409	7,792	8,617	9,448	8,676	9,997	10,845
TIME SEC (TEST 2)	4,671	7,771	7,773	7,971	8,622	7,958	9,539	9,405	8,913	15,048
TIME SEC (TEST 3)	3,936	8,424	7,838	9,935	7,751	9,469	6,559	10,481	8,055	11,312
MEDIAN	4,666333333	7,96633333	8,27733333	8,77166667	8,055	8,68133333	8,51533333	9,52066667	8,98833333	12,4016667

Nº CLIENTS	165	180	195	210	225	240	255	270	285	300
TIME SEC (TEST1)	13,29	13,765	13,185	19,425	19,27	22,421	22,255	21,416	27,165	25,483
TIME SEC (TEST 2)	12,658	15,244	15	15,911	18,564	21,771	21,553	22,616	25,651	24,182
TIME SEC (TEST 3)	9,858	11,853	15,004	20,514	18,156	19,367	20,083	18,566	20,99	21,49
MEDIAN	11,93533333	13,6206667	14,3963333	18,6166667	18,66333333	21,1863333	21,297	20,866	24,602	23,7183333

Nº CLIENTS	315	330	345	360
TIME SEC (TEST1)	30,108	31,232	37,909	34,446
TIME SEC (TEST 2)	21,835	34,988	33,586	49,465
TIME SEC (TEST 3)	31,577	34,409	29,775	44,801
MEDIAN	27,84	33,543	33,7566667	42,904

14.6.1.4.3 *WiMAx 003*

Identifier	WiMAX 003
NITOS lab infrastructure used	Servers hub: Node 047 Clients hub: Node 048
No. of server instances	30
Number of Clients	15–600
Repetitions	3
Objective	Time to serve all clients. Independent result and median value for the three of them together
Results	Failed

As it was expected, this experiment collapsed the nodes. There were too many clients defined for the experiment. The test failed to be completed.

14.7 Discussion

During the development of the experiment, there has been a catch, and that is the fact that NITLab was not able to work in the way we expected. Due to the fact that we received some errors, while trying to reserve nodes via jFed at first, and later, when trying to mount the image onto the node, we have decided to use two nodes, a server hub and a client hub, in the expectation that they will be useful to measure the architecture we have designed. Had we been able to reserve and use dozens of nodes, we may have been able to see a different outcome, so with that idea in mind, the experiment was repeated three times each, to get a median measure. We are now confident that more nodes would have not influenced the measures we have taken, because they would have acted in the same way, with some nodes acting as servers hubs, and others as clients hubs. The most interesting conclusion we have obtained is that, indeed, for large file sizes, LTE and WiMAX are much better than normal WiFi, but for less data, or a small file size, the overhead of many clients supersedes the benefits we could have obtained from a better network. With the tests that we have been able to perform the topology has not seemed critical for the performance of the platform.

Obviously, the architecture could be implemented, if we had enough machines, so we think the experiment proves that such an application could be indeed implemented and distributed, if the server architecture is powerful enough.

Our main motivation to run the experiment was the possibility to easily access infrastructures that otherwise would not be reachable for our company at this stage:

- Test in real environments for a 4-month period.
- Diversity of nodes that let us make different selections and configurations according to our needs.

In particular, the main asset was the access to LTE and WiMax nodes. Wi-Fi is in principle more reachable, but again, the dimension of the particular scenario that we are considering is not easily accessible for our company in a regular context.

From the experiment itself and the results obtained, the main value perceived is the knowledge that we have acquired to characterize the physical infrastructure that we would need to provide LIVEstats platform premium content over WiFi, LTE and WiMAX technologies to a high number of spectators on site on a sports stadium. We have been able to answer a number of questions and got insights about the relationship among parameters and characteristics of the different components of the platform and the wireless infrastructure that should be deployed.

Also, the preparation of the experiment required us to set up some developments and adaptations of the platform (in particular from Windows to Linux) that lead to improvement and fine-tune of the product, something that was not the main objective of the experiment, but has resulted in a very positive side effect.

The original aim of the experiment was to ascertain the amount of servers needed to be able to attend at least 40% of clients in a stadium in a timely fashion. According to the data we have obtained from the experiment, we extracted the following results:

14.7.1 Small File: From 0.5 to 2 Megabytes

1. According to our calculations, there is not a linear progression with the file size, meaning that a smaller file will most likely be served before than a bigger one. In a WiFi environment, considering 30 instances per machine, we would need between 30 and 50 machines to serve 25.000 clients, with a maximum waiting time of 10 seconds.
2. In an LTE environment, given that is at least 25% faster than a normal WiFi, we would be OK with maybe 30 machines, with 30 instances

per machine, for 25.000 clients. For a whole 100.000 people stadium, between 100 and 120 machines should be sufficient.

3. WiMAX is, in this case, very similar to LTE. Due to the small file size, the benefit would most likely be marginal, so we calculate around 100 machines for the whole stadium.

14.7.2 Normal File Size: From 8 to 12 Megabytes

1. For a WiFi connection, and between 8 and 12 megabytes of information to be transfered via WebSocket, we would need around 6000 server instances. Considering we could have as many as 30 instances per server, we would need around 200 different machines with WiFi connection to serve around 25.000 clients. To serve 100.000 clients (the maximum amount for our estimations), we would need around 800 machines, with 30 instances per machine, and the clients would be attended in less than 8 seconds.
2. For a LTE connection, and between 8 and 12 megabytes of information, we would need around 75% of the machines in the previous architecture. According to our calculations, 180 clients could be served, in less than 10 seconds, by a single machine with 30 server instances. Therefore, for 25.000 clients, around 120–135 machines would be needed. To serve 100.000 clients, we believe 500 machines would be sufficient.
3. For a WiMAX connection, and between 8 and 12 megabytes of information, we would need around 75% of the machines in the LTE architecture. That means, between 90 and 105 machines for 25.000 clients, and around 400, rounding, machines for a whole stadium with 100.000 clients trying to access the data.

14.7.3 Large File Size: From 30 to 50 Megabytes

1. As we stated before, there is not a linear progression in this case, so a larger file, in a WiFi environment, would need something between triple and quadruple the amount of machines for the same number of clients. This is due to the fact that, with a bigger file, the multiplexor will not be able to serve them so fast, the clients will require more time to download the information, so more servers would be advised in this particular case.
2. LTE gives a definite improvement over normal WiFi in large file sizes. We believe that with 150% of the servers used in the normal file size experiment, all the clients could be served in a timely manner, that is, no more than 10 seconds wait time for each one.

3. WiMAX is the most potent 4G wireless type of network there is, so we are quite confident that with between 125% and 150% of the machines specified in the normal file size experiment, we could serve all the clients in no more than 10 seconds per each client.

The aforementioned results assume that we are going to have all clients trying to try and obtain the server instance simultaneously, and that is why the number of machines/instances is so large in certain cases. If, assuming that no more than 20% of users are going to try to get the data at the same time, we are looking at something like 10% to 25% of the amount of server instances/machines that we would need for this architecture and serving times.

14.8 Conclusions

The results obtained brought great value for our company and our action plan: In a short time period we have had access to varied wireless technologies and infrastructures: Wi-FI, LTE and WiMAX in outdoor and indoor locations. Thanks to the testing we have gained knowledge for characterising the physical wireless infrastructure, so that we are now able to continue with the improvement and development of the business model for our product LIVEstats. This was the main objective of the experiment and we have met most of our expectations in this sense.

Conclusions extracted from our experiment have set the starting point that will let us define the real potential of our product. We will use and extrapolate results to model the service requirements and capabilities that the platform may offer:

- What seems to be the best wireless technology to implement our solution.
- What is the minimum required infrastructure required in order to be able to offer the service.
- What is the average number of users that we can serve with the platform as it is now in a concrete context.
- What is the average size of the video file that can be sent in order to assure an acceptable QoS.
- What is the minimum and maximum time to serve users in specific conditions.

Our next steps are to improve and redesign the business model according to the conclusions extracted from the experiment.

Apart from LIVEstats platform itself and its business model related aspects, the experiment brought value to our company in a very specific sense: what we have tested specifically during the experiment is the multiplexor module and its capacity to balance the load among server-client. This module is direct candidate to be integrated in more product developments of the company, and, in particular, in the field of Media and Entertainment, which is one of the main business lines of Planet Media. The company have consolidated clients in the area that require the latest innovations from us. In this sense, the experiment has given us the opportunity to test and validate the capacities of the module and gain knowledge about the possibilities of the tool, pros and cons, and experience to adapt results to further developments that may use it.

If we had not been allowed to make our experiment using Fed4FIRE facilities, we wouldn't have been able to progress on the development of the business model in the scenario for stadiums. We would have continued, at least for the moment, with the scenario of providing the service at the spectators' homes. We would have continued testing the platform and the multiplexor module with load and stress tests in the regular manner that we do with the rest of applications that are developed in our company. No wireless technologies (especially LTE and WiMAX) would have been tested at this stage, and for sure the dimension of the testing hypothesis and objectives would have been much more conservative, according to our test labs capabilities in our offices.

The knowledge acquired during the experiment on the performance of the platform using wireless technologies, and the conclusions extracted about the minimum requirements to serve a high number of clients constituted an enriching validation process: we have now a better understanding of the challenges and opportunities from now on to bring the product to a close to market stage.

15

An Experiment Description Language for Supporting Mobile IoT Applications

Kostas Kolomvatsos, Michael Tsiroukis and Stathes Hadjiefthymiades

Pervasive Computing Research Group,
Department of Informatics and Telecommunications,
National and Kapodistrian University of Athens,
Panepistimiopolis, Ilissia, 15784, Greece

Abstract

Mobile IoT applications consist of an innovative field where numerous devices collect, process and exchange huge amounts of data with central systems or their peers. Intelligent applications could be built on top of such information realizing the basis of future Internet. For engineering novel applications, experimentation plays a significant role, especially, when it is performed remotely. Remote experimentation should offer a framework where experimenters can efficiently define their experiments. In this chapter, we focus on the experiments definition management proposed by ***Road-, Air- and Water-based Future Internet Experimentation*** (RAWFIE). RAWFIE offers, among others, an experimentation language and an editor where experimenters can remotely insert their experiments to define actions performed by the nodes in a testbed. RAWFIE proposes the ***Experiment Description Language*** (EDL) that provides the elements for the management of devices and the collected data. Commands related to any aspect of a node behavior (e.g., configuration, location, task description) are available to experimenters. We report on the EDL description and the offered editors and discuss their key parts and functionalities.

15.1 Introduction

The ***Internet of Things*** (IoT) assumes the pervasive presence of numerous devices in the environment that are capable of performing simple processing tasks by involving multiple interactions among them. Such devices are wirelessly connected and adopt unique addressing schemes to be uniquely identified in the network. Objects make themselves recognizable and they obtain intelligence by taking or enabling context related decisions thanks to the fact that they can communicate information about themselves and they can access information that has been aggregated by other things, or they can be components of complex services [38]. The number of Internet-connected devices surpassed the number of human beings on the planet in 2011, and by 2020, Internet-connected devices are expected to a number between 26 and 50 billion. For every Internet-connected PC or handset there will be 5–10 other types of devices sold with native Internet connectivity [28]. Novel applications can be built on top of the vast network to improve the services offered to end users and, thus, to improve their quality of living. Mobile IoT involves devices capable of moving in the environment and record the ambient information. A typical example is the adoption of ***Unmanned Vehicles*** (UV). UVs can be categorized into: ***Unmanned Ground Vehicles*** (UGVs), ***Unmanned Aerial Vehicles*** (UAVs) and ***Unmanned Surface Vehicles*** (USVs). UVs act autonomously and can carry a set of sensors and communicate each other as well as with a central system where they can transfer the data that they record during their movement.

In this context, the research and technical challenges towards the development of a smart World are many. These challenges call for efficient solutions either *horizontally* (application neutral) or *vertically* (application dependent). Mobile IoT should overcome the vertical oriented legacy architectures and lead to open systems and integrated environments that will support intelligent applications on top of contextual knowledge collected/produced by autonomous nodes. The aim is to create innovative ecosystems of moving devices like UVs. An efficient means for building high quality applications is *remote experimentation*. Remote experimentation has already adopted in domains like education [1, 7, 13, 21]. It involves a real experiment with real equipment that is controlled remotely through the Internet. Remote experimentation can offer many advantages as physical experimentation is expensive, difficult to maintain and restricted to specific areas. Usually, *testbeds* are adopted to host a set of devices that will execute an experiment. A ***testbed*** is a platform/environment where hardware (e.g., a number of devices) is available to perform transparent and replicable testing of tools or technologies. Testbeds can be located around the World and host devices belonging to multiple types, if possible.

The ***Road-, Air- and Water-based Future Internet Experimentation*** (RAWFIE) platform comes to offer remote experimentation functionalities to the interested researchers and professionals. RAWFIE delivers a framework for interconnecting numerous testbeds over which remote experimentation will be realized. The RAWFIE platform originates in a European Union-funded (H2020 call: FIRE+ initiative) project which focuses on the mobile IoT paradigm and provides research and experimentation facilities through the ever growing domain of unmanned networked devices (vehicles). The purpose of the proposed structure is to create a federation of different network testbeds that work together to make their resources available under a common framework. Remote experimentation is realized on top of real devices. These devices have specific characteristics that may vary according to their type (e.g., UGVs, UAVs, USVs). Devices characteristics may vary even devices belong to the same category as they come from their manufacturers.

In remote experimentation, there is a gap between experimenters and devices realizing an experiment. Experimenters are researchers or professionals and may not be aware of the characteristics of the devices. Experimenters, likely, are not aware of the low level instructions that should be transferred to the devices during the execution of the experiments. To cover this gap and serve non experienced experimenters/professionals, RAWFIE offers an abstraction of the underlying functionalities. This abstraction is realized by a ***Domain Specific Language*** (DSL). A DSL is a language designed for a specific field of applications. Its aim is to solve problems related to a highly focused field of research. DSLs target to more specific tasks than classic programming languages. They provide expressions for describing parameters of a domain and they have a concrete syntax. A number of semantics are adopted to lead to the automatic generation of specific tools important for the creation of the final code [17]. The most significant advantage of the DSLs usage is that they provide the opportunity to non-experienced users to write more easily domain specific programs [20]. These programs are not dependent on the underlying platform, thus, providing an additional advantage. RAWFIE offers an ***Experiment Description Language*** (EDL), i.e., a DSL, and two editors (textual and visual) devoted to assist non-experienced users to easily define their experiments. A code generation component is responsible to translate each experiment expressed in the EDL into the information transferred to mobile nodes. Hence, RAWFIE efficiently interconnects experimenters coming from various domains with the nodes present in numerous testbeds.

The rest of the chapter is organized as follows. Section 15.2 is devoted to the description of the problem while Section 15.3 presents the related work. Section 15.4 discusses our approach for creating a DSL for abstracting the complexity of the UxVs characteristics and Section 15.5 reports on the technical details. Section 15.6 presents a case study where we create and launch an experiment with the proposed tools and Section 15.7 discusses our future research directions. Finally, in Section 15.8, we conclude our chapter.

15.2 Problem Statement

The definition of an experiment on top of a number of devices located in testbeds around the Globe involves the creation of a script containing commands that will be executed by the devices. Devices should receive the instructions defined in the script and move in the environment towards the execution of the experiment. For instance, an experiment may instruct a group of UVs to move around an area and collect data related to environmental conditions (e.g., temperature, humidity). In this scenario, experimenters should know the low level characteristics of the devices (e.g., commands for defining navigational instructions to UVs). However, this is not the usual case. It is difficult for experimenters to know the low level commands especially when they are working in a completely different domain. Imagine a researcher working in biomonitoring and the effects of environmental conditions in humans' health. The researcher does not have any technical knowledge on the functionalities provided by the UVs, however, he/she wants to instruct the devices to perform some processing tasks. The problem is more intense when we take into consideration that a testbed may incorporate many different devices with different characteristics. Experimenters cannot be aware of the different sets of commands to handle the heterogeneity of the devices. Due to the wide range of devices and technologies that testbeds could incorporate, a number of different commands could be adopted to instruct devices to execute experiments.

The above discussion shows the need for an abstraction in the underlying technologies. Such an abstraction will give the opportunity to experimenters to use the devices transparently and define commands for UVs in a more user friendly way. Hence, even non-experienced experimenters can use the provided platform and define their experiments that will be executed to any available testbed. In general, users, not having a lot of experience with programming languages, are not able to develop efficient software components like experiments for mobile IoT. In this case, *Model Driven Engineering* (MDE) can provide a lot of advantages not only to under-experienced programmers but also to proficient ones that are unfamiliar with the specific domain. MDE is a software development methodology for creating models for a specific domain. MDE technologies offer a promising approach to address the inability of the third generation languages to express domain concepts effectively [32]. The aim of MDE is to increase efficiency in developing applications. DSLs follow the principles of the MDE development and can provide a number of advantages in cases where domain programming knowledge is limited [22, 35]. DSLs target to more specific tasks than classic programming languages. They provide expressions for describing parameters of a domain of interest and they have a concrete syntax. A number of semantics are used in order to lead to the automatic generation of specific tools important for the creation of the final code [17].

RAWFIE offers the EDL, that provides a terminology for defining experiments for mobile IoT. The EDL offers an abstraction for any aspect of an experiment like the necessary metadata, statements, commands related to the devices, group of devices management and so on. The EDL terminology is invoked through the provided *Experiment Authoring Tool* (EAT). Two editors are provided: the ***textual*** and the ***visual***. Editors are built on top of the EDL and incorporate all the necessary functionalities like those originated in typical IDEs as well as functionalities related to the compilation and validation of the defined experiments.

15.3 Background and State of the Art

A number of research efforts deal with the devise and development of *Vehicular Ad Hoc Network* (VANET) testbeds for performing diverse applications (e.g., accident warning systems, traffic information control and prevention systems, pollution and weather monitoring, etc.). C-Vet [8] stands for the vehicular testbed developed in the University of California at Los Angeles (UCLA) campus offering both Vehicle to Vehicle (V2V) and Vehicle to Infrastructure (V2I) connectivity. The testbed is composed of 60 vehicles that circulate in the UCLA campus in order to support extended applications and services. CarTel [16] is a testbed developed by Massachusetts Institute of Technology (MIT) that has been active in Boston and Seattle. CarTel is comprised of six vehicles equipped with sensors and communications units that feature Wi-Fi (IEEE 802.11b/g) and Bluetooth. This testbed provides an important insight on how to handle intermittent connectivity, and how feasible this kind of connectivity is to explore a class of non-interactive applications. SAFESPOT [31] is a testbed that was run for 4 years in six cities across Europe. It uses vehicles equipped with OBUs, RSUs and Traffic Centres (communicating through Wi-Fi) to centralize traffic information and forward safety-critical messages. The project's goals were to: a) use the infrastructure and the vehicles as sources and destinations of safety-related information and develop an open, flexible and modular architecture and communication platform, b) Develop the key enabling technologies: ad-hoc dynamic network, accurate relative localisation, dynamic local traffic maps, c) Develop and test scenario-based applications to evaluate the impacts on road safety, d) Develop and test scenario-based applications to evaluate the impacts on road safety and e) Define a sustainable deployment strategy for cooperative systems for road safety, evaluating also related liability, regulations and standardisation aspects. HarborNet [2] is a real-world testbed for research and development in vehicular networking that has been deployed successfully in the sea port of Leixoes in Portugal. The testbed allows for cloud-based code deployment, remote network control and distributed data collection from moving container trucks, cranes, tow boats, patrol vessels and roadside units, thereby enabling a wide range of experiments and performance analyses.

DSLs have attracted a lot of attention in various application domains as they provide abstraction in the definition of applications oriented to a specific research field [14]. Every DSL has special characteristics and their size varies according the domain of application. Normally, DSLs are small in length and cover only the essential features and concepts of the domain under consideration [25], however, they are characterized by expressiveness [22]. This approach keeps the length of the notation small, thus, increasing the abstraction level. DSLs are more declarative or descriptive than legacy programming languages [36]. The design of a DSL involves the study of the domain under consideration and the identification of the most important concepts of that domain. The semantics of the domain should be implicit in the language notation [15].

For a theoretical survey in DSLs, the interested reader should refer in [25] while an empirical study on the use of a DSL in industry is presented in [12]. A number of contributions discuss the advantages of DSLs [18, 33, 34, 36] while a survey on the process for developing a DSL is described in [22]. In general, DSLs lead to easy maintenance of potential modifications, increase flexibility and productivity. DSLs are adopted to a set of research domains. In robotics, DSLs focus on increasing the level of automation, e.g., through code generation, to bridge the gap between the modeling of robotics and implementation. In [24], the authors survey the corresponding literature and classify a number of publications in the robotics field. DSLS are also adopted in banking [3], telecommunications [6], web services definition [12], autonomic computing [19]. DSLs are already adopted in a number of research projects like IPAC[1] and PoLoS[2]. The IPAC (Integrated Platform for Autonomic Computing) aims at delivering a middleware and service creation environment for developing embedded, intelligent, collaborative, context-aware services in mobile nodes. A DSL is implemented to support engineers to efficiently define applications that will be uploaded in mobile nodes. The PoLoS project aims to design specify and implement an integrated platform, which will cater for the full range of issues concerning the provision of *Location Based Services* (LBS). PoLoS proposes a DSL for 'annotating' LBSs that will be combined in the final workflow.

In [40], the authors demonstrate a framework to automate the generation of DSL testing tools. The presented framework utilizes Eclipse plug-ins for defining DSLs. Moreover, a set of tools concerning a translator, and an interface generator are responsible to map the DSL debugging perspective to the underlying *General Purpose Language* (GPL) debugging services. The aim is to present the feasibility and the applicability of debugging and testing information derived by a DSL in a friendly programming environment. A program transformation engine supporting the debugging process written in a DSL is described in [29, 39, 40]. The discussed approach concerns the methodology of generating a set of tools necessary to use a DSL from a language defined in a specific grammar. Such tools are: the editor, the compiler and the debugger [11]. This research effort focuses on issues related to the debugging support for a DSL development environment. The debugger is automatically generated by a language specification. Authors describe two approaches for weaving the debugger in conjunction with the *DSL Debugging Framework* (DDF) plug-in. The first approach is applicable when the aspect weaver is available for the generated GPL while the second approach involves the *Design Maintenance System* (DMS) [4] transformation and is applied when the aspect weaver is not available.

In [30], the authors describe a prototyping methodology for of *Domain Specific Modeling Languages* (DSMLs) on an independent level of the MDE architecture. They argue that the prototyping method should describe

[1]http://ipac.di.uoa.gr/

[2]http://polos.di.uoa.gr/

the semantics of the DSML in an operational fashion. For this, they use standard modeling techniques i.e., *Meta Object Facility* (MOF) [23] and *Query/View/Transformations* (QVT) Relations [27]. By combining this approach with existing metamodel-based editor creation technologies they enable the rapid and cost free prototyping of visual interpreters and debuggers. Authors utilize the *Eclipse Modelling Framework* (EMF) which is similar to MOF and using the Ecore metamodel of a DSML they can generate the DSML plug-in with EMF. The created plug-in provides the basis for the creation, access, modification, and storage of models that are instances of the DSML.

A logic programming based framework for specification, efficient implementation, and automatic verification of DSLs, is presented in [10]. Their proposal is based on Horn logic and, eventually, constraints to specify semantics of DSLs. The semantic specification serves as an interpreter or more efficient implementations of the DSL, such as a compiler, can be automatically derived by partial evaluation. The executable specification can be used for automatic or semi-automatic verification of programs written in a DSL as well as for automatically obtaining conventional debuggers and profilers. The syntax and semantics of the DSL are expressed through Horn logic. The Horn logic syntax and semantics are executable leading to the automatic definition of an interpreter. The authors in [10] present their approach and give some examples indicating the efficiency of the discussed methodology.

15.4 The Proposed Approach

15.4.1 The RAWFIE Platform

The purpose of the RAWFIE initiative is to create a federation of different testbeds that will be combined to make their resources available under a common framework. Specifically, RAWFIE aims at delivering a unique, mixed experimentation environment across the space and technology dimensions. RAWFIE will integrate numerous testbeds for experimenting in vehicular (road), aerial and maritime environments. The basic idea behind the RAWFIE effort is the automated, remote operation of a large number of robotic devices (UGVs, UAVs, USVs) for the purpose of assessing the performance of different technologies in networking, sensing and mobile/autonomic application domains. RAWFIE features a significant number of UVs for exposing to the experimenter a vast test infrastructure. All these items are managed by a central controlling entity which is programmed per case and fully overview/drive the operation of the respective mechanisms (e.g., auto-pilots, remote controlled ground vehicles). Internet connectivity will be extended to the mobile units to enable the remote programming (over-the-air), control and data collection. Support software for experiment management, data collection and post-analysis is virtualized to enable experimentation from everywhere in the world. The vision of *Experimentation-as-a-Service* (EaaS) is promoted through RAWFIE. The IoT paradigm is fully adopted and further refined for support of highly dynamic node architectures.

The RAWFIE architecture consists of tree tier design patterns. Each tier is separated in different software elements, each one providing a different functionality. The components are implemented with standard interfaces for safe interconnection between them. The discussed tiers are: i) the *front-end tier*, ii) the *middle tier* and iii) the *data tier*. The front end tier includes the services and tools that RAWFIE provides to experimenters to define and perform the experimentation scenarios. The RAWFIE *Web portal* provides to users, a web interface to federation resources and services. The user friendly environment of the portal makes experimenters creating straightforward successful experiment scripts. The front end tier has an authorization component, for checking the authorization of a user by his/her credentials. The *Testbed and Resource Discovery* component shows the availability of the testbed and the resources respectively while running. The *Experimentation Suite* is consisting of five tools and are the following: i) the *Monitoring tool* – it manages the presentation of the information needed for monitoring the status of the nodes and the data collected during the experiments; ii) the *Launching tool* – it is informed for the end of an experiment's execution to initiate the next booked scenario in the case of the entire use of a testbed or it is invoked (manually) to start an script that experimenters desire; iii) the *Booking tool* – it allows experimenters to book a spatiotemporal interval for running their experiments, thus, providing automatic coordination in the use of the testbed resources among experimenters; iv) the *Visualization tool* – it interconnects with the *Visualization Engine* of the middle tier receives the resource traces. The resource traces are graphically displayed to the web interface; v) the *Authoring tool* – it includes all the necessary mechanisms to allow access of the experimenters in the RAWFIE experimentation suite and the available EDL editors.

The RAWFIE middle tier is the layer that lies between the UVs testbeds and the experimenters (front-end tier). It provides the software interfaces needed, and includes useful software components related to security, trust, control and visualization aspects. This tier provides the infrastructure which facilitates the creation and integration of applications in the RAWFIE platform. It provides uniform, standard, high-level interfaces to the application developers and integrators so that the applications can be easily composed and reused. It will supply a set of common services to perform various general purpose functions in order to hide the distributed nature of the testbeds and facilitate the collaboration between different applications. The middle tier is consisted of the following modules: i) the *Experiment Validator* – it validates the experiment scenario to avoid syntactic and semantic errors. For instance, if the experimenter requests more resources than the available ones in the selected timeslot in the specified testbed site, the validator will avoid the execution of the experiment and send error message to the experimenter; ii) the *Experiment Controller* – it provides functionalities for the automatic control of the executed experiments according to the defined scripts; iii) the *Visualization Engine* – it is responsible for gathering sensing information from the UVs, processing the data and finally forwarding them to the visualization tool of the front-end tier; iv) the *Testbed directory* – it includes information relevant to the testbeds and resources (i.e., location, facilities) as well information on the capabilities of a particular resource and its requirements for executing experiments e.g., in terms of interconnectivity or dependencies; v) the *Data Collection and Analysis module* – it is responsible for the data collection and data the analysis-processing. Furthermore, it stores the measurement streams in the Data Storage components of the RAWFIE infrastructure. RAWFIE also provide a large, secure, cloud-based central repository in which collected data can be anonymized and made available to users; vi) the *Launching Service* – it provides functionalities related with the automatic and the manual launch of an experiment; vii) the *Booking Service* – it is adopted for performing bookings in the available testbeds and resources; viii) the *System Monitoring Service* – it secures that the platform works properly and identifies any potential error in the RAWFIE framework.

Finally, the data tier is in charge of ensuring data persistence. All the data elements and the code repos are stored to *Data Storage and Code Repositories* and servers to the Cloud, respectively.

15.4.2 The RAWFIE EDL

The Experiment Description Language (EDL) is a DSL for creating simple or more complex experimental scenarios for the IoT domain. The EDL is designed for the RAWFIE purposes aiming to help domain experts or non-experienced users (e.g., experimenters) to effectively create and handle

IoT remote experimentation. The major goal of the EDL is the provision of a high level of abstraction that shields experimenters from the complexities of the underlying implementation of the RAWFIE platform and the available devices. In the most interesting case, the EDL provides elements for handling resource requirements/configuration, location/topology information, task description, testbed-specific commands etc. Its syntax is simple and combines some common characteristics of well-known XML or legacy programming languages. The EDL is built with the help of the Xtext framework[3]. The following listing presents a small part of the proposed EDL grammar.

```
Experiment:
        'Experiment'
                metadata=MetadataSection
                (requirements=RequirementsSection)?
                (declarations=DeclarationsSection)?
                execution=ExecutionSection
        '~ Experiment'
;
/********** Metadata Section **********/
MetadataSection:
        'Metadata'
                met+=Metadata
        '~Metadata';
Metadata:
        'Name' name = ID
        (experimentVersion=Version)?
        (experimentDescritpion=Description)?
        (experimentDate=Date)?;
Version:
        'Version' ver=VER;
Description:
        'Description' name=ID;
Date:
        'Date' dat=DAT;
```

[3]https://eclipse.org/Xtext/

An experiment as realized through the EDL terminology is seen to have the following parts:

- ***Metadata section***. It contains generic information related to each experiment like the name, the date, etc. This information is important to define the necessary description for each experiment and, thus, to facilitate the efficient management of the available experiments.
- ***Requirements section***. It contains information related to the requirements of each experiment in terms of the testbed data, the location, the duration or the distance that the nodes should cover during the experiment execution. In addition, in this section, the experimenter should define the number of nodes that will be involved in the experiment and, thus, the RAWFIE platform is capable of knowing the needs for the experiments under consideration.
- ***Declarations section***. It concerns the necessary declarations like constants and variables declaration adopted to store data during the experiment execution. The discussed declarations are the key element to connect the experiment business logic with the data retrieved by UxVs and perform processing in a higher level than the device itself.
- ***Execution section***. It involves commands related to the management of the core business logic of each experiment. The EDL offers statements for the nodes or group of nodes management. Every aspect of nodes/groups behavior can be realized with specific terminology in the execution section. In addition, a number of statements are devoted to: (i) waypoints management; (ii) time line management (e.g., sequential or parallel execution); (iii) coordination management; (iv) control management (e.g., activation/deactivation of sensors); (v) configuration management (e.g., data management in each node); (vi) communication management (e.g., change in network interfaces).

It should be noted that 'typical' commands originated in legacy programming languages are also included in the EDL. Hence, assignments, conditionals statements (i.e., if, switch) and iterations (i.e., for, while) are also in place. In the following listing, we present a small part of an EDL script related to the definition of the behavior of a node. The 'Route' command instructs the node to follow a set of waypoints defined by multiple WP commands. Each waypoint is identified by three numbers: *time*, *x*, *y* and *z* coordinates. For instance, the command WP$<$3, 50, 22, 15$>$ instructs the node, at time 3, to be at the location (50, 22), at height/depth 15.

```
Node
        ID node1
        Route[
                WP<1, 10, 12, 12>
                WP<3, 50, 22, 15>
                WP<15, 84, 42, 15>
                WP<18, 36, 22, 15>
               ]
        DataManagement
                Time 14 Algorithm average(history = 10)
        ~DataManagement
        NodeCommunication
                NIC WiFi
        ~NodeCommunication
        DataManagement
                Time 25 Algorithm average(history = 5)
        ~DataManagement
~Node
```

15.4.3 The EDL Textual Editor

On top of the EDL terminology, RAWFIE provides two editors: the textual and the visual editors. Both editors are provided as a Web application in a common interface separated in two parts. Editors are responsible to provide the necessary functionalities to the experimenters towards the creation, update, compilation and validation of their experiments. Editors are a collection of tools for defining experiments and authoring EDL scripts through the RAWFIE Web portal. Rich editing facilities are supported in the textual editor together with an advanced content assist and checking mechanism at syntax time. The EDL keywords are highlighted with different color while the code folding (only in the standalone version of the textual editor) functionality enables blocks of code to be hidden or expanded at will. Some of the provided functionalities of the textual editor are: (i) syntax coloring; (ii) content assist; (iii) validation and quick fixes; (iv) code completion; (v) error checking. A set of additional tools for syntactic and semantic validation are also available. The textual editor gives 'access' to the EDL concepts through which an experiment will be defined. Editors are synchronized and experimenters have the opportunity to define nodes routes and other related information directly on the map of the area under consideration (in the visual editor)

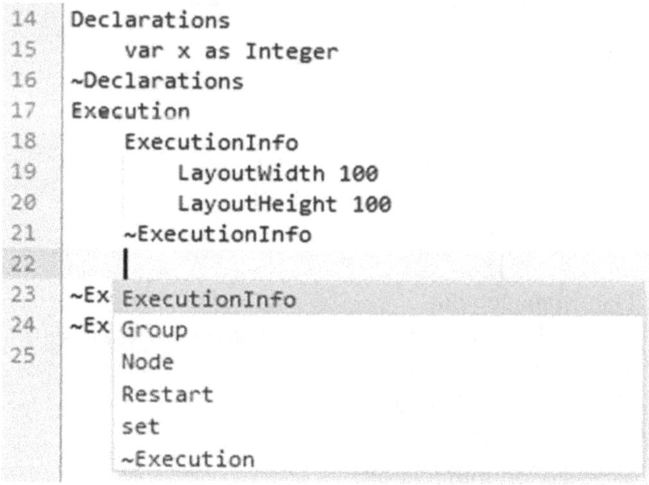

```
14 Declarations
15     var x as Integer
16 ~Declarations
17 Execution
18     ExecutionInfo
19         LayoutWidth 100
20         LayoutHeight 100
21     ~ExecutionInfo
22     |
23 ~Ex ExecutionInfo
24 ~Ex Group
25     Node
       Restart
       set
       ~Execution
```

Figure 15.1 The content assist functionality of the EDL textual editor.

and the list of waypoints is immediately transferred to the textual editor. In Figure 15.1, we see a snapshot of the provided textual editor where the content assist functionality gives us hints about the upcoming commands that should be inserted in an experiment.

15.4.4 The EDL Visual Editor

The visual editor is an innovative and powerful tool for creating experiments in the RAWFIE authoring tool. The main goal of the visual editor is to provide a user friendly environment that simplifies the creation of an experiment by adopting 'typical' GUI functionalities (e.g., mouse actions). Experimenters have the opportunity to define basic UVs actions (e.g., waypoint definition) directly on the map. A set of tools, in the form of buttons, are available to the experimenters. Each button has a specific orientation i.e., nodes management (e.g., addition, deletion), nodes behavior definition (e.g., activation of sensors, define data management algorithms) while with the use of the mouse, experimenters can define the route of each UV in the area. Every node is depicted in the map with a different color to avoid confusion in the cases where an experiment involves multiple nodes. In addition, the visual editor gives the opportunity to experimenters to define the time when an action/movement should take place maintaining the spatio-temporal aspect of the experiment. It should be noted that both editors are synchronized while the error messages and warnings are presented in the textual editor area.

15.4.5 The Validator and the Generator

The EDL validator is responsible for performing syntactic and semantic analysis on the provided EDL scripts. The validation is performed on top of the proposed EDL model that is based on the EDL grammar. The validator accesses the provided script and identifies any semantic errors that could jeopardize the execution of an experiment. Specific constraints should be fulfilled when the experiment workflow is defined. These constraints are continuously checked by the proposed editors and in case some of them are validated to be false, errors will be presented to the experimenters through various means (e.g., with red color). The main responsibilities of the validator are: (i) it provides syntactic and semantic validation of each experiment workflow; (ii) it applies a set of constraints that should be met in order to have a valid experiment; (iii) it is capable of applying semantic checking for nodes communication, spatio-temporal management, sensing and data management.

RAWFIE also offers a code generation component. When no errors are present, the component has the opportunity to generate specific files e.g., part of the final code to be uploaded in the UVs. The code generation component takes as input the experiment workflow in terms of EDL commands and transforms them in the appropriate target language. This component conveys design and implementation issues that need to be handled in such a way that will help experimenters to avoid errors and development problems. The module receives commands from the available editors, data from the underlying model (the terminology of the EDL as depicted by the Ecore model) to create the experiment code/files.

15.5 Technical Details

15.5.1 The EDL Grammar

The EDL and the provided editors are built by adopting the Xtext framework[4]. The Xtext is a framework for the development of DSLs. It offers functionalities that let engineers to define their language using a powerful grammar. The grammar is the most important part of the Xtext framework and, actually, it is DSL by itself. The grammar aims to provide functionalities for describing the concrete syntax of a DSL (e.g., the EDL) and how it is mapped to an in-memory representation. The in-memory representation of the EDL is the semantic model. The semantic model is produced during the experiment definition by

[4]https://eclipse.org/Xtext/

the parser. The definition of the EDL with the help of the Xtext involves the automatic creation of the corresponding *Ecore model* (i.e., a meta model of the EDL) that describes the structure of the EDL's *abstract syntax tree* (AST). The Xtext infers the Ecore model from the EDL grammar and adopts Ecore's EPackages to define the Ecore model. Ecore models are declared to be either inferred from the grammar or imported. By using specific directives, engineers instruct the Xtext to infer an EPackage from the grammar.

After the generation of the EDL meta-model (i.e., the Ecore model), we also get a set of tools and functionalities like the *parser*, the *linker*, the *type checker*, the *compiler* as well as editing support for Eclipse, IntelliJ IDEA and Web. The parser creates an in-memory object graph while experimenters define the script of each experiment. The object-graph is an instance of the EDL Ecore model. The parser is fed with a sequence of terminals and walks through the so-called *parser rules*. A parser rule produces a tree of non-terminal and terminal tokens i.e., the *parse tree*. Parser rules provide a building plan for the creation of EObjects that form the EDL semantic model (i.e., the AST). It should be noted that the EDL terminal rules are described using *Extended Backus-Naur Form*-like (EBNF) expressions.

15.5.2 The EDL Validator and Generator

The Xtext framework offers a set of automatic validation functionalities. Validation is very important to identify when the defined experiments are in 'agreement' with the EDL grammar. The first step of validation is performed by the available parser. The parser checks the syntactical correctness of any experiment while presenting error and warning messages. Such messages are automatically implied through the provided Xtext functionalities and show if an experiment complies with the terminology of the EDL grammar. In addition, the linker checks for broken cross-references between EDL concepts. The provided editors automatically validate all cross-links by navigating through the EDL model so that all the installed *Eclipse Modeling Framework* (EMF) proxies get resolved.

Apart from the automatic validation tools, RAWFIE EDL offers a set of custom tools adopted for validation purposes. The custom validator is written in the Xtend language[5] and adopts pure Java classes. The Xtend is a statically-typed programming language which translates to comprehensible Java source code. Syntactically and semantically, the Xtend has its roots in

[5]http://www.eclipse.org/xtend/

the Java programming language but is improved on many aspects. It offers extension methods for enhancing closed types with new functionalities while type inference and full support of generics offer compatibility with Java. Other advantages of the Xtend language are the operator overloading, powerful switch expressions, polymorphic method invocation, template expressions. The Xtend is very expressive, readable and any Xtend method can be invoked by Java classes in a transparent way.

The custom validator aims to define additional constraints for the defined experiments. In RAWFIE, the custom validator is adopted to define constraints in a semantic level for any experiment. The custom validator returns error or warning messages when violations in the experiment logic are present. The validator has access to the underlying database to get data related to the testbeds and UVs as well as to the experiments. Through this approach, RAWFIE platform can have full control of the defined experiments and forbid any action that cannot be performed by the nodes when the experiment will be realized. It should be noted that the custom validator is extended by adopting a Java class that includes the management of any check/functionality that is difficult to be handled by the Xtend language. In the following listing, we see a part of the validation script.

```
@Check
def checkDuration(RequirementsSection reqs) {
        if (Double.parseDouble(reqs.duration) <= 0)
      error("The experiment duration cannot be accepted!
            Please insert a positive number.", reqs,
            Literals.REQUIREMENTS_SECTION_DURATION, 101);
}
@Check
def checkMinDistance(RequirementsSection reqs) {
        if (Double.parseDouble(reqs.minDistance) <= 0)
      error("The experiment min distance cannot be accepted!
            Please insert a positive number.", reqs,
            Literals.REQUIREMENTS_SECTION_MIN_DISTANCE, 101);
}
@Check
def checkMaxDistance(RequirementsSection reqs) {
        if (Double.parseDouble(reqs.maxDistance) <= 0)
      error("The experiment max distance cannot be accepted!
```

```
                Please insert a positive number.", reqs,
                Literals.REQUIREMENTS_SECTION_MAX_DISTANCE, 101);
}
@Check
def checkAlgorithm(Algorithm users_algo) {
        if(!edlV.checkAlgorithm(users_algo.algName))
        error("Please type another algorithm. The " +
                users_algo.algName + " is not supported. Available
                algorithms: " + edlV.getAlgorithms(), users_algo,
                Literals.ALGORITHM_ALG_NAME, 101);
}
```

The Xtend language is also adopted for the creation of the EDL generator. The generator undertakes the responsibility of defining a set of files and code that will be executed directly by UVs. The generator is consisted of a set of Xtend files and multiple Java classes that depict each command defined by the experimenter into UVs commands. The Xtend can infer the types of variables, methods, closures, and so on and, thus, it can produce the mapping between EDL terminology and the target code.

15.5.3 The EDL Editors

The RAWFIE authoring tool offers two editors: the textual and the visual. Both editors offer their functionalities on top of the server part of the EDL. The server part is adopted to be the basis for building the Web version of the discussed editors. The EDL server is responsible to perform the validation (syntactic and semantic checking) as already described. All the backend Xtext functionalities are invoked with HTTP requests to the server-side component. The server immediately responds to any request and sends to the front end application data in the form of messages. The text content is either loaded from the Xtext server or provided through JavaScript. The Web integration of Xtext supports two operation modes: (i) *stateful mode* – in the stateful mode, an update request is sent to the server whenever the text content of the editor changes. With this approach a copy of the text is kept in the session state of the server, and many Xtext-related services such as AST parsing and validation are cached together with that copy; (ii) *stateless mode* – no update request is necessary when the text content changes, but the full text content is attached to all other service requests.

The client side of both editors is built through the adoption of JavaScript. A set of JavaScript files are responsible to visualize the proposed functionalities, accept experimenters commands and send the appropriate requests to the server-side component. The map presented in the visual editor is created with the adoption of OpenLayers[6]. OpenLayers is a pure JavaScript library for displaying map data in the most modern Web browsers, with no server-side dependencies. Experimenters have the opportunity to define in the graphical interface the routes and characteristics of the UxVs that they should perform during the execution of the experiment and, accordingly, the contents of both editors are synchronized. Hence, experimenters can easily switch from one editor to the other.

15.6 Case Study: Create and Launch an Experiment

In this case study, we show the steps required to define and launch an experiment. We assume that the experiment, initially, involves two (2) USV nodes. Assuming that the experimenter has booked the desired time for the experiment execution, he/she should login into the RAWFIE Web portal where he/she has access to the offered tools (Figure 15.3). There, the experimenter can access the authoring tool and use the provided editors. At the left, he/she can insert commands to the textual editor while at the right he/she can define nodes information in the visual editor.

For each editor, a set of buttons and menus are available. In Figure 15.3, we can see the available toolbars with a short description. Experimenters can use the available tools to insert the templates of specific commands. In Figure 15.4, we present an example where we insert the code templates for any basic part of an experiment.

In any step of the definition of an experiment, experimenter can use the provided content assist functionality to see the upcoming commands according to the EDL terminology (Figure 15.5). In addition, when an error is identified by the parser, the corresponding line of the experiment is marked with a red line and the error message is presented when the experimenter moves the mouse on the specific line (see Figure 15.6).

Nodes routes can be easily defined either in the textual or in the visual editor. As mentioned both editors are synchronized, thus, the experimenter can easily switch for the one to the other. In Figure 15.7, we see the routes for the two nodes under consideration. Experimenters can easily

[6]http://openlayers.org/

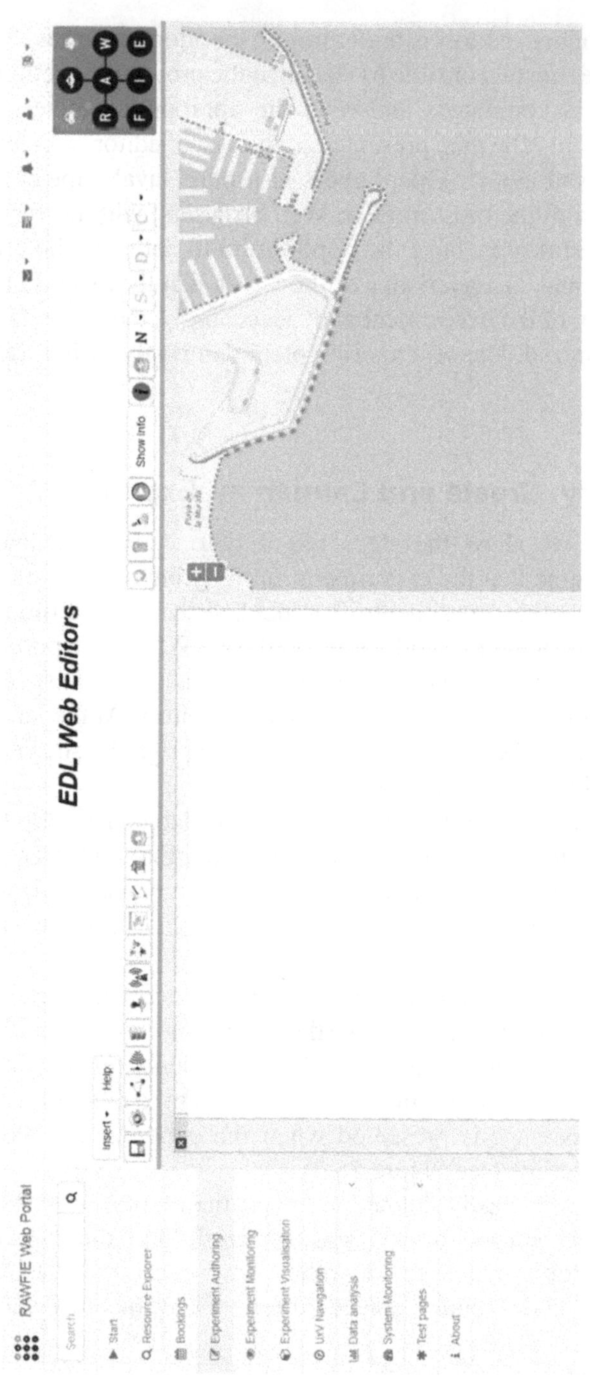

Figure 15.2 The RAWFIE Web portal and the authoring tool.

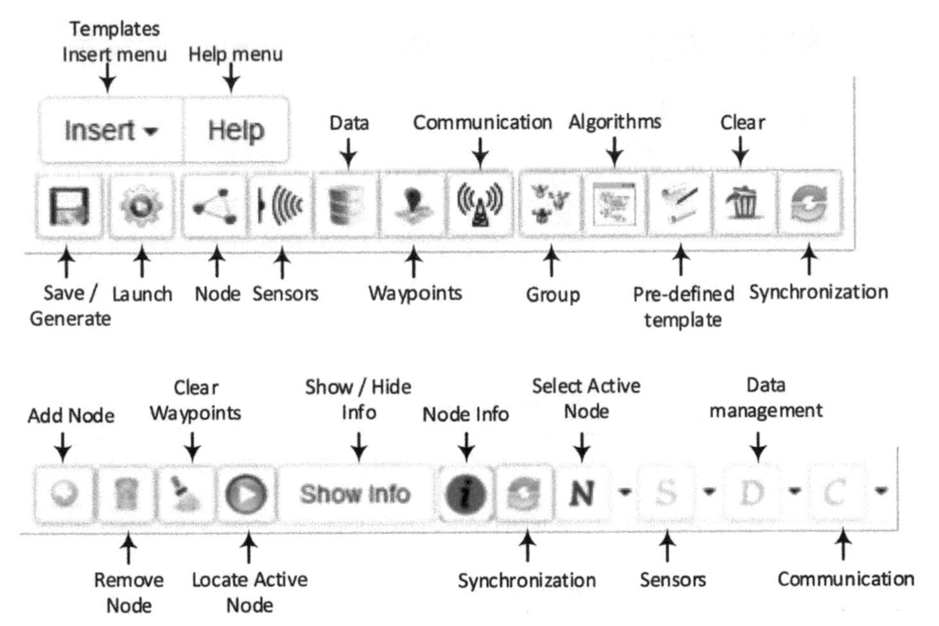

Figure 15.3 The editors' toolbars and buttons (above: the textual editor – below: the visual editor).

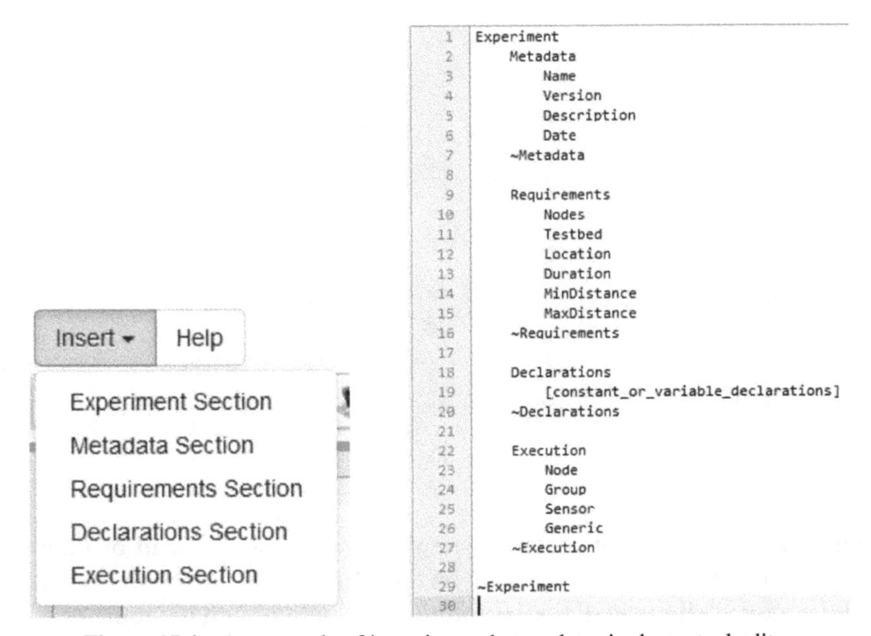

Figure 15.4 An example of inserting code templates in the textual editor.

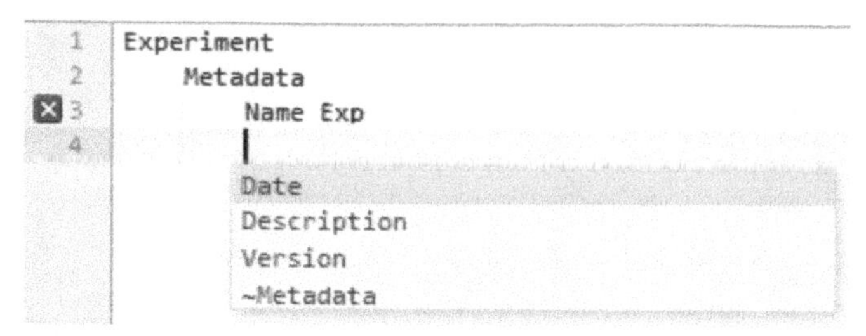

Figure 15.5 The content assist functionality.

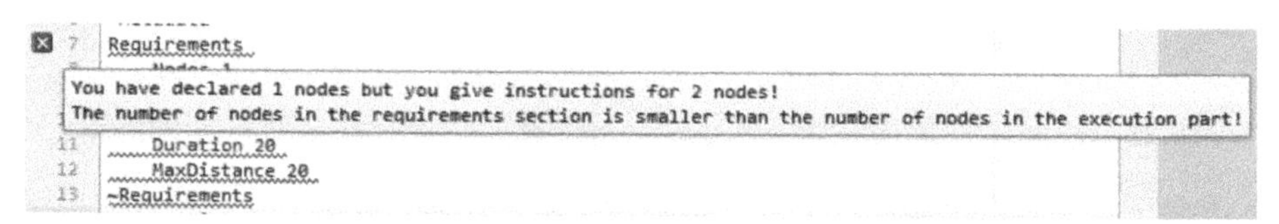

Figure 15.6 Error identification by the parser.

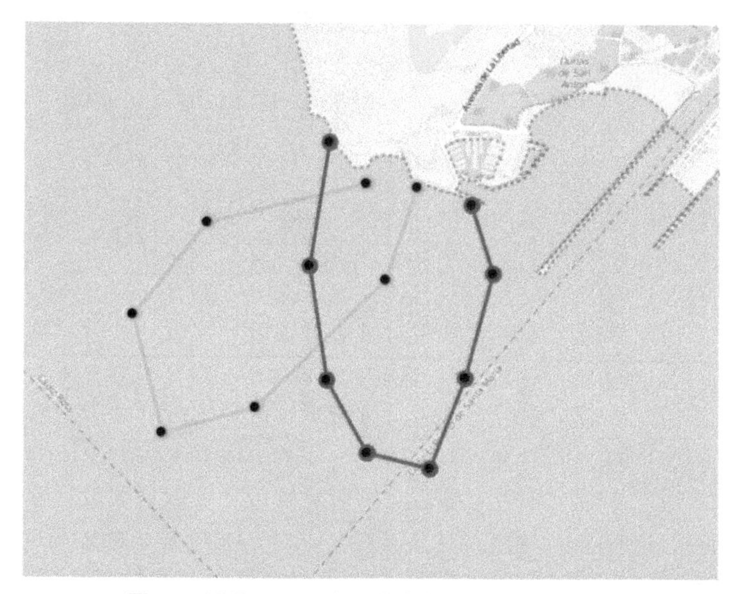

Figure 15.7 Waypoints definition for two USVs.

add more waypoints to each route by simply clinking on the map or they can move/remove waypoints by clicking on the mark (circle) of the waypoint that will be moved/eliminated. For performing any action with the route of a node, we should, first, select the corresponding layer as Figure 15.8 depicts.

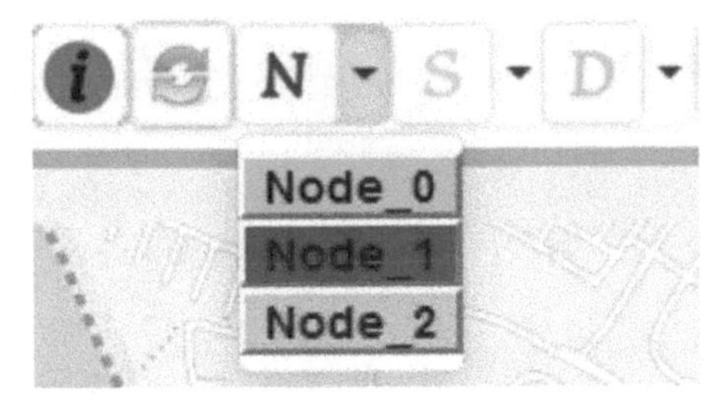

Figure 15.8 Node selection.

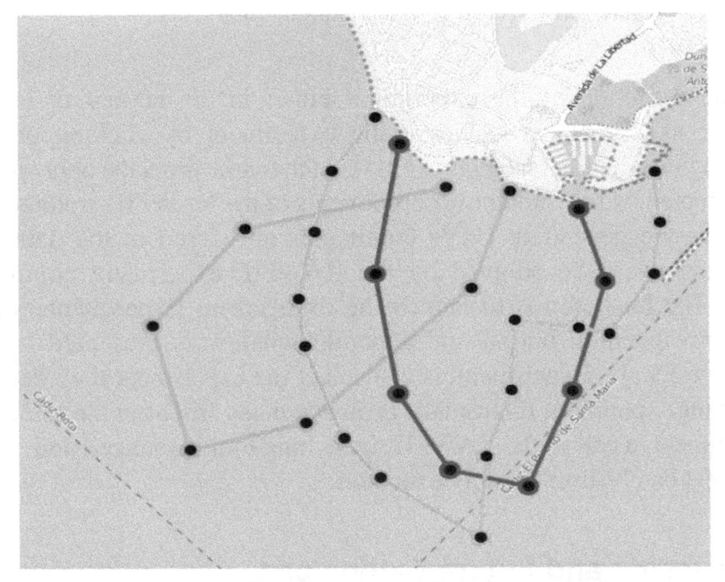

Figure 15.9 The addition of a node in the visual editor.

In Figure 15.9, we present the route of a third node that is added into our experiment.

For each node, we can also define the sensors, the data management algorithms or the communication interface that will be activated in specific time intervals during the execution of the experiment. In Figure 15.10, we present the popup window where the experimenter can manage the adoption of sensors for a USV. The specific example instructs a USV to activate the sonar from t = 4 to t = 10 and from t = 32 to t = 44. The same rationale stands for the invocation of data management algorithms and communication interfaces.

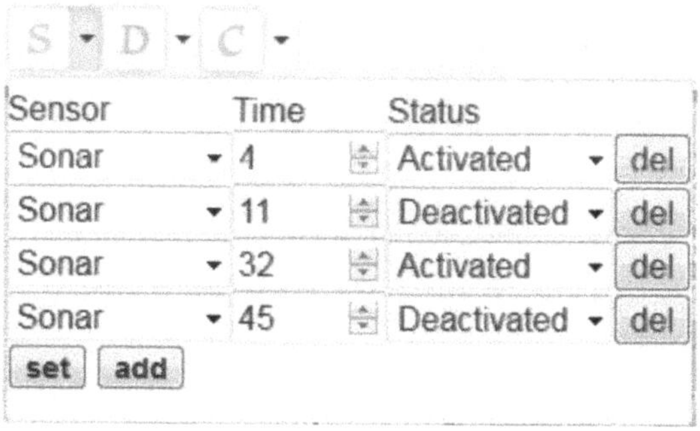

S D C

Sensor	Time	Status	
Sonar	4	Activated	del
Sonar	11	Deactivated	del
Sonar	32	Activated	del
Sonar	45	Deactivated	del

set add

Figure 15.10 A part of the custom validation script.

After the definition of the experiment either in the textual or in the visual editor, experimenters can save the experiment by clicking on the corresponding save button (see Figure 15.3). At the same time, the appropriate files to be adopted by the remaining components of the RAWFIE architecture and the UxVs are generated. UxVs commands are stored in the database and, accordingly, can be adopted by the RAWFIE experiment controller component. The final step is to launch the experiment. Experimenters can press the corresponding button and a popup window is presented in the screen (Figure 15.11). Experimenters can select the experiment they desired by selecting the experiment ID from the drop down list. Just after the selection of the experiment, a call to the RAWFIE launching tool is realized and, thus, the experiment can be immediately executed.

15.7 Discussion and Future Extensions

The RAWFIE EDL offers the necessary conceptual basis for efficiently creating and launching experiments for mobile IoT applications. The provided editors incorporate all the appropriate functionalities to assist experts as well as non-experienced users to define their experiments in a user friendly environment. In the first place of future research/development plans is the incorporation of the error messaging mechanism in the visual editor. Hence, the visual editor will become the appropriate tool for building experiments while the textual part of the RAWFIE EDL will remain as the place where experimenters can insert generic information for their scripts. Such information is related with experiments metadata or requirements. The vision is to have a fully graphical interface and only information that is difficult to be inserted in the visual editor will remain as part of the textual editor. Hence,

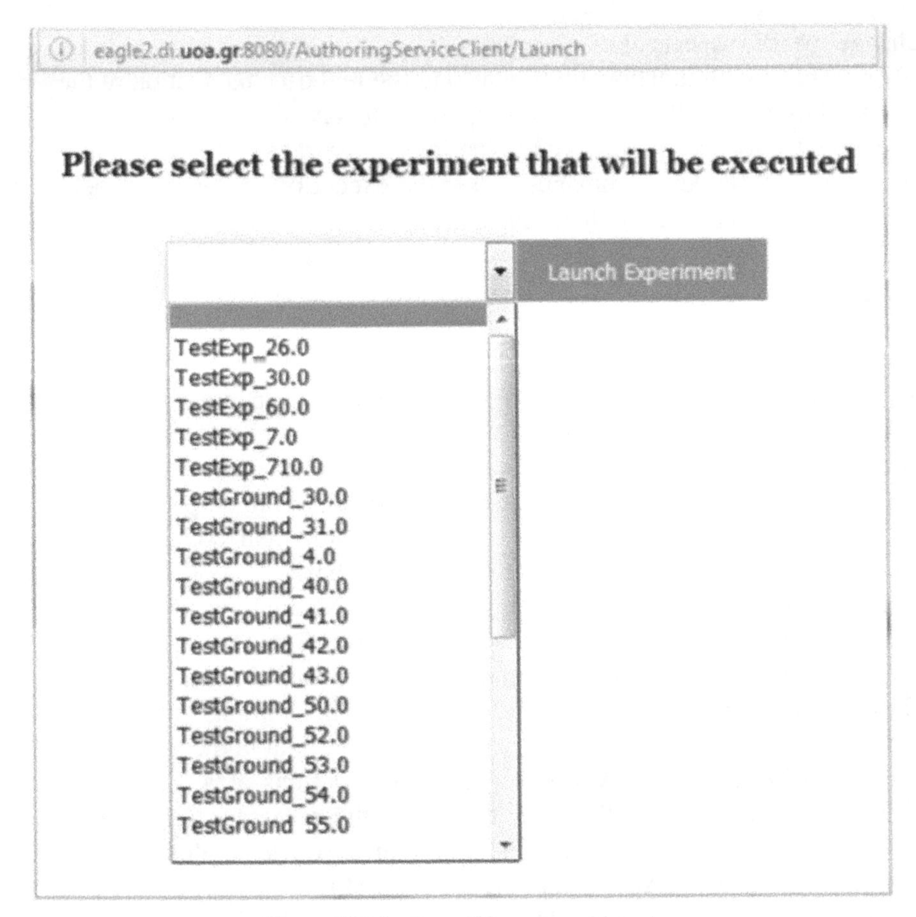

Figure 15.11 Launching an experiment.

errors related to possible collisions, semantic/syntactic violations, etc., will be depicted in the visual part of the provided editors through the adoption of specific icons and colors. Experimenters will be immediately informed about the presence of an error accompanied by suggestions for fixing the error as it already stands for the textual editor.

In addition, another extension is to combine the RAWFIE authoring tool with the experiment monitoring mechanism to get insights on the experiment execution in real time. The aim is to have the system proposing possible modifications in the experiment logic and depict the part of the experiment that is currently executed. This way, experimenters can see in real time the experiment workflow as it is executed by the nodes and decide if it is possible to change specific aspects of the script. For instance, the authoring tool can be easily enhanced with functionalities related to the real time navigation of the UVs and, thus, to be fully aligned with experimenters needs. Specific toolbars can be provided for such purposes and experimenters will have the opportunity to produce/generate new commands during the execution of the experiment and UVs will have to change their routes/actions.

15.8 Conclusions

Mobile IoT applications can play an important role to the development of techniques/tools/services for improving people's lives in the new era of the IoT. This can be done through the adoption of mobile nodes interacting with their environment to collect and process data. Remote experimentation can build on top of such autonomous devices and become the means for experimenting with novel technologies before they are applied into real conditions. In addition, remote experimentation can become the basis for collecting and processing information related to many domains and, thus, to provide the means for creating or improving new applications. The research project RAWFIE offers a platform where numerous devices can be used in remote experimentation activities. Due to the complexity in defining instructions by adopting commands immediately executed by the autonomous devices, the use of a DLS is an easy way to define instructions at a high level. RAWFIE proposes a DSL called EDL that offers the necessary terminology for efficiently defining experiments. A validator and a generator are also proposed to validate the experiments and transform the high level commands into commands immediately executed by the devices. In this chapter, we describe the EDL, the validator, the generator and the available editors. We also elaborate on technical details and provide a case study where we create and launch an experiment from scratch. Our aim is to show the efficiency of the proposed approach while describing future extensions that will improve the offered functionalities and increase the satisfaction level of potential experimenters.

References

[1] Albu, M. M., Holbert, K. E., Heydt, G. T., Grigorescu, S. D., Trusca, V., 'Embedding remote experimentation in power engineering education', in *IEEE Transactions on Power Systems*, Vol. 19, No. 1, pp. 139–143, Feb. 2004.

[2] Ameixieira, C., Cardote, A., Neves, F., Meireles, R., Sargento, S., Coelho, L., Afonso, J., Areias, B., Mota, E., Costa, R. A., Matos, R., Barros, J., 'HarborNet: A Real-World Testbed for Vehicular Networks', CoRR abs/1312.1920, 2013.

[3] Arnold, B. R. T., van Deursen, A., Res, M., 'An algebraic specification of a language for describing financial products', In Martin Wirsing, editor, *ICSE-17 Workshop on Formal Methods Application in Software Engineering*, 1995, pp. 6–13.

[4] Baxter I., Pidgeon C., Mehlich M., 'DMS: Program transformation for practical scalable software evolution', In *Proceedings of the International Conference on Software Engineering (ICSE)*, ACM Press, 2004, pp. 625–634.

[5] Blackwell, A., Britton, C., Cox, A., Green, T.R.G., Gurr, C., Kadoda, G., Kutar, M., Loomes, M., Nehaniv, C., Petre, M., Roast, C., Roe, C., Wong, A., Young, R., 'Cognitive dimensions of notations: Design tools for cognitive technology', In *Cognitive Technology: Instruments of Mind*, Springer-Verlag, 2001, pp. 325–341.

[6] Bonachea, D., Fisher, K., Rogers, A., Smith, F., 'Hancock: a language for processing very large-scale data', *SIGPLAN Notice*, vol. 35(1), 2000, pp. 163–176.

[7] de Lima, J. P. C., Rochadel, W., Silva, A. M., Simão, J. P. S., da Silva J. B., Alves, J. B. M., 'Application of remote experiments in basic education through mobile devices', *2014 IEEE Global Engineering Education Conference (EDUCON)*, Istanbul, 2014, pp. 1093–1096.

[8] Giordano, E., Tomatis, A., Ghosh, A., Pau, G., Gerla, M., 'C-VeT An Open Research Platform for VANETs: Evaluation of Peer to Peer Applications in Vehicular Networks', VTC Fall 2008: 1–2, 2008.

[9] Green, T. R. G., Blandford, A. E., Church, L., Roast, C. R., Clarke, S., 'Cognitive dimensions: achievements, new directions, and open questions', *Journal of Visual Languages & Computing*, vol. 17(4), 2006, pp. 328–365.

[10] Gupta, G., and Pontelli, E., 'Specification, Implementation, and Verification of Domain Specific Languages: A Logic Programming-Based Approach', Computational Logic: Logic Programming and Beyond, Essays in Honour of Robert A. Kowalski, Part I, 2002, pp. 211–239.

[11] Henriques P., Varanda Pereira M.J., Mernik M., Lenic M., Gray J., Wu H., 'Automatic generation of language-based tools using LISA', *IEE Proceedings Software*, vol. 152(2), 2005, pp. 54–69.

[12] Hermans, F., Pinzger, M., van Deursen, A., 'Domain-Specific Languages in Practice: A User Study on the Success Factors', Technical Report, Delft University of Technology, 2009.
[13] Herrera, O., Alves, G., Fuller, D., Aldunate, R., 'Remote Lab Experiments: Opening Possibilities for Distance Learning in Engineering Fields', chapter in Education for the 21st Century — Impact of ICT and Digital Resources, 2006.
[14] Hudak, P., 'Building domain-specific embedded languages', *ACM Computing Surveys*, vol. 28(4), 1996, pp. 196–202.
[15] Hudak, P., 'Modular domain specific languages and tools', In *Proceedings of the 5th International Conference on Software Reuse*, Washington, DC, USA, IEEE Computer Society, 1998.
[16] Hull, B., Bychkovsky, V., Zhang, Y., Chen, K., Goraczko, M., Miu, A., Shih, E., Balakrishnan, H., Madden, S., 'CarTel: a distributed mobile sensor computing system', SenSys: 125–138, 2006.
[17] Kelly, S., Tolvanen, J.-P., *'Domain-Specific Modeling Enabling Full Code Generation'*, John Wiley & Sons, Inc., 2008.
[18] Kieburtz, R. B., McKinney, L., Bell, J. M., Hook, J., Kotov, A., Lewis, J., Oliva, D. P., Sheard, T., Smith, I., Walton, L., 'A software engineering experiment in software component generation', In *International Conference on Software Engineering*, 1996, pp. 542–552.
[19] Kolomvatsos, K., Valkanas, G., Hadjiefthymiades, S., 'Debugging Applications Created by a Domain Specific Language: The IPAC Case', *Elsevier Journal of Systems and Software (JSS)*, vol. 85(4), 2012, pp. 932–943.
[20] Kosar T., Martínez López P. E., Barrientos P. A., Mernik, M., 'A preliminary study on various implementation approaches of domain-specific language', *Information and Software Technology*, vol. 50(5), 2008, pp. 390–405.
[21] Kozil, T., Marek, S., Preparing and managing the remote experiment in education, in *Proc. of the 15th International Conference on Interactive Collaborative Learning (ICL)*, 2012.
[22] Mernik, M., Heering, J., Sloane, A. M., 'When and How to Develop Domain-Specific Languages', *ACM Computing Surveys (CSUR)*, vol. 37(4), 2005.
[23] Meta Object Facility http://www.omg.org/spec/MOF/
[24] Nordmann, A., Hochgeschwender, N., Wrede, S., 'A Survey on Domain-Specific Languages in Robotics', Simulation, Modeling, and Programming for Autonomous Robots: 4th International Conference, SIMPAR 2014, Bergamo, Italy, October 20–23, 2014.

[25] Oliveira, N., Pereira, M. J. V., Henriques, P. R., da Kruz, D., 'Domain-Specific Languages: A Theoretical Survey', Faculdade de Ciências da Universidade de Lisboa, 2009.

[26] Pereira, M. J. V., Mernik, M., da Cruz, D., Henriques, P. R., 'Program comprehension for domain-specific languages', Computer Science an Information Systems Journal, Special Issue on Compilers, Related Technologies and Applications, vol. 5(2), 2008, pp. 1–17.

[27] Query/View/Transformation, http://www.omg.org/spec/QVT/

[28] Raymond James & Associates, 'The Internet of Things – A Study in Hype, Reality, Disruption, and Growth', online at http://www.vidyo.com/wp-content/uploads/The-Internet-of-Things-A-Study-in-Hype-Reality-Disruption-and-Growt....pdf, July 2016.

[29] Rebernak, D., Mernik, M., Wu, H., Gray, J., 'Domain-Specific Aspect Languages for Modularizing Crosscutting Concerns in Grammars', IET Software, vol. 3, Issue 3, 2009, pp. 184–200.

[30] Sadilek, D. A., and Wachsmuth, G., 'Prototyping Visual Interpreters and Debuggers for Domain-Specific Modelling Languages', in *Proc. of the 4th European Conference on Model Driven Architecture: Foundations and Applications*, Berlin, Germany, 2008, pp. 63–78.

[31] SAFESPOT, available at: http://www.safespot-eu.org/

[32] Schmidt D., 'Model-driven engineering', IEEE Computer vol. 39(2), 2006, pp. 25–31.

[33] Spinellis, D., 'Notable design patterns for domain-specific languages', Journal of Systems and Software, vol. 56, 2001, pp. 91–99.

[34] Spinellis, D., Guruprasad, V., 'Lightweight languages as software engineering tools', In *Proceedings of the Conference on Domain-Specific Languages*, 1997, pp. 67–76.

[35] Sprinkle, J., Mernik, M., Tolvanen, J.-P., Spinellis, D., 'What Kinds of Nails Need a Domain-Specific Hammer?', *IEEE Software*, vol. 26(4), 2009, pp. 15–18.

[36] van Deursen, A., Klint, P., 'Little languages: little maintenance', *Journal of Software Maintenance*, vol. 10(2), 1998, pp. 75–92.

[37] van Deursen, A., Klint, P., Visser, J., 'Domain-specific languages: an annotated bibliography', *ACM SIGPLAN Notices*, vol. 35, 2000, pp. 26–36.

[38] Vermesan, O., Friess, P., Guillemin, P., Sundmaeker, H., Eisenhauer, M., Moessner, K., Le Gall, F., Cousin, P., 'Internet of Things Strategic Research and Innovation Agenda', in *Internet of Things – Converging Technologies for Smart Environments and Integrated Ecosystems*, River Publishers, 2013.

[39] Wu, H., Gray, J. and Mernik, M., 'Grammar-driven generation of domain-specific language debuggers', *Software Practice and Experience*, Vol. 38, 2008, pp. 1073–1103.
[40] Wu, H., Gray, J., and Mernik, M., 'Unit Testing for Domain-Specific Languages', IFIP Conference on DSLs, LNCS 5658, 2009, pp. 125–147.

16

Recursive InterNetwork Architecture, Investigating RINA as an Alternative to TCP/IP (IRATI)

Eduard Grasa[1], Leonardo Bergesio[1], Miquel Tarzan[1], Eleni Trouva[2], Bernat Gaston[1], Francesco Salvestrini[3], Vincenzo Maffione[4], Gino Carrozzo[5], Dimitri Staessens[6], Sander Vrijders[6], Didier Colle[6], Adam Chappel[7], John Day[8] and Lou Chitkushev[8]

[1]Fundacio I2CAT, Spain
[2]University of Patras, Greece
[3]University of Firenze, Italy
[4]University of Pisa, Italy
[5]Nextworks, Italy
[6]University of Ghent, Belgium
[7]Interoute, UK
[8]Boston University, USA

16.1 Introduction

Driven by the requirements of the emerging applications and networks, the Internet has become an architectural patchwork of growing complexity which strains to cope with the changes. Moore's law prevented us from recognising that the problem does not hide in the high demands of today's applications but lies in the flaws of the Internet's original design. The Internet needs to move beyond TCP/IP to prosper in the long term, TCP/IP has outlived its usefulness.

The Recursive InterNetwork Architecture (RINA) is a new Internetwork architecture whose fundamental principle is that networking is only inter-process communication (IPC). RINA reconstructs the overall structure of the Internet, forming a model that comprises a single repeating layer, the DIF (Distributed IPC Facility), which is the minimal set of components required to allow distributed IPC between application processes. RINA supports inherently and without the need of extra mechanisms mobility, multi-homing and Quality of Service, provides a secure and configurable environment, motivates for a more competitive marketplace and allows for a seamless adoption.

RINA is the best choice for the next generation networks due to its sound theory, simplicity and the features it enables. IRATI's goal is to achieve further exploration of this new architecture. IRATI will advance the state of the art of RINA towards an architecture reference model and specifcations that are closer to enable implementations deployable in production scenarios. The design and implemention of a RINA prototype on top of Ethernet will permit the experimentation and evaluation of RINA in comparison to TCP/IP. IRATI will use the OFELIA testbed to carry on its experimental activities. Both projects will benefit from the collaboration. IRATI will gain access to a large-scale testbed with a controlled network while OFELIA will get a unique use-case to validate the facility: experimentation of a non-IP based Internet.

16.1.1 RINA Overview

RINA is the result of an effort that tries to work out the general principles in computer networking that applies to everything. RINA is the specific architecture, implementation, testing platform and ultimately deployment of the theory. This theory is informally known as the Inter-Process Communication "IPC model" [1, 2] although it also deals with concepts and results that are generic for any distributed application and not just for networking. RINA is structured around a single type of layer – called Distributed IPC Facility or DIF – that repeats as many times as needed by the network designer (Figure 16.1). In RINA all layers are distributed applications that provide the same service (communication flows between distributed applications) and have the same internal structure. The instantiation of a layer in a computing system is an application process called IPC Process (IPCP). All IPCPs have the same functions, divided into data transfer (delimiting, addressing, sequencing, relaying, multiplexing, lifetime termination, error check, encryption), data transfer control (flow and retransmission control) and layer management (enrollment, routing, flow allocation, namespace management, resource allocation, security management). The functions of an IPCP are programmable via policies, so that each DIF can adapt to its operational environment and to different application requirements.

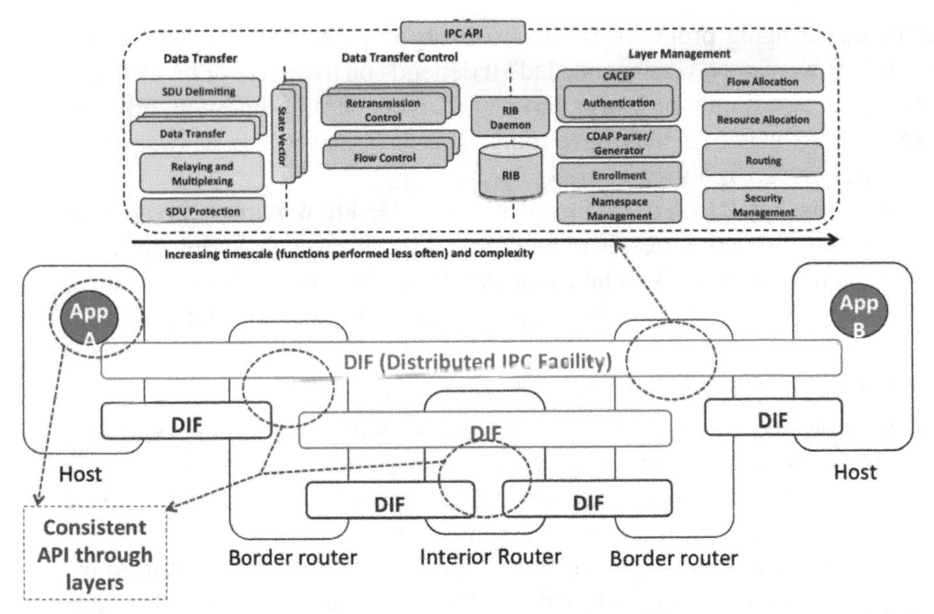

Figure 16.1 Illustration of the RINA structure: DIFs and internal organisation of IPC Processes (IPCPs).

The DIF service definition provides the *abstract description* of an API as seen by an Application Process using a DIF (specific APIs are system-dependant and may take into account local constraints; in some cases there may not be an API at all, but an equivalent way to have equivalent interactions). The Application Process might be an IPC Process, reflecting the recursive nature of RINA (a DIF can be used by any distributed application, including other DIFs). All DIFs provide the same service, called *flows*. A flow is the instantiation of a communication service between two or more application process instances.

In contrast with traditional network architectures in which layers have been defined as units of modularity, in RINA layers (DIFs) are distributed resource allocators [3]. It isn't that layers perform different functions; they all perform the same functions at different scopes. They are doing these functions for the different ranges of the environments the network is targeted at (a single link, a backbone network, an access network, an internet, a VPN, etc.). The scope of each layer is configured to handle a given range of bandwidth, QoS, and scale: a classic case of divide and conquer. Layers manage resources over a given range. The policies of each layer will be selected to optimize that range, bringing programmability to every relevant function within the layer [4]. How many layers are needed? It depends on the range of bandwidth, QoS, and scale: simple networks have two layers, simple internetworks, 3; more complex networks may have more. *This is a network design question, not an architecture question.*

One of the key RINA design principles has been to maximize invariance and minimize discontinuities. In other words, extract as much commonality as possible without creating special cases. Applying the concept from operating systems of separating mechanism and policy, first to the data transfer protocols and then to the layer management machinery (usually referred to the control plane), it turns out that only two protocols are required within a layer [1]:

- A single data transport protocol that supports multiple policies and that allows for different concrete syntaxes (length of fields in the protocol PDUs). This protocol is called EFCP – the *Error and Flow Control Protocol* – and is further explained in Section 4.2.
- A common application protocol that operates on remote objects used by all the layer management functions. This protocol is called CDAP – the Common Distributed Application Protocol.

Separation of mechanism and policy also provided new insights about the structure of those functions within the layer, depicted in Figure 16.1. The primary components of an IPC Process are shown in Figure 16.1 and can be divided into three categories: a) Data Transfer, decouple through a state vector from b) Data Transfer Control, decoupled through a Resource Information Base from c) Layer Management. These three loci of processing are characterized by decreasing cycle time and increasing computational complexity (simpler functions execute more often than complex ones).

- *SDU Delimiting*. The integrity of the SDU written to the flow is preserved by the DIF via a delimiting function. Delimiting also adapts the SDU to the maximum PDU size. To do so, delimiting comprises the mechanisms of fragmentation, reassembly, concatenation and separation.
- *EFCP, the Error and Flow Control Protocol*. This protocol is based on Richard Watson's work [5] and separates mechanism and policy. There is one instance of the protocol state for each flow originating or terminating at this IPC Process. The protocol naturally cleaves into Data Transfer (sequencing, lost and duplicate detection, identification of parallel connections), which updates a state vector; and Data Transfer Control, consisting of retransmission control (ack) and flow control.

- *RMT, the Relaying and Multiplexing Task.* It makes forwarding decision on incoming PDUs and multiplexes multiple flows of outgoing PDUs onto one or more (N – 1) flows. There is one RMT per IPC Process.
- *SDU Protection.* It does integrity/error detection, e.g. CRC, encryption, compression, etc. Potentially there can be a different SDU Protection policy for each (N – 1) flow.

The state of the IPC Process is modelled as a set of objects stored in the Resource Information Base (RIB) and accessed via the RIB Daemon. The RIB imposes a schema over the objects modelling the IPCP state, defining what CDAP operations are available on each object and what will be their effects. The RIB Daemon provides all the layer management functions (enrolment, namespace management, flow allocation, resource allocation, security coordination, etc) with the means to interact with the RIBs of peer IPCPs. Coordination within the layer uses the Common Distributed Application Protocol (CDAP).

16.2 IRATI Goals

The overarching goal of ARCFIRE is to contribute to the experimental research and development of RINA, investigating it as an alternative technology to functional layering and TCP/IP. This objective is divided into the following four goals:

1. **Enhancement of the RINA architecture reference model and specifications, focusing on DIFs over Ethernet**. The enhancement of the RINA specifications carried out within IRATI will be driven by three main forces: i) the specification of a DIF over Ethernet as the underlying physical media; ii) the completion of the specifications that enable RINA to provide a level of service similar to the current Internet (low security, best-effort) and iii) the project use cases targeting ambitious scenarios that are challenging for current TCP/IP networks (targeting features like multi-homing, security or quality of service). The industrial partners in the consortium will be leading the elaboration of the use cases, with the input of the External Advisory Board.

2. **RINA open source prototype over Ethernet for a UNIX-like OS**. This is the goal that can better contribute to IRATI's impact and the dissemination

of RINA. Besides being the main experimentation vehicle of the project, the prototype will provide a solid baseline for further RINA work after the project. By the end of the project the IRATI partners plan to setup an open source community in order to attract external interest and involve other organizations in RINA R&D.

3. **Experimental validation of RINA and comparison against TCP/IP**. This objective is enabled due to the availability of the FIRE facilities, which provide the experimentation environment for a meaningful comparison between RINA and TCP/IP. IRATI will follow iterative cycles of research, design, implementation and experimentation, with the experimental results retrofitting the research of the next phase. Experiments will collect and analyse data to compare RINA and TCP/IP in various aspects like: application API, programmability, cost of supporting multi-homing, simplicity, vulnerability against attacks, hardware resource utilization (proportional to energy consumption). The industrial partners in the consortium will be leading the choice of benchmarking parameters, with the input of the External Advisory Board.

4. **Provide feedback to OFELIA in regards to the prototyping of a clean slate architecture**. Apart from the feedback to the OFELIA [6] facility in terms of bug reports and suggestions of improvements, IRATI will contribute an OpenFlow controller capable of dynamically setting up Ethernet topologies to the project. IRATI will be using this controller in order to setup different topologies for the various experiments conducted during the project. Moreover, experimentation with a non-IP based solution is an interesting use case for the OFELIA facility, since IRATI will be the first to conduct these type of experiments in the OFELIA testbed.

16.3 Approach

The technical work of the IRATI Project comprises requirements analysis, design, implementation, validation, deployment and experimentation activities organized in three iterations. Such activities have been broken down in three technical work packages.

WP2 is the overarching work package that will define the scope of the use cases to be validated, propose a set of refinements and enhancements to the RINA architecture reference model and specifications, and elaborate a high level software architecture for the implementation. For each phase of the project, WP2 will:

- Elaborate the use cases to be showcased during the experimentation phases, analyze them and extract requirements. Use cases will try to focus at first on the availability/integration of core RINA functionalities in basic experimental setups; then, more complex scenarios that are challenging with the current Internet will be targeted to explore the full RINA functionalities and thus meet the expectations/take-up strategies of network operators and cloud service providers (like Interoute). The use cases will drive the experiment design and provide requirements for the completion/validation of the RINA architecture reference model and specifications.
- Analyse the RINA architecture reference model and specifications, identify holes in the mechanisms or missing policies, and propose enhancements/refinements.
- Based on the RINA architecture reference model and specifications on one side and the phase scenario and targeted platform on the other side, provide a high-level software architecture for the design and implementation of the prototype. This high-level software architecture will be the unifying document for the WP3 implementation tasks.

WP3 is the development work package of the project. Its overall objective is to translate the WP2 specifications and high-level software design into a set of prototypes that will be used by WP4 for its test-bed activities and experimentation. The main objectives of this WP are:

- to provide a common development environment
- to implement a RINA prototype over Ethernet for Linux/OS
- to integrate the various functionalities and components into a demon strable system (at node-level)

The architecture releases at the various project phases and the related functional decompositions delivered by WP2 are the starting point of work for WP3. Software prototypes are the major WP3 outcomes to be delivered to WP4. Moreover, it is expected that WP3 will produce a number of feedbacks on previous or concurrent activities, both internally (i.e. among tasks) and externally (i.e. towards other WPs). The feedbacks produced by WP3 to either internal or external tasks will have eventually an impact on the work produced by the target task, i.e. its deliverable. As a general rule, it is expected that major feedbacks on a task could lead to fix and reissue the deliverable(s) produced by that task previously.

WP4 is the experimentation and validation work package, responsible of the following goals:

- Design the experiments required to validate the use cases and deploy WP3 prototypes into the OFELIA facility for experimentation.
- Validate the correctness of the prototype with respect to its compliance with the use cases through experimentation.
- Compare and document RINA benefits against TCP/IP in different areas: application interface, multi-homing, support of heterogeneous applications, security and others identified by WP2.
- Based on the experiments result analysis, provide feedback to the RINA specifications enhancement and high-level software architecture design activities in WP2.

16.4 Discussion of Technical Work and Achievements

16.4.1 Enhancements of the RINA Specifications and Reference Model

16.4.1.1 Shim DIF over 802.1Q layers

This specification defines the shim IPC process for the Ethernet (IEEE 802.3) layer using IEEE 802.1Q [7]. Other Shim DIFs specifications will cover the use of Ethernet with other constraints. This type of IPC process is not fully functional. It presents an Ethernet layer as if it was a regular DIF. The task of a shim DIF is to put as small as possible a veneer over a legacy protocol to allow a RINA DIF to use it unchanged. In other words, because the DIF assumes it has a RINA API below, the Shim DIF allows a DIF to operate over Ethernet without change. The shim IPC process wraps the Ethernet layer with the IPC process interface. The goal is not to make legacy protocols provide full support for RINA and so the shim DIF should provide no more service or capability than the legacy protocol provides.

An Ethernet shim DIF spans a single Ethernet "segment". This means relaying is done only on 1-DIF addresses. Each shim DIF is identified by a VLAN (IEEE 802.1Q) id, which is in fact the shim DIF name. Each VLAN is a separate Ethernet Shim DIF. All the traffic in the VLAN is assumed to be shim DIF traffic.

Ethernet comes with the following limitations, which are reflected by the capabilities provided by the Ethernet shim DIF:

- It assumes that there is a single "network layer" protocol machine instance that is the only user of the Ethernet protocol machine at each system. The Ethertype field on the Ethernet header just identifies the syntax of the "network layer" protocol.
- Because it is only possible to distinguish one flow between each pair of MAC addresses, there is no explicit flow allocation.
- There are no guarantees on reliability.

These limitations impact the usability of the Ethernet shim DIF: it only provides enough capabilities for another DIF to be the single user of the Ethernet shim DIF. Therefore, the only applications that can register in an Ethernet shim DIF are IPC Processes. Moreover, since Ethernet doesn't provide the means to distinguish different flows nor explicit flow allocation, there can only be one instance of an IPC Process registered at each Ethernet shim IPC Process. The following figure illustrates an Ethernet frame as used by the Ethernet shim DIF.

Preamble (7)	Start of frame delimiter (1)	Destination MAC @ (6)	Source MAC @ (6)	802.1q tag (4)	Ethertype (2)	Payload (42-1500)	Frame check sequence (4)	Interframe gap (12)

- *Destination MAC address*: The MAC address assigned to the Ethernet interface the destination shim IPC Process is bound to.
- *Source MAC address*: The MAC address assigned to the Ethernet interface the source shim IPC Process is bound to.
- *802.1Q tag*: The DIF name.
- *Ethertype*: Although it is not strictly required to have a special Ethertype for the correct operation of the shim DIF (since all the traffic in the VLAN is assumed to be shim DIF traffic), it is handy to define an Ethertype for RINA (if, for no other reasons, to facilitate debugging). Therefore the Ethernet frames used within the shim Ethernet DIF will use the 0xD1F0 value for the Ethertype field.
- *Payload*: Carries the upper DIF SDUs. The maximum length of the SDU must be enforced by the upper DIF, since the Ethernet shim DIF doesn't perform fragmentation and reassembly functions. The only delimiting supported by this Shim DIF is "1 for 1," i.e. it is assumed the entire Payload is a single SDU.

Instead of using the RINA Flow Allocator, the Ethernet shim DIF reuses ARP in request/response mode to perform this function. ARP resolves a network layer address into a link layer address instantiating the state for a flow to this

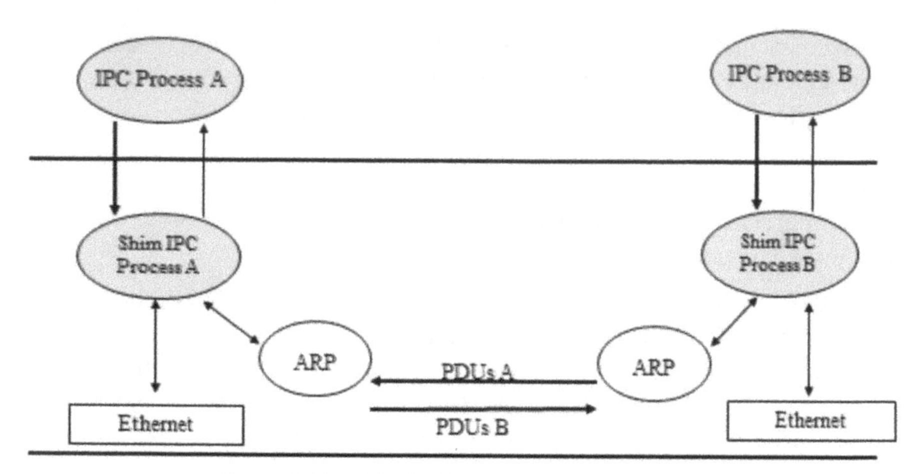

Figure 16.2 Relation between protocol machines.

protocol. In effect, in the context of the shim DIF, mapping the application process name to a shim IPC Process address and instantiating a MAC protocol machine equivalent of DTP, exactly the function that the Flow Allocator provides. The relation between these different protocol machines can be seen in Figure 16.2.

16.4.1.2 Shim DIF for hypervisors

In order to cope with the vast amount of virtualization technologies and hardware (HW) architectures, this specification aims at providing a generic schema in order to ease defining shim IPC Processes for Hypervisors (HV) [8]. Therefore, it does not target any particular HV/HW technology but describes basic mechanisms that shim IPC Process for specific virtualization technologies could inherit. It is expected that other specifications, focusing on specific HV/HW solutions, will be deriving from this one and will be defining the necessary low-level details.

Shim IPC Processes for HV are not fully functional processes, they just present their mechanisms for VM-to-VM intercommunication as if they were a regular DIF. The task of this shim DIF is to put as small as possible layering overhead on the VM-to-VM intercommunication mechanisms, wrapping them with the RINA API.

The scope of the communication mechanisms described is point-to-point and HV local, i.e. between guest and host VMs in the same HV system. Therefore, the host must rely on other shims IPC Processes to allow for inter-communications with other systems than the ones managed by the Hypervisor,

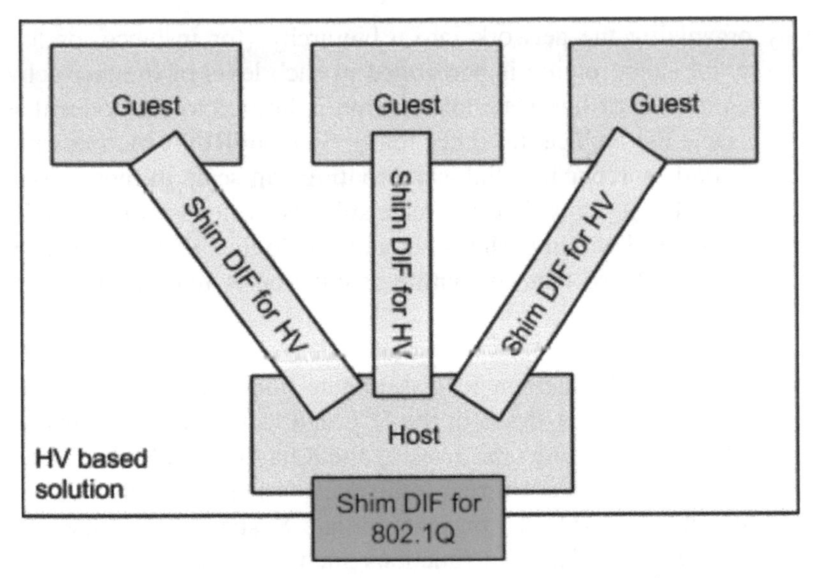

Figure 16.3 Environment of the shim DIF for Hypervisors.

for instance the shim IPC Process over Ethernet 802.1Q as depicted in Figure 16.3.

16.4.1.3 Link state routing policy

This specification describes a link-state routing based approach to generate the PDU forwarding table [9]. The PDU forwarding table is a component of an IPC Process. Routing in RINA is all policy; that is, each DIF is free to decide on the best approach to generate the PDU forwarding table for its environment. Link-state routing is just one of the possible policies. The ultimate goal of routing research in RINA is to create a framework for investigating different routing schemes in a recursive model. Since most DIFs will be of moderate size, starting with link-state routing seems a reasonable approach.

In its simplest form, link-state routing is based on the dissemination of link-state information among all the nodes in a network. Every node constructs the network connectivity graph based on its current view of the state of the links in the network, and applies an algorithm to this graph in order to compute the routes to every other node in the network. The routing table contains entries for the next hop to the shortest route to each other node (destination address) in the network. The forwarding table is generated by recording the mapping between each destination addresses and the corresponding outbound interface towards the next hop. Traditional link-state routing is made more scalable by organizing the network into a hierarchy (for instance, prefix-based), where link-state routing is performed in each level of the hierarchy (and the dissemination of link-state information is limited to the nodes that belong to the same hierarchical level or prefix). Since in RINA routing table sizes can be bound by recursing, link-state routing can scale in most cases. Even with large DIFs it is possible to create subnets of nodes within a DIF where link state is used within each subnet and topological addresses across the subnets of the DIF; so that no routing calculation is required between subnets [10].

The functions of the PDU forwarding table generator component are to collaborate with the RIB daemon to disseminate and collect information about the status of the N – 1 flows in the DIF and to use this information to populate the PDU forwarding table used by the Relaying and Multiplexing Task (RMT) to forward PDUs. Current link-state routing protocols, such as Open Shortest Path First (OSPF) or Intermediate System to Intermediate System (IS–IS), perform additional functions such as automatic neighbour discovery, adjacency forming or link-state failure detection. In an IPC Process these functions are performed by other components such as the Resource Allocator, therefore the PDU Forwarding Table generator doesn't need to do them (Figure 16.4).

The PDU Forwarding Table generator is one of the layer management components of the IPC Process. These components interact, using the RIB daemon, with their counterparts in neighbouring IPC Processes by exchanging Common Distributed Application Protocol (CDAP) messages over N – 1 flows. The CDAP messages perform remote operations (create/delete/start/stop/read/write) on one or more objects of the targeted IPC Process RIB. The PDU forwarding table generator is the handler of the RIB operations

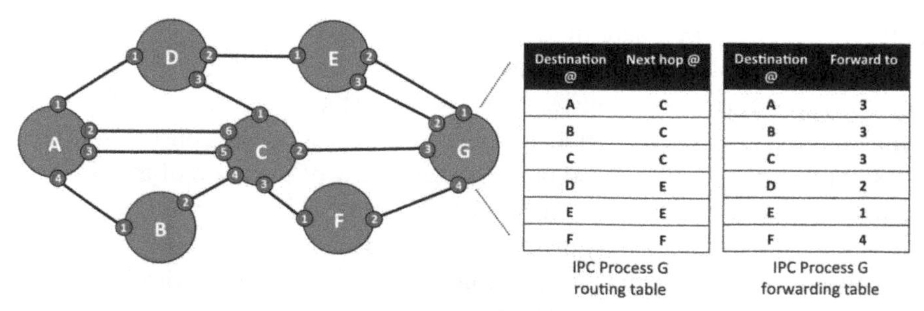

Destination @	Next hop @
A	C
B	C
C	C
D	E
E	E
F	F

Destination @	Forward to
A	3
B	3
C	3
D	2
E	1
F	4

Figure 16.4 A simple example of link-state routing.

targeting the objects related to link-state routing. To summarize, the important details for this specification are that:

- The Enrollment task explicitly notifies the PDU Forwarding Table Gene rator component when enrollment with a new neighbour IPC Process has completed successfully. Then the PDU Forwarding Table Generator can initiate the procedure to synchronize its knowledge about the state of N – 1 flows in the DIF with its new neighbour using the RIB daemon.
- The Resource Allocator component of the IPC Process will explicitly notify the PDU Forwarding table generator of events affecting the local N – 1 flows (local N – 1 flows are flows that have the IPC Process as source or target). Such events include the allocation, deallocation and changes in the status of these flows (N – 1 flow up / N – 1 flow down).
- The PDU Forwarding Table Generator component shares their view of the network using the RIB daemon to communicate with other IPC Processes. The RIB Daemon notifies the PDU Forwarding Table generator when it receives CDAP messages targeting these relevant objects (which are defined later in this document).
- The PDU Forwarding Table Generator propagates changes in the status of local or remote N – 1 flows by requesting the RIB Daemon to send CDAP messages to one or more neighbour IPC Processes. These CDAP messages cause operations on the relevant objects in the RIBs of neighbour IPC Processes.

Figure 16.5 illustrates these details, showing the inputs and outputs of the PDU Forwarding table component, and its relationship with the other components in the IPC Process.

16.4.2 RINA Implementation Activities

16.4.2.1 Implementation goals and major design choices

The IRATI implementation of RINA had to accomplish two main goals:

- **Become a platform for RINA experimentation**. The implementation has to be: i) flexible and adaptable (a RINA "node" can be configured as a host, border router or interior router); ii) the design must be modular so that it can be easily updated as new insights into RINA allow for simpler/better implementations; iii) it must be programmable so that researchers can experiment with different behaviours; iv) must be able to run over multiple lower layers (Ethernet, TCP, UDP, shared memory, USB, etc) and v) must be able to support native RINA applications.

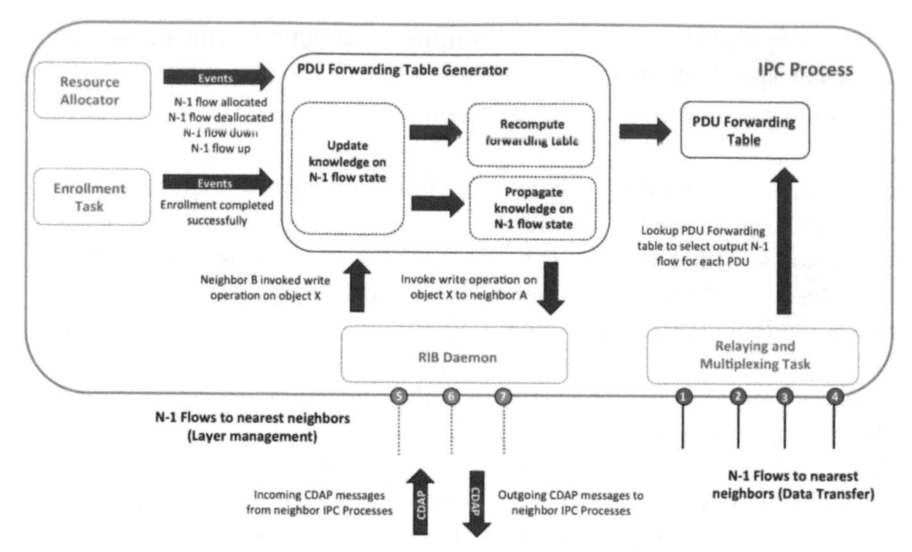

Figure 16.5 Inputs and outputs of the Link-state based PDU Forwarding Table generator policy.

- **Become the basis of future RINA-based products**. In order to achieve this goal, the implementation should: i) tightly integrate with the Opera ting System; ii) allow for performance-related optimizations; iii) enable hardware offload of some functions in the future; iv) seamlessly support existing applications and v) enable RINA to carry IP traffic – RINA as a transport of the IP layer (Figure 16.6).

Decision	Pros	Cons
Linux/OS vs other Operating systems	*Adoption, Community, Stability, Documentation, Support*	*Monolithic kernel (RINA/IPC Model may be better suited to micro-kernels)*
User/kernel split vs user-space only	*IPC as a fundamental OS service, access device drivers, hardware offload, IP over RINA, performance*	*More complex implementation and debugging*
C/C++ vs Java, Python, ...	*Native implementation*	*Portability, Skills to master language (users)*
Multiple user-space daemons vs single one	*Reliability, Isolation between IPCPs and IPC Manager*	*Communication overhead, more complex impl.*
Soft-irqs/tasklets vs. workqueues (kernel)	*Minimize latency and context switches of data going through the "stack"*	*More complex kernel locking and debugging*

Figure 16.6 Major design choices of the IRATI implementation.

With that goals and requirements in mind, the table displayed above shows a summary of the pros and cons behind the major design decisions taken in the IRATI implementation of the RINA architecture: Linux/OS was the chosen operating system; RINA functionalities were spread between the user-space and the kernel; C and C++ were chosen as the programming languages (however, bindings for other languages of the application API can be generated); the implementation contains multiple user-space daemons and the kernel infrastructure exploits the use of soft-irqs and tasklets in order to minimize latency and context switches in the kernel data-path.

Figure 16.7 illustrates the user-kernel split in terms of RINA components implemented in user-space or the kernel. Looking at the different tasks that form an IPC Process, the data transfer and data-transfer control ones – which have stringent performance requirements and execute at every PDU or every few PDUs – were implemented in the kernel. Layer management functions, which do not execute so often and can be much more complex, have been implemented in user-space. Shim IPC Processes have also been implemented in the kernel, since i) they usually need to access device drivers or similar functions only available in kernel space; and ii) the complexity of the functions performed by shim IPC Processes is relatively low.

16.4.2.2 Software architecture overview

The software architecture of the IRATI implementation is show in Figure 16.8. A more in depth description has been published in [11]. The main components of IRATI have been divided into four packages:

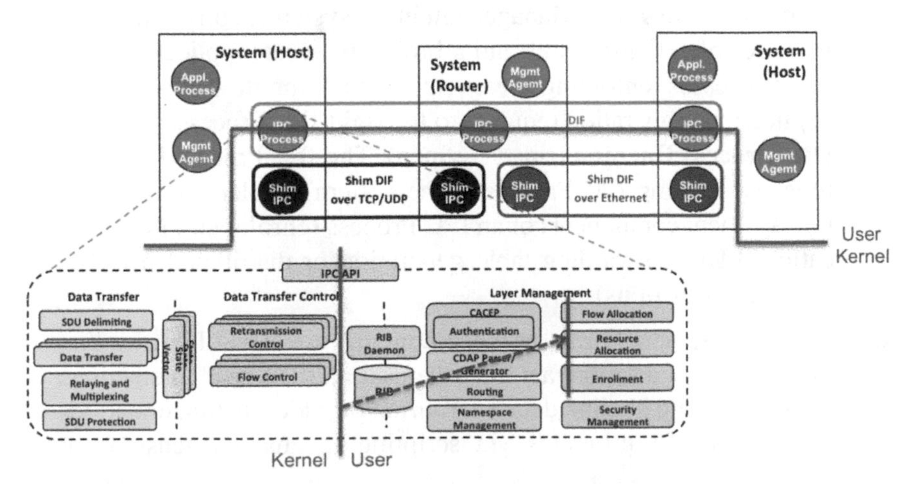

Figure 16.7 IPC Process split between the user-space and kernel.

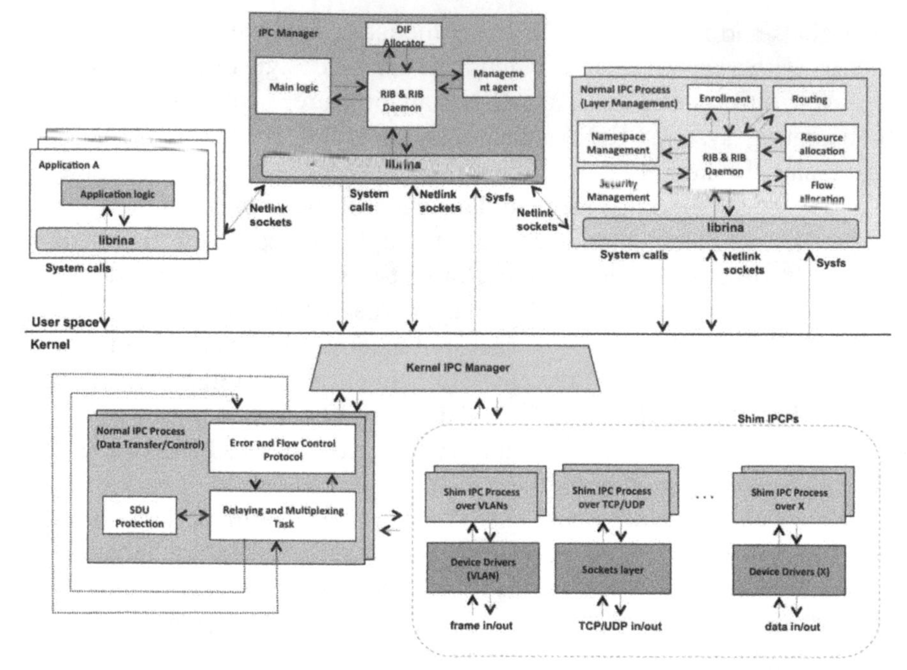

Figure 16.8 Software architecture of the IRATI RINA implementation.

1. **Daemons** (rinad). This package contains two types of daemons (OS Processes that run in the background), implemented in C++.
   - **IPC Manager Daemon** (rinad/src/ipcm). The IPC Manager Daemon is the core of IPC Management in the system, acting both as the manager of IPC Processes and a broker between applications and IPC Processes (enforcing access rights, mapping flow allocation or application registration requests to the right IPC Processes, etc.).
   - **IPC Process Daemon** (rinad/src/ipcp). The IPC Process Daemons (one per running IPC Process in the system) implement the layer management components of an IPC Process (enrollment, flow allocation, PDU Forwarding table generation or distributed resource allocation functions).
2. **Librina** (librina). The librina package contains all IRATI libraries that have been introduced to abstract from the user all the kernel interactions (such as syscalls and Netlink details). Librina provides its functionalities to user-space RINA programs via scripting language extensions or

statically/dynamically linkable libraries (i.e. for C/C++ programs). Librina is more a framework/middleware than a library: it has its own memory model (explicit, no garbage collection), its execution model is event-driven and it uses concurrency mechanics (its own threads) to do part of its work.

3. **Kernel components** (linux/net/rina). The kernel contains the implementation of the data transfer/data transfer control components of normal IPC Processes as well as the implementation of shim DIFs – which usually need to access functionality only available at the kernel. The Kernel IPC Manager (KIPCM) manages the lifetime (creation, destruction, monitoring) of the other component instances in the kernel, as well as its configuration. It also provides coordination at the boundary between the different IPC processes.
4. **Test applications and tools** (rina-tools). This package contains test applications and tools to test and debug the RINA Prototype. Right now the rina-tools package contains the rina-echo-time application, which can work both in "echo" (ping-like behavior between two application instances) or "performance" mode (iperf-like behaviour).

16.4.2.3 Open source

The IRATI software has been made available as open source. Interested users and developers can access the code at the IRATI github side [12], which includes documentation of the project, installation guides and tutorials on how to perform simple experiments. There is also a mailing list available to users and developers, in order to facilitate the interaction with the maintainers of the IRATI implementation.

16.4.3 Experimental evaluation of RINA on the FIRE infrastructure

16.4.3.1 Experimental evaluation of the shim DIF for hypervisors

In order to assess the possible gains from deploying the shim DIF for hypervisors in the DC, we measured the performance of the IRATI stack against the performance of the TCP/IP stack in Linux, when deployed to support VM networking. The full description of the experiment has been published in [13]. Note however up front that the IRATI stack is currently not optimized for performance yet. The tests reported in this section involve a single physical machine (the host) that acts as a hypervisor for one or two VMs. We performed two different test scenarios:

- Host-to-VM tests; where a benchmarking tool (rina-echo-time for IRATI tests and netperf for TCP/IP tests) is used to measure the goodput between a client running in the host and a server running on VM.
- VM-to-VM tests; where a benchmarking tool is used to measure the goodput between a client running on a VM and a server running on a different VM.

The measurements were taken on a processing system with two 8 core Intel E5-2650v2 (2.6 GHz) CPUs and 48 GB RAM. QEMU/KVM was chosen as the hypervisor, since it is one of the two hypervisors supported by the shim DIF for hypervisors provided by the IRATI prototype. For the host-to-VM scenario, three test sessions were executed. The first two tests sessions assess UDP goodput performance at variable packet size, therefore assessing the performance of traditional VM networking. The tap device corresponding to emulated NIC in the VM is bridged to the host stack through a Linux in-kernel software bridge.

The second test session makes use of a paravirtualized NIC model, the virtio-net device. Paravirtualized devices don't correspond to real hardware, instead they are explicitly designed to be used by virtual machines, in order to save the hypervisor from the burden of emulating real hardware. Paravirtualized devices allows for better performances and code reusability. The only difference between the first and the second test session is the model of the emulated NIC. In the first session, a NIC belonging to the Intel e1000 family is used, which is implemented in QEMU by emulating the hardware behaviour (full virtualization) – e.g. NIC PCI registers, DMA, packet rings, offloadings, etc. Despite being more virtualization-friendly than e1000 (or other emulated NICs like r8169 or pcnet2000), the guest OS still sees the virtio-net adapter like a normal ethernet interfaces, with all the complexities and details involved, e.g. MAC, MTU, TSO, checksum offloading, etc.

The third test session shows the performance of the shim DIF for hypervisors. A scenario comparable to the one deployed in the first and second test sessions involves a shim IPC process for hypervisors on the host and the corresponding one on the guest. No normal IPC processes are used, the applications can run directly over the shim DIF. This is a consequence of the flexibility of RINA, since the application can use the lowest level DIF whose scope is sufficient to support the intended communication (guest-to-host in this case) and that provides the required QoS.

The host runs our rina-echo-time application in server mode, while the guest runs rina-echo-time in client mode. Rina-echo-time is a simple RINA benchmarking application, which uses the IPC API to measure goodput. Each test run consists of the client sending to the server an unidirectional stream of PDUs of a specified size. Measurements have been taken varying the PDU size. The current maximum SDU size of the shim DIF for Hypervisors is the page size (4096 bytes on our machine). We repeated every measurement 20 times. The result of these goodput measurements for host-to-VM communication scenario are shown in Figure 16.9 – 95% confidence levels are also depicted, as well as a third degree polynomial regression line.

The shim DIF for hypervisors outperforms both e1000 and virtio-net NIC setups, which validates that a simpler and cleaner architecture allows for better performance, even with an unoptimized prototype. Next, similar goodput performance measurements were taken on the VM-to-VM scenario. Again, three test sessions were performed, the first two for traditional VM networking and the third one for IRATI stack. The setup of the first two sessions is very similar to corresponding one in the host-to-VM scenario. The VMs are given an emulated NIC, whose corresponding tap device is bridged to the host stack through a Linux in-kernel software bridge. Measurements are again performed with the netperf utlity, with the netperf server running on a VM and the netperf client running on the on the other VM.

In the case of the IRATI tests, point-to-point connectivity between host and VM is provided by the shim DIF for hypervisors. A normal DIF is

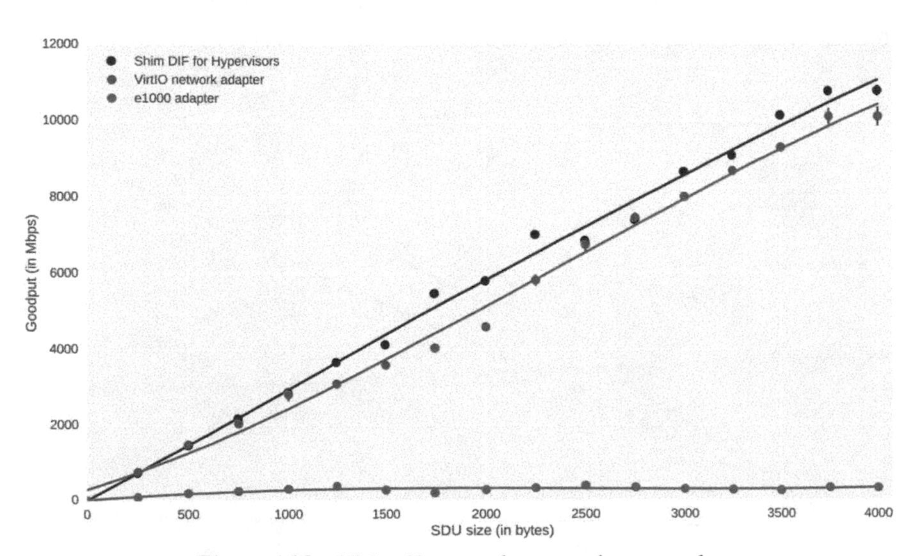

Figure 16.9 VM to Host goodput experiment results.

overlayed on these shim DIFs to provide connectivity between the two VMs. Tests are performed again with the rina-echo-time application, using a flow that provides flow control without retransmission control. Flow control is used so that receiver's resources are not abused. In TCP/IP, this kind of functionality – flow control without retransmission control – is not available. Hence we chose again UDP to perform the tests for the traditional networking solution, since its functionality is most similar. The result of these tests sessions are depicted in Figure 16.10. Full virtualization again performs poorly. The paravirtualized solution currently slightly outperforms the unoptimized IRATI stack. However, the IRATI prototype can still be optimized.

16.4.3.2 Evaluation of the link-state routing policy

The physical connectivity graph that we used for this experiment is shown in Figure 16.11. For each physical link, an instance of the shim DIF for 802.1Q was instantiated, after assigning a unique VLAN id to each link. A normal DIF was stacked on top of these shim DIFs. In this way, we show that routing works in a 1-DIF, a basic scenario. We performed tests with rina echo-time, an application that calculates the Round Trip Time (RTT) like the well-known ping tool. It calculates the time it takes for a client to send an SDU to the server and receive the same SDU back again. The server was running on node M. On every other node we ran the client and performed the test 50 times. The size of the SDU that was used was 64 bytes. We ran the application on top of a normal DIF, which runs on top of the shim DIF over 802.1Q.

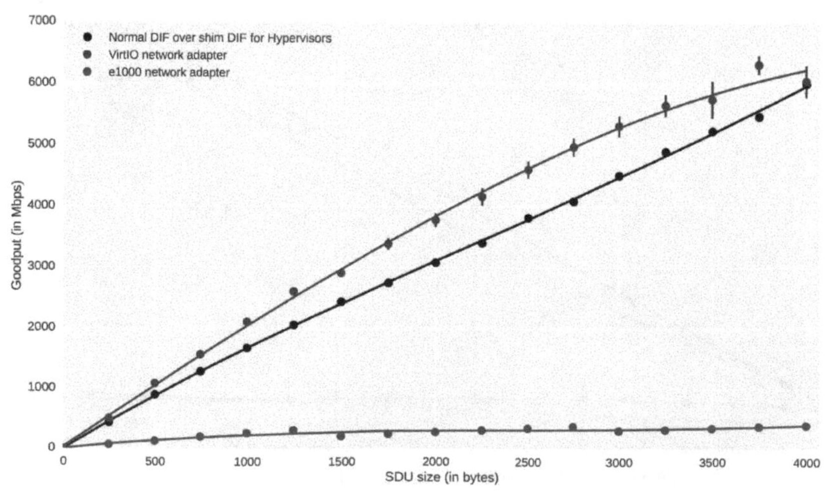

Figure 16.10 Host to Host goodput experiment results.

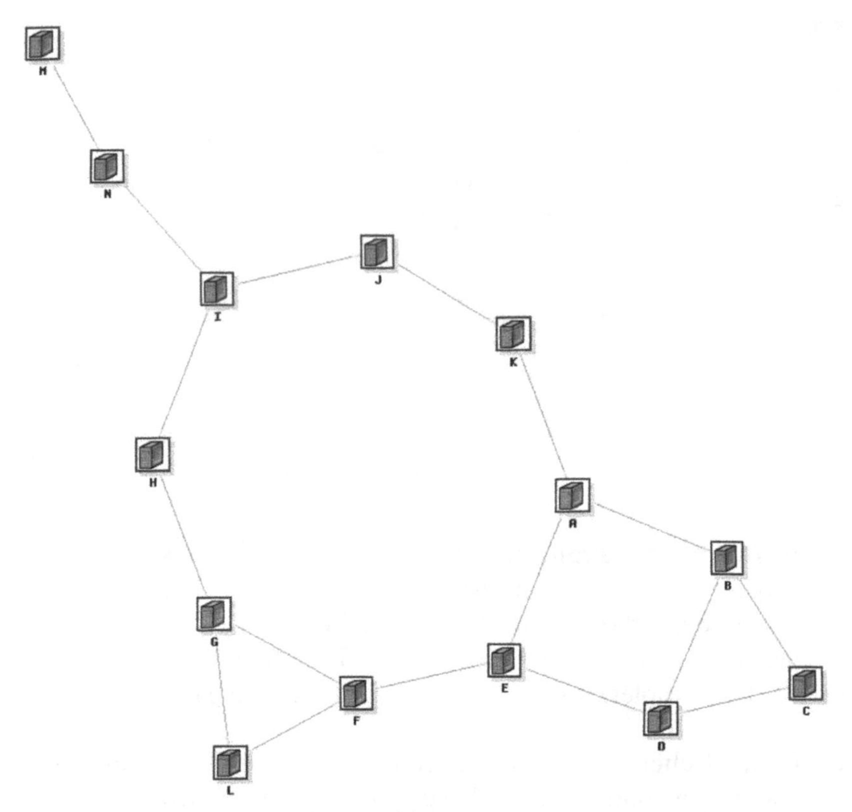

Figure 16.11 Physical connectivity graph for the routing experiments.

We did the experiment first with kernel-space debugging logs enabled, then we repeated the experiment with the debugging logs disabled (as these significantly stress the system) to get more accurate performance-oriented results. The results of the experiments can be seen in Figure 16.12. The data points on top represent the experiment with the kernel with logs enabled, while the data points at the bottom are the results of the experiment with the logs disabled. The further the distance from server M, e.g. the more hops needed to reach M, the longer the round trip time. In the case of a kernel with logs disabled, it takes 504.94 ± 68.10 μs to send and receive the same SDU again on node N, where no forwarding of SDUs was needed. Per extra node needed to forward the SDU, about 250 μs is added. Some nodes have the same distance to M, but their average round trip times differ somewhat from each other. In conclusion, the basic operation of the link-state routing policy has been verified.

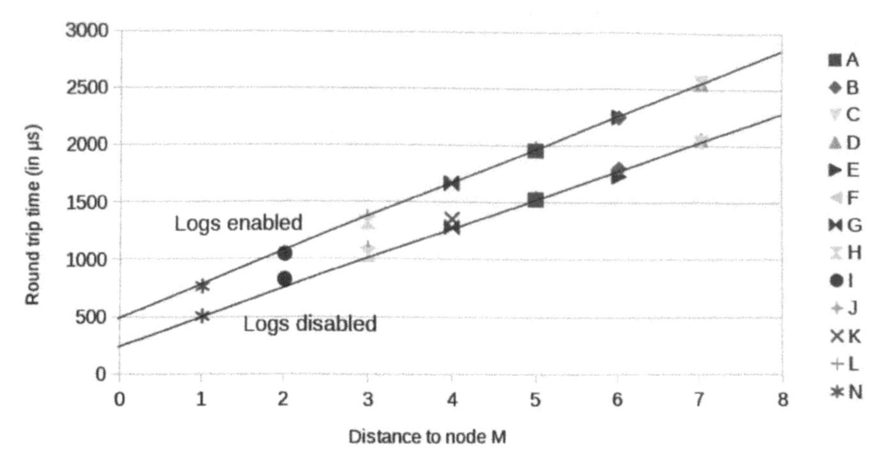

Figure 16.12 Results of the link-state routing experiments.

16.4.3.3 Performance evaluation on the iMinds OFELIA island

We carried out experiments to measure the performance of the phase 2 prototype. We executed them on the iLab.t Virtual wall, which is a controlled environment. The experiment depicted in Figure 16.13 was used to measure the performance. The complete results of these experiments have been published in [14].

The RINAperf client/server application is a RINA-native performance-measurement tool, measuring the available goodput between two application processes. We realised goodput measurements in the following 3 scenarios:

- The first scenario (Shim DIF + Application Process) runs the RINAperf application directly over the shim IPC process for 802.1Q. This scenario should be the fastest, but offers the least functionality.
- The second scenario (Shim DIF + Normal DIF A + Application Process) builds upon the first one by stacking a normal DIF on top of the shim IPC process for 802.1Q. A lot of functionality becomes immediately available since it is provided by the normal DIF (e.g. multihoming, QoS). This is a scenario that would be typically found in Local Area Networks (LANs), where the scope is the network.
- The third scenario (Shim DIF + Normal DIF A + Normal DIF B + Application Process) stacks another normal DIF on top of the previous. This scenario is added to show the influence of stacking multiple DIFs on top of each other and would the one be used in an internetwork: connecting together different networks.

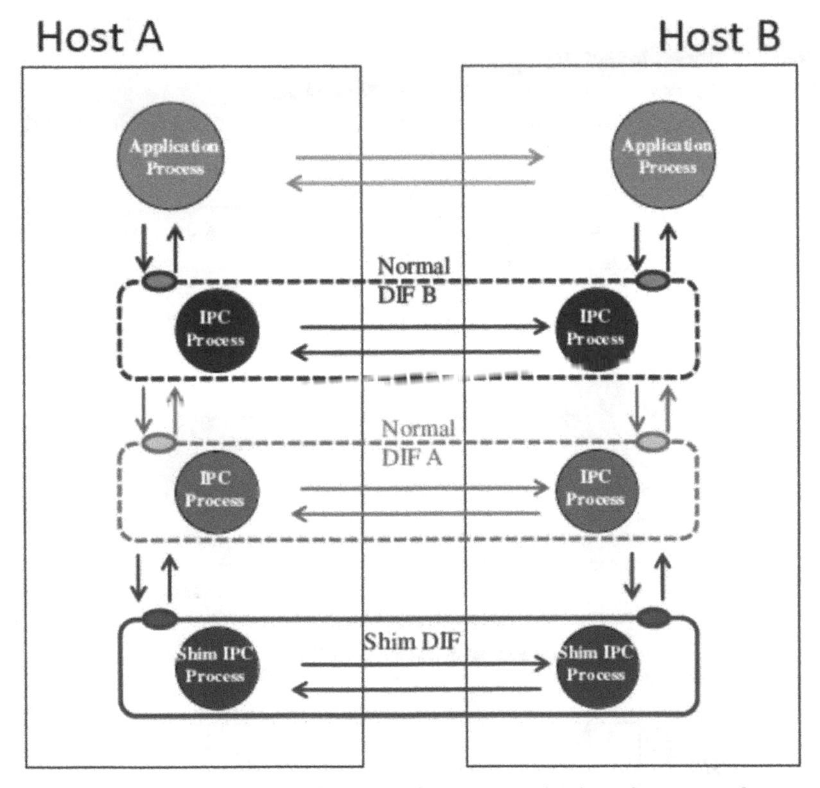

Figure 16.13 Scenario for the performance evaluation of prototype 2.

We executed experiments on 2 nodes from the iMinds OFELIA island iLab.t virtual wall aggregate. We used the RINAperf application to measure the maximum achievable goodput between two application processes given a certain SDU size. In all our experiments, we set the RINAperf timeout to 10 seconds and each one is repeated for different SDU sizes, ranging from 1 byte to the maximum SDU size for that scenario. The obtained goodput when these experiments were repeated in July can be seen in Figure 16.14. The values represent the mean of the goodputs obtained, together with their respective 95 percent confidence intervals (50 samples per interval).

As can be seen from Figure 16.14, the goodput increases as the SDU size increases, which is of course to be expected as the per-packet processing overhead due to the PCI headers is amortised over more bytes. The maximum throughput on the link when measured with iperf is 970 Mbit/s. Adding additional normal DIFs decreases the goodput, because of the extra processing

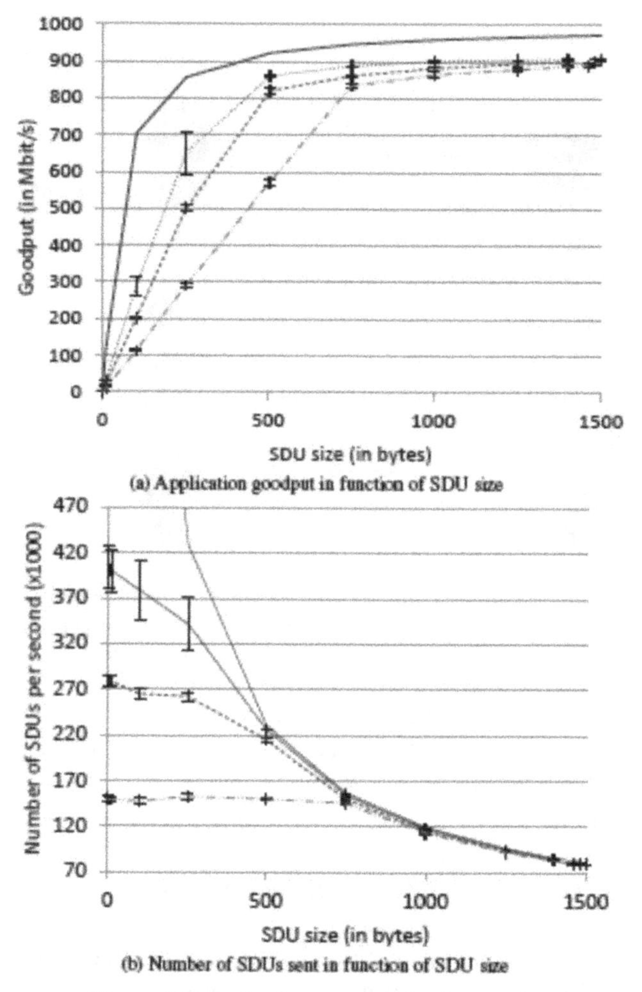

(a) Application goodput in function of SDU size

(b) Number of SDUs sent in function of SDU size

Figure 16.14 Performance evaluation results.

overhead and decreased maximum packet size. Tests executed with v0.8.0 at the maximum MTU allowed by the system recorded the mean goodput achieved as 907:67 ± 1:45 Mbit/s for scenario 1, 902:09 ± 9:32 Mbit/s for scenario 2, and 891:05 ± 6:91 Mbit/s for scenario 3. So each additional normal DIF incurred only a small performance penalty, and the overall goodput achieved was very close to line rate.

16.4.3.4 Validation of location-independence

First, the rina-echo-time server application registers with normal.DIF at System 3 (Figure 16.15). The "rina-echo-time" client application at System 1 allocates a flow to the server application, using the Application Name, sends a couple of SDUs and terminates. After that the rina-echo-time server at System 3 terminates, and an instance of the same application is registered at Sys tem 2. Then, the client application process at System 1 again allocates a flow to the server application process, sends a couple of SDUs and terminates again. Note that the application does not need any state updates. The DIF internally updates its Application-Name-to-IPC-Process address mappings in order to forward the flow allocation request to the right IPC process in both scenarios.

This section validated the location independence of Application Names and decoupling of connections and flows in the RINA architecture and the IRATI implementation. It shows correct operation when an application moves from on host (or, more accurately, from one IPCP process) to another. The directory of the underlying DIF is correctly updated, and the client application can reach the server application transparently without knowing it has moved location. This experiment illustrates one of the huge benefits RINA has with regards to application mobility due to it's location-independent naming scheme.

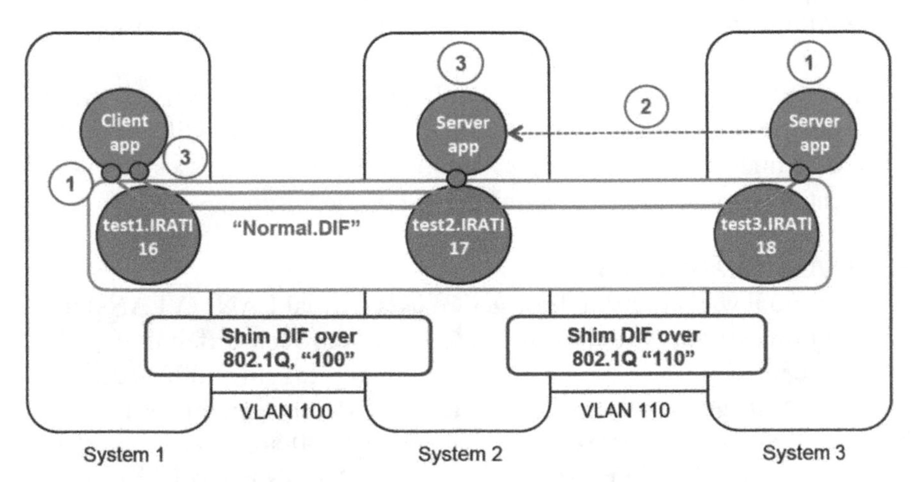

Figure 16.15 Experimental setup for application-location independence.

16.4.4 Feedback to the OFELIA Facility

16.4.4.1 IRATI VM image and XEN servers

Many of the modules of the RINA stack are located in kernel space. Since the Linux kernel versions used by the IRATI software are not always the same available in OFELIA resources, the VMs used for IRATI experimentation need to be upgraded in order to support the deployment of the RINA prototype. The VMs in OFELIA use the same unmodified kernel that the host server uses. This constraint was completely blocking for IRATI purposes, as previously exposed. Nevertheless, OCF (OFELIA Control Framework) is capable of supporting almost any kind of images to create VMs, using any of the described virtualization mechanisms. In order to address IRATI's requirements it was decided to create a new VM image based on IRATI's reference machine. This image has been installed on the servers of the I2CAT Island, and in later phases of the project it will be spread over the rest of the islands for the inter-island experiments. The new image template was updated to support OFELIA's authentication and access control features: authentication modules that allow users whose credentials are stored at the OFELIA LDAP directory direct access to the VM using SSH were added to the VM.

Finally, to fully integrate the new image template in the OFELIA testbed, some additions in the OCF, were required. The development done was integrated in OCF v0.7 and can be visualized in Figure 16.16. The following paragraph describes the main additions performed to the OCF:

- The Expedient web UI was modified to allow the creation of IRATI VMs.
- The Virtualization Aggregate Manager (VT AM) was modified to support the new VM.
- The OFELIA XEN Agent (OXA) running in the XEN servers was modified in order to handle the automatic creation and modification of HVM virtualized machines to provide them with IP addresses and ssh credentials in order to be accessible by the testbed users.

16.4.4.2 VLAN translator box

The iLab.t virtual wall emulab infrastructure uses Virtual LANs (VLANs) in the central Force10 hub switch to separate traffic between different experi ments. The central switch does not support double tagging (802.1ad), so 802.1Q VLAN-tagged frames cannot be used inside an experiment in this testbed environment (all frames with Ethertype 0x8100 are dropped by the central switch). The shim DIF over Ethernet uses a VLAN tag as DIF name, so we patched the Linux kernel and Network Interface Controller (NIC)

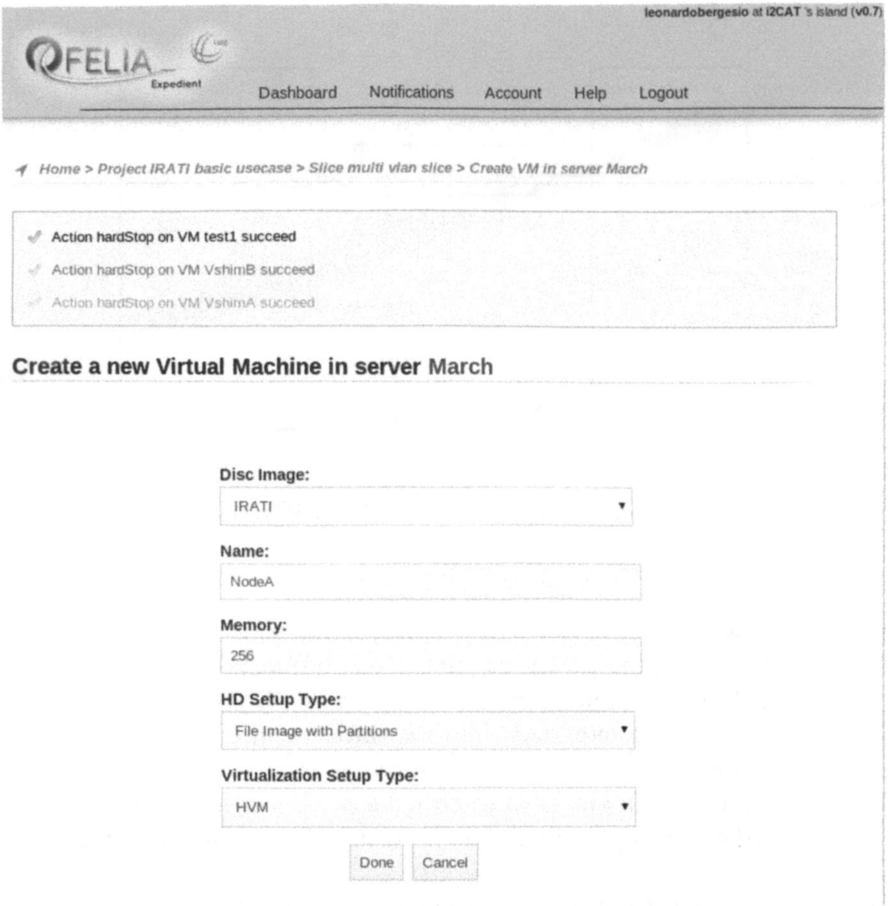

Figure 16.16 Creation of an IRATI VM with the OFELIA Control Framework.

device drivers of the physical machines to be used as RINA hosts. They use ethertype 0x7100 instead of 0x8100 for 802.1Q traffic, allowing transparent use of VLAN tags. In order to allow seamless operation within OFELIA, a translation needs to be done for traffic entering and leaving the iMinds virtual wall island. In order to translate the ethertype, we decided to implement a Linux Loadable Kernel Module (LKM) from scratch to do the translation. During this implementation it was found that the current kernel always untags incoming packets, stripping the VLAN header from the packet data and storing the VLAN header implicitly in a separate vlan tci field inside the socket buffer struct.

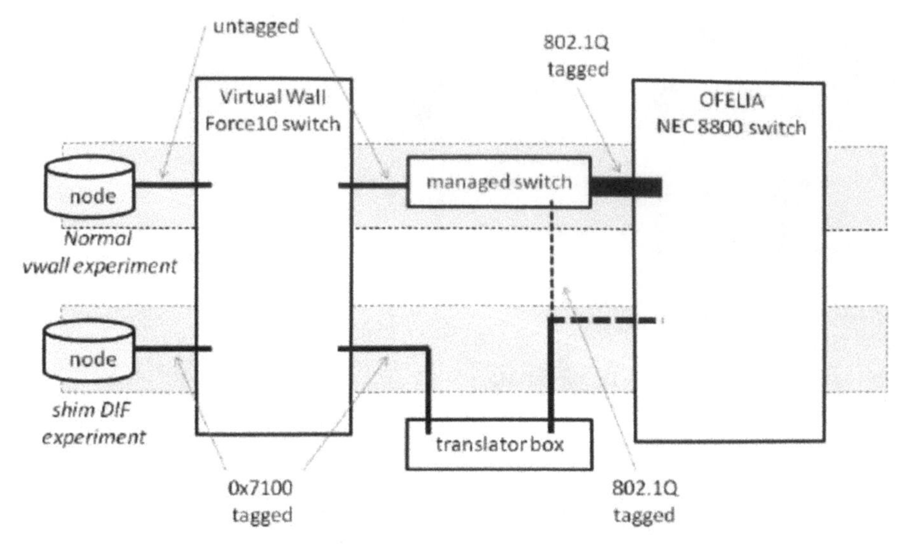

Figure 16.17 VLAN box in the OFELIA testbed.

The Linux LKM re-inserts the VLAN header (changing the ethertype to 0x7100) before sending the packet as "untagged" traffic. A server running this Linux kernel module is now integrated into the OFELIA infrastructure at iMinds. It may also be integrated into the infrastructure for Fed4FIRE. Some VMs in OFELIA do not support 1504 byte frames, so the Maximum Transmission Unit (MTU) will have to be reduced by 4 bytes to allow the VLAN header to be correctly inserted. This is not necessary between machines in the virtual wall.

Figure 16.17 shows (on top) the normal configuration for a virtual wall experiment. These use untagged traffic, and a managed switch is used to tag the traffic with the OFELIA-compatible VLAN tag, aggregate all traffic, and forward it over a 10 Gbit/s fiber to the central OFELIA switch (and vice versa for traffic coming from OFELIA). For traffic that is tagged (as explained, with a 0x7100 ethertype), the ethertype is translated to a normal 802.1Q 0x8100 one by the translator box, and then offered to either the managed switch for aggregation, or directly to a free port on the OFELIA switch.

16.5 Conclusions

In spite of being a small and relatively short research project, the impact of the work performed within the FP7 IRATI project is producing a long lasting impact. First of all IRATI produced the first kernel-based RINA implementation for the Linux/OS that can be overlaid on top of Ethernet, TCP/UDP, and shared memory mechanisms for Guest to Host communication in a Virtualized environment. Open sourcing this implementation has reduced the barriers of entry to researchers and innovators wanting to experiment with RINA. Second, initial experimental results of the RINA prototype over the FIRE testbeds have shown some of the RINA benefits in practice. Third, the IRATI project allowed the creation of a core group of European partners with expertise in RINA design and implementation. Last but not least, the dissemination and outreach activities of the project contributed to better position RINA in the radars of the industry, academia, funding bodies and Standard Development Organisations.

All these efforts have crystallized in the funding of three EC research projects which directly exploited FP7 IRATI results, in particular the open source RINA implementation: i) IRINA [15] studied the applicability of RINA in National Research Education Networks and GEANT; ii) PRISTINE [16] is working on bringing programmability to the IRATI implementation and studying/experimenting with policies in the areas of congestion control, resource allocation, routing, security and network management, and ARC-FIRE [17] is investigating the benefits of applying RINA to the design of converged operator networks and hardening the IRATI implementation to enable large-scale experiments on FIRE+ testbeds. Moreover, RINA is being considered for standardisation in the context of ETSI's Next Generation Protocols ISG [18] and ISO's SC6 Working Group 7 on Future Networks [19].

References

[1] J. Day, "*Patters in Network Architecture: A Return to Fundamentals*". Prentice Hall, 2008.

[2] J. Day, I. Matta, and K. Mattar. 2008. "*Networking is IPC: a guiding principle to a better Internet*". In *Proceedings of the 2008 ACM CoNEXT Conference (CoNEXT '08).*

[3] J. Day. "*About layers: more or less*". PSOC Tutorial, available online at http://pouzinsociety.org

[4] V. Maffione, F. Salvestrini, E. Grasa, L. Bergesio, and M. Tarzan, "A Software Development Kit to exploit RINA programmability". *IEEE ICC 2016, Next Generation Networking and Internet Symposium.*

[5] R. W. Watson. "Timer-based mechanisms in reliable transport protocol connection management". Book in innovations in Networking, pages 296–305. Artech House Inc.
[6] FP7 OFELIA Project website. Available online at http://www.fp7-ofelia.eu
[7] The IRATI consortium. "Specification of a Shim DIF over 802.1Q layers", available as part of IRATI's D2.3: http://irati.eu/deliverables-2/
[8] The IRATI consortium. "Specificafion of a Shim DIF for Hypervisors", available as part of IRATI's D2.3: http://irati.eu/deliverables-2/
[9] The IRATI consortium. "Specification of a Link-State based routing policy for the PDU Forwarding Table Generator"; available as part of IRATI's D2.4.: http://irati.eu/deliverables-2/
[10] J. Day, E. Trouva, E. Grasa, P. Phelan, M. P. de Leon, S. Bunch, I. Matta, L. T. Chitkushev, and L. Pouzin, "*Bounding the Router Table Size in an ISP Network Using RINA,*" Network of the Future (NOF), 2011.
[11] S. Vrijders, D. Staessens, D. Colle, F. Salvestrini, E. Grasa, M. Tarzan and L. Bergesio "Prototyping the Recursive Internetwork Architecture: The IRATI Project Approach", *IEEE Network,* Vol. 28, no. 2, March 2014.
[12] IRATI open source RINA implementation. Available online at https://github.com/IRATI/stack
[13] S. Vrijders, V. Maffione, D. Staessens, F. Salvestrini, M. Biancani, E. Grasa, D. Colle, M. Pickavet, J. Barron, J. Day, and L. Chitkushev 'Reducing the complexity of Virtual Machine Networking'. *IEEE Communications Magazine,* Vol. 54. No. 4, pp. 152–158, April 2016.
[14] S. Vrijders, D. Staessens, D. Colle, F. Salvestrini, V. Maffione, L. Bergesio, M. Tarzan, B. Gaston, E. Grasa; "Experimental evaluation of a Recursive InterNetwork Architecture prototype", *IEEE Globecom 2014,* Austin, Texas.
[15] IRINA Project website. Available online at http://www.geant.org/Projects/GEANT_Project_GN4/Pages/Home.aspx#IRINA
[16] FP7 PRISTINE Project website. Available online at http://ict-pristine.eu
[17] H2020 ARCFIRE Project website. Available online at http://ict-arcfire.eu
[18] ETSI NGP ISG website. Available online at http://www.etsi.org/technologies-clusters/technologies/next-generation-protocols
[19] ISO SC6 website. Available online at http://www.iso.org/iso/iso_technical_committee.html%3Fcommid%3D45072

17

FORGE: An eLearning Framework for Remote Laboratory Experimentation on FIRE Testbed Infrastructure

Alexander Mikroyannidis[1], Diarmuid Collins[2], Christos Tranoris[3], Spyros Denazis[3], DaanPareit[4], JonoVanhie-Van Gerwen[4], Ingrid Moerman[4], Guillaume Jourjon[5], Olivier Fourmaux[6], John Domingue[1] and Johann M. Marquez-Barja[2]

[1]Open University, UK
[2]TCD, Ireland
[3]University of Patras, Greece
[4]IMEC, Belgium
[5]CSIRO, Australia
[6]University Pierre & marie Curie LIP6, France

Abstract

The Forging Online Education through FIRE (FORGE) initiative provides educators and learners in higher education with access to world-class FIRE testbed infrastructure. FORGE supports experimentally driven research in an eLearning environment by complementing traditional classroom and online courses with interactive remote laboratory experiments. The project has achieved its objectives by defining and implementing a framework called FORGEBox. This framework offers the methodology, environment, tools and resources to support the creation of HTML-based online educational material capable accessing virtualized and physical FIRE testbed infrastructure easily. FORGEBox also captures valuable quantitative and qualitative learning analytic information using questionnaires and Learning Analytics that can help optimise and support student learning. To date, FORGE has produced courses covering a wide range of networking and communication domains. These are freely available from FORGEBox.eu and have resulted in over 24,000 experiments undertaken by more than 1,800 students across

10 countries worldwide. This work has shown that the use of remote high-performance testbed facilities for hands-on remote experimentation can have a valuable impact on the learning experience for both educators and learners. Additionally, certain challenges in developing FIRE-based courseware have been identified, which has led to a set of recommendations in order to support the use of FIRE facilities for teaching and learning purposes.

17.1 Introduction

The Forging Online Education through FIRE (FORGE)[1] FP7 project is focused on making practical and effective use of Future Internet Research and Experimentation (FIRE)[2] facilities by utilising them as eLearning resources for higher education institutions. FORGE offers engineering teachers and students access to world-class FIRE testbed infrastructure, while shielding them from the physical and sometimes political complexities of accessing and using experimentation equipment. This has the benefit of maximising the usage of expensive equipment to own, operate and maintain while simultaneously raising awareness of FIRE facilities among teachers, students and future researchers.

FORGE achieves its goals of experimentally driven research by complementing traditional classroom and online courses with interactive remote laboratory experiments. Our approach promotes the development of critical thinking and problem solving skills in students by turning them into active scientific investigators, equipped with world-class experimentation facilities (Marquez-Barja et al., 2014, Mikroyannidis et al., 2015, Jourjon et al., 2016).

FORGE acts as the glue that binds the eLearning and FIRE communities together (see Figure 17.1). This is achieved using the FORGEBox framework, which offers the environment, software components and resources to support the creation of HTML-based online educational material capable accessing virtualized and physical FIRE testbed infrastructure. FORGEBox is supported by the FORGE methodology, which helps course designers with establishing course requirements, identifying and integrating with suitable FIRE facilitates, authoring educational material and course deployment into interactive eBooks, Learning Management Systems (LMSs) and Virtual Learning Environments (VLEs). To support interoperability with existing LMSs and VLEs,

[1] http://ict-forge.eu

[2] http://cordis.europa.eu/fp7/ict/fire/

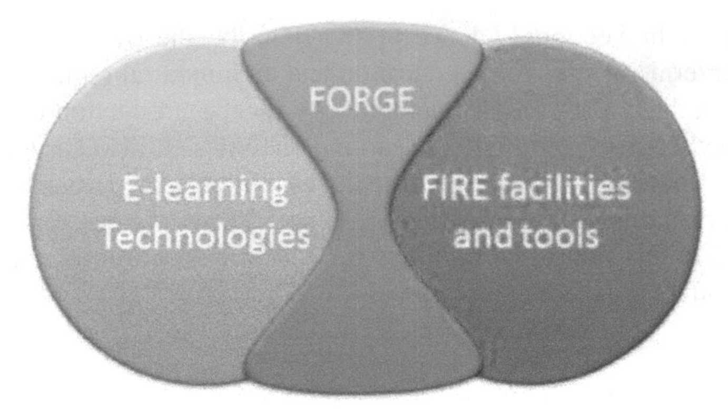

Figure 17.1 eLearning, FORGE and FIRE research facilities.

FORGEBox uses eLearning technologies such as the Learning Tools Interoperability (LTI) standard and SCORM. Additionally, FORGEBox captures valuable quantitative and qualitative learning analytic information based on the Experience API (xAPI) specification. This information can help optimise and support student learning and assist with course evaluation and future adaptation.

FORGE has produced experimentation courses covering a wide range of networking and communication domains, which have been undertaken by more than one thousand students across ten countries. Our research has shown that the use of remote high-performance testbed facilities for hands-on remote experimentation has had a valuable impact on the learning experience for both educators and learners. With the success of initial prototype courses, FORGE also created several advanced electrical engineering courses covering topics such as LTE and OFDM. The on-going FORGE open call courses such as the Internet Measurements MOOC[3], the partnership with Cisco[4] and the Go-Lab[5] projectalso prove its continuing progress. In spite of these successes however, there are several aspects that can be improved related to security, authentication, scalability and sustainability beyond project duration.

This chapter is organised as follows. In Section 17.2, we outline the problem statement in terms of online education and maximising FIRE testbed resources. This is followed by a synthesis of research into learning design theories and online labs for teaching telecommunications related content

[3]https://www.fun-mooc.fr/courses/inria/41011/session01/about

[4]PT Anywhere: http://pt-anywhere.kmi.open.ac.uk/

[5]FORGE widgets available via the Go-Lab project portal: http://www.golabz.eu/search/node/forge

in Section 17.3. In Section 17.4, we briefly describe the overall FORGE framework in terms of user roles, education and architectural requirements. Section 17.5 outlines the FORGE methodology for the production of FIRE testbed enabled courses and FORGE learning analytics. It also surveysfive post-graduate courses developed and deployed by project partners using the FORGE methodology. In Section 17.6, we discuss issues and challenges related to utilising FIRE facilities for educational purposes. Finally, Section 17.7 offers concluding remarks.

17.2 Problem Statement

Higher education is currently undergoing major changes, largely driven by the availability of high quality online materials, also known as Open Educational Resources (OERs). OERs can be described as "teaching, learning and research resources that reside in the public domain or have been released under an intellectual property license that permits their free use or repurposing by others depending on which Creative Commons license is used" (Atkins et al., 2007). The emergence of OERs has greatly facilitated online education (eLearning) through the use and sharing of open and reusable learning resources on the Web. Learners and educators can now access, download, remix, and republish a wide variety of quality learning materials available through open services provided in the cloud.

The OER initiative has recently culminated in MOOCs (Massive Open Online Courses) delivered via providers such as Udacity[6], Coursera[7] and edX[8]. MOOCs have very quickly attracted large numbers of learners; for example over 400,000 students have registered within four months in edX[9]. Also, in the four years since the Open University started making course materials freely available in Apple's iTunes U, nearly 60 million downloads have been recorded worldwide[10]. More recently, the Open University established FutureLearn[11] as the UK response to the emergence of MOOCs, in collaboration with premier British institutions such as the British Council, the British Library and the British Museum.

[6]http://www.udacity.com/
[7]https://www.coursera.org/
[8]https://www.edx.org/
[9]http://www.guardian.co.uk/education/2012/nov/11/online-free-learning-end-of-university
[10]http://projects.kmi.open.ac.uk/itunesu/impact/
[11]http://www.futurelearn.com/

These initiatives have led to widespread publicity and also strategic dialogue in the education sector. The consensus within education is that after the Internet-induced revolutions in communication, business, entertainment, media, amongst others, it is now the turn of universities. Exactly where this revolution will lead is not yet known but some radical predictions have been made including the end of the need for university campuses, while milder future outlooks are discussing ‘blended learning’ (combination of traditional lectures with new digital interactive activities). The consensus is however that the way higher education students learn is about to change radically.

The FIRE initiative holds the potential to contribute to these emerging trends in higher education, as it offers a wide range of experimentation facilities that can be used for teaching and learning online. FIRE’smission is to ensure that the European Internet Industry evolves towards a Future Internet containing European technology, services and values. Through the FIRE initiative and other similar regional and global initiatives a variety of facilities have been established to enable such experimentation. These facilities cover a plethora of different domains belonging to the Future Internet ecosystem, such as cloud computing platforms, wireless and sensor network testbeds, Software Defined Networking and OpenFlow facilities, infrastructures for High Performance Computing, Long Term Evolution (LTE) testbeds, smart cities and so on. However, the corresponding cost both for the establishment and operation of these infrastructures is not to be neglected. Hence, optimal usage of the facilities is desired by its owners, a goal which in general is not yet achieved today. To increase the usage, several steps can be taken.

One approach is to raise the awareness of the facilities within communities that are less familiar with the FIRE initiative. Another is to use the infrastructure not only for research and development, but also for other activities such as teaching through a constructivist approach. This means that students would be enabled to take certain initiatives in their learning, by setting up and conducting scientific experiments based on FIRE. In this way, using FIRE facilities for teaching computer science topics or other scientific domains would not only increase the usage of the facilities, it would also raise FIRE awareness in the long term since the students/experimenters of today are the researchers of tomorrow. And if educational materials were available that actually enable new types/areas of experimentation through FIRE, this would further lower the threshold for experimenters to explore new facilities and technologies.

The FORGE project offers a solution to this problem by adopting the latest trends in education in order to introduce the FIRE experimental facilities into the eLearning community. FORGE promotes the concept of experimentally-driven research in education by using experiments as an interactive learning and training channel for both students and professionals by raising the accessibility and usability of FIRE facilities. The goal is to create an open FORGE community and ecosystem where educational resources, collaborative tools and proposed experiments are offered and contributed for free.

17.3 Background and State of the Art

17.3.1 Learning Design

In this section we outline the various pedagogical theories associated with the process of designing courseware, or Learning Design as it is also known in the literature of Technology-Enhanced Learning (TEL). Learning Design (LD) is the act of devising new practices, plans of activity, resources and tools aimed at achieving particular educational aims in a given situation. LD should be informed by subject knowledge, pedagogical theory, technological know-how and practical experience. At the same time, it should also engender innovation in all these domains and support learners in their efforts and aims (Mor and Craft, 2012).

A learning design captures the pedagogical intent of a unit of study. It offers a broad picture of a series of planned pedagogical actions, rather than detailed accounts of a particular instructional event as might be described in a traditional lesson plan. As such, a learning design provides a model for intentions in a particular learning context that can be used as a framework for design of analytics to support faculty in their learning and teaching decisions (Lockyer et al., 2013).

The field of LD emerged in the early 2000s as researchers and educational developers saw the potential to use the Web to document and share examples of good educational practice. Smith and Ragan (2005) have proposed that LD might be more accurately described as Design for Learning. Some common definitions for LD are the following:

> *"A 'learning design' is defined as the description of the teaching-learning process that takes place in a unit of learning (e.g., a course, a lesson or any other designed learning event). The key principle in learning design is that it represents the learning activities and the*

support activities that are performed by different persons (learners, teachers) in the context of a unit of learning." (Koper, 2006).

"A methodology for enabling teachers/designers to make more informed decisions in how they go about designing learning activities and interventions, which is pedagogically informed and makes effective use of appropriate resources and technologies. This includes the design of resources and individual learning activities right up to curriculum-level design. A key principle is to help make the design process more explicit and shareable. Learning design as an area of research and development includes both gathering empirical evidence to understand the design process, as well as the development of a range of Learning Design resource, tools and activities." (Conole, 2012).

These definitions suggest two seemingly opposing approaches. However, Falconer et al. (2011) suggest that LD has two origins in TEL. The first one is the construction of computer systems to orchestrate the delivery of learning resources and activities for computer-assisted learning. The second is in the need to find effective ways of sharing innovation in TEL practice, providing an aid to efficiency and professional development for teachers. Therefore, Koper's definition represents the first TEL origin, while Conole's definition is derived from the second.

The most easily understood and adapted common elements within all learning designs include the following (Lockyer et al., 2013):

- A set of *resources* for the student to access, which could be considered to be prerequisites to the learning itself (these may be files, diagrams, questions, web pages, etc.).
- *Tasks* the learners are expected to carry out with the resources (prepare and present findings, negotiate understanding, etc.).
- *Support mechanisms* to assist in the provision of resources and the completion of the tasks; these supports indicate how the teacher, other experts, and peers might contribute to the learning process, such as moderation of a discussion or feedback on an assessment piece (Bennett et al., 2004).

Figure 17.2 provides an example learning design visual representation showing three common categories of resources, tasks, and supports.

In order to ensure that a learning design is sound, the learning outcomes should be in line with the assessment that is used to test for the achievement

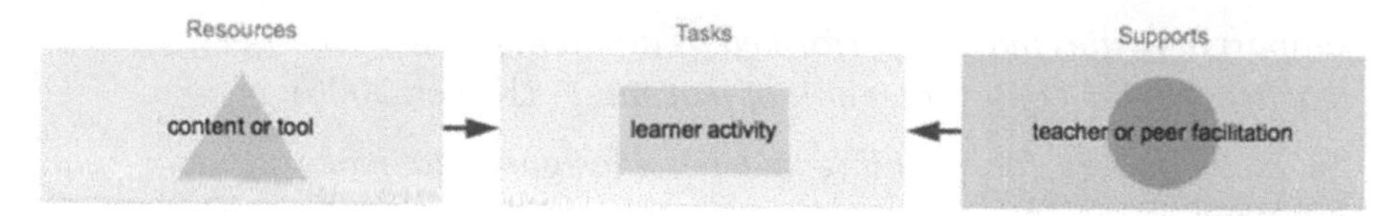

Figure 17.2 Learning design visual representation (Lockyer et al., 2013).

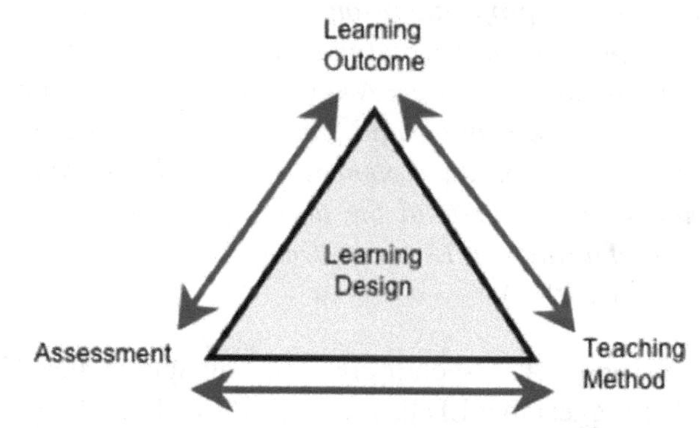

Figure 17.3 The instructional triangle of learning designs[12].

of learning outcomes. In addition, both learning outcomes and assessment should be aligned with the teaching method. Biggs (2011) refers to this as the "constructive alignment". The relationship between these three concepts can be represented as a triangle and it is often referred to as the "instructional triangle of learning designs", as shown in Figure 17.3.

With regards to the different skills that teachers need in order to implement a learning design successfully, Mishra and Koehler (2006) present the Technological Pedagogical and Content Knowledge (TPACK) model (see Figure 17.4). The TPACK model can be used as a foundation for analysing the pedagogical and technological elements of LD. The TPACK model puts emphasis on the intersections between Technological Knowledge, Pedagogical Knowledge and Content Knowledge, and proposes that effective integration of technology into the curriculum requires a sensitive understanding of the dynamic relationship between all three components.

The Instructional Management Systems (IMS) Learning Design (IMS-LD)[13] specification expresses a standardised modelling language for representing learning designs as a description of teaching and learning processes. The main objective of the IMS-LD specification is the provision

[12]http://www.open.edu/openlearnworks/course/view.php?id=1154

[13]http://www.imsglobal.org/learningdesign

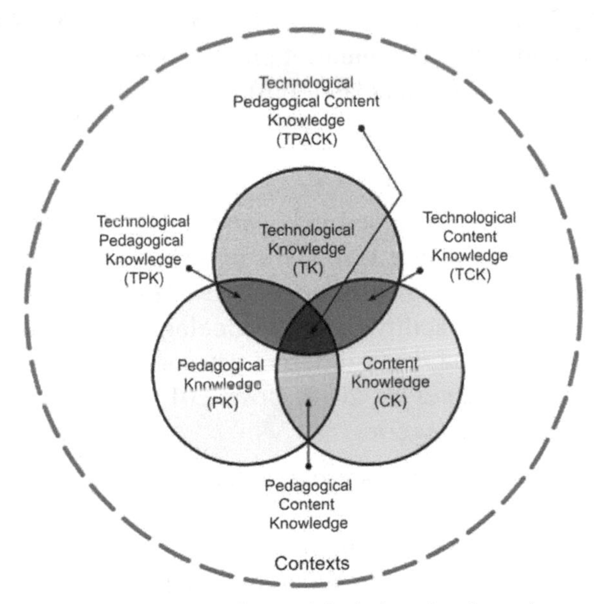

Figure 17.4 TPACK model of educational practice.

of a containment framework of elements that can describe any design of a teaching-learning process in a formal way. Thereby, the originally intended objectives of IMS-LD are (Koper, 2009):

- The standardised description of an adaptive learning and teaching process which takes place in a computer-managed course, i.e. these courses: are "developed" before they are used; can be used by different groups/classes of learners at different times (principle: "Develop once, run many times"); are managed by the computer (here: Runtime), not by the teacher; are designed to achieve certain learning outcomes for a given target group (prerequisites) as effective and efficient as possible for the individual learner.
- The support of all types of learning designs based on various pedagogical approaches.
- To have the learning and support activities at the centre, not the content.
- To provide an integrative framework for a large number of learning content such as IMS Common Cartridge (IMS-CC)[14], IMS Content Packaging (IMS-CP)[15], IMS Question and Test Interoperability Specification

[14]http://www.imsglobal.org/commoncartridge.html
[15]http://www.imsglobal.org/content/packaging/

(IMS-QTI)[16], Sharable Content Object Reference Model (SCORM)[17] as well as collaboration/communications services (e.g. audio/video conference, forum, and virtual classroom).

17.3.2 Online Labs

Online laboratories have been designed and operate under different themes in order to train students and enhance their skills in higher education programs (Harward et al., 2008). Depending on the methods used to access and to trigger the equipment at the backend facility and the technology used in the front-end graphical interface, from three to six different categories have been defined (Diwakar et al., 2013, Frerich et al., 2014, Bose, 2013). We can summarize these taxonomies into three categories:

1. Virtual labs, which are software-based laboratories, empowered by simulation tools.
2. Remote labs, based on remote experimentation on real lab equipment.
3. Hybrid labs, which combine the above two by processing output data from real measurements into simulation tools.

There are several works that describe the approaches that different universities and/or projects have applied to enable engineering-related online laboratories. Most of the approaches rely on simulation, providing virtual labs for teaching robotics (Abreu et al., 2013), electronic circuits (Bagchi et al., 2013), control systems (Diwakar et al., 2013) or a broad list of engineering disciplines (Bose, 2013).

Few approaches have been publicly proposed for teaching telecommunications related content in remote labs. Bose and Pawar (2012) proposed a remote lab called Virtual Wireless Lab where students can learn about the foundations of wireless signal, with concepts such as antenna radiation pattern, gain-bandwidth product of an antenna, cross polar discrimination and Signal Noise Ratio (SNR). The architecture proposed presents a front-end Adobe Flash enabled web page to access the back-end, which uses LabView to interface with telecommunications equipment such as spectrum analysers, oscilloscopes and signal generators. The eComLab supports a similar configuration and instruction by using a dedicated VNC-based virtual machines (VMs) managed by a gateway server that allows remote lab configuration and experimentation on Emona DATEx and NI ELVIS boards (Gampe et al., 2014). These VMs have direct access to the board hardware supporting

[16]http://www.imsglobal.org/question/
[17]http://www.adlnet.gov/scorm/

direct experiment control. A user can access these machines using a regular web browser with support for Flash and Java plugins. Due to the tools and equipment used by Virtual Wireless Lab and eComLab architectures, they do not support open easy to use interfaces for configuration, data collection, resource sharing, etc.

In contrast, the Smart Device specification (López et al., 2015), initiated by the Go-Lab EU FP7 project, advocates all devices (clients or servers) use common interfaces such as metadata, logging, data collection, configuration, and so forth to simplify communication between a remote labs, external services and applications. This is supported by; open protocols; WebSockets, which uses asynchronous bidirectional communication between client and server; and Swagger, a JSON-based description language for RESTful web services that easily integrates with WebSockets. Smart Device metadata is exposed on the Internet enabling applications, services and other devices to interact with the remote lab. Telecommunications courses, such as the oscilloscope lab available on Go-Lab project platform[18], can utilise the Smart Device specification to support design, integration and promote usability. These principles of openness and ease of use are similarly philosophies followed by the FORGEBox framework.

17.4 The FORGE Framework

The overall architectural approach of FORGE is displayed in Figure 17.5 and covers FORGE user roles (i.e. learners, course designers, instructors, and so forth) and requirements. The architectural approach is towards accomplishing our initial FORGE challenges include:

- To make the reservation of resources in (different) facilities easy for both teachers and learners;
- To allow easy fast experimentation control, from various devices and means, during the learning process;
- To know the identity of the user who is currently performing an experiment that was initiated from within a client web browser;
- To access resources that can only be reached over IPv6 or over a VPN;
- Avoid breaking the logical flow of an educational experiment when the user behaved unexpectedly;
- To allow multiple users sharing the same experiment;
- Handle a large number of simultaneous users.

[18]http://www.golabz.eu

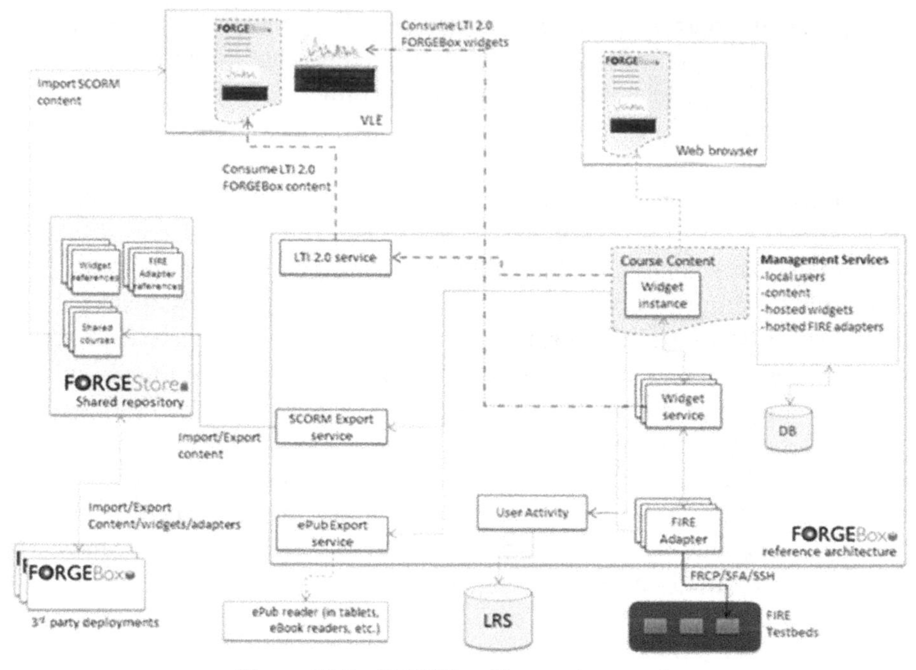

Figure 17.5 FORGE architectural approach.

It is also important to use existing eLearning technology and try to seamlessly integrate it with our FORGE artefacts. Thus, all developments made in our core entities, consider open and well known eLearning technologies. We investigated solutions of exploiting these eLearning technologies in two areas: interoperability and means to study user behaviour while learning on top of FIRE. These technologies are the Learning Tools Interoperability (LTI) standard, SCORM and the Experience API (xAPI), commonly known as the Tin Can API.

LTI adoption provides better integration between FORGE technology and existing LMSs and VLEs. LTI provides a seamless experience for learners while interacting both with the LMS/VLE content and the remote FIRE resource. Consequently, LTI makes it much easier for organizations to adopt and use the FORGE technologies and integrate them with their own already deployed learning systems. xAPI on the other hand allows instructors to study learners' behaviour while interacting with a facility.

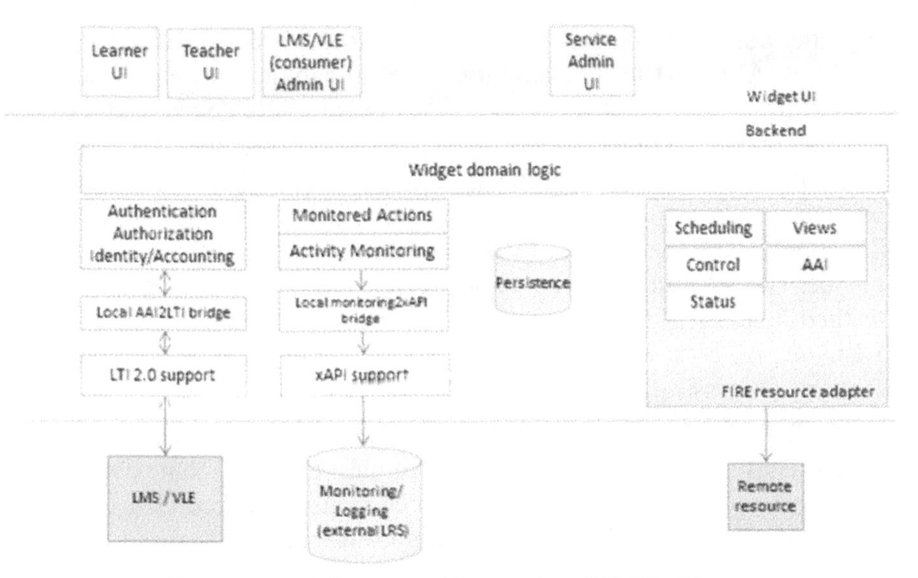

Figure 17.6 Reference architecture for a FORGE widget.

To address the above, we created a reference architecture for widgets and FIRE adapters that support interacting with remote facilities of FIRE through a VLE. Figure 17.6 displays our proposed reference architecture for a widget, with architectural components that a developer would need to implement in order to achieve the most desirable result of bridging learning with FIRE remote resource interactivity. Since widgets are web services hosted somewhere on the Internet ready to be consumed by other web content, the architecture defines both the widget UI as well as the backend domain logic and core architectural components. Next we discuss supported usage roles and each architectural component. These supported roles are:

- **Service Administrator**: the user responsible for the whole widget web service. Service Administrator can login to the host machine and administer the service that provides the widget to consumers. Service Administrator can also manage for example users, registrations etc. The use cases are specific to the capabilities that the widget service will offer. E.g. the administrator of the ssh2web widget can allow specific domains that can use the service.
- **LMS/VLE Administrator**: responsible to integrate the widget to the target learning system LMS/VLE or even in an eBook. He needs to

pay attention to the widget documentation, how it is delivered (i.e. as a URL), its API, it's LTI compatibility, etc. For example, an administrator responsible for a Moodle installation could visit FORGEStore and read the documentation of the widget. Then he could register the widget into the Moodle environment by using the LTI registration URL of the widget service.

- **Teacher/Instructor**: defines the behaviour and settings for a specific course. He can also use the interface to reserve resources or setup the testbed.
- **Learner**: interacts with the widget and the remote resource during the learning process.

The widget UI is the main component that a user uses to interact with the widget. To behave correctly, the Widget service must know the context that it works under, in order to properly display the equivalent UI according to the user role. Thus, if possible, the widget should be aware of:

- The consumer service into which it is hosted and operating (i.e. is it an LMS/VLE, the VLE URL, an eBook, etc.);
- The kind of consumer (i.e. its capabilities, browser, tablet etc.);
- The identity of the current user and his role;
- The current course (content or page reference).

All this information can be passed either through a widget API (e.g. passing URL parameters) or via more modern ways such as the LTI API. According to the user role there should be different UIs. Thus some first requirements for a widget service should be:

- An API to call the widget and pass user identity and context;
    - For this, LTI usage is encouraged
- Specific endpoints (URLs) that will service each user according to his role
    - E.g., service administrators visit http://www.mywidgeturl.org:8080/admin
    - E.g., a VLE admin visits http://www.mywidgeturl.org:8080/lti/register

It is not necessary for widgets to implement all these user interfaces. For example, the FORGE widgets of Teacher Companion Lab courses don't need to provide a Learner UI since they can be used only by Teachers.

17.5 Courseware and Evaluation

In this section the FORGE methodology is presented. It has been developed based an analysis of the state of the art in educational technologies with a specific focus on remote laboratories and online learning. We also outline the FORGE mechanisms for collecting learning analytics information based on a synthesis of available research and using existing technologies and standardisation efforts. Finally, we provide an overview and evaluation of five FORGE courses presented to over one thousand students.

17.5.1 The FORGE Methodology

One of the main goals of FORGE is to enable educators and learners to access and actively use FIRE facilities in order to conduct scientific experiments. We thus follow a constructivist approach to education where learning takes place by students creating artefacts rather than assuming the passive role of a listener or reader. Our approach is based on a wide range of studies that have shown that with the right scaffolding competent learners benefit greatly from constructivist or learning-by-doing approaches (De Jong, 2006, Hakkarainen, 2003, Kasl and Yorks, 2002). The experiment-driven approach of FORGE contributes to fostering constructivist learning by turning learners into active scientific investigators, equipped with world-class experimentation facilities.

From a learning technology perspective, FORGE is building upon new trends in online education. More specifically, in online educational platforms such as iTunes U, as well as in MOOCs, we see the large-scale take-up and use of rich media content. These include video in a variety of formats including webcasts and podcasts and eBooks, which can contain multimedia and interactive segments. In particular, eBooks provide a new level of interactivity since specific learning text, images and video can be closely integrated to interactive exercises[19]. In the context of the European project EUCLID[20] (EdUcationalCurriculum for the usage of Linked Data), we have been producing such interactive learning resources about Linked Data and delivering them in a variety of formats, in order to be accessed from a

[19]http://www.youtube.com/watch?v=KXCHKYsi1q8

[20]http://www.euclid-project.eu

variety of devices, both mobile (tablets and smartphones), as well as desktop computers. Building on this work, FORGE is producing interactive learning resources targeting a wide range of mediums and devices in order to maximise its impact on the eLearning community.

FORGE is enabling students to set-up and run FIRE experiments from within rich related learning content embedded as widgets inside interactive learning resources. Widgets are powerful software components that can be reused across different learning contexts and for different educational purposes. They offer a simple interface and can accomplish a simple task, such as displaying a news feed. They can also communicate with each other and exchange data, so that they can be used together to create mashups of widgets that complement each other. The portability of widgets as bespoke apps that can be embedded into a variety of online environments ensures that the FORGE learning solutions implemented as widgets have a high reusability factor across multiple learning domains and online learning technologies. Within FORGE, widgets enable educators and learners to access and actively use Future Internet facilities as remote labs in order to conduct scientific experiments. Learners and educators can setup and run Future Internet experiments from within rich related learning content embedded as widgets inside interactive eBooks and LMSs.

The FORGE methodology for the production of FIRE-enabled course consists of the following steps(Mikroyannidis et al., 2016):

- *Specifying course requirements.* In this step, the educator specifies the overall course requirements, including the learning objectives of the course, the required skills, the skills that will be acquired by learners after completing this course, the course timeframe, the number of learners and the method of delivery (online, face-to-face, or blended).
- *Identifying FIRE facilities.* In this step, the educator identifies the FIRE facilities that will suit the course requirements. These FIRE facilities will be selected based on their suitability for the learning objectives of the course and its associated skills. The number of learners and timeframe will also play a role in selecting a FIRE facility based on its availability. The first and most important task is to identify the *facility features* which match the intended course content. When someone, for example, wants to include experimental exercises using specially developed wireless transmission protocols, a facility should be chosen where one has permission to adapt the radio drivers or where one can

use cognitive radio devices, etc. A basic overview of the most prominent facility features covered in Fed4FIRE[21] portal.

- *Authoring educational content.* The educational content that will form the learning pathway of the course is authored in this step. Finding open educational resources that are suitable for the course is quite important, as these can be reused, adapted and repurposed to fit the course learning objectives and other requirements. These resources can have the form of text that describes the theory behind a specific exercise, questionnaires with multiple-choice options, videos with lectures, videos with instructions on how to conduct the exercise, images and diagrams about the architecture and topology of the required components, graphical representations of the desired results etc.
- *Integration of FIRE facilities and content.* In this step, the selected FIRE facilities and the educational content of the course are integrated in order to form the complete learning pathway. FIRE facilities are commonly integrated as widgets, which can be reused across different learning activities for different learning purposes.
- *Deployment.* The deployment of the course for delivery to learners is performed in this step. Depending on the course requirement for delivery (online, face-to-face, or blended), the educator can deploy the course within a LMS, a VLE, or as an interactive eBook.
- *Evaluation.* In this step, the educator evaluates the success of the course, based on qualitative feedback received from learners via surveys and questionnaires, or via quantitative data collected by Learning Analytics tools that track the interactions of learners with the course materials and with each other.
- *Reflection and adaptation.* By analysing the qualitative and quantitative data collected from the evaluation of the course, educators have the opportunity to reflect and draw some conclusions not only about potential adaptations and improvements to the course, but also, and most important, on the impact of the course on the students and their skills and knowledge acquired.

Figure 17.7 summarizes the FORGE methodology, showing the steps to be followed in order to deploy, create, use, and/or reuse a FORGE course. As depicted, two main phases should be considered: a) Course preparation, and b) Course deployment. In each phase, different processes are defined in order to

[21] http://www.fed4fire.eu

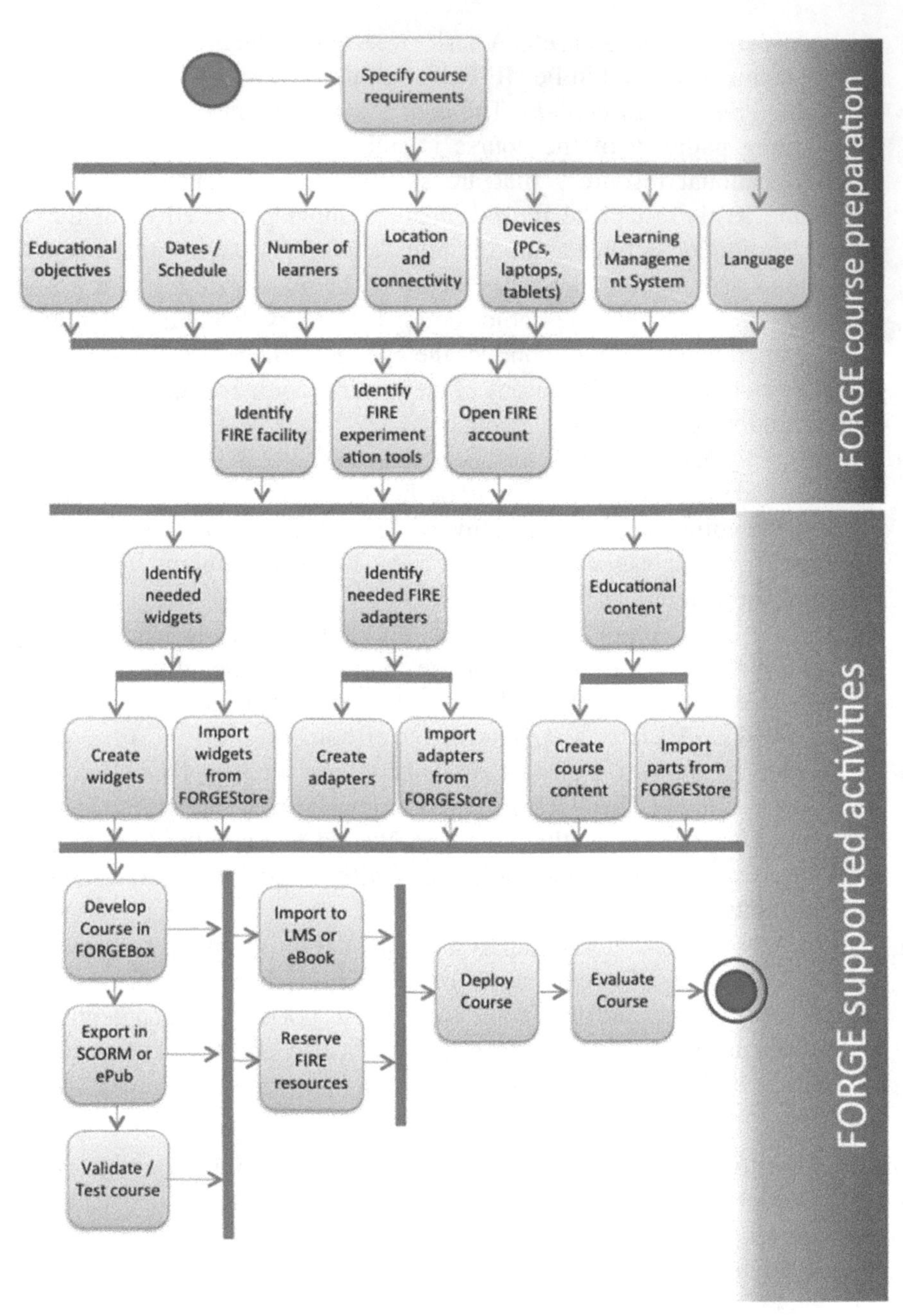

Figure 17.7 The FORGE methodology flowchart.

guide course developers and learners towards a successful course deployment and learning experience.

17.5.2 Learning Analytics

Learning Analytics can be described as the *"measurement, collection, analysis and reporting of data about learners and their contexts, for purposes of understanding and optimizing learning and the environments in which it occurs"*.[22] The field of Learning Analytics is essentially a *"bricolage field, incorporating methods and techniques from a broad range of feeder fields: social network analysis (SNA), machine learning, statistics, intelligent tutors, learning sciences, and others"* (Siemens, 2014).

Learning Analytics applies techniques from information science, sociology, psychology, statistics, machine learning, and data mining to analyse data collected during education administration and services, teaching, and learning. Learning Analytics creates applications that directly influence educational practice (Shum et al., 2012). For example, the OU Analyse[23] project deploys machine-learning techniques for the early identification of students at risk of failing a course. Additionally, OU Analyse features a personalised Activity Recommender advising students how to improve their performance in the course.

With Learning Analytics, it is possible to obtain valuable information about how learners interact with the FORGE courseware, in addition to their own judgments provided via questionnaires. In particular, we are collecting data generated from recording the interactions of learners with the FORGE widgets. We are tracking *learner activities*, which consist of interactions between a *subject* (learner), an *object* (FORGE widget) and are bounded with a *verb* (action performed). We are using the Tin Can[24] API (also known as xAPI) to express and exchange statements about learner activities, as well as the open source Learning Locker[25] LRS (Learning Record Store) to store and visualise the learner activities.

Figure 17.8 depicts the widget-based architecture adopted in FORGE. The FORGE widgets use LTI 2.0[26] for their integration within a LMS. The FIRE

---

[22] 1st International Conference on Learning Analytics and Knowledge – LAK 2011 https://tekri.athabascau.ca/analytics/

[23] https://analyse.kmi.open.ac.uk

[24] http://tincanapi.com/

[25] http://learninglocker.net/

[26] http://www.imsglobal.org/toolsinteroperability2.cfm

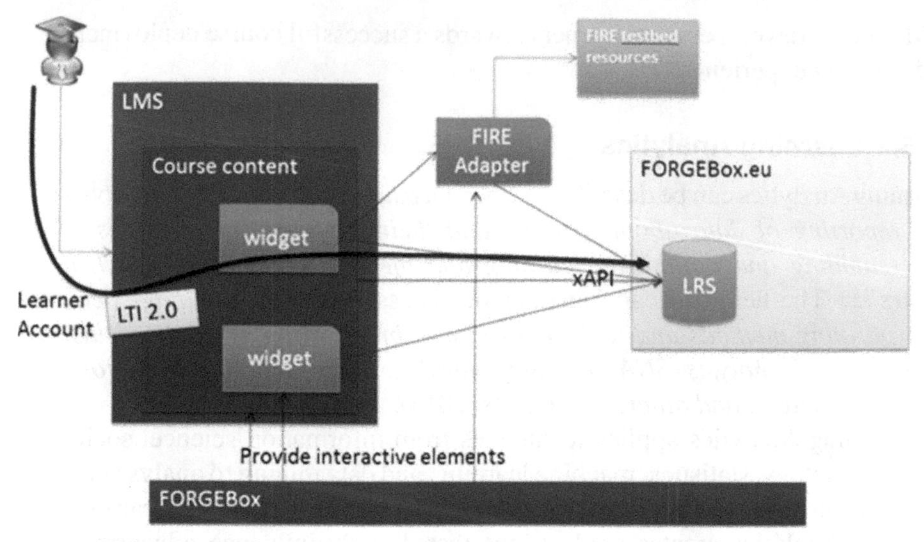

Figure 17.8 The widget-based FORGE architecture for learning analytics.

Adapters function as a middleware between the FORGE widgets and the FIRE facilities (testbeds), while the FORGEBox layer offers a seamless experience while learners are performing a course, reading content and interacting with FIRE facilities. All the interactions performed by users on the course content and the widgets are recorded and stored in the Learning Locker LRS using the xAPI.

Learner activities on the FORGE widgets typically include the initialisation of an experiment, setting the parameters of the experiment and, finally, completing the experiment. Therefore, the learner activities captured by the FORGE widgets use the following types of xAPI verbs:

- *Initialized*[27]: Formally indicates the beginning of analytics tracking, triggered by a learner "viewing" a web page or widget. It contains the (anonymised) learner id and the exercise/widget that was initialized.
- *Interacted*[28]: Triggered when an experiment is started by the learner, containing the learner id, the exercise and possible parameters chosen by the learner. These parameters are stored in serialized JSON form using the result object, as defined by the xAPI.

[27]http://adlnet.gov/expapi/verbs/initialized
[28]http://adlnet.gov/expapi/verbs/interacted

- *Completed*[29]: The final verb, signalling completion of an exercise by the learner. We can also include the duration that a learner took to perform the experiment, formatted using the ISO 8601 duration syntax following the xAPI specifications.

More specialised learner activities are also recorded by the FORGE widgets depending on the functionalities offered by each widget. For example, the PT Anywhere[30] widget, which offers a network simulation environment, records the following types of activities, reusing already defined vocabulary[31]:

- *Device creation, update and removal*: We use the verbs "create", "delete" and "update" from "http://activitystrea.ms/schema/1.0/".
- *Link creation and removal* (i.e., connecting and disconnecting two devices): The link creation and removal is expressed as a user creating a link that has its two endpoints defined as contextual information. Another alternative could have been to use non-existing connect/disconnect verbs to express that a user connects a device to another one (the latter should have been added as contextual information). However, we chose the first alternative because it reuses already existing verbs.

FORGE provides learners with Learning Analytics dashboards in order to raise their awareness of their learning activities by providing an overview of their progress or social structures in the course context. Learners are offered with detailed records of their learning activities, thus being able to monitor their progress and compare it with the progress of their fellow learners. Additionally, the Learning Analytics dashboards targeted to educators provide an in-depth overview about the activities taking place within their courses, thus making the educators aware of how their courses and experimentation facilities are being used by their students.

17.5.3 WLAN and LTE (iMinds)

iMinds has created two 'flipped labs' (for blended learning in a 'flipped classroom') for learners to better understand what is affecting the data throughput over two different types of wireless networks. One lab is using a Wireless Local Area Network (WLAN) network with Wi-Fi technology while the other lab is using a 4G cellular network with Long Term Evolution (LTE) technology. By changing parameters in web based 'widgets' with a cross-platform and

[29]http://adlnet.gov/expapi/verbs/completed
[30]http://pt-anywhere.kmi.open.ac.uk
[31]https://registry.tincanapi.com

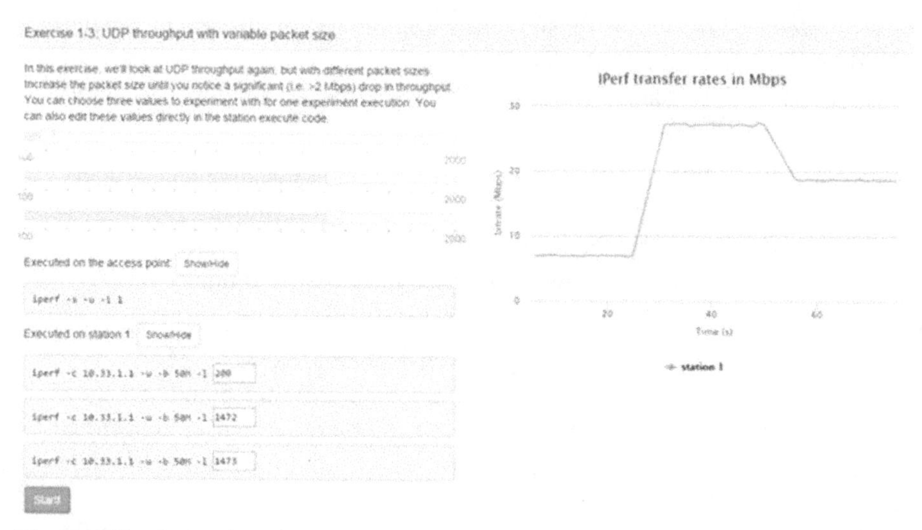

Figure 17.9 Screenshot of a web-based learner interface at the iMinds' WLAN and LTE lab.

easy-to-use interface (which is integrated in e.g. a LMS, eBook or any web page), learners can see the resulting throughput in a graph, based on the measurement results that are being collected from a real live experiment at the FIRE facilities of iMinds. Various back-end 'adapters' enable the communication between the front-end widgets and the actual resources at the FIRE facilities via the jFed CLI.

These labs were traditionally taught with local hardware, but were ported via the FORGE project to FIRE facilities. We now benefit from the resulting 'flipped labs' because the automating of the lab configuration simplifies lab sessions organisation at any given moment and at any given location. Furthermore, more advanced hardware of FIRE facilities can be used that would otherwise be unavailable locally.

The course is executed on two FIRE facilities, operated by iMinds: the w-iLab.t wireless testbed and the Virtual Wall (see Figure 17.10). The actual experimentation machines are located in the w-iLab.t testbed, where they are dynamically selected from 75 wireless nodes, depending on availability. These machines are controlled from a node of the Virtual Wall, a testbed consisting of 400 multi-core servers.

The Virtual Wall node contains the interactive course components and is responsible for controlling the wireless nodes with an "cOntrol and Management Framework" (OMF) Experiment Controller. All user interactions go

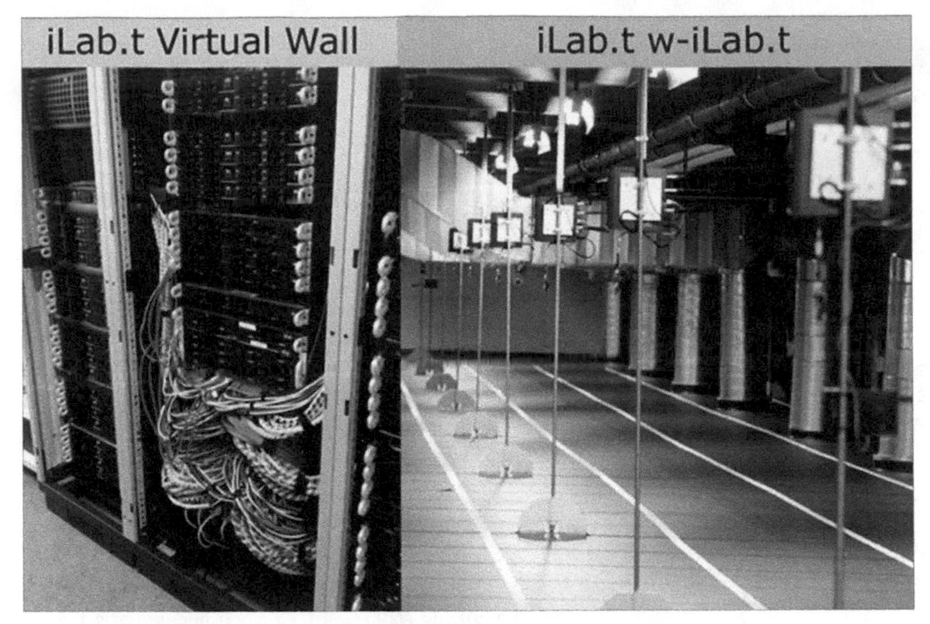

Figure 17.10 iMinds' iLab.t testing facilities used for the WLAN and LTE labs.

through this machine, which uses adapters and widgets, developed within FORGE, for executing and visualizing the experiments. Thanks to our extensions within the widgets and adapters, multiple learners can simultaneously access the interactive course and share FIRE resources.

The WLAN lab was executed for the first time for about 90 students in February 2015 at Ghent University (Belgium). Next additional executions took place at Trinity College Dublin (Ireland) in February 2015 for about 25 students and at Universidade de Brasil (Brazil) in May 2015 for about 20 students. During 2016, the WLAN course was executed a few more times: once again at Ghent University in March 2016 for about 90 students, once again at Trinity College Dublin in March 2016 for about 25 students, once again at Universidade de Brasil in May 2016 for 8 students and once at Universitat Politècnica de València (Spain) in December 2015 for 6 students. The LTE course was also deployed and executed in 2016: once at Ghent University in April–May 2016 for about 90 students and once at Universitat Politècnica de València (Spain) in December 2015 for 6 students.

Both teaching methods are possible for executing the WLAN or LTE lab: 'in-classroom' versus 'self-assessment'. When taught 'in-classroom', the lab

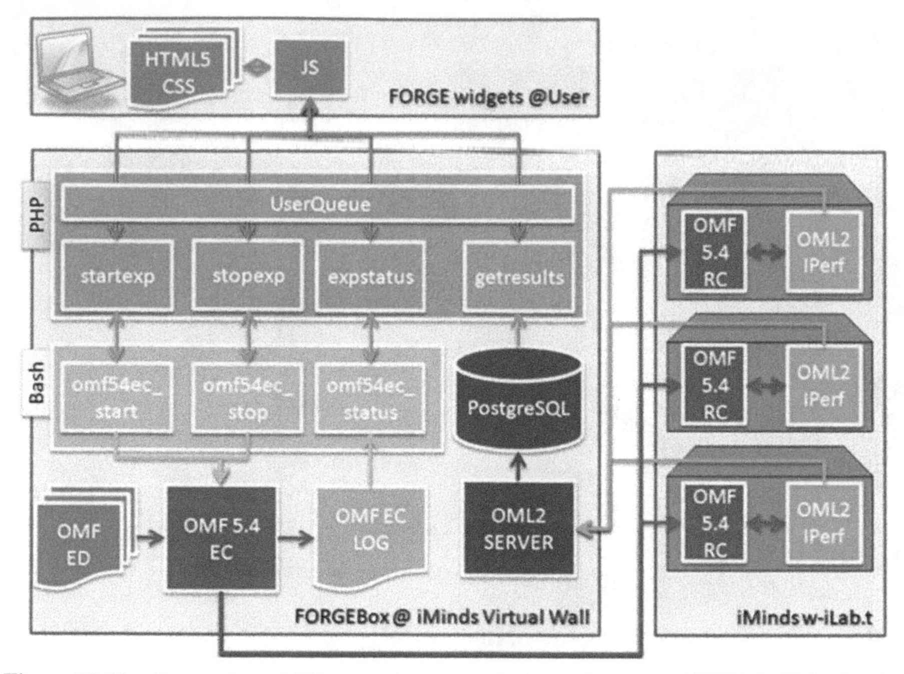

Figure 17.11 Interaction of different components between learner and FIRE facilities for the iMinds' WLAN and LTE labs.

was organized to last for 2–3 hours and took take place in a computer lab room where students could perform the FIRE experiments online with coaching of university staff members. The students answer the lab questions on paper or online and staff members corrected these afterwards. When taught as 'self-assessment home assignment', the students were given some time (typically about two weeks) to perform the lab individually at the time and the moment of their choice. The lab questions, which students had to answer, were converted to allow automated correction (i.e. multiple choice questions, numeric answers and 'fill-in-the-gap' questions) within a dedicated (Moodle-based) system to make this self-assessment possible and to provide immediate feedback to the students.

We also collected both qualitative and quantitative feedback from the students themselves. The qualitative feedback was collected via a survey, using 5-Likert-scale statements and some open questions. For quantitative feedback, learning analytics were applied using TinCan API, Learning Locker etc. In Figure 17.12, we have plotted the average score for each of the different qualitative survey questions, both for 2015 and 2016 for students at Ghent

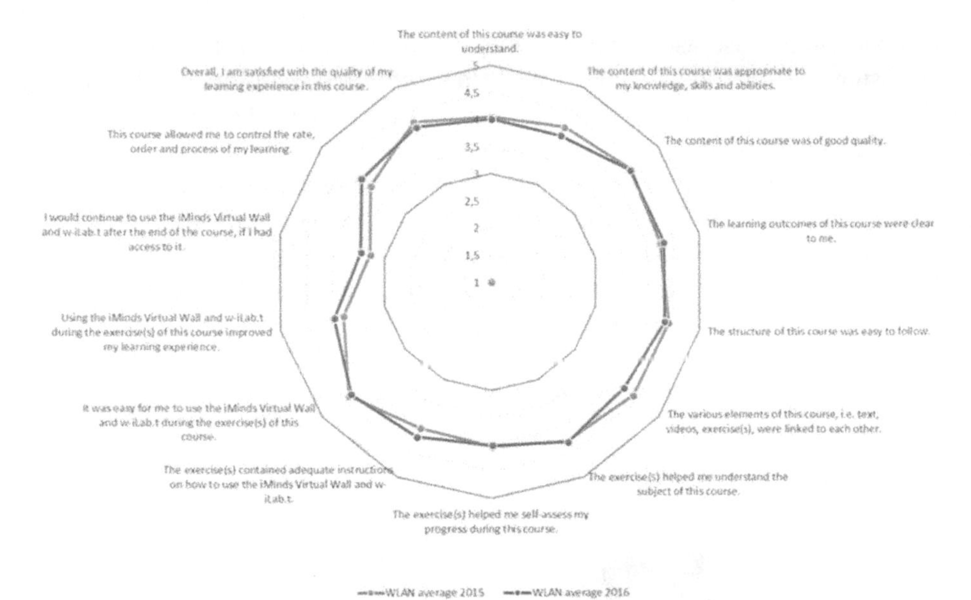

The content of this course was easy to understand.
The content of this course was appropriate to my knowledge, skills and abilities.
The content of this course was of good quality.
The learning outcomes of this course were clear to me.
The structure of this course was easy to follow.
The various elements of this course, i.e. text, videos, exercise(s), were linked to each other.
The exercise(s) helped me understand the subject of this course.
The exercise(s) helped me self-assess my progress during this course.
The exercise(s) contained adequate instructions on how to use the iMinds Virtual Wall and w-iLab.t.
It was easy for me to use the iMinds Virtual Wall and w-iLab.t during the exercise(s) of this course.
Using the iMinds Virtual Wall and w-iLab.t during the exercise(s) of this course improved my learning experience.
I would continue to use the iMinds Virtual Wall and w-iLab.t after the end of the course, if I had access to it.
This course allowed me to control the rate, order and process of my learning.
Overall, I am satisfied with the quality of my learning experience in this course.
5
4,5
3,5
3
2,5
2
1,5
1
WLAN average 2015
WLAN average 2016

Figure 17.12 Average score for the qualitative survey questions.

University. These had to be quoted on a 5-Likert-scale (1: strongly disagree, 3: neutral, 5: strongly agree). We notice very good scores for the different statements, averaging around ‘4’ (“agree” with the statement) and the scores are consistent for 2015 and 2016. This indicates we were able to implement a successful course concerning its quality, effectiveness and ease of use.

Some student quotes from the surveys, which represent the general tendency, are mentioned below:

- *“The iMinds wall was easy to use”*
- *“Hands on approach”*
- *“No configuration hassle”*
- *“Actually learned some interesting concepts”*
- *“Cool new concept”*
- *“I could “pause” the session whenever i wanted and resume when i had time.”*
- *“Modern and interactive learning environment”*
- *“I surprisingly enjoyed this Lab session a lot more than I thought I would”*
- *“The FORGE system is amazing!”*

The students were also asked whether they liked the overall concept of the home assignment or if they would have rather preferred traditional lectures and labs in-classroom. The results in Figures 17.13 and 17.14 show that more than 4 students out of 5 prefer this way of blended learning.

Based on the collected quantitative learning analytics information, teachers could analyse the most common mistakes students make and adapt their course to explain certain parts better. Furthermore, the activity of the different students could also be tracked and compared to their automated score within the self-assessment home assignment. This allows also to identify students who have cheated by extracting students who provide the correct answers to a question without having performed the necessary related experiment.

If you had the choice, what would you prefer? (68 responses)

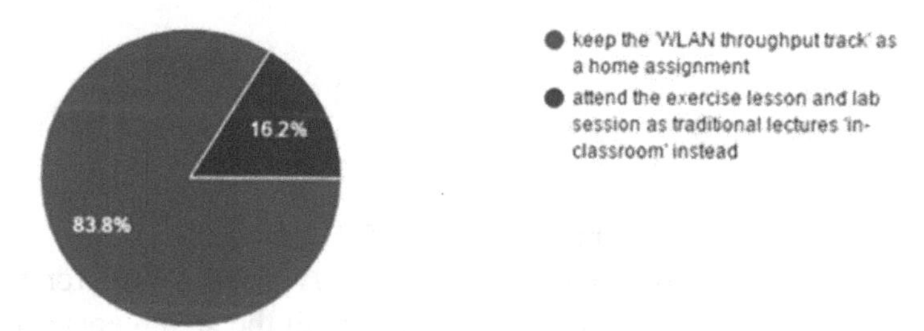

Figure 17.13 Preference of students (in 2016) for using the WLAN course as online home assignment versus teaching this via traditional in-classroom lectures.

If you had the choice, what would you prefer? (60 reacties)

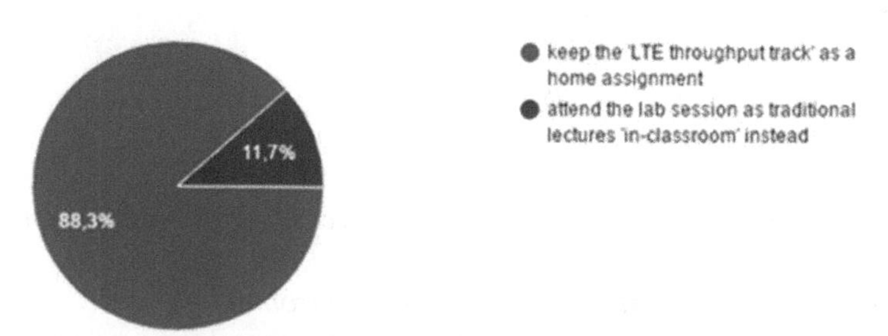

Figure 17.14 Preference of students (in 2016) for using the LTE course as online home assignment versus teaching this via traditional in-classroom lectures.

17.5.4 TCP Congestion Control and Metro MOOC (UPMC)

Université Pierre et Marie Curie (UPMC) run the PlanetLab Europe (PLE) Network Operation Center (NOC). Thanks to its experience, it invested in PLE related widgets and FIRE adapters for setting up a prototype course – UPMC TCP Congestion Control. This in turn supported the launch of an external course called METRO MOOC. The UPMC TCP Congestion Control prototype course focus on a fundamental mechanism of TCP. After few exercises illustrating the congestion control mechanisms, real traces of long distance traffic are performed on the PLE facility and are analysed with a packet analyser tool. The development of the course itself is in line with the methodology described by FORGE.

Concerning the execution of the course, the initial planning was to include the course inside a basic networking teaching unit, taught in French and in English for several kind of students (Classical students, part-time industry students and EIT-digital Master School students). The course took place in October 2014 and October 2015, each time for one week. The PLE resources reservation and the teaching team preparation were done the week before. The maximum number of student groups at the same time guided resource reservation (students work in pair in all tutorials works). The maximum number of students in a group at the same time was 2, and there was a total of 30 student groups. All groups perform the course during the same time. Therefore, we made reservation of 1 PLE slice with 66 nodes: 33 Clients (30 + 3 spare nodes) and 33 servers. The only resource where the pair must be alone is the client to generate a correct capture. The 33 servers can be shared and are supposed to serve 3 clients each. We used a dedicated tool to make the reservation on the PLE slice to generate all the configurations. Figure 17.15 shows an example of the configuration.

The UPMC prototype course was executed two times in the classroom with a web interface on the computer where students usually make their practical work. Groups of 30 students worked in pairs with one tutor. These course labs follow a 2-hour lecture about Congestion control theory. The labs last for 4 hours but the remote lab part is quite short and is only needed to get some remote traces to analyse locally with the tools commonly used by students. The qualitative feedback was collected via a survey. The demographic information concerning the student include the following:

- 2014 web based UPMC course: 168 students
  - 160 French speaking/8 English speaking (ICT Digital)
  - 23 female (14%)/124 male (86%)
  - Most of the student are in the 21–30 age slot (average 22.3)

Figure 17.15 The TCP congestion control widget.

- 2015 web based UPMC course: 150 students
    - 144 French speaking/6 English speaking
    - 26 female (17%)/124 male (83%)
    - Most of the student are in the 21–30 age slot (average 22.4)

In Figure 17.16, we have plotted the average score of each different question, both for 2014 and 2015.

UPMC also provided assistance with the creation of an external course called: "The Internet Measurements: a Hands-on Introduction" MOOC, which is offered by the French national e-learning platform France Université Numérique (FUN). This course has been developed with the METRO FORGE

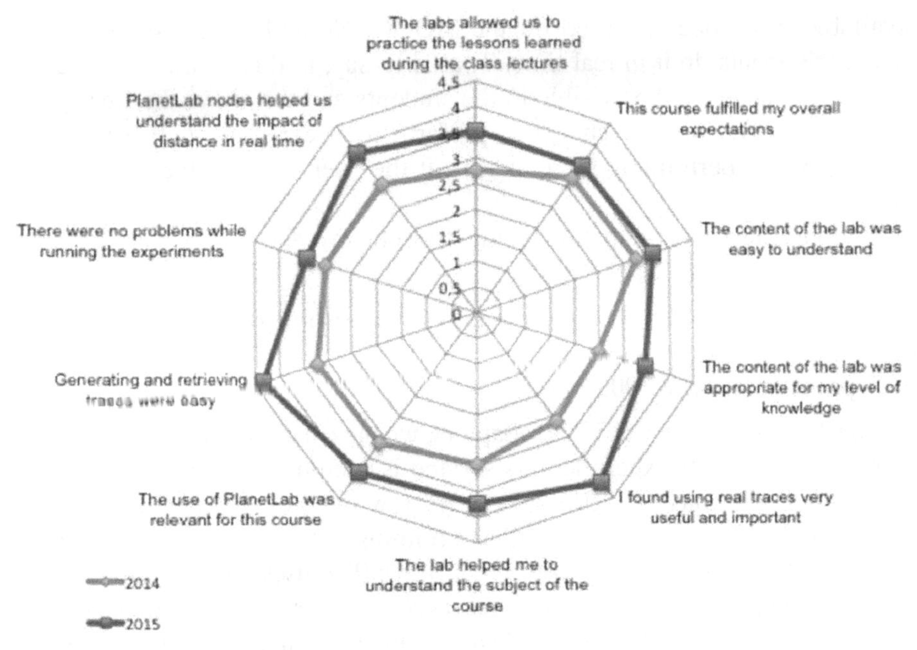

Figure 17.16 UPMC lab course questionnaire average score (2014 and 2015).

open call project proposed by INRIA. This MOOC is intended to attract as much as 5,000 students. It has been open to public since 23 May 2016 and also uses PLE testbed for experimentations such as ping, traceroute and iperf. There are potentially hundreds of requests coming to the PLE testbed every minute in the MOOC context along with the usual usage by the researchers and PLE members. It is quite obvious that PLE can't handle such a high volume of requests coming almost every minute. In order to not overload the testbed side, the MOOC developed a REST API based solution with job scheduling. All the MOOC experiment requests are stored directly in a NoSQL document based database at the beginning with a job status "waiting to be executed". There are several agents (threads) checking the new jobs as soon as they arrive and based on the job already in hand they schedule and process the new jobs. The agent also calculates an estimated time of executing the new jobs and informs the users to come back after a certain time period to check the results. When the agents are free to take new jobs they process them and change the job status to "executing" and then to "completed". In this way the agents don't overload the testbed by scheduling the time to process new jobs. It is worthwhile to mention that, all available PLE nodes are used for the MOOC. No nodes are reserved in advance, the agents do it in real time. That is to say, at this moment all the available PLE nodes will be visible to the students and they have the liberty to select any nodes of their choice to perform their experiment. If a node goes down while performing the experiment the user will receive an error message.

Some preliminary participation results from the MOOC:

- 1824 registration
- 1440 participants have 0% score (no exercise resolved)
- 155 certificate (score >50%)
- 27 participants have 100% score

In the latest edition of the FIRE PLE adapters, we have used all the PLE nodes available at the time of experiment. A service is running in the background checking the availability of PLE nodes every few minutes. By *available* we mean that all the PLE nodes that are up and running at the time of experiment not all the PLE nodes that are not reserved. Since PLE uses virtualisation for each reservation, a single node can be reserved by several users at the same time. Due to this powerful feature, the learners are given all the PLE nodes for experiment through a dropdown list. In order not to overload a specific node, we use a queuing mechanism. If multiple learners choose a specific node, we put them in the queue on a first come first serve basis. We also gave them an estimated time to completion of the experiment. For the moment, we're allowing only 2 experiments to run simultaneously on a single node. This can be scaled to more if the node is capable of handling it. Another reason for not allowing more than 2 experiments simultaneously is not to interrupt the usual usage of these nodes for the other PLE users. Figures 17.17–17.19 describe how the queuing system works for PLE.

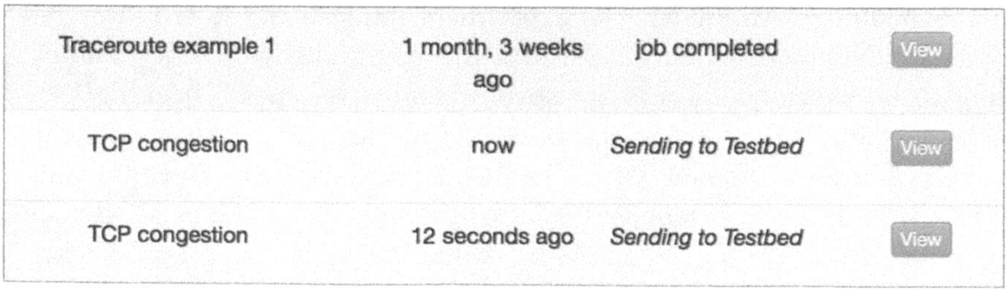

Traceroute example 1	1 month, 3 weeks ago	job completed	View
TCP congestion	now	*Sending to Testbed*	View
TCP congestion	12 seconds ago	*Sending to Testbed*	View

Figure 17.17 Experiments in queue before sending to PLE.

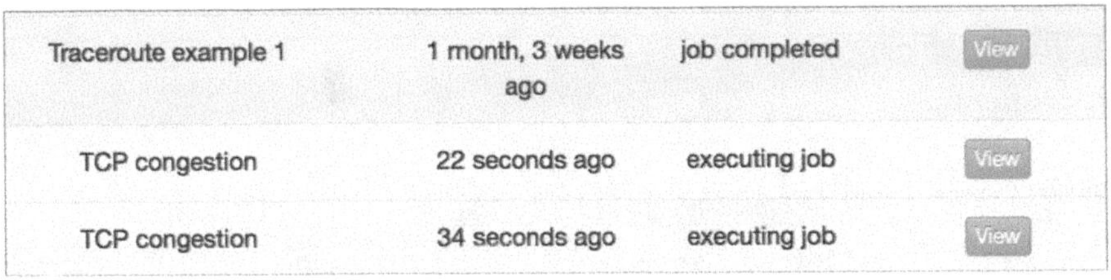

Traceroute example 1	1 month, 3 weeks ago	job completed	View
TCP congestion	22 seconds ago	executing job	View
TCP congestion	34 seconds ago	executing job	View

Figure 17.18 Experiment sent to PLE.

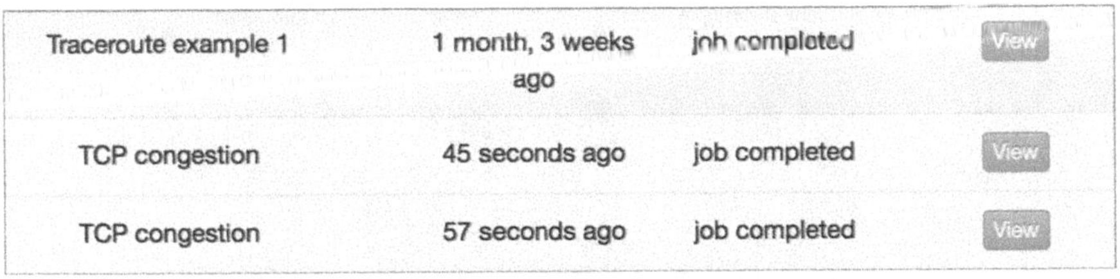

Traceroute example 1	1 month, 3 weeks ago	job completed	View
TCP congestion	45 seconds ago	job completed	View
TCP congestion	57 seconds ago	job completed	View

Figure 17.19 Experiment completed and result available.

17.5.5 OFDM (Trinity College Dublin)

In this course, students investigate how Orthogonal Frequency-Division Multiplexing (OFDM) wireless signals work by connecting students to advanced research hardware to investigate the sometimes troublesome digital multi-carrier modulation method as applied to wireless communications. Through this experience, students gain an appreciation of the most important factors in the use of OFDM for wireless experiments by exploring the configuration and use of real radios.

TCDs OFDM course runs completely on IRIS software defined radio (SDR) testbed facility. The IRIS testbed consists of 16 flexible radio USRP units each connected to a virtual machine that runs an SDR such as IRIS or GNU Radio. Resources are provisioned automatically by the gateway server, which also supports initialization of experimentation services such as measurement point data collection. A conceptual diagram of IRIS's virtualized cloud resources, radio hypervisor, user experiments and physical equipment is shown in Figure 17.20.

This course has been taught in the lab environment where students execute remote experiments on TCD's wireless testbed with support from a lab instructor. It has been presented nine times over the last twelve months in Brazil, Mexico and Ireland to a total of 148 students running at least 1,400 experiments. There are currently two versions of the course. The first version has been presented to approximately 132 students. To date, the second version

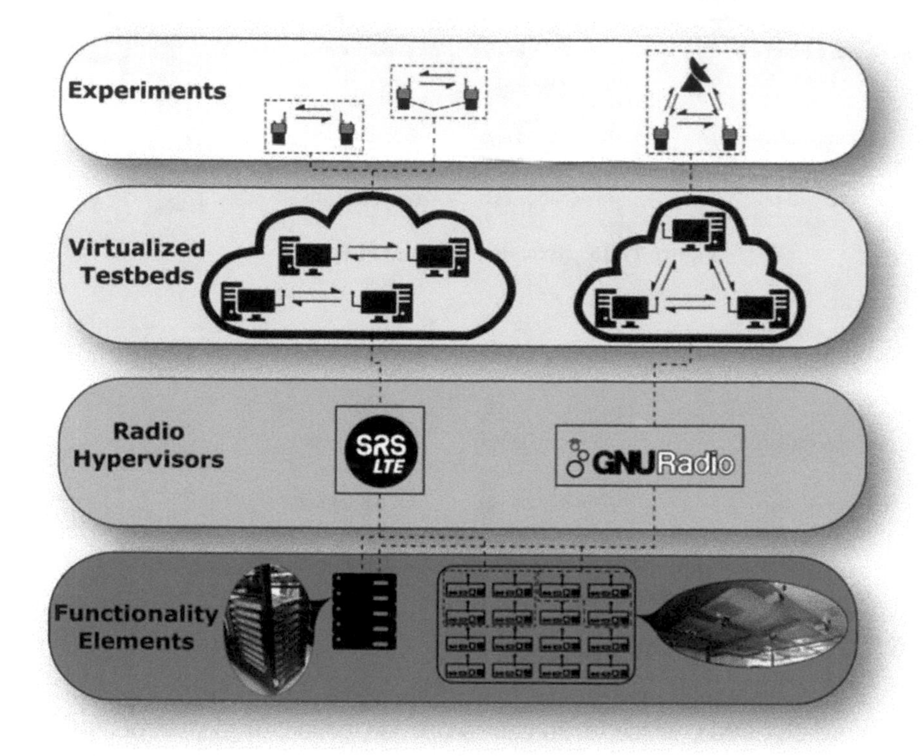

Figure 17.20 TCD's IRIS testbed.

of the course has been presented to 16 students running remote experiments from Brazil. At present, teachers reserve the testbed for use in a remote experimentation based lab.

Evaluation information was collected from students after execution of the first version of the course using the standard FORGE 5-Likert scale questionnaire template. A screenshot of sample summary feedback received from students using TCD's testbed is available in Figure 17.21. In general, over 90% of students who participated the lab agreed that the OFDM experiments helped them to understand the concepts being taught in the lectures. Additionally, over 90% agreed that they were aware the experiments were being run remotely on TCD's wireless testbed while over 76% agreed that the web interface helped to reduce the difficulty of the lab. Furthermore, over 90% of student agreed the lab helped them to self-assess their progress. Finally, almost 80% of students surveyed would like to use the testbed facility in the future if they had access to it. More detailed instructor feedback was also gathered, but in an informal manner via email. In general, the information received

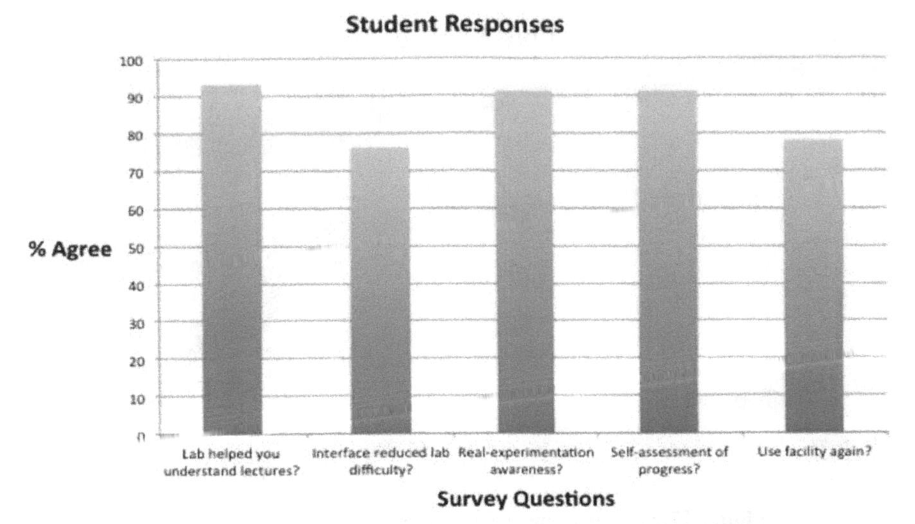

Figure 17.21 Responses from survey of first version of TCDs OFDM course.

about the course, content and structure was very positive from both students and teachers. However, several requirements to improve the course were also identified. These included the need for a more scalable, responsive, sustainable and reusable system with the ability to provide real-time information to end-users.

This feedback led directly to the course being redesigned and redeveloped to use GNU Radio, an open source SDR with a large user and developer community. This change was required as TCDs IRIS SDR system used in the first version of the course, which is flexible and adaptive for advanced wireless research and experimentation, was determined to have a much smaller development and support community than GNU Radio. It was decided that this would affect the long-term sustainability of the existing course and the development of future advanced SDR courses. Additionally, existing components developed by other users in GNU Radio could be easily incorporated into existing and new courses, which can reduce course development and testing time.

Furthermore, another change to the course involved collecting measurement data in real-time. This is now collected using OML and presented by JavaScript-based interactive widgets to students. A screenshot of sample real-time graphs displaying data received from TCD's IRIS testbed from the second version of the OFDM course supported by GNU Radio are available in Figure 17.22.

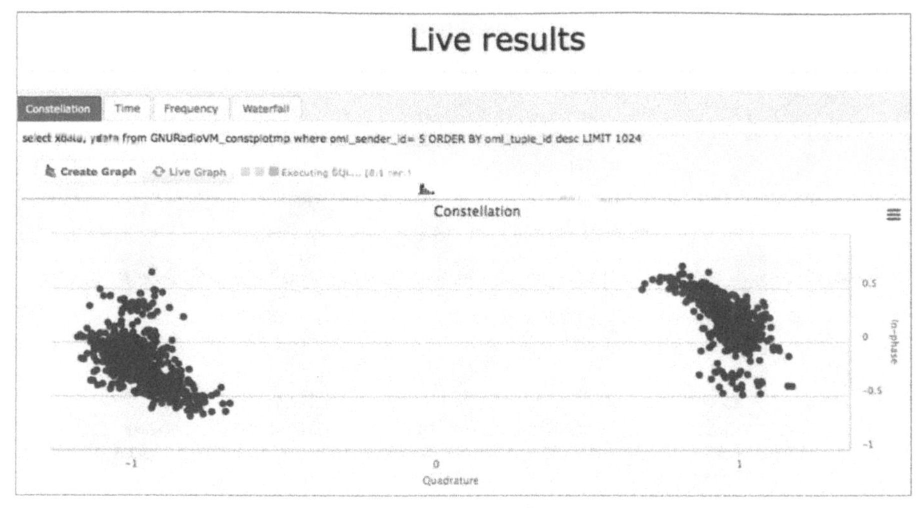

Figure 17.22 Real-time constellation measurements.

As the OFDM course architecture was almost completely redeveloped, we could only recycle the control component widget developed in the version 1 IRIS implementation. However, we reused and expanded a graphing widget developed by iMinds to support the generation of frequency, time, waterfall and constellation graphs from an SQLite database. Additionally, we utilized NICTA's OMF Measurement Library for the collection of real-time measurement point data during experiment execution. We also implemented some basic learning analytics, to help determine what commands were being executed by students primarily for technical support purposes. Furthermore, we developed an XML adapter to support users sending configuration parameters to GNU Radio in real-time.

GNU Radio is the most dependable, reliable, reusable and sustainable SDR platform to support remote experimentation on the IRIS testbed. This has been validated in a recent deployment of the OFDM course to 16 students in University of Brasilia, Brazil who were able to change OFDM parameters, send data packets and monitor USRP activity in real-time. Aside from some minor bugs experienced during course execution, positive feedback about the system stability and responsiveness and graph visualisation has been received from both students and teacher. Finally, the integration of learning analytics has helped the OFDM course implementers detect weaknesses in course design helping to further improve the quality of the online lab.

17.6 Discussion

FORGE has been investigating how FIRE facilities, which have been built primarily for research purposes, can be reused and adapted for teaching and learning purposes. The project has provided evidence that FIRE testbeds can function as world-class remote laboratories for educators and learners and can be used for online experimentation within a variety of learning contexts. However, the usage of FIRE for educational purposes also raises certain issues and challenges that the FIRE facilities have not encountered before (or not in a high degree).

A first challenge is security related. FORGE has created different web-based educational widgets that run on a web server, which can be part of the experiment itself. The experiments are thus executed and manipulated by the web server (via web based requests by the learner) rather than directly by the learner. The resources and accompanying widgets/adapters on the web server might furthermore not have been reserved by the learner himself/herself, and the learner might thus be controlling (via a web server) resources that were reserved by someone else (typically by the educator). This requires using a kind of 'proxy' or 'speaks-for' mechanism, securely allowing the sharing of resources amongst multiple FIRE accounts.

Another significant challenge lies in the fact that there is no common reservation system in place for all FIRE facilities. Depending on the scarcity of resources used by a lab, a certain reservation mechanism should be in place to guarantee the availability of the interactive exercises during a lab. When a group of learners (e.g. all students within the same classroom) are following the same course and executing the same experiments, a large number of FIRE resources will be required at the same moment of time. When the specific FIRE resources, which are needed, are scarce, a (very) high resource occupation will be imposed on the hosting FIRE facility. In order to still accommodate the experiments of the different learners while not overloading their own facility, FIRE facilities need to elaborate their policy strategy into different categories (e.g. 'best effort' or 'premium') to force a more well-thought usage of the facility by learners and experimenters alike. A FIRE facility would also need to provide some sort of reservation mechanism to guarantee resource availability to the learner in case of pre-planned lab sessions, while the FORGE widgets and adapters hide the specific reservation and scheduling mechanic for the learner. These policy strategies and associated business models are subject to the sovereignty of the different FIRE facilities. To limit the number of simultaneously used FIRE resources

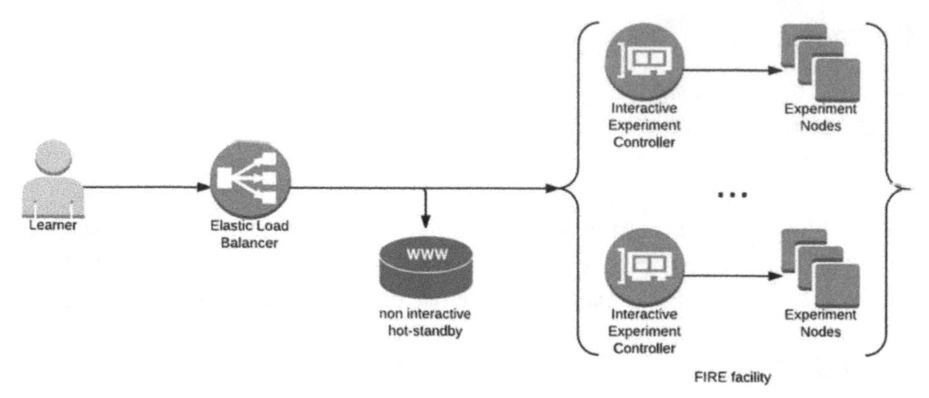

Figure 17.23 Load balancer with multiple experiment instances and graceful degradation via hot-standby.

by different learners, some of the FORGE adapters also add intermediate functionality by e.g. implementing a scheduling or queuing mechanism to allow multiple learners to share the same FIRE resources. A common reservation mechanism across FIRE testbeds would solve this additional complexity and would also provide an incentive and clear implementation path for FIRE facilities.

Since most FIRE facilities only offer 'best effort' resource availability, even with reservation, there is always the possibility of resource or total testbed failure. Even if there is no possible recourse to alleviate these kind of failures, a graceful degradation system can lessen the impact on the learner. A fall-back mechanism to a non-interactive version of the lab with a clear message to the learner can significantly increase the user experience. Ideally this fall-back mechanism would also allow to seamlessly switch back to the interactive version once connectivity is restored to the FIRE facility resources and retain any previous experiment results. This challenge can be solved by using existing load balancing techniques and software for redundant web services, as illustrated in Figure 17.23.

17.7 Conclusion

FORGE complements online learning initiatives with laboratory courses for an in-depth and hands-on learning experience. The constructivist approach of FORGE is based upon the notion of the experiment. FORGE allows students to create and conduct experiments using interactive learning resources within a comprehensive learning context. Towards this goal, the project has established a technological and pedagogical framework for remote labs and online experimentation, by defining a methodology for the design, delivery and evaluation of FIRE-enabled courseware.

FORGE has produced a wide range of networking and communication-courses, which have resulted in over 24,000 experiments undertaken by more than 1,800 students across 10 countries worldwide. FORGE has thus provided evidence that FIRE testbeds can function as world-class remote laboratories for educators and learners and can be used for online experimentation within a variety of learning contexts. Additionally, the project has identified certain challenges that have emerged from developing FIRE based courseware, leading to a set of requirements and recommendations for supporting the use of FIRE facilities for teaching and learning.

References

[1] Abreu, P., Barbosa, M. R. & Lopes, A. M. (2013) Robotics virtual lab based on off-line robot programming software. *2013 2nd Experiment @ International Conference (exp. at'13).* IEEE, 109–113.

[2] Atkins, D. E., Brown, J. S. & Hammond, A. L. (2007) A Review of the Open Educational Resources (OER) Movement: Achievements, Challenges, and New Opportunities. The William and Flora Hewlett Foundation, http://www.hewlett.org/uploads/files/Hewlett_OER_report.pdf

[3] Bagchi, D., Kaushik, K. & Kapoor, B. (2013) Virtual labs for electronics engineering using cloud computing. *Interdisciplinary Engineering Design Education Conference (IEDEC),* 2013 3rd. IEEE, 39–40.

[4] Bennett, S., Lockyer, L. & Agostinho, S. (2004) Investigating how learning designs can be used as a framework to incorporate learning objects. *Beyond the comfort zone: Proceedings of the 21st ASCILITE Conference*, 116–122.

[5] Biggs, J. B. (2011) *Teaching for Quality Learning at University: What the Student Does*. McGraw-Hill Education (UK).

[6] Bose, R. (2013) Virtual labs project: A paradigm shift in internet-based remote experimentation. *IEEE Access,* 1, 718–725.

[7] Bose, R. & Pawar, P. (2012) Virtual Wireless Lab—Concept, design and implementation. *Technology Enhanced Education (ICTEE), 2012 IEEE International Conference on.* IEEE, 1–7.

[8] Conole, G. (2012) *Designing for Learning in an Open World,* vol. 4. Springer Science & Business Media.

[9] De Jong, T. (2006) Scaffolds for computer simulation based scientific discovery learning. IN ELEN, J. & CLARK, R. E. (Eds.) *Dealing with Complexity in Learning Environments.* London, Elsevier Science Publishers.

[10] Diwakar, A., Poojary, S., Rokade, R., Noronha, S. & Moudgalya, K. (2013) Control systems virtual labs: Pedagogical and technological perspectives. *2013 IEEE International Conference on Control Applications (CCA).*

[11] Falconer, I., Finlay, J. & Fincher, S. (2011) Representing practice: practice models, patterns, bundles. . . . *Learning, Media and Technology,* 36(2), 101–127.

[12] Frerich, S., Kruse, D., Petermann, M. & Kilzer, A. (2014) Virtual labs and remote labs: practical experience for everyone. *2014 IEEE Global Engineering Education Conference (EDUCON).* IEEE, 312–314.

[13] Gampe, A., Melkonyan, A., Pontual, M. & Akopian, D. (2014) An assessment of remote laboratory experiments in radio communication. *IEEE Transactions on Education,* 57(1), 12–19.

[14] Hakkarainen, K. (2003) Emergence of Progressive-Inquiry Culture in Computer-Supported Collaborative Learning. *Science and Education,* 6(2), 199–220.

[15] Harward, V. J., Del Alamo, J. A., Lerman, S. R., Bailey, P. H., Carpenter, J., Delong, K., Felknor, C., Hardison, J., Harrison, B. & Jabbour, I. (2008) The ilab shared architecture: A web services infrastructure to build communities of internet accessible laboratories. *Proceedings of the IEEE,* 96(6), 931–950.

[16] Jourjon, G., Marquez-Barja, J. M., Rakotoarivelo, T., Mikroyannidis, A., Lampropoulos, K., Denazis, S., Tranoris, C., Pareit, D., Domingue, J. & Dasilva, L. A. (2016) FORGE Toolkit: Leveraging Distributed Systems in eLearning Platforms. *IEEE Transactions on Emerging Topics in Computing.*

[17] Kasl, E. & Yorks, L. (2002) Collaborative inquiry for adult learning. *New Directions for Adult and Continuing Education,* 2002(94), 3–12.

[18] Koper, R. (2006) Current research in learning design. *Educational Technology & Society,* 9(1), 13–22.

[19] Koper, R. (2009) IMS Learning Design: State-of-the-Art. *Keynote Speech at the 4th European Conference on Technology Enhanced Learning (ECTEL 2009) Workshop: Relating IMS Learning Design to Web 2.0 Technology.*

[20] Lockyer, L., Heathcote, E. & Dawson, S. (2013) Informing pedagogical action: Aligning learning analytics with learning design. *American Behavioral Scientist,* 0002764213479367.

[21] López, S., Carpeño, A. & Arriaga, J. (2015) Remote Laboratory eLab3D: A Complementary Resource in Engineering Education. *IEEE Revista Iberoamericana de Tecnologias del Aprendizaje,* 10(3), 160–167.

[22] Marquez-Barja, J., Jourjon, G., Mikroyannidis, A., Tranoris, C., Domingue, J. & Dasilva, L. (2014) FORGE: Enhancing elearning and research in ICT through remote experimentation. *IEEE Global Engineering Education Conference (EDUCON).* Istanbul, Turkey http://ict-forge.eu/wp-content/uploads/2013/11/educon.pdf, IEEE Education Society Publications.

[23] Mikroyannidis, A., Domingue, J., Pareit, D., Gerwen, J. V.-V., Tranoris, C., Jourjon, G. & Marquez-Barja, J. M. (2016) Applying a methodology for the design, delivery and evaluation of learning resources for remote experimentation. *IEEE Global Engineering Education Conference (EDUCON).* Abu Dhabi, UAE http://ict-forge.eu/wp-content/uploads/2013/11/EDUCON-2016-full-paper.pdf, IEEE Education Society Publications.

[24] Mikroyannidis, A., Domingue, J., Third, A., Smith, A. & Guarda, N. (2015) Online Learning and Experimentation via Interactive Learning Resources. *3rd Experiment@International Conference (exp.at'15).* Ponta Delgada, São Miguel Island, Azores, Portugal http://oro.open.ac.uk/43381/, IEEE Computer Society Publications.

[25] Mishra, P. & Koehler, M. J. (2006) Technological pedagogical content knowledge: A framework for teacher knowledge. *Teachers College Record,* 108(6), 1017.

[26] Mor, Y. & Craft, B. (2012) Learning design: reflections upon the current landscape. *Research in Learning Technology*, 20.

[27] Shum, S. B., Knight, S. & Littleton, K. (2012) Learning analytics. *UNESCO Institute for Information Technologies in Education. Policy Brief.* Citeseer.

[28] Siemens, G. (2014) The Journal of Learning Analytics: Supporting and Promoting Learning Analytics Research. *Journal of Learning Analytics,* 1(1), 3–5.

[29] Smith, P. & Ragan, T. (2005) Foundations of Instructional Design. In, Instructional Design. NJ: John Wiley & Sons Inc. pp. 17–37.

18

Triangle: 5G Applications and Devices Benchmarking

Almudena Díaz Zayas[1], Alberto Salmerón[1], Pedro Merino[1], Andrea F. Cattoni[2], Germán Corrales Madueno[2], Michael Diedonne[3], Frederik Carlier[4], Bart Saint Germain[4], Donald Morris[5], Ricardo Figueiredo[5], Jeanne Caffrey[5], Janie Baños[6], Carlos Cárdenas[6], Niall Roche[7] and Alastair Moore[7]

[1]University of Málaga, Andalucía Tech, Spain
[2]Keysight Technologies Denmark
[3]Keysight Technologies Belgium
[4]Quamotion, Belgium
[5]RedZinc, Ireland
[6]AT4 wireless, Spain
[7]University College London, UK

Abstract

The FIRE project TRIANGLE is building a framework to help app developers and device manufacturers in the evolving 5G sector to test and benchmark new mobile applications, devices, and services utilizing existing and extended FIRE testbeds. Connected apps will be a dominant software component in the 5G telco domain. Ensuring a correct and efficient behaviour of the applications and devices becomes a critical factor for the mobile communications market to meet the expectations of final users. While radio related certification of mobile devices has a strong standards based ecosystem there is still a lack of consensus on the benchmarking or testing methods at the apps level. The project will identify reference deployment scenarios, will define new KPIs and QoE metrics, will develop new testing methodologies and tools, and will design a complete evaluation scheme for apps and devices. At the same time the methodology to be used in the design and development of the TRIANGLE test framework will ensure that the testbed end user is not overwhelmed by the complexity of the overall testbed by providing an intuitive high level configuration layer for the experiments and a flexible framework architecture to incorporate new 5G networking topologies as they become available.

Keywords: LTE, 5G, Mobile Applications, Mobile Devices, Benchmarking.

18.1 Introduction

The focus of TRIANGLE project [1] is the development of a test framework that facilitates the evaluation of the QoE of new mobile applications and devices designed to operate in the future 5G mobile broadband networks. The framework will include testbeds which will comprise test equipment and test software, formal test specifications and test methodology and will exploit existing FIRE facilities adding new capabilities when necessary.

The project will identify reference deployment scenarios, will define new KPIs (Key Performance Indicators) and QoE metrics, will develop new testing methodologies and tools, and will design a complete evaluation scheme. The project will focus on the development of a framework to ensure users QoE in the new challenging situations, especially those due to heterogeneous networks and considering the important role software will have in the new 5G ecosystem.

The framework as value added will also provide the means to allow certification and quality mark for the applications compliant to the requirements and test specifications developed within the project. To ensure sustainability after the project the framework will be developed according to formal languages and methods and handover to key alliances. The formalization of the test scheme so that it can be used for certification will also be extensible to other FIRE test solutions. The outcome of the project will allow vendor differentiation, especially for start-ups and SMEs, in the current globalized and competitive markets and further visibility of FIRE facilities.

Moreover, it is expected that the proliferation of personal devices such as smartphones, tables, wearables and sensors will play a key role in health, safety, social and professional applications, areas in which testing is essential to guarantee performance and security issues under critical conditions such as mobility. In this respect TRIANGLE project will also focus on the testing of mobile devices.

The framework, methods and tools developed during the project will focus on providing the mechanisms to incorporate new wireless technologies and topologies envisaged in 5G and contribute to the new ecosystem.

The objectives of the project can be summarized as follows:

- Objective 1. Provision of a testing framework setting the pathway to test new applications and devices for the purposes of pre-normative benchmarking and ease the access of start-ups to a qualified testing environment.
- Objective 2. Development of networking infrastructures and measurement techniques and tools to pave the way for 5G scenarios.
- Objective 3. Foster collaboration between the FIRE community, certification bodies, testing houses, the research community and SMEs to maintain a strong competitive position of FIRE platforms in the industry and to improve the opportunities offered by FIRE to European technology organizations to build better devices and applications.

The project will be executed in a time frame where 4G mobile technologies mature and 5G is still in the requirements definition phase or early trials. Although there is an initial timeline and plan for IMT-2020 and technical discussion related to IMT2020 submission in RAN WGs will start from March 2016, a firm detailed architectural plan for 5G is not yet available. However 5G aspirations are well defined and the European industry is expected to significantly invest as we move towards 2020. Many new products and applications will be developed in the 4G world targeting the evolution towards 5G. For the success of a product it is very important to verify that it meets the standards and it functions close to the expectations of the final users before they become openly released both in the existing 4G scenario and in the targeted 5G uses cases. Being in a pre-normative, pre-standards phase for 5G, the approach within TRIANGLE is to work on the end-to-end testing setting the pathway towards the testing of fifth generation applications and their certification. Benchmarking against TRIANGLE test cases (TTC) will be provided to third parties (e.g., SMEs, app developers, devices vendors and network operators). An informal triangle mark will be provided based on the KPIs measured. Elements needed for a proper standalone certification scheme will be identified as well as the possible integration into existing mobile certification schemes such as GCF.

18.2 Motivation

The primary motivation of the TRIANGLE project is to promote the testing and benchmarking of mobile applications and devices in Europe as the industry moves towards 5G. This project will provide a pathway towards the verification of application level perceived performance in order to support qualified mobile developments in Europe, using FIRE testbeds as testing framework.

As shown in Figure 18.1 three distinct areas for testing and benchmarking are considered in the project: (i) applications and (ii) devices and (iii) mobile network operators. Applications are often provided by Small and Medium Enterprises (SMEs). Testing the performance of mobile application in the 5G uses cases defined by entities such as the Next Generation Mobile Network Alliance (NGMN) becomes critical due to the highly demanding requirements of these, which range from broadband access to low latency or higher user mobility. SMEs often find it difficult to gain access to testing processes under realistic network conditions; moreover it can be much harder for them to understand the requirements of standard bodies or even to know which testing scheme would be more appropriate for their products. In addition, the costs of testing (requiring specialised infrastructure) are high for small companies and start-ups, especially if the market share is small.

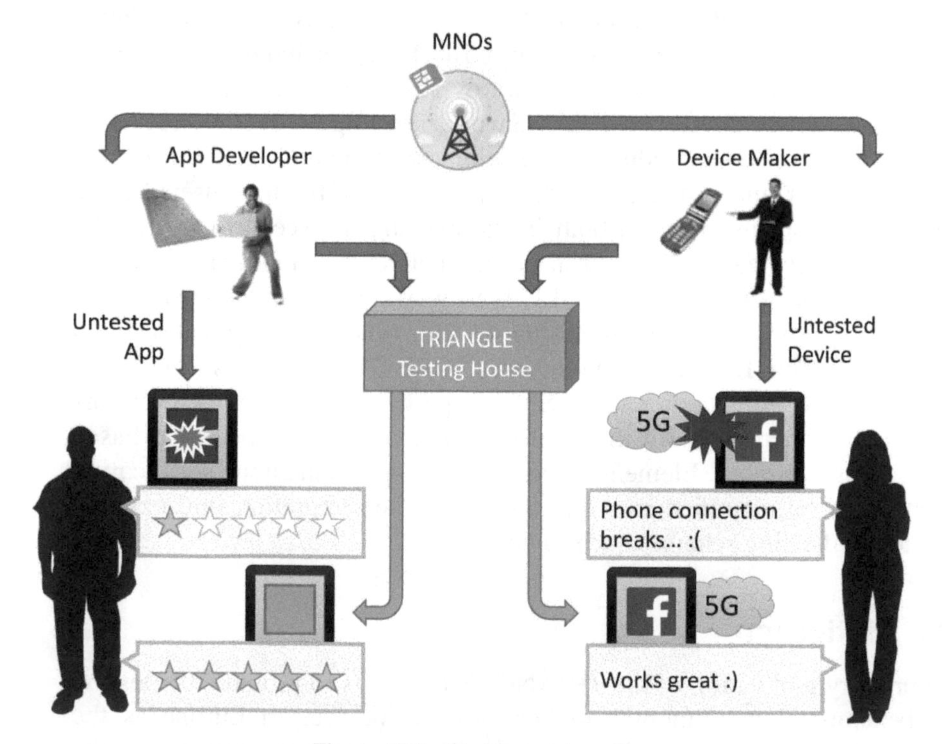

Figure 18.1 Problem statement.

The testing framework envisaged in this project can help to overcome these barriers. TRIANGLE project will provide not only a testing framework but also testing assessment through the provision of reference scenarios and Key Performance Indicators (KPIs) for a set of use cases covering areas defined for 5G by NGMN [2] and other standardization bodies [3–5].

18.3 Approach: Simplicity Operations for Testbed End Users

During the last years the estimated mobile traffic growth has been used as motivation for many wireless testbeds. Indeed this estimation is bigger year after year but has not reflected on the number of users of FIRE wireless testbeds. The main reason for this situation is the design of the FIRE testbeds themselves, which are network centric. Current testbeds are too focused on network configuration and have very complex and sophisticated configuration mechanisms, while the experimenters are not familiarized with the complex setup of the network resources and most of the time end up just using the default configuration. From experience obtained the federation of PerformNetworks testbed, it can said that most efforts were centered on providing access to all the low level parameters which have impact on the transport performance of the user traffic, however final users of the testbed do not know how to set up these parameters to generate a consistent experimentation scenarios, resulting in them.

The main idea that underpins the methodology to be used during the design and development of the TRIANGLE test framework is to ensure that the end user is not overwhelmed by the complexity of the overall testing testbed as a result of being exposed to its full set of details. In order to fully understand the testbed details the researcher will need multi-disciplinary knowledge (protocols, radio propagation, software, etc.). This is achieved by ensuring proper abstraction of underlying networking technologies by offering (see Figure 18.2): a) a high level configuration layer (personality) which calls on detailed scenarios definition, b) a flexible framework architecture to incorporate new 5G networking topologies when they become available The project will design a set of scenarios in the higher layer, scenarios that can be reproduced and whose final output is a Behaviour Indicator or Quality Mark, which defines how good the product (application or device) behaves when used in a realistic network, including energy consumption and model-based runtime checking of the apps and devices. Those scenarios can be modified by means of a scenario editor, which provides an API and a GUI to setup very

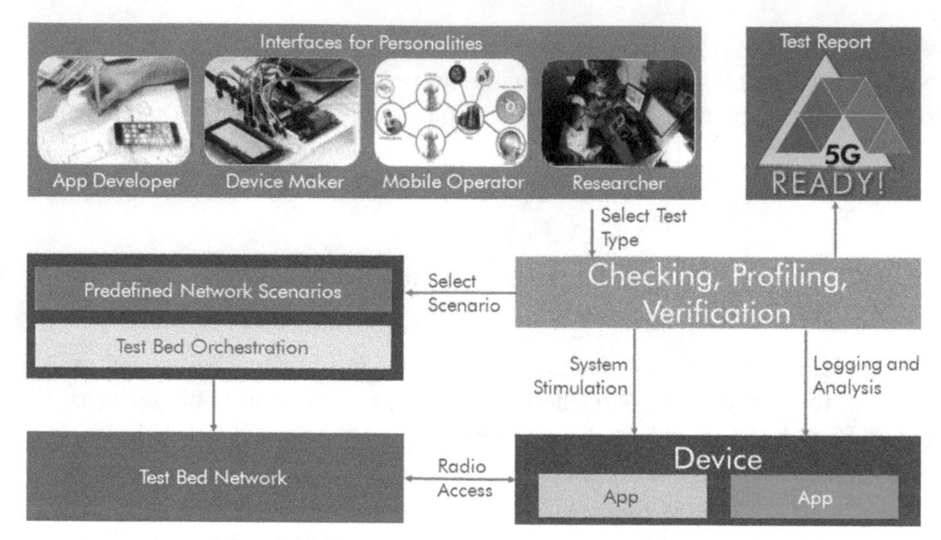

Figure 18.2 Triangle high level testing framework.

complex scenarios. This scenario editor can load pre-defined scenarios and can generate scenarios based on 5G use cases covering KPIs of interest for each one of them.

The project will provide a framework with different layers of abstraction using when possible commercial configuration interfaces as well as experi mentation standards, including those provided by Fed4FIRE [6]. Where needed, advanced users can be exposed to deeper configuration details and flexibility.

18.4 Technical Test Framework Approach and Methodology

18.4.1 TRIANGLE Components

TRIANGLE project will develop a testing framework based on existing testing know-how, and existing platforms: (i) the UXM Wireless Test Set from Keysight (ii) the UMA PerformNetworks testbed and tools, (iii) the AT4 Performance tool, (iv) the UCL app testbed, (v) the test automation tool from Quamotion and (vi) the RedZinc's virtual path slice solution VELOX. These existing platforms will be extended with new 5G requirements and functionalities.

An essential component in the testbed is an instrument capable of emulating multiple cellular networks in a controlled manner. To that end, TRIANGLE envisions the usage of the UXM Wireless Test Platform device by Keysight Technologies, which supports multiple radio access technologies (multi-RAT), including GSM/GPRS, UMTS and LTE-Advanced networks (i.e., 2G, 3G and 4G). The UXM features include intra-RAT and inter-RAT handovers, protocol debugging, IP end-to-end delay and throughput measurements, and performing RF conformance tests. Finally, it should be noted that the UXM also features an advanced fading engine with the main channels models defined by 3GPP.

The UCL App lab is a high level platform for distributing applications to a large-scale testbed for pre commercial testing and validation services. App Lab provides an app store offering in-the-wild user rapid field testing. App Lab collects valuable data for evaluation fast iteration of releases for app improvement cycles and includes audio and video capability as part of the government sponsored Innovate UK Digital Testbed. App Lab will allow the TRIANGLE partner organizations to have their own private mobile applications deployments for the pilot projects. As in standard app market-places (like Google Play or Apple Store), App Lab will allow developers to deliver mobile applications to pilot users (or a specific subset) after following an approval and publishing process. App Lab is comprised of client applications for Android and iOS mobile devices and a JEE server portal for application upload, management and distribution. The platform has a Web Management Console that carries out management tasks of the store: upload of new applications or updates (applications developed by the company, public applications or purchased applications), case definitions for testing, approval and publication of applications which will be adapted for the TRIANGLE pilot cases.

PerformNetworks (formerly PerformLTE) is a FIRE+ experimental platform, designed to offer a realistic experimentation environment covering LTE, LTE-A and future networks. The testbed is based on commercial off-the-shelf solutions (both in the radio and core network), software defined radio equipment and conformance testing equipment. The testbed offers a wide range of possibilities covering pilots, interoperability, performance evaluation, QoS, QoE and more. PerformNetworks is operated by the MORSE research group at the Universidad de Mlaga. The University of Mlaga also provides TestelDroid [10], a software tool that enables passive monitoring of radio parameters and data traffic in Android-based devices. Logging is implemented as an Android service that can be running in the background logging all the information while the application under test is being executed. This functionality enables monitoring of the traffic information generated by any application, which extends the testing to a very wide range of use cases. The parameters to be logged (network, neighbor cells, GPS, traffic) can be flexibly configured using the SCPI interface.

The AT4 Performance tool is composed of two components, Controller and Agents (data endpoints), and uses proprietary mechanisms to synchronize the Agents and provides accurate one-way measurements. This tool includes a built-in traffic generator with the capability of generating constant rates, ramps, loops and statistical traffic patterns which is something of utmost importance for setting up the desired environment in terms of varying traffic loads (e.g., for measuring LTE-U impact on Wi-Fi networks). Additionally, this tool has the ability to automate some mobile Apps on Android devices and measuring relevant QoE KPI such as YouTube buffering occurrences.

Quamotion Automate Mobile App Testing enables test automation, manual testing and exploratory testing of mobile apps. While RedZinc's virtual path slice solution enables that applications can demand to the network the prioritization of their traffic.

18.4.2 TRIANGLE's Components Orchestration

In order to orchestrate the components of the testbed and design repeatable test cases a control and management framework is also needed.

The the high level architecture of the TRIANGLE testbed, including this orchestration framework, is provided in Figure 18.3. On top of this architecture TRIANGLE will provide an online access Portal for people interested in running an experiment on the test bed, whether it is an app developer, a device maker, a telecom operator or in general a telecom engineer. The testbed Portal will be the main entry point for users wishing to have their applications tested/certified by TRIANGLE. It will provide an easy interface allowing users to request a testing campaign for an application, describe the scenarios which are part of the campaign and their parameters, check the execution of the tests, and obtain a report on the results.

Internally, the testbed Portal must be aware of the availability of resources in the testbed and must be capable of initiating the execution of testing campaigns. In order to reduce development time, TRIANGLE plans to use Labwiki as a the basis of the Portal. Labwiki is an existing web-based interface for OMF [8, 9] based testbeds, capable of running experiments and graphing its results.

Figure 18.4 provides a detailed description of the internal composition of the testbed. The architecture proposed is in-line with current tools promoted

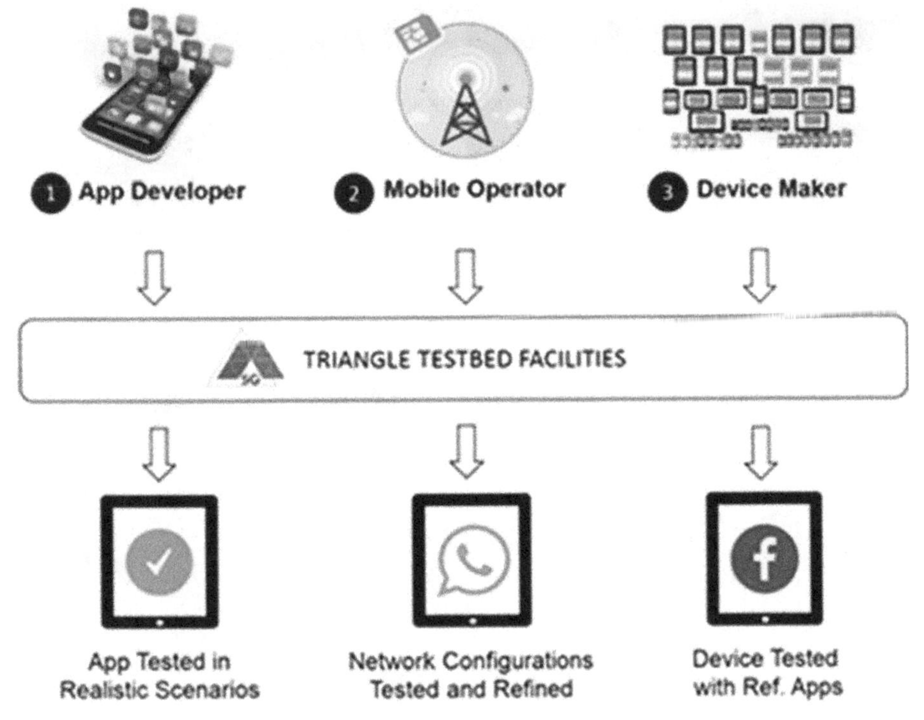

Figure 18.3 High level approach.

by Fed4Fire project and the FIRE community: OMF (Orbit Management Framework), OML (Orbit Measurement Libray), OEDL (OMF Experiment Description Language) and LabWiki [7].

OMF is the framework that manages experiment execution in Perform-Networks and other Fed4FIRE testbeds. It allows the definition of repeatable and automatable experiments thanks to the OEDL definition language. OEDL is a domain-specific language defined for the description of an experiment execution. OEDL provides a set of experiment-oriented commands and statements which can used to define the tests, the measurements and the graphical results. These OEDL scripts are interpreted by the Experiment Controller (EC), which orchestrates the resources present in the testbed during the execution of experiments. Each resource in the testbed is managed by a Resource Controller (RC). Thus, in the testbed, there are RCs for managing smartphones, network equipment, Quamotion tools, application servers, etc. Communication between the experiment controller and resource controllers is

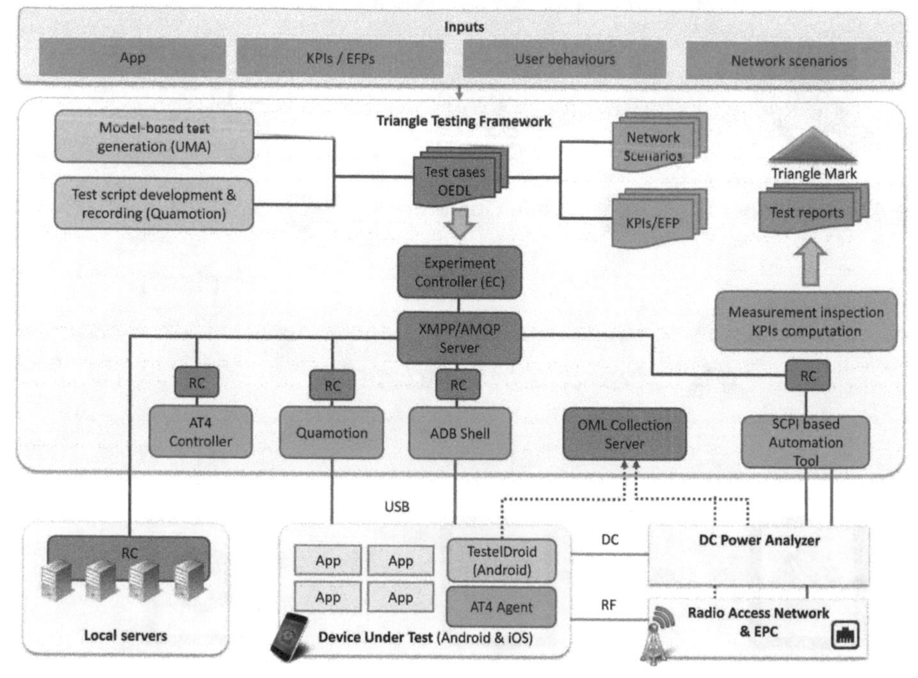

Figure 18.4 Testing framework architecture.

performed over the Advanced Message Queuing Protocol (AMQP). Measurement data obtained during experiment execution will be collected via OML in a central server. OML provides a programming library for easy application instrumentation, a collection point and a server which stores measurements in an experiment database. The instruments (eNodeB emulator and DC Power Analyser) are managed through the test automation platform provided by Keysight to manage SCPI (Standard Commands for Programmable Instruments) based instruments. The eNodeB emulator, is a generic platform used not only in conformance RF and signalling testing but also for design verification. In addition to LTE signalling and RF connection features, it also integrates channel emulation and digital generation of impairments such as AWGN, which is a critical feature for achieving high accuracy when setting Signal to Noise Ratio conditions. Standard multipath fading profiles defined by 3GPP are supported to emulate reference propagation conditions. The eNodeB emulator provides up to 3GPP release 10 and release 12 features, thus increasing the range of test possibilities with interesting network configurations. The DC Power Analyser is key to characterize power consumption in mobile devices.

Measurements collected by the tools and equipment available at the tesbed are sent to an OML collection server.

18.5 Testing Workflow Based on FIRE Technology

The application testing flow starts with the definition of the test cases. The test cases can be defined in several ways, supported by tools executed locally by the developer. The developer may write a script that contains the interactions that should be performed on the device (user behaviours), such as pressing buttons or entering text in text fields. Instead of writing the script by hand, this script may be generated by recording the interactions of the developer with a real device, which can then be replayed on the devices Both of these solutions will be provided by Quamotion. In addition, UMA will provide a model-based solution for app test case generation following the test specification methodology defined by AT4. The developer may provide a model of the possible user interactions with the app, which will then be used to automatically generate test cases. TRIANGLE users also have to indicate performance measurements (Key Performance Indicators (KPIs and Extra-Functional Properties) of interest from a list provided by the testbed. Finally the users of the testbed will select the network scenarios which are relevant for them: office, driving, pedestrian, Internet cafe, etc.

Once all the information is available the test cases can be completely defined and executed in the testbed. In order to coordinate the execution of the tools integrating the testbed each one will have associated a Resource Controller (RC). These RCs will allow the tools and the instruments to be controlled as part of the test, and receive commands from the experiment controller to execute a particular action. During the tests, the OML collection server will collect all the measurement results from all layers and measurement points present on the testbed. The results are passed to the Measurement inspection and KPI calculation, which will produce the final test report that leads to the "Triangle Mark".

18.6 Conclusion

TRIANGLE is the first FIRE project that provides a market oriented set of tools to perform a vendor-independent exhaustive analysis of a number of KPIs to qualify applications and devices in the pathway to 5G networks. TRIANGLE will provide a number of advances beyond the state of the art which includes to enable the testing with 5G networking features, to provide solutions for testing apps in the smartphone market, to provide apps oriented qualification of devices as a complement of radio access certification and to provide a sustainable business model involving stakeholders in certification and testing industry (including SMEs), research institutes (including FIRE) and apps developers.

References

[1] Andrea F. Cattoni et al., An End-to-End Testing Ecosystem for 5G, European Conference on Networks and Communications (EUCNC), Greece, 2016.

[2] 5G White Paper, Next Generation Mobile Networks (NGMN), 2015.

[3] Global Certification Forum (GCF), Key Performance Metrics, December 2014.

[4] GSM Association (GSMA), TS.09 Battery Life Measurement and Current Consumption Technique Version 7.6, June 2013.

[5] ETSI EG 202 810, Methods for Testing and Specification (MTS); Automated Interoperability Testing; Methodology and Framework, March 2010.

[6] Wim Vanderberghe et al., Architecture for the Heterogeneous Federation of Future Internet Experimentation facilities, Future Network & Mobile Summit 2013 Conference Proceedings, June 2013.

[7] G. Jourjon, T. Rakotoarivelo, C. Dwertmann, and M. Ott, International Conference on Computational Science, ICCS 2011 LabWiki: An Executable Paper Platform for Experiment-based Research, Procedia Comput. Sci., Vol. 4, pp. 697–706, 2011.

[8] T. Rakotoarivelo, M. Ott, G. Jourjon, and I. Seskar, OMF: A Control and Management Framework for Networking Testbeds, SIGOPS Oper. Syst. Rev., Vol. 43, No. 4, pp. 54–59, Jan. 2010.

[9] C. Dwertmann, M. Ergin, G. Jourjon, M. Ott, T. R. I. Seskar, and M. Gruteser, Mobile Experiments Made Easy with OMF/Orbit, Conference of the ACM Special Interest Group on Data Communication (SIGCOMM) on the applications, technologies, architectures, and protocols for computer communication, 2009.

[10] Andrés Álvarez, Almudena Díaz, Pedro Merino, F. Javier Rivas Tocado, Field measurements of mobile services with Android Smartphones, in *Consumer Communications and Networking Conference (CCNC)*. 2012 IEEE, pp.105–109, 14–17 Jan 2012.

PART IV

Research and Experimentation PROJECTS RECENTLY Funded

19

Recursive InterNetwork Architecture (ARCFIRE, Large-scale RINA benchmark on FIRE)

Sven van der Meer[1], Eduard Grasa[2], Leonardo Bergesio[2], Miquel Tarzan[2], Diego Lopez[3], Dimitri Staessens[4], Sander Vrijders[4], Vincenzo Maffione[5] and John Day[6]

[1]LMI
[2]Fundacio 12CAT, Spain
[3]TID
[4]University of Ghent, Belgium
[5]University of Pisa, Italy
[6]Boston University, USA

19.1 Introduction

The main goal of ARCFIRE is to bring RINA from the labs into the real-world. RINA, the Recursive InterNetwork Architecture, is an innovative "back-to-basics" network architecture that solves current limitations and facilitates full integration between distributed computing and networking. RINA addresses the challenges that drive the communications industry in moving from dedicated hardware to almost completely virtualised infrastructure.

New technologies such as 5G will change the communication industry even more significantly before 2020. Here, ARCFIRE contributes by providing experimental evidence of RINA's benefits, at large scale, in compelling and realistic business cases. This will motivate RINA adoption. ARCFIRE demonstrates – experimentally – RINA's key benefits integrating current European investment in advanced networks (IRATI, PRISTINE) and Future Internet testbeds (FIRE+) focusing on five goals:

1. Facilitate comparison of converged operator networks using RINA against operators' current network designs;
2. Produce a robust RINA software suite ready for Europe to engage in large-scale deployments and long-living experiments;
3. Provide relevant experimental evidence of RINA benefits to network operators, their equipment vendors, application developers and end-users;
4. Build on the current EU Future Internet community and raise the number of organisations involved in RINA development and deployment;
5. Enhance the FIRE+ infrastructure with ready to use RINA software.

ARCFIRE will have long-term sustainable impact on how we build infrastructure for the Networked Society. ARCFIRE's deployed software suite will enable equipment vendors to shorten their innovation life cycle, network operators to run advanced networks addressing their needs in a future-proof fashion, European SME's to find and exploit specialised markets, and application developers to explore unseen opportunities.

19.2 Problem Statement

The leitmotiv of ARCFIRE is to *experimentally demonstrate at large scale they key benefits of RINA, leveraging former EC investments in Future Internet testbeds (FIRE+) and in the development of the basic RINA technology (IRATI, PRISTINE).*
ARCFIRE's contribution is

1. showcase the benefits and viability of RINA via large-scale experimental deployments;
2. quantify those benefits by comparing RINA with current Internet technologies using different Key Performance Indicators (KPIs) and
3. motivate the academic and industrial computer networking research communities to engage in RINA research, development and innovation activities.

ARCFIRE addresses the following specific objectives:

1. *Compare the design of converged operator networks using RINA to state-of-the art operator network designs* – ARCFIRE analyses the design of current state of the art converged operator networks, carry out an equivalent design using RINA and compare both approaches using a set of KPIs.

2. *Produce a robust RINA software suite; mature enough for large-scale deployments and long-lived experiments* – IRATI, the most ambitious RINA implementation to date, is today mature enough to support short-lived experiments that allow only minor traffic variations in the range of a few hours (2–3) with a relatively small number of systems (up to 20), supporting only a couple of DIF levels. ARCFIRE improves the open source IRATI software suite so that it is possible to make large-scale experimental deployments with up to 100 nodes (physical or virtual), supporting tens to hundreds of DIFs, up to 5 levels deep, running experiments for up to a week. These metrics will allow the IRATI implementation to be used both for rich experimental research activities and for internal trial deployments by network operators.
3. *Provide relevant experimental evidence of RINA benefits for network operators, application developers and end-users* – ARCFIRE, via WP4, performs 4 extensive experiments with the goal of experimentally evaluating different aspects of converged RINA operator networks:
   - T4.2 looks at the benefits of RINA when managing multiple layers over multiple access technologies;
   - T4.3 assess how RINA improves the operation of resilient, virtualised services over heterogeneous physical media;
   - T4.4 analyses end-to-end service provisioning across multiple RINA network providers and
   - T4.5 studies the effectiveness of RINA against Distributed Denial of Service Attacks (DDoSs).

   All experiments will target large-scale deployments and run for relatively long periods of time (as defined in Objective 2).
4. *Raise the number of organisations involved in RINA research, development and innovation activities* – ARCFIRE implements a set of actions in order to raise the acceptance of RINA by the computer networking research community. These actions, refined as part of T5.1 activities, are designed to overcome two of the main reasons for the current low number of researchers involved in RINA R&D: facilitate the understanding of the RINA concepts and mechanics and disseminate experimental results that prove the benefits of RINA in high-impact scientific publications and conferences.
5. *Enhance FIRE+ as a platform for large-scale experimentation with RINA* – Facilitate experiments with the IRATI RINA implementation on the FIRE+ facilities by documenting all the experiments carried out

by the consortium using the FIRE+ infrastructure, publishing of all the configurations (and Virtual Machine) templates used in those experiments and adapting or extending the generic FIRE+ experiment provisioning, control and monitoring tools. ARCFIRE also provides feedback on these tools with respect to join FIRE+-GENI experiments.

19.3 Background and State of the Art

RINA, the Recursive InterNetwork Architecture is a "back to basics" approach learning from the experience with TCP/IP and other networking technologies in the past. To better understand the implications that this approach has uncovered and to explore its consequences, it is necessary to build systems that adhere to that architecture. Research results to date have found that many long-standing network problems are inherently solved by the structure of the resulting RINA theory of networking. Hence, additional mechanisms or, more commonly, the series of hacks and patches found with current technologies are not required.

Our back-to-basics approach reminded us that from its inception, networking was viewed as InterProcess Communication (IPC). Hence, RINA starts from the premise that networking is IPC and only IPC. Networking provides the means by which application processes on separate systems communicate, generalising the model of local IPC. Figure 19.1 shows a diagram of the RINA architectural model. In contrast to the fixed, five-layer model of the Internet, where each layer provides a different function, RINA is based on a single type of layer, which is repeated as many times as required by the network designer. The layer is called a Distributed IPC Facility (DIF), which is a distributed application that provides IPC services over a given scope[1] to the distributed applications above (which can be other DIFs or regular applications. These IPC services are defined by the DIF API, which provides operations to

1. allocate flows to other applications by specifying an application name and a set of characteristics for the flow such as delay, loss, capacity;
2. read and write data from the flows; and
3. de-allocate the flows and free the resources associated to them.

[1]Scope is the locus of distributed shared state that forms a layer, examples of these layers with different scopes may occur in point-to-point links, networks, networks of networks, virtual private networks, etc.

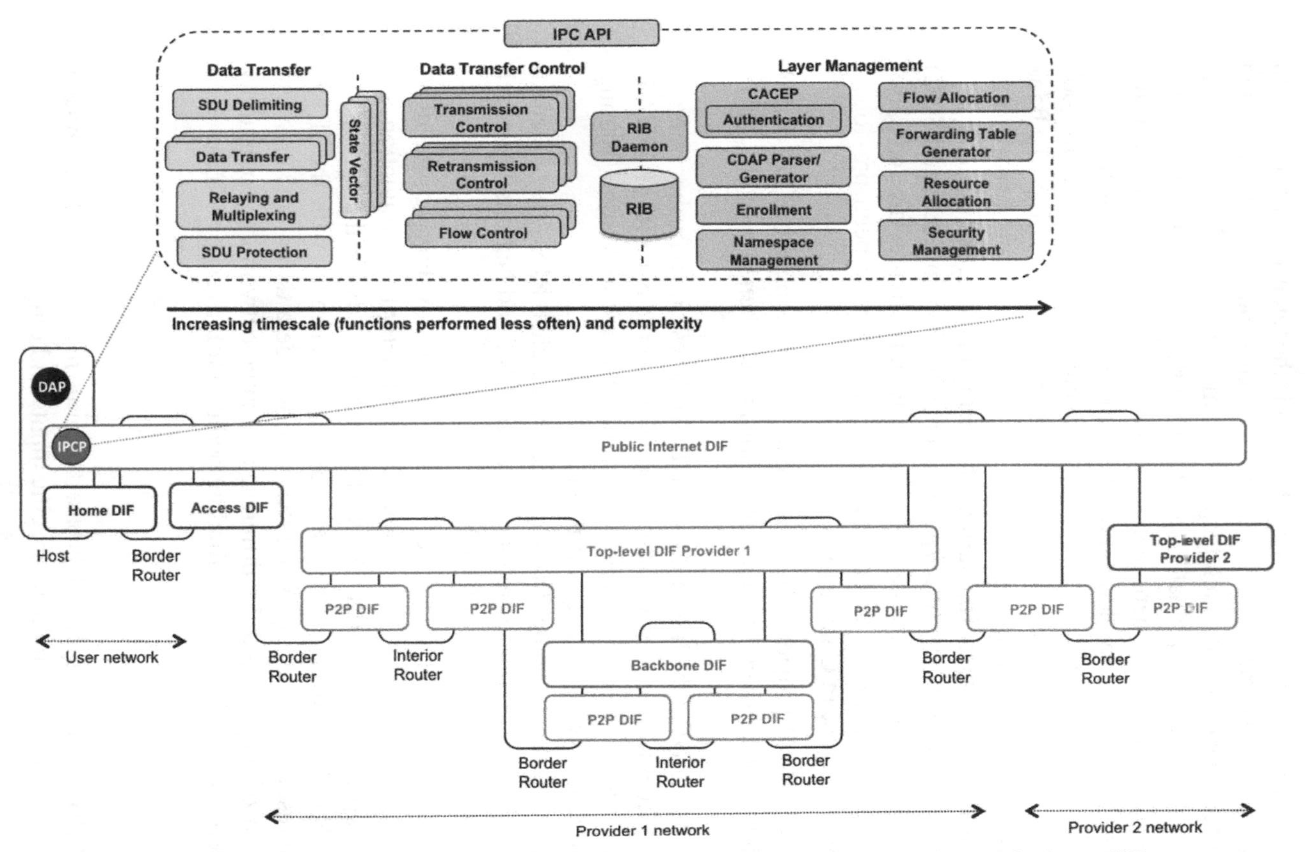

Figure 19.1 An example of the RINA architecture, with the same type of layer (DIF) repeated as required over different scopes. Different sets of policies customise each layer to its operational range.

All DIFs offer the same services through their API and have the same components and structure. However, not all the DIFs operate over the same scope and environment nor do they have to provide the same level of service. In RINA, invariant parts (mechanisms) and variant parts (policies) are separated in different components of the architecture. This makes it possible to customise the behaviour of a DIF to optimally operate in a certain environment with sets of policies for that environment instead of the traditional "one size fits all" approach or having to re-implement mechanisms over and over again.

The principles behind RINA, were first presented by John Day in his book "Patterns in Network Architecture: A return to Fundamentals". Since the book was published in 2008, several organisations have stated their interest in further researching RINA, as well as into turning the theory into practice by deploying RINA in the real world. The http://pouzinsociety.orgPouzin Society (PSOC) was formed in 2009 to coordinate all the international activities around RINA research and development. Three research initiatives have been previously funded by the European Commission:

- the FP7 IRATI project, which succeeded in developing the first RINA implementation over Ethernet and showing the benefits of the RINA structure;
- FP7 PRISTINE, which is building on IRATI's results to further improve this nascent technology and start demonstrating the benefits of RINA in specific areas such as congestion control, resource allocation, routing, security or network management; and
- the G3+ open call winner IRINA, which researched RINA as a potential alternative for the future architecture of GEANT and National Research and Education Networks (NRENs).

Both IRATI and PRISTINE have been/are very development-intensive projects, with a strong focus on implementation activities, simulation and low-scale experimentation. IRINA was focused on studying the benefits of RINA for NRENs and performing a small lab experiment with the IRATI stack. The results obtained by ARCFIRE will provide a definitive answer to the question of why should the different computer networking community stakeholders (academia, industry, funding bodies, etc) invest on RINA.

19.4 Approach

The ARCFIRE technical approach is based on three research and development activities, each of them with their specific inputs and outputs:

1. **Design of converged operator networks with RINA – WP2**
   This is the use case scenario of the project. ARCFIRE has chosen this scenario because it is the one that can best illustrate the benefits of RINA and allow for a greater diversity of experiments. ARCFIRE will take as an input the latest release of the RINA specification from the Pouzin Society, as well as the current state of the art in the design of converged networks (mainly contributed by ERICSSON and TID). The outputs of this design activity will be a set of new and enhanced RINA specifications (contributed back to the Pouzin Society). The specifications form the theoretical framework that will allow the design of the experiments and selection of KPIs and the set of new features in the RINA software suite that will need to be implemented by ARCFIRE.
2. **RINA software suite adaptation, enhancement and robustification – WP3**
   ARCFIRE will build on the results of FP7 IRATI, FP7 IRINA and FP7 PRISTINE and enhance the open source IRATI RINA software suite in order to make it compliant with the ARCFIRE requirements. This work will involve the design and development of new features required by the experimentation activities, but a good share of ARCFIRE's development activities will go into maturing the RINA stack, making it more stable and automated to enable large-scale deployments and long-running experiments. The resulting RINA software suite will be contributed back to the IRATI open source project, to make it available to the individuals and organisations interested in experimenting with RINA.
3. **Large-scale experimentation with the IRATI RINA implementation on FIRE+ facilities – WP4**
   The core activities of the project will exploit the large catalogue of Future Internet experimental facilities available in FIRE+, as well as GENI at the US. ARCFIRE will look at some of the key aspects in the operation of RINA-based converged operator networks to setup experiments that analyse and quantify some of the key benefits of RINA. These aspects include the management of multi-layer networks, provisioning of reliable services over heterogeneous physical media, end-to-end service provisioning across multiple network operators and effectiveness against DDoS attacks. Experimental activities will produce the key result of ARCFIRE: experimentally verification and quantification of RINA benefits over the current state of the art Internet architecture. A secondary but important output, which will be contributed back to FIRE+, are procedures and tools to facilitate large-scale RINA experimentation on FIRE+ experimental facilities.

19.5 Technical Work

Currently, operator networks suffer from a set of limitations and inefficiencies due to the design errors in the current Internet protocol suite and the series of patches that have been introduced to solve the problems caused by these errors. The greatest problem for the design and operation of service provider networks today is complexity. Today operators have to deploy separate, independently managed networks to support the public Internet and applications with strong quality guarantees, such as separate networks to carry voice to allow the usage of IP as a replacement of the SS7/TDM system in the Public Switch Telephone Network. The lack of a well-defined security architecture also fills the network with all sorts of equipment such as Network Address Translation boxes (NATs) or firewalls. The limited number of layers in the TCP/IP architecture, tied to the rigidness of their functions requires the introduction of pseudo-layers such as MPLS, VLANs, Q-in-Q, MAC-in-MAC, L2 and L3 tunnels and lately different network overlay/virtualisation technologies such as VXLAN or NGVRE; all of those requiring dedicated equipment and/or software to provide the new functionality. The poor support for mobility and multi-homing in IP networks requires a completely separate architecture for the mobile access network, with their own set of standards, protocols and equipment (GSM, UMTS or LTE as examples). Moreover, the lack of flexibility and adaptability in the Internet's transport layer makes it hard for operator networks to provide an optimal performance over heterogeneous physical media.

In contrast, the simple structure exhibited by RINA can be leveraged for designing simpler, more performant and predictable operator networks. Figure 19.2 shows a simplified example of an operator network (center), connected to a customer network (left) and to another operator network (right). If we focus at the border between the customer network and the operator 1 network, we can see that the customer gets access to one or more DIFs via the operator's top level DIF (called Operator 1 Metropolitan DIF in the Figure). These DIFs may be general-purpose DIFs with a large number of users (such as a public Internet DIF), or community or event application-specific DIFs with more specialised policies and tighter security. If the customer wishes to do so these set of DIFs can span to the customer's network (floating on the customer's internal layers with its own private addressing not visible from outside of the customer's network) and be exposed to individual applications running at the customer's premises: for instance, a "banking DIF" could be made available to a client application of an online banking tool, but hidden from other applications such as the web browser or the email client.

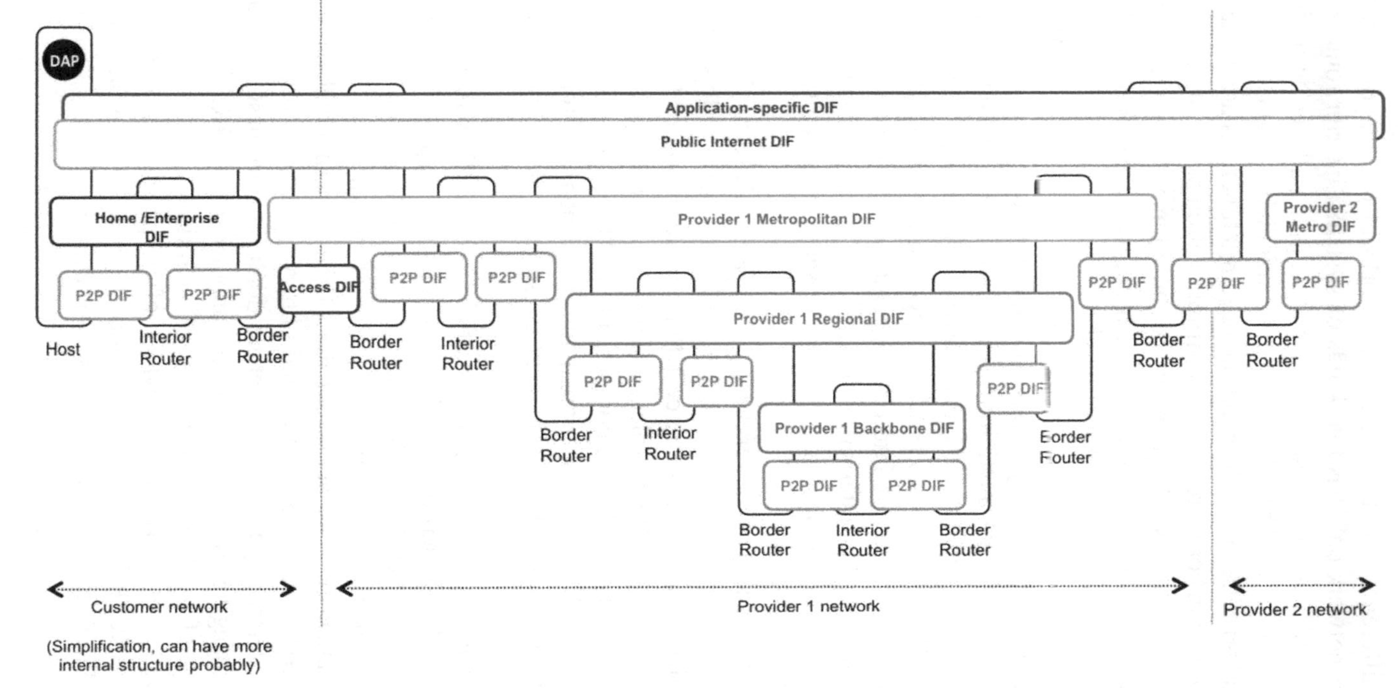

Figure 19.2 An example of a RINA operator network.

A crucial difference with the operator networks of today is the fact that RINA provider network only has two kinds of systems: interior routers and border routers (which is where recursion happens, going a layer up or down as shown in Figure 19.2). There is no need for middleboxes such as NATs, firewalls, tunnel termination devices: all the required functions are contained in interior routers or in border routers. The simplification potential in terms of management and operations is huge: networks built this way would be much simpler and cheaper to design, build and operate, which would make them easier to automate, more secure and predictable.

The designer of a RINA network can use as many layers (DIFs) as necessary to accomplish the design goals. The network designer can bound the size of the DIF, and just break a DIF into smaller DIFs if the first one becomes unwieldy. With RINA network designers have a structural tool to scale the network up: the DIF. There is no need to support layers than have to grow indefinitely such as in the current Internet. Also, interactions between DIFs are much more predictable than interactions between layers in the Internet, since all DIFs have the same interface and follow the same architectural patterns, albeit with different policies.

With regard to existing RINA software, ARCFIRE will use the IRATI open source RINA stack implementation, the Network Manager developed by the PRISTINE project and the open source RINA traffic generator (rina-tgen), initially developed by the IRINA project and now also part of the open source IRATI initiative. All these software components will be improved during the project lifetime. WP3 will be the work package responsible for adapting, improving and maintaining the RINA software suite for the purposes of the ARCFIRE project.

With regard to large scale experimentations, ARCFIRE will not focus only on a single FIRE+ facility; it will choose the most appropriate facility for the requirements of each individual experiment. This approach is today a realistic option due to the work the FIRE+ community has devoted to developing common tools for accessing the FIRE+ experimental facilities, deploying and controlling experiments and obtaining the experiments' data. By using this federated approach researchers don't have to master different toolsets when changing from one facility to another. FED4FIRE is leading the development of the FIRE+ federation of facilities, with 21 individual testbeds involved as of today, many of which offer open access programs to experimenters. ARCFIRE's preferred method to access FIRE+ facility will be via the FE4FIRE federation of testbeds. However, if a facility that is not member of FED4FIRE provides an interesting environment for the ARCFIRE experiments, the project will also consider its use.

19.6 Conclusion

ARCFIRE is guided by excellence in science and industrial leadership in integrated infrastructure with smart networking, focused on experimentally validating and benchmarking a breakthrough technology (RINA), thus bringing it closer to the market. In line with the H2020 objectives and the EU industrial policy goals, the impacts of ARCFIRE's results will contribute to the increase of European competitiveness and thought leadership in the area of networking and distributed computing, helping to create jobs and supporting growth. SMEs make up some $\frac{2}{3}$ of European industry's employment and a large share of EU industry's growth and jobs potential is to be found in its lively and dynamic SMEs. As such, the involvement of an SME in ARCFIRE NEXTWORKS – complemented by the participation of two industry players like Ericsson (the project coordinator) and Telefonica I+D is crucial for maximising the expected impact of ARCFIRE's actions. Last but not least, corporations Interoute – and SMEs in the External Advisory Board – TRIA Network Systems, Martin Geddes Consulting will provide a secondary exploitation path to ARCFIRE's outcomes.

20

ARMOUR

20.1 Project Objectives

Large-scale and distributed systems are good examples where the experimental approach is necessary. Such systems are built using very advanced features that have several multi-level interlinks/dependencies, which make it very difficult to analyse and predict the system overall behaviour. The runtime environments play a very important role in the overall performance and even different implementations of the same standard can impact their behaviour. Moreover, in large-scale distributed systems, the picture even more complex with the different resources potentially being heterogeneous, hierarchical, distributed or dynamic. Finally, failures, shared usages, etc. make the behaviour of large-scale distributed systems hard to predict[1]. The large-scale distributed Internet-of-Things (IoT) is a case where an experimentation (research) approach is required to have proper guarantee on its solutions.

The central goal of the ARMOUR project is then to perform large-scale experimentally-driven research as the way to provide properly tested Security & Trust solutions for large-scale IoT. "*Experiment is a test under controlled conditions that is made to demonstrate a known truth, examine the validity of a hypothesis, or determine the efficacy of something previously untried*"[2]. The ARMOUR experiments are aimed at determining the efficacy and performance of key Security & Trust methods in a large-scale distributed Internet-of-Things.

[1] Jens Gustedt, Emmanuel Jeannot, and Martin Quinson. Experimental Methodologies for Large-Scale Systems: A Survey. Parallel Processing Letters 2009 19:03, pp. 399–418.

[2] Experiment. (n.d.) American Heritage® Dictionary of the English Language, Fifth Edition. (2011). Retrieved April 10 2015 from http://www.thefreedictionary.com/experiment

ARMOUR identified 3 goals that define the approach being used to achieve the proposed Security and Trust solutions:

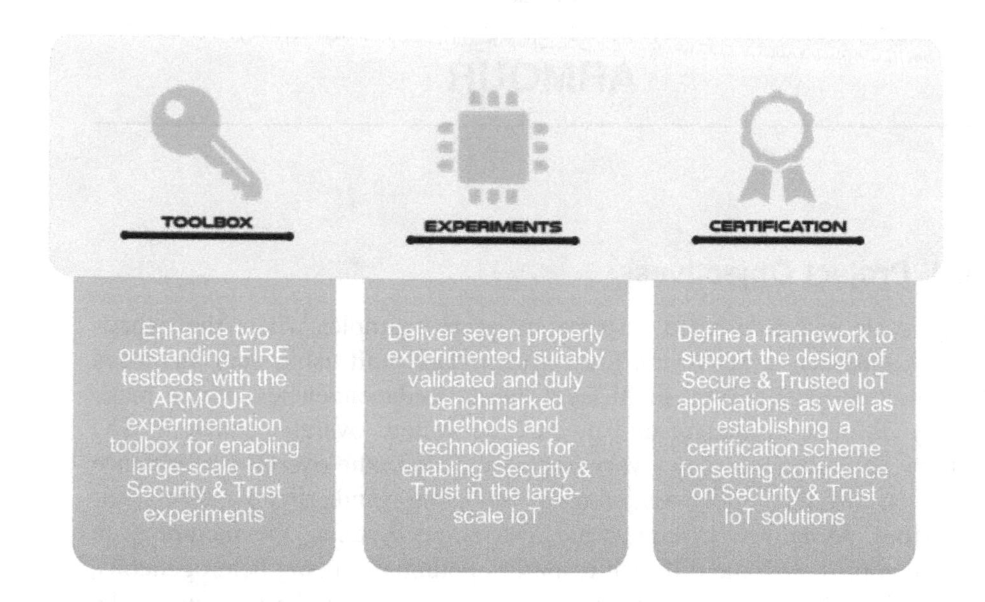

In testing and experimentation one often uses the term "large-scale" to denote an environment that exceeds the size, scope and capabilities of a laboratory environment. The notion of scale could not refer to the number of artefacts, whether these are switches, routers, computing nodes, sensors, cars, homes, etc. "Scale" can refer to the scope or extent of experimentation and "large" can imply heterogeneity based on the assumption that large-scale exceeds the borders of a single laboratory setting. "Large-scale" usually qualifies as in the upper thousands for sensors/small devices, or in the scope of routing nodes in the lower hundreds[3]. ARMOUR will perform large-scale experiments involving one-to-two thousand heterogeneous devices made available by a large-scale FIRE IoT facility – the FIT IoT-LAB testbed – that has been enhanced for supporting Security & Trusted experimentation.

Furthermore, a good experimentation implies verifying the repeatability, reproducibility, and reliability conditions in order to ensure generalisation of experimental results, and verifiability of their credibility. A proper experimentation methodology will be implemented, technologies subject to

[3]Experimentally driven research white paper, FP7-224524 FIRE Fireworks Support Action, April 2010.

experimentation will be benchmarked and even a new certification scheme will be designed for providing a "quality" label for large-scale IoT Secure & Trusted solutions. Also, applications of large-scale IoT Security & Trust will be studied and design guidance will be established for developing applications that are Secure and Trusted for the large-scale Internet-of-Things.

Finally, data and benchmarks from experiments will be properly handled, kept and made available via a FIRE data IoT facility – the FIESTA IoT/Cloud infrastructure. The FIESTA facility will be adapted and configured to hold the ARMOUR experimentation data and benchmarks. In this way, research data is properly preserved and made available to the research communities also making possible to compare results with experiments performed in other testbeds and/or also to confront results of disparate Security & Trust technologies.

20.2 Project Concept

In the following, we present a picture outlining the general main concepts subjacent to ARMOUR, and a brief description of the 7 concepts that shape the ARMOUR project. A detailed in-depth description of each of these concepts follows right after in the next sub-sections of this document.

- **Concept#1: ARMOUR Large-Scale IoT Security & Trust Testing Framework**. Security & Trust Experimentation on a large-scale Internet-of-Things brings some critical challenges for software testing techniques, concepts and tools in terms of business logic vulnerabilities, expected behaviour of IoT systems and the dimension, heterogeneity, compositionality and dynamicity of IoT systems. Presently no testing framework exists to cope with these challenges. ARMOUR will create a large-scale IoT Security & Trust testing framework that can be adapted easily to the various domains and experimented on the testbeds provided by the FIRE initiativc and beyond.

- **Concept#2: FIRE large-scale IoT testbed enabled for Security & Trust experimentations**. ARMOUR takes advantage of the unique FIRE FIT IoT-LAB for validating research results under large-scale real life conditions fostering the design and deployment of IoT Security & Trust solutions. For this, the IoT-LAB testbed will be enhanced with the ARMOUR Large-Scale IoT Security & Trust Testing Framework. The IoT-LAB testbed provides a unique open first class service to all IoT developers, researchers, integrators and developers: a large-scale experimental testbed allowing design, development, deployment and testing of innovative IoT applications. It offers a first class facility with thousands of wireless nodes to evaluate and experiment very large scale wireless IoT technologies ranging from low level protocols to advanced services integrated with Internet, accelerating the advent of ground-breaking networking technologies.
- **Concept#3: Large-Scale IoT Security & Trust Experiments**. With the FIRE FIT IoT-LAB ready for large- scale IoT Security & Trust experimentations it is possible then to perform the central goal of the project – the ARMOUR Large-Scale IoT Security & Trust Experiments. A set of experiments has been brought forward by the project partners (majorly by SMEs) based on their specific interests of technological performance improvement and/or innovation. These experiments will follow a well-defined methodology as to ensure reproducible, extensible, applicable and revisable experimentations. Experimentation process will be iterative in order to maximise solutions' efficacy.
- **Concept#4: Benchmarking Security & Trust technologies for the large-scale IoT**. It is a major necessity to provide tools for IoT stakeholders to evaluate the level of preparedness of their system to IoT security threats. Benchmarking is the typically approach to this and ARMOUR will be the first to establish a security benchmark for end-to-end security in the large-scale IoT. A new methodology for benchmarking Security & Trust technologies for IoT will be conceived (especially considering large-scale conditions) that will go beyond traditional approaches for security benchmarking by building up on the ARMOUR large-scale testing framework and process. And, the ARMOUR experiments will be benchmarked using the ARMOUR benchmarking methodology.

- **Concept#5: ARMOUR-tailored FIRE IoT/Cloud data testbed**. ARMOUR takes advantage of the FIESTA IoT/Cloud testbed to make experimentation and benchmarking data duly available, preserved, able to be inspected and visualised, and also making possible to compare data of experiments from disparate IoT testbeds. The FIESTA IoT/Cloud testbed provides access to and sharing of IoT datasets in a testbed-agnostic way and enables portability of IoT experiments across different testbeds, through the provision of interoperable standards-based IoT/cloud interfaces over disparate IoT experimental facilities. FIESTA implements a new first-of-a-kind meta-testbed that enables the execution of experiments that exploit data and resources from multiple underlying federated testbeds.
- **Concept#6: Certification/labelling scheme for Secure & Trusted IoT solutions**. Certification is a key element to support a specific level of trust on a (large-scale) IoT infrastructure/technology because the presence of non-certified IoT solutions/products could be open to vulnerabilities. ARMOUR will establish a rigorous certification scheme for labelling an IoT device/system with respect to (large-scale) Security & Trust. The ARMOUR benchmarking framework will be used as a basis for the certification activities so that IoT technologies and deployments could apply for a certificate to prove its security level toward third parties.
- **Concept#7: Secure & Trusted Internet-of-Things Applications**. It is fundamental to understand how security and privacy solutions are able to support the lifecycle of IoT applications. Particularly, how different security and privacy solutions or components, which are defined in their respective systems or contexts, can be used in a harmonised way to support the design and deployment of secure IoT applications. To this, ARMOUR will create procedures to test and validate the migration and the extendibility of IoT applications from the security and privacy viewpoints especially considering uses in a large-scale Internet-of-Things, e.g. considering the migration aspects (from one release to another of the IoT application) or the level of crypto-agility and flexibility, etc.

20.3 Project Approach

The ARMOUR project considers a large-scale experimentally-driven research approach. The large-scale distributed Internet-of-Things is a case where an experimentally-driven approach is required due to its high dimensionality, multi-level interdependencies and interactions, non-linear highly-dynamic behaviour, i.e. due to its complex nature. The large-scale experimentally-driven research approach makes possible to experiment and validate research technological solutions in large scale conditions and very close to real-life environments.

The ARMOUR large-scale experimentally-driven approach is realised by a well-established methodology for conducting good experiments that are reproducible, extensible, applicable and revisable. The methodology aims at verifying the repeatability, reproducibility, and reliability conditions to ensure generalisation of experimental results, and verifiability their credibility.

The general steps of the ARMOUR methodology encloses generically four steps: (1) Experimentation Definition & Supports; (2) Testbeds Preparation & Experimentation Set-up; (3) Experiments Execution, Analysis and Benchmark; (4) Certification/Labelling & Applications Framework. The figure below depicts the methodology.

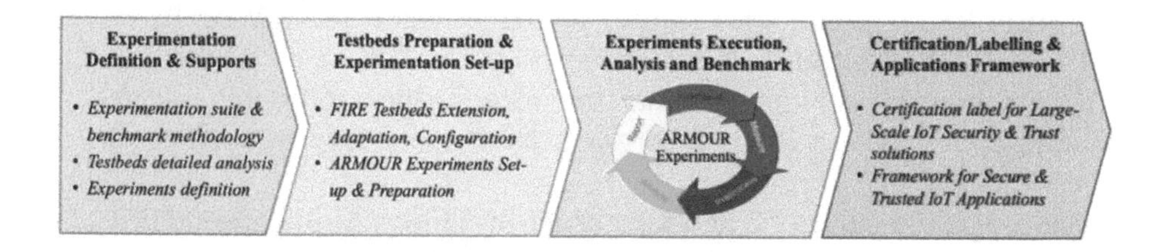

Step 1 – Experimentation Definition & Supports

The first step marks the start of the experimentation process and relates to the detailed definition of the ARMOUR experiments and the supports for experimentation (namely the Experimentation Suite and the Benchmarking Framework). This step basically involves:

- Definition of the IoT Security & Trust experiments (including defining testing scenarios, needed conditions, analysis dimensions) and the technological architecture for ARMOUR experimentations;
- Research and development of the ARMOUR technological experimentation suite and benchmarking methodology for executing, managing and benchmarking large-scale Security & Trust IoT experiments;
- Analyse the FIT IoT-LAB testbed and FIRE FIESTA IoT/Cloud testbed for assessing their specific composition, supports and services in view of the ARMOUR IoT Security & Trust experimentations.

Step 2 – Testbeds Preparation & Experiments Set-up

The second set of the experimentation methodology relates to establishing the proper conditions of conducting IoT Security & Trust experimentations using the selected testbeds and preparing the experiments. This step involves:

- Extending, adapting and configuring testbeds to enable IoT large-scale Security & Trust experiments:
  - Enhance FIT IoT-LAB testbed with the ARMOUR experimentation suite and, adjust/tune the testbed for multi-scenario large-scale IoT large-scale Security & Trust experiments;
  - Adapt and configure the FIRE FIESTA IoT/Cloud testbed for adequately supporting IoT Security & Trust experimentation data and benchmarks from ARMOUR large-scale experiments;
- Setting-up and preparing the ARMOUR experiments by specifying the security & trust test patterns for the experimentation that will then be used to execute and manage the experiments.

Step 3 – Experiments Execution, Analysis and Benchmark

The third step of the experimentation process relates to the actual execution of the ARMOUR experiments that represents the core research of the project to achieve the proven Security & Trust solutions for large-scale Internet-of-Things. This step takes the following sub-steps (iteratively):

- Configure – Install the scenario(s) for IoT large-scale Security & Trust experimentation;
- Measure – Do the measurements and collect the data from the ARMOUR experiments;
- Pre-process – Preform pre-processing and organisation of stored experimentation data;
- Analyse – Analyse data, do experiments benchmarking and compare performance;
- Report – Report on experimentation results (and possibly publish them even).

Step 4 – Certification/Labelling & Applications Framework

The final phase of the ARMOUR methodology relates to the creation of the certification label for large-scale IoT Security & Trust technologies and the establishment of a framework for Secure & Trusted IoT applications. These related especially with:

- A framework to define how different security & privacy solutions or components, defined in their respective systems or contexts, can be used to support the design/deployment of secure IoT applications;
- A new labelling scheme for high-dimensional Secure & Trusted Internet-of-Things solutions that provides the needed user and market confidence on their deployment, adoption and use.

21

Enabling a Mobility Back-End as a Robust Service (EMBERS)

Ricardo Vitorino[1], Margarida Campolargo[1] and Timur Friedman[2]

[1]Ubiwhere, Portugal
[2]UPMC, France

Several cities in Europe have started to provide real-life models of what a smart urban environment should be, through their deployments of innovative, sustainable, and efficient technologies. For instance, Vienna (Austria) is a leader in the domain of electric vehicles, with more than 400 charging stations in its streets, while also developing an open bike sharing system and car-share programs. Another example is Grenoble (France), with thorough testing of its Toyota electric vehicles fleet, integrated as part of the municipal transportation infrastructure, understood as a multimodal system, in which trips smoothly integrate the use of electric vehicles and the public transportation network. Many other cities are following these examples, establishing their own systems, with a strong emphasis on smart city mobility. The first generation of smart city mobility solutions tends to be "all-in-one" packages provided by a single vendor, which possibly comprises a diverse range of elements, such as:

- Sensors on the street (for instance for parked vehicles and traffic) and gateways that receive sensors' data via radio frequency communication and which pass the data on over fixed-line networks;
- A mobility back-end that receives and processes the data plus a management tool that controls the system;
- Information panels that display parking and traffic information in the city, apps that drivers use for navigation and for finding parking and websites that help guide users via the web.

The EMBERS project anticipates that a second generation of smart city technology will evolve to an entirely different solution. Ubiwhere, the company that is at the heart of EMBERS, is a provider of smart city solutions through its Citibrain brand, which realises that city managers, rightly, no longer wish to purchase single turnkey solutions. They would like to avoid locking their municipalities into a single vendor and become completely dependent upon that vendor for all of the system's components and maintenance. The same is true for urban mobility providers looking to upgrade their systems to a smart city ecosystem. As system assemblers, they seek the possibility to compose a multi-faceted system from multiple vendors, each one of which provides the best-of-breed component for the particular application domain that it serves.

With the purpose of allowing a municipality, or a system assembler working under contract, to put a set of diverse elements together into a single coherent system, each part must follow a common specification for interoperability, including, for instance, an open standard set of Application Programming Interfaces (APIs). With open specifications, the integration of third-party urban furniture, applications and other digital systems becomes an uncomplicated chore. Hence, cities are jumping at the possibility of more open systems, as exemplified by the Open & Agile Smart Cities initiative, in which 89 municipalities from 19 countries in Europe, Latin America and Asia-Pacific are adopting FIWARE's version of the Open Mobile Alliance's Next Generation Service Interfaces (OMA NGSI). By providing generic ways for a client and server to interact and change their state, the APIs that municipalities are embracing today mark an essential first step towards facilitating open data exchange. They are designed to promote interoperability across a full range of Internet of Things (IoT) devices, apps, and servers, but without going into individual application domains. The project's platform will go the next step by providing a free and open API that is specific for smart city mobility. It will allow developers of devices and apps to start immediately exchanging information about vehicles, roads, routes, parking spaces, drivers, and so forth, without having to reinvent this vocabulary and its rules or follow a closed, proprietary system.

EMBERS' first goal is to transform the Citibrain's platform into a market-ready solution. It is entirely more challenging to try to achieve interoperability with multiple third-party devices and applications by offering the platform via an open, public, mobility API for smart cities than to run it inside one's personal system, integrated with components developed by oneself, via a proprietary interface.

Much of EMBERS' effort is devoted to subjecting the solution to rigorous and extensive testing on FIRE+ testbeds to ensure its robustness. These testbeds are the FUSECO Playground, which offers a machine-to-machine (M2M) experimentation environment, and the FIT IoT-LAB facility, which allows testing with large numbers of sensor devices. While the first testbed has been used to test the machine-to-machine (M2M) interactions, the latter has been applied to test interactions of the solution with many different Internet of Things (IoT) devices.

EMBERS also aims at inviting designers of applications and devices to develop their interfaces to work with the smart city mobility API. Ubiwhere will offer this domain-specific interface via a full suite of lower-level APIs, to ensure the widest possible freedom of choice at a time when no single specification has emerged as a dominant standard. Participation will be encouraged both through hackathons and through an app contest, each offering prizes to the best entrants. Furthermore, three developers will receive funding via an open call, which will allow them to experiment with their applications in the same FIRE+ testing environments using the EMBERS platform. These activities focused on building third-party engagement intend to achieve EMBERS' second goal: to stimulate the smart city mobility ecosystem in Europe.

Finally, EMBERS third goal is to contribute back to FIRE+, by enhancing the experimentation capabilities of two of its principal facilities by the lessons learned. Furthermore, EMBERS will provide a valuable use case for efficient work with enterprises, which in turn is expected to motivate further usage in the future.

As referred before, municipalities are seeking alternatives to "all-in-one" closed smart city mobility systems, as they would like to be able to compose a system from multiple suppliers. With the possibility of combining the best-of-breed components from competing suppliers, city councils can serve better the interests of the citizens and the municipality. The key to enabling such freedom of choice is to choose systems that offer a set of standard and open APIs, which will allow the integration of multiple components.

In this context, EMBERS focuses on one central element, with which all of the other modules must operate: the mobility back-end, developed by the company Ubiwhere as a stand-alone product. In EMBERS, the deployment and testing in a set of varied scenarios using the FIRE+ facilities (FUSECO Playground and FIT IoT-LAB) will enable Ubiwhere to make critical design decisions about the product. Through dissemination events, such as the app challenge, hackathons, and ultimately an open call, EMBERS will bring in other companies that offer products in the smart city mobility field, allowing them to test the interoperability of their components against these same APIs and scenarios. In addition to helping the project's lead user, Ubiwhere, to move its product towards the market, EMBERS aims at transforming the smart city mobility ecosystem in Europe.

Three diagrams illustrate the overall concept: Figure 21.1 shows a fully integrated system of the most common type nowadays; Figure 21.2 shows the initial state of development of the system that EMBERS tests and Figure 21.3 illustrates the system as it will be at the end of the project.

In all three figures, there are devices and applications at the top and the mobility back-end at the bottom, which consists of a set of modules that handle all different aspects related to mobility in smart cities. The Parking Events module, for instance, tracks the occupancy of parking spaces and devices, such as road sensors, providing information to the module about vehicles arriving and leaving. An application might draw upon the module to inform users about the locations of available parking spaces and, similarly, devices such as

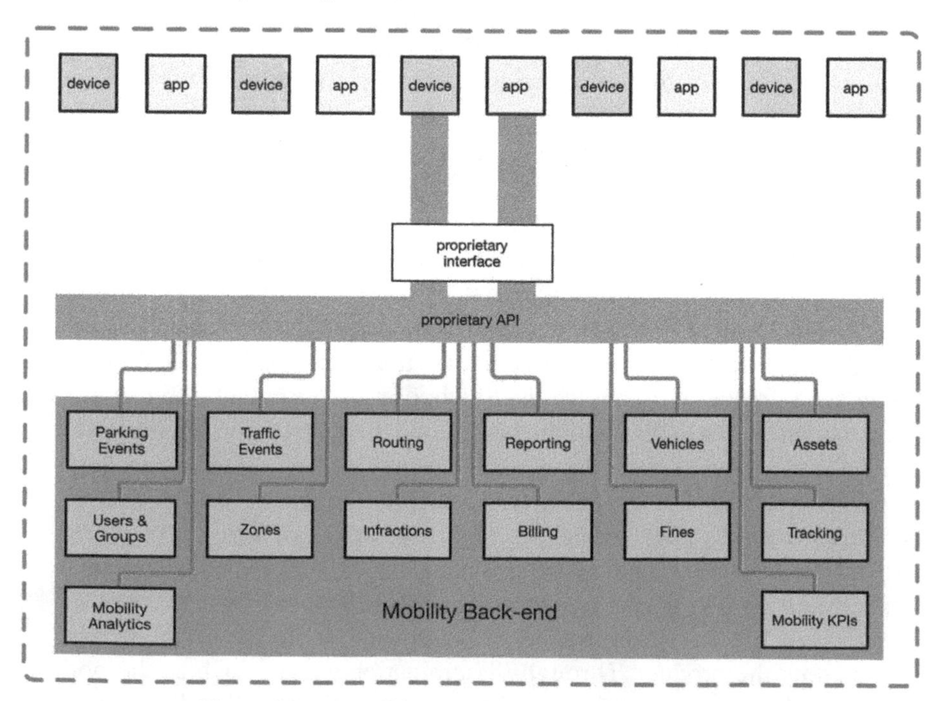

Figure 21.1 An all-in-one smart city mobility solution.

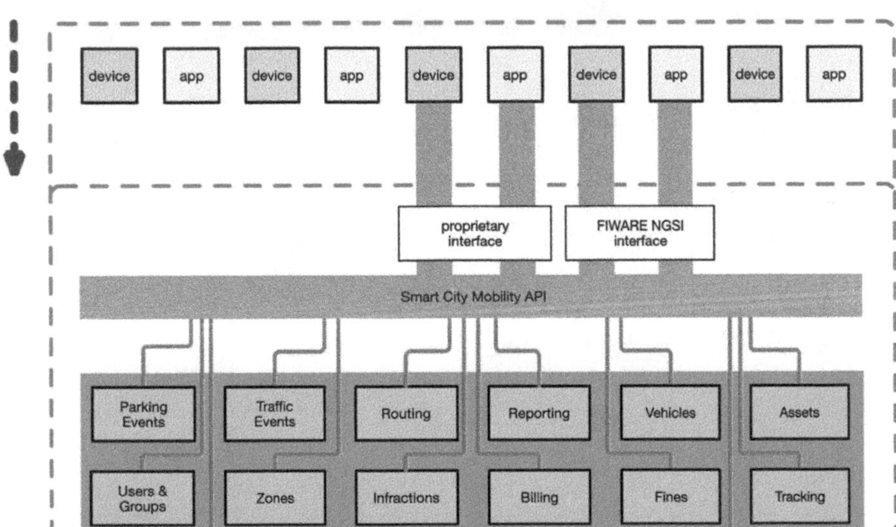

Figure 21.2 Positioning at the start of the EMBERS project.

roadside information panels can receive such data, supplied by applications as the identities of vehicles and their drivers. These exchanges take place via a domain-specific API, shown in grey, with each module providing a small part of the overall API. During Ubiwhere's first development of Citibrain's Smart Parking and Smart Traffic Management products, the domain-specific API was purely proprietary, just like the interface by which devices and apps connected to the back-end (shown in Figure 21.1 as a white box, above the domain-specific API). The dashed line around the entire system (see Figure 21.2) represents the product marketed, which is an "all-in-one" solution.

Figure 21.2 shows the changes devised and implemented as Ubiwhere prepared to break out the platform as a separate component. At this stage, the smart city mobility API became designed for external developers, structured around hardware and software development kits (SDKs). Alongside with the initial proprietary interface, a set of FIWARE core components is introduced, so that any devices or apps created for this interface can interact with the back-end in a standardised way. The aim, as shown by the dashed arrow, is to push down the boundary of the product (the dashed line), so that it no longer

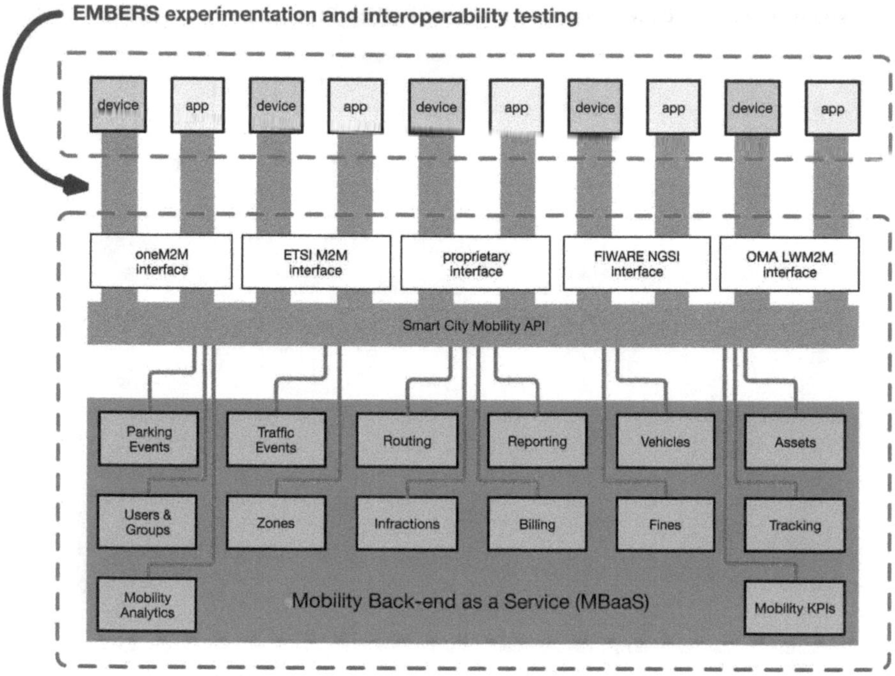

Figure 21.3 The EMBERS project and its result.

encompasses the devices and applications. In Figure 21.3, the work required from the EMBERS project can be better understood.

Alongside with FIWARE NGSI and Ubiwhere's proprietary interface, other generic interfaces will be available: oneM2M, ETSI M2M, and OMA LwM2M, as it is too early to tell which of these interfaces (if any) will ultimately emerge as a de facto standard. The EMBERS purpose is to lower the barriers to entry to the device and app developers' community (some of the developers may be writing to a specific type of interface while others support another interface) and, as such, the domain-specific smart city mobility API will be made available to them. Moreover, thanks to the FIRE+ testbeds, experimentation and interoperability testing will take place on the interactions between devices/apps and the platform, i.e. the interface between the two clearly separated boundaries.

With the hackathons, the app challenge, the open call experiments, and with some real-world pilots, such a platform, fully separated as a product, from the devices and apps, yet interacting with them via well-defined APIs, is perceived to be ready to market.

EMBERS will set up a federated M2M/IoT testing and experimentation infrastructure, based on the two FIRE+ testbeds that fit particularly well to this purpose: the FUSECO Playground in Germany, and FIT IoT-LAB in France. The first phase involves interconnecting the testbeds and enabling cross-domain testbed resource management (including user authentication, authorization, resource discovery, provisioning) and cross-domain experimentation (including experiment control and cross-domain monitoring) utilising the current standard FIRE+ tools and interfaces, such as FITeagle, MySlice, SFA, OMF, and OML. Afterwards, a standards-based (ETSI M2M, oneM2M) M2M platform, OpenMTC, is integrated into the federated infrastructure and the FIRE+ tools, mentioned above, for M2M/IoT resource management and experimentation. Setting up the core systems for testing allows the subsequent integration of Ubiwhere's Mobility Back-end as a Service into EMBERS' IoT/M2M test infrastructure. At this point, standards-based sensor/actuator communication (i.e., either via WebSockets, COAP, MQTT, RESTful interfaces) with the M2M/IoT platform, as well as standards-based interoperability between platform APIs and M2M/IoT applications (on top) is provided. Various sensor technologies, a) already existing in the two FIRE+ testbeds, b) provided by Ubiwhere, and c) provided by third party application developers or experimenters, are integrated into the EMBERS M2M/IoT experimentation environment. The whole physical infrastructure (including sensors and actuators), FIRE+ federation systems, and services for testing and resource management, M2M/IoT platform (OpenMTC integrated with Ubiwhere's MBaaS) becomes available for designing and scripting tests targeted at maturing and benchmarking the platform, repeatedly executed in different scenarios. Experiences and encountered issues while conducting end-to-end tests on EMBERS will be fed back to the M2M/IoT experimentation platform developers and used to improve the EMBERS facility, before, in the next phase, in the hackathon, app contest, and the open call, tests are being executed by third parties on top of the EMBERs M2M/IoT experimentation facility.

As for the city pilots, Ubiwhere will take the results of the experimentation in FIRE+ testbeds and translate them into a real life environment. Existing and new sensors for parking and traffic management will be deployed and integrated with the new version of the Mobility Platform. Ubiwhere will work with municipalities, local universities and innovation hubs to engage third parties who wish to create new applications for their cities, based on the capabilities at hand. Ubiwhere is also taking on the task of integrating transportation open and linked data from these regions to facilitate app development. Results of the pilots will be monitored, measured and validated against those collected in controlled FIRE+ infrastructure (and those obtained before the start of the project) so that the company and FIRE+ testbed owners can understand how the experimentation period allowed Ubiwhere to improve its product and increase its Technological Readiness Level (TRL).

22

F-Interop – Online Platform of Interoperability and Performance Tests for the Internet of Things

Sébastien Ziegler[1], Loïc Baron[2], Brecht Vermeulen[3] and Serge Fdida[2]

[1]Mandat International, Switzerland
[2]University Pierre & Marie Curie LIP6, France
[3]IMEC, Belgium

22.1 Introduction

The Internet of Things (IoT) will be massive and pervasive, with 50 to 100 Billion smart things and objects connected by 2020. It will impact diverse application domains, from smart cities to agriculture, from manufacturing to eHealth and energy. The success of this new technological revolution will require adequate standardization and interoperability. Since 1995, interoperability is recognized by the International Telecommunication Union (ITU) as the main obstacle to IoT development and adoption by the market.

F-Interop (www.f-interop.eu) [1] is a European research project addressing this challenge, by researching and developing an online platform of testing tools for the IoT, including:

- Interoperability tests
- Conformance tests
- Performance tests, including scalability, Quality of Service, and Privacy.

It intends to support and accelerate the development of emerging IoT standards and once the standard is stable enough to support SMEs to align and comply with such standards.

22.2 Context and Problematic

In order to be widely adopted, new technologies, products and solutions go through several steps:

- Standardization: stakeholders discuss and align their views to converge towards common standards and specifications.
- Conformance & Interoperability: stakeholders test and validate that their implementation is conform to the standard.
- Optimization: in terms of Quality of Service, scalability, energy consumption, etc.
- Market Launch: the solution is ready for roll-out into the market.

Each phase traditionally requires extensive testing, where different vendors meet face-to-face to test interoperability by going through an exhaustive list of "interoperability tests". The consequence is that:

- The current process is extremely labor-intensive, as engineers travel across the globe often only to find out what they need to make a minor fix;
- The cost associated with engineering time and travel expenses is often too high for SMEs;
- Time-to-market is unnecessarily stretched, giving vendors who want to adopt emerging standards a disadvantage compared to vendors who come to market with entirely proprietary solutions.

The concept of F-Interop is somehow to "dematerialize" the process of testing, exploiting the asset of the European FIRE research infrastructure. It aims to develop online and remote interoperability and performance test tools supporting emerging technologies from research to standardization and to market launch. The outcome will be a set of tools enabling:

- Standardization communities to save time and resources, to be more inclusive with partners who cannot afford travelling, and to accelerate standardization processes;
- SMEs and companies to develop standards-based interoperable products with a time-to-market cut by 6–12 months, and significantly lowered engineering/financial overhead.

22.3 Technical Approach and Outcomes

The goal of F-Interop is to extend FIRE+ with online interoperability and performance test tools supporting emerging IoT-related technologies from research to standardization and to market launch for the benefit of researchers, product development by SMEs, and standardization processes. Specifically, F-Interop will combine three complementary approaches:

22.3.1 Online Testing Tools

First and foremost, F-Interop is researching and developing online testing tools for the IoT, enabling to test interoperability, conformance, scalability, Quality of Service (QoS), the Quality of Experience (QoE), and energy efficiency of IoT devices and services.
Testbeds federations with a shared "Testbed as a Service"
F-Interop brings together 3 testbed federations and facilities, encompassing over 32 testbeds and thousands of IoT nodes, with:

- Fed4FIRE, which federates 24 FIRE+ testbeds, bringing together cloud, IoT, wireless, wireless mobile, LTE, cognitive radio, 5G, openflow, SDN, NFV and network emulation technologies.
- OneLab, which federates testbeds for the future Internet, including IoT, cognitive radio, wireless, cloud and overlay network technologies.
- IoT Lab, which federates IoT and crowdsourcing/crowd-sensing testbeds, including smart campus, smart building and smart office testbeds.

In order to support this integration, F-Interop is extending the testbeds federation architecture model with a new layer enabling shared services among several testbed federations. This approach enables to interface "Testbed as a Service" (TBaaS) with existing federations through a clearly specified API, enabling remote access and interaction with various experimental platforms.

As we can see, Fed4Fire, OneLab and IoT Lab were mostly providing access to raw resources (compute, storage, network, data). F-Interop will propose access to a higher, richer service focused on IoT testing and exploiting the resources of the federated underlying testbeds.

22.3.2 Support and to IoT Standardization and Industry

F-Interop works in close collaboration with several standardization bodies, and is directly contributing to three IoT standardization processes: oneM2M, 6TiSCH (IETF) and the Web of Things (W3C). It will also explore the possibility to support and enable new online certification and labelling mechanisms such as the IPv6 Ready logo. More generally, F-Interop intends to enable an easier participation of researchers and industry in standardization processes.

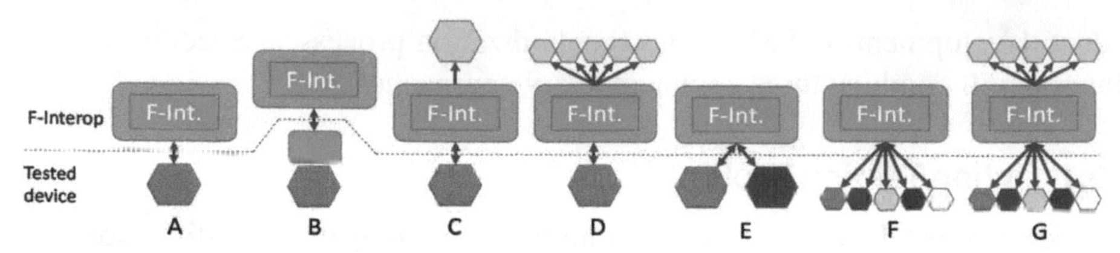

Figure 22.1 Multiple testing schemes.

It will also run an open call for SMEs and developers, inviting them to use and enrich the platform with additional modules and extensions.

22.3.3 Flexible Testing Schemes

F-Interop is researching and exploring various testing schemes and configurations, by interconnecting devices under tests with the server testing tools, resources provided by the F-Interop connected federations of testbeds, and resources provided by other users, as illustrated in Figure 22.1, where the salmon hexagon represents a device under test.

A. Tested Device using F-Interop Platform
B. Deported test with downloaded resource
C. Remote interop with 2 participants
D. Interop against testbed
E. Local interop with 2 participants
F. Remote interop with N participants
G. Remote interop with N participants and testbeds

22.4 Architectural View

22.4.1 F-Interop Platform and Test Tools

F-Interop offers a service allowing users to plug an IoT implementation and run interoperability, conformance and performance tests. The implementation under test can be a device owned by a user but it can also be hosted at a testbed among the three federations having a partnership with F-Interop: Fed4FIRE, OneLab and IoT Lab. This service is composed of the following building blocks:

- Implementations under test.
- Testbed federations (hosting implementations and the F-Interop service itself).

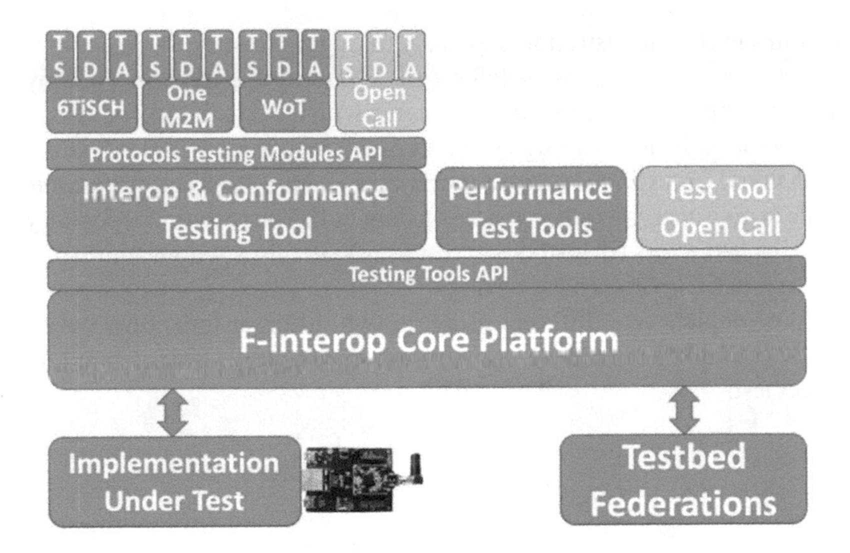

- F-Interop Core Platform is a bus of events ensuring the exchange of messages between devices and test tools.
- Test tools are responsible for orchestrating the execution of the tests.
- The tests must be defined in three steps: test specification (TS), test description (TD) and test analyzer (TD).
    - TS: test specification
    - TD: test description
    - TA: test analyzer issues verdicts about the results of the tests.

22.4.2 F-Interop Architecture

The basic architecture of F-interop answers three requirements:

- Allowing users to do remote interop, conformance and performance testing.
- Allow users to use devices in testbeds for this testing (varying from using only testbed resources such as IoT sensors, virtual machines, etc, to combining resources at the user premises with the testbed resources).
- Allow contributors to extend the F-interop framework with extra functionality and tests.

The figure below shows the basic architecture with two different users doing interop/conformance testing against the F-interop central servers. The graphical user interface (GUI) is the interaction point with the F-interop system (creating account, logging in, choosing test, start test, see results of tests).

The GUI talks to the orchestration and analysis engine to start, stop, analyze tests, while the orchestration engine talks to the F-interop agent running nearby the devices under test (e.g. on a laptop connected with USB to the IoT devices). All communication is done through secured AMQP messaging. After testing, results are stored in a central result repository (where a user can access his own results or give access to other people on demand). The resource repository lists all available devices (in testbeds and at user's premises who want to give remote access to their devices for testing).

The figure depicts two different users on two different locations, but of course these can be two users on the same location, or a single user, depending on the exact test.

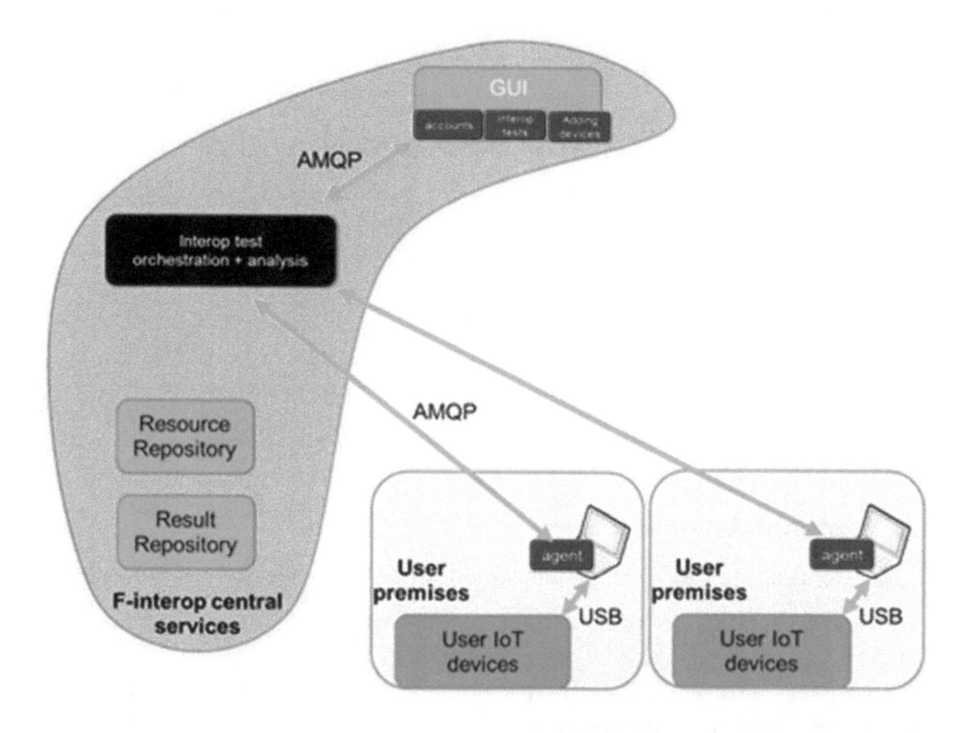

The next figure shows a similar architecture, but now resources at testbeds are involved. The figure depicts an example where a user tests devices at his premises against devices at a testbed, but of course alternatives such as running everything on a testbed, combining resources at multiple testbeds, etc are all possible. The components are merely the same, but as can be seen, there is now a TestBed as a Service Layer (TBaaS) which reserves and provisions resources at testbeds and then launches the test suites. Interesting here is that also the orchestration and analysis modules run on virtual machines on testbed and allow full access for users. This makes it possible for contributors of test and analysis software to deploy their own instantiation of part of the F-interop platform and as such they can extend it easily and contribute back to F-interop.

On top of the TBaaS layer, there is also an external API based on REST and based on the testbed standard resource description called an RSpec (XML based). A full blown test can be launched easily in this way.

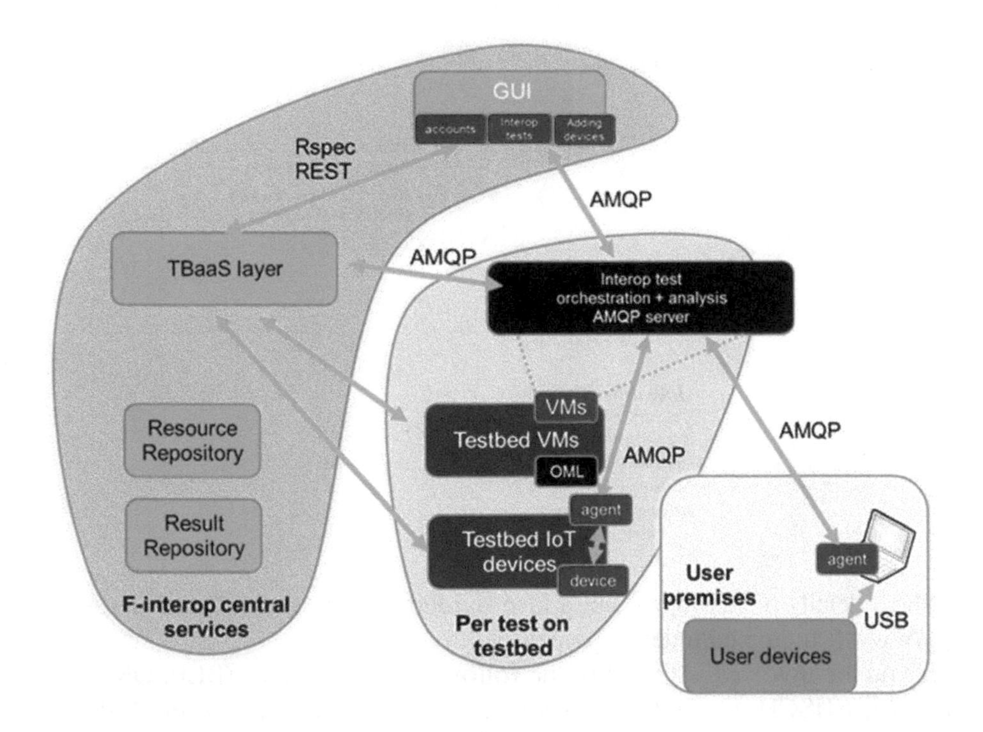

22.5 Open Call

F-Interop is already developing tools supporting a set of core protocols (e.g. 6TiSCH, One M2M, CoAP) and defining standard APIs to allow easy plug and extension with new tools and protocols. This will allow to inject events in the network in order to test protocols performance under different conditions and identify possible intervention points for modifications and improvements of standards.

To further growth the ecosystem of protocols, communities and devices the F-interop platform can support, thus cutting the time for interoperability tests of businesses operating in the IoT space, the project is now organising competitions to partner with external businesses (with funding ranging from 10 K to 100 K EUR depending on project nature) to extend the core platform with new tools and to exploit the current open nature of the designed architecture and APIs.

We are currently planning dissemination and communities engagement activities, in order to promote the project aims and to engage with third parties interested to apply for funding and to understand the possible tools that the platform might benefit from.

The timeline is provided below.

Now: Open Call and current architecture presentation
• EUCN, London Digital Catapult and other partners open F-Interop meetup
Until Jan 25th: Open call open to submission
• Target communities and promotion champions identified, continuous publicity
• Proposals submission
Until July 31st: Communities involvement and feedback collection/ review, final call shaping
• Protocols, tools and communities

More details on the open call and provided grants can be found here: http://www.f-interop.eu/index.php/open-call

If you are involved with one of the following communities (BLE, OMA LWM2M, 6TiSCH, CoAP/6LowPAN/RPL, IPv6 ready, Low Power Wide Area Network, Web of Things, Hypercat, COEL, 5G IoT) or new emerging ones (e.g. Internet of Underwater Things) in the IoT ecosystem, you might be interested in the F-interop open call.

22.6 Conclusion

F-Interop is a new project that strongly changes the potential usage of the FIRE testbeds. Indeed, the principle is to provide remote testing, conformance and performance test exploiting the large set of resources exposed by the underlying testbeds or federation of testbeds. In order to achieve this goal, F-Interop is developing architecture for TaaS as well as an adequate APIs to bridge the gap between what exists today, serving general-purpose experiments and our target related to IoT interoperability testing. Therefore, we can see that F-Interop will bring together two communities (FIRE and Test community) and expect to impact the way testing is done today, especially in the growing field of IoT where heterogeneity is key.

23

Q4Health: Mission Critical Communications Over LTE and Future 5G Technologies

Cesar A. Garcia Perez[1], Alvaro Rios[1], Pedro Merino[1], Kostas Katsalis[2], Navid Nikaein[2], Ricardo Figueiredo[3], Donal Morris[3], Terry O'Callaghan[3] and Pilar Rodriguez[3]

[1]Universidad de Málaga, Andalucia Tech, Spain
[2]Eurecom, Communication System Department, Sophia Antipolis, France
[3]Redzinc Services Limited, Dublin, Ireland

Abstract

Mission critical communications have been traditionally provided with proprietary communication systems (like Tetra), offering a limited set of capabilities, and mainly targeting voice services. Nevertheless, the current explosion of mobile communications and the need for increased performance and availability especially in mission critical scenarios, require a broad type of services to be available for these platforms. In this sense the LTE technology is very promising, as it provides mechanisms to enforce QoS, has standardized many useful functions in public safety scenarios (like group communications, positioning services, etc.), while it is being evolved to match future 5G requirements. The Q4Health project aims to prepare for market and optimize the BlueEye system, a video service platform for first responders. In our approach we use two FIRE+ platforms for demonstrations: OpenAirInterface and PerformNetworks. Q4Health is driving the optimization of the system with the execution of a set of experiments focusing on a different aspect of the network (core network, radio access and user equipment) and aims to cover current LTE standards, but also future 5G enhancements. The projects™ outcomes will be the optimization of the overall BlueEye system and the enrichment of the involved FIRE+ facilities with more services, functions and programmability.

Keywords: LTE, 5G, Mobile Communications, QoS, QoE.

23.1 Introduction

Q4Health project[1] [1] aims to improve two existing FIRE platforms PerformNetworks[2] [2] and OpenAirInterface[3] [3] in order to provide better innovation services to third parties. This is done by a use case provided by the company Redzinc Services Ltd. that provides BlueEye, a wearable real-time video application for first responders, and VELOX, a virtual path slice solution to enable QoS.

The project is based on a scenario in which first responders of a medical emergency (i.e. paramedics in an ambulance) have a wearable video equipment in the form of hands-free glasses with a dedicated LTE connection, and the objective is to guarantee the video transmission to a hospital where a doctor can monitor the condition of the patient in real time and suggest different treatments in its way to the emergency room. The main challenge is to achieve an interruption-free video broadcast while the ambulance pass through different LTE cells and available Wi-Fi hotspot, always within the accepted parameters defined for this type of traffic (under 150 ms for both audio and video transmissions) [4].

Through a series of experiments in several components of a mobile network infrastructure, all the components of the project will be improved. PerformNetworks will support 5G low latency prototypes and new environments to optimize heterogeneous handover, OpenAirInterface will be extended to test antenna performance and to provide an API for the eNodeB scheduler, BlueEye video service will be optimized to react better to channel and network conditions and VELOX will implement new drivers to expand its end to end QoS capabilities. The experiments will cover all the components and stacks of the network, from the physical layer in the eNodeB to a high level parameter optimization in the *Evolved Packet Core* (EPC). The optimization of all the platforms will improve the experimentation services offered by the two FIRE platforms and will accelerate the time-to-market of the BlueEye and VELOX systems. The experiments are designed to overcome the following challenges, that have been previously identified in field campaigns:

[1]http://www.q4health.eu/

[2]http://performnetworks.morse.uma.es/

[3]http://www.openairinterface.org/

- Trigger QoS enforcement procedures dynamically in the LTE network.
- Reduce the end to end latency of mission critical services.
- Improve the behavior of the eNodeB scheduling when attending real-time applications.
- Improve service coverage inside buildings by the introduction of heterogeneous handover and service adaptation.
- Support low latency group communications between adjacent peers.
- Optimize the service quality for the different components and services of an eHealth system (sensors with different criticallity, video, audio, etc.).

The experiments designed to overcome this challenges can be divided in two main groups, one focused on the radio access (described in Section 23.3) which include the base stations and the *user equipment* (UE), and another focused on the EPC (described in Section 23.4) in which new functionality is explored, such as recent 3GPP standards and also the evolution towards 5G.

23.2 Motivation

In most of the world the *Public Protection and Disaster Relief* (PPDR) services use proprietary technologies for its communication needs [5]. Although these technologies have visible advantages for this kind of services, namely the robustness of the communication and the constant availability, for some time now its users have been looking for alternatives that alleviate its shortcomings. One of the main problems of the systems used by first responders, like the *Terrestrial Trunked Radio* (TETRA) [6] used in Europe, is the cost of deploying and maintain a large network to provide coverage to only a few subscribers. This cost also prevents research and advancements in those systems because only large groups or companies can assume that monetary investment, and these groups are not willing to commit resources to upgrade services that have a very low number of customers. The end result of this lack of investment is that emergency communication services don't evolve with the rest of the technology. We can see an obvious example of the misalignment between the capacity of the systems available to emergency teams and those available to the general public looking at the situations of data transmission in mobile networks. While most of us have a theoretical capacity of transmit tens of megabits per second, the TETRA system mentioned earlier in its basic form has a limit of only 36 kbps.

For these reasons there is a push from both the network operators and the final users to upgrade these aging systems to new technologies like LTE [7, 8] which natively support many of the features required by mission-critical communications [9]. For the formers the main motivation is the cost reduction of providing premium services through already deployed networks, while the users can see a great benefit in the use of new technologies like image or video transmission to the operating centers in emergency situations, capability made unavailable in current systems simply due the lack of bandwidth.

With LTE as a basis of the future 5G mobile technology there is also a need to characterize and optimize data links in the search of low-latency and high priority communications [10]. This involves work in every level of the mobile network stack, from the analysis and optimization of the RF schedulers present in the radio station to the identification and dispatch of traffic from specific users or services in the core of the operator network. There is currently a trend that believes the performance of this classic division in layers and its corresponding architectural function (i.e. latency in the physical layers, throughput and network addressing in higher layers) can be greatly expanded by flattening the network structure with the introduction of *Software Defined Network* (SDN) functionality traversal to the architecture [11, 12].

With these additions new possibilities are suddenly available to optimize the communication paths for the traffic characteristics of the PPDR users. By introducing SDN technology in the access layers we can create new services like data broadcast for groups of users without adding new traffic in the backhaul network, or reducing E2E latency by spread geographically the functionality of some of the entities of the operator network converting them in new *Network Virtual Functions* (NVF) that can be executed in the node that needs such services.

23.3 Experiments Focused on the Radio Access

Different radio access equipment can be used in the project depending on the objective of the experiment. The equipment available consists of:

- Commercial equipment used to optimize the platform for the current deployments. This includes the commercial deployments, that can be used just for characterization purposes, and the proprietary indoor deployment available in PerformNetworks, that provides similar functionality but with the possibility of changing the configuration of all the elements.

- Conformance testing equipment. This type of devices is used by manufacturers on the design and verification process of LTE equipment. The conformance testing equipment can be configured to broadcast in any commercial frequency (which allows it to be used with devices of different countries). Furthermore it enables the fine configuration of many parameters on the LTE stack (even non commercially available configurations) and it also provides channel emulation, that can be used to test the UE under different channel conditions but maintaining the reproducibility (enabling exhaustive characterization of the terminals).
- *Software Defined Radio* (SDR) based equipment. This is the more flexible approach for research, as this equipment combined with available open source stacks can be used to test new concepts on the lower layers. This will be the approach for the experiments that required experimentation beyond market.

It is planned to evaluate different LTE dongles in order to have a comparison of their behavior under different channel conditions and their application level performance. This comparison will be mainly performed in the conformance testing equipment and could be used to select the most appropriate for the application. Besides the characterization of different LTE UEs there are additional experiments foreseen, focused in different parameters of the radio interface and covering the antenna and the MAC schedulers.

The modem of the BlueEye platform should be kept in a belt so the selection of an appropriate antenna is very important. The evaluation of the different antennas will be made by the execution of a series of experiments covering multiple topologies under different interfering conditions. Several metrics of the application layer will be considered such as transmission rate, error rate, etc. The setup to be used for these experiments is depicted in Figure 23.1.

OAI eNodeB will be used as base station as it can provide measurements on the power reported by the UE while being connected to the PerformNetworks EPC[4]. The evaluation scenarios include:

- Static UE (the UE is not moving). The UE will be located in different fixed locations from the eNodeB and the measures will be acquired with and without obstacles.
- Moving UE. The idea of this scenario is to test the antennas in mobility conditions. In the laboratory scenarios the mobility profile will be pedestrian but a vehicular one could be used when evaluating on external networks.

[4]Provided by Polaris Networks.

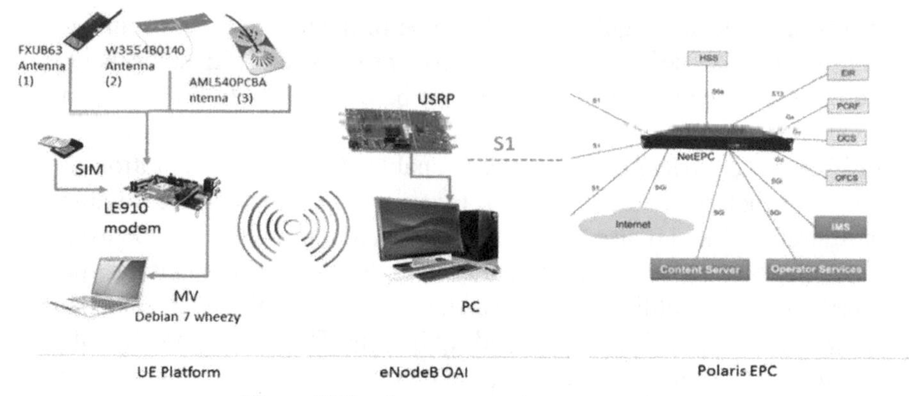

Figure 23.1 Antenna experiments overview.

The behavior of the scheduler is very important as it has a big impact on the performance of the applications. This is especially relevant in mission critical real-time communications that are very sensitive to the latency and traffic shaping introduced by the scheduler. Current schedulers algorithms do not differentiate their behavior based on the traffic type (this information is not available in the MAC layer) but introducing these parameters into the decision process could achieve gains in the performance of the applications. To to add them the OAI eNodeB will be modified in order to support passing of information to the scheduler during execution time.

The main objective of the scheduler experiments is to analyze what are the optimal cell-specific and protocol configurations available for a base station as wells as a scheduling policy that is able to consider traffic characteristics to meet the application requirements. The technical approach considers the introduction of programmable *Radio Access Network* (RAN) technologies, using the SDN design paradigm, on the OAI eNodeB. The data plane will be decoupled from the control plane, with a remote controller communicating with a local agent residing in the base station. Execution time decisions will be made regarding MAC scheduling and Resource Blocks allocation to facilitate real time prioritization of video traffic for the first responder. Figure 23.2 depicts the scheduler experiment setup. A new API will be included in the OAI eNodeB that could be accessed by third party applications.

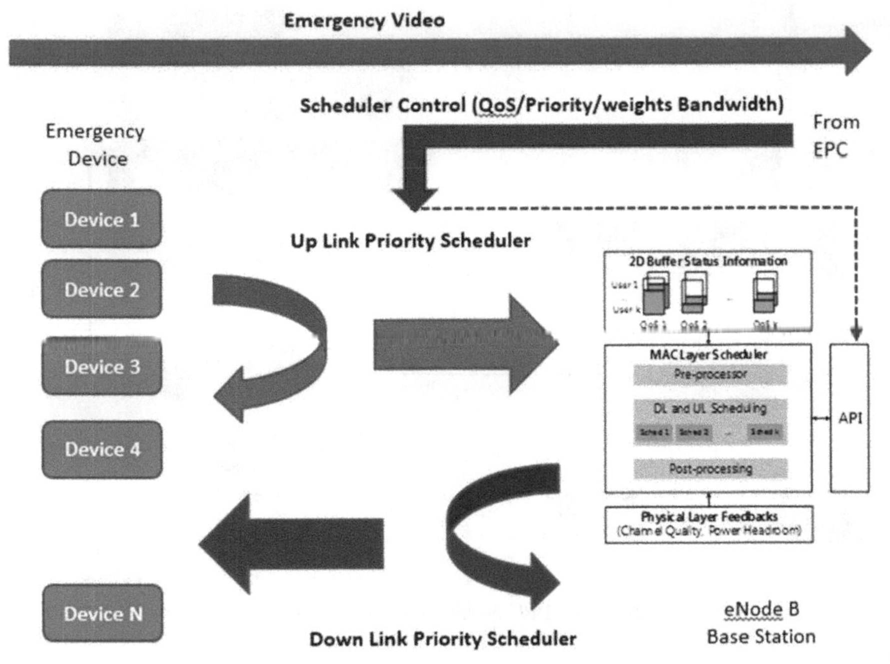

Figure 23.2 Scheduler overview.

23.4 Experiments Focused on the EPC

The experiments based on the EPC contemplates the measurements of metrics on standard procedures of the core network but also the evolution of it to integrate SDN techniques, *Mobile Edge Computing* (MEC) and *Over-The-Top* (OTT) QoS requirements of third party applications. Figure 23.3 depicts the target testing architecture for the core network. The following domain division has been assumed:

- Operator Domain is the operator private network and it is normally not available to third parties.
- Operator Domain for Third Party Network Applications is still part of the operator domain but it provide access to third parties to some functionality such as the *Rx* interface or a SDN controller.
- Internet Domains, this will be outside the operator domain and it comprises many different domains.

The components of the EPC are the standard ones plus the addition of an instance of *Open vSwitch* (OVS) in the backhaul. This OVS can be used to

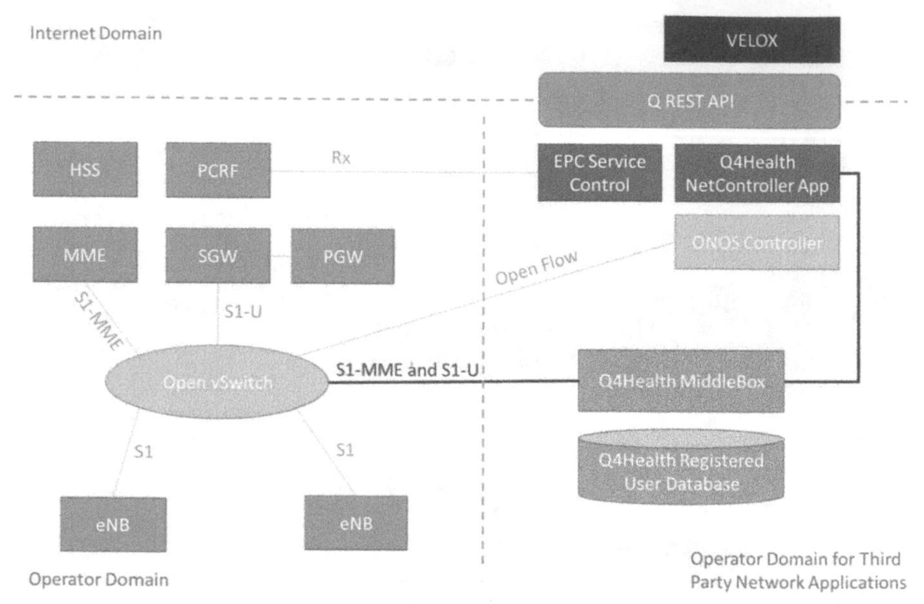

Figure 23.3 EPC architecture overview.

forward the control and data plane of the eNodeBs towards a Middlebox. The rules to do so will be injected via a network application running on top of an ONOS SDN controller. The Q REST API provides access to different functionality such as group video communications, low latency services and OTT QoS demand from third party applications.

The first set of experiments in the core network will acquire statistics about the signaling procedures of the control plane. Several tools to extract the success rates as well as the mean times of common procedures have been implemented. The statistics extracted with these tools can be used to compare different modems but also to evaluate the effect of the changes introduced in the network. Some of the signaling KPIs to be produced are:

- EPS Attach Success Rate (EASR), that will be used to determine how Q4HEALTH can improve the connectivity in indoor deployments.
- Dedicated EPS Bearer Creation Success Rate (DEBCSR), the QoS demands from the video application will trigger bearer creation on the core network.
- Dedicated Bearer Set-up Time by MME (DBSTM), the setup time is also important as it will determine the time to establish the QoS enforcement in the link, which is important in heavy-traffic scenarios.

- Service Request Success Rate (SRSR), which determines the success rate when the UE goes from idle to connected (for instance after a paging procedure.)
- Mean Active Dedicated EPS Bearer Utilization (MADEBU) determines the resources allocated by the UE, and this could be used as based to determine costs based on potential prices.
- Inter-RAT Handover Success Rate (IRATHOSR), to characterize the performance of seamless handover.

The main line of exploration of the handover procedure will be stress testing by constantly triggering the procedure on the network as well as the seamless heterogeneous handover to Wi-Fi that will be provided by the ANDSF and ePDG [13] components in the EPC. It will be studied under different scenarios including:

- Commercially available networks. The results on live networks will be used as a baseline. In this scenarios the handover will be studied from an RRC perspective by analyzing traces at that level of the stack, obtained with drive testing tools.
- Release 12 Emulator which can be used to test the handover procedure under different channel conditions and network configurations.
- A Small Cells scenario which will be use to test the heterogeneous handover by integrating them with a release 12 EPC with support for non-3GPP access networks.

Another important aspect of these systems is the possibility of configuring quality of service for their systems. This functionality is implemented by the VELOX engine, which supports different drivers to enable it. VELOX will implement a driver for the scheduling request experiments previously described but also drivers to enforce QoS between the different domains (by the insertion of rules in the transport switches) and a driver to trigger QoS demands via the Rx interface in the core network. With these drivers the system will be capable of enforcing a determined QoS in all the elements of the network, improving considerably the overall performance and reliability of the system.

The introduction of the Middlebox depicted in Figure 23.3 has two purposes, on one hand it can be used to provide low latency communications between peers geographically close. On the other hand it can be used to implement group video communications easily. For both functionality the OVS instance in the backhaul will be configured to copy and forward all the control plane traffic to the Middlebox. The Middlebox will analyze this traffic in order to produce a database of all the UEs connected as well as information about the *tunnel endpoint identifier* (TEID) of their tunnels, the endpoints addresses (the addresses of the eNodeB and the SGW) and details on the QoS configuration. The data plane will be redirected to the Middlebox (it will not be sent to the EPC) that will decide if it is able to process the traffic locally or if it has to forward it to the EPC. In the case that the final destination of the traffic is a user or a service registered in the Middlebox, the Middlebox will remove the GTP transport headers and redirect it to the appropriate peer. In the rest of the cases the traffic will be forwarded as usual to the EPC. With this approach the latency could be reduced up to 78% [14] as it reduce the transit times of the backhaul, EPC and transport networks.

23.5 Conclusion

These set of experiments designed in the Q4Health project will improve the performance and increase the functionality of the BlueEye project. Furthermore the project will produce an integrated experiment combining all the developments of the project. The objective of this final experiment is to showcase the platform to potential users of the testbed and the BlueEye system.

The combined experimental platform provides a very realistic end-to-end network, with access to the configuration of almost all of the levels of the stack. The functionality covered by the platform is focused in improving the innovation capabilities of the platform's users, but also with an eye in the latest communication trends trying to incorporate tools that enables the latest industry state of the art and best practices.

The development of 5G prototypes both of EPC components and enhanced LTE radio access will boost the number of users the platforms can attract and will also result in scientific contributions. BlueEye will be optimized to support current and future mobile communications in different markets which will boost the business opportunities of the platform.

References

[1] Cesar A. Garcia-Perez, Alvaro Rios, Pedro Merino, Kostas Katsalis, Navid Nikaein, Ricardo Figueiredo, Donal Morris, and Terry O'Callaghan. Q4Health: Quality of service and prioritisation for emergency services in the LTE RAN stack. In *Networks and Communications (EuCNC), 2016 European Conference on*, June 2016.

[2] A. Díaz Zayas, C. A. García Pérez, and P. Merino Gomez. PerformLTE: A testbed for LTE testing in the future internet. In *Wired/Wireless Internet Communications: 13th International Conference, WWIC 2015, Malaga, Spain, May 25–27, 2015, Revised Selected Papers*, volume 9071, page 46. Springer, 2015.

[3] Florian Kaltenberger, Raymond Knopp, Navid Nikaein, Dominique Nussbaum, Lionel Gauthier, and Christian Bonnet. OpenAirInterface: Open-source software radio solutions for 5G. *In EUCNC 2015, European Conference on Networks and Communications, 29 June–02 July 2015, Paris, France*, Paris, FRANCE, 06 2015.

[4] ITU. Performance and quality of service requirement for international mobile telecommunications-2000 (IMT-2000) access networks. ITU Recommendation ITU-R M.1079-2, International Telecommunication Union, 2003.

[5] T. Doumi, M. F. Dolan, S. Tatesh, A. Casati, G. Tsirtsis, K. Anchan, and D. Flore. LTE for public safety networks. *Communications Magazine, IEEE*, 51(2):106–112, February 2013.

[6] A. P. Avramova, S. Ruepp, and L. Dittmann. Towards future broadband public safety systems: Current issues and future directions. In *Information and Communication Technology Convergence (ICTC), 2015 International Conference on*, pages 74–79, Oct 2015.

[7] R. Ferrus and O. Sallent. Extending the lte/lte-a business case: Mission- and business-critical mobile broadband communications. *Vehicular Technology Magazine, IEEE*, 9(3):47–55, Sept 2014.

[8] A. Kumbhar and I. Guvenc. A comparative study of land mobile radio and lte-based public safety communications. In *SoutheastCon 2015*, pages 1–8, April 2015.

[9] A. Diaz Zayas, C. A. Garcia Perez, and P. M. Gomez. Third-generation partnership project standards: For delivery of critical communications for railways. *IEEE Vehicular Technology Magazine*, 9(2):58–68, June 2014.

[10] NGMN Alliance. NGMN 5G White Paper. Technical report, Next Generation Mobile Networks, February 2015.

[11] P. Ameigeiras, J. J. Ramos-munoz, L. Schumacher, J. Prados-Garzon, J. Navarro-Ortiz, and J. M. Lopez-soler. Link-level access cloud architecture design based on sdn for 5G networks. *IEEE Network*, 29(2):24–31, March 2015.

[12] T. Chen, M. Matinmikko, X. Chen, X. Zhou, and P. Ahokangas. Software defined mobile networks: concept, survey, and research directions. *IEEE Communications Magazine*, 53(11):126–133, November 2015.
[13] 3GPP; Technical Specification Group Services and System Aspects. Architecture enhancements for non-3GPP accesses. TS 23.402, 3rd Generation Partnership Project (3GPP), December 2015.
[14] Cesar A. Garcia-Perez and Pedro Merino. Enabling low latency networks in LTE. In *Foundations and Applications of Self* Systems (FAS*), 2016 IEEE International Conference on*, September 2016.

PART V

INTERNATIONAL COLLABORATION on Research and Experimentation

24

WAZIUP: Open Innovation Platform for IoT-Big Data in Sub-Sahara Africa

Abdur Rahim Biswas

Corentin Dupont and Congduc Pham

24.1 Introduction

ICT developments in Sub-Saharan Africa has the potential to cut across traditional sectors: notable examples are the introduction of micro-health insurance and health-savings accounts through mobile devices; index-based crop insurance; crowd-sourcing to monitor and manage the delivery of public services. These innovative applications recognize and leverage commonalities between sectors, blur traditional lines, and open up a new field of opportunities. The opportunity for ICT intervention in Africa is huge especially of IoT and big data: those technologies are promising a big wave of innovation for our daily life. The era of IoT can connect billions of sensors, devices, equipment, systems. In turn, the challenge is about driving business outcomes, consumer benefits, and the creation of new value. The new mantras for the IoT era is the collection, convergence and exploitation of data. The information collection involves data from sensors, devices, gateways, edge equipment and networks. This information allows increasing process efficiency through automation while reducing downtime and improving people productivity. The WAZIUP project will show:

Potential of IoT and Big data in Africa: Over the last several years there has been a lot of discussion and research on IoT to understand the reference architecture, what is IoT and how it can impact our daily life. It is not a question any more on whether IoT and big data will come or not: most of the companies have defined internal business activities to go along with this global move. According to the EC nearly five billion things will be connected by 2015, reaching 25 billion by 2020, helping citizens save energy, reduce traffic jams, increase comfort, and get better healthcare and increased independence. Revenues in the European Union from IoT are estimated to represent €400 million in 2015 and are set to increase to more than €1 trillion by 2020. However, countries in Sub-Saharan Africa are still far from being ready to enjoy the full benefit of IoT. This is because of many challenges, such as lack of infrastructure, high cost platforms and complexity in deployment. At the same time, it is very urgent to promote IoT worldwide: WAZIUP will contribute by reducing part of the technology gap between EU and Africa. Thus, the goal of WAZIUP is to deploy and validate real-life IoT and big data pilot cases with several Sub-Saharan African countries.

There are two key reasons why IoT and Big Data should be addressed now in Africa.

- *For EU*: to create critical mass within the IoT innovation ecosystem and facilitate co-creation of products and services in open ecosystems; to this respect, it is necessary to be cooperative with Africa. As the continent has full of young talent (more than 40% of the population in sub-Saharan countries is younger than 15 years old), the cooperation on IoT and Big data with Africa will boost the economy for both continent;
- *For Africa*: There are many challenges to the adoption of IoT and Big Data in Africa. This is why WAZIUP is conceived as a pathfinder project for Africa. We believe that the technological landscape in Africa can move very fast, it is hence urgent to promote the WAZIUP technologies in Africa and to harmonize with global IoT and Big Data movement in order to better prepare for the upcoming ICT wave.

The reason why WAZIUP targets the rural community in Sub-Saharan Africa is because about 64% of the population is living outside cities. The region will be predominantly rural for at least another generation. The pace of urbanization here is slower compared to other continents, and the rural population is expected to grow until 2045. The majority of rural residents manage on less than few Euros per day. Rural development is particularly imperative in sub-Saharan Africa, where half of the rural people are depending on the agriculture/micro and small farm business, other half faces rare formal employment and pervasive unemployment. For rural development, technologies have to support several key application sectors like living quality, health, agriculture, climate changes, etc. To reach WAZIUP goal, one has to overcome both technical challenges as well as economical challenges. WAZIUP project consider how to best design and deploy the IoT-Big Data technology considering cost and energy challenges in the first place. WAZIUP will target the removal of three major barriers:

- *Rural Access to Technology*: Vast distances and poor infrastructure isolate rural areas, leaving those who live there poorly integrated into modern ICT ecosystems. WAZIUP will offer long-range IoT communication network to connect rural communities: the software service platform will offer highly innovative monitoring, recommendation, notification services based on the data coming from multiple rural application sectors.
- *Cost of hardware and services*: Power consumption and deployment costs are the two most important issues for devices: the first issues are universal for IoT and the later one is more specific to Sub-Shahara Africa. High delivery and infrastructure costs discourage service providers from reaching the countryside. The potential of IoT, in Sub-Saharan Africa, can only be realized if the cost is resolvable as most of the rural population in the Africa is at the poverty level. WAZIUP will take this challenge as the main one to be addressed. WAZIUP will also consider power consumption: devices must reduce the overall power consumption. However the deployment challenges cannot only be realized by reducing the devices as well as service costs, let alone to reduce the joint cost of the devices and service: there has to be an innovative business model. We envision mostly spin-offs enterprises for micro-small scale services that could afford to rent the devices to farmers and provide them services. In this case the cost of services must be also affordable. Hence WAZIUP will have a dedicated effort to design a viable exploitation model which may lead to the creation of small-scale innovative service companies.
- *Quality of service*: the technology of WAZIUP can be used to overcome a structural problem in the work market in Africa: very often communities located in isolated areas are left behind in the innovation process not because they are unwilling to benefit from changes in the technology, but rather because by definition those areas attract fewer and less qualified professionals, civil servants, skilled workers, and innovators than urban centers. Having a technology which offers remote assistance and control indeed greatly mitigates such effect of marginalization. Furthermore, some of the advanced intelligent services, e.g., those qualified as "watch-dog" applications – as in the case of cattle-rustling prevention – have the role of increasing security and/or reliability in remote locations and thus

have the potential to increase the general quality of experience in the usage of ICT solutions.

Beside the cost and power consumption, the robustness of hardware is a core requirement: hardware has to be robust enough so as to require lower maintenance and handle environmental and deployment threats as well. WAZIUP will collect the grand challenge: reduce costs, reduce power consumption but at the same time increase the robustness of the hardware. WAZIUP will bring in existing IoT-Cloud and big data platform developed in several EU as well as industrial projects. The technologies will address the specific needs and conditions of the business use case identified in the projects.

24.2 Objective

WAZIUP is a H2020 international cooperation action. The project is driven by a consortium of 5 EU partners and of 7 partners from 4 sub-Saharan African countries. Furthermore, it has support from multiple African stakeholders with the aim of defining new innovation space to advance the African Rural Economy. It will do so by involving end-users communities in the loop, namely rural African communities of selected pilots, and by involving relevant public bodies in the project development. WAZIUP will accelerate innovation in Africa by coupling with current EU innovation in the sector of IoT and Big Data: this EU technology will be specialized to generate African cost effective technologies with an eye to preparing the playground to the future technological waves by solving concrete current needs. WAZIUP will deliver a communication and big data application platform and generate locally the know how by training by use case and examples. The use of standards will help to create an interoperable platform, fully open source, oriented to radically new paradigms for innovative application/services delivery. WAZIUP is driven by the following visions:

- *Empower the African Rural Economy.* Develop new technological enablers to empower the African rural economy now threatened by the concurrent action of rapid urbanization and of climate change. WAZIUP technologies can support the necessary services and infrastructures to launch agriculture and breeding on a new scale;
- *Serve the Wealth Growth of Rural Communities.* Create innovation across a dated agribusiness/agriculture/rural sector: increasing agriculture's value and by adding to sub-saharan countries economical growth, such innovation contributes towards poverty reduction of communities living in the rural areas;

- *Innovate Agro-Industry Processes.* Increase efficiency of production and processing in small-scale agro-industry SMEs, catalyze better yields and advance agribusiness;
- *Improve work conditions.* WAZIUP technology aims at improving work and living quality by affordable and available specific IoT services tailored for African rural communities;
- *Tailored IoT and Big-data Technology.* Offer smart sensor and data-driven applications and services addressing the end-users needs and requirements (understanding users requirements and preference delivering towards more personalized and easy users interfaces and applications)
- *Value-added cost and energy efficiency.* IoT application and services based on WAZIUP open IoT-Big data platform will focus on ease of maintenance and low cost of solutions;
- *Lower Entry Level.* Provide to application developers a mature platform, as well as tools and standards that are inexpensive, easy and relevant.

In order to achieve the above aims, a strong dissemination and exploitation effort of the project will be dedicated to a) strengthening linkages of end-users with industries, b) engage innovation space and living labs to accelerate innovation coaching/training/start-up activities (e.g., community-driven development paradigms), c) promote value-addition to business outputs, d) challenge the value-chain of African agribusiness through technology for value increase.

The proposed solutions will be tested for a set of real-life use cases covering several countries. At higher level, WAZIUP will implement a regional innovation platform, where SMEs could continue to develop/plug-in solutions using the technical elements and the open data provided in the project. The ultimate target is to create large African industries, SMEs ecosystem, and induce a network-effect.

The above objectives require tackling several challenges which we enlist below:

- *Challenge 1: Innovative design of the IoT platform for the Rural Ecosystem.* Low-cost, generic building blocks for maximum adaptation to end-applications in the context of the rural economy in developing countries.
- *Challenge 2: Network Management.* Facilitate IoT communication and network deployment. Lower cost solutions compared to state of the art technology: privilege price and single hop dedicated communication networks, energy autonomous, with low maintenance costs and long lasting operations.

- *Challenge 3: Long distance.* Dynamic management of long range connectivity (e.g., cope with network & service fluctuations), provide devices identification, abstraction/virtualization of devices, communication and network resources optimization.
- *Challenge 4: Big-data.* Exploit the potential of big-data applications in the specific rural context.

From a technical standpoint, WAZIUP introduces innovation by constructing on the following pillars of IoT/Big Data technology, specifically tailored for the rural ecosystem:

- *Privacy and security*: through attention to all related privacy and security aspects with specifics addressing the involved communities (farmers, developers);
- *Personalized and user friendliness*: models will receive requirements from users' needs and will ensure compliance with all most common usability standards (e.g., Web Accessibility Initiative – WAI or ISO/TR 16982:2002);
- *An Open interoperable platform*: through open standard and protocols from the Geospatial Consortium (OGC), W3C, IEEE from the European SDOs (CEN, CENELEC and ETSI, etc.) for all its key technology;
- *Continued Openness:* through the release of open specification and open software components and/or algorithms;
- *Low-cost and low-energy consumption*: through the design of innovation hardware (sensors/actuators), and of IoT communication & network infrastructure.

24.3 Technical Solution

In WAZIUP the challenges outlined above will be tackled using an open IoT-big data platform with affordable sensors connected through an IoT-Cloud open platform. This platform will also make use of mobile phones and real-time processing to empower users and deliver the needed services. Hereafter a compact list of core technical functionalities encompassed by the platform:

- Cloud-based real-time data collection combined with analytics and automation software: thus, the platform will offer cost-effective solutions for aggregating different machines and sensor types to engender efficiency, smart automation and optimization in the rural context.

- Intelligent analytics of sensor and device data: studied in order to optimize for performance of the rural workplace, detect potential outages, and finally reduce overall maintenance costs.
- Integration to 3rd parties' platform: enables customers' benefit of scaling fast and easy.
- PaaS (Platform-as-a-Service) provider: WAZIUP will provide to business clientele with independently maintained platform upon which their web application, services and mobile applications can be built.

The set of value-added services to be delivered by WAZIUP:
Long-distance real-time/near real-time monitoring and users notification: WAZIUP project enables the comprehensive monitoring of a product's condition, operation, and external environment through sensors and external data sources. By data processing, a product can alert users or others of changes in circumstances or performance. Monitoring also allows companies and customers to track a product's operating characteristics and history and to better understand the product's usage history. Usage history, in turn, may deeply depend on the specific rural communities involved.

Long-distance control of the system and devices: WAZIUP devices and platforms can be controlled through remote commands or algorithms that are built into the device or reside in the product cloud. Modern control techniques act through software embedded in the product or deployed in the cloud. This allows the customization of product performance to a degree that previously was not cost effective or often even possible. The same technology also enables users to control and personalize their interaction with the product in many new ways.

Optimization and big data analytic application: The rich flow of monitoring data from connected sensors/products, coupled with the capacity to control product operations, allows companies (SMEs, NGO) to optimize product performance in numerous ways, many of which have not been previously possible. WAZIUP project can apply algorithms and analytics to in-use or historical data to dramatically improve output, utilization, and efficiency of processes in the rural context. Real-time monitoring data on product condition and product control capability enables firms to optimize service by performing preventive maintenance when failure is imminent; they can also accomplish repairs remotely, thereby reducing product downtime and the need to dispatch repair personnel to remote rural areas.

24.4 Applications Cases

We present a detailed analysis of the different application cases selected for WAZIUP project, these are precision agriculture, cattle rustling, logistic & transport, fish farming and urban waste management.

24.4.1 Precision Agriculture

The goal of this application cases is to improve the working conditions and yield in the agricultural field by giving precise information on the ground status. To achieve this, we will gather data on the environmental conditions with dedicated sensors, analyze data and make optimized and personalized predictions for the farmers.

24.4.2 Cattle Rustling

Cattle rustling is a serious problem observed in African countries, particularly in Senegal. This is a recurring phenomenon that causes a lot of problems to farmers. Cattle's stealing is extremely expensive; it represents millions for farmers but also for the state annually. Faced with this problem, the famer is often helpless.

24.4.3 Logistics and Transport, Saint-Louis, Senegal

Whether by air, ground or sea, transportation and logistics are essential components to many enterprises' productivity, and access to real-time data is critical. Many industries and business sectors are struggling to grasp the possibilities of data-driven technology, but companies in transport and logistics are way ahead. By their very nature, the logistics providers that move objects by air, sea, rail, and ground have widely distributed networks and rely on rapid information about those networks to make decisions. As a result, they were quick to see the benefits of new sensor and connection technology, placing them at the forefront of the transition to a connected world.

24.4.4 Fish Farming, Kumasi, Ghana

In order to increase the management efficiency of the fish farms, this pilot will deploy a network of sensors to monitor remotely and in real time the water situation and quality within the fish ponds.

24.4.5 Environment and Urban Agriculture

African cities have the fastest urbanization speed of the world. Some cities like Kinshasa will have its population tripled by 2050. Thus, the African urbanity becomes the perfect experimental field to test urban smart systems. The most important challenges are the household living conditions improvement of food security, appropriate waste management and digitalization of the different sectors.

Table 24.1 is summarizing the uses cases that will be validated in the project.

24.5 WAZIUP Platform as a Service (PaaS)

Platform as a Service (PaaS) is a category of the cloud computing service that provides a platform allowing customers to develop, run, and manage applications; without the complexity of building the infrastructure typically associated with developing and deploying applications. Typically, a PaaS framework will compile an application from its source code, and then deploy it inside lightweight virtual machines or containers. This compilation and

Table 24.1 Pilot use cases

Application Domain	Use Cases
Precision agriculture	• UC1: Monitor soil • UC2: Field Weather Situation • UC3: Storage Moisture and Temperature
Cattle rustling	• UC1: Real-time position and itinerary of the cattle's herd • UC2: Ability to receive notification in critical situations
Logistics and transportation	• UC1: Track operations and remote monitoring • UC2: Real-time visibility across the supply chain • UC3: Check the integrity, identification, authentication and traceability of goods
Fish farming	• UC1: Fish Pond Water Quality • UC3: Cost-Efficient Feeding System
Environment and Urban agriculture	• UC1: Indoor/small scale farming automation • UC3: Confirm emptied waste bins

deployment is done with the help of a file called the manifest, which allows the developer to describe the configuration and resource needs for the application. The manifest file will also describe the services that the application requires and that the platform will need for provision. Furthermore, PaaS environments usually offer an interface for applications to scale up or down, or to schedule various tasks within the applications.

The idea of WAZIUP is to extend the paradigm of the PaaS to IoT. Developing an IoT Big Data application is a complex task. A lot of services need to be installed and configured, such as databases and complex event processing engines. Furthermore, it requires an advanced knowledge of the various communication protocols, the programming of embedded devices, the storage, processing and analysis of the data in a distributed fashion and finally the programming of GUIs and user interactions. The promise of the PaaS extended to IoT is to abstract away this work to a large extent.

Figure 24.1 shows the PaaS deployment in WAZIUP. Traditional PaaS environments are usually installed on top of IaaS (in blue in the picture). The blue boxes are physical servers, respectively the Cloud Controller and one Compute node. The PaaS environment is then installed inside the IaaS VMs, in green in the picture. We use Cloud Foundry as a PaaS framework. It comes with a certain number of build packs, which and programming languages compilers and runtime environments. It also provides a certain number of preinstalled services such as MongoDB or Apache Tomcat. The manifest file,

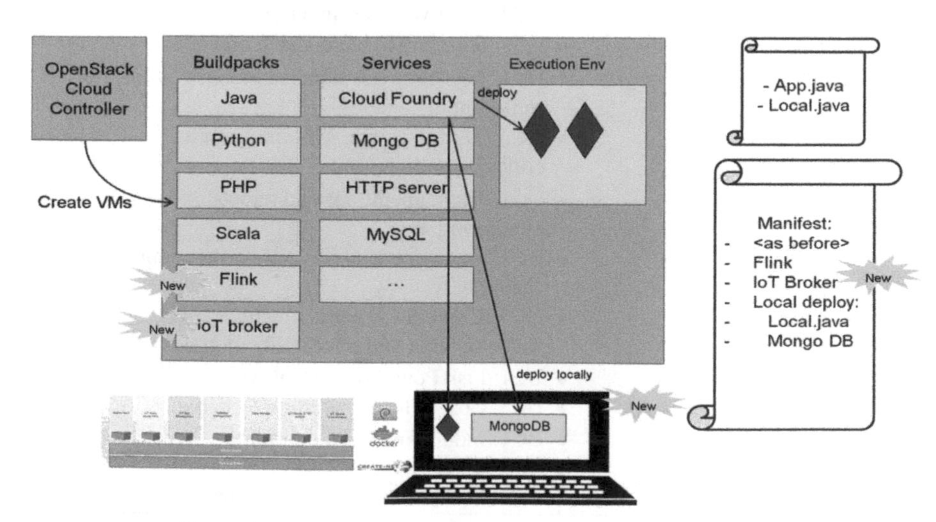

Figure 24.1 PaaS deployment extended for IoT in WAZIUP.

showed on the right hand side, provides a high-level language that allows describing which services to instantiate. We propose to extend this language to IoT and big data services such as:

- Data stream and message broker
- CEP engines
- Batch processing engines
- Data visualization engines

Furthermore, we propose to include in the manifest a description of the IoT sensors that are required by the application. This query includes data such as the sensor type, location and owner. The manifest also includes the configuration of the sensors. The application will then be deployed both in the global Cloud and in the local Cloud.

24.5.1 Local and Global Clouds

The WAZIUP project defines two different types of "Cloud": the local Cloud and the global Cloud. A local Cloud is an infrastructure able to deliver services to clients in a limited geographical area. The local Cloud replicates some of the features provided by the traditional Cloud. It is used for clients that may not have a good access to the traditional Cloud, or to provide additional processing power to local services. In order for such an infrastructure to be considered as a local Cloud it must support a virtualization technology. In the case of WAZIUP, the local Cloud comprises the end user or service provider PC and IoT Gateway. The local Cloud characteristics are:

- Existence of IoT devices attached
- Can have geographical characteristics
- Must support virtualization
- Must support local cloud components
- Has an identifiable administrator/owner
- Has certain regulations/privacy considerations for data access and treatment

The global Cloud, on the other end, is a "backbone infrastructure" which increases the business opportunities for service providers and allows services to access a virtually infinite amount of computing resources. In order for such an infrastructure to be considered as a global Cloud it must support a virtualization technology and be able to host the global cloud components of the WAZIUP architecture.

24.6 WAZIUP Architecture

This section provides the details of the WAZIUP architecture. A functional overview is given, followed by the actor definition, the components and finally the sequence diagrams.

24.6.1 Functional Overview

This section presents the functional view of the architecture.

Figure 24.2 shows the functional overview of WAZIUP. The topmost block represents the Cloud platform, the middle one is the network connectivity while the bottom one is the local deployment, including gateway and sensors. Table 24.2 shows the functional domains that have been identified, with a description for each of them.

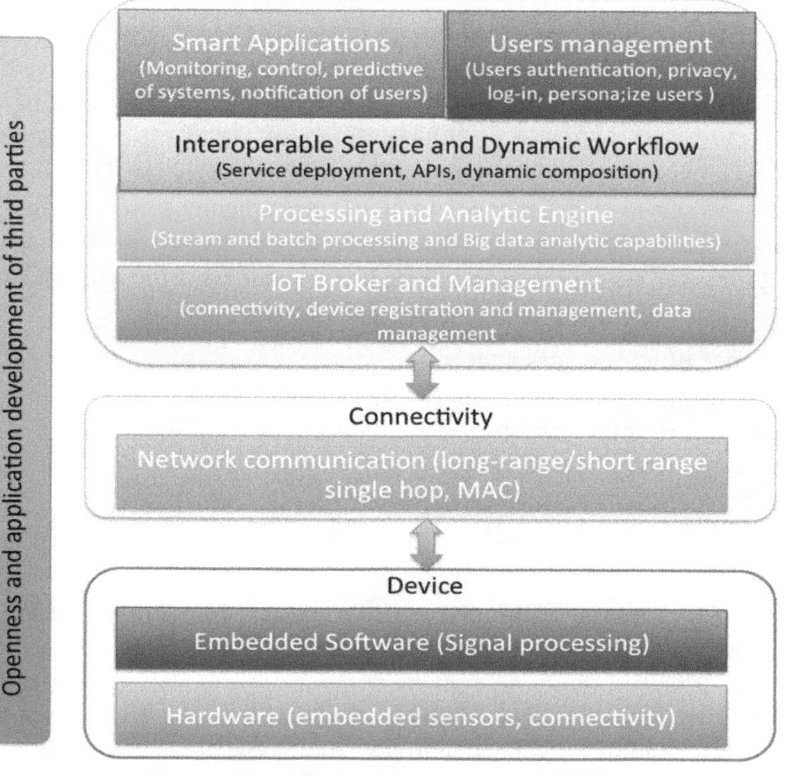

Figure 24.2 Functional overview of WAZIUP.

Table 24.2 Functional domains

Functional Domains	Description
Application platform	Application writing, deploying, hosting and execution.
IoT platform	The connectivity of IoT devices, the sensors data and metadata.
Stream and data analytic	Data brokering, stream processing and data analytics.
Security and privacy	Management of the identification, roles and connections of users. Also includes data anonymisation of the data and securisation of the transmissions.
Platform Management	Status of the components, deployment of the platform.

24.6.2 Components

Figure 24.3 presents the full WAZIUP architecture. It shows the four functional domains: Application Platform, IoT Platform, Security and Privacy and finally Stream & Data Analytic. The Application Platform involves the development of the application itself and its deployment in the Cloud and in the Gateway. A rapid application development (RAD) tool can be used, such as Node-Red. The user provides the source code of the application, together

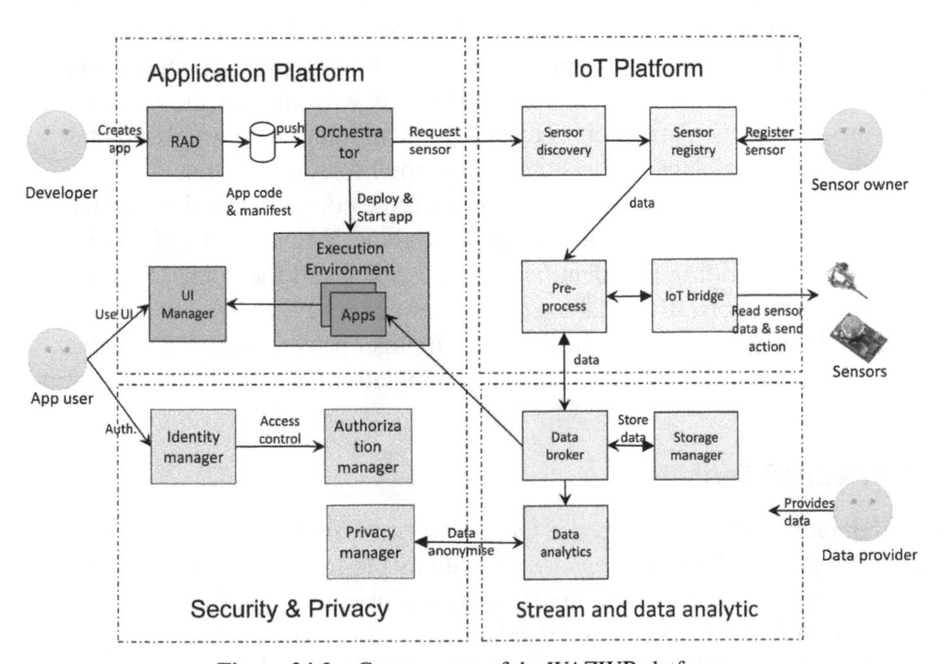

Figure 24.3 Components of the WAZIUP platform.

with the manifest. As a reminder, the manifest describes the requirements of the application in terms of:

- computation requirements (i.e. RAM, CPU, disk)
- references to data sources (i.e. sensors, internet – sources ...)
- big data processing engines (i.e. Flink, Hadoop ...)
- configuration of sensors (i.e. sampling rate)
- local and global application deployment

The application source code, together with the manifest, is pushed to the WAZIUP Cloud platform by the user. The orchestrator component will read the manifest and trigger the compilation of the application. It will then deploy the application in the Cloud Execution Environment. It will also instantiate the services needed by the application, as described in the manifest. The last task of the orchestrator is to request the sensor and data sources connections from the IoT components of the architecture. The sensor discovery module will be in charge of retrieving a list of sensors that matches the manifest description. On the left side of the diagram, the sensor owners can register their sensors with the platform. External data sources such as Internet APIs can also be connected directly to the data broker. The sensors selected for each application will deliver their data to the data broker, through the IoT bridge and pre-processor. This last component is in charge of managing the connection and configuration of the sensors. Furthermore, it will contain the routines for pre-processing the data, such as cleaning, extrapolating, aggregating and averaging. Historical data can be stored using the Storage manager.

The Security and Privacy domain contains three components: the Identity Manager, the Authorization Manager and the Privacy Manager. The first one is in charge of providing the identification, the roles and the connections of the users. The Authorization Manager provides the access policy for each of the WAZIUP resources. Finally the Privacy Manager provides services for the privacy of communication and also the anonymization of data.

24.7 WAZIUP Test-Beds

We will deploy a network test-bed at Gaston Berger University (UGB), Saint-Louis, Senegal, to validate the sensor and gateway platforms and to test various sensor settings in various rural environments. Computing facilities at UGB will host the WAZIUP platform to test advanced sensor and data management. The Internet access will also enable the small-size, single-application scenarios where public IoT data clouds will be used.

Figure 24.4 Deployment of sensor nodes around a gateway use case integration.

As can be seen in Figure 24.4, UGB has high buildings for the LPWAN antennas installation. By deploying LPWAN devices we can build a test-bed allowing LPWAN connectivity of IoT devices within a range of more than 15Kms in LOS in typical rural areas. In addition, the geographic location of UGB perfectly suit our needs as it is located within LPWAN radio range of the downtown Saint-Louis city as well as within range of many typical rural areas for test diversity, such as small villages, crop field and farms.

An important feature provided by WAZIUP is the possibility to run the sensor-gateway system in an autonomous manner, without Internet connectivity nor access to dedicated servers. The gateway can therefore also store data locally and make them available through local computing facilities (e.g. laptop, smartphones, tablets) for standalone surveillance applications. Figure 24.5 illustrates the various scenarios that WAZIUP will support: (top) gateway will Internet connectivity provided by cellular technologies, (middle) gateway will Internet connectivity provided by a WiFi (possibly ADSL-based)

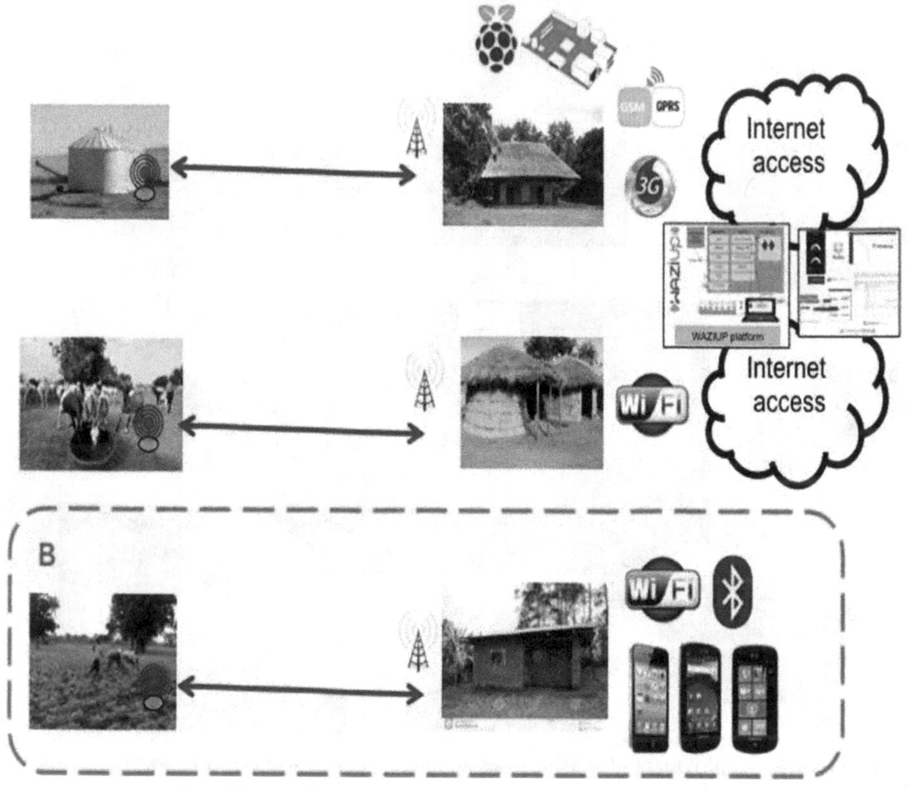

Figure 24.5 WAZIUP deployment scenarios.

access, (bottom) gateway without Internet connectivity, providing connectivity (short range) to local computing facilities (e.g. laptop, smartphones, tablets).

After the integration of the sensors and the gateway, the WAZIUP platform will be ready to receive and process information coming from the sensors.

24.8 Conclusion

With IoT, Sub-Saharan African countries can dramatically improve their productivity by enabling rapid and cost-effective deployment of advanced and real-time monitoring. However, deploying an IoT platform for Africa comes with many challenges. Among them, the most important are supporting low cost, low power, low bandwidth, and intermittent connection from Internet.

Furthermore, widely accessible technologies such as SMS and voice calls need to be supported to reach the maximum users. In this chapter, we proposed an architecture for the WAZIUP IoT Big Data platform. The concepts that underpin the WAZIUP platform are three: the PaaS approach to IoT, the data processing capacity inspired from Big Data techniques and finally the local and global Cloud. The idea of extending the PaaS approach to IoT is to propose a platform dedicated to IoT developers that can reduce the time-to-market for an application by cutting the development costs. The Big Data techniques enable the processing of the huge amount of data produced by sensors. Finally the local and global Clouds address the intermittent connection challenge: when Internet is not available, the user can still access some IoT functionalities from the local Cloud. The project will develop several applications use cases to validate its concepts. The application cases selected are precision agriculture, cattle rustling, logistic & transport, fish farming and urban waste management. Each use cases will be developed and deployed in one of our test-beds in Africa.

25

Understanding the Challenges in the Optical/Wireless Converged Communications

Federated Union of Telecommunications Research Facilities for an EU-Brazil Open Laborntory (FUTEBOL)[1]

Johann M. Marquez-Barja[1], Nicholas Kaminski[1], Paulo Marques[2], Moises R. Nunes-Ribeiro[3], Juliano Wickboldt[4], Cristiano B. Both[5] and Luiz A. DaSilva[1]

[1]Centre for Future Networks and Communications (CONNECT), Trinity College Dublin, University of Dublin, Ireland
[2]Instituto de Telecomunicacoes (IT), Aveiro, Portugal
[3]Software Defined Networks Research Group, Department of Informatics Federal University of Espirito Santo (UFES), Av. Fernando Ferrari, 514 Vitoria (ES), Brazil
[4]Institute of Informatics, Federal University of Rio Grande do Sul (UFRGS), Av. Bento Goncalves, 9500, Porto Alegre (RS), Brazil
[5]Federal University of Health Sciences of Porto Alegre (UFCSPA), Brazil Av. Sarmento Leite, 245, Porto Alegre, (RS), Brasil

Abstract

Current wireless trends (cell densification, coordinated communication, massive MIMO) pose a new set of challenges that require the joint consideration of optical and wireless network architectures. These problems are of direct impact to emerging economies such as Brazil, with highly heterogeneous infrastructure capabilities and demand, as well as to more established markets such as the EU, which aims to regain its leadership in the next generation

[1]www.ict-futebol.eu

of telecommunication technologies. FUTEBOL composes a federation of research infrastructure in Europe and Brazil, develops a supporting control framework, and conducts experimentation-based research in order to advance the state of telecommunications through the investigation of converged optical/ wireless networks. Also FUTEBOL establishes the research infrastructure to address these research challenges through innovation over this infrastructure, with a consortium of leading industrial and academic telecommunications institutions.

Keywords: Wireless, Optical, Converged Networks.

25.1 Introduction

There have been approaches to improve telecommunications by linking up wireless networks with optical ones, such as Radio over Fiber solutions. However, it is still a fact that wireless and optical network research problems are often treated in isolation of each other [Beas13, Saskar07]. Current wireless trends (cell densification, coordinated communication, massive MIMO, millimetre-wave) pose a new set of challenges that require the joint consideration of optical and wireless network architectures. These problems are of direct impact to emerging economies such as Brazil, with highly heterogeneous infrastructure capabilities and demand, as well as to more established markets such as the European Union, which aims to regain its leadership in the next generation of telecommunication technologies. The Federated Union of Telecommunications Research Facilities for an EU-Brazil Open Laboratory FUTEBOL project develops the research infrastructure to address these research challenges and conducts research over this infrastructure.

Europe and Brazil have a long research cooperation tradition in the area of Science & Technology culminating in the signing in 2004 of the Agreement for Scientific and Technological Cooperation[2]. The Agreement identifies some priority areas for future cooperation including Information and Communications Technology (ICT). EU-Brazil research cooperation in the area of ICT has been developing since the launch of the first coordinated call in 2011 and addresses some topics dealing with Future Internet, micro-electronics and micro-systems, cloud computing, technologies and applications for a smarter

[2]http://ec.europa.eu/world/agreements/prepareCreateTreatiesWorkspace/treatiesGeneralData.do?step=0&redirect=true&treatyId=2041

society and e-infrastructures. It is supported by an EU-Brazil Dialogue on Information Society with specific working groups in some areas addressing not only research and innovation matters but also ICT policy and regulatory aspects. EU-Brazil research cooperation in the area of ICT, including cloud computing, is also regarded as having a crucial strategic value and high societal impact. The current call further realises some of the objectives of the cooperation agreement by focusing on Advanced CyberInfrastructure including **Experimental Platforms** federating network resources in Europe and Brazil building on FIRE (Future Internet) and FIBRE (Future Internet Brazilian Environment for Experimentation) developments. Joint work on the areas above is expected to be continued in the work-programme 2016–17 of Horizon 2020.

The overall objective of the FUTEBOL project is to **develop and deploy research infrastructure**, and an associated **control framework for experimentation**, in Europe and Brazil, that enables **experimental research** at the **convergence point between optical and wireless networks**.

Considerable progress has been made in the past few years on the development of federated telecommunications research infrastructure in Europe, through the FIRE[3] program. More recently, the FIBRE project enabled optical fibre interconnection of research facilities in Europe and Brazil. However, telecommunications research remains largely segregated between optical networks and wireless systems and rarely do researchers cross the boundary between the two. We argue that the needs of future telecommunication systems, be it for high data rate applications in smart mobile devices, machine-type communications and the Internet of Things (IoT), or backhaul requirements brought about from cell densification, require the co-design of the wireless access and the optical backhaul and backbone. FUTEBOL aims at developing a converged control framework for experimentation on wireless and optical networks and to deploy this framework in federated research facilities on both sides of the Atlantic.

Within this chapter we present an overview of the research targeted by the FUTEBOL project. FUTEBOL is in an early stage of execution, and the research questions regarding optical/wireless converged networks are starting to be answered by understanding the challenges the such networks. However, the approach to tackle such challenges is described along this chapter.

[3]http://www.ict-fire.eu

25.2 Problem Statement

As of May 2014, Brazil was one of the largest internet markets in the world. Recent industry data projects the current online penetration rate of 49.3 percent to grow to 59.5 percent in 2017, which at that point will equal more than 123 million internet users across all online devices. Recent statistics suggest that out of the current 84 million internet users, 43 million are mobile internet users, 41 million of these accessing the internet via a mobile phone. The strong presence of mobile internet correlates with a recent Brazilian survey stating that 52 percent of Brazilian consumers do not have any internet access at home[4]. It is also estimated that in Brazil there may be up to ten times more mobile phones per base station than in Europe[5]. Notably, Brazil is characterized by enormous environmental diversity, ranging from mega-cities to wide expanses of low population density; this creates challenges to providing high level of communication services to all citizens. The lack of investment in research infrastructure is a limiting factor for digital inclusion in Brazil, relevant to democratize quality access to a broad range of internet-enabled services.

From the European mobile operator point of view, a key requirement of future mobile networks is that significant additional network capacity has to be added at lowest possible cost, to combat the current trend of stagnant revenues while traffic grows exponentially. Approximately 24% of a network operator's costs come from OPEX, including the cost of network operation and maintenance, training and support, energy, site rental. The experimental research infrastructure enabled by FUTEBOL will demonstrate how wireless/optical convergence will support future traffic growth and new mobile services, while limiting the CAPEX/OPEX required to deploy and maintain the network.

FUTEBOL's focus is on converged optical/wireless experimentation. On the wireless side, new spectrum access modalities such as Licensed Shared Access (LSA) will soon open new bands for mobile broadband, and more spectrum also means that less investment in infrastructure would be needed. The proliferation of small cells increases frequency reuse and is responsible for a major proportion of the gains in mobile network capacity. On the optical network side, network function virtualization and the concept of software defined networks are revolutionizing the way that network resources are managed. We view virtualization on the optical side and densification

[4]http://www.statista.com/topics/2045/internet-usage-in-brazil

[5]http://wireless.ictp.it/tvws/book/5.pdf

and capacity increase on the wireless access as major game changers in future networks that should be co-designed and experimented with together. FUTEBOL creates the infrastructure that enables academic and industrial researchers in Europe and Brazil to experiment at the convergence points between wireless and optical networks.

25.3 Background and State-of-the-Art

Converged wireless-optical network architectures started to appear in the second half of the last decade. Due to the high cost of developing a highly capillary fibre access network, these solutions considered increasing the broadband coverage by deploying a wireless network, mainly in dense urban areas, backhauled by a fibre access networks, typically implemented as a Passive Optical Network (PON). Typically the wireless access network was implemented as a mesh of WiFi enabled nodes. Architectures such as those dubbed CROWN [Kazowsky07], MARIN [Shaw07], or WOBAN [Sakar07] were developed to address the challenge of optical/wireless integration. All such models employed the PON as a mechanism to backhaul the wireless network, i.e. providing connectivity to the general Internet, while also integrating dynamic routing in the wireless mesh.

Example challenges addressed by prior architectures include: load balancing both on the wireless mesh and PON access points, and integrated routing algorithms between the wireless mesh and a multi-wavelength PON network. In particular the WOBAN architecture addressed additional issues such as optimization of node location and fault tolerance while also providing an implementation of a laboratory prototype [Chowdhury09], where a number of Optical Line Terminals (OLTs), Optical Network Units (ONUs), and WiFi access points were interconnected in order to measure their overall integrated performance. As the first highly integrated optical-wireless prototype, WOBAN allowed the authors of [Chowdhury09] to measure throughput, packet loss, and jitter of the integrated system in various applications, such as file transfer or VoIP, under varying background load conditions.

Over time, as LTE started being developed together with concepts of small and femtocells, the attention moved towards integration of optical access and mobile networks. Due to the increasing number of mobile cells in a small or femtocell deployment, current practice of backhauling a base station with point-to-point fibre links becomes highly uneconomical, as the cost of the backhaul network can easily exceed that of the mobile infrastructure. Backhauling base stations with a Passive Optical Network becomes thus an interesting alternative, as the infrastructure cost can be shared with other services, such as residential broadband.

In [Milosavljevic13], a novel network architecture has been proposed that addresses the challenges posed by the emergence of mobile cloud computing. A comprehensive solution of optical-wireless convergence has been described, that takes into consideration not only the aggregated bandwidth of the ONU/eNBs, but also the individual bandwidth requirements of wireless users, allowing better resource management, reduced delay, and scalable optical links. This is achieved via long-reach TDWM-PON, as well as a combination of centralized and decentralized backhauls. The solution tackles bandwidth outage with dynamic routing at the intersection of the presented heterogeneous backhaul networks (CRAN and IP backhauling), as well as with redundant links. These links are also considered in order to provide path for transmission in case of fibre failure. The proposed solution is able to constantly adjust to the conditions of the RAN in terms of channel estimation, as well as bandwidth demands; the network becomes aware of the conditions that are static throughout time in its coverage (e.g. buildings and terrain anomalies), in addition to those that are dynamic (e.g. congestion in specific areas during specific hours of the day).

Access networks are responsible for a significant part of overall telecom network energy consumption, and their demand for energy also increases rapidly with the ever-growing traffic volume they carry. Sustainability necessitates energy conserving solutions which also carefully limit the negative effects on other system qualities. It is expected that future access networks are based on a converged optical/wireless architecture. The work in [Ladanyi14] examines a hybrid small cell LTE and PON network. The authors analyse the impact of serving the user population with a reduced number of active cells, either due to failure or selective switch-off of chosen cells. Multiple optical topologies are considered for connecting the cells of the wireless network. Extensive simulations are used to quantify the interdependence of energy consumption, network availability and the QoS experienced by the consumer. In [Ladanyi14b] one investigates the trade-off between serving the user population with a reduced number of active cells and the quality of network services.

The transmission of a 3GPP LTE signal over a seamlessly integrated radio-over-fibre and millimetre-wave wireless link at 90-GHz band has been theoretically analysed and experimentally demonstrated in [Dat13]; one successfully transmitted and demodulated all the test signal models defined by 3GPP for LTE eNB over the proposed system. The measured error vector magnitude for all test signal transmission is well under the limit threshold defined by the standard. The proposed system can be realized as an attractive means for a high capacity backhauling network for high speed mobile networks using small cell and/or carrier frequency at millimetre-wave.

In [Vall-llosera14] the authors have discussed a strategy for increased capacity based on improving macro cells, increasing density of the macro layer, and adding small cells for indoor users. The authors show the differences between 3GPP RoF solutions and WiFi and conclude that a 3GPP coordinated indoor small cell is the best solution to provide the best mobile broadband experience. One has shown a new architectural all-optical solution, the fibre radio-dot, that brings the macro layer indoors. This solution is an upgrade of the current radio dot system (a radio dot system is an alternative macro base station architecture based on detaching the antenna radio head from the radio units), and uses logical point-to-point connections to the antenna, therefore enabling higher bandwidths and no eavesdropping. Because it is a macro cell solution it offers frequency reuse, low latency, interference management, cell selection and network management, features that non-coordinated solutions cannot provide. The authors have tested a full fibre radio-dot system in the lab, and concluded that their analogue radio over fibre solution meets mobile broadband requirements while using a cost-effective link technology.

The work in [Yamada14] proposes a new type of wireless network named Virtual Single Cell (VSC) network. The VSC network is a small cell network which allows smooth packet transfer to moving terminals as if they stayed in a single cell. Each terminal is closed in an LMC (Logical Macro Cell) which consists of a few numbers of adjacent cells around the cell with the target terminal, and LMC is handed over to follow the moving terminal. If the interruption due to cell to cell move in an LMC is small enough, the total network can be a virtual single cell. The authors of [Yamada14] examine the 3G-LTE based handover procedures for the small cell network under the conditions that the total network is synchronously operated considering the cell size is very small, and cells are contiguously placed. PON with multicast functionality contributes to configuring LMC.

A number of solutions have specifically targeted machine type communications (MTC) communications and the Internet of Things (IoT), where a multitude of heterogeneous access networks are emerging and the integration of them in a single platform ensuring seamless data-exchange with Data-Centres is of major importance. In [Orphanoudakis14] one describes HYDRA (HYbriD long-Reach fibre Access network), a novel network architecture that overcomes the limitations of both long-reach PONs as well as mobile backhauling schemes, leading to significantly improved cost and power consumption figures. The key concept is the introduction of an Active Remote Node (ARN) that interfaces to end-users by means of the lowest cost/power consumption technology (short-range xPON, wireless, etc.) whilst on the core network side it employs adaptive ultra-long reach links to bypass the Metropolitan Area Network. The proposed architecture can enhance performance while supporting network virtualization and efficient resource orchestration based on Software Defined Networking (SDN) principles and open access networking models.

From an optical transmission perspective backhauling solutions are categorized into:

- Pure Backhauling, where the signal is processed entirely at the base station and cells connected at an IP level.
- Radio-over-Fibre (RoF), where the optical carrier is modulated by the RF signal.
- Fronthauling, where the RF signal is sampled and the I/Q samples transmitted digitally over fibre.

There has been over the past few years, in the research community, a trend of moving from backhauling towards fronthauling solutions, as this enables better efficiency and cost reduction from a mobile network perspective. Fronthauling allows reducing processing equipment at each mobile site and centralizing it in one location that can handle all processing required for a number of base stations. Besides enabling lower power consumption and better sharing of processing resources it is also more suitable to implement advanced functionalities such as Coordinated Multi-Point (CoMP) or Inter Cell Interference Coordination (ICIC).

Although fronthauling does have clear advantages from a mobile network viewpoint it does bring some important issue from an optical transport perspective [Pizzinat15], as it poses very strict requirements on the maximum latency budget (lower than 400 μs) and it increases the required capacity by over an order of magnitude compared to backhauling (the typical increase being a factor of 16). While studies have attempted to reduce the capacity requirement by adopting compression techniques [Park14], [Lorca13], and cope with the latency requirements [Tashiro14], it is uncertain whether such technology will prove adequate. The fact that the optical access network needs to be carefully design to accommodate the fronthauling requirements could largely increase its cost and thus offset many of the economic advantages it brings on from a mobile network perspective.

Recently a new research area, dubbed Split Base Station Processing aiming at harnessing benefits from both backhauling and fronthauling has emerged, looking at intermediate solution between the two. The idea is to add some additional signal processing at the base station, compared to the fronthauling solution, to relax latency requirements and reduce capacity occupation. The authors of [Dotsch13] provide an excellent study of different processing splitting options, highlighting benefits and constraints.

From the work carried out over the past decade it is clear that there is much to gain in integrating the mobile and optical access domains and large economic benefits can be achieved by converging multiple services on top of the same optical access infrastructure. There are however challenges to be overcome in assuring tight integration between the two worlds in order to maintain low latency and end-to-end high quality of service, across the two domains.

FUTEBOL is exploring these aspects of network convergence enabling easy access to integrated optical/wireless research infrastructures.

Notable European projects in this domain include: the FP7 project COMBO constructing a fixed-mobile convergence testbed, looking at LTE as mobile technology [Baldo14]; the FP7 project ACCORDANCE which investigates the introduction of OFDMA into a PON architecture offering optical backhauling for wireless and copper based networking[6]; the FP7 project iJOIN which designs an open access and backhaul network architecture for small cells based on cloud networks[7].

25.4 FUTEBOL Approach

The infrastructure and control framework created in FUTEBOL is being federated according to principles developed in the FIRE program and facilities in the two continents interconnected through infrastructure deployed by the FIBRE project, as shown in Figure 25.1.

As mentioned before the main goal of FUTEBOL is is to enable experimental research at the convergence point between optical and wireless networks through the development of research infrastructure between Europe and Brazil. Nevertheless, the following objectives will be also addressed during the project's lifespan:

[6]www.ict-accordance.eu

[7]www.ict-ijoin.eu

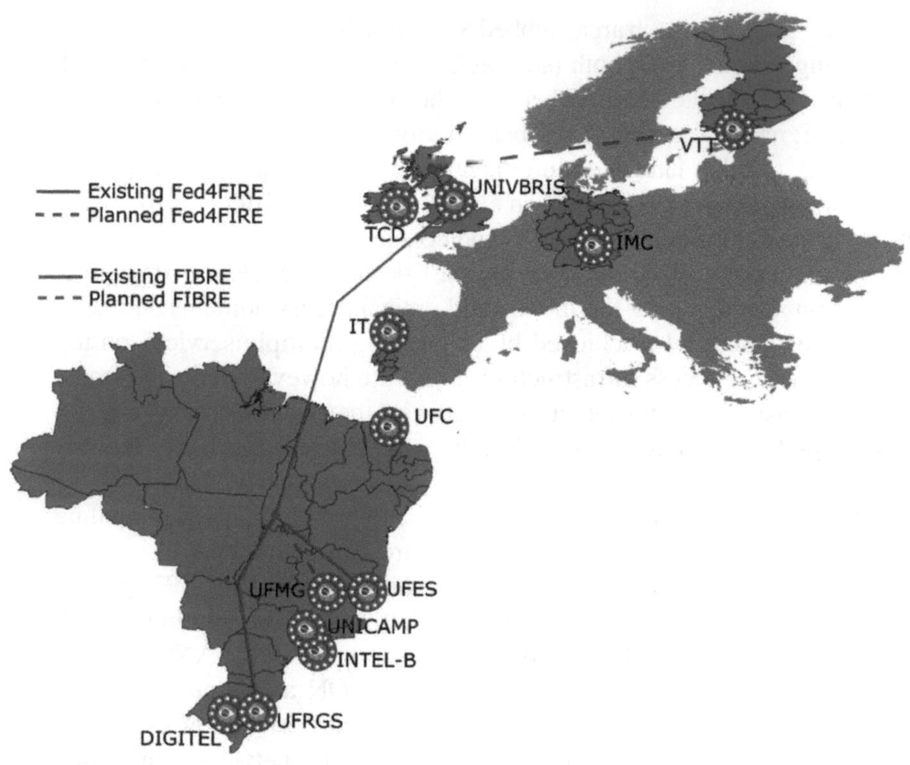

Figure 25.1 FUTEBOL consortium geographically distributed in Brazil and Europe.

- To deploy facilities in Europe and Brazil that can be accessed by external experimenters for experimentation that requires integration of wireless and optical technologies.
- To develop and deploy a converged control framework for experimentation at the optical/wireless boundary, currently missing in FIRE and FIBRE research infrastructure.
- To conduct industry-informed experimental research using the optical/wireless facilities.
- To create a sustainable ecosystem of collaborative research and industrial/academic partnerships between Brazil and Europe.
- To create education and outreach materials for a broad audience interested in experimental issues in wireless and optical networks.

In order to reach the above objectives, FUTEBOL composes a federation of research infrastructure, develops a supporting control framework, and conducts experimentation-based research advancing the state of telecommunications through the investigation of the optical/wireless boundary of networks. Figure 25.2 illustrates the layer natured of FUTEBOL: the end-user driven advancement of telecommunications relies on the development of the FUTEBOL converged control framework, which, in turn, requires the composition of federated research infrastructure. Through this approach,

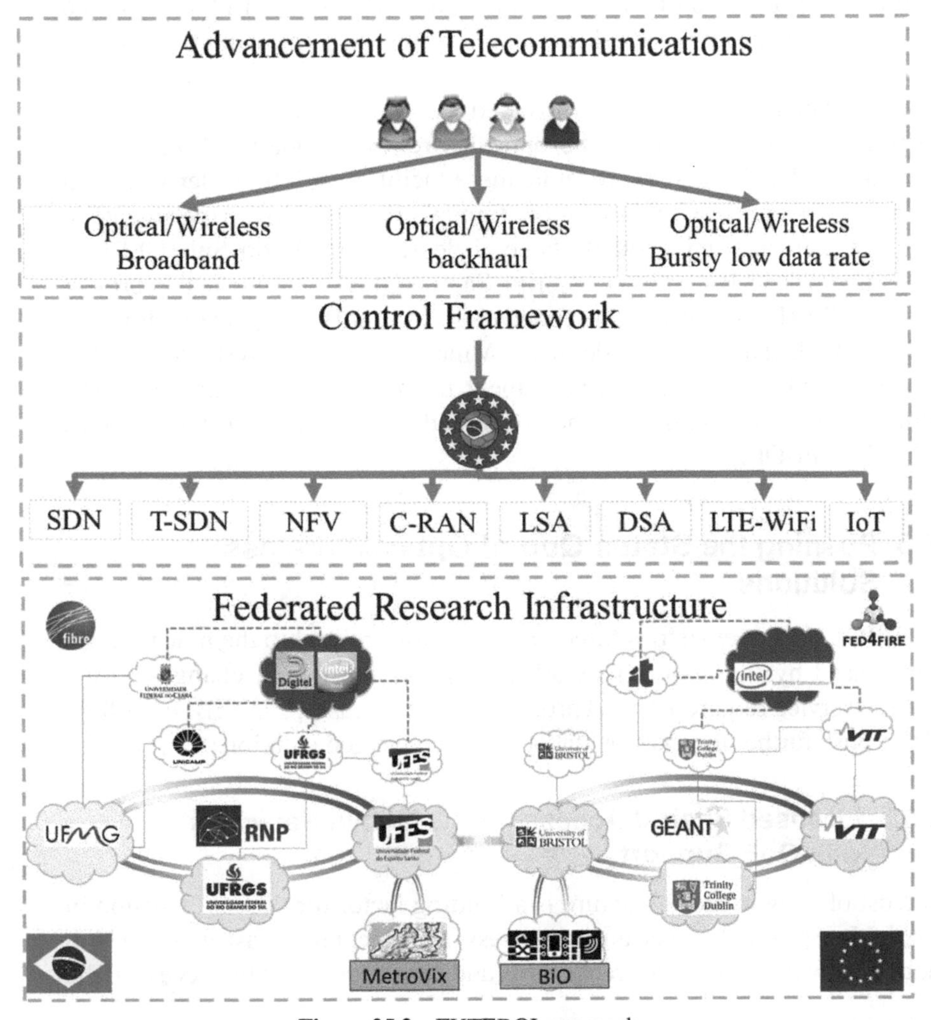

Figure 25.2 FUTEBOL approach.

FUTEBOL provides a complete, top-down development of research infrastructure, tailored to the needs of end-users throughout Brazil and Europe. Moreover, the combination of leading academic and industrial partners within the FUTEBOL consortium provides the key ingredients required to connect the broader telecommunications community to the advancements achieved through research.

European partners, Trinity College Dublin (TCD), University of Bristol (UNIVBRIS), and Teknologian Tutkimuskeskus VTT Oy (VTT), bring to the project mature, proven optical and wireless research infrastructure. While these facilities provide advanced capabilities in both optical and wireless experimentation, there is no converged control framework that enables integrated experimentation at the optical/wireless boundary. FUTEBOL is providing such a framework, federate these facilities with tools derived from the Fed4FIRE[8] project, and open this framework to external experimenters. Brazilian partners, Universidade Federal do Rio Grande do Sul (UFRGS) and Universidade Federal do Espírito Santo (UFES), bring to the project existing FIBRE[9] islands, as a foundation for further expansion through FUTEBOL. Universidade Federal de Minas Gerais (UFMG) will deploy such an island through their involvement in the project. All three of these partners will adopt and deploy the converged control framework developed within FUTEBOL.

25.5 Pushing the Status Quo of Optical/Wireless Solutions

FUTEBOL project envisions three main use cases related to the new requirements faced by the optical network to support substantial changes in the wireless service requirements. Through such showcases, FUTEBOL will be able to push further the current status of the converged solutions.

25.5.1 Licensed Shared Access for 4G Mobile Networks with QoE Support

The cost of network infrastructure is a limiting factor for digital inclusion in Brazil. More available spectrum means less investment in infrastructure which can be a relevant factor to decrease costs and democratize quality access to the

[8]www.fed4fire.eu
[9]www.fibre.org.br

internet in Brazil. Licensed Shared Access (LSA) is becoming an emerging model in Europe that promises to open new spectrum for mobile operators. In European, a 100 MHz frequency band in 2.3–2.4 GHz are expected for LSA to be used by cellular stakeholders on a shared basis. In this context, this experiment will exploit the 4G/LTE LSA trial environment developed by VTT in Finland to demonstrate the viability of LSA as a way to increase capacity with limited infrastructure investment in spectrum bands of common interest for Brazil and Europe.

It is expected that use of LSA will provide more spectrum bandwidth capacity to support a larger number of end-users in certain areas. However, the use of LSA will also increase the network handovers for entering and vacating the LSA band. In this sense, is required to evaluate the impact of LSA on the end-users Quality of Experience (QoE). The performance impact of using LSA will be assessed concerning how it affects the QoE for end-users on the shared spectrum. Additionally, this experiment will evaluate the system performance, functionality and incumbent characteristic of LSA under the latest ETSI specifications.

The main aim of this experiment is to evaluate the latest ETSI specifications about LSA and quantify its performance regarding QoE. Further, this experiment will demonstrate the wireless/optical co-work, through the dynamic reconfiguration of the optical backhaul to deal the rapid increase of wireless capacity due to the use of LSA spectrum opportunities in small cell scenarios.

25.5.2 The Design of Optical Backhaul for Next-Generation Wireless

Extreme cell densification and data rates expected in wireless networks create unprecedented demand on the optical backhaul both in terms of delivering capacity to the appropriate point in the network and in support tight delay constraints for applications reliant on the tactile internet. Following such trends, it becomes apparent that convergence among wireless and optical networks needs to be considered throughout the communications protocol stack, from the physical layer to the network layer and the interface with the service layer. In fact, operators in Europe and Brazil are currently struggling to satisfy the requirements for the 4G wireless access networks using advanced front-/back-hauling techniques. Their first approach was to keep existing wireline and wireless architectures as unchanged as possible. Wireless base-stations have been connected to the core network exclusively via IP. So, in this case, the backhauling network simply needs to provision for tunnels for transporting S1 (i.e. LTE protocol between the base-station and the core-network) and X2 (i.e. LTE protocol for connecting base-stations) packets. The second approach emerged with the cloud radio access network (C-RAN) concept, embedding the wireless subsystem into the wireline network. An architecture capable of supporting the massive deployment of remote radio heads, while providing different degrees of centralized processing, allows for the development of more advanced virtualization solutions at the access segment. A hyper dense deployment of radio elements will open up new research challenges in terms of interference management and mobility and will fundamentally change the needed point of attachment for the fibre optical infrastructure. From the mobile operator point of view, a key requirement of future mobile networks is that this additional network capacity has to be added at the lowest possible cost, in contrast to the current trend of decoupling revenues and traffic. C-RAN will allow for a significant reduction of costs for operators, because part of the RAN computation complexity is moved to the cloud infrastructure.

25.5.3 The Interplay between Bursty, Low Data Rate Wireless and Optical Network Architectures

Machine Type Communication (MTC) is a use case mainly characterized by a very large number of connected devices that typically transmit relatively low volume of data. Most of these services are non-delay-sensitive (e.g., utility metering). Devices are required to be simple and cheap, and have a very long battery life. This use case addresses the needs of a massive deployment of ubiquitous machine-to-machine communication, involving devices ranging from low complexity to those that are more advanced. Ubiquitous devices will sometimes communicate in a local context, which means that the traffic pattern and routes may be different than in cloud or traditional human-centred communication. To integrate the ubiquity of communication in a unified optical/wireless network is a challenge e.g. for applications combining information from different types of sources. Another challenge lies in how to manage the signalling overhead created by the high number of devices. Many devices frequently exchange short bursts of data with their network-side application. When there are no other communication needs, the devices have only a small amount of data to send but nevertheless have to go through a full signalling procedure to transmit the data. This wastes battery life, spectrum and network capacity. To handle this type of transaction more efficiently, the network needs to support a truly connectionless mode of operation, where devices can simply wake up and send a short burst of data. Upon reception of the short burst, device and application-related state information can be retrieved from a controller function and resources to handle the packet allocated accordingly.

25.6 Conclusions

The experimental research infrastructure enabled by FUTEBOL will demonstrate how wireless/optical convergence will support future traffic growth and new mobile services, while limiting the CAPEX/OPEX required to deploy and maintain the network. Moreover, FUTEBOL envisages the creation of a federated control framework to integrate testbeds from Europe and Brazil for network researchers form academia/industry with unprecedented features. Our major goal is to allow the access to advanced experimental facilities in Europe and Brazil for research and education across the wireless and optical domains. To accomplish this goal, we are developing a converged control framework to support optical/wireless experimentation on the federated research infrastructure from all associated partners/institutions. This way, industry-driven use cases can be deployed to produce advances in research at the optical/wireless boundary.

Acknowledgement

FUTEBOL has received funding from the European Union's Horizon 2020 for research, technological development, and demonstration under grant agreement no. 688941 (FUTEBOL), as well from the Brazilian Ministry of Science, Technology and Innovation (MCTI) through RNP and CTIC.

References

[1] **[Baldo14]** N. Baldo et al., "A Testbed for Fixed Mobile Convergence Experimentation: ADRENALINE-LENA Integration." *European Wireless Conference* 2014.

[2] **[Beas13]** J. Beas, G. Castanon, I. Aldaya, A. Aragon-Zavala and G. Campuzano, "Millimeter-Wave Frequency Radio over Fiber Systems: A Survey," in *IEEE Communications Surveys & Tutorials*, vol. 15, no. 4, pp. 1593–1619, Fourth Quarter 2013.

[3] **[Chowdhury09]** P. Chowdhury et al., "Hybrid Wireless-Optical Broadband Access Network (WOBAN): Prototype Development and Research Challenges", *IEEE Network*, vol. 23, no. 3, pp. 41–48, May 2009.

[4] **[Dat13]** P. T. Dat et al., "Performance Evaluation of LTE Signal Transmission Over a Seamlessly Integrated Radio-Over-Fiber and Millimeter-Wave Wireless Link", *IEEE Global Communications Conference (GLOBECOM)*, pp. 748–753, December 2013.

[5] **[Dotsch13]** U. Dotsch et al., "Quantitative Analysis of Split Base Station Processing and Determination of Advantageous Architectures for LTE", *Bell Labs Technical Journal*, vol. 18, no. 1, pp. 105–128, January 2013.

[6] **[Ladanyi14]** A. Ladanyi et al., "Impact of optical access topologies onto availability, power and QoS", *International Conference on the Design of Reliable Communication Networks (DRCN)*, April 2014.

[7] **[Lorca13]** J. Lorca et al., "Lossless compression technique for the fronthaul of LTE/LTE-advanced cloud-RAN architectures", *IEEE International Symposium on a World of Wireless, Mobile and Multimedia Networks (WoWMoM)*, June 2013.

[8] **[Ladanyi14b]** A. Ladanyi et al., "Tradeoffs of a converged wireless-optical access network", *International Telecommunications Network Strategy and Planning Symposium (NETWORKS)*, September 2014.

[9] **[Milosavljevic13]** M. Milosavljevic et al., "Self-organized Cooperative 5G RANs with Intelligent Optical Backhauls for Mobile Cloud Computing", *IEEE International Conference on Communications (ICC)*, pp. 900–904, June 2013.

[10] **[Orphanoudakis14]** T. G. Orphanoudakis et al., "Next Generation Optical Network Architecture Featuring Distributed Aggregation, Network Processing and Information Routing", *European Conference on Networks and Communications (EuCNC)*, June 2014.

[11] **[Park14]** S.-H. Park et al., "Inter-Cluster Design of Precoding and Fronthaul Compression for Cloud Radio Access Networks", *IEEE Wireless Communications Letters*, vol. 3, no. 4, pp. 369–372, August 2014.

[12] **[Pizzinat15]** A. Pizzinat et al., "Things You Should Know About Fronthaul", *IEEE/OSA Journal of Lightwave Technology*, vol. 33, no. 5, pp. 1077–1083, March 2015.

[13] **[Saskar07]** S. Sarkar, S. Dixit and B. Mukherjee, "Hybrid Wireless-Optical Broadband-Access Network (WOBAN): A Review of Relevant Challenges," in *Journal of Lightwave Technology*, vol. 25, no. 11, pp. 3329–3340, Nov. 2007.

[14] **[Shaw07]** W.-H. Shaw et al., "MARIN Hybrid Optical-Wireless Access Network", *Optical Fiber Communications Conference (OFC)*, March 2007.

[15] **[Tashiro14]** T. Tashiro et al., "A novel DBA scheme for TDM-PON based mobile fronthaul", *Optical Fiber Communications Conference (OFC)*, March 2014.

[16] **[Vall-llosera14**] G. Vall-llosera et al., "Small cell strategy: meeting the indoor challenge", *IEEE International Conference on Communications (ICC)*, pp. 392–396, June 2014.

[17] **[Yamada14]** T. Yamada and T. Nishimura, "Possibility of a new type of wireless network – Connected small cells with 3G-LTE signaling primitives and PON for the backhaul", *International Telecommunications Network Strategy and Planning Symposium (NETWORKS)*, September 2014.

26

ECIAO: Bridging EU-China Future Internet Common Activities and Opportunities

26.1 Introduction

The ECIAO project (August 2013 to July 2015) supported EU-China co-operation on activities related to Future Internet Experimental Research (FIRE) and IPv6 deployment.

In particular, the project explored EU-China co-operation on the following topics:

- Strengthening EU-China joint research efforts on the Future Internet by facilitating the trialling of interoperable solutions and common standards, and exchanging best practices on the federation of testbeds.
- Reinforcing academic and industrial co-operation on Future Internet experimental research, through a better networking between European and Chinese actors. The ECIAO web portal at http://www.euchina-fire.eu/, linked also to leading social networks and – with a dedicated helpdesk service – offered an efficient Co-operation Platform stimulating collaboration between EU and China researchers.
- Exchanging best practices for IPv6 deployment and supporting the creation of interconnected IPv6 pilots between Europe and China.

26.2 Problem Statement

Among the salient problems which existed prior to ECIAO were the differences in experience and contexts regarding testbeds and their federation between Europe and China. This was coupled with the added factor of technology constantly evolving at a fast pace.

Therefore, a common action on testbed federation was seen as a necessary requirement. ECIAO looked at both global and EU/China research on federation, taking into account the existing work being done on the Chinese testbeds through PTDN (Public Packet Telecom Data Network) and the Internet Innovation Union (IIU), and in EU projects such as OFELIA, BonFIRE, OpenLab and Fed4FIRE.

Interoperability was seen as another important step to bring key technical areas of interest for both Europe and China from research to the market.

Contributions to standards in the domain of Future Internet was an area in which there was no optimal communication between EU and Chinese standardisation experts. In China, a main focus of standardisation was on PTDN, but on the EU side, the equivalent topics were being followed in ETSI by the AFI (Autonomic Future Internet) NTECH Technical Committee.

The massive adoption of IPv6 in China requires experience from complex deployments including IPv4 and IPv6 transitions. This expertise is available in Europe.

26.3 Background

In 2009, the first EU-China Information Society Dialogue between the DG Information Society and the Chinese Ministry for Industry and Information Technology (MIIT) took place. This Dialogue was in addition to the EU-China Information Society Dialogue that started one year earlier. The new Dialogue focused on topics of policy-making in the area of the telecommunications and information society framework.

One of the key topics that arose from this Dialogue in 2010 was the need to exchange experiences and plans on further developing the overall policy framework for the Information Society. The European Commission was working on the policy framework to succeed the i2010 programme ("A European information society for growth and employment"), and was calling for input from experts and stakeholders, domestically and abroad, whilst at the same time the Chinese government was conducting research and internal consultations on Information Society-related elements of its forthcoming 12th 5-Year-Plan to guide the Chinese industrial and social policy over the next years.

During this period of important strategic decisions, both for the EU and China, the Information Society Policy Dialogue sought to improve mutual understanding of the respective approaches and support the development of global strategies for global information and communication networks.

To help provide inputs related to the "Internet of the Future" for the Dialogue, an EU-China expert group was established in 2010. The expert group met twice – in July and September 2010 – and had regular exchanges which led to recommendations in important areas including IoT, FIRE and IPv6.

The ECIAO project was designed to follow-up on the recommendations from this expert group.

26.4 Approach

The project targeted five actions:

- The analysis of Future Internet research topics in Europe and China and the identification of common topics for co-operation. These were: IPv6, SDN, NFV, IoT, 5G, Cloud and AFI-PTDN.
- The identification and documentation of common ongoing technical collaborations that were ready to move to the stage of interoperability testing between Europe and China. The interoperability testing of IoT and IPv6 was facilitated.
- The facilitation of joint contributions to standards in the domain of the Future Internet (AFI-PTDN).
- The exchanging of best practices in IPv6 deployment between Europe and China, including the setting up of a common pilot.
- Ensuring a better networking and enhanced co-operation between European and Chinese organisations, through the creation of an interactive web portal, the provision of helpdesk services and the organisation of dedicated events.

This approach was accompanied by a solid dissemination strategy that comprised three consecutive phases:

- *Awareness-oriented phase*: At the start of project, this phase raised public, industry and research community awareness about the project and the problems that it aimed to address. During this phase of the dissemination, the tasks involved the setting up of the basic marketing materials and awareness-raising presentations at various related events.
- *Result-oriented phase*: During this phase the results of the project were published to promote these to stakeholders in EU and China.

- *Exploitation-oriented phase*: Specific activities were undertaken in order to improve the online Co-operation Platform and to increase the co-operation in standardisation, testbed federation and resource sharing, IPv6 education and best practices, etc.

26.5 Achievements

The success of the ECIAO project manifested itself in several important achievements:

An online *Co-operation Platform* helping EU and Chinese organisations network together and discuss co-operation on Future Internet topics.

An online *Support Desk* for EU and Chinese researchers and industries looking for background information in the area of FIRE, or for support in establishing contacts in this field.

9 articles on EU-China Future Internet co-operation sent by experts from both Europe and China in response to the project call for articles.

Two conferences supporting EU-China co-operation on IPv6 and Future Internet testbeds and strengthening collaboration on AFI-PTDN standardisation.

Two webinars imparting knowledge on Public Packet Telecom Data Network (PTDN) and AFI.

Two workshops discussing AFI standardisation and SDN, NFV and IPv6 impacts, as well as EU-China testbeds federation, including a *Fed4FIRE tutorial* on how to use worldwide federation techniques.

Improvement of interoperability between European and Chinese developments due to successful *interoperability events in IPv6 and SDN* organised or supported by the project.

A successful *EU-China IPv6* pilot set up between partners sites in Beijing (the Beijing Internet Institute and the Beijing University of Post and Telecommunications) and Paris (Mandat International and France Telecom/Orange).

Increased awareness on *testbed federation techniques* preparing the ground for worldwide federation between GENI (USA), Fed4FIRE (Europe) and CENI (China).

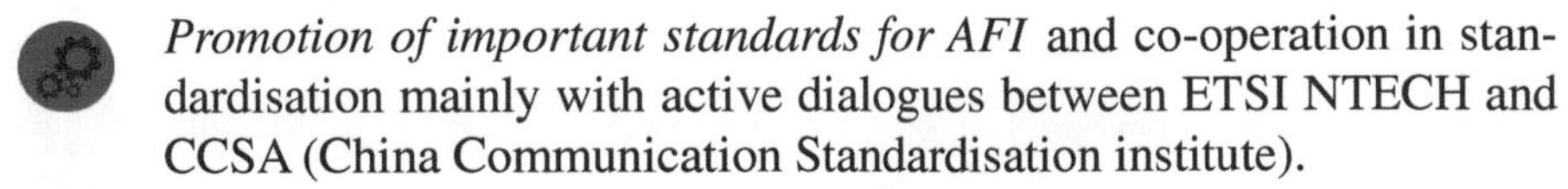

Promotion of important standards for AFI and co-operation in standardisation mainly with active dialogues between ETSI NTECH and CCSA (China Communication Standardisation institute).

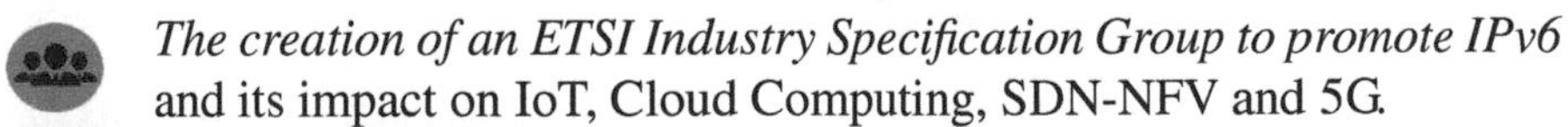

The creation of an ETSI Industry Specification Group to promote IPv6 and its impact on IoT, Cloud Computing, SDN-NFV and 5G.

Important contribution to the Fed4FIRE architecture implementation.

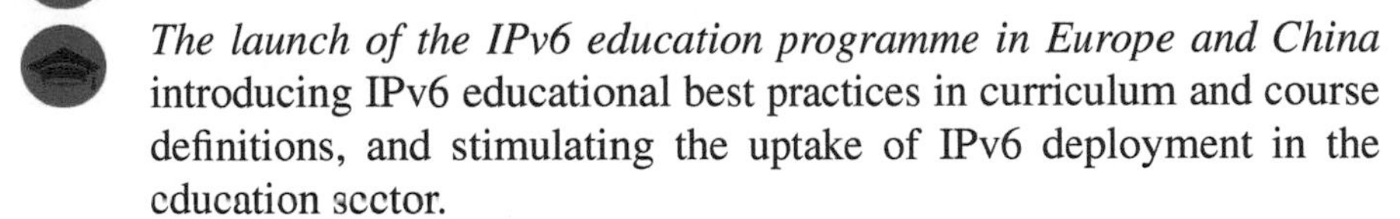

The launch of the IPv6 education programme in Europe and China introducing IPv6 educational best practices in curriculum and course definitions, and stimulating the uptake of IPv6 deployment in the education sector.

Two eBooks on IPv6 best practices written by experts in the field offering deployment recommendations (1st eBook) and providing an IPv6 Roadmap exploring the transition process (2nd eBook).

A virtual community of stakeholders supporting EU-China co-operation on FIRE on LinkedIn, Twitter and Weibo.

26.6 Conclusions

The ECIAO project successfully built a sustainable partnership between European and Chinese organisations in order to foster co-operation in the domain of Future Internet research experimentation and IPv6. This was achieved by facilitating the development of interoperable solutions and common standards, reinforcing academic and industrial co-operation, the exchanging of good practices for IPv6 deployment and supporting the creation of interconnected IPv6 pilots between Europe and China.

ECIAO exploited work done in several past and on-going FIRE projects and took into account various experiences from them.

Six Chinese testbeds began the process of federating with EU testbeds, using the Fed4FIRE specifications.

The impact of ECIAO reaches not only into the scientific and technical communities but also fosters social and political information exchanges and co-operation between EU and China. Research organisations, regulators, policy-makers, enterprises and consortia have taken advantage to build on our work, which will be continued through a similar initiative in H2020, focussing on EU-China collaboration for IoT and 5G.

27

EU-US Collaboration in FIRE

Brecht Vermeulen, Tim Wauters, Peter Vandaele and Piet Demeester

IMEC, Belgium

27.1 History

In the FIRE domain, there is a long standing history of collaboration between EU and the US. E.g. the Onelab project worked closely together as early as 2006/2007 with the US Planetlab community to build a federated European Planetlab (PLE, Planetlab Europe) which is still operational and federated till today. Several research institutes collaborate since long on a bilateral/open source base as well. Several EU-US workshops have been organized in the last 10 years as well with the goal to bring researchers closer together. In the remainder of this part on EU-US collaboration, we will focus on the last 4 years of collaboration guided from with the FedFIRE project, that started in 2012.

27.2 Liaison – Mission Statement

Since 2012, the Fed4FIRE project, as a largely collaborative and federation project, is responsible from the EU side for the organization of the GENI-FIRE collaboration that has been further enhanced. The mission of this collaboration was established in a "Joint Statement of Interest": "*The EU and US research communities wish to perform collaborative research, on the basis of equality and reciprocity, in areas of mutual interest, which may be characterized as (a) investigations of the research infrastructures suitable for hosting at -scale experimentation in future internet architectures, services, and applications, and (b) use of such infrastructures for experimental research. We envision that our collaboration will encompass joint specification of system interfaces, development of interoperable systems, adoption of each other's tools, experimental linkages of our testbeds, and experimentation that spans our infrastructures. We further envision that students and young professors from the US and EU will visit each other and collaborate deeply in these activities, in hopes of sparking friendships and life-long research collaborations between the communities.*"

The goal of this liaison was to bring (young) researchers together in workshops and give them the possibility to collaborate by funding travels and visiting stays.

27.3 GENI-FIRE Collaboration Workshops

On 14 and 15 October 2013, Fed4FIRE hosted a workshop to stimulate collaborations between GENI and FIRE researchers. The workshop took place in Leuven, Belgium, and was attended by 38 participants. Next to members of the organization, each community was represented by a group of 15 carefully selected participants. Invitations were based on relevant expertise. The following topics were discussed: experiments, general aspects (architecture, API's, terminology), resource description, policies, data plane and education. Several concrete opportunities for collaboration were identified and initiated (joint definition of an ontology for resource description, joint specification of a new version of the SFA Aggregate Manager API, etc.). The discussions also allowed the participants to identify specific opportunities for collaborations based on visits of US researchers at specific EU partners (which can be funded by NSF), or by setting up joint experiments remotely.

Jointly with the partners from GENI, Fed4FIRE has organized the second GENI/FIRE Collaboration workshop on 5 and 6 May 2014 in Boston, following the successful workshop that Fed4FIRE hosted in Leuven (Belgium) on 14 and 15 October 2013. 35 top-level experts, equally balanced between EC and US, were invited to this closed workshop. The overall goal of the workshop was twofold: on one hand the goal was to report on actual EU-US collaborations that were initiated at the first workshop in Leuven, and to stimulate and facilitate their continuation. On the other hand, the workshop also scheduled some time to discuss new topics which have not been touched before, but which are believed to have great potential for seeding additional collaborations between the communities at each side of the Atlantic. These topics included wireless networking, software-defined networks, instrumentation and measurement, control and operability and experimenter support. Through joint funding mechanisms between Fed4FIRE and US (SAVI grants), up to nine travels have been organized for meetings for GENI/FIRE collaboration.

After previous workshops in Leuven and Boston, Fed4FIRE organized a third GENI-FIRE collaboration workshop in Paris, hosted by UPMC, on November 20–21, 2014. The meeting was attended by 44 invited experts from EU and US. The participants engaged in a lively discussion on the future of Internet-wide testbed federations. The workshop's agenda was structured into five sessions. A first general session provided the highlights of ongoing funded collaborations and future plans. The following sessions focused on specific topics, such as Wireless testbeds (both in FIRE and GENI, dealing with LTE and WIFI technologies), Ontologies (lessons learned from Exo-GENI and CREW, accomplishments in jFed and Open Multinet), Federation aspects (including interconnectivity, policies and common federation APIs) and Clouds (ongoing initiatives, common cloud APIs, tools and resource representations).

The fourth workshop organized by GENI and Fed4FIRE took place in Washington on September 17–18, 2015. It was attended by 35 invited experts from EU and US that discussed in six sessions on the following topics: a general session on reporting, demonstrations and discussion on funded travels for collobaration. A second session discussed Cloud topics, followed by a session on wireless and one on ontologies. The fifth session was on federation (global federation, policies, SDX, connectivity), while the final session was on monitoring.

For the next workshop, organized in Brussels (18–20 April 2016) just before the Net Futures 2016 event, it was decided to extend the workshop also to partners from Brazil and Japan, besides EU and US. This GEFI workshop ("Global Experimentation for the Future Internet") had 11 US participants, 5 Brazilian participants, 3 Japanese and 12 EU participants and organizers. Six sessions were organized on the following topics: overview by the funding organisations on the goals of the workshops, Federation/software defined infrastructure and connectivity, cloud and big data, wireless/cognitive radio and convergence, 5G/NFV and SDN and a last session on the Internet of Things (IoT). On April 20th, there was also a session organized in the Net Futures conference with speakers of the 4 countries/continents. Each session was chaired by two people from a different country/continent.

27.4 FIRE-GENI Summer Schools (FGRE)

Besides the above mentioned workshops, till now also three summer schools were co-organized between GENI and Fed4FIRE with the goal to bring tutors and students from EU and US together and let them collaborate.

GENI pays travel funding for the US students and tutors. Fed4FIRE foresees travel funding only for Fed4FIRE students and tutors.

The 1st Fed4FIRE-GENI Research Experiment Summit (FGRE 2014) event has been held in Ghent, Belgium on July 7–11, 2014. It consisted of keynote speeches, presentations on experiences, tutorials, hands-on experiments and team projects. The summit provided participants with opportunities to learn and use the various resources and tools available in the shared Fed4FIRE and GENI experimentation environments.

More than 50 applicants and tutors, including undergraduate and graduate students, faculty member at different-level colleges and researchers from industry (both SME and large companies) have participated at the event and collaborated in a lively atmosphere. The scheme below (of the 2016 summer school) shows that the summer school starts with 2.5 days of tutorials followed by 2.5 days of team projects where about 20 students work really together on small projects they define themselves.

In July 2015, the 2nd FGRE summer school was organised with 37 participants from EU, US and South-Korea, and 14 tutors from EU, US and South-Korea. 14 students took part in the project teams.

In July 2016, the 3rd FGRE summer school was organised in Gent with 35 participants from EU and US, and again 14 tutors from EU and US. Interesting to know is that only one participant came from a Fed4FIRE partner while the others were from US (10) and from other companies and institutes not taking part in Fed4FIRE. For this summer school we also organised the

<table>
<tr><th></th><th>Monday</th><th colspan="3">Tuesday</th><th colspan="2">Wednesday</th><th>Thursday</th><th>Friday</th></tr>
<tr><td>Block I (8:30-10)</td><td>Welcome
Intro to Fed4FIRE
Intro To GENI</td><td colspan="2">Bigdata introduction</td><td>Wireless/sensor introduction</td><td>LTE GENI</td><td>Openflow OVS</td><td>Team Project</td><td>Team Project</td></tr>
<tr><td>Block II (10:30 – 12:30)</td><td>Getting started tutorial (Lab zero)</td><td colspan="2">Hadoop</td><td>Wireless + robots</td><td>LTE FLEX</td><td>Openflow hardware</td><td>Team Project</td><td>Team Project presentations</td></tr>
<tr><td>Block III (1:30 – 3:30)</td><td>More advanced tutorial (Lab one)
Openflow introduction</td><td>Spark</td><td>Smart city</td><td>Cloudlab</td><td>Phantomnet</td><td>Team Project</td><td>Team Project</td><td></td></tr>
<tr><td>Block iV (4-5.30)</td><td>Emulation, scaling up, stitching</td><td>Stream processing</td><td>IoT Sensors</td><td>Automate – genilib ansible chef</td><td colspan="2">Team Project</td><td>Team Project</td><td></td></tr>
<tr><td>Evening</td><td></td><td colspan="3">Social event (19:00)</td><td colspan="2"></td><td></td><td></td></tr>
</table>

tutorials in tracks of a single day or half day on the following topics: big data, Wireless/IoT/SmartCity, Cloud, LTE and Openflow, each with two or more tutorials on the same subject.

All the tutorials of the summer schools are available at http://doc.ilabt.iminds.be/fgre/

27.5 Dissemination at the Geni Engineering Conferences (GEC)

Most of the collaboration results were shown at Geni Engineering Conferences organized three times a year, both at demo nights or in particular discussion or tutorial sessions where we could give tutorials with EU developed tools (e.g. jFed).

In particular at GEC22 in Washington, April 2015, a team that started working together in the FIRE-GENI Summer school of 2014, won the best demo award with the work they started in 2014. At the same GEC, a joint EU-US plenary demo was demonstrated as well. On exactly the same moment, the same demo was shown at the Net Futures 2015 conference as well.

27.6 Standardization

The relevant APIs that have been defined, implemented and used in both GENI and Fed4FIRE (and beyond) for allocating and provisioning resources on testbeds are:

- The AM (Aggregate Manager) API, currently at version 3. This API was mainly defined by GENI with some input from others.
- The SA (Slice Authority) and MA (Member Authority) API, currently at version 2. These APIs were defined together by US GENI, EU Fed4FIRE and some non-Fed4FIRE EU partners.

These APIs have now been moved to Github (https://github.com/open-multinet) where everyone can contribute through issue creation/pull requests. A compiled version of the API documentation can be found at https://fed4fire-testbeds.ilabt.iminds.be/asciidoc/federation-am-api.html

In GENI, a reference implementation was created for these APIs (both client and server side), that can be found at https://github.com/GENI-NSF. In Fed4FIRE, a full test and monitoring framework was developed to test the compliance to these APIs. jFed (http://jfed.iminds.be) is used as test and experimenter interface, while the Fed4FIRE monitoring is done for Fed4FIRE testbeds (https://flsmonitor.fed4ifre.eu) and non-Fed4FIRE test-beds (GENI, South-Korea, Japan, Cloudlab, . . .): https://flsmonitor.fed4fire.eu/fls.html?testbedcategory=international_federation&hideinternalstatus&showlogintests

27.7 Some Technical Highlights from the EU-US Collaboration

During the intense collaboration of GENI and Fed4FIRE, the following technical highlights were reached amongst others:

- Agreement on APIs for AM API, MA, API, SA API needed for secure testbed access in a uniform way.
- Development of reference tools and frameworks for these APIs.
- Development of compliance testing and monitoring tools for these APIs.
- A way for provisioning end-to-end layer 2 connectivity through VLAN stitching (also between US and EU and beyond).
- Exchange of hands-on tutorials for the testbeds between EU and US: without any change, EU or US students can use tutorials to learn specific things.

- Successful federation of testbeds and authorities between EU and US.
- Prototype for a joint ontology base for testbed resources.

27.8 Conclusion

In this section, we described the collaboration between EU FIRE and US GENI going on for more than 10 years and during the last 4 years specifically. A combination of successful workshops on invitation, funded travels of students, summer schools and technical collaboration on e.g. tools and standards have led to an excellent understanding and collaborative mood. It is clear that a federation of testbeds and authorities and the accompanying developments lead to stronger research and research outcome on both sides of the ocean. The Fed4FIRE project runs till the end of 2012, but in the successor project Fed4FIRE+ the international collaboration is even more important, so we are quite confident that this collaboration will even get stronger in the future.

28

FESTIVAL: Heterogeneous Testbed Federation Across Europe and Japan

Martino Maggio[1], Giuseppe Ciulla[1], Roberto Di Bernardo[1], Nicola Muratore[2], Juan R. Santana[3], Toyokazu Akiyama[4], Levent Gurgen[5] and Morito Matsuoka[6]

[1]Engineering Ingegneria Informatica SpA, Palermo, Italy
[2]Demetrix Srl, Palermo, Italy
[3]Universidad de Cantabria, Santander, Spain
[4]Kyoto Sangyo University, Kyoto, Japan
[5]Commissariat à l'énergie atomique et aux énergies alternatives, Grenoble, France
[6]Osaka University, Osaka, Japan

Abstract

FESTIVAL is an H2020 EU-Japan collaborative project that aims to federate heterogeneous testbeds, making them interoperable and building an "Experimentation as a Service" (EaaS) model. Going beyond the traditional nature of experimental facilities, related to computational and networking large scale infrastructures, FESTIVAL testbeds have heterogeneous nature and in order to be federated they have been clustered in four categories: "Open Data" (i.e. open datasets), "IoT" (i.e. sensors and actuators), "IT" (i.e. computational resources) and "Living Labs" (i.e. people). Considering that every testbed category provides specific resources, the main challenge for FESTIVAL is to develop a platform that can allow experimenters to access very different assets in an homogeneous and transparent way, supporting them in the phases of the experiments. The FESTIVAL architecture, based on a multi-level federation approach, proposes a solution to this problem providing also a set of functionalities to manage and monitor the experiments. FESTIVAL tools, also, include the possibility to access FIWARE Generic Enablers allowing to deploy predefined components to address specific needs in the experimentation (e.g. data analysis, big data management etc.). The FESTIVAL platform will be tested on three different smart city domains across Japan and Europe: smart energy, smart building and smart shopping.

28.1 Introduction

There have been long years of research work in Europe and Japan on federation of testbeds. More recently this research also involved Internet of Things (IoT) paradigm that is radically changing the way we interact with daily life objects at various environments such as home, work, transportation and city. FESTIVAL [1] is a H2020 European-Japanese collaborative project that aims to federate heterogeneous (IoT) testbeds, making them interoperable and building an "Experimentation as a Service" (EaaS) model. The objective of the project is also to facilitate the access to those testbeds to a large community of experimenters allowing them to perform their experiments taking benefit of various software and hardware enablers provided both in Europe and in Japan. FESTIVAL testbeds will connect cyber world to the physical world, from large scale deployments at a city scale, to small platforms in lab environments and dedicated physical spaces simulating real-life settings.

This chapter will describe the results achieved in the first 18 months of the FESTIVAL project in terms of platform design and development and use case and experiments definition. The first section of the chapter introduces the experimental testbed involved in the project that will be federated through the FESTIVAL platform: the testbeds have heterogeneous nature and, in order to be federated, they have been clustered in four categories: "Open Data" (i.e. open datasets), "IoT" (i.e. sensors and actuators), "IT" (i.e. computational resources) and "Living Labs" (i.e. people). The Section 28.2 presents the FESTIVAL EaaS approach, description of the main components of the architecture and the different federation layers: from the federation of the same resource types to the homogenous representation of all the heterogeneous resources in a common data model. The third section describes the technical details of the refe rence implementation of the architecture composed by both existing software components and new ones specifically developed for FESTIVAL. Section 28.4 presents the typical experiment workflow that can be performed by the experimenters using the FESTIVAL functionalities through a dedicated experimentation portal. The Section 28.5 describes the use case domains and some specific experiments that will be executed during the project. The last section presents some conclusions related to the work performed, also indicating the next steps to be achieved in the following phases of the FESTIVAL project.

28.2 FESTIVAL Experimental Testbeds

FESTIVAL project aims to federate distributed and heterogeneous testbed among Europe and Japan. The different testbeds involved due to their heterogeneous nature have been classified in four different categories: Open Data, IoT, IT and Living Labs. In this document the specific testbeds included in these four clusters are presented.

28.2.1 Open Data Oriented Testbeds

Open Data oriented testbeds provide the experimenters with (Linked) Open Datasets. Open Data oriented testbeds consist in Open Data Management systems, specific web platforms, based on different technologies, designed to manage open datasets in various formats. In FESTIVAL there are four main platforms: Santander Datos Abiertos [2], Metropole of Lyon's Open Data [3], FIWARE Lab Data Portal [4] and FESTIVAL Japanese open data portal [5].

Santander Datos Abiertoss provided by Santander City Council, is a CKAN [6] based platform that includes Open Data about transport (e.g. buses, taxis, traffic information), urban planning and infrastructures (e.g. parks and gardens location, municipality buildings), shops, demography (current and historic census), society and well-being, culture and leisure (labour calendar and cultural programming). Metropole of Lyon's Open Data is related to the municipal territory of Lyon and provides wide-ranging access to public data such as the land register map for the conurbation, the surface area taken up by greenery, the availability of shared bikes or the locations of automatic car-sharing stations, real-time traffic data, highway events and traffic history. FIWARE Lab Data Portal is a CKAN based platform that provides a huge amount of data from different smart cities collected by FIWARE generic enablers instances [7]; FESTIVAL Japanese open data portal represents the CKAN node that will be used to collect the open data results of the use case experimentations and applications deployed in Japan.

28.2.2 IoT Oriented Testbeds

IoT oriented testbeds are physical places in which sensors or other smart objects are deployed, these can be part of experiments in order to gather data or to perform actions. IoT oriented testbeds in FESTIVAL are: iHouse [8]

(Ishikawa, Japan) an experimental smart house facilities built by ISICO (Ishikawa Sunrise Industries Creation Organization); Connectivity Technologies Platform (PTL) [9] (Grenoble, France), a facility to test connectivity technologies in real environments, SmartSantander [10] (Santander, Spain), an experimental test facility for the research and experimentation of architectures, key enabling technologies, services and applications for the Internet of Things in the context of a city; ATR Data Center [11, 12], an experimental data center facility that aims to reduce total energy consumption using specific monitoring technologies and prediction algorithms; Maya & Kameoka Stations [13–15] two Japanese railways smart stations equipped with an environmental monitoring sensors network (pollen, PM2.5, vibration, acceleration, noise, temperature).

28.2.3 IT Oriented Testbed

IT oriented testbeds provide virtualised IT resources such as virtual machines or virtual networks that can be used by the experimenter to deploy and execute software components and applications for their experiments. In particular, in FESTIVAL the two main IT oriented testbeds are JOSE platform and the Engineering FIWARE-Lab.

JOSE (Japan) [16] provides a Japan-wide open testbed, which consists of a large number of wireless sensors, SDN capabilities and distributed "CLOUD" resources. The facilities of JOSE are connected via high-speed network with SDN features.

The FIWARE-Lab (Italy) [17] is an instance of FIWARE based on a cloud environment allowing users to deploy, configure and execute a set of Generic Enablers. The cloud infrastructure is based on OpenStack [18], an open source software for creating cloud platforms. This specific FIWARE-lab instance is will provide specific computational resources dedicated to the FESTIVAL project.

28.2.4 Living Lab Testbed

Living Lab testbeds are represented by the living labs participating in FESTIVAL project, that provide services, people and physical places to experimenters in order to perform different activities (e.g. service co-design with user involvement, event organisations, expert consultant etc.). Living labs and services they provide have an important role in experiments because they allow the active participation of people and experts with specific skills in the activities related to experiments. The two Living Labs involved in the

FESTIVAL project are The Lab [19] (Osaka, Japan) and TUBA [20] (Lyon, France). The Lab constitutes a space where general public (such as researchers, creators, artists, students, citizens) can experience the latest technologies and have interactions with exhibitors. Communicators are the specialised staff who bring visitors together to The Lab with other people, things, and information and play the role of gathering the comments and reactions of visiting members of the public, and feeding this information back to companies, researchers, and other event organizers.

TUBA, which stands for "Experimentation Urban Test Tube", is operated by Lyon Urban Data: an association based on a mixed consortium of public and private entities. TUBA is a 600 m^2 place dedicated to experimenting new services and helping developing new projects (from start-ups to SMEs and large companies) based on data.

28.3 EaaS Model and FESTIVAL Federation

The main concept behind the design of the FESTIVAL architecture is the definition of an Experimentation as a Service (EaaS) model and its realization through a multi-level federation approach.

The adoption of the EaaS model aims to implement a platform that allows experimenters to create replicable and scalable experiments and to access very different testbeds in a homogeneous and transparent way, supporting them in all the phases of the experiments.

Starting from a set of homogeneous access APIs, namely FESTIVAL EaaS APIs, the platform supports experimenters in performing and managing multi-domain experiments. It provides discovery functionalities to find resources matching requirements for their experiments and the ability to analyse results collected during the execution of experiments.

This objective is achieved by structuring the FESTIVAL architecture with two levels of federation: the first one federates the resources of the same categories ("resource-based" federation) and the second one provides unified access to them ("experiment-based" federation).

The Figure 28.1 depicts the high level overview of FESTIVAL architecture that presents the EaaS Model and the different federation levels.

At the lowest layer, the figure shows all the testbeds involved in FESTIVAL, and classified according to a specific typology: Open Data, IoT, IT and Living Lab. The first federation level, "resource-based federation", provides functionalities for the integration and harmonization of each types of resource, using ad-hoc components called *Aggregators*. Each Aggregator

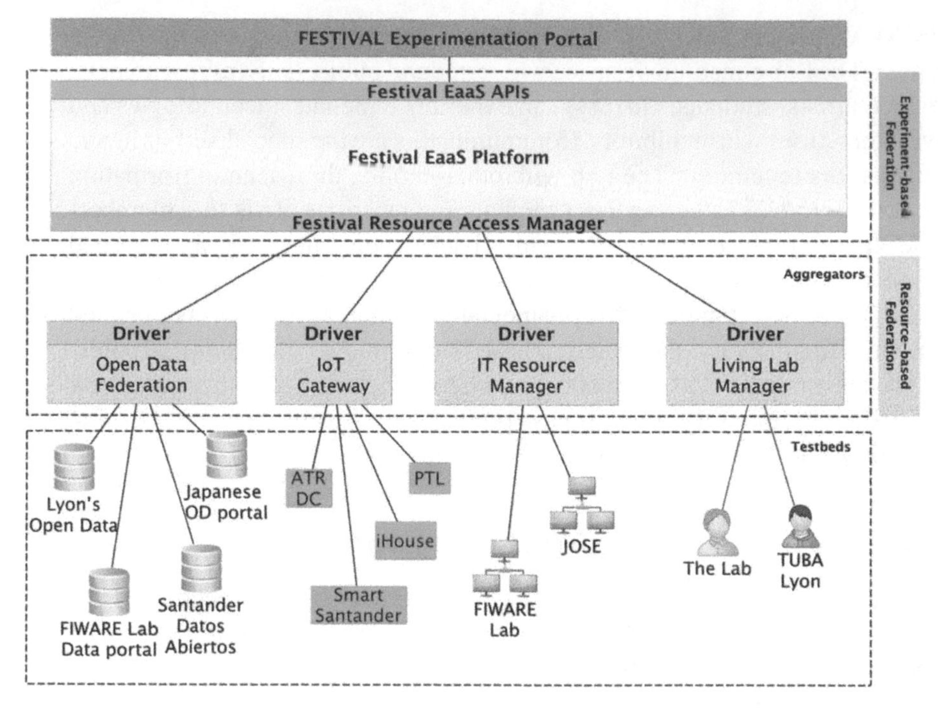

Figure 28.1 FESTIVAL architecture.

is independent from the FESTIVAL platform and federates resources of the same type, providing a common data model describing them in a uniform way.

FESTIVAL supports the following aggregators:

- **Open Data Federation**: it is the component in charge of federating different Open Data Management Systems (ODMS), providing functionalities to search generic open data and to perform queries on Linked Open Data (e.g. RDF).
- **IoT Gateway**: this component provides a uniform access to different IoT devices such as sensors or more generally smart objects, supporting different IoT protocols.
- **IT Resource Manager**: this component realizes the federation of computational resources, and in particular it is in charge of managing all the aspects related to the reservation and access of the virtual machines.
- **Living Lab Manager**: this component is able to federate living labs considering their services, methodologies used and expertise of its members as resources that can be involved in an experiment. The component allows to manage all of these living lab assets in a uniform way.

FESTIVAL platform is designed to be flexible and scalable: it is able to federate new types of resources, in addition to the already supported ones, adding new instances of aggregators of the same categories or completely new ones. (e.g. support more IoT devices adding more IoT gateway aggregators).

The aggregators are connected to the FESTIVAL platform through their specific "drivers" which are in charge of translating the heterogeneous resources in the "FESTIVAL resource data model" that will be used in the upper layer. FESTIVAL resource data model is designed to represent heterogeneous types of resources in the same way and at the same time to maintain their specific peculiarities.

Translation performed by drivers completes the federation process, allowing to the **FESTIVAL EaaS layer** to implement the "experiment-based federation", managing the resources without taking into account their different nature. The **FESTIVAL EaaS layer** contains the core components of the platform and manages the resources in all the phases of an experiment, executing the business logic to create and define the experiment, to access and reserve the resources and finally to execute and monitor the experiment. All the EaaS layer functionalities are accessible by external applications through a set of APIs (FESTIVAL EaaS API). FESTIVAL in particular provides the **Experimentation portal**, a web application designed to facilitate experimenters in the managing of their experiments and in the discovery of resources provided by the different federated testbeds.

28.4 FESTIVAL Reference Implementation

The following section presents some of the most relevant technologies and components used to implement the FESTIVAL architecture already presented in the previous sections:

28.4.1 Aggregators

The aggregators represent the components that federate the resources of a specific category of testbeds. The concrete components that implements the aggregators in the FESTIVAL reference implementation are the following:

Federated Open data platform: it is the component that implements the open data federation. It is an open source server-side JEE based components that is able to connect, to retrieve, search and visualize datasets of different types coming from open data repositories using specific connectors. It supports connectors for CKAN and SOCRATA [21] portals, but new open data nodes based on different technologies can join the federation by implementing specific REST API (Federation API). The platform includes also the Federated Open Data Catalogue, a web interface allowing end users to access directly to the Open Data Federation functionalities.

SensiNact Platform [22] is the gateway responsible for the aggregation of the IoT resources. SensiNact supports different protocols and API layers such as MQTT, REST API, JSON-RPC and Bluetooth 4.0, among others. In order to aggregate the IoT FESTIVAL testbeds, new bridges have been developed, which communicate with the testbeds resources: KNX Bridge connects SensiNact to PTL testbeds, using the KNX Worldwide Standard for Home and Building Control [23]; Orion Bridge connects SensiNact to SmartSantander testbeds, using the Orion Context Broker [24], a Generic Enabler provided by FIWARE; EchoNet Bridge: this bridge connects SensiNact to EchoNet Lite devices. EchoNet Lite [25] is a communication protocol developed in Japan focused in devices for SmartHome; OpenHab Bridge/Discovery: this bridge connects SensiNact to OpenHab [26] (a platform to control Smart Objects) running instances in the local network and provides access and discovery capabilities to items configured in OpenHab as devices into SensiNact.

SFA aggregator: it aims at federating IT virtual resources, that in the projects are provided by JOSE and FIWARE-Lab testbeds. In the perspective of reusing solutions already implemented in other FIRE initiatives, the approach followed is the implementation of a SFA [27] (Slice-based Federation Architecture) component for OpenStack, a solution implemented in other international research projects and more recently in the Fed4Fire European project [28]. Different implementation of this framework are available and among them, SFAWrap [29] was chosen as basis for FESTIVAL IT Aggregator. SFAWrap uses a specific driver to communicate with OpenStack services via Python API.

Living Lab Manager is the component in charge of managing the living labs federated in FESTIVAL. It has been implemented as a Java EE application that provides a set of APIs to add, update and delete living labs data in the federation.

The Living Lab Manager manages living labs and their resources according to a specific data model designed in the FESTIVAL project; this data model defines resources of living labs in terms of their main assets. The resources identified are:

- Services: identifies which activities can be performed in the Living Lab.
- Methodologies: each service can be composed by one or more tasks that can be executed by using specific methodologies.
- Expertise: special skills and knowledge of the members of the Living Lab (e.g.: legal expertise, social science, user-centred design, etc...).
- LL communities/stakeholder groups: from the experiment point of view it is important to define some features that the people involved in the LLs task should have, by identifying specific stakeholder groups.

The interactions between the experimenter and the Living Labs have to be managed differently than other aggregators because a Living Lab involves human resources, so the process cannot be completely automated. Living Lab manager provides functionalities to manage requests coming from experimenters in order to define and confirm participation in the experiment, through an asynchronous communication process.

FESTIVAL platform provides a set of specifications, Driver APIs, to federate Aggregators. Each aggregator has to implement this set of APIs, using their specific technologies exposing them as REST services. This mechanism allows to federate any kind of aggregator with minimal effort, because only the implementation of the corresponding driver is required.

28.4.2 FESTIVAL Resource Model

In order to integrate and harmonize the different types of resources that will be available and accessed through FESTIVAL platform, a common representation of them was investigated and a specific model was defined: FESTIVAL Resource Model.

The FESTIVAL Resource Model is composed of two sections: a general section containing a list of attributes to provide generic information in order to allow the identification of the resources in the system. In particular, a combination of three IDs (IDs of the resource, the testbed and the aggregator as assigned by the driver) identifies univocally each resource available in FESTIVAL platform. The second one is a custom section containing a list of attributes allowing experimenters to know the right way to manage the resource in the experiment: the actions that an experimenter can perform on the resource, additional information useful to manage the resource and a state variable that informs the experimenter about the possibility or not to lock the resource for a period of time for an exclusive access.

28.4.3 FESTIVAL EaaS Platform

The FESTIVAL Platform provides to the experimenters the core functionalities to create, manage and monitor experiments using the resources of the federated testbeds: the main EaaS platform modules are:

Platform Administration: it is in charge of providing the specific functionalities to manage the platform, such as assignment of given roles to the platform experimenters and performing actions on existent experiments such as extending the duration or deleting an experiment.

Experiment Control: this module is responsible for taking care of all activities linked to the experiment execution: execution control, resource allocation and failure control.

Resource Access Manager: it manages the access to the resources. The module provides the business logic to perform the corresponding requests to the different drivers.

Experiment Monitoring: this module will provide experimenter with the possibility to monitor the experiments in the platform. To manage collecting of measures, FESTIVAL platform provides an OML server [30].

Storage Service: the role of this module is to normalize the data access and manage the connection to the databases. Every EaaS module usee the storage service to store and retrieve platform data.

Analysis and Software Tools Repository: it provides a repository of several software and analysis tools to support experiments; in addition to this, the module manages some predefined VM templates that include specific libraries or services useful for the experimenter (e.g. instances of FIWARE GE).

EaaS KPI Monitoring: this module interacts with other EaaS modules to centralize the KPI measurement features. The general approach for KPI measurement collection is to implement a "probe" to related EaaS module that raises the necessary data using EaaS API or other technologies (e.g. sniffer). This data is then sent to the EaaS KPI Monitoring module to be analysed.

Security: it is the module in charge of providing the Authentication, Authorization and Accounting for the FESTIVAL platform. The implementation follows an approach based on PeP Proxy [31] using Keyrock [32] as authentication system.

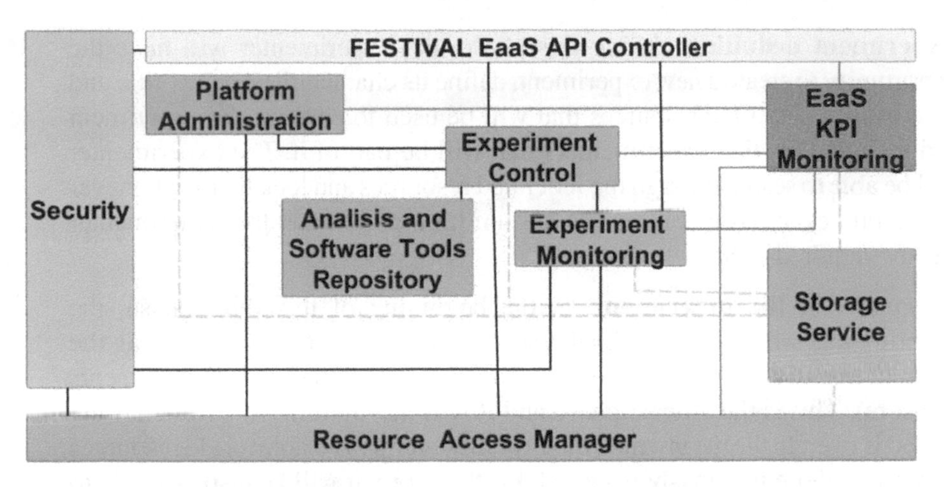

Figure 28.2 FESTIVAL EaaS platform modules.

FESTIVAL platform provides a REST interface represented by EaaS APIs to enable experimenters and in general any client application to access and to interact with FESTIVAL platform and to manage experiments and resources involved in their execution. The EaaS Controller manages the request to the API performing security access and redirecting it to the underlying EaaS modules.

28.5 FESTIVAL Portal and Experiment Workflow

The FESTIVAL experimenter portal is the main access point to the platform for new experimenters. Its main purpose is to provide a graphical user interface to the experimenter in order to use the EaaS platform functionalities. In particular, the user will be able to perform generic experiments that are based on a typical workflow composed by four different phases, as explained below:

Account creation and documentation: firstly, the experimenter will access to the experimentation portal to learn about the resources offered in the platform, as well as the testbeds that have been federated. Afterwards, if he/she wants to continue with the experiment in FESTIVAL, he/she will be requested to create an account to obtain the credentials to access the experimenter API. Finally, the experimenter will have access to the documentation of the EaaS API and to access the resources.

Experiment definition: during this phase, the experimenter will have the opportunity to create a new experiment, define its characteristics, (e.g. title and description) select the resources that will be used to perform the experiment and include the other experimenters that will be part of it. The experimenter will be able to search through the federated resources and lock them, whenever the resources are available, in the chosen time period and the experimenter has the required rights.

Execution of the experiment: at the beginning of this third phase, the experimenter will trigger the final reservation of the resources, including the instantiation of the VMs and the subscription to different information elements (sensors). This is the longest phase and it will last until the experiment time period is over. In the meanwhile, the experimenter will be able to add resources or remove those previously reserved. Furthermore, it will be also possible to communicate with the living labs managers to request the involvement, in the experiment, of real end-users if needed. During this phase the monitoring of the experiments is also performed in terms of resource availability and collection of results.

End of the experiment: the last phase of an experiment over the EaaS FESTIVAL platform is the conclusion of the experiment and obtaining results. The experiment can terminate when: 1) the experimenter considers that the experiment being performed is finished even if the time planned is not over; 2) the defined time period for the experimenter is finished and it is automatically stopped. Once an experiment is considered as finished, the platform will release the involved resources, which entails closing all the remaining subscriptions, deleting the VMs that were instantiated, and stopping the FESTIVAL functionalities used by the experimenter.

28.6 FESTIVAL Use Case Experiments

As one of the main pillars of the FESTIVAL initiative, the project envisions the development and research of several experiments in three domains, which are described below.

- **Smart Energy**: this domain includes the experiments that deal with energy management, and energy savings related to the modification of different actuators. The experiments consider the automation of energy distribution and consumption based on the measurements provided by IoT devices.

- **Smart Building**: this domain includes all the related applications in building automation in different building facilities, such as houses or big buildings. Automation processes are performed by interpreting values returned for IoT devices using different techniques, such as machine learning.
- **Smart Shopping**: this last domain refers to all applications relating to shopping areas. The applications found in this domain envision to improve the relationship of the customers and the shopping area, by taking different measurements such as customer position or environmental parameters. In addition, it aims at improving direct communication of customers and shop owners through new channels, such as smartphones.

Within the framework of the different domains, the experiments serve also as an example of what can be done using the EaaS FESTIVAL platform. Among others, the description of some of the experiments being carried out in FESTIVAL can be found below.

In the Smart Energy domain, the "xEMS" project aims at reducing the power consumption of data centres. This is carried out implementing machine learning techniques to modify the local energy management system depending on the conditions of the data centre, which are provided by the deployed sensors. The "SNS-like EMS", or social networking system for energy management, is also another example of the use of different sensors that are deployed in several locations within the same building to reduce power consumption. In this case, data is gathered not only from the sensors but also from the users that shares their perception through chat-based messages. Both experiments take advantage of the existing facilities in FESTIVAL, such as the Knowledge capital living lab or the ATR data centre. Furthermore, the platform also provides the infrastructure to deploy the applications to process the algorithms.

Regarding the Smart Building domain, there are several experiments being carried out in the facilities of FESTIVAL. The "Smart Camera" experiments take advantage of the PTL facility for the tests, where modified cameras gather specific image features, avoiding privacy issues, to perform statistical analysis, such as counting people or providing signals to actuators. Other experiment related to the Smart Building domain is the "Smart Station". This experiment consists in the deployment of sensors in some railways stations in Japan, in order to retrieve environmental parameters and other data, such as occupancy or traffic information for multimodal transportation.

Finally, there are several projects under the Smart Shopping paradigm. One of them, the "Connected Shop" experiment intends to analyse the Probe Request packets sent by the IEEE 802.11 interfaces, found in all the smartphones nowadays, to locate customers and predict their behaviour. Other parameters, such as humidity and temperature are also considered in this experiment, which take advantage of the EaaS FESTIVAL platform to access to the sensor data and make use of the storage capabilities. In this sense, also using the processing capabilities from the IT testbed federated in FESTIVAL to perform highly demanding processing capabilities, the "Smart Shopping System and Recommendation Analysis" applies big data techniques to Smart Shopping applications. The goal of this experiment is to obtain the customers' preferences and the best configurations for the exhibitors to increase the customer attention and hence the the sales.

28.7 Conclusions

This chapter has shown the progresses of the FESTIVAL project, in particular in terms of design and implementation of the testbed federation platform. As shown, the project tried to propose a solution in the field of testbed federation with a specific focus on IoT: the novelty of the solution proposed by FESTIVAL is specifically related to the problem of the heterogeneity of the testbeds involved, going beyond the traditional federation of estbed that provides virtual computational resources. FESTIVAL defines an architecture that harmonises very different resources that include open data, IoT devices, cloud IT resources and human resources.

The FESTIVAL reference implementation, developed using new technical solutions and existing relevant Japanese and European assets, will provide to the experimenters with the possibility to access to heterogeneous large scale testbeds to perform experiments in different domains (e.g. Smart Energy, Smart Building, Smart shopping). The platform will be released for project members by October 2016 and will be freely accessible for external experimenters by January 2017: in this second part of the project all the technical achievements described in this chapter will be validated by concrete use cases. This will demonstrate the impact of the FESTIVAL EaaS approach, on the number and quality of the experimentations that are run on the testbeds, thus presenting both small and large scale trials over various application domains.

Acknowledgments

This work was funded in part by the European Union's Horizon 2020 Programme of the FESTIVAL project (Federated Interoperable Smart ICT Services Development and Testing Platforms) under grant agreement no. 643275, and by the Japanese National Institute of Information and Communications Technology.

References

[1] FESTIVAL project, [Online]. Available: http://www.festival-project.eu
[2] Santander Datos Abiertos, [Online]. Available: http://datos.santander.es/
[3] Metropole of Lyon's Open Data, [Online]. Available: http://data.grand lyon.com
[4] FIWARE Lab Data Portal, [Online]. Available: https://data.lab.fiware.org
[5] FESTIVAL Japanese open data portal, [Online]. Available: http://festival.ckp.jp
[6] CKAN open-source data portal platform, [Online]. Available: http://ckan.org
[7] The FIWARE catalogue, [Online]. Available: http://catalogue.fiware.org
[8] Smart House Testbed for both Physical and Cyber Entities, 2011. [Online]. Available: https://www.naefrontiers.org/File.aspx?id=29971
[9] LINKING TECHNOLOGIES R&D PLATFORM, [Online]. Available: http://www.irtnanoelec.fr/linking-technologies/
[10] SmartSantander project, [Online]. Available: http://smartsantander.eu
[11] T. Deguchi, Y. Taniguchi, G. Hasegawa, Y. Nakamura, N. Ukita, K. Matsuda e M. Matsuoka, Impact of workload assignment on power consumption in software-defined data center infrastructure, *Proceedings of IEEE CloudNet 2014,* pp. 440–445, 2014
[12] S. Tashiro, Y. Tarutani, G. Hasegawa, Y. Nakamura, K. Matsuda e M. Matsuoka, A network model for prediction of temperature distribution in data centers, *Proceedings of IEEE CloudNet 2015,* pp. 251–256, 2015
[13] Opening of a new Smart Station used by FESTIVAL in Japan, [Online]. Available: http://www.festival-project.eu/en/?p=871
[14] Maya Station, [Online]. Available: https://en.wikipedia.org/wiki/Maya_Station
[15] Kameoka Station, [Online]. Available: https://en.wikipedia.org/wiki/Kameoka_Station

[16] JOSE: Japan-wide orchestrated smart ICT testbed for future smart society, [Online]. Available: http://cordis.europa.eu/fp7/ict/future-networks/documents/eu-japan/stream-c-teranishi.pdf
[17] Engineering FIWARE-Lab, [Online]. Available: http://fiware.eng.it
[18] OpenStack, [Online]. Available: https://www.openstack.org
[19] The Lab, [Online]. Available: http://kc-i.jp/en/facilities/the-lab
[20] TUBA, [Online]. Available: http://www.tuba-lyon.com
[21] Socrata data platform, [Online]. Available: http://www.socrata.com/
[22] L. Gürgen, C. Munilla, R. Druilhe, E. Gandrille e J. B. D. Nascimento, sensiNact IoT Platform as a Service, in *Enablers for Smart Cities1: Foundations*, Wiley-ISTE, 2016
[23] KNX, [Online]. Available: http://www.knx.org/knx-en/index.php
[24] Publish/Subscribe Context Broker – Orion Context Broker, [Online]. Available: http://catalogue.fiware.org/enablers/publishsubscribe-context-broker-orion-context-broker
[25] H. Kodama, The ECHONET Lite Specifications and the Work of the ECHONET Consortium, [Online]. Available: https://www.ituaj.jp/wp-content/uploads/2015/04/nb27-2_web-02-spec.pdf
[26] OpenHab, [Online]. Available: http://www.openhab.org/
[27] L. Peterson, R. Ricci, A. Falk e J. Chase, Slice-based federation architecture Working draft, Version 2.0, 2010
[28] Fed4Fire project, [Online]. Available: http://www.fed4fire.eu
[29] SFAWrap, [Online]. Available: https://sfawrap.info/
[30] OML – Measurement Library, [Online]. Available: https://oml.mytestbed.net/
[31] PEP Proxy – Wilma, [Online]. Available: http://catalogue.fiware.org/enablers/pep-proxy-wilma
[32] Identity Management – KeyRock, [Online]. Available: http://catalogue.fiware.org/enablers/identity-management-keyrock

29

TRESCIMO: Towards Software-Based Federated Internet of Things Testbeds across Europe and South Africa to Enable FIRE Smart City Experimentation

Nyasha Mukudu[1], Ronald Steinke[2], Giuseppe Carella[3], Joyce Mwangama[1], Andreea Corici[2], Neco Ventura[1], Alexander Willner[3], Thomas Magedanz[2], Maria Barros[4] and Anastasius Gavras[4]

[1]University of Cape Town, South Africa
[2]Fraunhofer FOKUS, Germany
[3]Technical University of Berlin, Germany
[4]Eurescom GmbH, Germany

Abstract

Smart Cities are able to offer efficient services for creating more sustainable environments. However, this will require stakeholders to have access to suitable testbed infrastructure for prototyping new services. This chapter describes the underlying federated testbed infrastructure, designed and implemented within the EU TRESCIMO project, to support Smart City experimentation. The focal point of this chapter is to share the technical challenges on the design and implementation decisions made to provide a sustainable state of the art federated testbed infrastructure between the Technical University of Berlin (TUB) and the University of Cape Town (UCT). The starting point for this testbed implementation is the layered TRESCIMO Smart City reference architecture, onto which existing standard compliant and SDN/NFV based testbed toolkits have been mapped. The chapter describes these toolkits and their integration within the context of TRESCIMO. Furthermore, it will outline how this SDN-based infrastructure setup can be utilised in future FIRE SDN projects, such as SoftFIRE.

Keywords: Experimentation, Internet of Things, Network Function Virtualisation, Smart City, Software Defined Networks.

29.1 Introduction

Many cities around the world are presently confronted with the challenge of providing services that address economic, social and environmental requirements faced by residents. A drastic increase in urbanisation is expected to result in two-thirds of the world population residing in cities [5]. This will result in an unprecedented demand for services such as energy management, clean water, healthcare and transportation just to name a few. This has driven many cities to investigate the adoption of Smart Cities services that utilise Internet of Things (IoT) generated data to make informed decisions which result in enhanced city services. Smart Cities have the potential to intelligently manage available resources in order to create integrated, habitable and sustainable urban environments [1]. The realisation of the Smart City vision will require the integration of various domains within a common framework.

Machine-to-Machine (M2M) communication technology will enable various physical objects within cities to be connected. Smart City services should have the capability to analyse the data and provide instant real-time solutions for many challenges faced by cities. Additionally, many different technical and non-technical service requirements must be considered. Consequently, large-scale experimentation is required to provide the necessary critical mass of experimental data required by businesses and end-users to evaluate the readiness of M2M and other IoT technologies for market adoption.

In this chapter, we describe a prototyping environment implemented to address the need for testing facilities for Smart Cities. This testing environment, developed as part of the Testbeds for Reliable Smart City Machine-to-Machine Communications (TRESCIMO) project, includes participants from Europe and South Africa, and allows for the ability to test Smart City use cases from both a developed and developing world context. Many partners were involved in the testbed deployment and experimentation aspects of the project: Technical University of Berlin (TUB) in Germany; University of Cape Town (UCT) in South Africa; Fraunhofer FOKUS Institute (FOKUS) in Germany; Council for Scientific and Industrial Research (CSIR) in South Africa; and Fundaci i2CAT (i2CAT) in Spain.

29.2 Problem Statement

The need for large-scale testbeds for Smart Cities has been recognised by industry and academia in order to develop a reference implementation model for Smart Cities [9]. However, creating an experimental facility capable of coping with the diverse nature of Smart City services and the number of connected devices remains a challenge. In addition, the expected growth in demand for services makes it imperative for testbeds to allow service providers to create strategies to cope with these increases.

In the TRESCIMO project, we created a federation of testbeds that allows for experimentation which makes use of enabling technologies, standardised platforms and applications for Smart Cities. These Smart Cities can have different needs and requirements thus allowing for flexible configurations depending on contexts and use cases. Experimental tools are incorporated as an effective option to study the behaviour of integrated software and hardware before implementing real deployments.

The TRESCIMO facility is based on a virtualised standardised M2M platform Open Machine Type Communications Platform (OpenMTC) and an open-source, Slice-based Federation Architecture (SFA)-compatible framework for managing and federating testbeds, Future Internet Teagle (FITeagle). The facility consists of three interconnected sites located in Berlin – Germany, Cape Town – South Africa and Pretoria – South Africa. Additionally, the ability to federate with other testbeds will allow the sharing of resources (i.e. sensors, actuators and data) among different services and users regardless of their location.

29.3 Background and State of the Art

In order to evaluate new protocols and architectures that will enhance Smart City deployments, testbeds that incorporate a wide range of heterogeneous resources are needed. Additionally, considerations should be made on the variability of resources, the size of traffic or data generated, and the operational complexity they introduce. Although there are several existing wireless and wired testbeds that offer researchers the ability to perform Smart City related experimentation, many of these testbed do not cater from some important user specific requirements. Some of these requirements include ease of user management, experiment control and connectivity.

Different approaches have been developed to overcome these issues. The focus of the SmartSantander project was to create a European experimental test facility for architectures, key enabling technologies, services and applications for smart cities [15]. The facility had to support both experimenters and real world end users using sensors deployed within the city of Santander in Spain. This was achieved using separate planes for IoT data and testbed management data. In addition, the suitability of IoT technology for supporting real world smart cities was demonstrated.

In the GEneralized architecture for dYnamic infrastructure SERvice (GEYSERS) project, cloud-based infrastructure is provided to experimenters as an on demand service [2]. This approach results in a more seamless and coordinated provisioning of virtual infrastructures composed of network and IT resources. Adopting a similar approach, TRESCIMO was able to cope with the diverse range of deployment scenarios found in smart cities. The Federated E-Infrastructure Dedicated to European Researchers Innovating in Computing Network Architectures (FEDERICA) project is used to create virtual versions of physical testbed resources [3].

29.4 Smart City Testbed Design

The TRESCIMO research facility was designed to allow for experimentation making use of enabling technologies, standardised platforms and applications for Smart Cities with different configurations. This section provides details on key requirements for Smart City experimentation facilities identified using use cases from various Smart City application domains. Based on these requirements, the architecture of the federated TRESCIMO testbed was developed.

29.4.1 Design Considerations

Key requirements for testbeds to adequately support IoT experimentation have been identified [9]. However, Smart City services impose additional requirements on testbeds due to the diverse range of possible usage scenarios. As a result, use cases for Smart City experimentation were selected from the energy management, environmental monitoring, healthcare, safety and education domains. This wide range of use cases were analysed, and common experimentation requirements were identified.

In the TRESCIMO facility, a collection of services and infrastructures are used to create an environment capable of meeting the needs of Smart City experimenters. The following subsections highlights key requirements and considers how they are addressed within the TRESCIMO testbeds.

29.4.1.1 Federation

Federation with other testbeds is required to allow experimenters access to a greater pool of available resources [9]. Resources that can be shared include data produced by sensors, control of actuators and integrated M2M/IoT components deployed in the individual testbeds. The TRESCIMO facility combines the CSIR, TUB and UCT testbeds to provide Future Internet Research and Experimentation (FIRE) users with access to resources from the individual testbeds. Furthermore, this creates a facility capable of conducting experiments in both a developing African and a developed European context [4].

29.4.1.2 Heterogeneity

Smart Cities consist of various domains (e.g. healthcare and energy), that utilise a wide range of applications, platforms and devices. Consequently, it is necessary to adopt integrated cloud-oriented architectures of networks, services, interfaces and data analysis tools to meet the needs of future smart cites [11]. TRESCIMO uses a standardised M2M platform to provide support for a wide range of IoT devices and technologies. Furthermore, experimenters are able to provision various virtual smart city infrastructures, to enable the deployment of various experimental scenarios.

29.4.1.3 Scale

Smart City services will have to cope with a large amount of data from sources such as IoT nodes, other services and people. In order to facilitate large-scale experimentation, current IoT testbeds such as SmartSantander offers access to thousands of nodes [15]. The TRESCIMO facility uses a combination of physical and virtual IoT devices to allow experimenters access to a larger number of devices.

29.4.1.4 Reliability

The testbed facility is intended to allow users to run their experiments to completion without any unexpected interruptions. Consequently, it is necessary for testbed operators to utilise monitoring tools to ensure that the facility and its infrastructure are operational. In the case of TRESCIMO, experimenters are able to receive monitoring data for infrastructure used in a particular experiment.

29.4.1.5 Resource management

In order to manage the connected resources in the testbed, it has to support resource discovery, resource reservation and resource provisioning. In the

FIRE context this functionality is provided to FIRE experimenters via a standards-based federation interface.

29.4.1.6 Flexibility

The testbed is designed to be extendible to meet future requirements or additional use cases. This allows for the realisation of a wide range of users, use cases and complex infrastructures. In the TRESCIMO facility, a modular approach was adopted to ease the process of adding new platforms and devices. This high level of flexibility will allow for experimenters to develop new services and devices.

29.4.2 Architecture Overview

The TRESCIMO architecture was designed to provide IoT testbed Smart City use cases. Additionally, the architecture can be partly or fully mapped to trials that verify the functionality in real world scenarios. In order to provide flexibility and accelerated setup for the different use cases and experiments, an Software-Defined Networking (SDN)/Network Functions Virtualisation (NFV) approach was selected. This enabled the capability to provide testbed components as virtualised resources that could be instantiated on-demand as required. In addition to virtualised resources, physical real-world resources and devices were also provided and integrated for dynamic and flexible utilisation.

Figure 29.1 illustrates the overview architecture. The left side of the diagram highlights the general reference architecture while the right side shows the extracted elements from the general architecture concentrating on the elements implemented or deployed at the various geographical locations. In the lower layer of the diagram, the different devices used for the different use cases are shown. These include the Smart Energy trial devices developed by Eskom and CSIR, the environmental monitoring devices deployed by i2CAT. Devices were connected to the rest of the architecture by using simple area networks such as LAN or WLAN, and when this was not possible they were connected via a Delay Tolerant Network (DTN).

Common to all locations is the use of an M2M middleware layer. The middleware exposes gateways that allow for interconnection of devices to the backend of the middleware. The M2M middleware, OpenMTC, developed by FOKUS was adopted. Optionally, the Smart City Platform (SCP) developed by CSIR, could be deployed to provide an enriched depiction or analysis of the

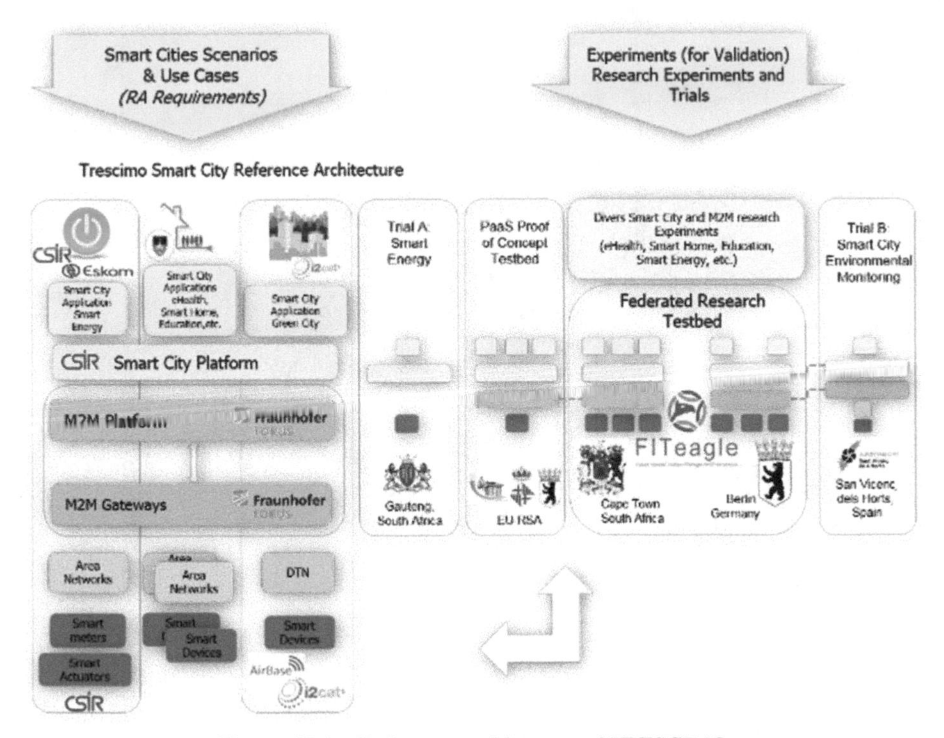

Figure 29.1 Reference architecture of TRESCIMO.

aggregated data exposed by the M2M middleware platform. This SCP provides interfaces for the applications that were used and tested in the experiments and trials. The complete stack of the architecture was federated and managed by FITeagle from TUB, which also provides the standardised SFA interface for experimenters.

The final deployment in Figure 29.2 shows the federation with all involved sites connected via the Internet. The TUB testbed (on the left) implements all the components of the TRESCIMO architecture stack. The TUB testbed is able to integrate physical IoT devices and can also emulate virtual devices. The UCT testbed (on the right) implements all the components except for the SCP. This testbed also integrates physical devices. The CSIR Proof-of-Concept testbed (in the middle) implements a small subset of the architecture in order to demonstrate the use case of the energy management trial within the TRESCIMO stack. In this trial, due to regulations and restrictions, only the SCP and the Active Devices could be deployed. All three testbeds are

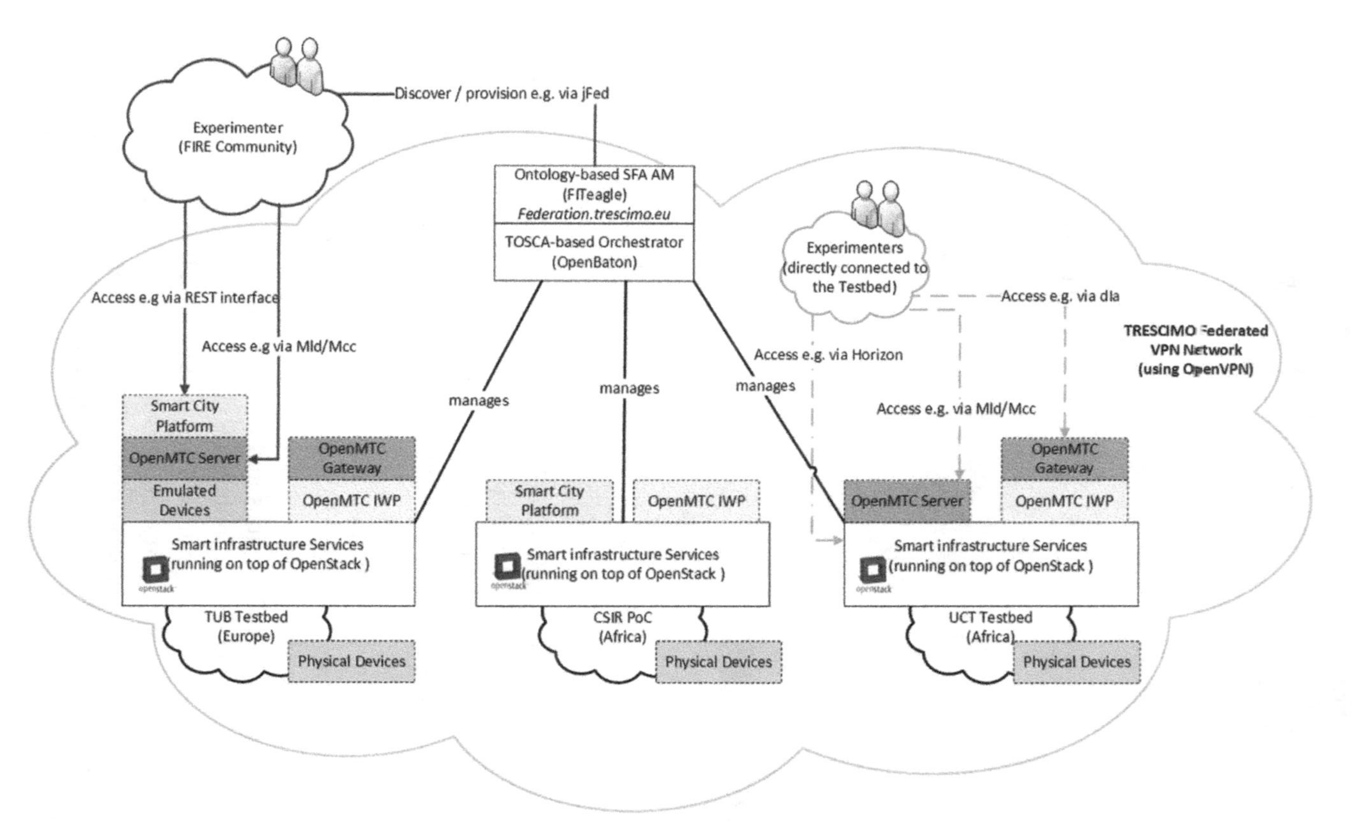

Figure 29.2 Current TRESCIMO federated testbed architecture (based on [4]).

based on an OpenStack setup in order to deploy the needed parts of the architecture. These OpenStack instances are managed by FITeagle with the help of OpenBaton. FITeagle provides the interface to the experimenter in order to setup the experiments. It returns the established endpoints to the instances that are created so that the experimenter is able to connect to the created testbed resource.

29.5 Technical Work/Implementation

Several software tools were utilised to realise the designed reference architecture. To enable access to external experimenters, FITeagle was used as it provides a standardised SFA interface. FITeagle and OpenStack were further interworked with OpenBaton which allowed for enhanced orchestration features of the testbed resources. OpenBaton provides a Topology and Orchestration Specification for Cloud Applications (TOSCA) interface to FITeagle. For the transportation of M2M data a combination of the toolkits OpenMTC and Open5GMTC were used.

29.5.1 Cloud Management – OpenStack

Smart city infrastructure is hosted and managed using OpenStack based cloud servers[1]. An OpenStack server is deployed at each of the individual testbed sites in the TRESCIMO federation as shown in Figure 29.3. Each site hosts a selection of the TRESCIMO Smart City stack and is controlled by a shared NFV Orchestrator (OpenSDNCore). The OpenSDNCore communicates with the OpenStack controllers at each site in order to launch the required components of the TRESCIMO smart city stack.

Each of the individual testbed sites utilises a unique addressing scheme for locally available resources. To meet the requirements of smart city experimenters, it was necessary to enable experimenters to remotely provision and access components in the TRESCIMO smart city infrastructure. To support the provisioning process, the individual OpenStack controllers are interconnected to the OpenSDNCore orchestrator. Public IP addresses are assigned to the relevant endpoints of the provisioned infrastructure. Furthermore, private IP addresses are used to allow sensors to communicate with provisioned M2M gateways and servers. In addition, communication among provisioned virtual machines was achieved using a common IP range (trescimo-net).

[1] https://www.openstack.org/

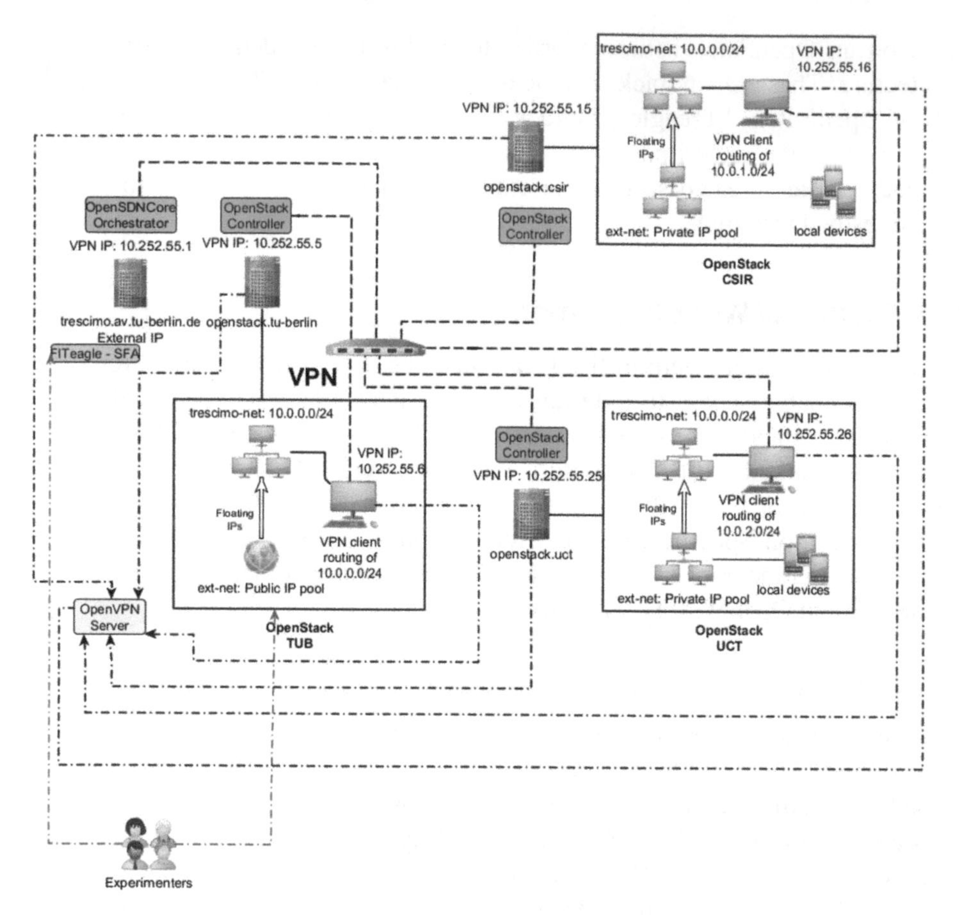

Figure 29.3 Testbed interconnections.

29.5.2 Experimentation Management – FITeagle

TRESCIMO uses FITeagle [16], the first implementation of a semantic-based Slice-based Federation Architecture (SFA) Aggregate Manager (AM), to dynamically provision Software Defined Infrastructures (SDI) for the Smart City context. These SDIs are made available to the Global Environment for Network Innovations (GENI) [14] and FIRE community. For the TRESCIMO testbeds, the FITeagle framework offers interfaces to the FIRE community and handles all aspects of the experiment life cycle including Authentication (AuthN) and Authorization (AuthZ). This is based on X.509 certificates signed by trusted Certificate Authorities (CAs), such as from Federation for FIRE[2] or PlanetLab Europe (PLE)[3].

While physical devices are managed by dedicated resource adapters deployed within FITeagle, requests for virtualised services are forwarded to the OpenSDNCore Orchestrator framework via its TOSCA interface. Depending on the selected location, the relevant OpenStack sites are then contacted to instantiate the requested services. The underlying workflow with references to the related SFA AM method calls is depicted in Figure 29.4.

Due to the fact that the TRESCIMO facility uses the OpenSDNCore Orchestrator to provisionVirtualised Network Function (VNF)s internally, compatibility with the offered TOSCA [13] interface was improved. For this, the Open-Multinet (OMN)[4] ontology and translator, developed within the World Wide Web Consortium (W3C) Federated Infrastructures Community Group[5], were extended to provide the required properties and mappings to GENI v3 Resource Specifications (RSpecs) [12].

Figure 29.5 highlights the resource adapters implemented to enable FITeagle to interact with virtual and physical infrastructure for realising the required API functionality. These resource adapters are:

- **TOSCA Resource Adapter (TOSCA RA)** is responsible for communicating with the OpenSDNCore for orchestration of virtual infrastructure. OpenBaton receives a Resource Description File (RDF)-style topology file and converts it into a TOSCA-compatible topology for use by the OpenStack controller. This will result in the provisioning of the required Smart City infrastructure and feedback on the created endpoints will be provided to the experimenter in the form of an RDF response.
- **Interworking Proxy Resource Adapter (IWP RA)** handles interaction with the interworking proxy and the devices it manages. This resource adapter provides experimenters with details on the connected devices and options for setting the URIs of the M2M endpoints used in a particular experiment. As a result, experimenters can instruct the interworking proxy to create resources inside the desired M2M endpoints. Connected devices can be addressed by using the resource URIs or can be found via discovery on a specified M2M gateway. Devices can be connected to the federated testbed using a wired connection or a wireless area network as shown in Figure 29.5.

---

[2]www.fed4fire.eu

[3]https://www.planet-lab.eu

[4]http://open-multinet.info

[5]https://www.w3.org/community/omn

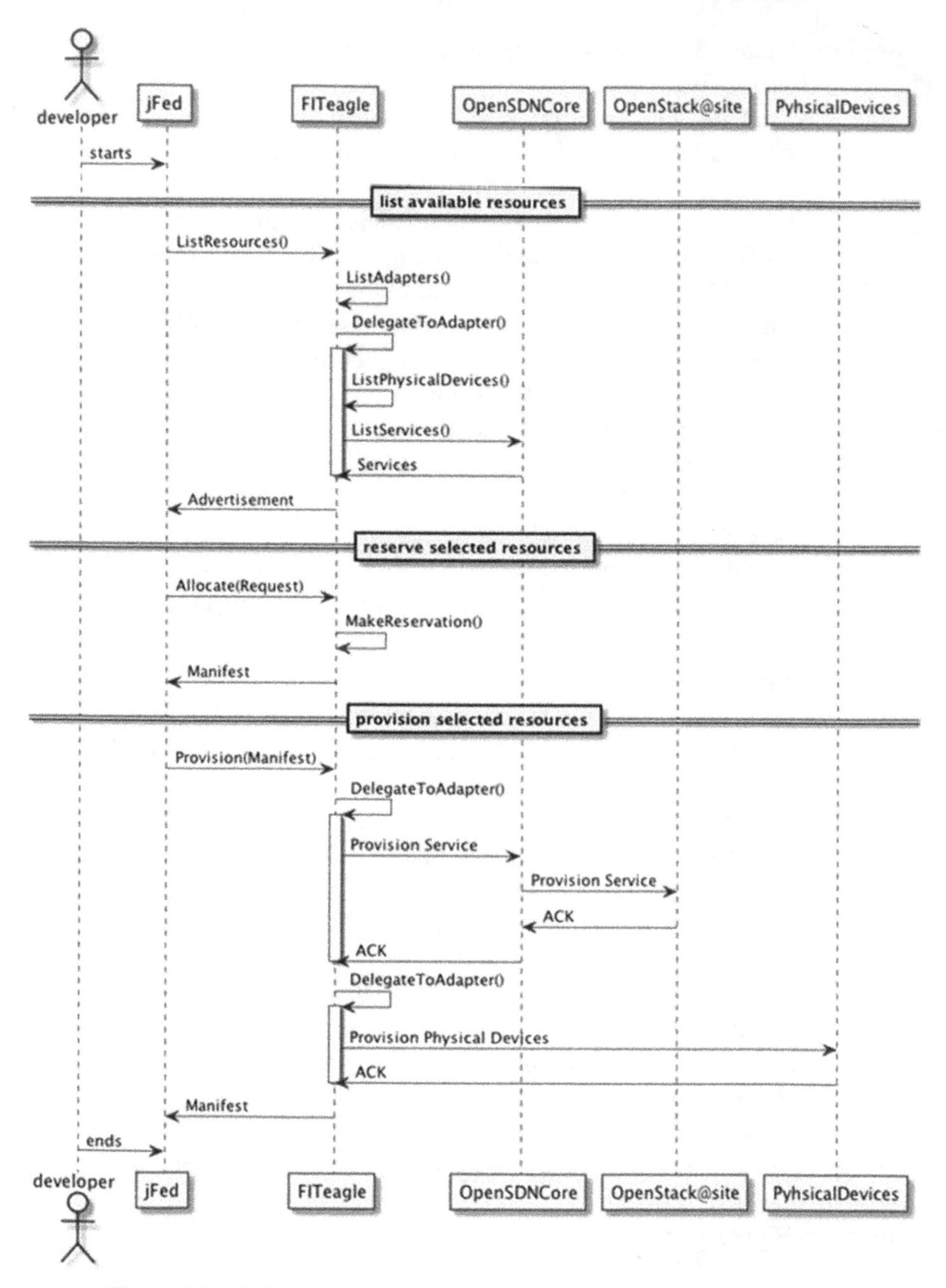

Figure 29.4 Message flow between FITeagle, OpenSDNCore and devices.

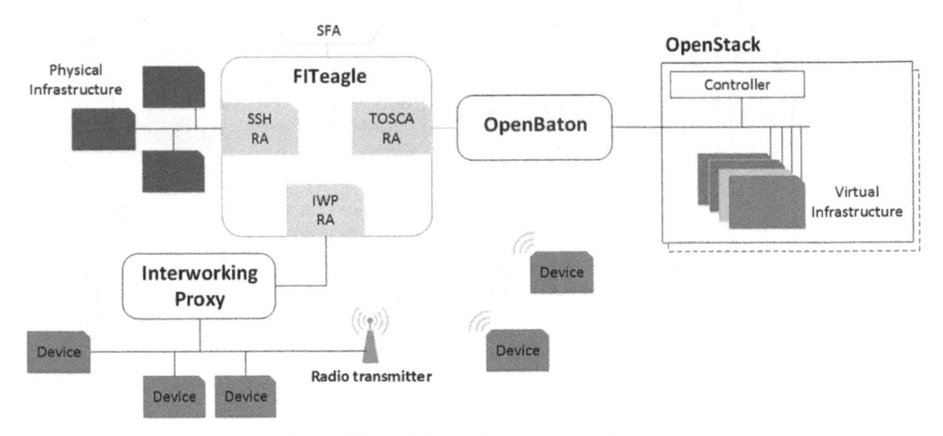

Figure 29.5 FITeagle resource adapters.

- **Secure Shell Resource Adapter (SSH RA)** focuses on creating specific user profiles for physical hosts connected to the federated testbed. It allows experimenters secure access to specific hardware resources using a SSH server and provided SSH keys.

29.5.3 NFV Management and Orchestration (MANO) – OpenBaton

With the increasing acceptance of NFV and SDN technologies, Telco Operators are modifying their traditional network infrastructures in order to provide more flexibility enabling new business opportunities. One major change introduced by those technologies is the capability of deploying on demand different network topologies on top of the same physical infrastructure.

The aim of the European Telecommunications Standards Institute (ETSI) NFV Industry Specification Group (ISG) was to provide a set of guidelines that can be used by different software-based network functions providers for providing their solutions as a service. In particular, the ETSI NFV Management and Orchestration (MANO) domain defines an architecture for managing those VNFs on top of a common NFV Infrastructure. In TRESCIMO, the NFV MANO capabilities have been required specifically for the management of the M2M VNFs on top of the federated testbed. OpenSDNCore, a Fraunhofer FOKUS toolkit, has been selected as NFV Orchestrator for the TRESCIMO testbed. Fraunhofer FOKUS decided to open source its MANO platform (part of the OpenSDNCore project) under the name of OpenBaton. Figure 29.6 shows the OpenBaton architecture.

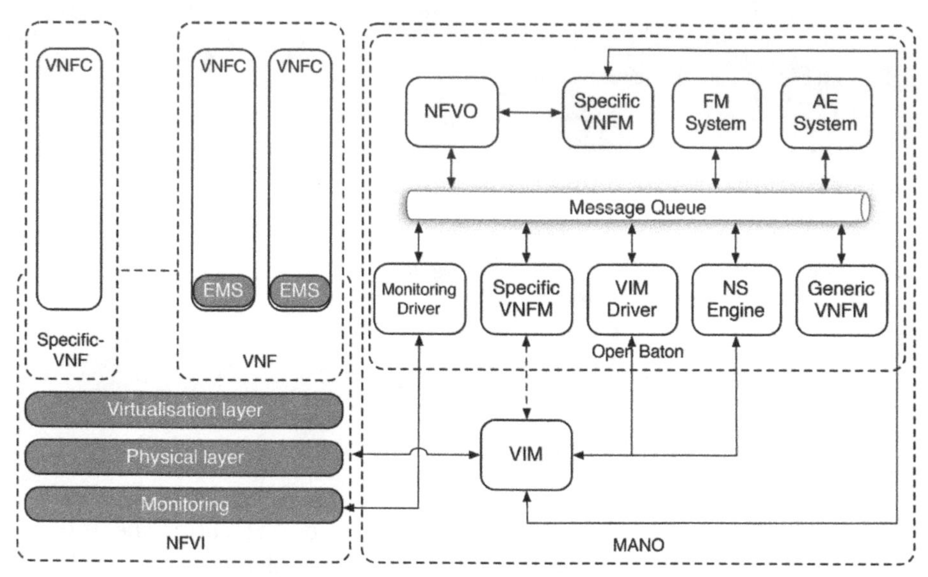

Figure 29.6 OpenBaton functional architecture.

OpenBaton is based on a very modular and extensible architecture. Internally each functional entity is implemented as an independent component, and the communication is based on micro-services principles using a messaging system. The Advanced Message Queuing Protocol (AMQP) is used as messaging protocol between the different components. The main entry point is represented by the Network Function Virtualisation Orchestrator (NFVO) providing a OASIS TOSCA Northbound API which can be consumed by FITeagle for requesting the instantiation of Network Services (NSs) as composition of different M2M VNFs. The NFVO interacts with the Generic Virtual Network Function Manager (VNFM) for managing the lifecycle of those VNFs. The instantiation of the virtual compute and network resources is done via the Virtualised Infrastructure Manager (VIM) Driver. Considering that OpenStack represents the standard de-facto implementation of the VIM, for TRESCIMO has been employed and integrated the OpenStack Plugin[6]. OpenBaton provides the capability of interacting with a multi-site NFV Infrastructure. Basically, each testbed has been registered as an independent Point of Presence (PoP) using the OpenBaton dashboard. Each different VNF can be configured in order to be deployed on a specific PoP accordingly.

[6]https://github.com/openbaton/openstack-plugin

29.5.4 M2M Platform – OpenMTC/Open5GMTC

The M2M platform was used to transport the data from the devices to the SCP or to the applications directly. In fact two toolkits were used. OpenMTC, a reference implementation of the ETSI M2M and the OneM2M standard, was used to transport the data. Open5GMTC, which utilises the OMA Lightweight M2M Standard and CoAP as the transportation protocol, was used to configure the gateway and the applications so that they can register with the correct endpoint and deliver the data to the specified destination.

The OpenMTC framework is separated in a gateway and a backend component. The gateways are registered with the backend and the applications are registered with either the gateway or the backend. The devices can act like an application or a so called Interworking Proxy (IWP) that mediates between the specific devices and the framework. In doing so, the devices working as actuators or sensors will receive a digital representation in the system. The IWP is then an application from the perspective of the gateway or the backend. The applications can access historic data and also subscribe to receive new incoming data. By using access rights, other applications within different operating domains or use cases can be granted permission to view the same data if required.

The Open5GMTC, which is an enhancement of the OpenMTC, is also divided into a server instance and corresponding clients. In the TRESCIMO project the device management (DM) capabilities were used in order to configure some components of the architecture. From Figure 29.7 one can see the used example in the Proof of Concept implementation. Here the M2M DM Server which is a component of Open5GMTC was deployed statically, as well the Active Gate IWP and the DTN Gateway with their connected devices. The emulated devices, the M2M Platform and the SCP were deployed dynamically and configured to establish connections between them. The M2M Platform was connected with the M2M DM Server so that the latter can configure the static instances of the Active Gate IWP and the DTN Gateway to connect to the M2M Platform when it is available.

29.6 Results and/or Achievements

In TRESCIMO the following objectives were achieved: the architecture to fulfil the use cases was defined and successfully setup; and the toolkits were integrated and the various TRESCIMO stack components were deployed within the testbeds and the trials. It was shown that with the incorporation of

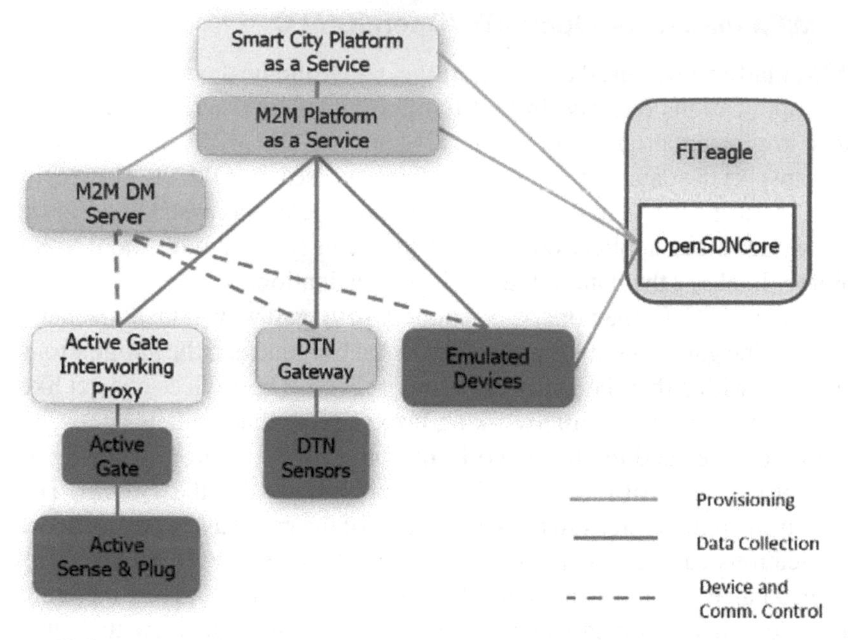

Figure 29.7 Architecture used in the PoC to illustrate the cooperation of the M2M frameworks.

SDN and NFV, the TRESCIMO testbeds were able to provide the necessary environment for IoT experimentation to be carried out.

29.6.1 Integration of the Toolkits

The integration of the various toolkits was an important part of the project as it allowed for the validation of the designed architectural framework. In the second prototype version of the TRESCIMO stack, illustrated in Figure 29.8, one can see all of the integrated components and their interconnection. The M2M Platform is connected on the Southbound interface to various devices either via different Interworking Proxies (IWP) or directly through M2M Gateways. The M2M Platform is linked to the Device Management (DM) Server. On the Northbound interface the M2M Platform interfaces with the SCP which supports heterogeneous applications. These applications can also be directly connected to the M2M Platform bypassing the SCP. On the right seats FITeagle which uses the orchestrator of the OpenSDNCore OpenBaton. The combination is used to deploy and orchestrate the components of the heterogeneous (emulated) devices, the M2M Gateway and Platform, the SCP and the heterogeneous applications.

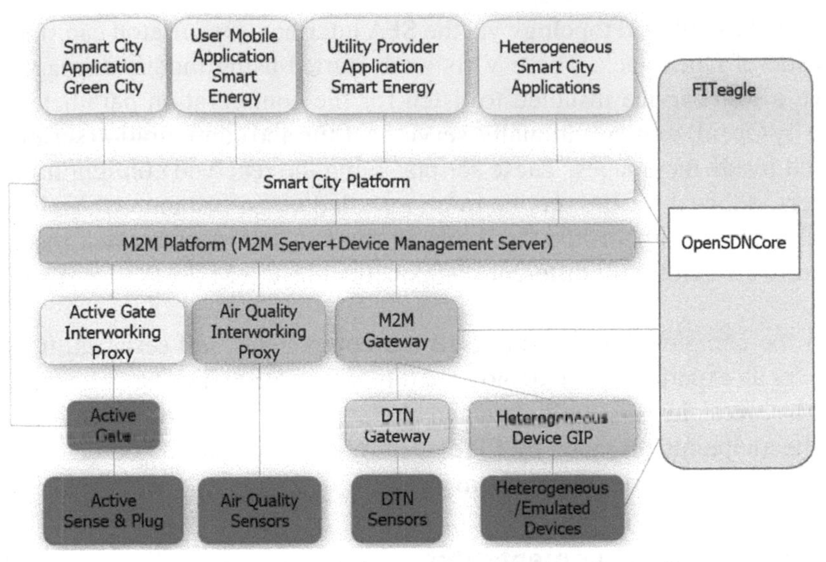

Figure 29.8 Detailed prototype architecture version 2.

The TRESCIMO stack of the components were deployed inside an OpenStack cloud computing environment. Some of these components are statically deployed and thus always available. This was the case for FITeagle, OpenBaton and the M2M DM Server. The integration of the components can be divided into two types: IoT stack components integration and infrastructure or management components integration.

The central element of the IoT architecture was the M2M framework. In order to enable the different devices to connect and communicate with the framework IWPs were developed specific to each device. This allowed for the data generated by the devices to be accessible on the M2M platform by the relevant high-level applications. In order for these applications to access this data, the SCP enabled the discovery of new devices and functionality to subscribe to the data generated by the devices. Although the communication from the devices to the SCP was done by the M2M framework a data model was necessary to link the data to the real world.

The integration of the management part of the architecture was performed by linking the different management tools to each other. The FITeagle framework, which provided the SFA interface to the experimenter, was connected to the OpenBaton VNF orchestrator via a TOSCA-interface. To achieve this a TOSCA-adapter in FITeagle was added and TOSCA was added to the API of OpenBaton. Additionally some RSpecs were added or modified to be able to define the to be deployed topology via the SFA interface. OpenBaton can start VMs inside of OpenStack. These VMs were started using modified images that have a web service installed to listen for the configuration parameters received by OpenBaton. Additionally, service startup and configuration scripts are placed inside the images. These scripts, using the received configuration parameters, allowed for the service to be installed on an instantiated VM at its launch time. To support the flexibility of the IoT architecture stack these scripts were created for all different components like the M2M Platform and the SCP.

With the complete integration of all the components and services, it is possible for an experimenter to deploy a setup of the TRESCIMO architecture stack with several devices configured and also spread over different locations. Due to the endpoints provided by FITeagle, the experimenter could connect to these endpoints and run the experiment.

29.6.2 Smart City Experimentation

The main aim of the TRESCIMO facility is to allow for research activities in the areas of M2M, IoT and Smart Cities. This will require experimenters to be able to provisioning connected M2M nodes and services in order to meet the requirements of a particular experiment. A few of the smart city experimentation use cases, to validate the functionality of the testbed, are presented in the following sections. In addition, the key observations for each use case are discussed.

29.6.2.1 Smart buildings

Connected homes are equipped with various IoT devices and applications that are capable of communicating with remote users [8]. This enables services such as remote control of appliances to be deployed. In the case of TRESCIMO, the possibility of reusing existing smart home devices to provide new services for residents is investigated. This involved using low powered devices to monitor energy consumption and control model appliances. To cope with the limitations of these devices, an M2M application (Interworking Proxy) that abstracts the complexities of the devices and provides data to external applications via an OpenMTC gateway was created. This application was deployed in conjunction with OpenMTC gateways for a number of experimental scenarios. We observed that this approach enabled smart energy applications to utilise data from home automation application without having to know any of the procedures necessary to collect the data. However, these

smart energy applications still needed to have knowledge of the data structure used by the Interworking Proxy to store data. Extending the IWPs to utilise a semantic description generator may solve this problem [10].

29.6.2.2 Energy management

The adoption of smart metering technology will enable energy providers to use demand side management (DSM) techniques to monitor and manage energy consumption in homes. In particular, energy providers are interested in implementing demand response (DR) and making consumers more energy aware [6]. This will enable energy providers to send residents requests to reduce current energy consumption based on the state of then national power grid. This technique is used to ensure the stability of the electricity grid. To facilitate DR actions, it is necessary for energy providers and consumers to agree on the appliances available for control and incentives for the consumer [7]. For demand response applications, it is imperative that the energy provider be able to receive real time data on the energy consumption of individual appliances from IoT devices deployed in the home. This data will be used to generate DR requests for residents. We conducted DR experiments within the TRESCIMO facility in order to test the ability of the smart city infrastructure to support time sensitive applications. An energy manager application was created to generate DR messages based on the current state of electricity grid. These messages were delivered to consumers prompting them to reduce their current energy consumption. The developed energy applications were able to send messages and carry out required actions. An example of the messages sent is shown in Figure 29.9. In this example, the residents allows an energy application to switch off devices on their behalf. We discovered that it was possible to implement a privacy aware DR solution by storing consumer information about energy used by specific appliances on a residential M2M gateway.

29.6.2.3 Education

This experiment scenario focuses on enabling university lecturers to provide a laboratory for their classes, thus enabling students to experiment with state of the art technology. The teacher utilises the jFed experimenter client[7] developed in the scope of the Fed4Fire project[8]. This client utilises the SFA interface of the TRESCIMO testbed provided by FITeagle. After successfully

[7]http://jfed.iminds.be/
[8]http://www.fed4fire.eu

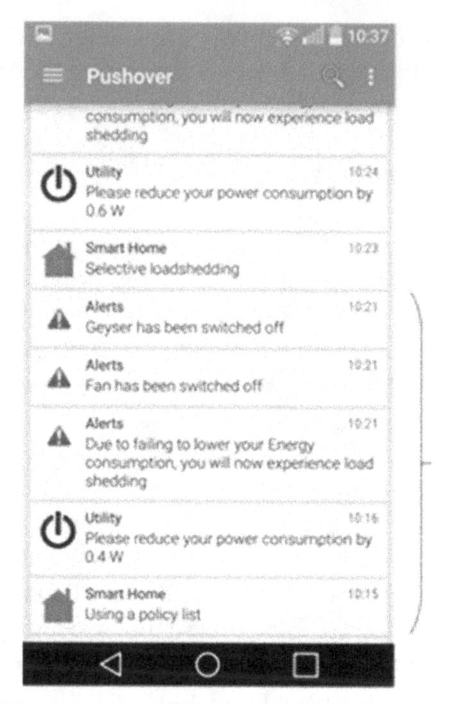

Figure 29.9 Example of DR related messages sent to a resident.

logging in with a valid X509 certificate the teacher is able to create an M2M topology using the graphical interface and by providing an RSpec detailing the required infrastructure. Figure 29.10 shows a sample topology created using the jFed client. After successfully creating the topology, the lecturer then provides the students details of the accessible resources and their endpoints. In TRESCIMO, it was possible for the lecturer to create topologies that use resources from multiple testbed sites. For the topology shown in Figure 29.10, energy measuring devices and an M2M gateway was deployed at UCT in South Africa, while the M2M server was deployed at TUB in Germany. Students were able to access both the gateway and server within an acceptable level of latency.

29.7 Discussions and Conclusions

We have shown that an intercontinental Infrastructure-as-a-service IoT testbed may be setup consisting of multiple sites connected via VPN connection, capable of supporting several different experimental use cases. The testbeds

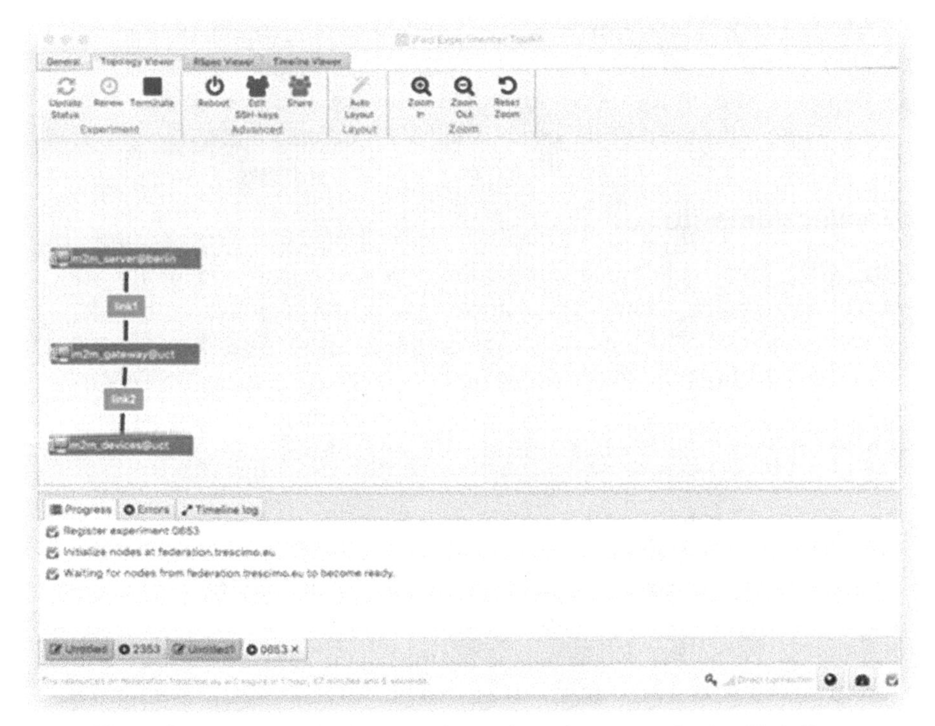

Figure 29.10 Example of an experimental topology created using jFed client.

are federated by providing the FIRE SFA interface. This enables FIRE and GENI experimenters to utilise resources deployed at any of the three testbed sites.

Further short-term work includes the complete replacement of the old OpenSDNCore orchestrator by OpenBaton and the integration of more generic devices to support use cases in the areas of healthcare and security. Based on feedback provided by experimenters, we are investigating the possibility of providing some of the applications developed during the smart city experimentation process, as on demand services for experimenters. This will hopefully ease the process of getting started with smart city experimentation.

According to FIRE's definition of testbed federation, each testbed site should host a FITeagle instance that supports FIRE tools such as the jFed client. In TRESCIMO, the testbeds are interconnected via a Virtual Private Network such that they can be viewed, in FIRE terms, as one distributed multi-site testbed. FIRE experimenters can thus access TRESCIMO testbed resources by first communicating with the FITeagle instance hosted at the

TUB testbed site. The long-term objectives are for each individual testbed site to be independently federated. That is, a FITeagle and OpenBaton instance will be deployed at each location and enable instances to communicate with each other without the need of the established VPN connection.

Acknowledgements

The TRESCIMO project has received funding from the European Union's 7th Programme for research, technological development and demonstration under grant agreement no. 611745, as well as from the South African Department of Science and Technology under financial assistance agreement DST/CON 0247/2013. The authors thank all the TRESCIMO partners for their support in this work.

References

[1] J. M. Barrionuevo, P. Berrone, and J. E. Ricart. Smart Cities, Sustainable Progress. *IESE Insight*, (14):50–57, Third Quarter 2012.

[2] Bartosz Belter, Juan Rodriguez Martinez, Jos Ignacio Aznar, and Jordi Ferrer Riera. The {GEYSERS} optical testbed: A platform for the integration, validation and demonstration of cloud-based infrastructure services. *Computer Networks*, 61:197–216, 2014. Special issue on Future Internet Testbeds Part I.

[3] M. Campanella and F. Farina. The FEDERICA infrastructure and experience. *Computer Networks*, 61:176–183, 2014.

[4] L. Coetzee, A. Smith, A. E. Rubalcava, A. A. Corici, T. Magedanz, R. Steinke, M. Catalan, J. Paradells, H. Madhoo, T. Willemse, J. Mwangama, N. Mukudu, N. Ventura, M. Barros, and A. Gavras. TRESCIMO: European Union and South African Smart City contextual dimensions. In *Internet of Things (WF-IoT), 2015 IEEE 2nd World Forum on*, pages 770–776, Dec 2015.

[5] DESA United Nations. *United Nations, Department of Economic and Social Affairs, Population Division: World Urbanization Prospects, the 2014 Revision: Highlights (ST/ESA/SER.A/352)*. UN publications, 2014.

[6] Karen Ehrhardt-Martinez, Kat A. Donnelly, and John A. Laitner. Advanced Metering Initiatives and Residential Feedback Programs: A Meta-Review for Household Electricity-Saving Opportunities, Report E105. Technical report, American Council for an Energy – Efficient Economy, 2010.

[7] ETSI. TR 102 691 V1.1.1 (2010-05) Machine-to-Machine communications (M2M); Smart Metering Use Cases. Technical report, ETSI, May 2010.

[8] ETSI. TR 102 857 V1.1.1 (2013-08) Machine-to-Machine communications (M2M); Use Cases of M2M applications for Connected Consumer. Technical report, ETSI, August 2013.

[9] A. Gluhak, S. Krco, M. Nati, D. Pfisterer, N. Mitton, and T. Razafindralambo. A survey on facilities for experimental internet of things research. *IEEE Communications Magazine*, 49(11):58–67, November 2011.

[10] A. Gyrard, C. Bonnet, K. Boudaoud, and M. Serrano. Assisting iot projects and developers in designing interoperable semantic web of things applications. In *2015 IEEE International Conference on Data Science and Data Intensive Systems*, pages 659–666, Dec 2015.

[11] R. R. Harmon, E. G. Castro-Leon, and S. Bhide. Smart cities and the Internet of Things. In *2015 Portland International Conference on Management of Engineering and Technology (PICMET)*, pages 485–494, August 2015.

[12] G. Klyne, J. J. Carroll, and B. McBride. *Resource description framework (RDF): Concepts and abstract syntax, W3C, W3C Recommendation.* 2004.

[13] D. Palma and T. Spatzier. *Topology and orchestration specification for cloud applications (TOSCA) version 1, Nov.* 2013.

[14] L. Peterson, R. Ricci, A. Falk, and J. Chase. Slice-based federation architecture. Technical report, GENI, 2010.

[15] Luis Sanchez, Luis Muoz, Jose Antonio Galache, Pablo Sotres, Juan R. Santana, Veronica Gutierrez, Rajiv Ramdhany, Alex Gluhak, Srdjan Krco, Evangelos Theodoridis, and Dennis Pfisterer. SmartSantander: IoT experimentation over a smart city testbed. *Computer Networks*, 61: 217–238, 2014.

[16] Alexander Willner, Daniel Nehls, and Thomas Magedanz. FITeagle: A Semantic Testbed Management Framework. In *Proceedings of the 10th EAI International Conference on Testbeds and Research Infrastructures for the Development of Networks & Communities*, pages 1–6, Vancouver, Aug 2015. ACM.

30

Federated Experimentation Infrastructure Interconnecting Sites from Both Europe and South Korea (SmartFIRE)

Kostas Choumas[1], Thanasis Korakis[1], Jordi Ordiz[2], Antonio Skarmeta[2], Pedro Martinez-Julia[3], Taewan You[4], Heeyoung Jung[4], Hyunwoo Lee[5], Ted "Taekyoung" Kwon[5], Loic Baron[6], Serge Fdida[6], Woojin Seok[7], Minsun Lee[7], Jongwon Kim[8], Song Chong[9] and Brecht Vermeulen[10]

[1]University of Thessaly
[2]University of Murcia
[3]National Institute of Information and Communications Technology (NICT)
[4]Electronics and Telecommunications Research Institute
[5]Seoul National University
[6]University Pierre and Marie Curie
[7]Korea Institute of Science and Technology
[8]Gwangju Institute of Science and Technology
[9]Korea Advanced Institute of Science and Technology
[10]iMinds

30.1 Introduction

The main achievement of SmartFIRE [1, 2] is the design and implementation of a shared experimental facility spanning different testbeds, which are platforms located in Europe (EU) and South Korea (KR) and offered for conducting transparent and replicable testing of networking protocols and technologies. The SmartFIRE user is able to exploit a *federation* of testbeds, meaning that the testbeds are capable to operate both individually as well as in a unified and collaborative manner. Before SmartFIRE, the participating testbeds in the federation were able to provide diverse resources for experimentation, including WiFi, WiMax and LTE enabled nodes, as well as virtual and physical OpenFlow switches and other SDN devices. However, the user of these infrastructures was not able to access them simultaneously and experiment on heterogeneous network environments that leverage on all the aforementioned resources. Now, the SmartFIRE testbeds have been significantly extended and federated in the experimental, as well as the control plane.

The federations on the experimental and control plane respectively mean the development and operation of simplified and user friendly interfaces, which take charge of the orchestration of the experimentation over all testbeds (experimental plane), as well as the reservation and provision of the resources of all testbeds (control plane). Both directions are supported by the leading experimental and control frameworks adapted by most global testbeds, which are the cOntrol and Management Framework (OMF) [3, 4] and the MySlice tool [5] that is based on the Slice Federation Architecture (SFA). The OMF framework, initially supporting control and experimentation in wireless testbeds, has been expanded in order to support SDN experimentation with OpenFlow switches and Click Modular Routers [7, 8], thus integrating wireless with OpenFlow testbeds [9]. Moreover, unique features, only existing in the KR testbeds have been integrated into OMF [10], unleashing the hidden potential of experimenting with novel resources. On the other hand, the developed SFA extensions enable the federation in the control plane allowing the assignment of multiple heterogeneous resources under a single slice, meaning an isolated set of resources that could be used together in one experiment.

The interconnection of the aforementioned EU sites takes advantage of the GEANT [11] network, while the respective KOREN/KREONET [12, 13] is utilized by the KR sites. The two disjoint networks are interconnected via the Trans-Eurasia Information Network (TEIN) [14] and the Global Ring Network for Advanced Application Development (GLORIAD) [15]. Last but not least, SmartFIRE really showcases its potential with the implementation of two representative use cases, designed to demonstrate the power of the EU-KR shared Future Internet experimental facility. The first experiment explores different mobility scenarios based on the Mobile Oriented Future Internet (MOFI) architecture [16], while the second experiment shows the benefits of an Information Centric Networking (ICN) architecture that achieves efficient content delivery. Finally, the large scaled federated facility is a significant promotion on the joint experimentation among EU and KR researchers, encouraging them to conceive and implement innovative protocols, able to take advantage of the current leading network technologies.

30.2 Problem Statement

The KR testbeds of SmartFIRE (OF@TEIN testbed of GIST, KREONET-Emulab of KISTI, MOFI testbed of ETRI, ICN-OMF testbed of SNU and Open WiFi+ testbed of KAIST) are geographically distributed in multiple sites throughout KR, including most of the well-known local facilities and sharing a minimum set of common SDN features, which are exploited towards the support of new networking protocols and architectures. Although they are able to offer individually an enriched environment for SDN experimentation, they do not share the same enhanced set of wireless experimentation capabilities with the EU testbeds. On the other hand, the EU testbeds of SmartFIRE (NITOS testbed of UTH, w-iLab.t testbed of iMinds and GAIA testbed of UMU) offer different aspects of wireless access networks, which are being controlled and operated in a common manner, participating also in other federations such as Fed4FIRE [17] and OpenLab [18]. However, they could improve significantly their experimentation diversity if they could be federated with the KR testbeds.

More specifically, the experimentation capabilities of all SmartFIRE testbeds, which are depicted in Figure 30.1, are presented below:

- Gwangju Institute of Science and Technology (GIST) offers OF@TEIN, which is an aggregated OpenFlow island consisting of 7 racks, located over 7 international sites. In the OF@TEIN testbed, similar to the GENI racks, a unique rack is designed and deployed to promote the international SDN research collaboration over the intercontinental network of TEIN. OF@TEIN aims at a) the design and verification of the racks (with domestic-vendor OpenFlow switch), b) the site installation and verification of the OF@TEIN network, and c) the design and development of

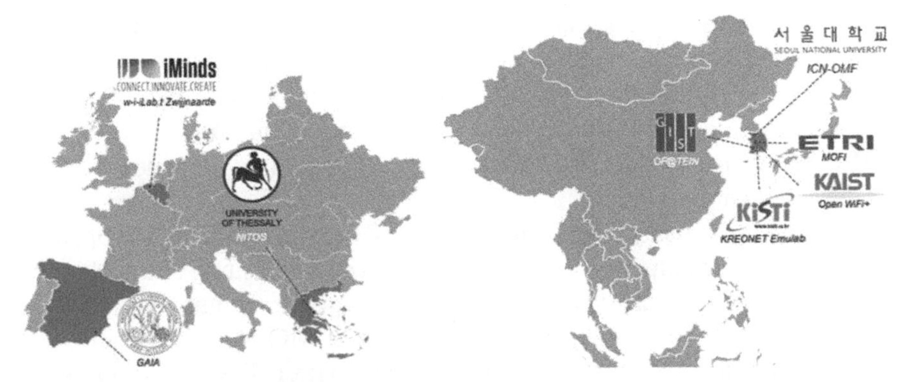

Figure 30.1 SmartFIRE testbeds.

the OF@TEIN experimentation tools. GIST provides also a cloud service based on OpenStack, offering virtualized resources.

- Korea Institute of Science and Technology Information (KISTI) offers an emulation based network testbed in the KREONET domain. It is called KREONET-Emulab and provides the opportunity for evaluation of several network protocols. Many network protocols, which cannot perform over KREONET due to unexpected hazard, can be freely tested in KREONET-Emulab. It consists of 42 powerful servers, each of them equipped with 5 network interfaces, one for the control and four for the experimentation. Each server can work as a router with 4 paths, and each network interface can be configured up to 1 Gbps.
- Electronic and Telecommunications Research Institute (ETRI) proposes the network architecture of MOFI. Following a completely different approach from the current IP networking, MOFI enables the development of networks with Future Internet support of mobile intrinsic environments. The evaluation of the MOFI architecture relies on the OpenFlow-based mobility testbed of ETRI. The mobility testbed is an aggregation island, consisting of four interconnected South Korean domain networks. Their interconnection is based on the KOREN networking infrastructure.
- Seoul National University (SNU) proposes the ICN-OMF architecture, as a result of the research on the development of content centric networking applying it to the OMF framework. In particular, ICN-OMF is the architecture for the development of ICN-based networks using Open-vSwitch and CCNx over virtual machines. It provides functionalities for in-network caching, as well as for their name-based forwarding. Additionally, SNU operates the ICN-OMF testbed which enables the experimentation on ICN.
- Korea Advanced Institute of Science and Technology (KAIST) provides a wireless mesh network, named Open WiFi+, which is a programmable testbed for experimental protocol design. It is located at the campus of the KAIST University and it consists of 56 mesh routers, 16 of them being deployed indoors and 40 outdoors, each of them equipped with three IEEE 802.11 b/g/n WiFi cards. Moreover, 50 sensor nodes are deployed at the same campus.
- University of Thessaly (UTH) provides the NITOS facility, which is open to the research community 24/7 and it is remotely accessible. The testbed consists of 100 powerful wireless nodes, each of them equipped with 2 WiFi interfaces, some of them being 802.11n MIMO cards and the rest 802.11a/b/g cards. Several nodes are equipped with USRP/GNU-radios,

cameras and temperature/humidity sensors. The nodes are interconnected through a tree topology of OpenFlow switches, enabling the creation of multiple topologies with software-defined backbones and wireless access networks. The testbed features programmable WiMAX and LTE equipment, fully configurable with an SDN backbone.

- iMinds supports the generic and heterogeneous w-iLab.t facility. It consists of two wireless sub testbeds: the w-iLab.t office and w-iLab.t Zwijnaarde. The w-iLab.t office is deployed in a real office environment while the testbed Zwijnaarde is located at a utility room. There is little external interference at the Zwijnaarde testbed as no regular human activity is present and most of its walls and ceiling are covered with metal. The majority of devices in w-iLab.t are embedded PCs equipped with WiFi interfaces and sensor nodes. Since the Zwijnaarde testbed was deployed more recently, the devices in this testbed are more powerful in terms of processing power, memory and storage.
- Universidad de Murcia (UMU) offers the research and experimentation infrastructure of GAIA. GAIA comprises several network nodes interconnected with different technologies. On the one hand, they are connected to the campus network through Gigabit Ethernet switches and thus they form the point of attachment to the Internet. On the other hand, they are connected to a CWDM network, which acts as backbone/carrier network and can be adapted to different configurations, depending on the specific requirements of each experiment. GAIA has also a wide wireless and WiMAX deployment along the campus. This, together with other smaller wireless deployments, allows the experimentation with many local and wide-range wireless technologies, including mobility and vehicle (V2V) communications.

Based on the individual experimentation capabilities of these testbeds, the main SmartFIRE goal was to operate an extended federated facility in KR and EU that could combine the strong points of the independent testbeds towards a joint infrastructure that could efficiently provide more research abilities for experimentation. The federated function and collaborative operation of all these testbeds could allow the experimentation with novel applications, under highly automated conditions, easily operated and managed. For example, the OpenFlow-based SDN is an emerging technology, which was supported by few testbeds before SmartFIRE, enabling the experimentation with content-centric architectures and protocols focusing only on the wired networking. SmartFIRE improved the capabilities of this experimentation environment by enabling the wireless support for all OpenFlow-enabled testbeds.

Except for the lack of diversity in the experimentation environments, coming from the fact that the SmartFIRE testbeds were not federated, another problem in the status before SmartFIRE was the adoption of different control and experimentation frameworks from these testbeds. The use of different frameworks made difficult the repeatability in the experiments description and execution. Before SmartFIRE, the researchers had to become familiar with the special tools used in each isolated testbed, in order to succeed conducting their experimentation. If they wanted to repeat their experimentation in another testbed with the same resources, they had to translate their experiment description to another accessible language for the tools of the other testbed. After SmartFIRE, the included testbeds are handled by common and widely adopted tools and frameworks that release the users from time-consuming learning curves and translations of experiment descriptions.

30.3 Background and State of the Art

Future Internet is the breakthrough innovation that changes our society in terms of economic, social, entertaining, informational, governmental or daily aspects. Many organizations and institutions are working for improving and upgrading the status of the nowadays Internet, facing a variety of inherent problems related to scalability, suitability, sustainability, energy efficiency, security, etc. Their efforts include the participation in various projects that introduce the fundamental redesign of the Internet architecture and protocols. These projects mostly support the development of large-scale experimental facilities and testbeds that provide easy experimentation in proposed theoretical formulations.

The Future Internet Research and Experimentation [19, 20] (FIRE) program, funded by the European Commission, and the Global Environment for Network Innovation [21] (GENI), funded by the National Science Foundation (NSF), are two leader projects that concentrate on the creation and functionality of large-scale testbeds that will provide insights and directions for the Future Internet evolution. Under these programs there are various testbeds that seek to provide researchers with well-dimensioned computing, storage, sensor and network resource slices.

The FIRE initiative in Europe aims at experimental research and funds projects to produce Future Internet research and experimentation facilities, like OpenLab and OFELIA [22]. OpenLab provides an open, both large-scale and sustainable federated testbed, including PlanetLab Europe, the NITOS wireless testbed and other federated testbeds like PlanetLab Korea and PlanetLab Central. OFELIA is another program that provides OpenFlow-based experimentation capabilities to experimenters and researchers, spanning multiple OpenFlow-based islands in Belgium, Germany, Spain, Switzerland, UK, Italy and Brazil. Fed4FIRE is the newest effort to create a federation of all FIRE testbeds, providing easy access to them through a powerful and well accepted set of tools, which is elaborated in synergy with GENI.

OMF6 is the latest version of OMF, which is a generic framework included in the FIRE and GENI adopted tools and allows the definition and orchestration of experiments using shared resources from different federated testbeds. OMF6 is the successor of OMF5, which was originally developed for single wireless testbed deployments, but now it is extended to support multiple deployments and various features. Its architecture is modular consisting of different components endowed with the operation of the experiment orchestration and the resource control. As it is depicted in Figure 30.2, using a simple human readable experiment definition, OMF6 is supporting the whole experiment lifecycle, cooperating also with its accompanying framework, OMF Measurement Library (OML). The experimenter submits a simple script to the OMF6 Experiment Controller (EC) and the underlying functionality is responsible for setting up the resources, running the defined applications and collecting the results in an organized way.

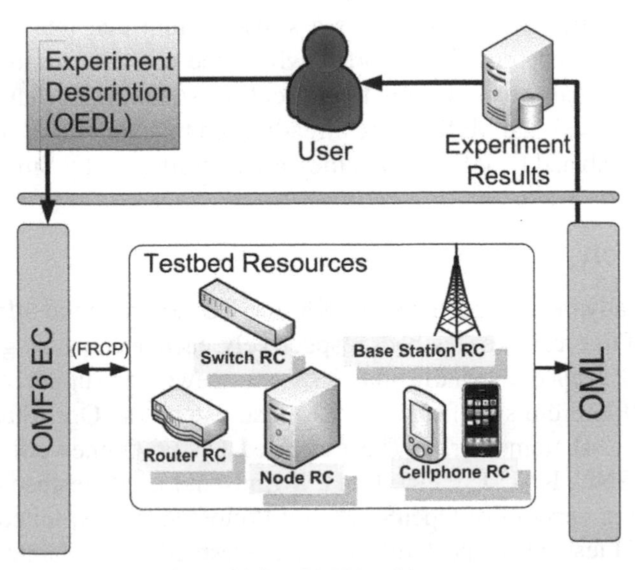

Figure 30.2 OMF6 architecture.

First, OMF6 provides a domain-specific language based on an event-based execution model to fully describe even complex experiments (OEDL). OMF6 also defines a generic resource model and concise interaction protocol (FRCP), which allows third parties to contribute new resources as well as develop new tools and mechanisms to control an experiment. It uses a standardized sequence of messages sent by the EC to the Resource Controllers (RCs) and vice versa. The RC is a daemon that behaves as a proxy between the EC and the resource, translating the messages of the EC to executable commands for the resource and vice versa. Testbed operators are able to use this flexible protocol to extend the experimenter's control on new testbed resources or even establish federations of testbeds, thus enhancing the experimentation ecosystem. In SmartFIRE, we took advantage of this feature in order to extend OMF6 support to OpenFlow and SDN resources, such as Click Modular Routers. In this way, there is one framework in the experimental plane that is able to coordinate the experimentation over network topologies with both wireless and SDN resources.

In the control plane, both FIRE and GENI have designed the Slice Federation Architecture (SFA), defining the interface that should provide each testbed want to be federated. The testbeds resources have to be described in RSpecs (Resource Description) and managed by a SFA Aggregate Manager (AM), which provides a SFA compatible interface. This interface enables the discovery, reservation, provisioning and releasing of all testbed resources in a unified and common way. MySlice is a software tool that interacts with the SFA interfaces and provides a portal, where the user is able to see the testbeds, as well as information about their location, status and resources (see Figure 30.3). The SFA and MySlice are already used by many FIRE projects, including OpenLab and Fed4FIRE, and they are also adopted by SmartFIRE.

30.4 Approach

As we have already mentioned, the set of KR and EU experimental infrastructures consists of five and three testbeds respectively, each one featuring unique characteristics. Except of expanding the OMF6 framework to support all these features, SmartFIRE did significant work to also extend the OpenStack [23] and ProtoGeni [24] frameworks. The extended OMF6 framework is now utilized by the SNU, KAIST, ETRI, UTH, iMinds and UMU testbeds, while the corresponding versions of OpenStack and ProtoGeni are exploited by the GIST and KISTI testbeds respectively. All these testbeds are federated in the control plane with use of the SFA based software of MySlice.

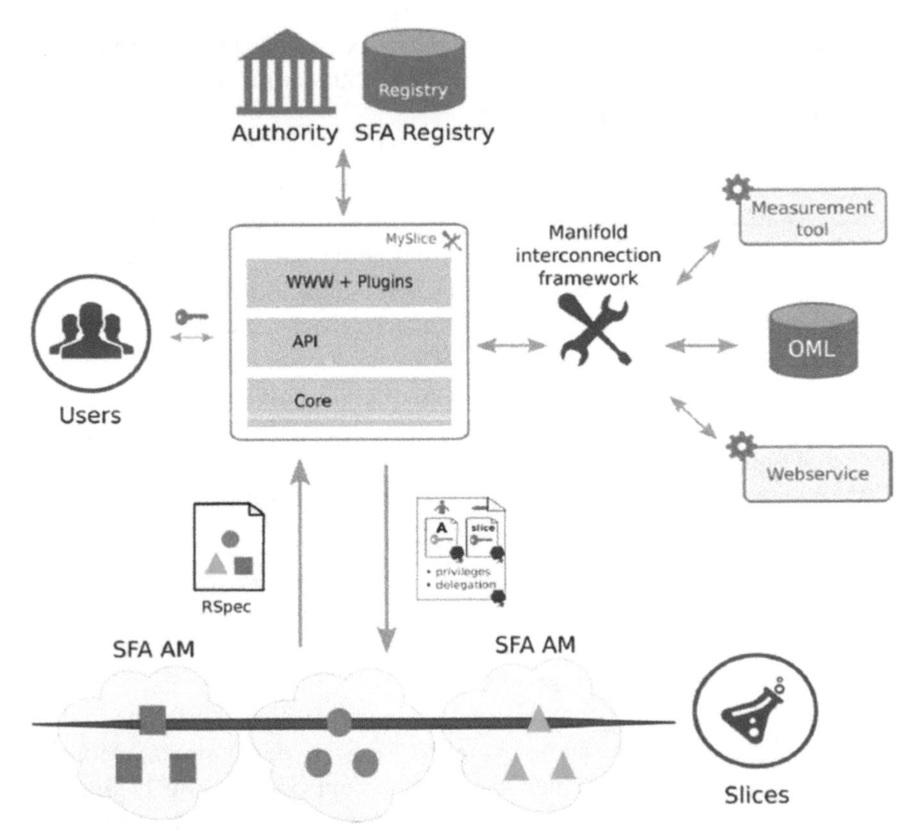

Figure 30.3 MySlice and SFA.

a) The contributions on the OMF6 framework are summarized below:

- **OMF6 extensions for ICN experimentation support**. ICN recently attracts much attention from researchers of Future Internet Architecture, due to its novel communication model, distributing/retrieving the contents by its name (i.e., "what") rather than accessing the location the contents resides ("where"). "Cisco's Visual Networking index: Global Mobile Data Traffic Forecast Update, 2013–2018" reported that the communication behavior of Internet has been shifted to publisher/subscriber model, which is more optimistic for content distribution/retrieval. However, the current IP-based Internet architecture is not designed to accommodate this communication model. With this motivation, ICN proposes to remedy the problems the Internet encounters (i.e., an inefficient communication model for contents distribution and retrieval) exploiting the capabilities of SDN. Appropriate OMF6 extensions are required for enabling the configuration of ICN topologies with parameterized number of content publishers and subscribers.

- **Enabling experimentation on Mobility-based communication using OMF6**. GENI has noted that wireless/mobile will be the major access means for Future Internet. MOFI effectively realizes a seamless mobility architecture in the Future Internet. It utilizes Host Identifiers (HID) that represent "who is the user (human) or user's equipment (host)?" and Locators (LOC) that represent "where is the user or the equipment?". The HID is decoupled from the LOC. The enhanced OMF6 framework of SmartFIRE is able to control and manage domain networks that contain various Open-vSwitch (OvS) [25] resources as access routers and follow the MOFI architecture.
- **OMF6 support for new software and hardware resources, like**:
    - The OvS software that enables the creation of virtual switches with use of Linux operated computers. Although Open-vSwitch has been initially developed for managing wired networks by creating virtual switches, it can be efficiently used for managing wireless interfaces that are parts of such a switch.
    - The Click Modular Router is another long established software tool that its capabilities can be exploited for SDN development. More specifically, Click enables the development of Software Defined Routers with use of Linux operated computers. In [26], the authors present how they utilize Click and OMF6 to experiment with distributed loading shedding schemes.
    - The FlowVisor [27] software that enables the slicing of OpenFlow switches. FlowVisor will be used as the network virtualization layer, allowing for the physical network to be sliced by the control framework, and for each slice to be controlled by the OpenFlow controller associated with this slice. This feature is very crucial for testbed facilities with slicing mechanisms, enabling the simultaneous use of the included resources from multiple experimenters.
    - The Linux operated computers named M-Boxes, which enable the experimentation on virtual wireless topologies with use of Virtual Machines (VM) and OvS.
    - Wireless Access Points (AP) that are extended in order to be controlled even in terms of the utilized transmission power. The experimentation with wireless APs illustrates the big difference between the theoretical and actual performance of the protocols

for wireless networking. In the experimental research, the use of actual radio resources is becoming more and more important and the need for a real wireless environment testbed is increasing.

b) The contributions on SFA and MySlice are:

- **The development of the SmartFIRE portal**, which provides a graphical user interface that allows users to register, authenticate, browse all SmartFIRE testbeds resources, and manage their slices. This work was important to provide a unified and simplified view of many hidden components to the experimenter.
- **The definition of new RSpecs for the new SmartFIRE resources**. In this way the new resources can be viewed in the SmartFIRE portal. Of course, the RSpecs are reproducible and it is not needed to be defined again for the same type of resources used in another testbed.

c) Last but not least, the physical interconnection of the EU and KR testbeds and their federation in a unified experimentation platform that enables their control and management through a single framework.

30.5 Technical Work

30.5.1 ICN-OMF

Although there are lots of proposals investigating on architecture of ICN, their methodologies to evaluate and validate ideas still stay at unrealistic simulation or small-scale emulation. However, to become a new protocol deploying at the production networks, it should be validated and evaluated on real physical testbed, providing scalability, configurability, and low-cost to researchers. Thus, if there is a formulaic testbed for ICN, the experimenters can only focus on their own experiments without concerning cumbersome learning curves. SmartFIRE developed and deployed ICN-OMF, leveraging and extending OMF6 to control and manage globally dispersed ICN nodes (i.e., publishers, subscribers, or routers).

The experimenter is able to use OEDL to describe the ICN experiment, while the final ED is given to the EC which communicates with the ICN-RCs, as it is illustrated in Figure 30.4. The whole experimentation process is the same with the OMF6 one that was described before, with the only difference that all related parts are enhanced to support ICN experimentation. In particular, the ICN-RC is responsible for configuring the physical nodes by creating virtual switches and virtual machines behaving as ICN nodes on demand.

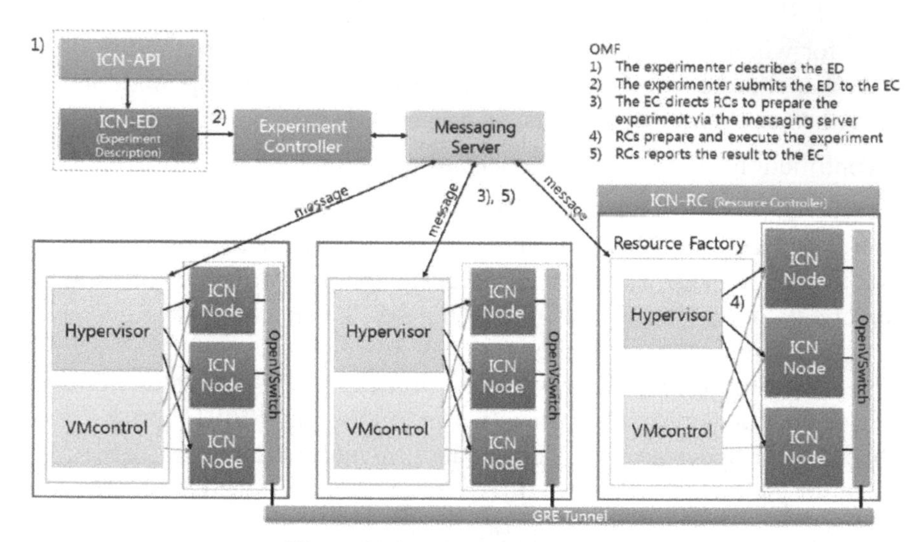

Figure 30.4 ICN-OMF framework.

Since ICN is a clean-slate network architecture that is not compatible with the current IP-based network, ICN networks are built as overlay networks. The CCNx open-source software is used, since it is one of the most well known candidates for ICN and it is widely used. SmartFIRE experimenters are able to utilize this framework in order to validate and evaluate their ideas in a convenient way.

30.5.2 MOFI-OMF

ETRI builds the MOFI testbed which consists of multiple domains interconnected through a global backbone network, such as Internet. All communication entities have one or more global HIDs that are necessary for the development of end-to-end communication channels. In the MOFI architecture, each domain network has multiple Access Routers (ARs) that take care of attached communication entities and one or more Gateways (GWs) that interconnect the domain networks through the backbone network. This network architecture is implemented with use of OpenFlow and SDN technologies. The ARs are based on the Open-vSwitch software and they are controlled by OpenFlow NOX controllers. These controllers are responsible for the domain networks, as well as the GWs that support the inter-domain communication.

In order the MOFI testbed to be operated and managed by OMF6, SmartFIRE extended OMF6 to control and manage the MOFI domain networks consisting of various resources such as the OvS based ARs and GWs.

As it is depicted in Figure 30.5, MOFI network domains are created with use of OvS instances which are configured by the corresponding RCs. The OvS instances are either the points of attachment for the hosts or the GWs that interconnect the domains. Both RCs and EC are able to support the description and control of specific MOFI components, which are identified and used by the experiments without the need to deploy the full networking stack on a generic resource. During the configuration phase of resources, OMF6 deploys specific links among the components, thus the switches and routers are connected to each other and to the hosts. Moreover, OMF6 manages the specification of low-level software components, such as kernel modules. This is addressed during the resource set-up phase so the general deployment of the networking element into the resource is sped up and the experiment definition is simplified. Network links should be supported as a special resource, which is not supported now, since the links are associated to the resources they are connected.

30.5.3 Open-vSwitch (OvS)

The creation of virtual OpenFlow switches relies on OvS, which is used in multiple commercial products and runs in many large scale production networks. Although OvS has been initially developed for managing wired networks by creating virtual switches, it can be efficiently used for managing

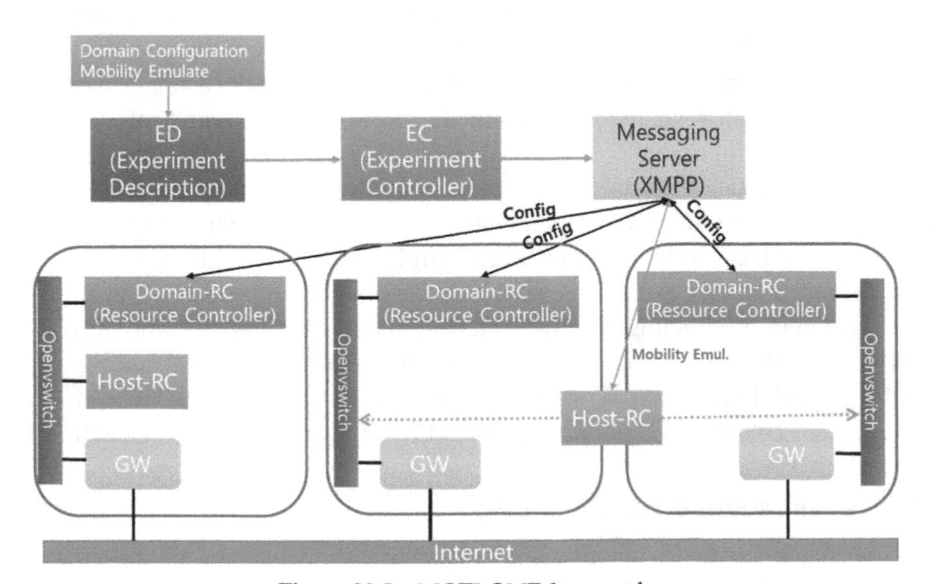

Figure 30.5 MOFI-OMF framework.

wireless interfaces that are parts of a switch. If such an interface is placed in OvS, the experimenter has the ability to intercept the traffic that is exchanged over the wireless interface as Ethernet based frames (since the wireless header is removed upon each packet reception by the wireless driver). Although this seems to be a time saving advantage for the researcher, it also poses many questions regarding the controllability of the SDN enabled wireless switch. To this aim, SmartFIRE enables a Simple Network Management Protocol (SNMP) agent process on the wireless nodes, which allows us to remotely configure the wireless interfaces in a software defined manner.

Based on these processes for creating wireless OpenFlow switches, SmartFIRE developed the corresponding extensions to the OMF6 framework that allow this functionality. The OMF RC entity is significantly extended, in order to be in charge of receiving the proper configuration messages and applying the corresponding settings to the resources. All the existing commands of the OvS API are supported by our extensions. Complementary to this, SmartFIRE developed the appropriate exchange messages among the OMF6 entities for instructing the RC to send the appropriate snmp-set commands for configuring the wireless interfaces, and the snmp-get commands for retrieving their status. Accordingly, the OMF EC entity, which is in charge of sending the appropriate messages to the RC based in the experiment description submitted by the user, has been extended to support this functionality.

The messages exchanged are based on the FRCP standardized by the OMF6 research community. With our extensions, the experimenter can now use the testbed framework to transparently create and configure virtual switches, combining even wireless resources, in large scale using a user friendly and human readable experiment description. An example of such a setup is the configuration of an OvS instance consisting of the wireless interface of a node, the initiation of an OpenFlow controller to control this switch, and multiple wireless clients to connect and generate traffic accordingly on the wireless network, in a single experiment definition. The aforementioned OMF6 extensions have been developed and evaluated using virtual switches combining several heterogeneous wireless technologies. Namely, our extensions support the concurrent operation and configuration of Atheros and Intel based Wi-Fi interfaces, Intel and Teltonika WiMAX interfaces and Huawei LTE interfaces in a single OpenFlow wireless switch.

30.5.4 Click Modular Router (Click)

Click Modular Router is another long established software tool that its capabilities can be exploited for SDN development. More specifically, Click enables the development of Software Defined Routers with use of Linux operated computers. In Roofnet [28], Click developers investigated wireless connectivity issues and proposed a routing algorithm named after it. Their framework is extensible and well documented, enabling the implementation of many routing algorithms with significantly low effort. The alternative option for packet forwarding in a wireless mesh is the 802.11s protocol, that relies on a similar approach called path selection. Nonetheless, Click is much more flexible and extensible than the existing 802.11s implementations. Many testbeds utilize Click to implement wireless mesh networks, enforcing their computer resources to behave as wireless software defined routers.

Our extensions to support the Click framework have not been so straightforward as for the OvS framework. Since Click is a highly configurable tool, with many users being able to develop their own extensions of the supported functionality using Click elements or modules (as they are called by the supporting community). However, SmartFIRE follows a different approach in order to support as many as possible configurations. The corresponding RC is only responsible for executing the Click router in the user-space level with the appropriate arguments. The experimenter submits to the EC a configuration file that describes the desired Click settings. With this approach, the experimenters can now define new elements, which did not exist at the time that our developments took place, and use them to orchestrate their experimentation in large scale mesh networks. SmartFIRE has moved one step beyond in the extension of our framework and enabled OML support in the core Click system, responsible for capturing the output of Click execution and injecting the measurements in an OML database. Using our provided hooks in the Click version 1.8, the experimenter can easily support new measurements from the lately released elements.

30.5.5 FlowVisor

FlowVisor behaves as a network hypervisor, which enables the concurrent usage of an OpenFlow switch by multiple experimenters. FlowVisor is nothing more than a special purpose OpenFlow controller, which acts as a transparent proxy between any OpenFlow switch and multiple experimentation specific OpenFlow controllers. From the perspective of the OpenFlow switch, FlowVisor is its controller. It isolates parts of the underlying hardware switch and provides access to these subparts to experimentation specific controllers. Slicing might depend on several attributes of the switch, like for example the number of ports used, the physical switch memory or processing power utilized per controller instance. The slicing may also be based on the packet flow characteristics, like the IP/MAC source and destination addresses or the VLAN tagging.

SmartFIRE extended OMF6 to control the FlowVisor process and allocate completely isolated OpenFlow switch slices upon a user's request. The slices are isolated based on the switch's physical ports, thus preventing each experimenter to interact with the traffic intended for another slice. When a user reserves testbed nodes attached to a physical switch, OMF6 transparently creates an OpenFlow slice, consisting of the ports where the reserved nodes are attached. As it is illustrated in Figure 30.6, with only the OMF6 functionality without the SmartFIRE extensions, each user reserves two nodes that share a wireless connection using an idle or non interfering with other users' wireless frequency. Our extensions take place at the wired OpenFlow enabled backbone connection of the nodes, and upon the node reservation set up the appropriate FlowVisor instance which abstracts the testbed switch that the two depicted nodes connect to.

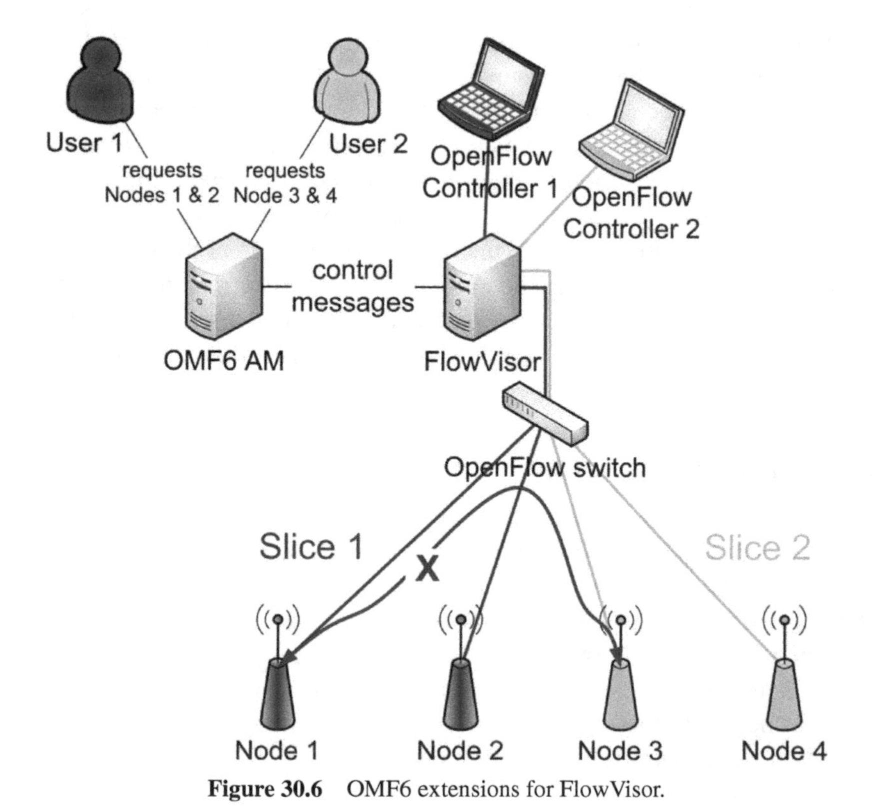

Figure 30.6 OMF6 extensions for FlowVisor.

30.5.6 Open WiFi+

SmartFIRE uses both commercial and open source APs to implement an experimentation environment close to the actual environment. Firstly, Open Wi-Fi+ was using commercial APs which can be found in the market. Secondly, by deploying the APs in a real office, SmartFIRE developed a Wi-Fi environment for experimentation under real world settings. Now, Open WiFi+ supports 8 APs, and the user is able to control each AP's transmission power. By using the power control, the user can change the wireless topology of his/her experiment.

30.5.7 SFA and MySlice

SFA has been designed as an international effort, originated by the GENI framework, to provide a secure common API with the minimum possible functionality to enable a global testbed federation. It provides a secure API that allows authenticated and authorized users to browse all the available resources and allocate those required to perform a specific experiment, according to the agreed federation policies. Therefore, SFA is used to federate the heterogeneous resources belonging to different administrative domains (authorities) to be federated.

The federation architecture adopted in SmartFIRE project is composed of 3 main components:

- Registry
- Aggregate Manager
- MySlice portal

The SFA Registry holds the certificate of the root authority of the federation. Its database is responsible for storing the users and their slices with the corresponding credentials. The project partners have decided to use a central Registry for the SmartFIRE federation. This component can also be federated with other Registries by exchanging certificates as a proof of trust relationship. The SmartFIRE federation is structured as a hierarchy of partner institutions. Each institution is responsible of its users and must validate the requests of new users belonging to their institution.

Another component of the SFA layer is the AM, which is required in each SFA-compliant testbed. The AM is responsible for exposing an interface that allows the experimenters authenticated by the Registry to browse and reserve resources of a testbed. The SFA AM exports a slice interface that researchers interact with to set up, control, and tear down their slices. Each testbed has

Table 30.1 AM Software used by SmartFIRE testbeds

Institution	Testbed	AM Software
UTH	NITOS	OMF-SFA Broker
UMU	GAIA	OMF-SFA Broker
SNU	ICN-OMF	OMF-SFA Broker
iMinds	w-iLab.t	OMF-SFA Broker
KISTI	KISTI-Emulab	ProtoGeni
KAIST	Open WiFi+	OMF-SFA Broker
ETRI	MOFI	OMF-SFA Broker
GIST	OF@TEIN	SFA Wrap

an AM, which relies on different software as shown in Table 30.1. But they all expose an SFA compliant API, allowing users to reserve resources across different testbeds.

MySlice was introduced by UPMC as a mean to provide a graphical user interface that allows users to register, authenticate, browse all the testbeds resources, and manage their slices. This work was important to provide a unified and simplified view of many hidden components to the experimenter. At the same time, it provides an open environment for the community to enrich the portal through various plugins specific to each testbed or environment. The basic configuration of MySlice consists on the creation of an admin user and a user to whom all MySlice users could delegate their credentials for accessing the testbed resources. In order to enable MySlice to interact with heterogeneous testbeds, MySlice has to be able to generate and parse different types of RSpecs; this task is performed by plugins. MySlice has been widely adopted by the community and is currently an international effort. As of today MySlice has been adopted by the following testbeds (or Projects): FIT (France), F-Lab (France), Fantaastic (EU), Fed4Fire (EU), Openlab (EU), FIBRE (Brazil), FORGE (EU), CENI (China), SmartFIRE (Korea) and III (Taiwan).

30.6 Results and/or Achievements

30.6.1 Multi-Domain, ID-Based Communications and Seamless Mobility with MOFI

One of the use cases proving the value of SmartFIRE platform is the mobility use case, which shows the service continuity using a video streaming application, when host moves and connects to different Access Routers (AR) and to different Gateways (GW) (as it is depicted in Figure 30.7. According to

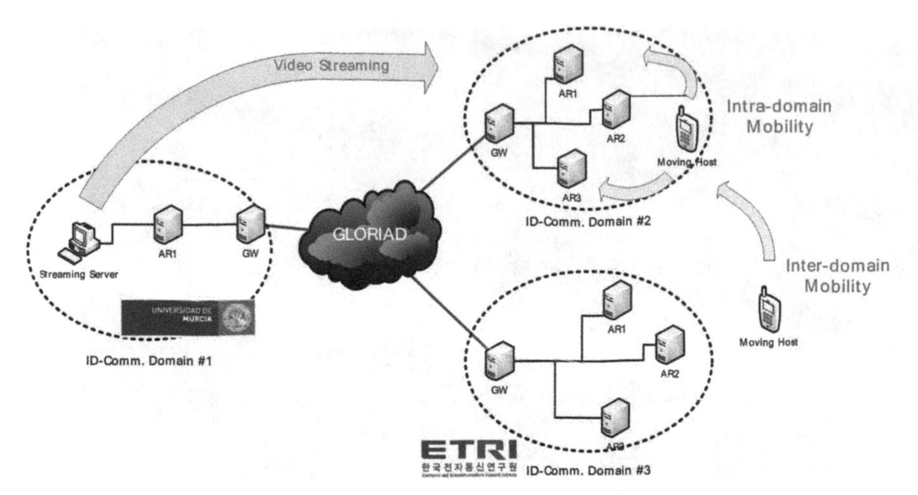

Figure 30.7 Seamless mobility scenario.

this use case, a video streaming server is deployed at UMU's domain network (Domain #1), while a video streaming client, which is a moving host, is located at one of the two ETRI's domain networks (Domain #3). The connection of both domains utilizes a dedicated Layer 2 intercontinental virtual link between UMU and ETRI. The video streaming application, which is installed in a server located in Domain #1, starts streaming to the client located in Domain #3, while the client moves to another AR whether in the same domain or to a different domain network (Domain #2). We have seen that video streaming service is provided continuously under this mobility scenario.

As we evaluate the above mobility scenario, the experimenter is able to continuously check the service by observing the experimentation messages, as it is depicted in Figure . This Figure 30.8 shows a connection between client and server. When the client moves to another domain, then the connection is lost for a while. After moving into another domain, the client registers its own Identifier to the new domain gateway by starting the HBR (Host Binding Request) process. The connectivity is resume soon and although video streaming is stopped for a while, after one second the client is able to deliver the streaming data smoothly.

In addition to the aforementioned mobility experiment, an extra experiment featuring multi screen streaming over the ID-based communication architecture of the MOFI testbed has been implemented. This experiment is designed to showcase the capabilities of the ID-based communication architecture for seamless service mobility. In particular, it focuses on a video

Figure 30.8 Experimentation messages for handover.

streaming service that is dynamically directed to different mobile hosts. The demonstrated service is called multi-screen streaming service and shows the opportunities of the Service ID (SID) concept (SID is used to uniquely identify an upper-layer application or service that is running on the host).

This experiment is depicted in Figure 30.9 and was also demonstrated at ICT 2015. A MOFI domain network is consisted of an AR, a GW, and various screens (such as Smart Phone, Tablet and PC monitor) equipped with USB

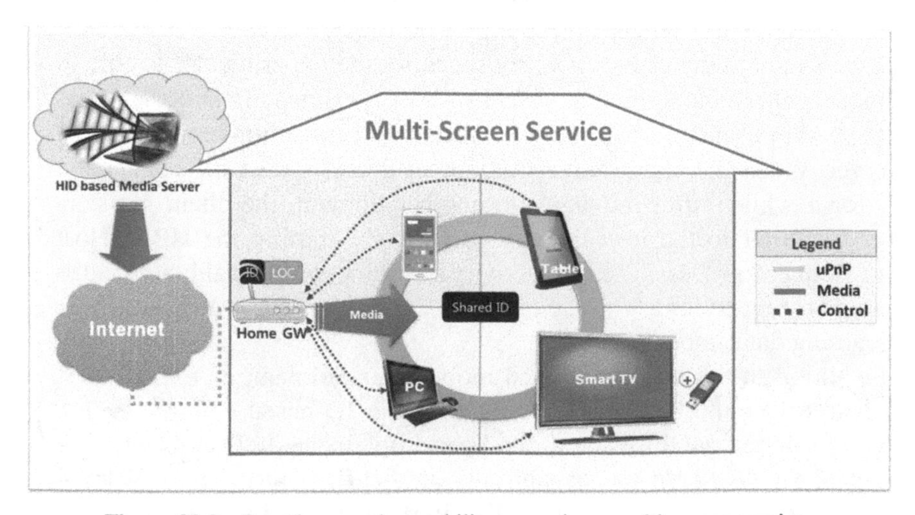

Figure 30.9 Seamless service mobility scenario as multi-screen service.

dongles that provide the processing power. In the service mobility scenario, when these screens are already connected to the MOFI domain network, they start negotiating for a common SID (Service IDentifier). During this experimentation, a daily scenario is emulated, where a user is coming back to home from his/her work. The video is first streamed to the smart phone. When the user enters the house, a new ID-based domain is available to the phone and various screens are discovered after the end of the SID negotiation process. These screens share the same SID, which corresponds to the original smart phone's HID. At this point, the smart phone selects another destination screen that will receive the ongoing video stream. Since the MOFI GW maintains a table that maps each screen's HID to its location, the selected screen is assigned the same HID and as a consequence, the video is directly forwarded to the desired screen through the GW.

SmartFIRE showcases two types of mobility scenario, named server mobility and service mobility. In server mobility scenario, the video stream is initiated between a server located in Spain and a video client located in ETRI, and the experimenter is able to observe the media data being streamed. In the intra-domain server mobility event (meaning the event when the server moves to another access router located within the same domain) there is no observable impact in the streaming session. On the other hand, in the inter-domain server mobility event (meaning the event when the server moves to another access router located in a different domain), it is a challenge to avoid the connection break due to the change of the destination address (which is the locator in the ID-based communication networks). In order to provide seamless streaming service, the SDN controllers that are in charge of mapping each host's HID to a specific locator, they update this mapping information and create the flows that will forward the traffic to the new location. After updating the mapping information, streaming data is forwarded to the new domain network to which the client is now attached. Thanks to the mapping the streaming service is provided in a non-disruptive way.

Secondly, the seamless service mobility scenario works in a pretty similar way to the above scenario without the procedure of synchronization for same SID. The video is initially streamed from the video server to the controller (smart phone). Then, the controller gets information that there are other screens within the Domain through uPnP. Discovered screens (tablet and PC monitor) are listed in the contoller. At the same time, the SID synchronization process is triggered in order to assign the smartphone's original HID forming a SID as group-ID. After that, the smart phone is able to choose a screen among a list of available screens to which the video streaming can be forwarded and displayed at the screen. This is achievable thanks to the MOFI ID-based architecture and to the usage of SDN with which the same HID (ipv6 address in our case) can be assigned to multiple devices.

30.6.2 Content-Based Video Communications on Wireless Access Network

In this scenario, a content-based architecture is utilized, that is implemented using SDN technology (leveraging physical OpenFlow switches and nodes with virtual OpenFlow and Click Modular Router instances) on top of the UTH, SNU, KAIST and GIST testbeds. All these resources are controlled and managed with the support of the OMF6 tools developed by this project. The utilized resources are interconnected including Layer 2 intercontinental virtual links, based on the GEANT-GLORIAD-KREONET services. Wireless devices laying on UTH testbed are connected to a Content-based network on SNU, GIST and KAIST, where the IP addressing scheme is replaced by a novel one, based on content identifiers. The goal of this innovation is to use identifiers that specify only the content and not the location of this content, as the IP addresses do. Each piece of content is placed on multiple sides on the Content-based network developed on the aforementioned South Korean testbeds. The target of the Content-based architecture is the forwarding of the content from the most appropriate side to the requesting wireless device, while the streaming over the UTH wireless mesh is based on a Backpressure routing scheme.

We expected and proved that the time performance is much better when the data is cached. The in-network caching, which is an inherent feature of ICN, improves significantly the end-to-end delay of the video streaming from distributed South Korean sites to UTH. The interesting part of our experimentation is the trade-off between the time spent for the caches management and the reduction of the time delay in the packet forwarding. We showcase that the appropriate design and deployment of the Content-based and wireless access networks is fundamental for significant gains in terms of end-to-end delay in video streaming. In the following Figure 30.10, you can see the topology we created in the SmartFIRE testbed and we demonstrated in ICT 2015.

The results of our experimentation are very promising, since we observed significant improvements in terms of end-to-end delay in the video streaming. We showcase that the video streaming lasts for shorter time interval when more and more devices ask for the same video content, since the new devices

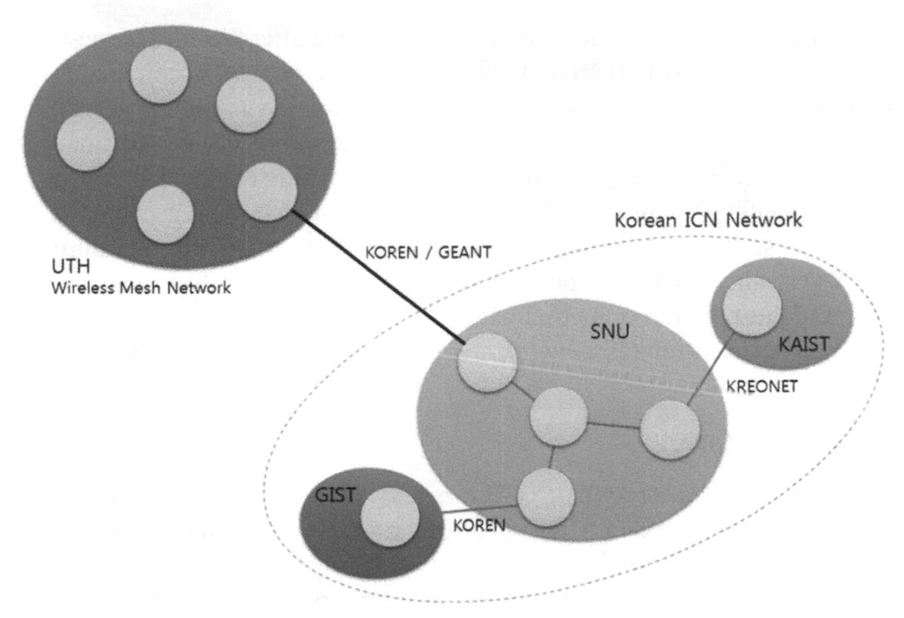

Figure 30.10 Topology of the experimentation on a content-based communication network.

requesting the same video are able to download from closer geographically caches. We measured the end-to-end delay for the video streaming by collecting and subtracting the timestamps of the video packets when they are generated and when they are delivered.

Figure 3.11 shows the end-to-end delay of the last request for a specific video content, when there is one WiFi device that requests for the same content. The requests are sent every 15 seconds. The y-axis shows the delay measured in milliseconds, while the x-axis shows the number of the requests that device has done.

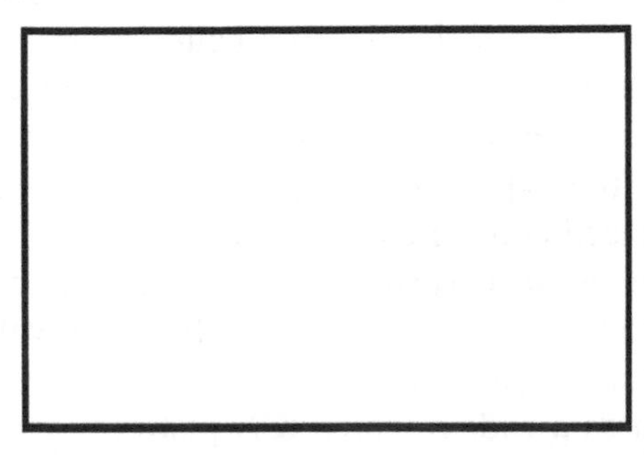

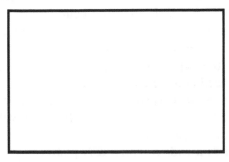

Figure 30.11 End-to-end delay in the content-based communication network.

From the graph, we can see that the delay time is stable after the third request. It shows that after almost 30 seconds (2 requests), the content is completely cached in the ICN gateway, which is later used as a video streaming server. The time before, the content was being cached and could not be streamed from the closer ICN gateway, but it had to be downloaded from the remote video server. Having in mind that the length of the content (video) used in this experiment is seven seconds, we observe that the delay time after the third request is very short (less than one second). It is much lower than the delay of the first request, which is approximately 30 seconds. The graph illustrates the significant effect of the in-network caching in the end-to-end delay.

30.7 Discussion

The showcases presented as part of the results of the SmartFIRE project demonstrate the usefulness of a federated solution that interconnects multiple isolated and heterogeneous testbeds. We have shown how experiments can reinforce the research results obtained by two network approaches that have totally different natures.

On the one hand, the ID-based communications is a key research field in current networks. In fact it has been sometimes associated with the raise of SDN, which treats underlying network identifiers as such, identifiers, instead of using them to build some kind of addresses. The experiments showcased during the development of the SmartFIRE platform have supported the definition of specific requirements in terms of new interfaces to the OMF/SFA infrastructures and the specific support for heterogeneous resources, since MOFI resources have been exposed as network-level elements while other resources were exposed as lower-level elements. Moreover, the mobility requirements of the ID-based approach has supported the outcomes of setting up the wireless (and WiMAX) technologies within the SmartFIRE infrastructure. This has been translated into a better set of features and the extension of the kind of experiments supported by the final federated platform, enriching the FIRE ecosystem at the same time.

On the other hand, the experiment targeting content-based video communications has introduced another current research field, the ICN, and has related it to specific wireless scenarios. This has been the key to retrieve more requirements to apply to the SmartFIRE platform but, with valuable research results for the ICN community, it has also established the basement for a wider research initiative towards a wide research ICN platform for researching on video communications and their implications in wireless networks.

30.8 Conclusions

The current research efforts to study, analyze, find problems, and finally improve the Internet have to deal with the lack of real scenarios and infrastructures to experiment with. This makes it really difficult to achieve production ready solutions with a high degree of confidence. In order to improve this ecosystem, the experimentation driven research has been pushed to the network research field. It states the aspects that good research methodology [30] and results should meet in order to ensure they provide enough evidences to the research community, as well as the companies that will translate those results to products that impact to final users. In addition, the FIRE initiative has responded to this problem by establishing the objective of building a framework to support such kind of research. The SmartFIRE project has been incepted and developed with this objective in mind.

The federation of the SmartFIRE testbeds is the most outstanding result of the project's contributions. The physical interconnection of the testbeds, as well as the development of a common framework for controlling, managing, provisioning and reserving the testbeds' resources, enables the heterogeneous and large-scale experimentation in a unified and human-friendly platform.

This chapter has described the execution and final results obtained from the integration of the proof of concept experiments within the infrastructure provided by the SmartFIRE project. They have been used to get the most relevant requirements for such infrastructure in terms of resources and experimentation tools, and they have been used to improve even more the results of the project. This way, the showcases have also been used to evaluate, validate, and demonstrate the benefits provided by the resulting infrastructure. Also, they have shown enormous interest from the research community, especially from South Korea, so their execution and results have had good reception among them.

References

[1] SmartFIRE: K. Choumas, T. Korakis, H. Lee, D. Kim, J. Suh, T. Kwon, P. Martinez-Julia, A. Skarmeta, T. You, L. Baron, S. Fdida and J. Kim, "Enabling SDN Experimentation with Wired and Wireless Resources: The SmartFIRE facility", *Proceedings of CloudComp 2015*, Daejeon, South Korea, October 2015, (Best paper Award).

[2] SmartFIRE-link: SmartFIRE: Enabling SDN Experimentation in Wireless Testbeds exploiting Future Internet Infrastructures in South Korea and Europe, http://eukorea-fire.eu (accessed July 2016)

[3] OMF: Thierry Rakotoarivelo, Maximilian Ott, Guillaume Jourjon and Ivan Seskar, “OMF: a control and management framework for networking testbeds”, *SIGOPS Oper. Syst. Rev.*, vol 43, no. 4, pp. 54–59, January 2010.
[4] OMF link: OMF: cOntrol and Management Framework, http://mytestbed.net (accessed July 2016)
[5] MySlice: MySlice: https://www.myslice.info (accessed July 2016)
[6] SFA: SFA: Slice-base Federation Architecture v2.0, http://groups.geni.net/geni/wiki/SliceFedArch (accessed July 2016)
[7] Click: E. Kohler, R. Morris, B. Chen, J. Jannotti and M. F. Kaashoek. “The Click Modular Router”, ACM Trans. on Computer Systems, vol. 18, no. 3, pp. 263–297, August 2000.
[8] Click-link: Click: Click Modular Router, http://read.cs.ucla.edu/click/click (accessed July 2016)
[9] CNERT: K. Choumas, N. Makris, T. Korakis, L. Tassiulas and M. Ott, “Testbed Innovations for Experimenting with Wired and Wireless Software Defined Networks”, *Proceedings of CNERT workshop of IEEE ICDCS 2015*, Columbus, Ohio, USA, June–July 2015.
[10] ICN-OMF: H. Lee, D. Kim, J. Suh and T. Kwon, “ICN-OMF: A Control, Management Framework for Information-Centric Network Testbed”, *Proceedings of ICOIN 2015*, Siem Reap, Cambodia, January 2015.
[11] GEANT: GEANT: Pan-European Research and Education Network, http://www.geant.net/ Pages/default.aspx (accessed July 2016)
[12] KREONET: KREONET: Korea Research Environment Open NETwork, http://www.kreonet.re.kr/en_main (accessed July 2016)
[13] KOREN: KOREN: Korea Advanced Research Network, http://www.koren.kr/koren/eng/index.html (accessed July 2016)
[14] TEIN: TEIN: The Trans-Eurasia Information Network, http://www.tein.asia/tein4/index.do (accessed July 2016)
[15] GLORIAD: GLORIAD: Global Ring Network for Advanced Applications Development, http://www.gloriad.org/gloriaddrupal/index.php (accessed July 2016)
[16] MOFI: H. Jung, S. Koh, and W. Park, “Towards the Mobile Optimized Future Internet”, *Proceedings of ACM CFI 2009*, Seoul, Korea, 2009.
[17] Fed4FIRE: Fed4FIRE: Federation for Future Internet Research and Experimentation, http://www.fed4fire.eu (accessed July 2016)
[18] OpenLab: OpenLab: http://www.ict-openlab.eu/home.html (accessed July 2016)

[19] FIRE: FIRE: Future Internet Research and Experimentation, https://www.ict-fire.eu (accessed July 2016)

[20] A. Gavras, A. Karila, S. Fdida, M. May and M. Potts, "Future internet research and experimentation: the FIRE initiative", *ACM SIGCOMM*, vol. 37, no. 3, pp. 88–92, July 2007.

[21] GENI: GENI: Exploring Networks of the Future, https://www.geni.net (accessed July 2016)

[22] OFELIA: OFELIA: OpenFlow in Europe: Linking Infrastructure and Applications, http://www.fp7-ofelia.eu (accessed July 2016)

[23] OpenStack: OpenStack: https://www.openstack.org (accessed July 2016)

[24] ProtoGeni: ProtoGeni: http://www.protogeni.net (accessed July 2016)

[25] OVS: Open vSwitch: http://openvswitch.org/ (accessed July 2016)

[26] K. Choumas, G. Paschos, T. Korakis and L. Tassiulas, "Distributed Load Shedding with Minimum Energy", *Proceedings of IEEE INFOCOM 2016*, San Francisco, CA, USA, April 2016.

[27] FlowVisor: Rob Sherwood, Glen Gibb, Kok-Kiong Yap, Guido Appenzeller, Martin Casado, Nick McKeown and Guru Parulkar, "Can the production network be the testbed?", *Proceedings of OSDI 2010*, Berkeley, CA, USA, 2010.

[28] John Bicket, Daniel Aguayo, Sanjit Biswas and Robert Morris, "Architecture and evaluation of an unplanned 802.11b mesh network", *Proceedings of ACM MobiCom 2005*, New York, NY, USA, 2005.

[29] A. Gavras, (ed.), "Experimentally driven research, white paper, On the existence of experimentally-driven research methodology, Version 1 (April 2010)", http://www.ict-fireworks.eu/fileadmin/documents/Experimentally_driven_research_V1.pdf (downloaded July 2012).

[30] A. Gavras, A. Bak, G. Biczók, P. Gajowniczek, A. Gulyás, H. Hrasnica, P. Martinez-Julia, F. Németh, C. Papagianni, S. Papavassiliou, M. A. Skarmeta, "Heterogeneous Testbeds, Tools and Experiments – Measurement Requirements Perspective, Measurement Methodology and Tools", *Lecture Notes in Computer Science*, vol. 7586, pp. 139–158, 2013.

Printed in the USA
CPSIA information can be obtained
at www.ICGtesting.com
LVHW020523090923
756175LV00005B/81